COMMON ELEMENTS

Name	Symbol	Approx. at. wt.	Common ox. nos.	Name	Symbol	Approx. at. wt.	Common ox. nos.
aluminum	Al	27.0	+3	magnesium	Mg	24.3	+2
antimony	Sb	121.8	+3, +5	manganese	Mn	54.9	+2, +4, +7
arsenic	As	74.9	+3, +5	mercury	Hg	200.6	+1, +2
barium	Ba	137.3	+2	nickel	Ni	58.7	+2
bismuth	Bi	209.0	+3	nitrogen	N	14.0	−3, +3, +5
bromine	Br	79.9	−1, +5	oxygen	O	16.0	−2
calcium	Ca	40.1	+2	phosphorus	P	31.0	+3, +5
carbon	C	12.0	+2, +4	platinum	Pt	195.1	+2, +4
chlorine	Cl	35.5	−1, +5, +7	potassium	K	39.1	+1
chromium	Cr	52.0	+2, +3, +6	silicon	Si	28.1	+4
cobalt	Co	58.9	+2, +3	silver	Ag	107.9	+1
copper	Cu	63.5	+1, +2	sodium	Na	23.0	+1
fluorine	F	19.0	−1	strontium	Sr	87.6	+2
gold	Au	197.0	+1, +3	sulfur	S	32.1	−2, +4, +6
hydrogen	H	1.0	−1, +1	tin	Sn	118.7	+2, +4
iodine	I	126.9	−1, +5	titanium	Ti	47.9	+3, +4
iron	Fe	55.8	+2, +3	tungsten	W	183.8	+6
lead	Pb	207.2	+2, +4	zinc	Zn	65.4	+2

COMMON IONS AND THEIR CHARGES

Name	Symbol	Charge	Name	Symbol	Charge
aluminum	Al^{+++}	+3	lead(II)	Pb^{++}	+2
ammonium	NH_4^{+}	+1	magnesium	Mg^{++}	+2
barium	Ba^{++}	+2	mercury(I)	Hg_2^{++}	+2
calcium	Ca^{++}	+2	mercury(II)	Hg^{++}	+2
chromium(III)	Cr^{+++}	+3	nickel(II)	Ni^{++}	+2
cobalt(II)	Co^{++}	+2	potassium	K^{+}	+1
copper(I)	Cu^{+}	+1	silver	Ag^{+}	+1
copper(II)	Cu^{++}	+2	sodium	Na^{+}	+1
hydronium	H_3O^{+}	+1	tin(II)	Sn^{++}	+2
iron(II)	Fe^{++}	+2	tin(IV)	Sn^{++++}	+4
iron(III)	Fe^{+++}	+3	zinc	Zn^{++}	+2
acetate	$C_2H_3O_2^{-}$	−1	hydrogen sulfate	HSO_4^{-}	−1
bromide	Br^{-}	−1	hydroxide	OH^{-}	−1
carbonate	CO_3^{--}	−2	hypochlorite	ClO^{-}	−1
chlorate	ClO_3^{-}	−1	iodide	I^{-}	−1
chloride	Cl^{-}	−1	nitrate	NO_3^{-}	−1
chlorite	ClO_2^{-}	−1	nitrite	NO_2^{-}	−1
chromate	CrO_4^{--}	−2	oxide	[illegible]	−2
cyanide	CN^{-}	−1	perchl	[illegible]	−1
dichromate	$Cr_2O_7^{--}$	−2	perma	[illegible]	−1
fluoride	F^{-}	−1	peroxi	[illegible]	−2
hexacyanoferrate(II)	$Fe(CN)_6^{----}$	−4	phospl	[illegible]	−3
hexacyanoferrate(III)	$Fe(CN)_6^{---}$	−3	sulfate	[illegible]	−2
hydride	H^{-}	−1	sulfide	[illegible]	−2
hydrogen carbonate	HCO_3^{-}	−1	sulfite	SO_3^{--}	−2

HOLT, RINEHART AND WINSTON, PUBLISHERS
New York · London · Toronto · Sydney · Mexico City · Tokyo

H. Clark Metcalfe
P.O. Box V2, Wickenburg, Arizona, 85358; formerly teacher of chemistry at Winchester-Thurston School, Pittsburgh, Pennsylvania, and Head of the Science Department, Wilkinsburg Senior High School, Wilkinsburg, Pennsylvania.

John E. Williams
Formerly teacher of chemistry and physics at Newport Harbor High School, Newport Beach, California, and Head of the Science Department, Broad Ripple High School, Indianapolis, Indiana.

Joseph F. Castka
Formerly Assistant Principal for the Supervision of Physical Science, Martin Van Buren High School, New York City, and Adjunct Associate Professor of General Science and Chemistry, C. W. Post College, Long Island University, New York.

The Holt Modern Chemistry Program

Modern Chemistry (Student Text)
Modern Chemistry (Teacher's Edition)
Laboratory Experiments in Modern Chemistry
Exercises and Experiments in Modern Chemistry
Exercises and Experiments in Modern Chemistry (Teacher's Edition)
Tests in Modern Chemistry (Duplicating Masters)
Problems for Modern Chemistry
Teacher's Resource Book for Modern Chemistry

Cover: Photography by Joe Salenetri, courtesy of Squibb Corporation.

Photo and art credits appear on pages 727–728.

Printed in the United States of America

ISBN 0-03-001274-0

678 032 98765

PREFACE

The 1986 edition of MODERN CHEMISTRY continues the nearly 70-year tradition of an accurate, up to date, easily understood textbook that fulfills the various curriculum requirements of an introductory course in chemistry. Teachers will find ample material in MODERN CHEMISTRY for an outstanding college preparatory program. For those students who do not plan to go further in science, there is sufficient elementary theory and descriptive information for a complete general course of study. The Teacher's Edition of MODERN CHEMISTRY gives suggestions for implementing both types of chemistry courses and suggestions for an advanced course as well.

In MODERN CHEMISTRY the subject matter of chemistry is organized in a logical, workable sequence. Descriptive and theoretical topics are alternated to provide classroom variety and a well-correlated laboratory program. Special attention has been given to the introduction of technical vocabulary words. Each of these words is printed in ***boldface italics*** and carefully defined when first introduced. Numbered Sample Problems and Solutions are used to illustrate quantitative relationships. Practice Problems with answers reinforce the understanding of these concepts.

Significant content changes in this edition of MODERN CHEMISTRY include a description of the production of elements 108 and 109, an explanation of atomic structure exclusively in terms of energy levels, a definition of atomic weight that allows a more logical development of the concepts of atomic weight and the mole, a new explanation of how electrons constitute a stable electron cloud about a nucleus, revised data for electron affinity, a different method of calculating percentage of ionic character from electronegativity data that gives results more in agreement with the actual nature of the compound, the introduction of carbyne as a third allotropic form of carbon, a clearer explanation of factors affecting equilibria, the consistent treatment of oxidation and reduction as changes in oxidation state, the revision of definitions of anode and cathode to conform to their specialized use in electrochemistry, an increased coverage of early discoveries in radioactivity, and a more detailed discussion of controlled fusion reactions. As usual many details of descriptive chemistry have been updated as to sources, preparations, physical and chemical properties, and uses of compounds. The Questions and Problems sets have been expanded and extensively rewritten. Some questions about social and environmental problems related to chemistry have been included. The Teacher's Edition of *Exercises and Experiments in Modern Chemistry* includes a very detailed and reproducible list of qualitative and quantitative **Performance Objectives** for each chapter.

At the beginning of each chapter there is a brief list of Goals. Marginal Notes are strategically placed throughout the text to reinforce concepts, provide supplementary information, and assist the student in developing productive study techniques. At the end of each chapter there is a Summary, a Vocabulary review list, and suitable Questions and Problems. The Questions, which are based on the text itself, are graded according to difficulty into Groups A and B. The Problems are similarly graded. The average student should master all the Group A Questions and Problems; the better student will be able to do both Group A and Group B exercises. Except for review material, each Question and Problem is keyed to the detailed list of **Performance Objectives** for each chapter. In the back of the book are an Appendix containing Data Tables, a Glossary, and an index.

Because of their clarity and great learning value, line drawings are used extensively. Design coordination was contributed by Claudia DePolo and Brian Molloy. The text is also illustrated with many fine photographs researched by Yvonne Gerin. New photographs involving chemistry laboratory equipment were taken for this edition by Yoav Levy with coordinating and supervising assistance by Sheila Kogan. The chemistry laboratory facilities of Wickenburg High School were made available for these photographs by Dr. James M. Randall, Superintendent, Wickenburg Unified School District, Arizona, and Robert Kilker, Principal, Wickenburg High School. Pamela I. Rhoda, Teacher of Chemistry and Physical Science, Wickenburg High School, assisted in preparing the laboratory set-ups. Dudley G. Smith, Teacher of Chemistry, Newport Harbor High School, Newport Beach, California, helped provide special materials and equipment. There are Photo Essays on careers in chemistry, oxygen, crystals, and steel production. Caption copy for the chapter opener nobelists was researched and written by Diane Khoury.

The text was written by H. Clark Metcalfe and John E. Williams. Joseph F. Castka was responsible for the preparation of EXERCISES AND EXPERIMENTS IN MODERN CHEMISTRY, LABORATORY EXPERIMENTS IN MODERN CHEMISTRY, and TESTS IN MODERN CHEMISTRY, with editorial assistance from Florence Anderson. Mr. Dean M. Hurd, chemistry teacher at Carlsbad High School in Carlsbad, California prepared additional materials for the laboratory experiments. Mr. Hurd also prepared the manuscript for PROBLEMS FOR MODERN CHEMISTRY. The Teacher's Edition and the detailed lists of Performance Objectives were prepared by Zoe A. Godby Lightfoot, Teacher of Chemistry and Head of the Science Department, Carbondale Community High School, Carbondale, Illinois. Research assistance on references and resources in the Teacher's Edition was provided by Dr. Billie L. Perkins.

The authors wish to thank the following persons who have assisted in the preparation of this revision by providing information, helpful criticism, and advice: Wayne A. Fithian, Agronomist, Ogallala, Nebraska; Frances P. Gray, Teacher of Chemistry, Manchester West High School, Manchester, New Hampshire; Jerry L. Green, Chairman of the Science Department, Miamisburg Senior High School, Miamisburg, Ohio; Dr. George R. Gross, Teacher of Chemistry, Union High School, Union, New Jersey; Sister Mary Realino Lynch, B.V.M., Formerly Teacher of Physical Sciences and Chairman of the Science Department, Holy Family High School, Glendale, California; Lester M. Mack, Science Consultant, Pontiac School District, Pontiac, Michigan; Dr. Vance B. Miller, Family Physician, Wickenburg, Arizona; Richard L. Phillips, Agricultural Entomologist, Holdrege, Nebraska; and Jack T. Yaxley, Teacher of Chemistry, South Plantation High School, Plantation, Florida.

H. Clark Metcalfe   John E. Williams   Joseph F. Castka

CONTENTS

HOW TO USE THIS BOOK

Major Goals help the student focus on the important concepts of the chapter.

Subtopic and numbered **Section** headings point out the main ideas of the section.

The twelve **Units** are divided into thirty-one **Chapters.**

Numerous **Marginal Notes** reinforce concepts, offer additional information, and assist student development of effective study techniques.

Otto Hahn (German) was awarded the 1944 Nobel Prize in chemistry for discovering the fission of heavy nuclei. Atoms are composed of a positive nucleus surrounded by negative electrons. Hahn found that when the nucleus of uranium-235 is hit with slow neutrons, it splits into two new atoms of approximately equal mass, such as barium and krypton. A great deal of energy is released when the structure of matter changes in this way. The fission process is utilized in nuclear reactors and in the atomic bomb.

Matter and Its Changes

2

GOALS
In this chapter you will gain an understanding of: • heterogeneous and homogeneous matter • mixtures, compounds, and elements • physical and chemical changes • endothermic and exothermic processes • reaction tendencies in nature: energy-change tendency, entropy-change tendency

COMPOSITION OF MATTER

2.1 Classes of matter Chemistry has been described as the study of the structure and composition of matter, of changes in its composition, and of the mechanisms by which these changes occur. The methods of chemistry involve metric measurements and often require computations with measurement data. To pursue this study, it is essential that some basic concepts of matter be recognized.

Matter is commonly *described* as anything that occupies space and has mass. It includes all materials found in nature. Some materials are made up of parts that are not alike. The dissimilarities may be subtle, or they may be readily apparent. *Matter that has parts with different properties is said to be* ***heterogeneous*** (het-er-oh-*jee*-nee-us). Granite, a common rock, is heterogeneous because it is composed of different minerals, each having characteristic properties. Distinctly different parts are easily observed.

Other materials appear uniform throughout; all parts are alike. The properties of any one part are identical to the properties of all other parts. *Matter that has identical properties throughout is* ***homogeneous*** (hoh-muh-*jee*-nee-us). Sugar and ordinary table salt are examples of homogeneous materials.

Matter is either heterogeneous or homogeneous.

The many different kinds of matter throughout the world are the materials with which chemists work. It would be difficult and time-consuming to study these materials without first organizing them into similar groups. One method of classifying matter is according to its three different phases: solid, liquid, and gas. Matter is also divided into three general classes on the basis of its properties: *elements, compounds,* and *mixtures.* These and other ways of classifying matter make the study of chemistry simpler. Figure 2-1 will be helpful in the study of Sections 2.2 and 2.3.

41

Chemistry Nobelists featured with each chapter relate chemistry to a variety of other topics, which are historical in context and can be applied to present- day chemistry.

New terms, definitions, and principles are printed in *italics* and ***boldface italics*** for emphasis and rapid review.

Unit 3 Chemical Formulas and Equations

SAMPLE PROBLEM 7-4

Hydrogen peroxide is found by analysis to consist of 5.9% hydrogen and 94.1% oxygen. Its molecular weight is determined to be 34.0. What is the correct formula?

SOLUTION

1. The empirical formula, determined from the analysis by the method described in Section 7.12, is

HO

2. The molecular formula determined from the empirical formula and molecular weight is

$$(\text{HO weight})_x = 34.0$$
$$(1.0 + 16.0)_x = 34.0$$
$$x = 2$$
$$\text{molecular formula} = (HO)_2 \text{ or } H_2O_2$$

Figure 7-4. Compounds formed by the same two elements combined in different proportions may have very different chemical properties. In (top photo), water, H_2O, is unreactive with the green felt. In (bottom photo), hydrogen peroxide, H_2O_2, bleaches the green dye in the felt.

Practice Problems

1. A certain hydrocarbon compound is found by analysis to have the empirical formula CH_3. Its molecular weight is 30.0. What is the molecular formula? *ans.* C_2H_6

7.10 Law of multiple proportions It is not uncommon for two or more compounds to be composed of the same two elements. Hydrogen and oxygen form a series of two compounds, water (H_2O) and hydrogen peroxide (H_2O_2). The constituent elements, hydrogen and oxygen, are present in each of these two compounds in unvarying but different *mass* proportions.

For water, the composition is about 2 parts hydrogen to 16 parts oxygen, or 1 part hydrogen to 8 parts oxygen. For hydrogen peroxide, the composition is 1 part hydrogen to 16 parts oxygen. Considering quantities of these two compounds *that have the same masses of hydrogen*, the relative masses of the second element (oxygen) are related *as a ratio of small whole numbers*.

Consider, for example,

H_2O	1 g of H	and	8 g of O
H_2O_2	1 g of H	and	16 g of O

The *masses of hydrogen* in this series, 1 g and 1 g, are in the constant ratio of 1 : 1. The *masses of oxygen*, 8 g and 16 g, are in the

168

Sample Problems and **Solutions** illustrate important quantitative relationships.

Practice Problems, with answers, provide reinforcement of the understanding of basic concepts.

Photographs and **Diagrams** enhance and illustrate ideas presented in the chapter.

Important equations are printed in **boldface** and sometimes in color.

Chapter 7 Chemical Composition

Table 7-2
METALLIC ION NAME EQUIVALENTS

Old system		New system	
chromic	Cr^{+++}	chromium(III)	Cr^{+++}
cobaltous	Co^{++}	cobalt(II)	Co^{++}
cobaltic	Co^{+++}	cobalt(III)	Co^{+++}
ferrous	Fe^{++}	iron(II)	Fe^{++}
ferric	Fe^{+++}	iron(III)	Fe^{+++}
cuprous	Cu^{+}	copper(I)	Cu^{+}
cupric	Cu^{++}	copper(II)	Cu^{++}
mercurous	Hg^{+}	mercury(I)	Hg_2^{++}
mercuric	Hg^{++}	mercury(II)	Hg^{++}
plumbous	Pb^{++}	lead(II)	Pb^{++}
plumbic	Pb^{++++}	lead(IV)	Pb^{++++}
stannous	Sn^{++}	tin(II)	Sn^{++}
stannic	Sn^{++++}	tin(IV)	Sn^{++++}

SUMMARY

The formulas for common ionic compounds can be written using the ion-charge method. The ion notations are written in the order used in the name of the compound. The number of each names of the constituent ions. If the metallic ion present has different possible oxidation states, Roman numerals are added to the name of the ion to indicate the oxidation state. A system of

VOCABULARY

Avogadro number
binary compound
empirical formula
gram-atomic weight
gram-formula weight
gram-molecular weight
mole
molecular weight
percentage composition

QUESTIONS

GROUP A

1. To what characteristics of the Group I metallic elements can the ease with which their atoms form +1 monatomic ions be attributed? *A4a*

GROUP B

16. Name each of the following binary compounds: (*a*) SO_2; (*b*) SO_3; (*c*) CO; (*d*) CO_2; (*e*) CS_2; (*f*) CCl_4; (*g*) $SbCl_3$; (*h*) As_2S_3; (*i*) As_2S_5; (*j*) $BiBr_3$. *A2e*

PROBLEMS

GROUP A

1. What is the formula weight of potassium chlorate, $KClO_3$? *B3a*
2. Determine the formula weight of acetic acid, $HC_2H_3O_2$. *B3a*

GROUP B

21. What is the percentage composition of the drug chloramphenicol, $C_{11}H_{12}N_2O_5Cl_2$? *B3b*
22. One compound of platinum and chlorine is

171

Graphs and **Tables** present data in a clear, concise manner. Important **Tables** are contained in the **Appendix.** A **Glossary** provides definitions of important words. The **Index** gives page references for key terms.

End-of-chapter **Summary** is a brief review of the major concepts.

Key words covered in the chapter are reviewed in the **Vocabulary.**

Questions and **Problems,** graded according to difficulty, are designed to check your progress in understanding chapter material. They are keyed to **Performance Objectives** found in the *Teacher's Resource Book.* If there is no objective code listed, don't worry. It means that you have mastered the material in a previous chapter.

Marie Sklodowska Curie (French) received the Nobel Prize in chemistry in 1911 for discovering the radioactive elements radium and polonium (named after her native Poland) and for isolating radium and studying its compounds. In 1903 she shared the Nobel Prize in physics with her husband, Pierre Curie, and A. Henri Becquerel for their studies of the radiation phenomena. She became the first woman professor at the Sorbonne in Paris, where the Radium Institute was organized in 1914 for her research. She died in 1934 from the effects of radium poisoning.

Measurements in Chemistry

1

GOALS

In this chapter you will gain an understanding of: • the nature of chemistry • careers in chemistry • the scientific process • matter and energy • measurement accuracy and precision • computation techniques

INTRODUCTION

You are about to begin your study of chemistry. Perhaps you are wondering what this science is about and how it is important to you. Chemistry is the study of materials, their composition and structure, and the changes they undergo. Chemistry is a body of systematized knowledge gained from observation, study, and experimentation—as opposed to guesswork and opinions. By studying chemistry you will be able to understand the nature of the materials around you and the changes they undergo.

Numerous practical benefits accrue from an understanding of chemistry. The food, clothing, building materials, and medicines that are necessary to life are products of the application of chemical knowledge. Many of the materials used in such fields as architecture, engineering, and nuclear science result from the work of chemists. Without these and other innovations, our way of life would be radically different.

You are beginning your study of chemistry at a time when there is great concern about complex problems affecting the quality of life. Many of these problems relate to the contamination of our environment. Chemical research has been responsible for the production of many synthetic materials, but such processes often contribute to environmental pollution. Mining coal and metal ores scars the landscape. Burning coal and smelting metal ores foul the atmosphere. Industrial wastes often disrupt natural environmental processes, upsetting established balances in nature. However, the study of chemistry may also reveal solutions to these and other complex problems.

Chemistry is both an experimental and an intellectual science. It can take you to the core of many different kinds of problems. It can also help you gain a deeper and more satisfying understanding of your environment, enabling you to recognize

Figure 1-1. A stream fouled by the illegal dumping of industrial wastes.

and appreciate order in nature. The Greek scholar Aristotle referred to humans as "creatures who delight to know." If you are curious and wish to know more about natural processes, about water and solutions, and about the minerals of the earth and the gases of the atmosphere, the activities of chemistry beckon to you.

Chemical knowledge of a practical nature dates far back in history. Prehistoric people produced watertight pottery by shaping, baking, and glazing certain clays. About 5000 years ago, fire was found to be useful in preparing copper, bronze, and iron. Nearly 2500 years ago, the Egyptians found methods for manufacturing glass and extracting medicines from plant sources. Each of these early discoveries was a chemical operation that greatly influenced the course of civilization. Progress in understanding the underlying causes of these and other chemical transformations was slow in coming, however. Although the Greeks attempted to analyze materials and describe their various properties as early as 300 B.C., it was not until 2000 years later that adequate chemical explanations began to appear. Since then, tremendous progress has been made in both understanding and applying the marvels of chemistry. Frontiers still remain, however. As chemical research continues to unravel the complexity of materials, chemists will continue to make valuable contributions to human progress. Research is an ongoing process in all the sciences, including chemistry. Chemical research promises a brighter future for all. To be a chemist is to be an architect of the future.

Although scientists are interested in learning facts, a mere collection of facts about the physical world does not constitute a science. A major activity of scientists is using such facts to build models or provide explanations that relate the facts to each other and that can be used to predict future observations. Another activity of scientists is testing the models and explanations. Testing confirms the validity of scientific concepts.

If, over a long time, experimental evidence supports an explanation, it is accepted as valid. If even one observation does not follow the behavior predicted, however, the explanation is no longer valid and must be revised or discarded. Explanations and models are never certain, and scientists must learn to accept and work with a degree of uncertainty.

Scientists are naturally curious people who attempt to find interrelationships among their many observations. Scientists are patient people, for nature reveals its secrets slowly. Scientists are careful people. They constantly test their conclusions, knowing full well that scientific explanations are at best imperfect and often temporary. Finally, scientists are optimistic people who have great faith in the orderliness of nature. What scientists hope to accomplish is appropriately expressed in *An Introduction to Mathematics* by the famous English philosopher Alfred North

Whitehead: "The progress of Science consists in observing interconnections and in showing with a patient ingenuity that the events of the ever-shifting world are but an example of a few general relations, called laws. To see what is general in what is particular, and what is permanent in what is transitory, is the aim of scientific thought."

The Organization of Chemistry

1.1 Chemistry: a physical science A convenient way to organize the various sciences is to group them into two general categories: (1) *biological sciences,* which are concerned with living things, and (2) *physical sciences,* which deal with the relationships in nature.

Chemistry is a physical science. It is a body of systematized knowledge gained from observation, study, and experimentation, as opposed to guesswork and opinion. ***Chemistry** is the science of materials, their composition and structure, and the changes they undergo.*

The scope of chemistry is very broad. To make the study of chemistry easier, chemical knowledge is arranged in separate systems or branches. Six main branches of chemistry are listed as follows:

1. *Analytical chemistry* is concerned with the separation, identification, and composition of materials.
2. *Organic chemistry* is the chemistry of carbon compounds.
3. *Inorganic chemistry* is the chemistry of materials other than those classed as organic.
4. *Physical chemistry* involves the study of the physical characteristics of materials and the mechanisms of their reactions.
5. *Biochemistry* includes the study of materials and processes that occur in living things.
6. *Nuclear chemistry* involves the study of subatomic particles and nuclear reactions.

Each of these branches of chemistry is further divided into more specialized areas.

Chemistry is a tool for investigating the materials that constitute the environment. Knowing the composition and structure of materials provides a basis for understanding the changes these materials undergo in natural processes and in planned experiments.

Figure 1-2. Dr. Jennie R. Patrick is the first black woman in the United States to earn a doctoral degree in chemical engineering. Her research led to new insights into the behavior of vapor explosions of superheated liquid mixtures. Her work promises to result in energy conservation and pollution control.

1.2 Careers in chemistry Because chemistry covers a broad scope of both intellectual and experimental activities, various opportunities exist for those who wish to pursue careers as chemists, chemical engineers, or chemical technicians. In chemistry, as in other fields of endeavor, education, ability, and interest determine a person's career opportunities.

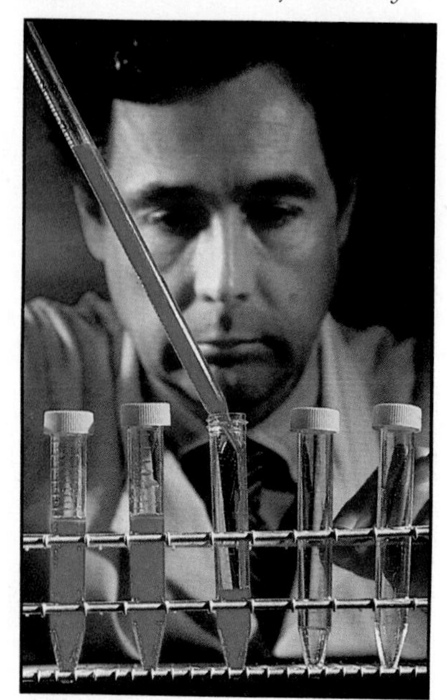

Figure 1-3. Dr. Michael Parvin of the Center for Disease Control in Atlanta checks the progress of an experiment evaluating the properties of a virus.

Chemists pursue gratifying careers in a wide range of scientific activities. They conduct basic research in the quest for new knowledge and understanding. They are involved in the technology of applied chemistry—the application of chemistry to those processes that supply material needs. They perform chemical analyses of substances, supervise production, develop manufacturing processes, or manage organizations that make, use, or sell chemicals. They do technical writing, work in sales and marketing, or become consultants. They teach chemistry in universities, colleges, or high schools. Chemists sometimes earn law degrees and become patent attorneys that have special technical qualifications.

Chemical engineers are chemists with engineering training. They are usually concerned with large-scale industrial processes. They are involved in the design, construction, and operation of equipment required in chemical processes. Frequently they specialize in the marketing aspects of industry.

Chemical technicians usually work under the supervision of a professional chemist. The chemist may design and plan procedures for the technician to carry out in the laboratory. Chemical technicians are trained to use instruments and to perform various laboratory operations.

1.3 The modern era of chemistry Centuries ago skilled workers undoubtedly used some practical chemical knowledge in working with bronze, in recovering iron from its ore, and in making pigments and pottery. The practice of alchemy in the Middle Ages was the forerunner of chemistry. Seeking a means to change metals into gold, the alchemists conducted their work in an atmosphere of mystery and magic. However, by the seventeenth century, the transition from alchemy to chemistry had begun; the use of reliable quantitative measurements challenged the beliefs of the alchemists.

A plausible theory of combustion, called the *phlogiston* (flo-*jis*-tun) *theory,* was introduced in 1669. According to this theory, all combustible materials contain a mystical substance called *phlogiston* that is released in the form of heat and flame during the combustion process. The phlogiston theory required that a body lose weight during combustion because the phlogiston escaped.

Although the phlogiston theory was easily understood and applied, some early scientists doubted its correctness. They argued that if it is true a material gives off phlogiston during combustion, the ash that remains must always weigh less than the original material. These doubters challenged the phlogiston theory experimentally. They showed that the "ashes" of burned metals weighed more than the metals themselves. Despite such strong challenges, the phlogiston theory of the nature of fire dominated all chemical thought for more than a century. It was the last important generalization of the alchemy era.

The initial concept of modern chemistry, that elements are the stuff of which things are made, was developed in Europe at about the time of the American Revolution. Oxygen was first prepared in 1771 and was quickly recognized as the part of the air actively involved in ordinary combustion processes. These events set the stage for a basic understanding of chemical reactions and a determined search for chemical elements.

Through the efforts of many investigators, about 90 elements were eventually recognized. By 1920 most of these elements had been isolated and their properties studied. As in other areas of science, research—creatively conceived and skillfully performed—is primarily responsible for the refinement and growth of chemical knowledge to its present state.

Figure 1-4. An engineer may specialize in processing problems, marketing, or production. This biochemical engineer is gathering data from a cell concentrator.

1.4 The scientific method Some important scientific discoveries have come about quite by accident. Others have been the result of brilliant new ideas. Most scientific knowledge, however, is the result of carefully planned work carried on by trained scientists. Their technique, known as the ***scientific method,*** *is a logical approach to the solution of problems that lend themselves to investigation.* The scientific method requires honesty, the ability to withhold a decision until all the evidence is gathered, and a desire for knowledge.

Scientists believe that there is order in nature, that natural events occur in an orderly way, and that the rules governing these events can be discovered and described. Chemists, like other scientists, strive to explain a large number of related observations in terms of broad principles or *generalizations.* Basic scientific research is devoted to the discovery of these principles. *The generalizations that describe behavior in nature are called* ***laws*** *or* ***principles.*** Unlike laws that restrict behavior, scientific laws describe what occurs in nature. Laws of science may be expressed by concise statements or by mathematical equations.

One of the distinguishing qualities of scientists is their curiosity. It prompts them to ask two important questions: "What occurs?" and "How does it occur?" When an event or situation in nature, called a *phenomenon,* is observed, the answers to these questions are sought by conducting systematic, disciplined, and persistent investigations.

There are four distinct phases in applying the scientific method: *observing, generalizing, theorizing,* and *testing.*

1. *Observing.* The scientist accumulates as much reliable data as possible about an observed phenomenon. The initial interest is the discovery of *what* actually occurs. These data may come

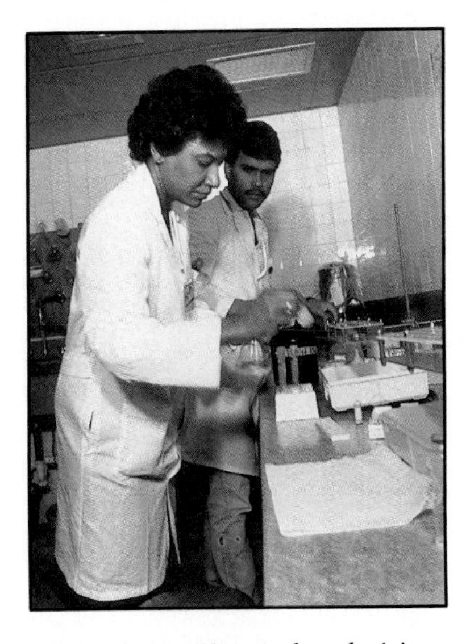

Figure 1-5. Chemical technicians test for drugs possibly present in urine and blood specimens of athletes participating in the 1983 Pan American Games held in Caracas, Venezuela.

from direct observations, from a search for information previously reported in scientific literature, and from well-planned and skillfully performed experiments.

Chemists know that observing is most productive when the conditions that affect the observations are brought under their control. Thus, observing is generally done in the *laboratory,* where conditions can be controlled by the observer. *A sequence of observations carried out under controlled conditions is called an* ***experiment.*** Experimentation provides the foundation upon which modern science is built.

2. *Generalizing.* The scientist organizes the accumulated data and looks for relationships among them. Relationships that are discovered may enable the scientist to formulate a broad generalization describing what occurs. When well established by abundant supporting data, this generalization may be recognized as a new law or principle that describes the behavior.

3. *Theorizing.* When the scientist knows *what* occurs, it becomes possible to move on to the more stimulating task of determining *how* it occurs. The scientist attempts to develop a plausible explanation and construct a simple physical or mental model that relates the observed behavior to familiar and well-understood phenomena. *A plausible explanation of an observed natural phenomenon in terms of a simple model with familiar properties is called a* ***theory.***

4. *Testing.* Once a satisfactory theory is developed, it is tested and retested to establish its validity. Scientists continually test observational and experimental data and predictions by subjecting them to new experiments. A theory is retained only as long as it is useful. It may be discarded or modified as a result of new experimentation. A theory that stands up under scientific testing is valuable because it serves as a basis for predicting behavior not previously investigated. This testing and predicting is the heart of the scientific method and the stimulus for the growth of scientific knowledge.

The term "theory" is often used by chemists to refer to the laws and experimental evidence that make up a body of related knowledge. Some examples of broad chemical theories concerning the behavior of matter that you will study in later chapters are the *atomic theory,* the *kinetic theory* of gases, and the *theory of ionization.*

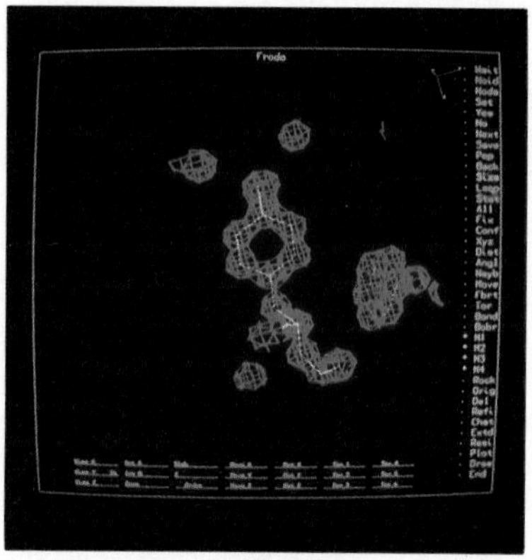

Figure 1-6. A physical model is often used by chemists to help them understand the behavior of matter. On this computer screen, graphics are used to show the molecular structure of an amino acid side chain.

MATTER AND ENERGY

1.5 The concept of matter All materials consist of matter. Through the senses of sight, touch, taste, and smell, various kinds of matter can be recognized. This book, your desk, the air you breathe, the water you drink are all examples of matter.

Some kinds of matter are easily observed. A stone or a piece of wood can be seen and held in the hand. Water, even in a quiet pool, is easily recognized. Other kinds of matter, such as the air, are recognized less readily. You ride on compressed air in automobile tires. You know of the tremendous damage caused by the rapidly moving air in a hurricane or tornado.

Matter *is described as anything that occupies space and has mass.* Matter may be acted upon by *forces* that change its motion or position. Matter possesses *inertia,* a resistance to change in position or motion. The concept of inertia as a property of matter is important in the study of physics and chemistry. Imagine a basketball being used as a substitute for a bowling ball. Its effect on the pins would not be the same as that observed in Figure 1-7. Although approximately the same size, the bowling ball contains more matter (and of a different kind) than the basketball. The inertia of the bowling ball is correspondingly greater than that of the basketball. Thus its tendency to remain in motion is also greater.

Figure 1-7. A basketball or a hollow bowling ball would not produce the results shown in this photograph.

Although these comments are descriptive of matter, they do not give a completely satisfactory definition of matter. Scientists, even with great knowledge of the properties and behavior of matter, are not able to define it precisely.

1.6 Mass and weight *The quantity of matter that a body possesses is known as its* ***mass.*** If an object is at rest and you move it, you notice that it resists your effort. If the object is moving and you stop it, you notice that it resists this effort also. The object's *mass* is the measure of its resistance to change in position or motion. This property has been identified as *inertia.* Thus, ***mass*** *is the measure of the inertia of a body* and is responsible for it.

Mass is also responsible for the weight of a body. ***Weight*** *is the measure of the earth's attraction for a body.* An object weighs less at a high altitude and at the bottom of a deep mine shaft than it does at sea level. Although the weight of the object varies, its mass remains unchanged. *The mass of a body is constant.*

The weight of an object can vary from place to place.

Mass is usually measured by comparison with known masses, as shown in Figure 1-8. If the masses of two objects are the same, they have equal weights while in the same location. Thus the mass of an object, when determined by "weighing" it on a platform balance, is sometimes referred to as its "weight." The measurement of mass by "weighing" is common in chemistry and is not confusing if the meanings of the terms *mass* and *weight* are understood. Chemists are primarily concerned with measurements of mass, and the term *mass* will be used here with its proper meaning.

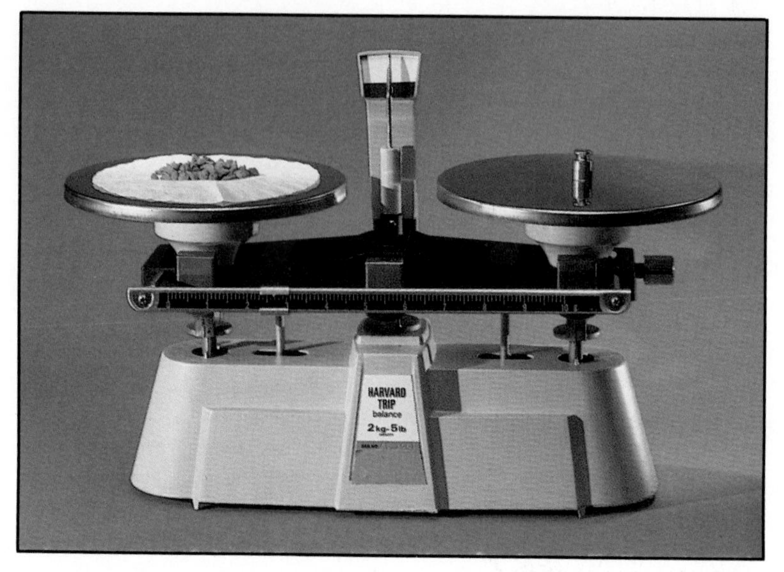

Figure 1-8. The mass of the materials on the left pan is determined to be 12.7 g because it is counterbalanced by the combination of the known 10-g mass on the right pan and the 2.7-g mass indicated by the slider position on the graduated beam.

Figure 1-9. The densities of materials vary. The cork floats because it is less dense than water. The metal key sinks because it is more dense than water. Both mass and volume measurements are required to express density.

1.7 Density Matter occupies space and therefore has *volume*. From everyday experiences, you recognize that materials have different masses. Lead can be described as "heavy" and cork as "light." Such statements have little meaning unless *equal volumes* of lead and cork are being compared. *The mass of a unit volume of a material is called its* ***density.*** Density is expressed by the equation

$$D = \frac{m}{V}$$

Where m is the mass of a material and V is its volume, D is its density. A comparison of the masses of equal volumes of different materials indicates that the densities of different kinds of matter vary. Thus density reveals the *concentration* of matter. See Figure 1-9.

1.8 Phases of matter A block of ice is a *solid*. When it melts, it forms a *liquid*. When liquid water evaporates, it forms a *gas*. Iron is another familiar solid that melts and becomes a liquid if its temperature is high enough. At a temperature near that of the surface of the sun, liquid iron boils to form a gas, iron vapor. Materials exist either in the *solid*, *liquid*, or *gas* phase and may undergo changes of phase under suitable conditions.

Gas, liquid, and solid are also referred to as the "states" of matter.

A block of wood placed on a table retains its shape and its volume. To change either its shape or its volume, considerable external force would have to be applied. A solid does not require lateral (side) support in order to retain its shape. A solid resists compression. ***Solids*** *have both definite shape and definite volume.*

A "phase" is a uniform part of a system separated from other uniform parts by boundary faces.

Suppose water is poured onto a table. It flows out over the surface because a liquid is not rigid. To retain water, lateral support must be provided. If you attempt to pour a quart of milk into a pint bottle, you observe that the volume of milk doesn't change. The pint bottle holds only half of the milk. ***Liquids** have definite volume, but not definite shape.* A liquid assumes the shape of its container.

Ice in water represents a two-phase system, a solid phase and a liquid phase.

When an automobile tire is inflated with air, the air assumes the shape of the tire, which is its container. The tire is actually full of air. If a blowout occurs, however, the air escapes and thus expands in volume. A pint of gas expands and fills all the space when placed in a truly empty quart bottle. ***Gases** have neither definite volume nor definite shape.* A confined gas has both the volume and shape of its container. Models of solid, liquid, and gas phases of matter are shown in Figure 1-10.

From your experience with toy balloons, bicycle tires, and volleyballs, you know that the volume of a gas is affected by heat and pressure. Because of this behavior, a measurement of volume of a gas has meaning only if the temperature and pressure of the gas are specified.

Both liquids and gases are known as *fluids.* Liquid and gaseous materials flow and require vessels to contain them. Although solids are considered rigid, none is perfectly rigid. Given two different solids of the same size and shape, one is likely to be more flexible (less rigid) than the other. Similarly, there are no perfectly fluid materials. Molasses, water, and carbon dioxide all flow as fluids, but certainly at different rates.

Because a liquid has a definite volume, it may have a *free surface.* A free surface is one that is not confined or supported by its container. Thus water can be contained in an open vessel. In ordinary situations, gases must be confined on all sides of their containers. A gas is a fluid that does not have a free surface.

Fluids that cannot exist as liquids having a free surface at ordinary conditions of temperature and pressure are correctly called *gases.* ***Vapor** is the term used for the gaseous phase of materials that normally exist as liquids or solids.* Thus the correct descriptions are *oxygen gas* and *water vapor*.

The structure of a solid is usually well-ordered. Particles of the material are arranged in a regular pattern. They are closely packed and rigidly bound in fixed positions. This gives the solid its definite size and shape.

The particles of a liquid are also closely packed. This explains why the volume of a solid changes very little when the solid melts. The particles of a liquid, however, are not bound in fixed positions. A liquid has fluidity; its structure is less orderly than that of a solid, and it is without shape.

The particles of a gas are widely dispersed (separated) in a random, disorganized fashion. Under ordinary conditions of temperature and pressure, the attraction between such particles is so small that it can be disregarded. A gas is a fluid of very low density.

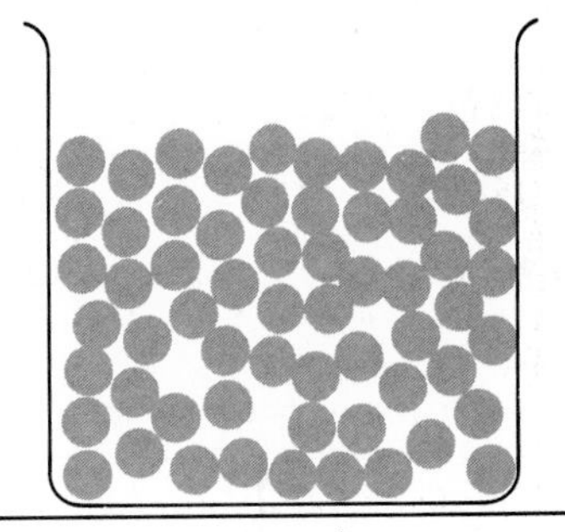

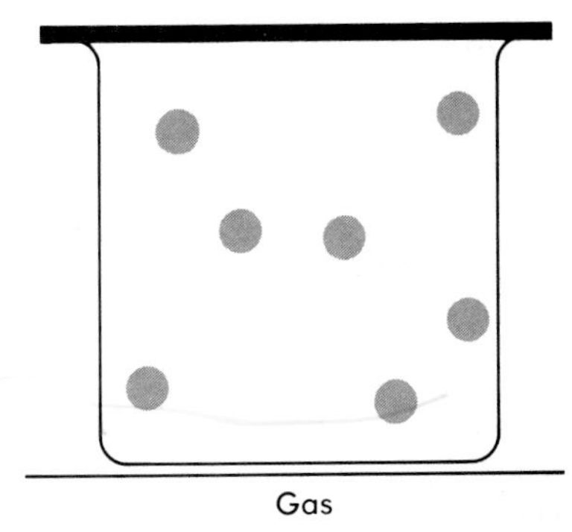

Figure 1-10. Models of solid, liquid, and gas phases of matter.

1.9 Properties of matter Matter can be identified by studying its properties. Many liquids, like water, are colorless. Some colorless liquids have distinctive odors; water is odorless. Water freezes at 0 °C, boils at 100 °C, and has a density of 1 gram per cubic centimeter at 4 °C. Because no other liquid exhibits exactly these properties, they can be used to help identify liquid water. *Properties of matter that are useful in identifying it are called* ***specific*** *(or characteristic)* ***properties.*** The most useful specific properties are those that can be measured quantitatively and expressed by a number of measurement units.

When determining the identity of matter, it is helpful to distinguish between *physical* properties and *chemical* properties. ***Physical properties*** *are those that can be determined directly without altering the identity or composition of a material.* They include *color, odor, solubility, density, hardness, melting* and *boiling points,* and *crystalline* or *amorphous* (noncrystalline) *form.* These properties do not apply equally to all phases of matter. For example, hardness and crystalline form are not properties of fluids. Similarly, odor is of little value in describing many solids. **CAUTION:** *The odor of a nonpoisonous or nonirritating vapor can be tested safely by wafting. The taste test is never used in chemistry.*

Chemical properties *are those that describe the behavior of a material in processes that alter its identity.* Chemical properties include *chemical activity* or behavior, which results in a change of identity. Some materials are chemically *active,* reacting vigorously with certain other materials. Some are *inactive.* They do react with other materials, but not very readily. Still others, said to be *inert,* do not react with other materials under the ordinary conditions for chemical reactivity. Information about whether a material burns and how it reacts with the gases of the air, with water, and with acids is valuable to the chemist.

1.10 The concept of energy Scientists experience the same difficulty in defining energy as they do in defining matter. They know a great deal about energy and how it can be used and controlled, but they cannot define it precisely. ***Energy*** *is described as the capacity for doing work.* It is an action quantity associated with changes in matter. Energy is related to matter, but is not a form of matter. Matter possesses energy.

Work, in a physical sense, involves the application of a force through a distance. Work = force × distance.

The most familiar forms of energy are *mechanical energy* and *heat energy.* Mechanical energy can be of two types: *potential energy* or energy of position, and *kinetic energy* or energy of motion. Water held behind a dam has potential energy because of its elevated position. The potential energy of a given mass of water increases with the height of its position. When water is released from the dam, its potential energy is converted to kinetic energy as it falls to a lower level. The kinetic energy of the mass of falling water increases with the *square* of its speed.

When you double the speed of your automobile, its destructive capacity (its kinetic energy) increases fourfold.

Energy is released as heat when fuels are burned. The heat

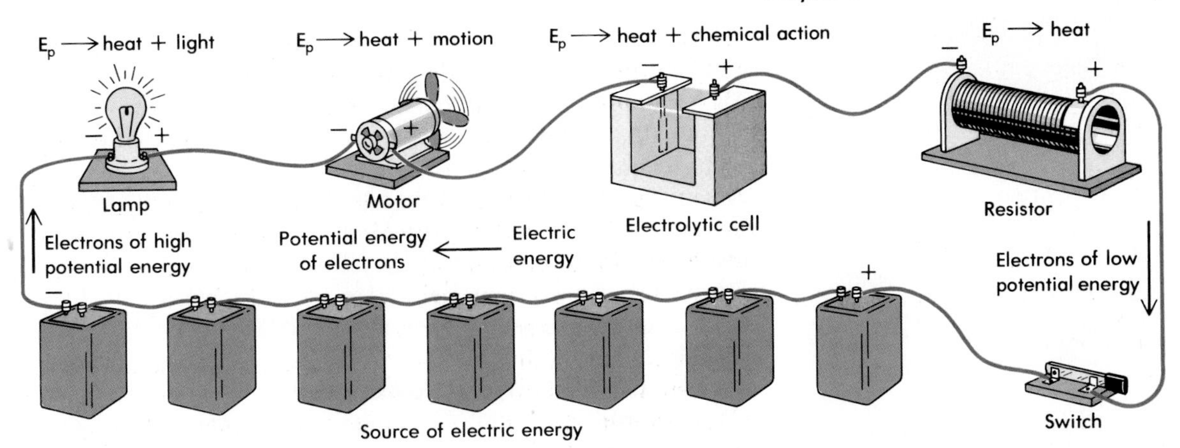

Figure 1-11. Examples of energy transformations.

energy of burning fuels and the kinetic energy of falling water are major sources of industrial and domestic energy.

Other forms of energy are *electric* energy, *chemical* energy, *radiant* energy, and *nuclear* energy. Chemical energy is a basic concern of chemistry. Radio waves, infrared and ultraviolet radiations, visible light, X rays, and gamma rays are examples of radiant energy. Solar energy is a familiar example of radiant energy. The role of solar energy and nuclear energy as basic energy sources is of increasing importance today.

Chemical energy can be thought of as a kind of potential (stored) energy.

One form of energy can be transformed into other forms of energy. The transformation of potential energy to kinetic energy for falling water has already been mentioned. As coal or oil is burned, some of the chemical energy of the fuel and of the oxygen from the air is released as heat energy. The heat energy may be transferred to water and the water converted to steam. The steam then drives a turbine that produces mechanical (rotational) energy. The turbine drives a generator producing electric energy. This electric energy can be transformed into heat and light (radiant energy) in an incandescent lamp. The electric energy can also be transformed into mechanical energy in an electric motor that drives a clock or a locomotive. It is the *transformation* of energy that is usually observed, and the *energy change* during the transformation is what is measured. Some familiar energy transformations are illustrated in Figure 1-11.

1.11 Conservation of matter and energy In 1905, Albert Einstein (1879–1955) suggested that matter and energy are related. His famous equation

$$E = mc^2$$

represents this relationship. In the equation, E is the amount of energy, m is the amount of matter, and c is a constant equal to the speed of light. Many experiments have established the validity of this equation.

Figure 1-12. Solar collectors at Sandia Base, New Mexico and on the roof of a New York City apartment building provide energy.

Matter can be converted to energy and energy to matter. The conversion factor, c^2, has a very large numerical value, because the value of c is of the order of 3×10^8 meters per second. Consequently, a very small amount of matter is converted to a very large amount of energy. This conversion suggests that matter and energy are not two independent physical quantities that can be defined individually. Instead, *matter and energy may represent two different forms of a single, more fundamental, physical quantity.* From their studies of matter and energy, scientists have formulated a law of nature known as ***the law of conservation of matter and energy:*** *Matter and energy are interchangeable, and the total matter and energy in the universe is constant.*

Chemical reactions are always accompanied by energy changes. Energy is either released or acquired. Only in nuclear reactions, involving tremendous quantities of energy, does the amount of matter transformed into energy become significant. Changes in mass are not measurable in ordinary chemical reactions. This fact leads to the following generalization: *In ordinary chemical reactions, the total mass of the reacting materials is equal to the total mass of the products.*

Measurements in Chemistry

1.12 The metric system Progress in science depends on a precise system of measurements based on a set of *universal standards.* The English system used in daily activities presents some problems in scientific use. In a sense, it is a system that just grew. The units have practical size for common use, and the unit subdivisions are based on the convenient practice of halving and quartering. For example, the pound is subdivided into half-pounds, quarter-pounds, and so on. However, the chief disadvantage of the English system is that no single, simple numerical relationship exists between different units of measure. In linear measure, for example, 12 inches = 1 foot, 3 feet = 1 yard, and $5\frac{1}{2}$ yards = 1 rod.

The *metric system,* developed in France near the end of the eighteenth century, is used in scientific work throughout the world. It is in general use in practically all countries except the United States, Great Britain, and other countries with English-speaking backgrounds.

Many American industries that participate in foreign markets have adopted the metric system. This move was necessary to keep them competitive in markets that are almost entirely metric. By the Metric Conversion Act of 1975, the United States was formally committed to encourage, but not to require, the change to metric measurements.

The metric system is a *decimal* system that has a single, simple numerical relationship between units. Computations with mea-

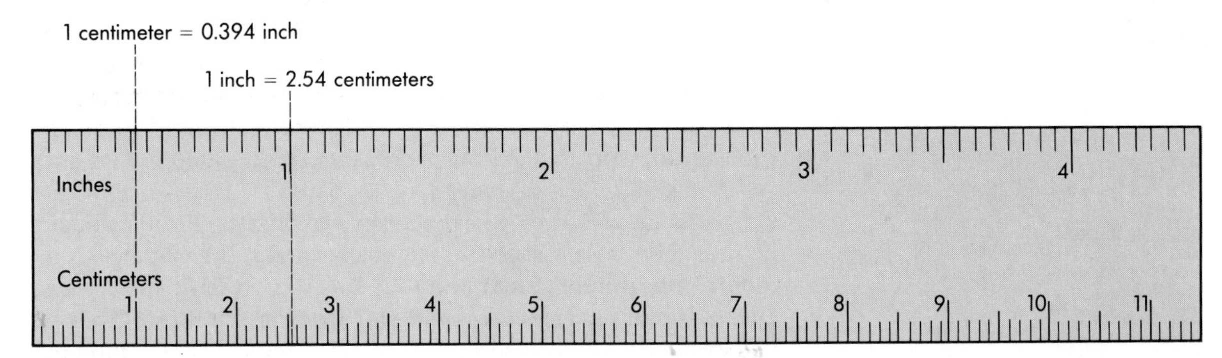

Figure 1-13. The common English/metric rule illustrates one relationship between English and metric measurements.

surements are easily performed. The disadvantage of the metric system in everyday usage is that metric units do not have the practical sizes of English units.

A standard set of prefixes is used to expand the basic metric system and provide a simple, uniform structure for expressing measurements of very large and very small magnitudes. Latin prefixes identify descending values. Examples are *deci-*, 0.1, *centi-*, 0.01, and *milli-*, 0.001. Greek prefixes identify ascending values. Examples are *deka-*, 10, *hecto-*, 100, and *kilo-*, 1000. Common prefixes, their symbols, and factors are listed in Table 1-1. A more extensive list is provided in Appendix Table 1.

Metric measurements are used exclusively in chemistry. You will be concerned mainly with measurements of *length*, *mass*, and *capacity* (volume of liquids and gases). These measurements are expressed in terms of the *meter* (m), the *gram* (g), and the *liter* (L), respectively. The metric prefixes most commonly used are *milli-*, *centi-*, and *kilo-*. Examples are milliliter (mL), centimeter (cm), and kilogram (kg). The relationships shown in Table 1-2 will be used throughout your study of chemistry. You will find it helpful to memorize them.

When the metric system was originally established, it was based on natural standards. The meter was the fundamental unit

Table 1-1
COMMON METRIC PREFIXES

Factor	*Prefix*	*Symbol*
10^6	mega-	M
10^3	kilo-	k
10^2	hecto-	h
10^1	deka-	da
10^{-1}	deci-	d
10^{-2}	centi-	c
10^{-3}	milli-	m
10^{-6}	micro-	μ (Greek mu)

Table 1-2
COMMON METRIC EQUIVALENTS

Length	
10 millimeters (mm)	= 1 centimeter (cm)
100 centimeters (cm)	= 1 meter (m)
1000 meters (m)	= 1 kilometer (km)
Capacity	
1000 milliliters (mL)	= 1 liter (L)
1000 liters (L)	= 1 kiloliter (kL)
Mass	
1000 milligrams (mg)	= 1 gram (g)
1000 grams (g)	= 1 kilogram (kg)

of measure from which all other units were to be derived. Today, for all but the extreme precision of measurements required in very sophisticated scientific research, *the* **standard meter** *is the distance between two parallel lines engraved on a special platinum-iridium bar maintained at 0 °C in the International Bureau of Weights and Measures in Sèvres, France.*

This standard meter bar is a physical instrument that could be damaged, lost, or destroyed. Accordingly, the General Conference on Weights and Measures meeting in Paris in October 1983 defined an *indestructible* meter that can be reproduced in any properly equipped laboratory. This reproducible standard meter is based on the speed of light in a vacuum, which is now defined as a *constant*, exactly 299,792,458 meters per second. *The* **meter** *is now defined as the distance light travels in a vacuum during* 1/299,792,458 *second*. This indestructible standard meter is important for scientists requiring precise distance measurements. There are those, however, for whom it may be enough to know that the meter is a little longer than a yard.

One gram (1 g) was intended to be the mass of one cubic centimeter (1 cm^3) of water measured at its temperature for maximum density. Because the gram is an inconveniently small unit of mass, the *standard* unit of mass today is the *kilogram*. *The* **gram** *is now defined as one-thousandth of the standard kilogram preserved at the International Bureau of Weights and Measures.*

$$1 \text{ g} = 0.001 \text{ kg}$$

The **liter (L)** *is a special name for a cubic decimeter.* One cubic decimeter (1 dm^3) is equal to one-thousand cubic centimeters

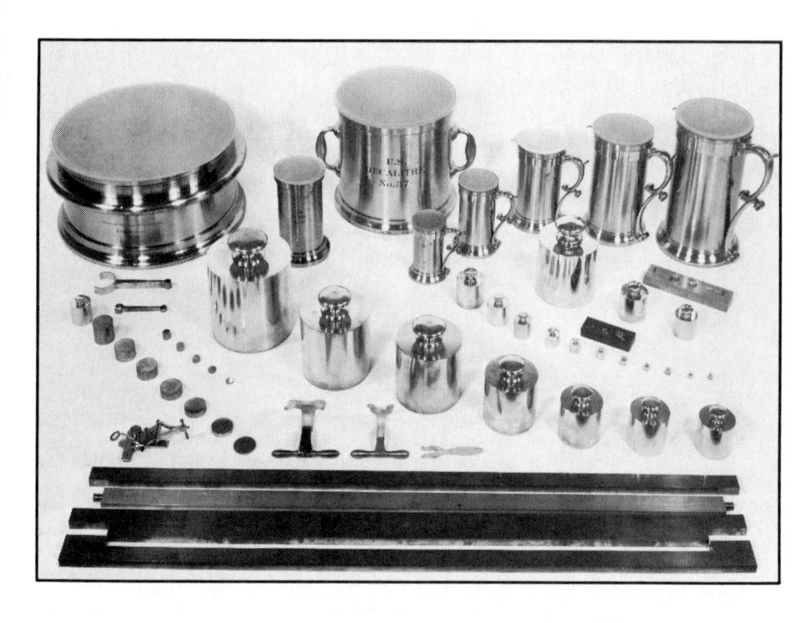

Figure 1-14. To the right are prototype standards of metric length, mass, and volume distributed to the states at the direction of Congress in 1836 and 1866. Prototype kilogram K-20 is pictured above. Between 1965 and 1978, the National Bureau of Standards issued new modern standards to the states to replace these nineteenth century artifacts.

Table 1-3
METRIC-ENGLISH EQUIVALENTS

Metric to English	*English to Metric*
1 cm = 0.3937 in = 0.03281 ft	1 in = 2.54 cm = 0.0254 m
1 m = 39.37 in = 3.281 ft = 1.094 yd	1 ft = 30.5 cm = 0.305 m
1 km = 3281 ft = 0.6214 mi	1 yd = 91.4 cm = 0.914 m
1 cm^3 = 0.0610 in^3 = 0.0000353 ft^3	1 mi = 1609 m = 1.609 km
1 L = 1.06 qt = 0.265 gal = 0.0353 ft^3	1 qt = 946 mL = 0.946 L
1 g = 0.0353 oz = 0.00220 lb	1 oz = 28350 mg = 28.35 g
1 kg = 2.20 lb = 0.00110 tn	1 lb = 453.6 g = 0.4536 kg
1 metric tn (10^3 kg) = 2200 lb = 1.10 tn	1 tn = 907 kg = 0.907 metric tn

(1000 cm^3). Study the cube shown in the schematic diagram in Figure 1-15. Therefore, a one-liter flask holds 1000 cubic centimeters of a fluid when filled. One milliliter (1 mL) is then equivalent to one cubic centimeter.

$$\mathbf{1\ L = 1\ dm^3 = 1000\ cm^3}$$

$$\mathbf{1\ mL = 0.001\ dm^3 = 1\ cm^3}$$

Representative metric units and their English equivalents are listed in Table 1-3.

Fluids are usually measured in flasks and other containers graduated in capacity units. Consequently, the volumes of liquids and gases are most conveniently expressed in milliliter (mL), liter (L), and kiloliter (kL) units. This practice is common in chemistry and will be followed generally in this book. Because water is a universal standard, the remarkable simplicity of the metric system can be recognized in the relationships shown in Table 1-4.

Table 1-4
VOLUME-MASS RELATIONS FOR WATER

1 L of water has a volume of 1000 cm^3 and a mass of 1 kg
1 mL of water has a volume of 1 cm^3 and a mass of 1 g

Figure 1-15. Some comparisons between the English and metric systems.

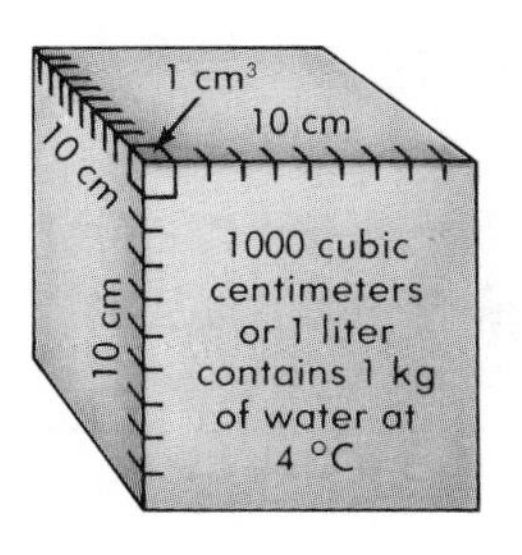

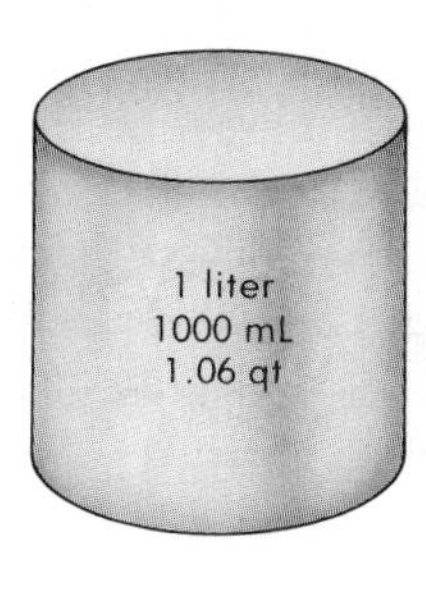

1.13 Measurement dimensions A *physical quantity* is the description of a measurement process. *Length* is a physical quantity. The measurement process is indicated by the definition of ***length*** *as a measured distance or dimension.* The measurement procedure determines the magnitude, or size, of the physical quantity. The magnitude *is described by both a number and a unit,* because the number of measurement units counted depends on the unit used. *The number without the unit is worthless.*

Basic physical quantities are measured in *arbitrary units* established by international agreement. These quantities are the *fundamental quantities* of the measurement system, and their measurement units are *fundamental units. Length (l), mass (m),* and *time (t)* are examples of fundamental physical quantities. Their standardized measurement units are the *meter* (m), the *kilogram* (kg), and the *second* (s), respectively. (Recall that the gram unit of mass is now defined in terms of the standard kilogram.) These quantities, together with *temperature (T),* measured in degrees, and *amount of substance,* measured in moles, are the important fundamental quantities in chemistry.

Symbols for physical quantities are printed in ***italic typeface.***

Symbols for measurement units are printed in **roman typeface.**

The degree is discussed in Section 1.15; the mole is discussed in Section 3.15.

All other physical quantities are *defined* in terms of fundamental quantities. They are known as *derived quantities.* Thus, the magnitudes of many different physical quantities can be determined by the measurements of a few fundamental quantities.

Volume is a derived quantity. The volume (V) of a cube is defined as the product of the lengths (l) of its three-dimensional sides.

$$V = l \times l \times l = l^3$$

The volume may be expressed in km^3, m^3, dm^3, or cm^3, depending upon the unit of length used. All are derived volume units and represent the length unit cubed, l^3. Volume, as a derived quantity, has the dimensions of *length*^{3}, or l^3.

Recall that a cubic decimeter (dm^3) *is called a liter* (L).

Density is also a derived physical quantity. Density (D) is defined in Section 1.7 as the mass (m) per unit volume (V) of an object or a material.

$$D = \frac{m}{V}$$

The density of a material can be determined by measuring its mass and its volume. The densities of solids are expressed in grams per cubic centimeter or kilograms per cubic meter. The densities of liquids are expressed in grams per cubic centimeter or grams per liter (cubic decimeter). The densities of gases are expressed in grams per liter. These derived units—g/cm^3, kg/m^3, g/mL, and g/L—represent a mass unit divided by a length unit cubed. Density has the dimensions of *mass/length*^{3}, or m/l^3.

The mass of 10.0 cm^3 of iron is 78.7 g and the mass of the same volume of mercury is 135 g. The mass of 6.50 L of oxygen is

9.29 g. Their respective densities are

$$D_{\text{iron}} = \frac{78.7 \text{ g}}{10.0 \text{ cm}^3} = 7.87 \text{ g/cm}^3$$

$$D_{\text{mercury}} = \frac{135 \text{ g}}{10.0 \text{ cm}^3} = 13.5 \text{ g/cm}^3$$

$$D_{\text{oxygen}} = \frac{9.29 \text{ g}}{6.50 \text{ L}} = 1.43 \text{ g/L}$$

Density can be expressed only in derived units that represent the dimensions of mass/length^3.

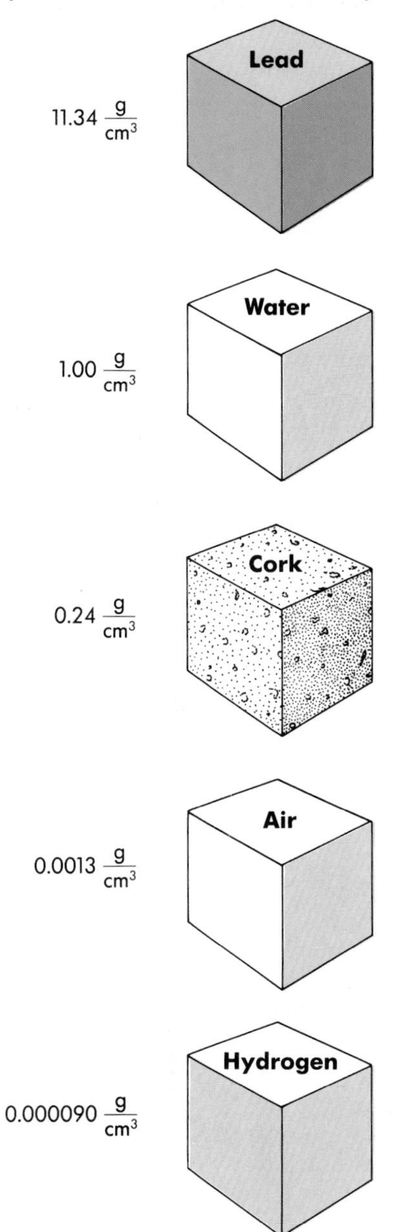

Figure 1-16. Density is a derived physical quantity expressed in units of measure that have the dimensions of mass/length^3.

1.14 Temperature and heat Temperature and heat are different, but related, physical quantities. It is important to understand the subtle distinction between them. Just as you can push an object to estimate its mass or lift it to estimate its weight, you can touch an object to determine its *hotness* or *coldness*. You then describe the sensation by using a term such as hot, warm, cool, or cold. Your sense perceptions of hotness and coldness are used to assign a property called *temperature* to the object.

Temperature sense, while generally useful, may be unreliable under some conditions. If you place one hand in cold water and then in cool water, the cool water feels warm. But if your hand had been in hot water first, the cool water would feel cold.

This experiment suggests that the temperature sensation depends on the transfer of heat energy to the hand or away from it. If a system, a body of matter, has a higher temperature than its surroundings, energy flows away from the system. If the temperature of the system is lower than its surroundings, energy flows to the system. The energy that is *in transit* is called heat. ***Heat** is the energy transferred between two systems as a consequence of the difference in temperature between the systems. The **temperature** of a system is a measure of its ability to transfer heat to or acquire heat from other systems.*

When two systems with different temperatures are in contact, heat energy flows from one to the other. *Temperature* is the property that determines the direction of the heat transfer. The warmer system cools as it gives up heat, and the cooler system warms as it acquires heat. This is true as long as neither system experiences a change of phase. When the temperatures of the two systems are equal, the tendencies for heat to flow between them are equal, and no further net transfer of heat occurs. The two systems are now in *thermal* (heat) *equilibrium*. It follows that *systems in thermal equilibrium have the same temperature.*

At equilibrium, opposing tendencies are equal.

Heat and temperature are different physical quantities that can be sensed qualitatively. They can be determined quantitatively, however, only in terms of measurable quantities that are independent of sense perceptions. Heat is measured as a *quantity of energy,* whereas temperature is measured as the *heat intensity* of matter. A burning match and a campfire might both be at the same *temperature,* but the quantities of *heat* given up are quite different. A small warm object will transfer heat to a large cool object whose total heat content is greater than that of the small warm object.

Heat flows from a region of high heat intensity to a region of low heat intensity.

1.15 Measuring temperature Some properties of matter vary with temperature and, therefore, can be used to measure temperature. For example, most materials expand when warmed and contract when cooled. The most familiar instrument for measuring temperature, the mercury thermometer, is based on the nearly linear expansion and contraction of liquid mercury with changing temperature.

In Section 1.13, temperature was described as a fundamental physical quantity. Its measurement unit, the *degree,* is an arbitrary unit determined by international agreement.

Mercury thermometers for scientific use are calibrated in the *Celsius* temperature scale. This scale was devised by the Swedish astronomer Anders Celsius (1701–1744). Celsius established his thermometer scale by defining two *fixed points:* the normal freezing point of water, the *ice point,* as 0 °C, and the normal boiling point of water, the *steam point,* as 100 °C. He divided the interval between the ice point and the steam point into 100 equal parts. Each part represents a temperature change of 1 °C. By extending the same scale divisions beyond the two fixed points, he could measure temperatures below 0 °C and above 100 °C, as illustrated in Figure 1-17.

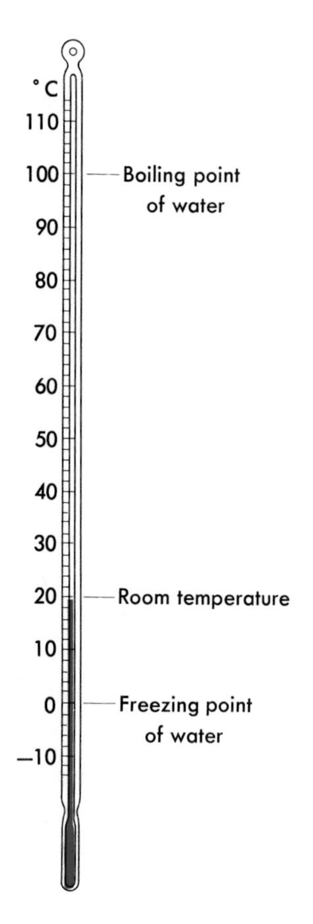

Figure 1-17. The Celsius thermometer scale is used in science. Compare the labeled temperatures with those on the common Fahrenheit thermometer scale.

In modern thermometry, an absolute (thermodynamic) temperature scale is constructed by defining a single standard fixed point. This temperature scale, called the *Kelvin scale,* is described in Section 10.10.

1.16 Measuring heat As heat is transferred *to* a material, the temperature of the material *increases;* as heat is transferred *from* the material, its temperature *decreases.* This is true only if neither heat-transfer process is accompanied by a change in phase. Heat quantities are measured during the heat-transfer processes associated with *thermal* properties of matter. Some important thermal properties of matter are *heat of solution* (Chapter 13), *heat of combustion, heat of formation,* and *heat of reaction* (Chapter 20). The measurement of thermal properties is called *calorimetry.*

Thermal: pertaining to heat and also to temperature.

The unit of heat energy commonly used in calorimetry is the *calorie. The* ***calorie* (cal)** *is traditionally defined as the quantity of heat required to raise the temperature of 1 gram of water through 1 Celsius degree.*

The calorie is a very small unit of heat and is sometimes inconvenient to use. A larger unit, the *kilocalorie*, is often used in calorimetry measurements. *The* **kilocalorie (kcal)** *is the quantity of heat required to raise the temperature of 1 kilogram of water through 1 Celsius degree.*

$$1 \text{ kcal} = 10^3 \text{ cal}$$

The kilocalorie is the "large Calorie" used for calorie counting in dietetics.

Dietetics: the application of principles of nutrition.

The calorie, as traditionally defined, lacks good measurement precision because the heat required to raise the temperature of one gram of water through one Celsius degree varies slightly with the water temperature. The calorie is now defined as an auxiliary heat unit in terms of a dimensional unit of energy, E.

The kinetic energy, E_K, of a moving body is determined by the mass of the body and its velocity (speed) according to the equation

$$E_K = \frac{1}{2} mv^2$$

Speed is determined by measuring the distance (length) traveled and the time required. Kinetic energy then has the dimensions of mass × (length ÷ time)^2, or ml^2/t^2. Other forms of energy can be shown to have the same dimensions as kinetic energy.

If mass is measured in kilograms, length in meters, and time in seconds, the dimensional unit for energy is kg m^2/s^2. Generally, complex dimensional units are assigned special names. This dimensional unit, kg m^2/s^2, is called a *joule* (J).

It is easier to say or write "10 joules" than "10 kilogram meter square per second square."

$$1 \text{ J} = 1 \frac{\text{kg m}^2}{\text{s}^2}$$

For example, if

$$E = 15.2 \text{ kg m}^2/\text{s}^2,$$

then

$$E = 15.2 \text{ J}$$

is an equivalent expression.

The calorie can now be defined as an auxiliary unit of heat energy in terms of the joule. *One* **calorie** *is equivalent to 4.19 joules of heat energy.*

Look up the value of the mechanical equivalent of heat in Appendix Table 3.

$$1 \text{ cal} = 4.19 \text{ J}$$

$$1 \text{ kcal} = 4.19 \text{ kJ}$$

In Table 1-5 the energy values of some common foods are given in both kilocalories (nutritional Calories) and kilojoules. A comparison of the kcal and kJ columns may be helpful in visualizing the relative magnitudes of these two heat units.

Table 1-5
AVERAGE ENERGY VALUES OF SOME COMMON FOODS

Food	*Measure*	*Energy* (kcal)	(kJ)
apple	1 large	100	419
bacon	2 slices	95	398
banana	1 medium	130	545
bread	1 slice	60	250
cottage cheese	25 cm^3	93	390
cupcake	1 average	200	838
custard	100 mL	125	524
hamburger	113 g (¼ lb) patty	420	1760
hot dog	1 average	138	578
honey	10 mL	67	280
ice cream	100 g	185	775
orange juice	200 mL	83	350
peach	1 medium	50	210
peanut butter	32 g (1 tbsp)	190	796
potato	1 medium	100	419
skim milk	200 mL	74	310
sugar (sucrose)	4 g (1 tsp)	16	67
whole milk	200 mL	133	557

Heat energy measurements are converted from calories to joules and from joules to calories as follows:

$$23.8\ \cancel{\text{cal}} \times \frac{4.19\ \text{J}}{\cancel{\text{cal}}} = 99.7\ \text{J} \qquad 25.1\ \cancel{\text{kcal}} \times \frac{4.19\ \text{kJ}}{\cancel{\text{kcal}}} = 105\text{kJ}$$

$$99.7\ \cancel{\text{J}} \times \frac{\text{cal}}{4.19\ \cancel{\text{J}}} = 23.8\ \text{cal} \qquad 105\ \cancel{\text{kJ}} \times \frac{\text{kcal}}{4.19\ \cancel{\text{kJ}}} = 25.1\ \text{kcal}$$

1.17 Uncertainty in measurement The experimental process is an essential part of chemistry, and reliable measurement information is an essential part of the experimental process. Unfortunately, *the measurement of any physical quantity is subject to some uncertainty*. If a measurement is to have much worth, it must include some indication of its *reliability*. Consequently, the complete expression of a measured quantity must include the *number value*, the *measurement unit* used, and some indication of *how reliable the number is*.

Corollary: No measurement of a physical quantity is absolutely certain.

Contributions to the uncertainty in measurements of physical quantities accrue from *limitations in accuracy* and *limitations in precision* inherent in all measurement processes. In common usage, *accuracy* and *precision* are practically synonymous terms. In regard to measurements, however, they have distinctly different meanings. If you wish to become competent in making measurements and interpreting measurement data, you must understand this distinction.

1. **Accuracy** *denotes the nearness of a measurement to its accepted value.* It refers to the correctness of measurement data. Accuracy is expressed in terms of *absolute* or *relative error.*

An *absolute error,* $\mathbf{E_a}$, is the difference between an *observed* (measured) value, **O,** and the *accepted* value, **A,** of a physical quantity.

$$E_a = O - A$$

In laboratory experiments, absolute errors are referred to as *experimental errors.*

If the absolute error, $\mathbf{E_a}$, is compared to the accepted value, **A,** the resulting ratio is called the *relative error,* $\mathbf{E_r}$. It is ordinarily expressed as a percentage.

$$E_r = \frac{E_a}{A} \times 100\%$$

In laboratory experiments, relative errors are referred to as *percentage errors.*

Observe that the accepted value of a measured quantity is used to determine both the absolute error and the relative error. Thus, the accuracy of a measurement can be determined *only* if the accepted value of the measurement is known.

2. **Precision** *is the agreement between the numerical values of a set of measurements that have been made in the same way.* Precision refers to the reproducibility of measurement data or to the amount of measurement detail. It relates to the ± uncertainty in a set of measurements. *Precision conveys nothing about accuracy.* Good measurement precision can be obtained from an instrument even though it may introduce appreciable error in the measurement. The distinction between measurement error and measurement precision is illustrated in Figure 1-18.

The precision of measurement data is expressed in terms of *absolute* or *relative deviation.* One simple form of deviation, the *absolute deviation,* of a set of identical measurements shall be considered as follows:

An *absolute deviation,* $\mathbf{D_a}$, is the difference between an observed value, **O,** and the arithmetic mean, **M,** for a set of several identical measurements.

$$D_a = O - M$$

A set of three *identical* samples of potassium chlorate are decomposed to determine the mass of oxygen in each sample. The results are recorded in Table 1-6. Assuming that there is an equal chance for the individual mass measurements to be either high or low, the average (mean) mass for the set is taken as the "best" value.

The absolute deviations of the individual values from the average value are then calculated using the expression for $\mathbf{D_a}$

The value of **O** *can be either higher or lower than that of* **A.**

Poor accuracy/Good precision

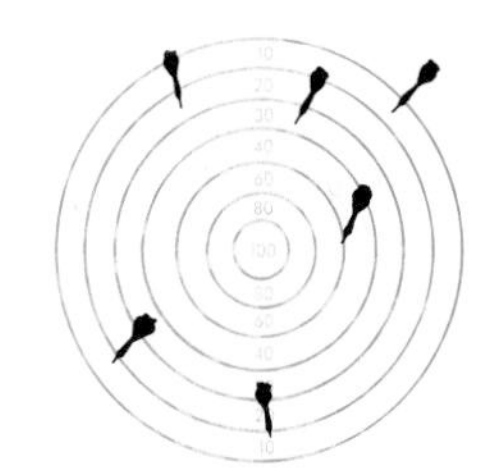

Poor accuracy/Poor precision

Good accuracy/Good precision

Figure 1-18. An illustration of the distinction between accuracy and precision.

TABLE 1-6
AVERAGE DEVIATION OF A SET OF MEASUREMENTS

Sample	*Mass of oxygen*	*Deviation*
1	3.92 g	0.02 g
2	3.97 g	0.03 g
3	3.93 g	0.01 g
Average	3.94 g	0.02 g

previously given. The average of these deviations, $\mathbf{D}_{a(av)}$, provides a measure of the precision of the experiment. The uncertainty in measurement is ±0.02 g. The mass of oxygen derived from the experimental data can be expressed as 3.94 ± 0.02 g.

An understanding of the distinction between measurement accuracy and measurement precision is necessary for using measurement instruments properly and processing measurement data correctly.

Measurement detail relates to good precision.

You will use instruments for measuring length, mass, time, temperature, etc. These instruments can provide measurement detail within their design limits. Instruments of good quality that provide large measurement detail (good precision) are also designed for small error tolerance (good accuracy). Such instruments are periodically recalibrated against dependable standards to maintain their proficiency.

Tolerance: the maximum allowable error in the scale graduations of a measuring instrument.

1.18 Significant figures In Section 1.17 you learned that all measurements are subject to some uncertainty. The mass of oxygen recovered from each of three identical samples of potassium chlorate was shown to be 3.94 g. The uncertainty in this measurement set was expressed as a deviation of ±0.02 g, and the mass of oxygen per sample was expressed as 3.94 ± 0.02 g.

The concept of uncertainty in measurements gives meaning to the method of significant figures.

This deviation of ±0.02 g indicates that there is uncertainty in the *second decimal place* of the recorded measurement. The digit occupying this place could be as low as 2 and as high as 6. That is, the actual mass of oxygen lies between 3.92 g and 3.96 g. Furthermore, *all digits to the left of this uncertain digit are certain*. Of course, any digit carried to the right of this uncertain digit makes no contribution to the measurement information. Its presence would, instead, create a false impression of the measurement reliability.

For all ordinary uses of measurement data in chemistry, it is sufficient to recognize that *the last digit of the measurement expression is uncertain*. In the example given, the magnitude of this uncertainty, ±0.02 g, can be omitted and the measurement can be recorded simply as 3.94 g *with the knowledge that the last digit, 4, is uncertain*. This measurement, 3.94 g, consists entirely of digits that have *physical significance;* they are called *significant figures*. **Significant figures** *in a measurement expression comprise all digits that are known with certainty, plus the first digit that is uncertain*. The position of the decimal point is irrelevant.

The method of significant figures provides a simple and convenient way of expressing measured quantities. The ± uncertainty in the last significant digit is not required and, therefore, does not have to be calculated. It is necessary only to keep track of the uncertain digit in calculations involving measurement data and to recognize that this uncertain digit is the last digit retained in the result of the calculations.

Figure 1-19. Typical new-generation laboratory balance that is readable to 1 cg. Just place an item on the pan and read its mass to 1 cg in an instant.

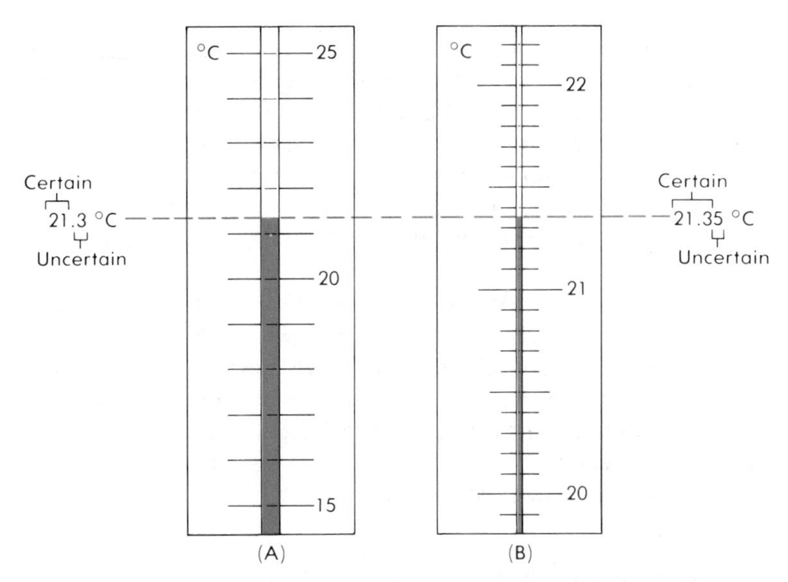

Figure 1-20. Sections of two thermometer temperature scales. Scale (A) has 1-degree graduations and scale (B) has 0.1-degree graduations. The temperature readout from scale (A) has three significant figures and the readout from scale (B) has four significant figures. Observe that the estimated digit in each readout makes a significant contribution to the temperature information.

In your use of measuring instruments, the last digit (the uncertain digit) of a measurement can ordinarily be estimated one place beyond the smallest graduation of the readout scale. For example, the temperature readout from thermometer **A** of Figure 1-20 consists of 3 significant figures; that from thermometer **B** consists of 4 significant figures. Observe that the uncertain digit in each readout *represents an estimate beyond which no further significant contribution to the measurement precision is possible.* The significant measurement detail is limited to the one estimated place.

Suppose you wish to determine the volume of a metal block. Your measuring instrument is a centimeter rule having 1-mm scale divisions, as shown in Figure 1-13. You find the sides to be 3.5**4** cm, 4.8**5** cm, and 5.4**2** cm, estimating the value of the last digit in each case. Observe that the last digit in each measurement represents a reasonable estimate to the nearest 0.1 mm. You can have no idea of the digit that should occupy the next decimal place. Each measurement consists of two *certain figures* and one *uncertain figure,* thus *three significant figures.*

All uncertain digits in this series of calculations are shown in ***bold typeface.***

The area of one surface is

$$3.5\mathbf{4}\text{ cm} \times 4.8\mathbf{5}\text{ cm} = 17.\mathbf{1690}\text{ cm}^2$$

The product of any number multiplied by a doubtful figure is also doubtful. Therefore, only one such doubtful figure can be carried. The result is rounded off to 17.**2** cm^2.

The volume of the block is then calculated.

$$17.\mathbf{2}\text{ cm}^2 \times 5.4\mathbf{2}\text{ cm} = 93.\mathbf{224}\text{ cm}^3$$

Again, the result is properly rounded off to 93.**2** cm^3, the volume of the metal block. Had all of the doubtful figures been retained throughout the computations, the volume would have been expressed as 93.**055980** cm^3. Obviously, precision to a millionth of a cubic centimeter cannot be obtained from a centimeter scale having 0.1-cm graduations. *Assuming more decimal places does not improve the measurement accuracy or precision.*

In the example given above, the number of significant figures in both the measurements and computation results are easily recognized because all figures are nonzero digits. It is not as easy to determine *when* zeros in a measurement expression are significant. For example, the mean distance to the moon to *six* significant figures is 384,558 km. The distance is more commonly expressed as 385,000 km, precise to *three* significant figures. The three zeros that follow the digit 5 merely serve to locate the (understood) decimal point. Similarly, a measured length of 0.00531 cm is precise to *three* significant figures. Here again the zeros are used to locate the decimal point. However, the measurements 104.06 m and 100.60 m each contain *five* significant figures. The question naturally arises: When are zeros significant?

A person using a measuring instrument knows, of course, whether a zero in the readout represents a significant contribution to the measurement information or a method of properly locating the decimal point. This information can be communicated to others by following accepted rules for identifying zeros as significant figures in measurement data. The rules concerning zeros in a measurement expression are stated and illustrated in Table 1-7. These rules are followed throughout this book.

Uncertainty is inherent in any measurement procedure. The accuracy of a measurement can be determined if the accepted value is known. The precision can always be expressed by the proper use of significant figures. Judgments based on precision alone must be considered with caution.

Table 1-7
ZEROS IN MEASUREMENT EXPRESSIONS

Rule	*Measurement expression*	*Significant figures*
1. All nonzero digits **are** significant.	127.34 g	5
2. All zeros between two nonzero digits **are** significant.	120.007 m	6
3. Zeros to the right of a nonzero digit, but to the left of an understood decimal point, **are not** significant **unless** specifically indicated as significant by a bar placed above the rightmost such zero that is significant.	109,000 km	3
	109,$\bar{0}$00 km	4
	109,0$\bar{0}$0 km	5
	109,00$\bar{0}$ km	6
4. All zeros to the right of a decimal point but to the left of a nonzero digit **are not** significant.*	0.00406 kg	3
5. All zeros to the right of a decimal point and to the right of a nonzero digit **are** significant.	0.04060 cm	4
	30.0000 s	6

*The single zero conventionally placed to the left of the decimal point in such an expression is never significant.

CAREERS IN CHEMISTRY

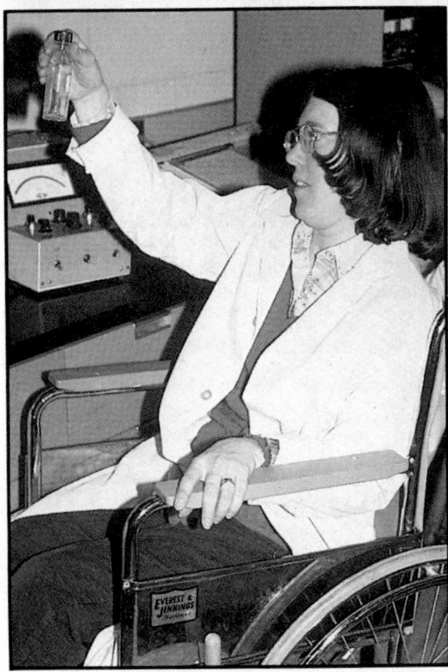

The people pictured here are applying the principles and concepts of chemistry in their daily work as research chemists, biochemists, organic chemists, industrial chemists, environmental chemists, agricultural chemists, laboratory technicians, and teachers. Their efforts will increase our understanding of the world and supply the research and technology vital to our future progress.

Practice Problems

In each of the following measurements, *(a)* determine the number of significant figures, and *(b)* identify the applicable rules (Table 1-7).

1. 1.030 cm *ans.* 4; 2, 5
2. 2,074,000.0 s *ans.* 8; 2, 5
3. 0.00080 kg *ans.* 2; 4, 5
4. 0.00001 g *ans.* 1; 4
5. 367.52 g *ans.* 5; 1
6. 0.00320 m *ans.* 3; 4, 5
7. 601,500 km *ans.* 4; 2, 3
8. $1,570,50\bar{0}$ cm/s *ans.* 6; 2, 3
9. $47,0\bar{0}0$ kg *ans.* 4; 3
10. 0.1020 L *ans.* 4; 2, 5

1.19 Scientific notation Scientific work often involves the use of numbers that are extremely large or exceedingly small. The speed of light is approximately 30,000,000,000 cm/s. The mass of the earth is approximately 6,000,000,000,000,000,000,000,-000,000 g. An electron's mass is 0.000,000,000,000,000,000,-000,000,000,910,9534 g. The wavelength of yellow light is about 0.000059 cm. These numbers are inconvenient to write, difficult to read, and cumbersome to use in calculations.

To express such numbers in a simpler way, they can be written as powers of 10. This *exponential notation,* commonly referred to as *scientific notation,* has the form

$$M \times 10^n$$

where M is a number between 1 and 10 (having one digit to the left of the decimal point) and n is an integer. For example, the speed of sound in water is about 1500 m/s. Written in scientific notation, it has the form

The set of integers consists of positive integers, negative integers, and zero.

$$1500 = 1.5 \times 1000 = 1.5 \times 10^3$$

and

$$1500 \text{ m/s} = 1.5 \times 10^3 \text{ m/s}$$

The unusual quantities given above in positional (ordinary) notation are written in scientific notation in Table 1-8. To change a number from positional notation to scientific notation:

1. Determine M by moving the decimal point in the original number to the left or the right so that only one nonzero digit is to the left of it.

2. Determine n by counting the number of places the decimal point has been moved. If moved to the left, n is positive; if moved to the right, n is negative.

The laws of exponents apply in computations involving numbers expressed in scientific notation.

Laws of exponents are reviewed in Section 1.20.

When a measurement is written in the form $M \times 10^n$, all digits, zero and nonzero, expressed in M *are significant*. For exam-

Table 1-8
NUMBERS IN SCIENTIFIC-NOTATION FORM

30,000,000,000 cm/s	$= 3 \times 10^{10}$ cm/s
6,000,000,000,000,000,000,000,000,000 g	$= 6 \times 10^{27}$ g
0.000,000,000,000,000,000,000,000,000,910,9534 g	$= 9.109534 \times 10^{-28}$ g
0.000059 cm	$= 5.9 \times 10^{-5}$ cm
$100\bar{0}$ mL	$= 1.000 \times 10^3$ mL
1000 mL	$= 1 \times 10^3$ mL
$10\bar{0}0$ mL	$= 1.00 \times 10^3$ mL
10050 mL	$= 1.005 \times 10^4$ mL

ple, the distance $402{,}5\bar{0}0$ m is precise to five significant figures. Expressed in scientific notation, it becomes 4.0250×10^5 m. If rounded off to three significant figures, it becomes 4.02×10^5 m. The distance from the earth to the sun is 149,740,000 km, or 1.4974×10^8 km. Rounded off to three significant figures, the distance is $15\bar{0}{,}000{,}000$ km, or 1.50×10^8 km. Because *only significant figures* comprise *M*, the implied precision of the measurement can be determined at a glance.

See Table 1-9 for rounding-off rules.

Practice Problems

1. Express each measurement to *3 significant figures* in scientific notation.
   (*a*) 9,454,500,000,000 km — *ans.* 9.45×10^{12} km
   (*b*) 22.4136 L/mole — *ans.* 2.24×10^1 L/mole
   (*c*) 0.0032416 m — *ans.* 3.24×10^{-3} m
   (*d*) 0.0000000140 cm — *ans.* 1.40×10^{-8} cm
   (*e*) $2{,}210{,}\bar{0}00{,}000$ beats — *ans.* 2.21×10^9 beats
2. Express each measurement in positional notation.
   (*a*) 6.022045×10^{23}/mole
   ans. 602,204,500,000,000,000,000,000 mole
   (*b*) 1.05602×10^7 kg — *ans.* 10,560,200 kg
   (*c*) 2.7316×10^2 K — *ans.* 273.16 K
   (*d*) 2.2×10^4 m/hr — *ans.* 22,000 m/hr
   (*e*) 1.4×10^{-8} cm — *ans.* 0.000000014 cm

1.20 Operations with significant figures The results of calculations involving measurements of physical quantities can be no more precise than the measurements themselves. Certain rules must be observed when performing calculations so that the results do not imply greater precision than the measurements. The following rules are valid for most purposes and should be used unless otherwise indicated.

"Rules of thumb" are generally useful, but they do not cover all cases.

1. *Addition and subtraction.* Recall that the rightmost significant figure in a measurement is uncertain. Therefore, *the rightmost significant figure in a sum or difference occurs in the leftmost place at which the doubtful figure occurs in any of the measurements involved.* The following addition exercise demonstrates this rule.

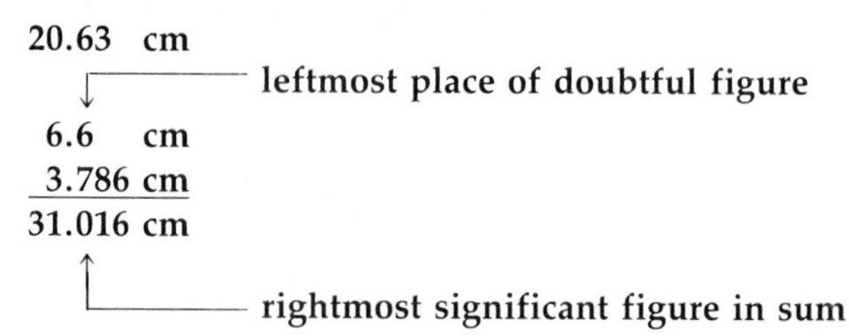

The sum 31.016 cm is then rounded off to allow only the first doubtful figure and is recorded as 31.0 cm. This is the best expression for the sum in this addition problem. *Before adding or subtracting measurements expressed in scientific notation, all terms must be adjusted to the same power of ten.*

2. *Multiplication and division.* Three points discussed in Section 1.18 must be remembered in multiplication and division operations involving measured quantities: *(a)* The rightmost significant figure in a measurement is uncertain. *(b)* The product of any digit multiplied by an uncertain digit is also uncertain. *(c)* Only one uncertain digit is retained in the result. Therefore, *the product or quotient is precise to the number of significant figures contained in the least precise factor.* The result in either operation is rounded to the same number of significant figures contained in the factor having the least number of significant figures.

In each of the following examples, typical computation results obtained by slide rule and by calculator are given. These results are then adjusted to the proper number of significant figures representing the numerical product or quotient. The conventions for rounding off computation results are illustrated in Table 1-9.

Multiplication: 9.25 m × 0.52 m × 11.35 m
By slide rule, the product is 54.6 m^3.
By calculator, the product is 54.5935 m^3.
Observe that the factor 0.52 m has only two significant figures. The product can have no greater precision than this least precise factor. Therefore, the result is properly rounded off to two significant figures.

$$9.25 \text{ m} \times 0.52 \text{ m} \times 11.35 \text{ m} = 55 \text{ m}^3$$

Division: 69.48 m ÷ 3.62 s
By slide rule, the quotient is 19.2 m/s
By calculator, the quotient is 19.19337 m/s
The result is properly expressed to three significant figures.

$$69.48 \text{ m} \div 3.62 \text{ s} = 19.2 \text{ m/s}$$

Table 1-9
ROUNDING OFF THE RESULT OF A COMPUTATION

Rule 1. Choose the required number of significant figures that is closest to the number to be rounded off.
Rule 2. When two choices are equally close, choose the one ending in an even digit.

Computation result	*Rounded-off result*		
	4 sig fig	*3 sig fig*	*2 sig fig*
1.6666 g	1.667 g	1.67 g	1.7 g
27.155 s	27.16 s	27.2 s	27 s
2.3145 kg	2.314 kg	2.31 kg	2.3 kg
150.33 °C	150.3 °C	150 °C	150 °C
8.5455 g/cm^3	8.546 g/cm^3	8.55 g/cm^3	8.5 g/cm^3
30.050 cm	30.05 cm	30.0 cm	30 cm

In multiplication or division operations with measurements expressed in scientific notation, the expressions for M are handled according to the rule above. Keep in mind that all digits are significant. The laws of exponents govern the multiplication and division of 10^n terms. In multiplication, the exponents are added.

$$10^3 \times 10^4 = 10^{3+4} = 10^7$$
$$10^6 \times 10^{-2} = 10^{6+(-2)} = 10^4$$
$$10^4 \times 10^{-6} = 10^{4+(-6)} = 10^{-2}$$

In division, the exponent of the divisor is subtracted from the exponent of the dividend. (Simply change the sign of the exponent of the divisor and add exponents.)

$$10^3 \div 10^2 = 10^{3-2} = 10^1$$
$$10^4 \div 10^{-3} = 10^{4+3} = 10^7$$
$$10^{-5} \div 10^2 = 10^{-5-2} = 10^{-7}$$

Number expressions that are not measurements should not be interpreted as having limited significance. When such exact or defined values are included in a computation with measurements, they have no influence on the number of significant figures in the result. For example, the freezing point of water is *defined* as 0 °C. It is exactly zero degrees. A thermometer readout, on the other hand, is a measurement and is subject to some uncertainty.

A defined number value is not a measurement.

1.21 Operations with units You will recall from Section 1.13 that a system of measurement is based on a few selected physical quantities—such as *length, mass, time,* and *temperature*—that are measured in arbitrary units. All other quantities are derived from the fundamental quantities and are measured in derived units. Speed (or velocity) is a derived quantity. The average speed of a moving object is defined as the ratio of the distance (length) traveled to the elapsed time of travel.

$$\textbf{speed} = \frac{\textbf{length}}{\textbf{time}} = \frac{l}{t}$$

Suppose an object travels 576 m in 20.0 s. Its average speed will be

$$\textbf{speed} = \frac{l}{t} = \frac{\textbf{576 m}}{\textbf{20.0 s}} = \textbf{28.8 m/s}$$

Speed is expressed in dimensions of length/time.

Based on the definition of speed, its *dimensions* are length/time, or l/t. Observe that the dimensions of speed are also specified in terms of the derived unit m/s. Speed *can be expressed* in any derived unit *that represents its dimensions of length/time.* Speed *cannot be expressed* in any derived unit *that does not represent its dimensions of length/time.* This concept is fundamental in computations involving the dimensions of physical quantities.

A measurement is expressed as a significant number of some kind of dimensional unit: 12.5 g, 6.72 cm, 10.0 s, 42.1 °C, 0.09 g/L, 28.8 m/s. For a mathematical equation to represent the correct solution to a problem, the equation must be *dimensionally correct.* That is, *both sides of the equation must have the same dimensions.* For example, density is defined as the ratio of mass/volume and is shown in Section 1.13 to have the dimensions of mass/length^3. The dimensional unit can be g/cm^3. If,

$$D = \frac{m}{V} \qquad \text{then,} \quad \frac{\mathbf{g}}{\mathbf{cm^3}} = \frac{\mathbf{g}}{\mathbf{cm^3}}$$

$$m = D \times V \quad \text{then,} \quad \mathbf{g} = \frac{\mathbf{g}}{\cancel{\mathbf{cm^3}}} \times \cancel{\mathbf{cm^3}} \quad \text{and} \quad \mathbf{g} = \mathbf{g}$$

$$V = \frac{m}{D} \qquad \text{then,} \quad \mathbf{cm^3} = \frac{\mathbf{g}}{\mathbf{g/cm^3}}$$

Terms common to both numerator and denominator are "canceled" by dividing both by the common term.

$$\text{or,} \quad \mathbf{cm^3} = \cancel{\mathbf{g}} \times \frac{\mathbf{cm^3}}{\cancel{\mathbf{g}}} \quad \text{and} \quad \mathbf{cm^3} = \mathbf{cm^3}$$

Observe that dimensional units *common to both numerators and denominators* are "canceled" and removed from the expressions in the same familiar way as numerical factors.

Only *identical* units cancel, as the above operations illustrate. If mass is given in gram units in the numerator of an expression and in kilogram units in the denominator, the units do not cancel directly. For the mass units to cancel, either the measurement in grams must be converted to kilograms, or the measurement in kilograms must be converted to grams. A valid *conversion factor* that can be introduced into the expression for the purpose of making the desired unit conversion is needed.

Metric units are related as powers of 10 by the system of prefixes. One kilogram is *exactly* one thousand grams; one centimeter is *exactly* one one-hundredth meter; one second is *exactly* one thousand milliseconds. Such a unit relationship can be expressed as a fraction and inserted into an equation as a conversion factor. The form of this conversion factor must allow for the cancellation of the unit to be converted and the retention of the unit required.

Exact conversion factors derived from defined relationships are not involved in significant-figure operations.

Suppose a length l is measured to three significant figures as 1.30 m. How is l expressed in centimeters? By definition, 1 m is exactly 100 cm. A conversion factor with the cm unit in the numerator (unit required) and the m unit in the denominator (unit to be canceled) can be assembled.

(correct) $$l = 1.30\ \cancel{\text{m}} \times \frac{100\ \text{cm}}{\cancel{\text{m}}} = 13\bar{0}\ \text{cm}$$

The result, $13\bar{0}$ cm, remains precise to three significant figures.

Had the conversion factor been assembled in the inverted form, the meter units would not cancel, the answer unit would not be in centimeters, and the *numerical result would not be a correct answer.*

The conversion factor, 100 cm/m, does not have limited significance.

(incorrect) $$l = 1.30\ \text{m} \times \frac{\text{m}}{100\ \text{cm}} = \frac{0.0130\ \text{m}^2}{\text{cm}}$$

Compare these two examples. Both forms of the conversion factor are valid expressions. The second form, m/100 cm, however, is the conversion factor for converting centimeters to meters. Observe that

$$\cancel{\text{m}} \times \frac{\text{cm}}{\cancel{\text{m}}} = \text{cm} \quad \text{and} \quad \cancel{\text{cm}} \times \frac{\text{m}}{\cancel{\text{cm}}} = \text{m}$$

The conversion factor used in the first (correct) example not only yields the intended unit in the answer *but also sets up the arithmetic computation that gives the correct numerical answer.*

These simple examples of the use of conversion factors reveal an important strategy for solving problems involving measurements and their dimensional units. *In any numerical equation, the units associated with the various quantities are treated algebraically and are canceled, combined, etc., just like factors in the equation.*

Measurement units are treated as factors in equations.

When the numerical expression for the solution to a problem has been assembled, the units in the expression should be "solved" for the answer unit *before* the arithmetic is done. If the answer unit is dimensionally correct for the physical quantity required, such as g/cm^3 for density or m/s for speed, this indicates that the arithmetic computation should yield the correct numerical answer. On the other hand, unit operations that yield an incorrect answer unit signal that the expression is incorrect and cannot give a correct answer to the problem.

The factor-label method uses unit operations to set up the arithmetic in problem solutions.

This *first* unit-operations step is a simple and rapid analysis of the dimensional character of the solution setup for the problem. It is the key step in the problem-solving technique called *dimensional analysis*. This technique is often called the *factor-label* method. It is the method used in this book. Later, the factor-label method will be extended to include chemical formulas in the unit-operations step of problem solving. The following Sample Problems illustrate this factor-label method.

SAMPLE PROBLEM 1-1

A chemistry student was asked to determine the density of an irregularly shaped sample of lead, but was not supplied with any measurement data.

SOLUTION

The student measured the mass of the lead on a "centigram" balance as 49.33 g. Recalling from General Science that a solid displaces its own volume in a liquid, the student immersed the lead in water contained in a cylinder graduated in 0.1-mL divisions. The sample displaced 4.35 mL of water, the 0.05 mL being estimated.

Because the volume of a solid is normally expressed in cubic measure, the measured volume was converted to cubic centimeters.

Since 1 mL = 1 cm^3,

$$4.35 \cancel{\text{mL}} \times \frac{1 \text{ cm}^3}{\cancel{\text{mL}}} = 4.35 \text{ cm}^3$$

By definition:

$$D = \frac{m}{V} = \frac{49.33 \text{ g}}{4.35 \text{ cm}^3} = 11.34 \text{ g/cm}^3$$

The multiplication/division rule for significant figures indicates an answer to 3 significant figures in this computation. Therefore, the result is rounded off to 11.3 g/cm^3.

A more efficient solution setup for this problem is

$$D = \frac{m}{V} = \frac{49.33 \text{ g}}{4.35 \cancel{\text{mL}} \times 1 \text{ cm}^3/\cancel{\text{mL}}} = 11.3 \text{ g/cm}^3$$

SAMPLE PROBLEM 1-2

Determine the concentration of table salt (sodium chloride), in grams of salt per gram of solution, when $40\bar{0}$ mg of the salt is dissolved in $10\bar{0}$ mL of water at $6\bar{0}$ °C.

SOLUTION 1

(Two solution methods are demonstrated.)

The problem requires that the solution concentration be expressed in grams of salt per gram of solution. Therefore, the solution concentration is derived from the ratio: mass salt/mass solution. The mass of the solution is the sum of the mass of salt dissolved and the mass of water used. The volume of water at $6\bar{0}$ °C is known, but its mass is required. Density relates the mass and volume of a material. The density of water at $6\bar{0}$ °C is listed in Appendix Table 10. Locate this table and verify that the density of water at $6\bar{0}$ °C is 0.983 g/mL.

By definition: $$D = \frac{m}{V}$$

Solving for m: $$m = D \times V$$

Substituting: $$m = 0.983\ \frac{\text{g}}{\cancel{\text{mL}}} \times 10\bar{0}\ \cancel{\text{mL}}\ \text{water} = 98.3\ \text{g water}$$

Observe that the unit operations yield the correct answer unit for the mass of water.

The mass of salt is given in milligrams. As milligrams and grams cannot be added, milligrams of salt must be converted to grams.

By definition: $$1\ \text{mg} = 0.001\ \text{g}$$

$$40\bar{0}\ \cancel{\text{mg}}\ \text{salt} \times \frac{0.001\ \text{g}}{\cancel{\text{mg}}} = 0.400\ \text{g salt}$$

Observe that the factor 0.001 g/mg is exact.

$$\text{mass of solution} = 0.400\ \text{g salt} + 98.3\ \text{g water}$$

$$\text{mass of solution} = 98.7\ \text{g solution}$$

$$\text{solution concentration} = \frac{\text{mass salt}}{\text{mass solution}} = \frac{0.400\ \text{g salt}}{98.7\ \text{g solution}}$$

$$\text{solution concentration} = 0.00405\ \text{g salt/g solution}$$

SOLUTION 2

In this example, the complete solution setup is developed in literal terms (without number values) as an algebraic equation. Measurement values are then substituted, and the indicated unit operations are performed to verify the correct answer unit—grams salt per gram solution. It is an efficient method and a productive approach to the problem-solving process.

$$\text{solution concentration} = \frac{\text{mass salt}}{\text{mass solution}} = \frac{m_s}{m_{soln}}$$

But, $$m_{soln} = \text{mass salt} + \text{mass water} = m_s + m_w$$

And, $$m_w = \text{density of water} \times \text{volume of water} = D_w \times V_w$$

Therefore,

$$\text{soln conc} = \frac{m_s}{[m_s + (D_w \times V_w)]_{soln}}$$

$$\text{soln conc} = \frac{40\bar{0}\text{ mg salt} \times 0.001\text{ g/mg}}{(40\bar{0}\text{ mg} \times 0.001\text{ g/mg} + 0.983\text{ g/mL} \times 10\bar{0}\text{ mL})_{soln}}$$

$$\text{soln conc} = \frac{0.400\text{ g salt}}{(0.400\text{ g} + 98.3\text{ g})\text{ soln}} = \frac{0.400\text{ g salt}}{98.7\text{ g soln}}$$

$$\text{soln conc} = 0.00405\text{ g salt/g soln}$$

Practice Problems

1. Using a platform balance, the mass of an irregular block of iron is found to be 280.2 g. The iron was immersed in water and found to displace 35.6 mL of the water. Determine the density of the iron. *ans.* 7.87 g/cm^3
2. Determine the concentration of a solution in grams of salt per gram of solution when the solution is prepared by dissolving 10.2 g of a certain salt in 50$\bar{0}$ mL of water at 5$\bar{0}$ °C. *ans.* 0.0202 g salt/g soln

1.22 Proportions 1. *Direct proportions.* Two variables are said to be *directly proportional* to one another if their *quotient* has a constant value. If the quantity y is proportional to the quantity x, then y varies directly as x, and their quotient (or ratio) has the constant value k.

$y \propto x$ and $\frac{y}{x} = k$ where $\propto$ is a proportionality sign, and k is the proportionality constant.

Then,

$$y = kx$$

The density of a material has been defined as the ratio of its mass to its volume.

$$D = \frac{m}{V} \quad \text{then,} \quad m = DV$$

The expression $m = DV$ has the form $y = kx$ when D is kept constant.

At 4 °C the density of water, D_w, has the constant value of 1.00 g/mL, and the mass of water, m_w, varies as its volume, V_w.

$$m_w = D_w V_w$$

At 4 °C,

$$m_w = \frac{1.00\text{ g}}{\text{mL}} \times V_w$$

Table 1-10
VARIATIONS OF MASS WITH VOLUME OF WATER AT 4 °C

Volume (ml)	*Mass* (g)
$1\bar{0}$	$1\bar{0}$
$2\bar{0}$	$2\bar{0}$
55	55
75	75
95	95
110	110

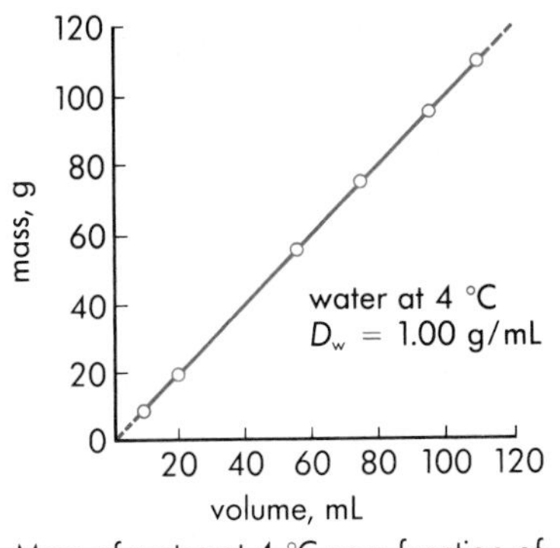

Mass of water at 4 °C as a function of its volume

Figure 1-21. A graph of direct proportion is a straight line.

From this expression it is apparent that the mass of water varies directly with its volume at a constant water temperature. The mass and volume of water at a constant temperature are examples of variables that are directly proportional to one another. Representative variations of mass with volume are tabulated in Table 1-10.

The relation between the variables in Table 1-10 can be summarized by means of a graph in which mass m_w is plotted as a function of volume V_w. The assigned values of V_w, called the *independent variable,* are plotted horizontally on the x-axis. The determined values of m_w, called the *dependent variable,* are plotted vertically on the y-axis. This is shown in Figure 1-21. Observe that the graph of a direct proportion is *linear* (straight-lined).

V_w, with assigned values, is the independent variable; m_w, with determined values, is the dependent variable.

2. *Inverse proportion.* Two variables are said to be *inversely proportional* to one another if their *product* has a constant value. If the quantity y is inversely proportional to the quantity x, then y varies *inversely* as x and their product has the constant value k, the proportionality constant.

$$xy = k \quad \text{and} \quad y = \frac{k}{x}$$

From your experience, you know that the *time* required to travel a *certain distance* is a function of the *average speed* maintained over the distance. For a constant distance, d, the time, t, varies *inversely* as the speed, v.

Distance is measured in length units.

$$vt = d \quad \text{and} \quad t = \frac{d}{v}$$

Observe that the expression $t = d/v$ has the form $y = k/x$ when d is a constant distance.

The data listed in Table 1-11 illustrate the variations in time required to traverse the same distance at different speeds. The relation between the variables can be summarized by a graph of these data in which time t is plotted as a function of speed v. The graph is shown in Figure 1-22. Observe that the graph of an inverse proportion is a *hyperbola.*

Plot the independent variable horizontally and plot the dependent variable vertically.

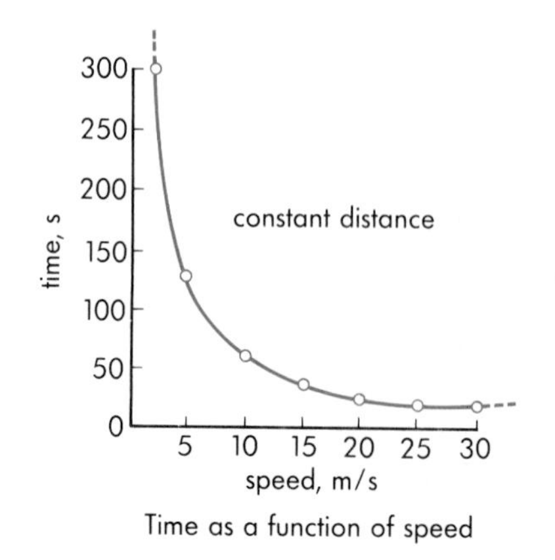

Figure 1-22. A graph of an inverse proportion is a hyperbola.

Table 1-11
VARIATIONS OF TIME WITH SPEED OVER A CONSTANT DISTANCE

Speed (m/s)	*Time* (s)
2.0	300
5.0	120
10	60
15	40
20	30
25	24
30	20

SUMMARY

Chemistry is the science dealing with materials, their composition, and the changes they undergo. The scientific method is a logical approach to the solution of problems under investigation. Four steps are usually involved in the application of the scientific method: observing, generalizing, theorizing, and testing.

Matter is anything that occupies space and has mass. Matter has inertia. The mass of an object is a measure of its inertia. Weight is the measure of the earth's gravitational attraction for an object. A material's mass is determined indirectly by comparing its weight with that of a known mass in the same location on earth.

The three phases of matter are solid, liquid, and gas. These phases are largely a function of temperature. The properties of matter are classified as either physical or chemical.

Energy is the capacity for doing work. The law of conservation of matter and energy states that matter and energy are interchangeable and the total matter and energy in the universe is constant.

The metric system is a decimal system with a simple numerical relationship between units. A single set of prefixes extends the basic system of measurement. A few fundamental quantities are measured in terms of arbitrary units. All other physical quantities are derived from these fundamental quantities.

Heat is defined in terms of the energy transfer associated with a difference in temperature between two systems. Temperature is an indication of heat intensity within a system. It is a fundamental quantity measured in an arbitrary unit, the °C. Heat is a form of energy commonly measured in calories. Heat energy is also measured in joules. The calorie is defined in terms of the joule.

There is some uncertainty inherent in the measurement of any physical quantity. Accuracy and precision are used to express the nature and degree of measurement uncertainty. A measurement expression includes a number magnitude, a dimensional unit, and some indication of the measurement reliability. The measurement reliability is indicated by the method of significant figures.

Computations involving measurement information follow the factor-label method of operations. Measurement units are treated algebraically as factors in the mathematical expressions of the solutions to problems.

Measured quantities that vary with one another are directly proportional if their quotient is constant. They have a linear relation, and their graph is a straight line. Measured quantities that vary with one another are inversely proportional if their product is constant. Their graph is a hyperbola.

VOCABULARY

accuracy
calorie
centi-
chemical property
chemistry
density
dimensional analysis
direct proportion
energy
gram
heat
hyperbola
inertia
inverse proportion
joule
kilo-
kinetic energy
law
liter
mass
matter
meter
milli-
physical property
potential energy
precision
scientific method
scientific notation
significant figure
temperature
theory
weight

QUESTIONS

Group A

1. Define chemistry. *A1*
2. *(a)* Identify six main branches of chemistry. *(b)* Why is chemical knowledge organized into separate branches? *A1*
3. What was the phlogiston theory? *A1*
4. What important contributions to chemistry are attributed to Marie Curie? *A3b*
5. Scientists who investigate natural phenomena are seeking the answers to what two important questions? *A2b*
6. Describe the scientific method. *A2b*
7. Define: *(a)* inertia, *(b)* mass, in terms of inertia, *(c)* weight. *A1*
8. What distinguishes *(a)* a solid from a liquid? *(b)* a liquid from a gas? *A2f*
9. Compile a list of properties of materials that are classified as physical properties. *A2i*
10. Compile a list of properties of materials that are classified as chemical properties. *A2i*
11. State the law of conservation of matter and energy. *A2n*
12. Which metric prefixes are most commonly used in chemistry? *A2o*
13. In what metric units would you express: *(a)* the area of the cover of this book? *(b)* your family's daily milk consumption? *(c)* the length of the eye of a darning needle? *(d)* your own mass? *(e)* the speed of a moving automobile? *A2o*
14. Which of the units required in Question 13 are derived units? *A2r*
15. Name five fundamental quantities that are important in chemistry and give the standard unit for each. *A2r, A2s*
16. *(a)* What is the distinction between heat and temperature? *(b)* In what units will each be measured? *A2t*
17. Distinguish between accuracy and precision in measurements. *A2v, A2w*
18. Define significant figures as they relate to measurement information. *B1b*
19. To change a measurement recorded in positional notation to scientific notation, the decimal point is moved seven places to the right. What is the exponent of 10? *B3c*
20. What relationship exists between two variables *(a)* whose product has a constant value? *(b)* whose quotient has a constant value? *B1d*

Group B

21. Having defined chemistry in response to Question 1, in your own words state briefly the meaning the term "chemistry" conveys to you. *A1*
22. Of what importance is the study of chemistry in the attainment of your educational goals? *A1, A2a*

23. What fact enables you to determine the mass of a body by comparison with a known mass, as illustrated in Figure 1-18? *A2c*
24. Accepting the fact that an object weighs less at a high altitude above the earth than it does at the surface, *(a)* can you develop an argument that supports this fact? *(b)* Will your argument, or a modified version of it, support the fact that the object also weighs less at the bottom of a deep mine shaft than it does at the surface? *A2c*
25. Define the term "free surface" as used in the following statement: "A liquid has a single free surface." *A2f, A2g*
26. Why are both liquids and gases classified as fluids? *A2g*
27. What determines whether a certain property of a material is classed as physical or chemical? *A2i*
28. Volume is a property of a material. *(a)* Is it a specific property? *(b)* Is mass a specific property? *(c)* Is the ratio of mass to volume a specific property? Explain. *A2e, A2i*
29. Explain why the expression of the magnitude of any measurement requires both a number and a unit. *A2g, A2r*
30. Acceleration is defined as the change in velocity (speed) per unit of time. *(a)* What are the dimensions of acceleration? *(b)* What is the derived unit of measure for acceleration if distance is measured in meters and time is measured in seconds? *A2r, A2s*
31. A platform balance has scale graduations to 0.1 g and is known to be sensitive to 0.01 g. Using this balance, a student records the mass as 70.14 g. *(a)* How was the digit in the hundredth-gram place determined? *(b)* The recorded measurement contains how many significant figures? *B1b*
32. Your laboratory partner was given a meter stick graduated in millimeters and the task of measuring the length of a block with as much measurement detail as possible. The following set of independent measurements was reported: 12.95, 12.9 cm, 12.95 cm, 129.55 mm, 13 cm. *(a)* State which of these reported measurements you accept, giving the reason. *(b)* Give your reason for rejecting unacceptable measurements. *B1b*
33. Copy each of these measurements and underscore all significant figures in each. *(Do not mark in this book.)* *(a)* 127.50 km; *(b)* 1200 m; *(c)* 90027.00 cm^3; *(d)* 0.0053 g; *(e)* $67\bar{0}$ mg; *(f)* 0.0730 g; *(g)* 43.053 L; *(h)* 300900 kg; *(i)* 147 cm; *(j)* 6271.9 cm^2. *B1c*
34. The speed of light in vacuum, an important physical constant, is listed in Appendix Table 3. *(a)* How many significant figures are included in the constant expression? *(b)* Convert the expression to positional notation, retaining all significant figures given in Table 3. *(c)* Round off your expression for *(b)* to four significant figures. *(d)* Round off your expression in *(c)* to three significant figures. *(e)* Round off your expression in *(d)* to one significant figure. *(f)* What conversion factor would you use to convert your constant to one expressed in km/s? *(g)* Express the constant in scientific notation to three significant figures with the dimensional unit of km/s. *(h)* Express the constant in scientific notation to one significant figure with the dimensional unit of cm/s. *B1a, B1c, B3c*
35. Explain the meaning of the following statement: "For an equation to represent the correct solution to a problem, the equation must be dimensionally correct." *A3e*
36. When solving a chemistry problem by the factor-label method, the units in the solution expression are first solved for the answer unit. What advantage accrues from this factor-label technique? *B3g*

PROBLEMS

Group A

1. Using the appropriate conversion factors, demonstrate the conversion of the following measurements to millimeters: *(a)* 1 cm, *(b)* 1 m, *(c)* 1 km. *A2o, B1a*
2. By the method of Problem 1, demonstrate

the conversion of *(a)* 1.00 ft to centimeters, *(b)* 2.0 m to centimeters, *(c)* 2.0 m to inches. (Note: use Table 1-3 as required.) *A2o, B1a*

3. Convert *(a)* 1.5 L to milliliters, *(b)* 10 L to cubic centimeters, *(c)* 1 m^3 to liters. *A2o, B1a*

4. Calculate the number of milligrams in *(a)* 0.425 kg, *(b)* 1.15 lb. *(c)* How many grams are there in 2.65 kg? *A2o, B1a*

5. *(a)* What is your height in centimeters? *(b)* What is your mass in kilograms? *A2o, B1a*

6. A Florence flask has a capacity of 2.5×10^2 mL. *(a)* What is its capacity in liters? *(b)* How many grams of water will the flask hold? *A2o, B1a*

7. Mars revolves around the sun at an average distance of 141,500,000 mi in a period of 687 days. *(a)* Express the distance in scientific notation as kilometers. *(b)* Express the period in years. (Assume a 365-day year.) *A2o, B1a, B3c*

8. The thickness of an oil film on water is 0.0000005 cm. Express this thickness in scientific notation. *B3c*

9. The distance light travels through space in one year is called a light-year. Using the speed of light in vacuum listed in Appendix Table 3 rounded off to 3 significant figures and assuming a 365-day year, determine the distance of a light-year in kilometers. *B3c, B3d*

10. In a calorimetry experiment, the quantity of heat released during the combustion of a sample was measured as 736.2 cal. *(a)* How many joules of heat energy were released? *(b)* Express this quantity of heat in kilojoules. *B3a, B3b*

Group B

11. Express the distance 152.20 cm in each of the following units, showing the conversion computation in each case: *(a)* meters, *(b)* millimeters, *(c)* kilometers, *(d)* inches. *B1a, B3a*

12. A cubic box holds $100\bar{0}$ g of water. *(a)* What is the capacity of the box in milliliters? *(b)* What is the length of one side in centimeters? *(c)* What is this length in meters? *B1a, B3a*

13. A laboratory test tube is 125 mm long and 25.0 mm in diameter. *(a)* Neglecting the fact that the bottom of the test tube is rounded, calculate its capacity in milliliters. *(b)* How many grams of water will it hold? *B1a, B3a*

14. A 1.00-L graduated cylinder has an inside diameter of 8.24 cm. There is a 52-mm ungraduated portion at the top. What is the total height of the cylinder in centimeters? *B3a, B3g*

15. Each member of a class of 24 students needs 8.600 g of sodium chloride for an experiment. The instructor sets out a new 1.000-lb jar of the salt. What mass, in grams, of the salt remains at the end of the laboratory period? *B3a*

16. The density of mercury, to three significant figures, is 13.5 g/mL. *(a)* What is the mass of 8.20 mL of mercury? *(b)* What volume would $12\bar{0}$ g of mercury occupy? *B2b, B3d, B3e, B3g*

17. A platform balance showed the mass of a molded glass figurine to be 247.6 g. When immersed in water, the figurine displaced 83.9 mL. What is the density of the glass? *B2b, B3d, B3e, B3g*

18. A mass of 575 mg of a certain salt is dissolved in $15\bar{0}$ mL of water at 80.0 °C. Determine the concentration of the solution in terms of grams of salt per gram of solution. *B3d, B3e, B3g*

19. Chemists have determined that 18.0 g of water consists of 6.02×10^{23} molecules. Assuming that a teaspoon holds 3.70 mL of water, determine the number of water molecules the teaspoon can hold. *B3c, B3d, B3e, B3g*

20. Suppose you are able to remove individual molecules of water from the teaspoon of Problem 19 at the rate of 1 molecule per second. How many years would be required to empty the spoon?

Otto Hahn (German) was awarded the 1944 Nobel Prize in chemistry for discovering the fission of heavy nuclei. Atoms are composed of a positive nucleus surrounded by negative electrons. Hahn found that when the nucleus of uranium-235 is hit with slow neutrons, it splits into two new atoms of approximately equal mass, such as barium and krypton. A great deal of energy is released when the structure of matter changes in this way. The fission process is utilized in nuclear reactors and in the atomic bomb.

Matter and Its Changes

2

GOALS

In this chapter you will gain an understanding of: • heterogeneous and homogeneous matter • mixtures, compounds, and elements • physical and chemical changes • endothermic and exothermic processes • reaction tendencies in nature: energy-change tendency, entropy-change tendency

COMPOSITION OF MATTER

2.1 Classes of matter Chemistry has been described as the study of the structure and composition of matter, of changes in its composition, and of the mechanisms by which these changes occur. The methods of chemistry involve metric measurements and often require computations with measurement data. To pursue this study, it is essential that some basic concepts of matter be recognized.

Matter is commonly *described* as anything that occupies space and has mass. It includes all materials found in nature. Some materials are made up of parts that are not alike. The dissimilarities may be subtle, or they may be readily apparent. *Matter that has parts with different properties is said to be* ***heterogeneous*** (het-er-oh-*jee*-nee-us). Granite, a common rock, is heterogeneous because it is composed of different minerals, each having characteristic properties. Distinctly different parts are easily observed.

Other materials appear uniform throughout; all parts are alike. The properties of any one part are identical to the properties of all other parts. *Matter that has identical properties throughout is* ***homogeneous*** (hoh-muh-*jee*-nee-us). Sugar and ordinary table salt are examples of homogeneous materials.

Matter is either heterogeneous or homogeneous.

The many different kinds of matter throughout the world are the materials with which chemists work. It would be difficult and time-consuming to study these materials without first organizing them into similar groups. One method of classifying matter is according to its three different phases: solid, liquid, and gas. Matter is also divided into three general classes on the basis of its properties: *elements, compounds,* and *mixtures.* These and other ways of classifying matter make the study of chemistry simpler. Figure 2-1 will be helpful in the study of Sections 2.2 and 2.3.

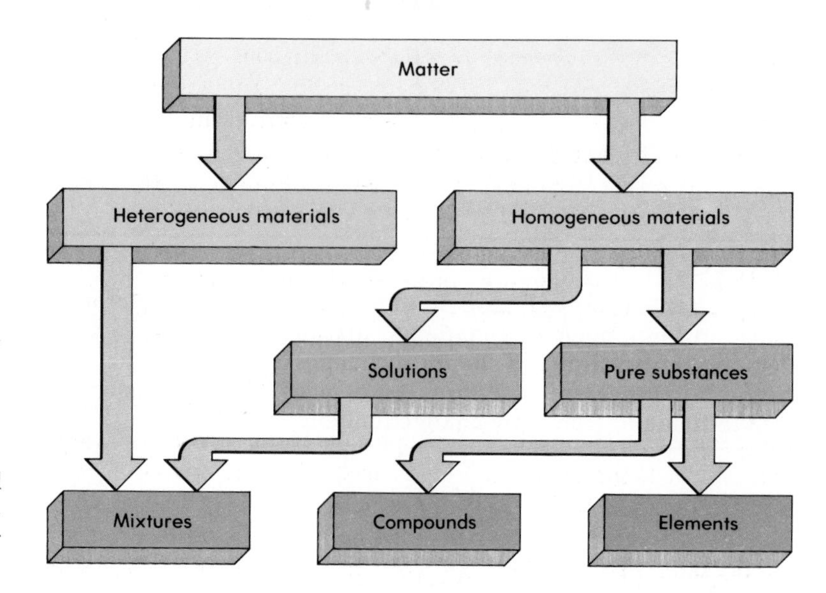

Figure 2-1. All matter is divided into three general classes: mixtures, compounds, and elements. Solutions are homogeneous mixtures.

2.2 Mixtures Suppose you examine a piece of granite closely with a hand lens. Three different crystalline materials can be seen. They are quartz, feldspar, and mica. The properties of each differ greatly. Granite is a heterogeneous material, having parts with different properties. It is a *mixture*. One part of a piece of quartz has the same properties as every other part. This fact is also true of feldspar and mica. Each of these components of granite is a homogeneous material. Heterogeneous materials are mixtures of homogeneous materials.

Heterogeneous materials are mixtures.

A heterogeneous mixture has no set of unique properties. Instead, its properties are a combination of the properties of its homogeneous parts. All mixtures are not heterogeneous, however. When sugar is dissolved in water, the resulting solution has similar properties throughout. Thus the solution is homogeneous. The amount of sugar or water may be increased, but the solution remains a homogeneous mixture of the two materials. The solution has the sweet taste of the sugar it contains. The water may be removed by evaporation and the sugar recovered in its original form. Solutions are homogeneous mixtures.

Some homogeneous materials are mixtures.

Solutions are mixtures that are homogeneous.

Air is a solution of gases. It is a mixture composed principally of nitrogen and oxygen. Other gases present in air are argon, carbon dioxide, and water vapor. Each gas displays its own unique properties. Alloys, primarily combinations of metals, are usually solid solutions. *A **mixture** is a material consisting of two or more kinds of matter, each retaining its own characteristic properties.*

2.3 Substances include compounds and elements Materials with identical properties throughout are homogeneous. In chemistry, such homogeneous materials are called *pure sub-*

stances, or simply *substances*. *A* **substance** *is a homogeneous material consisting of one particular kind of matter.* Both the sugar and the water of a sugar-water solution are substances in this sense. Unlike granite, which has properties attributable to quartz, feldspar, and mica, sugar has properties attributable to the sugar itself, properties that stem from its particular composition. *A substance has a definite chemical composition.*

Mixtures are not substances in a chemical sense. Elements and compounds are substances.

Suppose a small quantity of sugar is heated in a test tube over a low flame. The substance melts and changes color. Finally a black residue remains in the bottom of the test tube. Drops of a clear, colorless liquid appear around the cool, open end. The black substance has the properties of carbon and the liquid has the properties of water. The properties of the sugar no longer exist. In fact, the sugar no longer exists. Instead, two different substances, carbon and water, are observed.

The decomposition of sugar by heat is an example of a chemical process.

Each time the experiment is repeated, the sugar decomposes in the same way to yield the same proportions of carbon and water. Sugar is recognized as a complex substance showing a constant composition. It is an example of a *compound*. *A* **compound** *is a substance that can be decomposed into two or more simpler substances by ordinary chemical means.*

Water can be decomposed into two simpler substances, hydrogen and oxygen. Thus water is also a compound. Chemists have not succeeded in decomposing carbon, hydrogen, or oxygen into simpler substances. The conclusion is that they are elemental substances or *elements*. **Elements** *are substances that cannot be further decomposed by ordinary chemical means.*

Figure 2-2. The components of a mixture can be separated by using differences in their physical properties. In the top photo, the mixture of two solids is separated by shaking it in a liquid because the components settle out at different rates. In the bottom photo, the mixture is separated by filtration after shaking it in a liquid in which one component (but not the other) is soluble.

2.4 The known elements

One of the fascinating facts of science is that all known matter is composed of approximately 100 elements. A few elemental substances, such as gold, silver, copper, and sulfur, have been known since ancient times. During the Middle Ages and the Renaissance, more elements were discovered. Through the years, improved research techniques have enabled scientists to add to the list of elements.

The list of known elements has grown to a total of 109 at the time of this writing. The 92 elements ranging from hydrogen to uranium are traditionally known as *natural elements*. With a few exceptions, they are found in nature in either free or combined form. These elements make up the pre-Atomic Age list of known elements. The elements beyond uranium are called the *transuranium elements*. They are "artificial" elements prepared from other elements during *synthesis* procedures.

In the decade before World War II, a great experimental study of atomic structure was undertaken. Enrico Fermi, an Italian theoretical physicist, stated that it should be possible to prepare the ninety-third and ninety-fourth elements from uranium.

Figure 2-3. A vial containing the compound berkelium trichloride.

Element 93 was first produced in the laboratories of the University of California, Berkeley, in 1940. It was named *neptunium* for the planet Neptune. This planet is beyond the planet Uranus just as element 93 is beyond uranium (element 92) on the list of elements. Later, element 94 was produced in the same laboratories. It was given the name *plutonium* for the planet Pluto, which is beyond Neptune.

These triumphs were followed by the production of *americium* (am-er-*ih*-see-um), named for America; *curium* (*ku*-ree-um), named in honor of Marie and Pierre Curie; *berkelium* (*berk*-lee-um), named for Berkeley (a site of the University of California); and *californium*, named for the university and the state. More recently, *einsteinium*, named for Albert Einstein, *fermium*, named for Enrico Fermi, and *mendelevium* (men-del-*ev*-ee-um), named for Dmitri Mendeleev, have brought the total to 101.

In 1957, a team of American, British, and Swedish scientists working at the Nobel Institute in Sweden announced the discovery of element 102. They suggested the name *nobelium*. Careful experiments by other scientists, however, failed to confirm their discovery.

The following year, a research group at Lawrence Radiation Laboratory of the University of California produced element 102 and identified it by chemical means. They retained the name *nobelium* to honor Alfred Nobel, who made a significant contribution to the advancement of science through his establishment of the Nobel Prizes.

Cyclotron: a particle accelerator. An apparatus that accelerates charged particles to high speeds for nuclear bombardment experiments.

In 1961, element 103 was produced by scientists at the Lawrence Radiation Laboratory. The name *lawrencium* has been assigned to element 103 in honor of Ernest O. Lawrence, the inventor of the cyclotron and founder of the laboratory in which the element was first produced.

In 1964 and the decade that followed, both American and Russian scientists reported the synthesis of elements 104, 105, and 106. In 1976 the Russian scientists reported the production of element 107. A group of West German scientists also produced and identified element 107 in 1981. This same group reported the synthesis of element 109 in 1982 and the synthesis of element 108 in 1984. At the time of this writing, the production of neither element 108 nor element 109 has been verified experimentally by other scientists.

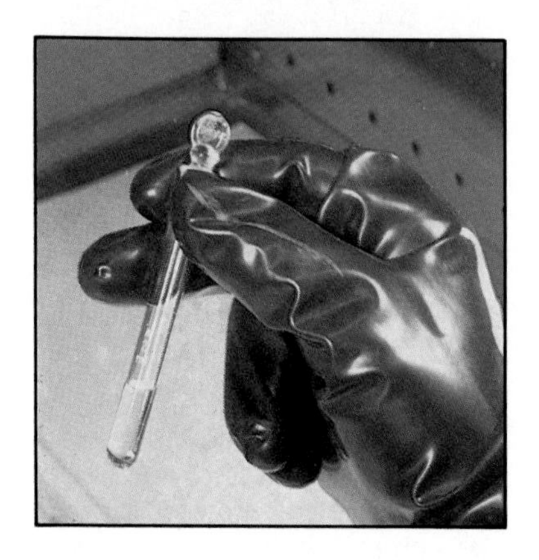

Figure 2-4. A vial containing californium, transuranium element number 98.

Traditionally the "discoverer" of a new element proposes its name, but the element is not officially named until its discovery has been clearly established by the scientific community. At the time of this writing none of the new elements beyond lawrencium (element 103) has an official name.

To fill a need for names for these elements, as well as others that may be predicted but not yet produced, the *International Union of Pure and Applied Chemistry* has devised a systematic

Table 2-1
SYSTEMATIC NOMENCLATURE FOR ELEMENTS

1. The systematic name is derived from the element number using the following numerical roots:

0 = nil		
1 = un	4 = quad	7 = sept
2 = bi	5 = pent	8 = oct
3 = tri	6 = hex	9 = enn

2. The roots are assembled in the order of the digits in the element number and terminated by "ium" to spell out the element name—with the following exceptions:
   (*a*) The final "n" of "enn" is omitted when it occurs before "nil."
   (*b*) The final "i" of "bi" and of "tri" is omitted when it occurs before "ium."
3. The symbol of the element is composed of the initial letters of the numerical roots that make up the name.
4. In the element name each root is pronounced separately. The root "un" is pronounced to rhyme with "moon."

Element number	*Element name*	*Element symbol*
101	Mendelevium	Md
102	Nobelium	No
103	Lawrencium	Lr
104	Unnilquadium	Unq
105	Unnilpentium	Unp
116	Ununhexium	Uuh
127	Unbiseptium	Ubs
238	Bitrioctium	Bto
349	Triquadennium	Tqe
490	Quadennilium	Qen

Pure and Applied Chemistry, February 1979, pp. 383–384. *International Union of Pure and Applied Chemistry.* Used by permission.

nomenclature for elements beyond 100. It is intended that the systematic names be used for such elements until official names are established. The systematic name is composed of numerical roots as listed in Table 2-1. The roots are assembled in the order of the digits in the element number and are followed by the suffix "ium." Using this scheme, the systematic name for element 104 is un-nil-quad-ium, or *unnilquadium* (oon-nill-kwawd-ee-um). The name for element 105 is *unnilpentium;* for 106, *unnilhexium;* for 107, *unnilseptium;* for 108, *unniloctium;* and for 109, *unnilennium*. The systematic name for predicted element 118 would be *ununoctium*. An element such as number 201 would have the systematic name *binilunium* (by-nill-oon-ee-um).

2.5 General classes of elements Each element has its own set of properties. General similarities among the properties of large

Figure 2-5. Mercury (top photo) is the only metallic element that is a liquid at room temperature. Sulfur (bottom photo) is a nonmetallic element. At room temperature its stable form has a rhombic crystalline structure.

groups of elements provide one way of classifying them. In this sense, chemists recognize two general classes of elements, *metals* and *nonmetals*.

Metals. Many familiar metals are recognized because of their appearance. They have a *metallic* luster; that is, they shine like silver. Copper and gold have this metallic luster and they also have distinctive colors. Metals are good reflectors of heat and light. They are good conductors of heat and electricity. Metals are *ductile;* they can be drawn out into fine wire. They are *malleable;* they can be hammered (or rolled) into thin sheets. They are *tenacious;* they can resist being stretched and pulled apart. Metals have these metallic properties of ductility, malleability, and tenacity (tensile strength) to varying degrees.

Familiar metals include gold, silver, copper, aluminum, mercury, magnesium, tin, and zinc. At ordinary temperatures mercury is a liquid metal, as shown in Figure 2-5. Less familiar metals include platinum, sodium, potassium, calcium, cobalt, titanium, gallium, tungsten, and uranium. Gallium is a liquid metal at body temperature, as illustrated melting in the palm of a hand in Figure 2-6.

Nonmetals. Nonmetallic elements are usually poor conductors of heat and electricity. As solids, they are brittle and are neither ductile nor malleable. Sulfur is a typical solid nonmetal. Crystalline sulfur is shown in Figure 2-5. Other common nonmetals that are solids at room temperature are iodine, carbon, and phosphorus. Bromine is a liquid nonmetal. Nonmetals such as oxygen, nitrogen, chlorine, and hydrogen are gases.

A few elements are not accommodated entirely by either of these two general classes of elements. For example, the elements helium, neon, argon, krypton, xenon, and radon are gases at ordinary temperatures. As such they are nonmetallic. However, unlike nonmetallic elements, they are essentially without chemical reactivity. Some form no known compounds; others form only certain compounds under unusual conditions. Ordinarily these elements are not included in the general classes of metals and nonmetals, but are classified as *noble* gases.

A few other elements have some properties midway between those of metals and nonmetals. Silicon and germanium, for example, are usually described as nonmetals. However, both are much better conductors of electricity than sulfur, but much poorer conductors than silver or copper. They are called *semiconductors.* Silicon, germanium, and other elements such as boron, arsenic, and antimony are traditionally called *metalloids*. They are solids under normal conditions and have semiconducting properties to varying degrees. They are basically nonmetals with some metallike properties. As semiconductors, silicon and germanium are important in the production of transistors.

2.6 Chemical symbols Ancient alchemists apparently recognized a need for symbols to represent the various substances known to them. For example, the outline of a crescent moon was commonly used as a symbol for silver. An inverted triangle, a short wavy line, or a combination of the two was the symbol for water. The symbols used today to represent chemical elements evolved from this ancient shorthand.

In 1808, John Dalton (1766–1844) introduced a framework for the symbols of the known chemical elements. He used a circle as the symbol for oxygen, a circle with a dot at its center for hydrogen, and a circle with a particular inscribed design or letter for each of the remaining elements, as shown in Figure 2-7. Observe that he represented compounds by combining the symbols of elements.

Jöns Jakob Berzelius (1779–1848), a Swedish chemist, was the first to use letters as symbols for elements. Instead of Dalton's small circles with identifying marks, Berzelius used the first letter of the name of an element as its symbol. For example, he used the letter **O** as the symbol for oxygen and the letter **H** as the symbol for hydrogen.

As the number of known elements increased, Berzelius added a second letter whose sound was conspicuous when the name of the element was pronounced. For example, the symbol for carbon is **C;** for calcium, **Ca;** for chlorine, **Cl;** for chromium, **Cr;** and for cobalt, **Co.** The first letter of a symbol is *always* capitalized, but the second letter of a symbol is *never* capitalized. For example, **Co** is the *symbol* for cobalt: **CO** is the *formula* for the *compound* carbon monoxide, which is composed of the elements carbon, **C,** and oxygen, **O.** Compare Dalton's and Berzelius' representations of water shown in Figure 2-7 with the familiar H_2O used today.

In several instances the symbol for an element is derived from the Latin name of the element. For example, the symbol for

Figure 2-6. (Top photo) The metallic element gallium occurs as gray-black rhombic crystals. (Bottom photo) Gallium melts in your hand. Its melting point is about 30 °C, and normal body temperature is 37 °C. Its boiling point is nearly 2000 °C.

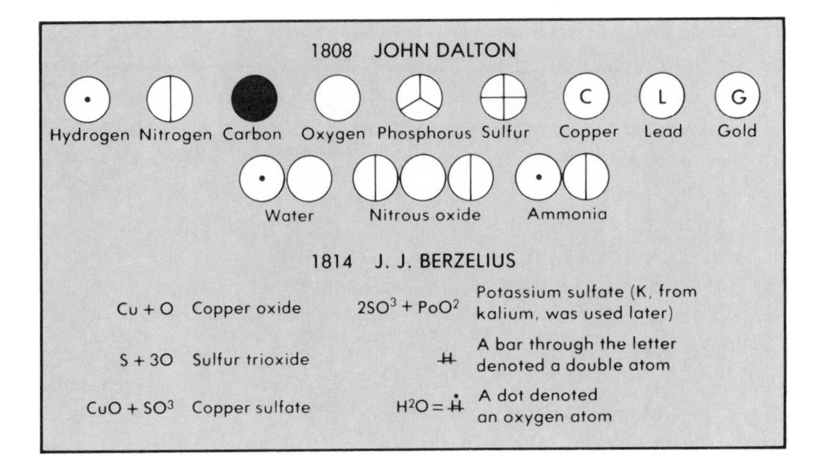

Figure 2-7. John Dalton developed symbol notations for elements and used them to describe the composition of compounds. His notations simplified the system of astrological symbols used by the early alchemists. Berzelius adopted Dalton's system but replaced his graphic symbols with the first letter or letters of the Latin names of the elements.

The first letter of a symbol is ***always*** *capitalized.*

The second letter of a symbol is ***never*** *capitalized.*

iron is **Fe,** from the Latin *ferrum*. The symbol for *lead* is **Pb,** from the Latin *plumbum*. The symbols for silver, **Ag,** sodium, **Na,** and potassium, **K,** are derived from the Latin *argentum, natrium,* and *kalium,* respectively. Some familiar elements and their symbols are listed in Table 2-2.

A chemical symbol is more than an abbreviation; it has quantitative significance. The symbol **K** not only represents the element potassium, but also *one atom* of potassium. Similarly, **Fe** represents 1 atom of iron. Atoms, the basic units of elements, combine with other elements to form chemical compounds. The symbols for the elements will acquire additional significance as your study of chemistry progresses.

Table 2-2
COMMON ELEMENTS AND THEIR SYMBOLS

Name	*Symbol*
aluminum	Al
antimony	Sb
arsenic	As
barium	Ba
bismuth	Bi
bromine	Br
calcium	Ca
carbon	C
chlorine	Cl
chromium	Cr
cobalt	Co
copper	Cu
fluorine	F
gold	Au
hydrogen	H
iodine	I
iron	Fe
lead	Pb
magnesium	Mg
manganese	Mn
mercury	Hg
nickel	Ni
nitrogen	N
oxygen	O
phosphorus	P
platinum	Pt
potassium	K
silicon	Si
silver	Ag
sodium	Na
strontium	Sr
sulfur	S
tin	Sn
titanium	Ti
tungsten	W
zinc	Zn

2.7 The earth's elemental composition Approximately 90 elements are known to occur in measurable amounts in either a free or combined state in the earth's crust. The atmosphere consists almost entirely of two elements, nitrogen and oxygen. Water, which covers a great portion of the earth's surface, is a compound of hydrogen and oxygen. Natural water also contains many dissolved substances.

Only about 30 elements are fairly common. Table 2-3 shows the relative mass distribution of the 10 most abundant elements in the earth's crust, atmosphere, lakes, rivers, and oceans.

Although the solid crust of the earth is the foundation of human existence, it constitutes only about 0.4% of the total mass of the earth and less than 1% of its volume. The mantle accounts for about 67.2% and the core 32.4% of the earth's total mass.

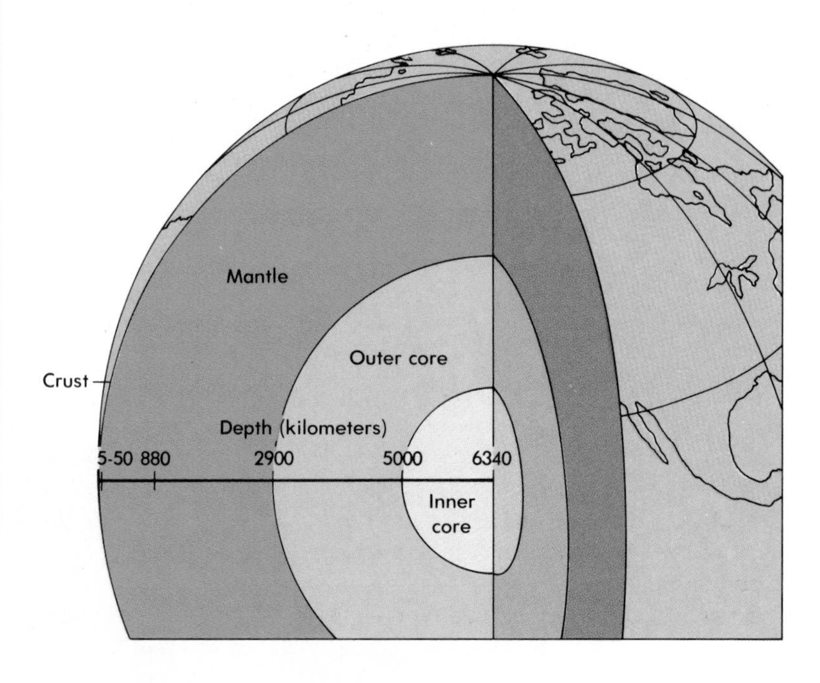

Figure 2-8. Regions of the interior of the earth.

The two most abundant elements in the solid crust are oxygen and silicon. Together they account for almost 75% of the mass of the continental crust. Eight elements make up over 98% of the mass of the continental crust. They are listed in Table 2-4. These eight common elements, along with the less common ones, are combined in many ways to produce the more than 2000 different minerals found in the earth's solid crust.

Scientists at the Smithsonian Institution have estimated that five elements account for more than 94% of the mass of the solid earth, including the continental crust, mantle, and core. These elements are listed in Table 2-5.

The mantle is believed to consist almost entirely of compounds of four elements—magnesium, iron, silicon, and oxygen. Earth scientists believe that the earth's core contains about 85% elemental iron, about 7% nickel, and 8% of a mixture of silicon, cobalt, and sulfur.

Table 2-3
COMPOSITION OF SURFACE ENVIRONMENT

Element	*Mass distribution*
oxygen	49.5%
silicon	25.8%
aluminum	7.5%
iron	4.7%
calcium	3.4%
sodium	2.6%
potassium	2.4%
magnesium	1.9%
hydrogen	0.9%
titanium	0.6%
all other elements	0.7%

2.8 Compounds differ from mixtures When matter is made up of two or more elements, the elements are either mixed mechanically or combined chemically. The material is either a *mixture* or a *compound*, depending on what has happened to the elements. If the material is a mixture, the properties of each element present will persist. If the elements are chemically combined, the material is a complex substance with its own characteristic properties.

Suppose powdered sulfur and powdered iron are mixed thoroughly on a sheet of paper. There is no evidence of a chemical reaction; neither light nor heat is produced. The two substances may be mixed in any proportion. It is possible to use a large portion of iron and a small portion of sulfur or a large portion of sulfur and a small portion of iron.

As the paper containing this mixture is moved back and forth over a strong magnet, the iron particles separate from the sulfur. When a small portion of the mixture is added to a solution of hydrochloric acid, the iron reacts with the acid and disappears from view, leaving the sulfur unaffected. When another portion of the mixture is added to liquid carbon disulfide, the sulfur dissolves, leaving the iron unchanged.

In each of these tests, the properties of iron and sulfur persist. This is typical of a mixture; the *components* do not lose their identity. They may be mixed in any proportion without showing any evidence of chemical activity.

It is possible to cause the iron and sulfur to react chemically and form a compound. Suppose these two elements are mixed in the ratio of 7 g iron to 4 g sulfur, and the mixture heated strongly in a test tube over a Bunsen burner flame. With a rise in temperature the mixture begins to glow. Even after its removal from the flame, the mixture continues to react and the whole mass

Table 2-4
COMPOSITION OF THE CONTINENTAL CRUST

Element	*Mass distribution*
oxygen	46.6%
silicon	27.7%
aluminum	8.1%
iron	5.0%
calcium	3.6%
sodium	2.8%
potassium	2.6%
magnesium	2.1%

Table 2-5
COMPOSITION OF THE SOLID EARTH

Element	*Mass distribution*
iron	34.6%
oxygen	29.5%
silicon	15.2%
magnesium	12.7%
nickel	2.4%

Table 2-6
DIFFERENCES BETWEEN MIXTURES AND COMPOUNDS

Mixture	*Compound*
1. In a mixture, the components can be present in any proportion.	1. In a compound, the constituents have a definite proportion by mass.
2. In the preparation of a mixture, there is no evidence of any chemical action taking place.	2. In the preparation of a compound, evidence of chemical action is usually apparent (light, heat, etc.).
3. In a mixture, the components do not lose their identity. They can be separated by physical means.	3. In a compound, the constituents lose their identity. They can be separated only by chemical means.

The process of combining two or more substances to form a single more complex substance is called ***synthesis.***

becomes red hot. *Both heat and light are produced during the chemical reaction in which sulfur and iron combine to form a compound.*

After the reaction has ceased and the product has been removed, careful examination shows that it no longer resembles either the iron or the sulfur. Each element has lost its characteristic properties. The iron cannot be removed by a magnet. The sulfur cannot be dissolved out of the product with carbon disulfide. In the original mixture, hydrochloric acid reacted chemically with the iron, and odorless hydrogen gas was produced. Hydrochloric acid also reacts chemically with this new product, and a gas is again produced. However, in this reaction, the gas is distinctly different from that formed in the previous reaction. This gas has an odor. It is *hydrogen sulfide,* a poisonous gas that is notorious for its rotten-egg odor. This different product gives evidence that a new substance with a new set of properties is formed during the reaction between the iron and sulfur. It is a *compound* composed of iron and sulfur.

Analysis *is the process of determining the identity and quantity of each constituent element in a compound.*

By chemical analysis, the new compound is found to consist of iron, **Fe**, and sulfur, **S**, in a definite mass relationship. It is an *iron sulfide* compound consisting of 63.5% Fe and 36.5% S. This composition represents a mass ratio Fe : S of 7 : 4. Had the starting mixture for the reaction been in the ratio of 8 g Fe : 4 g S, 1 g of iron would remain as an unused surplus after the reaction was complete. *A compound is always composed of the same elements in a definite mass relationship.* Significant differences between mixtures and compounds are summarized in Table 2-6.

2.9 Law of definite composition Louis Proust (1755–1826), a French chemist, was one of the first to observe that elements combine with one another in a definite mass ratio. About 50 years later, Jean Servais Stas, a Belgian chemist, performed a series of precise experiments that confirmed this observation. Proust's observation is now known as the ***law of definite composition:*** *Each compound has a definite composition by mass.*

Using the law of definite composition, a manufacturer of chemical compounds can determine precisely the quantity of each constituent required to prepare a specific compound.

Figure 2-9. A mixture of powdered iron and sulfur can be separated by a magnet because iron, but not sulfur, is attracted by the bar magnet.

2.10 Common examples of mixtures and compounds In Section 2.2, air was described as a mixture. Its composition varies somewhat in different localities. Other familiar examples of mixtures are baking powders, concrete, and various kinds of soil. There is practically no limit to the number of possible mixtures. They may be made up of two or more elements, of two or more compounds, or of both elements and compounds. For example, brass is a mixture of two elements, copper and zinc. Common gunpowder is a mixture of two elements, carbon and sulfur, with a compound, potassium nitrate. A solution of common salt is a mixture of two compounds, sodium chloride and water.

A large dictionary may define almost a half-million words. All of these words are formed from one or more of the 26 letters of our alphabet. Try to imagine the number of compounds possible from different combinations of 100 or more elements. Of course, some elements do not react readily with others to form compounds. Helium, **He**, exists only in the free state. It forms no known compounds. Enough elements do combine, however, to make the several million compounds known to chemists. Water, table salt, sugar, alcohol, baking soda, ether, glycerol, cellulose, nitric acid, and sulfuric acid are some examples of common compounds.

The simplest compounds consist of two different elements chemically combined. Iron sulfide is such a compound. It is composed of iron and sulfur. Carbon dioxide is composed of carbon and oxygen. Hydrogen and oxygen combine to form water. Table salt consists of sodium combined with chlorine. Sodium is an active metallic element that must be protected from contact with air and water. Chlorine is a poisonous gas. However, when combined chemically, the two elements form common table salt or *sodium chloride.*

Many compounds are composed of three different elements. Carbon, hydrogen, and oxygen are the constituent elements in sugar. These same three elements occur combined in different proportions in many other compounds having decidedly different properties. Examples are acetic acid, alcohol, ether, and formaldehyde.

Changes in Matter

2.11 Physical changes Solids melt, liquids boil, gases condense, liquids freeze, and many solids, liquids, and gases dissolve in water. A length of platinum wire glows when heated in a Bunsen flame. As the platinum wire cools, its characteristic

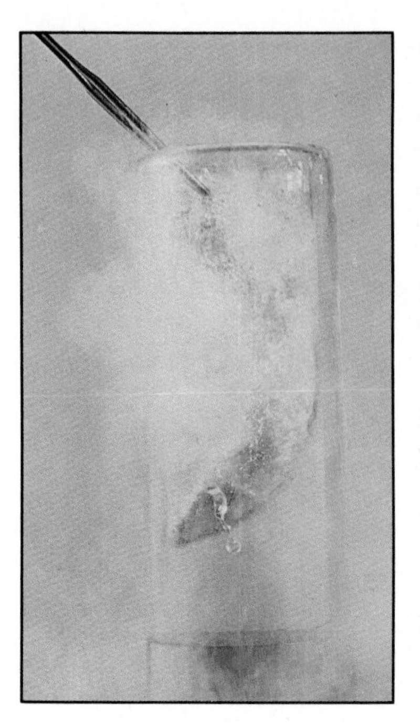

Figure 2-10. After a banana is dipped in liquid nitrogen at −195.8 °C, it can be used as a hammer to drive the nail into the block of wood.

metallic luster is observed to have remained unchanged. In each of these actions, matter undergoes a change that involves some quantity of energy. The form of matter may be different, or a change of phase may have occurred. However, the *identity* of matter has not changed. A reversal of the action that caused a change may restore the material to its original form and the original identifying properties may again be recognized. For example, water is rapidly converted to water vapor when heated to its boiling point. As the vapor loses heat, it condenses to water, and the properties of liquid water are again recognized.

These examples together with Figure 2-10 illustrate *physical changes.* Only physical properties are altered during physical changes. No new substances with new properties are formed. ***Physical changes*** *are those in which certain physical properties of substances change and their identifying properties remain unchanged.*

Modern concepts of the solution process suggest that some types of physical changes may involve intermediate processes that are not physical in nature. These considerations will be examined in Chapter 13.

2.12 Chemical changes Wood burns, iron rusts, silver tarnishes, milk sours, plants decay, and acids react with metals. These reactions are examples of processes in which the identifying properties of the original substances disappear as new substances with different properties are formed. Changes occur that alter the composition of matter. ***Chemical changes** are those in which different substances with new properties are formed.*

Chemical reactions may involve the formation of compounds from elemental substances. Complex substances may be broken down into simpler compounds or into the constituent elements. Compounds may react with other compounds or elements to form new and different substances.

Chemistry is concerned specifically with the chemical changes substances undergo and with methods of controlling the chemical reactions in which the changes occur. To recognize these processes and control them, chemists need to know how the reactions occur. They observe changes in the properties of substances in bulk and try to interpret these changes in terms of the behavior of particles of the substances.

2.13 Chemical reactions involve energy Chemical changes are always accompanied by energy changes. Substances possess energy because of their composition and structure. It is a kind of potential energy that chemists refer to as *chemical energy*.

The products of chemical reactions differ in composition and structure from the original reactants. Thus they have larger or smaller quantities of chemical energy than those of the original reactants. If the product has a *smaller* quantity of energy, energy is *liberated* during the reaction—usually as heat and sometimes light or electric energy. If the product has a *larger* quantity of energy, energy is *absorbed* during the reaction.

*Energy is **absorbed** in endothermic reactions.*

Calcium carbide, a compound formed of the elements calcium and carbon, is produced in the intense heat of an electric furnace. The compound carbon disulfide is formed when hot sulfur vapor is passed over white-hot carbon in an electric furnace. Heat energy is absorbed continuously while these chemical reactions proceed. *Any process that absorbs energy as it progresses is said to be **endothermic**.*

*Energy is **released** in exothermic reactions.*

Some chemical reactions are important because of their products. Others are important because of the energy that is released. When fuels are burned, large amounts of heat energy are released rapidly. Many similar reactions occur in nature but take place so slowly that the release of heat is not noticed. *Any process that liberates energy as it proceeds is said to be **exothermic**.* The majority of chemical reactions that occur spontaneously in nature are exothermic. One notable exception is the photosynthesis process in green plants. This process is endothermic. Solar energy is stored in the product of photosynthesis (a sugar) as chemical energy.

As fuels are burned, light energy usually accompanies the release of heat. A photoflash lamp is designed to release a maximum amount of energy as light. The final proof of a chemical

Figure 2-11. The elements sulfur and zinc mixed in the evaporating dish at room temperature (top photo) do not react chemically until the temperature of the mixture is raised, as in the bottom photo. Once started, the reaction is exothermic.

A Celsius temperature readout of 10 degrees is expressed as 10 °C.

A Celsius temperature interval or temperature difference of 10 degrees is expressed as 10 C°.

reaction rests with the analysis of the products. However, *the evolution of heat and light* usually indicates a chemical reaction, as illustrated in Figure 2-11.

The explosion of dynamite is a violent exothermic chemical reaction that is ordinarily used to produce mechanical energy. Similarly, the combustion of a mixture of gasoline vapor and air in the cylinder of an automobile engine produces mechanical energy.

The zinc cup in a flashlight cell acts as the negative electrode in a chemical reaction when the cell is in use. During the reaction, chemical energy is converted to electric energy. The resulting *electric current* gives evidence of the chemical reaction by heating the lamp filament to incandescence.

The production of a gas usually indicates that a chemical reaction is taking place. However, a boiling liquid, a dissolved gas escaping from a solution, or gas escaping from the surface of a solid should not be mistaken as evidence of chemical reaction.

The formation of a precipitate, an insoluble solid formed when one solution is added to another, provides evidence that a chemical reaction occurred as two solutions were mixed.

Chemists make use of various agents to initiate chemical reactions or to control reactions already started. Some form of energy is often used.

1. *Heat energy.* A match is kindled by rubbing it over a rough surface to heat it by friction. If paper is held in the flame of the burning match, the paper ignites and begins to burn. Heat from the burning match is required to start this combustion; once started, however, it proceeds as an exothermic reaction. It is not necessary to continue supplying heat from the burning match in order to keep the paper burning.

Many chemical reactions that occur in the preparation of foods are endothermic. Heat is supplied to sustain these reactions. As a rule, increasing the temperature increases the rate of chemical reaction. *Each increase in temperature of 10 C° approximately doubles the rate of many chemical re.ctions.*

2. *Light energy.* Photosynthesis, the endothermic reaction by which green plants manufacture food, requires light energy. In photography, a camera shutter is opened for a fraction of a second and light falls on the light-sensitive film. This light energy triggers a chemical reaction in the film, and a "latent" (invisible) image is formed. From this image a picture can be developed at some future time.

3. *Electric energy.* If a direct current of electricity is passed through water (to which a little acid has been added to improve its conductivity), the water is decomposed to hydrogen and oxygen. Similar procedures are used for recharging storage batteries, electroplating one metal on the surface of another, recovering aluminum and other metals from their ores, and purifying some metals. Electric energy is also used to produce heat for

thermal processes. The use of an electric furnace in the production of calcium carbide and carbon disulfide was mentioned earlier in this section.

4. *Solution in water.* Baking powder is a mixture of two or more compounds. No chemical reaction occurs as long as the mixture is kept *dry*. However, when water is added to baking powder, a chemical reaction begins immediately, and a gas is released. Many chemicals that do not react in the *dry* state begin to react as soon as they are dissolved in water.

5. *Catalysis* (kuh-*tal*-uh-sis). Some chemical reactions are promoted by *catalysts* (*kat*-uh-lists). These are specific agents that promote reactions that would otherwise be difficult or impractical to carry out. A catalyst does not start a chemical reaction that would not occur of and by itself. For example, oxygen can be prepared in the laboratory by heating a mixture of potassium chlorate and manganese dioxide. Without the manganese dioxide, the preparation would have to be carried out at a higher temperature. Also, the gas would be produced more slowly. The manganese dioxide aids the reaction by its presence. It can be recovered in its original form at the conclusion of the experiment. *A* ***catalyst*** *is a substance or combination of substances that increases the rate of a chemical reaction without itself being permanently changed.*

Many chemical processes, such as the production of vegetable shortening, the manufacture of synthetic rubber, and the preparation of high-octane gasoline, depend on specific catalytic agents for their successful operation.

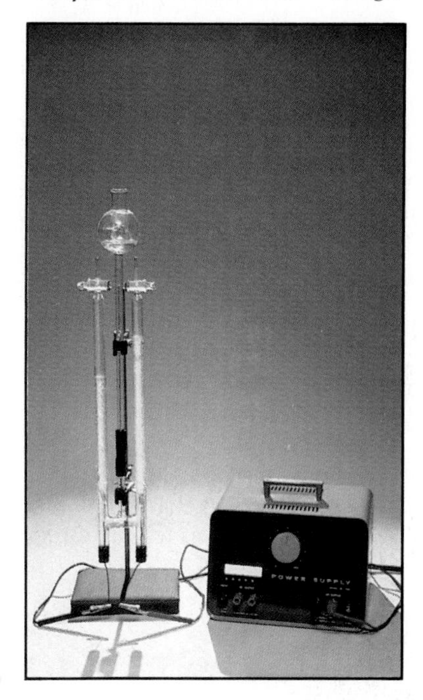

Figure 2-12. Electric energy can be used to decompose water. Oxygen gas is collected above the positive electrode and hydrogen gas is collected above the negative electrode. Observe that the volume ratio of hydrogen gas to oxygen gas is 2:1.

2.14 Reaction tendencies It is not surprising to see a ball roll down an incline unaided; it is just what would be expected. Potential energy is given up in this process, and the ball attains a more stable state at a lower energy level. It would be surprising, however, if the ball rolled up the incline by and of itself. From such experiences with nature, a basic rule for natural processes becomes apparent: *There is a tendency for processes to occur that lead to a lower energy state.* This tendency in nature is toward greater *stability* (resistance to change) in a system.

The great majority of chemical reactions in nature are exothermic. As the reactions proceed, energy is liberated and the products have less energy than the original reactants. With the above rule in mind, one would expect exothermic reactions to occur spontaneously; that is, they have the potential to proceed without the assistance of an external agent.

Spontaneous chemical reactions are self-acting because of inherent properties of the system.

It should follow that endothermic reactions, in which energy is absorbed, would not occur spontaneously but would proceed only with the assistance of an external agent. Certainly energy must be expended on the ball to roll it up the incline. At the top of the incline the ball's potential energy is high and its stability is low.

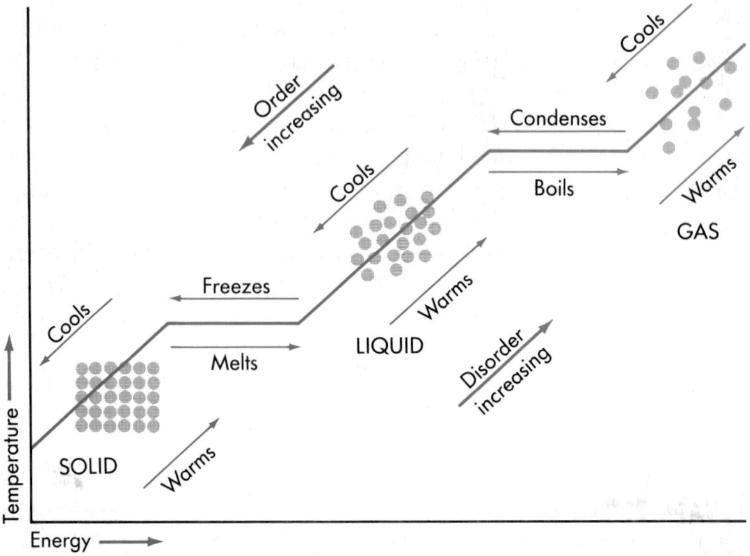

Figure 2-13. Change of phase. Observe that each change of phase occurs at a constant temperature.

There are several processes that occur spontaneously with the absorption of energy. When steam is passed over hot carbon, carbon monoxide gas and hydrogen gas are produced and heat is absorbed. The products are in a higher energy state than the reactants. This spontaneous reaction is endothermic, and the rule appears to have failed. Apparently the tendency of a reaction to proceed spontaneously is not determined exclusively by the energy-change rule.

An ice cube melts spontaneously at room temperature as heat is transferred from the warm body of air to the cool ice and to the ice water formed. The well-ordered structure of the ice crystal (the solid phase) is lost, and the less orderly liquid phase of higher energy content is formed. This phase-change mechanism is illustrated in Figure 2-13.

What is observed here is a tendency for the ice to move into the less orderly liquid phase. This observation suggests a second basic rule for natural processes: *There is a tendency for processes to occur that lead to a less orderly or a more disordered state.* This tendency in nature is toward greater *disorder* in a system. A disordered system is one that lacks a regular arrangement of its parts. *That property which describes the disorder of a system is called* ***entropy.*** The more disordered or more random the system, the *higher* is its entropy. Liquid water has higher entropy than ice.

Entropy is a measure of the disorder of a system.

Thus, processes in nature are driven in two ways: toward *lowest* energy and toward *highest* entropy. When these two oppose each other, the dominant factor determines the direction of the spontaneous change. In the steam-plus-hot-carbon reaction, the temperature is high enough that the entropy factor overcomes the unfavorable energy-change factor. The spontaneous endothermic reaction occurs.

If the ice cube is subjected to a temperature below 0 °C, it does not melt. Liquid water placed in this environment freezes. The temperature is low enough that the energy-change factor overcomes the entropy-change factor. Heat is given up by the water (lower energy), and the well-ordered ice crystal is formed (lower entropy). These examples suggest that an entropy change is favored by high temperatures and an energy change is favored by low temperatures. Entropy as a factor in reaction systems will be considered in more detail in Section 20.7.

2.15 Nuclear changes New substances are produced during a chemical reaction by rearranging the atoms of the original substances. In a *nuclear change*, new substances with new properties are also produced. There is a difference, however. *In a* ***nuclear change,*** *the new substances are formed by changes in the identity of the atoms themselves.*

In nature, some nuclear changes take place spontaneously. Radium atoms disintegrate in successive stages, finally becoming lead. Scientists are able to bring about many important nuclear changes. The elements beyond uranium named in Section 2.4 are products of nuclear changes. Nuclear reactions, both natural and artificial, will be discussed in Chapter 31.

SUMMARY

Matter is classified as a mixture or a pure substance. Pure substances are homogeneous and are either compounds or elements. Compounds can be decomposed into two or more simpler substances by ordinary chemical means. Elements cannot be further decomposed by ordinary chemical means. Mixtures are composed of two or more kinds of matter, each retaining its own characteristic properties.

Of the 109 known elements, the 92 ranging from hydrogen to uranium represent the pre–Atomic Age list of elements. Approximately 90 of these elements occur free or combined in the earth's crust. Only about 30 elements are fairly common. Based on their properties, elements are recognized as metals or nonmetals. Some elements having certain properties characteristic of metals and others of nonmetals are called metalloids, or semiconductors.

Symbols are used to represent elements. The symbol of an element stands for one atom of that element. The atom is the smallest particle of an element that can enter into a combination with other elements. Compounds are composed of two or more elements. Every compound has a definite composition by mass.

Physical changes are those in which certain identifying properties of a substance remain unchanged. Examples are freezing, boiling, and dissolving. Chemical changes are those in which different substances with new properties are formed. Reactions that absorb energy as they proceed are endothermic. Reactions that liberate energy are exothermic. The rates of some chemical reactions can be accelerated by catalysts.

There is a basic tendency for processes to occur in nature that lead to a lower energy state. A second basic tendency is for processes to occur that lead to a more disordered or more random state. Entropy is the property that describes the disorder of a system. The more disordered the system, the higher is its entropy.

VOCABULARY

catalyst
chemical change
chemical symbol
compound
ductile
element
endothermic
entropy
exothermic
formula
heterogeneous
homogeneous
malleable
metal
metalloid
mixture
nonmetal
nuclear change
physical change
precipitate
semiconductor
spontaneous reaction
substance
symbol

QUESTIONS

Group A

1. What are the three general classes of matter? *A1, A2a*
2. Distinguish between matter and a substance. *A1, A2a*
3. Distinguish between a complex substance and an elemental substance. *A1, A2a*
4. *(a)* What are the two general classes of elements? *(b)* Do all elements fit definitely into one of these classes? *A1, A2e*
5. Distinguish between a compound and a mixture. *A1, A2c, A5a*
6. What are the five most abundant elements by mass in the earth's surface environment? *A3d*
7. *(a)* What are the properties of metals? *(b)* of nonmetals? *A2e*
8. *(a)* How many elements are known? *(b)* How many were known prior to the Atomic Age? *A3b*
9. What is the meaning of a chemical symbol? *A3c*
10. *(a)* List five familiar substances that are elements. *(b)* List five that are compounds. *(c)* List five familiar mixtures.
11. What is the difference between a physical change and a chemical change? *A2f*
12. How can a chemist usually increase the speed of a chemical reaction? *A2g*
13. If two or more elements have symbols beginning with the same letter, how are they distinguished? *A3c*
14. If a symbol has two letters, *(a)* what is always true of the first letter? *(b)* what is always true of the second letter? *A3c*
15. Write the names of the elements represented by these symbols: *(a)* S; *(b)* Zn; *(c)* K; *(d)* N; *(e)* Ni; *(f)* Co; *(g)* Ba; *(h)* Fe; *(i)* Cl; *(j)* Cr; *(k)* Mg; *(l)* Mn; *(m)* As; *(n)* Pb; *(o)* Na. *A4*
16. Write the symbols for the following elements: *(a)* aluminum; *(b)* tungsten; *(c)* mercury; *(d)* carbon; *(e)* bromine; *(f)* silicon; *(g)* tin; *(h)* hydrogen; *(i)* gold; *(j)* silver; *(k)* fluorine; *(l)* strontium; *(m)* calcium; *(n)* phosphorus; *(o)* bismuth. *A4*
17. Write the systematic name and symbol for: *(a)* element 101, *(b)* element 102, *(c)* element 103, *(d)* element 213, *(e)* element 290. *A6a*
18. Distinguish between exothermic and endothermic processes. *A2g*

Group B

19. What difference in the properties of white sand and sugar would enable you to separate a mixture of the two substances? *A2f*
20. How would you carry out the separation and recovery of both components of the sand-sugar mixture of Question 19? *A2f*
21. Why is a solution recognized as a mixture? *A2a*
22. What is the meaning of the phrase "definite composition by mass"? *A5b*
23. Why is the law of definite composition important to chemists? *A5b*
24. Consult the list of known elements in Appendix Table 4 and compile a list of those about which you already have some knowledge. Give the name, symbol, and pertinent information in column form. *A2b, A3d*
25. Given two liquids, one a solution and the other a compound, distinguish the solution from the compound. *A2c, A5a*

26. Suppose you heat three different solids in open vessels and then allow them to cool. The first gains mass, the second loses mass, and the mass of the third remains unchanged. How can you reconcile these facts with the generalization that in a chemical reaction the total mass of the reacting materials is equal to the total mass of the products? *A2f, A6b*
27. How can you explain the fact that gold, silver, and copper were known long before such metals as iron and aluminum?
28. Suppose you were given a sample of iodine crystals, a sample of antimony, and a sample of a mixture of iodine and antimony that had been ground together to form a fine powder of uniform consistency. Look up the physical and chemical properties of both iodine and antimony. Then list the properties of each element that you believe would be useful in separating and recovering them from the mixture. On the basis of these properties, devise a procedure to separate the two elements from the mixture and recover the separate elements. *A2f*
29. Which of these changes are physical and which are chemical? *(a)* burning coal; *(b)* tarnishing silver; *(c)* magnetizing steel; *(d)* exploding gunpowder; *(e)* boiling water; *(f)* melting shortening. *A2f*
30. Which of the chemical changes listed in Question 29 are also exothermic? *A2g*
31. Show by example how each form of energy produces chemical changes: *(a)* heat energy; *(b)* light energy; *(c)* electric energy. *A6b*
32. What evidences usually indicate chemical reaction? *A6b*
33. Can you suggest a reason why iron and sulfur unite in definite proportions when iron sulfide is formed?
34. How can you determine whether a particular change is physical or chemical? *A2f, A6b*
35. What two basic tendencies in nature appear to influence reaction processes? *A5c*
36. An ice cube melts at room temperature and water freezes at temperatures of 0 °C and below. From these facts, what can you infer concerning the relationship between the temperature of a system and the influence of the entropy factor on the change that the system undergoes? *A5c*

PROBLEMS

Group B

(Show problem solutions by factor-label operations. Refer to Section 1.21.)

1. Determine the volume of an iron sphere that has a mass of 4.05 kg. (See Appendix Table 16 for properties of elements.)
2. If 1.25 L of water absorbs 45.5 kcal of heat, how much does the temperature of the water rise?
3. The temperature of $45\bar{0}$ g of water is raised from 21.0 °C to 100.0 °C. Determine the quantity of heat energy required to effect this change in temperature, *(a)* in kilocalories, *(b)* in kilojoules.
4. The Einstein equation $E = mc^2$ shows that 1.0 g of matter is the equivalent of 2.1×10^{10} kcal of energy (Section 1.11). When 2.0 g of hydrogen reacts with 16 g of oxygen to form 18 g of water, 68 kcal of heat energy is given up. This quantity of energy is attributed to what loss of mass in the reaction system?
5. Ordinary table salt is the compound sodium chloride, NaCl. According to the law of definite composition, this compound, NaCl, always consists of 39.3% sodium and 60.7% chlorine. (A percentage composition may be interpreted as *grams of element per $10\bar{0}$ g of compound*.) A chemist carries out a reaction between 10.0 g of sodium and 20.0 g of chlorine to produce sodium chloride. *(a)* Which reactant element is present in excess? (Justify your answer.) *(b)* What quantity of sodium chloride is produced? *(c)* What quantity of the element in short supply would have been required to react with all of the element present in excess? *A5b*

The 1943 Nobel Prize in chemistry was given to George de Hevesy (Hungarian) for his use of isotopes as tracers in chemical research. He prepared phosphorus-32, a radioactive isotope containing one more neutron than the more common phosphorus-31. He then incorporated this isotope into various organic compounds and studied physiological processes by following the route of phosphorus-32 through the body. He also calculated the relative abundance of the chemical elements and helped to discover the element hafnium.

Atomic Structure

3

GOALS

In this chapter you will gain an understanding of: • the atomic theory • the structure of the atom • atomic number, mass number, and atomic mass • isotopes • atomic weight and gram-atomic weight • the relationship between a mole and the Avogadro number

3.1 Particles of matter For a very long time people have believed that matter is made up of simple, indivisible particles. As early as 400 B.C. some Greek thinkers had the idea that matter could not be destroyed. They believed that matter could be divided into smaller and smaller particles until a basic particle of matter that could not be divided further was reached. Such basic particles were thought to be the smallest particles of matter that existed. Democritus called them *atoms*. The word "atom" is from the Greek word meaning "indivisible."

Democritus (deh-mock-writ-us) (460–370 B.C.) was a Greek thinker. He believed that the hard atoms of the four primitive elements (earth, air, water, and fire) moved in a vacuum. The shape and size of these atoms explained some of their properties. For example, the atoms of fire were tiny spheres. Because of their smooth surfaces, they did not link with the atoms of the other elements. The atoms of earth, air, and water had shapes that enabled them to connect with each other and form visible matter.

When you crush a lump of sugar, you can see that it is made up of many small particles of sugar. You may grind these particles into very fine powder, but each tiny piece is still sugar. Now suppose you dissolve the sugar in water. The tiny particles disappear completely. Even if you look at the sugar-water solution through a microscope, you cannot see any sugar particles. However, if you taste the solution, you would know that the sugar is still there.

If you open a laboratory gas valve, you can smell the escaping gas. Yet you cannot see gas particles in the air of the room, even if you use the most powerful microscope. These observations and many others like them have led scientists to believe that the basic particles of matter must be very, very small.

3.2 The atomic theory Invention of the chemical balance gave chemists a tool for studying the composition of substances quantitatively. Chemists showed, in the half-century prior to the year 1800, that substances they investigated are chemical combinations of a fairly small number of elements. They learned much about how elements combine to form compounds and how compounds can be broken down into their constituent elements.

From this knowledge they formulated several quantitative laws of chemical combination, such as Proust's law of definite composition (see Section 2.9).

John Dalton was an English schoolmaster. He was interested in the composition and properties of the gases in the atmosphere. He kept a daily record of the weather from 1787 until 1844.

John Dalton's atomic theory, first conceived in 1803 and published in 1808–1810, postulated the existence of a different kind of atom for each element. His theory was ultimately accepted because it satisfactorily explained these laws. It forms the basis of modern atomic theory.

A theory is a plausible explanation of an observed natural phenomenon in terms of a simple model with familiar properties. See Section 1.4(3).

In science, a theory is never secure. Each new observation tests it. If the theory does not explain the new observation satisfactorily, the theory is either rejected or revised. With increased knowledge, Dalton's original theory has been revised. For example, Dalton thought that atoms of the same element were identical in all respects, particularly mass. But evidence now shows that all atoms of the same element do not have exactly the same mass. Generally, the atomic theory has stood the test of time. It has satisfactorily explained observations and laws in many different fields. It now includes information about:

1. the structure and properties of atoms;
2. the kinds of compounds that atoms form;
3. the properties of compounds that atoms form;
4. the mass, volume, and energy relations of reactions between atoms.

*An **atom** is the smallest unit of an element that can exist either alone or in combination with atoms of the same or different elements.* It is the smallest individual structure of an element that retains the properties of the element. The chemical and physical properties of matter lead scientists to make the following statements about atoms and their properties:

1. All matter is made up of very small structures called *atoms*.
2. Atoms of the *same element* are *chemically alike;* atoms of *different elements* are *chemically different*.
3. Individual atoms of an element may not all have the same mass. However, *the atoms of an element,* as it occurs naturally, *have,* for practical purposes, *a definite average mass that is characteristic of the element.*
4. Individual atoms of different elements may have nearly identical masses. However, *the atoms of different naturally occurring elements have different average masses.*
5. Atoms are not subdivided in *chemical reactions*.

Right now statements 3 and 4 may be a bit puzzling. In Section 3.14, their meaning will be made clearer.

3.3 The structure of the atom For nearly 100 years, scientists have been gathering evidence about the structure of atoms. Some of this evidence has come from the study of radioactive elements such as radium and uranium. Particle accelerators, the mass spectrograph, the X-ray tube, the spectroscope, and a variety of other electronic devices have provided additional information. The field-ion microscope and the transmission electron microscope have made possible photographs of arrays of atoms. See Figures 3-1 and 3-2. From all this information scientists have developed a theory of atomic structure. This model of atomic structure will be described in the following sections of this chapter and in Chapter 4. As you read, remember that this explanation is based on the best, present understanding of experiments on atomic structure. Future experiments may necessitate changes in the model.

At the present time, scientists know that atoms are not simple indivisible particles. Instead, they are composed of several different kinds of still smaller particles that are arranged in a complex way.

An atom consists of two main parts. *The positively charged central part is called the **nucleus**.* It is very small and very dense. Its diameter is about 10^{-12} cm. Its density is about 20 metric tn/cm^3.

A more useful unit for atomic sizes is the angstrom.

$$\textbf{1 angstrom (Å)} = \mathbf{10^{-8}}\ \textbf{cm}$$

To get some idea of the extreme smallness of the angstrom, consider the fact that 1 cm is the same fractional part of 10^3 km (about 600 miles) as 1 Å is of 1 cm.

The diameter of the nucleus is about 10^{-12} cm, or about 10^{-4} Å. This is about one ten-thousandth of the diameter of the atom itself, since atoms range from 1 Å to 5 Å in diameter.

The idea that an atom has a nucleus is the result of experiments conducted about 1910 by the English physicist Ernest Rutherford (1871–1937). See the schematic diagram in Figure 3-3. Rutherford used an evacuated tube (a tube from which the gas has been pumped). In it, a beam of high-speed, positively charged particles was aimed at a thin sheet of gold. Most of the particles passed straight through the gold. A few were slightly deflected (turned away from a straight course) as they passed through the gold. A very few were greatly deflected back from the gold. These very few great deflections were explained by assuming that the positively charged particles bounced back if they approached a positively charged atomic nucleus head-on. The few slight deflections were explained by assuming that the particles were turned from their paths in near misses of nuclei. But most of the particles passed straight through the gold foil. Therefore, the experimenters reasoned, most of an atom must consist of space through which such particles could move readily. In fact, since a vast majority of the particles went undeflected, the nucleus must occupy a very, very small portion of the volume of an atom.

Figure 3-1. The point of a tungsten needle as viewed through a field-ion microscope. This instrument magnifies objects up to 3,000,000 diameters and shows the regular arrangement of tungsten atoms in the metal.

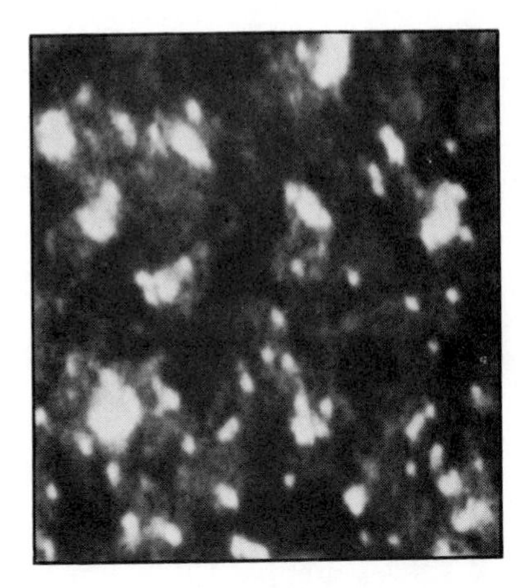

Figure 3-2. Single uranium atoms magnified more than five million times by a transmission electron microscope. The small bright spots are single uranium atoms. The larger white spots are clusters of uranium atoms. The background is a thin carbon filament, about six atoms thick, that supports the uranium atoms.

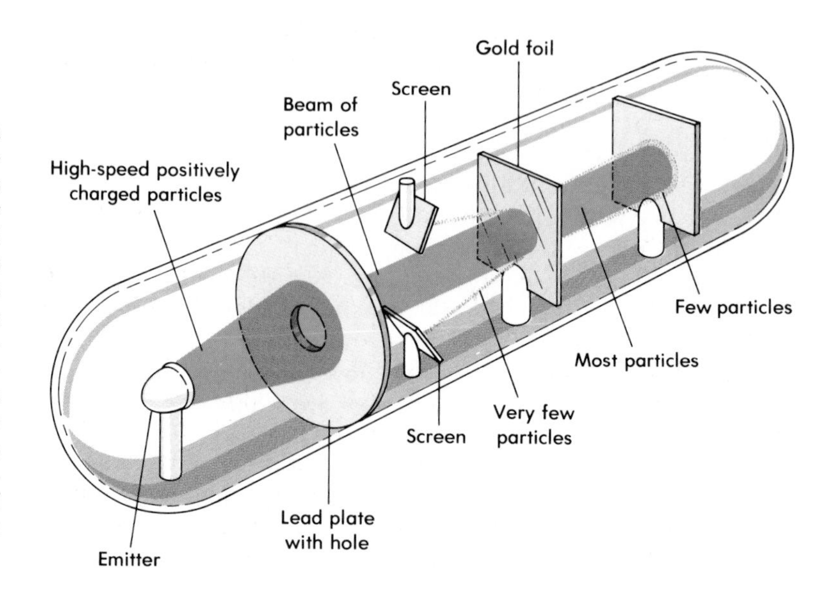

Figure 3-3. A schematic diagram of Rutherford's experiment. High-speed positively charged particles are given off in many directions by the emitter. The hole in the lead plate allows a beam of particles moving toward the gold foil to pass through. The solid portion of the lead plate absorbs the other particles. Most of the high-speed positively charged particles in the beam passed through the gold foil. A few were slightly deflected, while a very few were very greatly deflected. The screens emit a flash of light when they are struck by a charged particle.

The other part of an atom lies outside the central nucleus. It is made up of negatively charged particles and is called the *electron cloud.* The electron cloud gives an atom its volume and keeps out other atoms.

Each atom is electrically neutral. The electric neutrality indicates that the total positive charge of the nucleus must equal the total negative charge of the electron cloud.

3.4 The electron cloud An electron cloud is made up of negatively charged particles called *electrons.* Electrons move about the nucleus with different energies. Electrons having similar energies are said to be in the same *energy level.*

About 1913, the Danish scientist Niels Bohr (1885–1962) compared the movements of electrons about the nucleus of an atom with the revolution of the planets around the sun. The motion of the electrons is now believed to be much less definite than the orbits of the planets. Electrons move about the nucleus of an atom much as bees move about in the area near their hive. Sometimes the electrons are near the nucleus; sometimes they are farther away. While electrons do have some properties of particles, it is not possible for scientists to determine the paths along which they move. The electrons just seem to occupy the relatively vast empty space around the nucleus. The positions of the electrons are best visualized as an electron cloud about the nucleus.

Electrons were discovered in 1897 by the English scientist J. J. Thomson (1856–1940).

***Electrons** are pointlike particles having a unit negative electric charge and a mass of* 9.110×10^{-28} *g.* Regardless of the atom of which an electron is a part, it is believed that *all electrons are* *identical.* The mass of an electron is $\frac{1}{1837}$ of the mass of the most common type of hydrogen atom. The most common type of hydrogen atom has the simplest structure and the least mass of any atom.

3.5 The nucleus of the atom Except for the simplest type of hydrogen atom, a nucleus is made up of two kinds of particles, *protons* and *neutrons*. ***Protons*** *are positively charged particles with a mass of 1.673×10^{-24} g*. This mass is $\frac{1836}{1837}$ of the mass of the simplest type of hydrogen atom. This atom consists of a single-proton nucleus with a single electron moving about it. Most of the mass of the simplest type of hydrogen atom is due to the mass of the proton. The diameter of a proton is about 10^{-5} Å. The electric charge on the proton has the same magnitude as that on an electron, but is positive in sign. In any atom, the number of electrons and protons is equal. Since protons and electrons have equal but opposite electric charges, an atom is electrically neutral.

Protons were discovered in the early years of this century. Today there is evidence that protons have an internal structure. One theory supposes that protons are mostly empty space containing three, hard, fundamental pointlike units called quarks.

Neutrons *are neutral particles with a mass of 1.675×10^{-24} g*, which is about the same mass as a proton. They have no electric charge. The name, mass, and charge of each atomic particle is given in Table 3-1.

Neutrons were discovered by the English scientist James C. Chadwick (1891–1974) in 1932. The experiment during which this discovery was made is described in Section 31.11. Neutrons are also believed to have an internal structure consisting of quarks.

Particles that have the same electric charge generally repel one another. Nevertheless, as many as 100 protons can exist close together in a nucleus. This close existence of protons in the nucleus can occur when up to about 150 neutrons are also present. When a proton and a neutron are very close to each other, there is a strong attraction between them. Proton-proton attractive forces and neutron-neutron attractive forces also exist when these pairs of particles are very close together. These short-range proton-neutron, proton-proton, and neutron-neutron forces hold the nuclear particles together. They are referred to as *nuclear forces*.

In nuclear changes, particles in addition to protons and neutrons come from the nucleus. We shall describe some of these in Chapter 31. We do not need to know about these particles now to understand atomic structure and chemical changes.

The nuclei of atoms of different elements are different. The amounts of positive charge are always different. The nuclei of atoms of different elements also have different masses, although the difference in mass is sometimes very slight.

Table 3-1
PARTICLES IN AN ATOM

Name	*Mass*	*Atomic mass* (See Section 3.13)	*Mass number* (See Section 3.7)	*Charge*
electron	9.110×10^{-28} g	0.0005486 *u*	0	−1
proton	1.673×10^{-24} g	1.007276 *u*	1	+1
neutron	1.675×10^{-24} g	1.008665 *u*	1	0

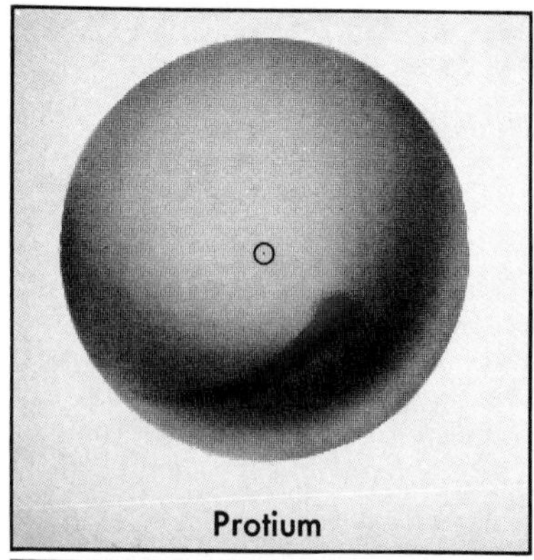

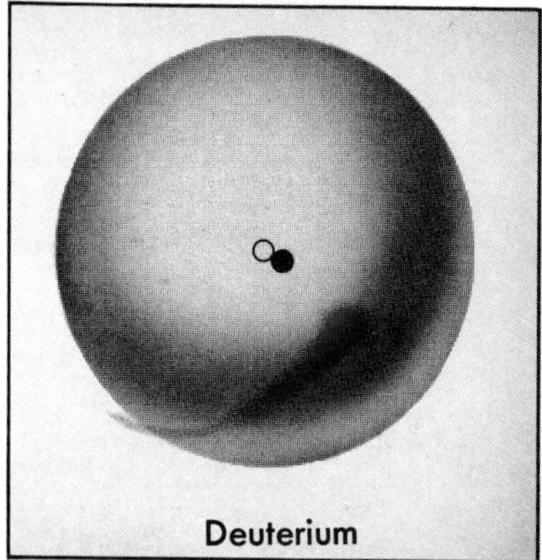

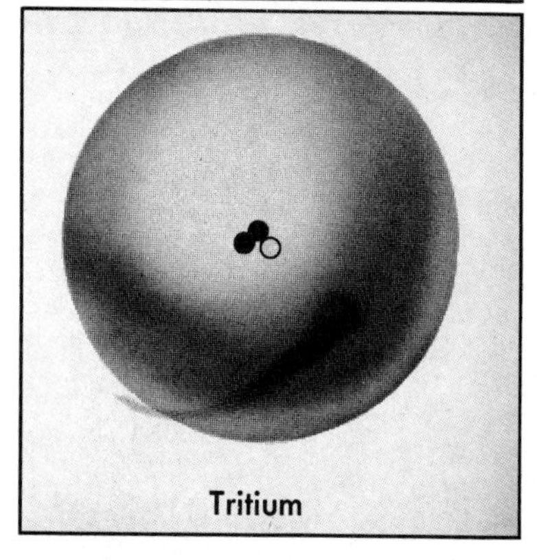

Figure 3-4. Tentative models of three isotopes of hydrogen: protium, deuterium, and tritium. Each has one proton in the nucleus and one electron moving about the nucleus in the 1st energy level. The only structural difference between them is the number of neutrons in the nucleus of each atom. The nuclei are enlarged in proportion to the size of the atom to show their composition.

3.6 Hydrogen atoms The most common type of hydrogen is sometimes called *protium* (*proht*-ee-um). Its atoms have the simplest possible composition. The nucleus of a protium atom consists of *one proton*. This proton has one electron moving about it. This electron could most probably be found at a distance from the nucleus corresponding to the lowest energy level that an electron can occupy. This energy level is called the ***1st energy level.*** An illustration may help you to better understand the sizes and distances between the particles of the protium atom. Picture the nucleus (a proton) as being 1 cm in diameter (about the diameter of the end of your little finger). Comparatively, the much smaller electron is on the average about 48 m from the nucleus. (The 48-m distance is about one-half the length of a football field.) The electron can be thought of as moving about the nucleus and effectively occupying the space surrounding it.

In addition to protium, which makes up 99.985% of naturally occuring hydrogen, there are two other known forms of hydrogen atoms. One of these is *deuterium* (dyou-*tir*-ee-um), which occurs to the extent of 0.015% in nature. Each deuterium atom has a nucleus containing *one proton* and *one neutron*, with one electron moving about it.

The third form of hydrogen is *tritium* (*trit*-ee-um). Tritium is a radioactive form. It exists in nature in very small amounts but can be prepared artificially by a nuclear reaction. Each tritium atom has a nucleus composed of *one proton* and *two neutrons*, with one electron moving about it. See Figure 3-4.

3.7 Atomic number and mass number *The **atomic number** of an atom is the number of protons in the nucleus of that atom.* An element consists of atoms, all of which have the same number of protons in their nuclei. Hence, all atoms of the same element have the same atomic number. (All unexcited neutral atoms of an element have the same arrangement of electrons about their nuclei, too.) The atoms of the element hydrogen, whether they be protium, deuterium, or tritium atoms, have one proton in each nucleus. Their atomic number, therefore, is 1. Any atom having the atomic number 1 contains one proton in its nucleus and is a hydrogen atom.

However, because protium, deuterium, and tritium nuclei contain different numbers of neutrons, these atoms have different masses. *Atoms of the same element that have different masses are called **isotopes.***

All elements have two or more isotopes. Isotopes of an element may occur naturally, or they may be prepared artificially. While isotopes have different masses, they do not differ significantly in chemical properties. See Appendix Table 2 for a list of natural and radioactive isotopes of some of the elements.

Each different variety of atom as determined by the number of protons and number of neutrons in its nucleus is called a ***nuclide.*** Nuclides having *the same number of protons* (the same atomic number) are *isotopes.* The three hydrogen isotopes are the nuclides protium, deuterium, and tritium.

An element is a substance all the atoms of which have the same number of protons; that is, they have the same atomic number. This statement is a more precise definition of an element than that given in Section 2.3.

In addition to their names, hydrogen nuclides may also be distinguished by their *mass numbers. The* ***mass number*** *of an atom is the sum of the number of protons and neutrons in its nucleus.* The mass number of protium is 1 (1 proton + 0 neutron). That of deuterium is 2 (1 proton + 1 neutron). That of tritium is 3 (1 proton + 2 neutrons). Sometimes these isotopes are named hydrogen-1, hydrogen-2, and hydrogen-3, respectively.

3.8 Elements in atomic number order At the time of this writing, 109 different elements are known or reported to exist. Their atomic numbers range from 1 to 109. The elements may be arranged in the order of increasing atomic number. This arrangement simplifies the understanding of atomic structure. If the elements are arranged in this way, the nuclei of the atoms of one element differ from those of the preceding element by one additional proton per nucleus. (The number of neutrons per nucleus may or may not change from atom to atom.)

3.9 Helium atoms The second element in order of complexity is helium. Since each helium nucleus contains two protons, the atomic number of helium is 2. Natural helium exists as a mixture of two isotopes. Helium-3 occurs to the extent of $1.34 \times 10^{-4}\%$. Helium-4 accounts for practically 100% of natural helium. These helium nuclides contain 1 neutron and 2 neutrons per nucleus, respectively. (Note that the number of neutrons in the nucleus of an atom may be determined by subtracting the atomic number from the mass number.) Moving about each helium nucleus are two electrons, both in the ***1st energy level.*** The chemical properties and spectrum of helium indicate that the 1st energy level may contain a *maximum* of two electrons. Thus, helium atoms have a *filled* 1st energy level. A model of a helium-4 atom is shown in Figure 3-5.

The atoms of hydrogen have one 1st-energy-level electron. The atoms of helium have two (the maximum number). Hydrogen and helium, then, form the first *series* of elements.

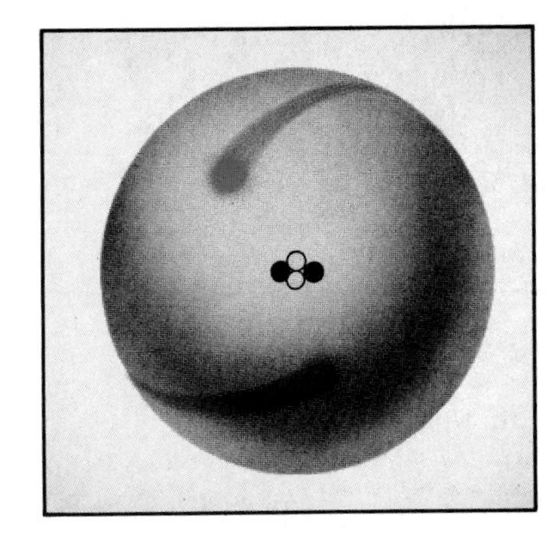

Figure 3-5. Tentative model of a helium-4 atom. Its nucleus consists of two protons and two neutrons. Its two electrons move about this nucleus and completely fill the 1st energy level of the atom.

3.10 Lithium atoms Lithium exists in nature as two isotopes. Each atom of one isotope contains 3 protons, 3 neutrons, and 3 electrons. Each atom of the other isotope contains 3 protons, 4 neutrons, and 3 electrons. By the composition of these atoms, the atomic number of lithium is known to be 3 (3 protons). The

Figure 3-6. Tentative model of a lithium-7 atom. This atom has a nucleus consisting of three protons and four neutrons. Two electrons are in the 1st energy level and one is in the 2nd energy level. Lithium is the first element in the second series.

mass number of the first isotope is 6 (3 protons + 3 neutrons). The mass number of the second isotope is 7 (3 protons + 4 neutrons). These isotopes are lithium-6 and lithium-7.

The chemical properties of lithium help chemists determine its electron arrangement. Two of the three electrons in a lithium atom move in the 1st energy level. The third moves about the nucleus at a greater distance and with higher energy than the other two. It moves in the next higher energy level. This is called the ***2nd energy level.*** The 1st energy level contains no more than two electrons. When it has this maximum of two electrons, additional electrons occupy higher energy levels at greater distances from the nucleus. See Figure 3-6.

3.11 Other atoms of the second series The element with atomic number 4 is beryllium. Naturally occurring beryllium consists of only one nuclide, beryllium-9. Beryllium nuclei consist of four protons and five neutrons. The four electrons are arranged with two in the 1st energy level and two in the 2nd energy level. Next in order of atomic structure are the elements boron, carbon, nitrogen, oxygen, fluorine, and neon. The atoms of each successive element have one additional proton per nucleus and may have one or two additional neutrons per nucleus. Each successive element has one additional electron in the 2nd energy level of each of its atoms. The atoms of the element neon have eight electrons in the 2nd energy level. The chemical properties and spectrum of neon indicate that the 2nd energy level may contain a maximum of eight electrons. So neon completes the second series of elements.

Table 3-2 shows the composition of the nuclei and electron configurations of atoms in the first and second series.

3.12 The atoms of the third series The elements in the third series are sodium, magnesium, aluminum, silicon, phosphorus, sulfur, chlorine, and argon. The atoms of these elements have a filled 1st energy level of two electrons and a filled 2nd energy level of eight electrons. Successive electrons occupy the ***3rd energy level.*** This level contains eight electrons in argon atoms.

Atomic masses are measured using a mass spectrograph. This device was developed by F. W. Aston, an English scientist, in 1919.

3.13 Atomic mass The actual mass of a single atom is very small. An atom of oxygen-16 has a mass of 2.65×10^{-23} g. A hydrogen-1 atom has a mass of 1.67×10^{-24} g. These numbers are not very easy to use in chemical arithmetic problems. Therefore, a system for expressing the masses of atoms in numbers on a relative scale has been worked out. These relative numbers are easier to use.

Table 3-2
NATURALLY OCCURRING NUCLIDES
(First and Second Series of Elements)

Name of nuclide	*Abundance*	*Atomic number*	*Mass number*	*Composition of nucleus*		*Electron configuration*	
				Protons	*Neutrons*	*1st energy level*	*2nd energy level*
hydrogen-1 (protium)	99.985%	1	1	1	0	1	
hydrogen-2 (deuterium)	0.015%	1	2	1	1	1	
helium-3	0.00013%	2	3	2	1	2	
helium-4	99.99987%	2	4	2	2	2	
lithium-6	7.42%	3	6	3	3	2	1
lithium-7	92.58%	3	7	3	4	2	1
beryllium-9	100%	4	9	4	5	2	2
boron-10	19.78%	5	10	5	5	2	3
boron-11	80.22%	5	11	5	6	2	3
carbon-12	98.892%	6	12	6	6	2	4
carbon-13	1.108%	6	13	6	7	2	4
nitrogen-14	99.635%	7	14	7	7	2	5
nitrogen-15	0.365%	7	15	7	8	2	5
oxygen-16	99.759%	8	16	8	8	2	6
oxygen-17	0.037%	8	17	8	9	2	6
oxygen-18	0.204%	8	18	8	10	2	6
fluorine-19	100%	9	19	9	10	2	7
neon-20	90.92%	10	20	10	10	2	8
neon-21	0.257%	10	21	10	11	2	8
neon-22	8.82%	10	22	10	12	2	8

A relative scale consists of numbers without units or with specially defined units. These relative-scale numbers must be directly proportional to the magnitude of some property of matter that can be measured. An example will make this definition clearer. Suppose you wish to set up a relative scale of weights of the members of your class. First, the weight of one member of the class is selected and given a simple numerical value. In theory, it does not make any difference whose weight is selected or what numerical value it is assigned. In practice, choices should be made on the basis of convenience and usefulness.

Direct proportion is explained in Section 1.22(1).

Suppose the weight of a 125-lb (pound) pupil is selected and assigned a value of 5.00 units. Now the *relative* weights of the other members of the class can be calculated by comparing their *actual* weights with that of the 125-lb pupil. A 15$\bar{0}$-lb pupil has a

Zeros to the right of a nonzero digit, but to the left of an understood decimal point, are not significant unless specifically indicated to be significant. The rightmost such zero that is significant is indicated by a bar placed above it.

weight that is $\frac{15\bar{0}\text{ lb}}{125\text{ lb}}$ or 1.20 times that of the 125-lb pupil. Since the 125-lb pupil has an assigned weight of 5.00 units on the relative weight scale, the $15\bar{0}$-lb pupil will have a relative weight of 1.20 × 5.00 units = 6.00 units. This relative weight could have been calculated in one step by using the expression $\frac{15\bar{0}\text{ lb}}{125\text{ lb}} \times 5.00$ units = 6.00 units. Similarly, the weight of a $20\bar{0}$-lb pupil will have a value of $\frac{20\bar{0}\text{ lb}}{125\text{ lb}} \times 5.00$ units = 8.00 units on the relative scale. The weight of a $10\bar{0}$-lb pupil will have a value of $\frac{10\bar{0}\text{ lb}}{125\text{ lb}} \times$ 5.00 units = 4.00 units on the relative scale. These relative weights, 5.00 units, 6.00 units, 8.00 units, and 4.00 units, are directly proportional to the actual weights of 125 lb, $15\bar{0}$ lb, $20\bar{0}$ lb, and $10\bar{0}$ lb, respectively. If it is the *relationship* between these weights that is important, then it does not matter whether the actual weights or the relative weights are used. In the example given, the actual weights and the relative weights are probably equally convenient to use. But the very small numbers that express the actual masses of atoms in grams are not convenient to use.

In order to set up a relative scale of masses of atoms, one atom is chosen and assigned a relative mass value. The masses of all other atoms are then expressed in relation to this defined relative mass. Such a system of relative masses was set up by the world organizations of chemists and physicists. In this system, the carbon-12 atom was chosen and assigned a relative mass of exactly 12 units. The units of this relative mass scale are called *atomic mass units* and are abbreviated *u*. One atomic mass unit, 1 *u*, is exactly $\frac{1}{12}$ the mass of a carbon-12 atom, or $1.6605655 \times 10^{-24}$ g. The *atomic mass* of a carbon-12 atom is exactly 12 *u*. *The mass of an atom expressed in atomic mass units is called the* ***atomic mass*** *of the atom.*

The hydrogen-1 atom has a mass about $\frac{1}{12}$ that of the carbon-12 atom. So it has a relative mass (atomic mass) of about $\frac{1}{12}$ of 12 or about 1 *u*. The accurate value for the atomic mass of hydrogen-1 atoms is 1.007825 *u*.

The mass of a hydrogen-2 atom is about $\frac{1}{6}$ that of a carbon-12 atom. Its atomic mass has been measured to be 2.014102 *u*. An oxygen-16 atom has about $\frac{4}{3}$ the mass of a carbon-12 atom. Careful measurements show its atomic mass to be 15.994915 *u*. The mass of a magnesium-24 atom is found to be slightly less than double that of a carbon-12 atom. Its accurate atomic mass is 23.985042 *u*. In the same way, the atomic mass of any nuclide is determined by comparison with the mass of a carbon-12 atom. Atomic masses are very accurately known, as the values given above indicate. You have learned that the *mass number* is the total number of protons and neutrons in the nucleus of an atom. You can now see that the *mass number* is also *the whole number closest to the atomic mass.*

The masses of the subatomic particles may also be expressed on the atomic-mass scale. The atomic mass of the electron is 0.0005486 *u*. That of the proton is 1.007276 *u*, and that of the neutron is 1.008665 *u*. See Table 3-1.

3.14 Atomic weight Naturally occurring elements usually exist as a mixture of several isotopes. The percentage of each isotope in the naturally occurring element is nearly always the same, no matter where the element is found. Naturally occurring hydrogen consists of 99.985% hydrogen-1 atoms, atomic mass 1.007825 *u*; and 0.015% hydrogen-2 atoms, atomic mass 2.014102 *u*. What will be the average atomic mass of the atoms in this mixture? From the data given, 99.985% or 0.99985 of the average atomic mass is 1.007825 *u*, while 0.015% or 0.00015 of the average atomic mass is 2.014102 *u*. Hence,

$$(0.99985 \times 1.007825\ u) + (0.00015 \times 2.014102\ u) = 1.00797\ u,$$

the average atomic mass of naturally occurring hydrogen atoms.

Recall that you convert a percentage to a decimal by dividing the percentage by 100%. This does not change the quantity because 100% = 1.

Naturally occurring oxygen consists of 99.759% of oxygen-16, atomic mass 15.99491 *u*; 0.037% of oxygen-17, atomic mass 16.99913 *u*; and 0.204% of oxygen-18, atomic mass 17.99916 *u*. The average atomic mass of the atoms in this mixture will be

$$(0.99759 \times 15.99491\ u) + (0.00037 \times 16.99913\ u) + (0.00204 \times 17.99916\ u) = 15.9994\ u.$$

The ratio of the average atomic mass of an element to $\frac{1}{12}$ of the mass of an atom of carbon-12 is the ***atomic weight*** *of the element.* The average mass of naturally occurring hydrogen atoms is 1.00797 *u*. From Section 3.13, $\frac{1}{12}$ the mass of an atom of carbon-12 is exactly 1 *u*. The ratio of these masses

The use of the word "weight" in the expression "atomic weight" is, of course, not correct since the quantity so named is a "mass." However, "atomic weight" has been used historically and still is used today for this concept.

$$\frac{1.00797\ \cancel{u}}{1\ \cancel{u}} = 1.00797$$

is the *atomic weight* of naturally occurring hydrogen. Note that *the atomic weight is the numerical portion of the average atomic mass.*

Similarly for oxygen, the average mass of naturally occurring oxygen atoms is 15.9994 *u*. Hence, the atomic weight of naturally occurring oxygen is 15.9994.

Atomic weights are important to the chemist because they indicate relative mass relationships between reacting elements. They enable the chemist to predict the quantities of materials that will be involved in chemical reactions.

Atomic weights appear in the periodic table that is on the inside back cover of this book and in Appendix Table 4. They include the most recent accurate figures. Atomic weights are occasionally revised when new data become available. You need not memorize them. Approximate atomic weights are given on the inside of the front cover and in Appendix Table 5. These values are accurate enough for use in solving problems in high school chemistry. However, for more advanced chemical work the accurate atomic weights in the periodic table or Appendix Table 4 must always be used.

If the atomic weight is expressed in gram units, we derive another important quantity in chemistry, the *gram-atomic weight.* The gram-atomic weight of hydrogen is 1.00797 g of hydrogen. For oxygen, the gram-atomic weight is 15.9994 g of oxygen.

3.15 The Avogadro number and the mole You have learned (Section 3.13) that the atomic mass of a carbon-12 atom is exactly 12 u. But atomic mass units are so very small that they are impractical for laboratory mass determinations. A chemist would find it much easier to weigh out exactly 12 g of carbon-12 (to the precision limit of the balance). How many atoms are there in exactly 12 g of carbon-12?

$$12.00000\ \text{g C} \times \frac{\cancel{u}}{1.6605655 \times 10^{-24}\ \cancel{\text{g}}} \times \frac{1\ \text{atom}}{12\ \cancel{u}} = 6.022045 \times 10^{23}\ \text{atoms C}$$

The number of atoms in the atomic mass of a nuclide in grams is an important unit of measure in chemistry. It is the number of atoms in exactly 12 grams of carbon-12, 6.022045×10^{23} atoms. It is also the number of atoms in 1.00797 g of naturally occurring hydrogen.

$$1.00797\ \text{g H} \times \frac{\cancel{u}}{1.6605655 \times 10^{-24}\ \cancel{\text{g}}} \times \frac{1\ \text{atom}}{1.00797\ \cancel{u}} = 6.022045 \times 10^{23}\ \text{atoms H}$$

It is also the number of atoms in 15.9994 g of naturally occurring oxygen. You should recognize that the number of atoms in these three cases is identical. This is true because the mass taken in grams is directly proportional to the atomic mass of a nuclide or to the average atomic mass of the atoms of each naturally occurring element. In the calculation, these numbers cancel out.

To help you understand the enormous number of units in one mole, imagine this situation. Suppose everyone living today on the earth (4.8 billion people) were to help count the atoms in one mole of an element (copper, for instance). If each person counted continuously at the rate of one atom per second, it would require about 4 million years for all the atoms to be counted.

This quantity, 6.022045×10^{23}, is so important in science that it has been given a special name. It is called the ***Avogadro number,*** honoring the Italian chemist and physicist Amedeo Avogadro (1776–1856). This constant is quite useful and should be remembered to at least three significant figures: 6.02×10^{23}.

The amount of substance containing the Avogadro number of any kind of chemical unit is called a ***mole*** *of that substance.* Thus, exactly 12 g of carbon-12 is a mole of carbon-12 atoms; and 1.00797 g of

1 mole of lead atoms
6.02×10^{23} atoms or 207.2 g Pb

1 mole of copper atoms
6.02×10^{23} atoms or 63.5 g Cu

1 mole of carbon atoms
6.02×10^{23} atoms or 12.0 g C

Figure 3-7. One gram-atomic weight of a naturally occurring element contains one mole of atoms, 6.02×10^{23} atoms, of the element.

naturally occurring hydrogen atoms is a mole of hydrogen atoms; and 15.9994 g of naturally occurring oxygen contains a mole of oxygen atoms. Note that *mole* is the name of the quantity containing a convenient number (6.02×10^{23}) of chemical units. Here the chemical unit is the atom. *For naturally occurring elements, one gram-atomic weight of the element contains one mole of atoms of the element. For individual nuclides, one gram-atomic mass* (the atomic mass in gram units) *of the nuclide contains one mole of atoms of the nuclide.* See Figure 3-7.

"Mole" is used by chemists as a counting unit the way a grocer uses "dozen" or "case," or a stationer uses "gross" or "ream." A dozen eggs is the quantity 12 eggs; a gross of pencils is the quantity 144 pencils; a ream of paper is the quantity 500 sheets of paper; a case of soft drink bottles is the quantity 24 bottles. A mole of atoms is the quantity 6.02×10^{23} atoms. The mole is a very important unit of measure in chemistry. It will be used throughout this text. See Table 3-3.

Table 3-3
MOLAR QUANTITIES

Atoms	*Atomic mass*	*Molar quantity*	*Number of atoms*	*Gram-atomic mass*	*Gram-atomic weight*
C-12	12 *u* exactly	1 mole	6.02×10^{23}	12 g exactly	—
H-1	1.007825 *u*	1 mole	6.02×10^{23}	1.007825 g	—
H-2	2.014102 *u*	1 mole	6.02×10^{23}	2.014102 g	—
H	1.00797 *u* wtd avg	1 mole	6.02×10^{23}	—	1.00797 g
O–16	15.99491 *u*	1 mole	6.02×10^{23}	15.99491 g	—
O–17	16.99913 *u*	1 mole	6.02×10^{23}	16.99913 g	—
O–18	17.99916 *u*	1 mole	6.02×10^{23}	17.99916 g	—
O	15.9994 *u* wtd avg	1 mole	6.02×10^{23}	—	15.9994 g

SAMPLE PROBLEM 3-1

What is the mass in grams of 3.50 moles of copper atoms?

SOLUTION

The atomic weight of copper from the table inside the back cover or from Appendix Table 4 is 63.546. Therefore, the gram-atomic weight of copper, or the mass of one mole of copper atoms, is 63.546 g. The mass of 3.50 moles of copper atoms is

$$3.50 \text{ moles Cu} \times \frac{63.546 \text{ g}}{\text{mole}} = 222 \text{ g Cu}$$

Since the number of moles of copper atoms is expressed in the problem to only three significant figures, the approximate atomic weight of copper found in the table inside the front cover or on Appendix Table 5, 63.5, could have been used in the solution. The same answer is obtained.

$$3.50 \text{ moles Cu} \times \frac{63.5 \text{ g}}{\text{mole}} = 222 \text{ g Cu}$$

Practice Problems

1. What is the mass in grams of 5.75 moles of copper atoms? *ans.* 365 g Cu
2. What is the mass in grams of 5.75 moles of silver atoms? *ans.* $62\bar{0}$ g Ag

SAMPLE PROBLEM 3-2

How many moles of atoms are there in 6.195 g of phosphorus?

SOLUTION

Since the mass of phosphorus is given to four significant figures, the atomic weight of phosphorus given to at least four significant figures in the table inside the back cover must be used. Rounded to four significant figures, 30.97376 is 30.97. Thus, there is one mole of phosphorus atoms in 30.97 g of phosphorus. Then 6.195 g of phosphorus contains

$$6.195 \text{ g P} \times \frac{\text{mole}}{30.97 \text{ g}} = 0.2000 \text{ mole P}$$

Significant-figure rules permit an answer calculated to four significant figures.

Practice Problems

1. How many moles of atoms are there in 18.58 g of phosphorus? *ans.* 0.5999 mole
2. How many moles of atoms are there in 18.58 g of nitrogen? *ans.* 1.326 moles

SUMMARY

The idea that matter consists of simple, indivisible, indestructible particles called atoms was proposed by Greek thinkers as early as 400 B.C. John Dalton, between 1803 and 1808, was the first to realize that an understanding of the nature and properties of atoms could explain the law of definite composition. It could also explain the way and the proportions in which substances react with one another.

An atom is the smallest unit of an element that can exist either alone or in combination with atoms of the same or different elements. The atomic theory states that

1. All matter is made up of very small structures called atoms.
2. Atoms of the same element are chemically alike; atoms of different elements are chemically different.
3. Individual atoms of an element may not all have the same mass. The atoms of an element, as it occurs naturally, have a definite average mass that is characteristic of the element.
4. Individual atoms of different elements may have nearly identical masses. The atoms of different naturally occurring elements have different average masses.
5. Atoms are not subdivided in chemical reactions.

An atom consists of a positively charged central part called the nucleus and negatively charged particles called electrons, which move about the nucleus in energy levels. Atoms are electrically neutral. Positively charged protons and neutral neutrons are held together in the nucleus by very short-range forces.

The hydrogen atom has a nucleus made up of one proton. One electron moves about this nucleus in the 1st energy level. Three isotopes of hydrogen (protium, deuterium, and tritium) are possible with nuclei containing, respectively, no neutron, one neutron, and two neutrons. Isotopes are atoms of the same element that have different masses. The atomic number of an atom is the number of protons in the nucleus of that atom. Therefore, the atomic number of the hydrogen atom is 1.

Each different variety of atom as determined by the number of protons and number of neutrons in its nucleus is called a nuclide. The mass number of an atom is the sum of the number of protons and neutrons in its nucleus.

The 1st energy level may contain a maximum of two electrons. The 2nd energy level may contain a maximum of eight electrons. Argon, the final element in the third series, has eight electrons in the 3rd energy level.

The mass of an atom expressed in atomic mass units is called the atomic mass of the atom. One atomic mass unit is exactly one-twelfth the mass of a carbon-12 atom. The whole number closest to the atomic mass is the mass number of the atom.

The ratio of the average atomic mass of an element to $\frac{1}{12}$ of the mass of an atom of carbon-12 is the atomic weight of the element. The atomic weight of an element expressed in gram units is the gram-atomic weight of the element.

The amount of substance containing an Avogadro number of any kind of chemical unit is called a mole of that substance. The Avogadro number is 6.02×10^{23}. For naturally occurring elements, one gram-atomic weight of the element contains one mole of atoms (6.02×10^{23} atoms) of the element.

VOCABULARY

angstrom
atom
atomic mass
atomic mass unit
atomic number
atomic theory
atomic weight
Avogadro number
electron
electron cloud
energy level
gram-atomic weight
isotope
mass number
mole
neutron
nucleus
nuclide
proton

QUESTIONS

Group A

1. What observations lead scientists to believe that the basic particles of matter are very small? *A3a*
2. What kinds of information are now included in the atomic theory? *A3a*
3. List five statements scientists make about atoms and their properties. *A3a*
4. *(a)* Name the two main parts of an atom. *(b)* Name the three types of particles found in an atom. *(c)* In which part of the atom is each particle found? *(d)* What is the mass and charge of each particle? *A1, A2a, A5*
5. Compare the diameter of a nucleus with the diameter of an atom. *A2b, A3d*
6. Define: *(a) energy level, (b) electron cloud. A1, A2b*
7. Describe the manner in which electrons move about the nucleus of an atom. *A2b, A3c, A5*
8. *(a)* Name the isotopes of hydrogen. *(b)* What is the structure of each? *A2d*
9. *(a)* How is the atomic number of an atom related to its structure? *(b)* For a neutral atom, what is the relationship between its atomic number and the number of electrons? *A2c, A3g*
10. Define: *(a) nuclide, (b) isotope. A1*
11. How is the mass number of an atom calculated from the number and kinds of particles composing it? *A2c*
12. An atomic nucleus that contains 7 protons and 8 neutrons is surrounded by 7 electrons: 2 in the 1st energy level and 5 in the 2nd energy level. *(a)* What is the atomic number of this nuclide? *(b)* What is its mass number? *(c)* What is the name of this nuclide? *A2c, A3g, A3h, A4, B1*
13. *(a)* Among the first ten elements, which two exist naturally as a single nuclide? *(b)* How many artificial nuclides of these two elements are there? *A4*
14. Define: *(a) atomic mass unit, (b) atomic mass. A1, A2f, A3i, A5*
15. Define: *(a) atomic weight, (b) gram-atomic weight. A1, A2f*
16. Atomic weights are very important to chemists. Why? *A2f, A3i*
17. From the Table of Atomic Weights, find the atomic number and atomic weight of: *(a)* iron; *(b)* aluminum; *(c)* iodine; *(d)* tin; *(e)* radium. *A4*
18. *(a)* How many atoms are there in exactly 12 g of carbon-12? *(b)* What name is given to this quantity? *(c)* What do scientists call the amount of a substance containing this number of chemical units? *A2g*
19. What is the mass in grams of: *(a)* 3.00 moles of helium atoms; *(b)* 7.50 moles of boron atoms; *(c)* 0.600 mole of neon atoms; *(d)* 0.350 mole of magnesium atoms; *(e)* 0.225 mole of silicon atoms? Use the Table of Atomic Weights and follow the rules for significant figure calculations. *A4, B6*
20. How many moles of atoms are there in each of the following:
(a) 25.0 g of lithium; *(b)* 10.0 g of sodium; *(c)* 50.0 g of argon; *(d)* 8.50 g of sulfur; *(e)* 90.0 g of tungsten? *A4, B6*

Group B

21. *(a)* Give two statements included in Dalton's atomic theory. *(b)* Do we believe these statements are still true? *(c)* Why? *A3a, A5*

22. In one of the Rutherford experiments, it was found that 1 high-speed positively charged particle in 8000 was deflected by 90° or more when directed at a thin sheet of platinum. What does this observation indicate about the structure of platinum atoms? *A3b, A5*

23. If the elements are arranged in order of increasing atomic number, how do atoms of successive elements differ in *(a)* number of protons? *(b)* number of electrons? *(c)* number of neutrons? *A2c, A3g*

24. How are the electrons arranged in the atoms of the elements of the second series? *A2b, B3*

25. Copy and complete the table below *on a separate sheet of paper.* *B2, B3*

26. Copper exists in nature as copper-63, atomic mass 62.9298 *u*, and copper-65, atomic mass 64.9278 *u*. Its atomic weight is 63.546. What must be the approximate abundance in nature of these two isotopes? *A2f*

27. The elements sodium, aluminum, and phosphorus have only one naturally occurring nuclide. How will the atomic mass of this nuclide and the atomic weight of the element compare? *A2f*

28. *(a)* What is the relationship between an atom containing 12 protons, 12 neutrons, and 12 electrons, and one containing 12 protons, 13 neutrons, and 12 electrons? *(b)* What is the relationship between an atom containing 12 protons, 13 neutrons, and 12 electrons, and one containing 13 protons, 12 neutrons, and 13 electrons? *A2e*

29. How can you calculate the mass in grams of a single atom of a nuclide if you know the atomic mass of the nuclide? *A2f, A2g*

30. If you divide the gram-atomic weight of an element by the Avogadro number, what is the meaning of the quotient? *A2f, A2g*

31. Calculate the atomic weight of magnesium. The naturally occurring element consists of 78.70% Mg-24, atomic mass 23.98504 *u*; 10.13% Mg-25, atomic mass 24.98584 *u*; and 11.17% Mg-26, atomic mass 25.98259 *u*. *B4*

32. Imagine you have an Avogadro number of cubes exactly 1 cm on a side. On a broad level surface, you use some of the cubes to make a square one-cube-thick layer exactly 100 km on a side. On this base you then pile the remaining cubes, layer on layer. When you have finished, how many km high is the pile of cubes? *A2g*

Name of nuclide	*Atomic number*	*Mass number*	*Composition of nucleus*		*Electron configuration* (Energy level)		
			Protons	*Neutrons*	*1st*	*2nd*	*3rd*
sodium-23	11						
magnesium-24	12						
aluminum-27	13						
silicon-28	14						
phosphorus-31	15						
sulfur-32	16						
chlorine-35	17						
argon-40	18						

Kenichi Fukui (Japanese) received the 1981 Nobel Prize in chemistry for studying how the electronic structures of molecules influence the chemical reactions that occur. Fukui proposed that the highest molecular orbital containing electrons on one molecule interacts with the lowest unoccupied molecular orbital on a second molecule. A new orbital and bond form as the reaction occurs. Fukui shared the Nobel Prize with Roald Hoffmann (American) who developed a theory based on orbital symmetry to explain why some chemical reactions occur as they do.

Arrangement of Electrons in Atoms

4

GOALS

In this chapter you will gain an understanding of: • electromagnetic radiation and spectra • quantum numbers • the electron configurations of the atoms • orbital notation, electron-configuration notation, and electron-dot notation

4.1 The nucleus and moving electrons In Chapter 3, you learned about the particles that make up atoms. The atom was described as a nucleus, containing protons and usually neutrons, surrounded by electrons. Although the electrons in an atom move about the nucleus, it is not possible to determine their paths. The regions the electrons occupy are called energy levels; the part of the atom where the electrons may most probably be found is called the electron cloud. Now you are ready to learn more about how electrons are arranged and held within the atom.

In Section 3.4 it was stated that Bohr compared the movements of electrons about the nucleus with the movement of planets around the sun. Let us examine this comparison further. There are two kinds of actions that operate at the same time to determine the path of a planet about the sun. One of these actions is gravitation, which, if a planet were not moving, would cause it to be attracted directly toward the sun. The other action, which occurs because the planet is moving, is its tendency, if gravitation did not operate, to continue to move through space in a straight line. This motion would carry the planet away from the sun. The balance between these two actions causes the planet to move in a path around the sun.

Now let us see how this analogy might apply to electrons moving about a nucleus. Recall that the nucleus has a positive charge because of its protons. A neutral atom contains an equal number of positively charged protons and negatively charged electrons. We would expect electrons to be drawn toward the nucleus by the attraction between oppositely charged particles. This attraction is similar to the action of gravitation in the orbiting of a planet around the sun.

When we attempt further comparison with the movement of a planet, however, the analogy begins to break down. Scientists

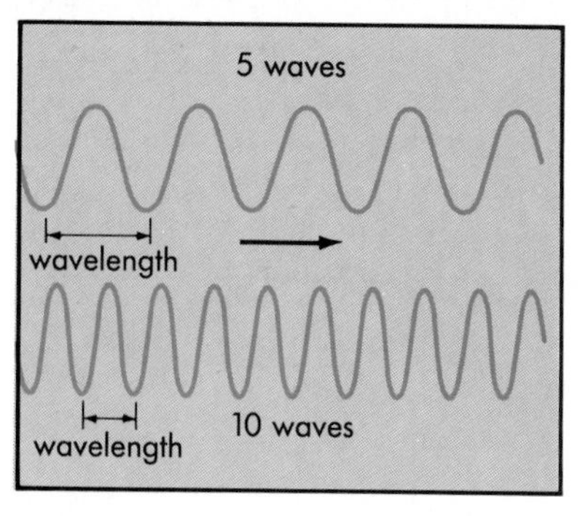

Figure 4-1. Both waves are traveling toward the right at the same speed. Assume they travel the distance shown in 1 s. Then the frequency of the top wave is 5 waves/s and the frequency of the bottom wave is 10 waves/s. The top wave has a wavelength twice that of the bottom wave. The wavelength is inversely proportional to the frequency. The top wave has a lower frequency and a longer wavelength. The bottom wave has a higher frequency and a shorter wavelength.

have observed that electrically charged particles moving in curved paths give off energy. If an electron moving about a nucleus continually gives off energy, it should slow down. It should gradually move nearer to the nucleus and eventually fall into it. This behavior would be like the slowing down of a satellite by friction with the earth's upper atmosphere. As this slowing down occurs, the satellite falls toward the earth and eventually burns up in the earth's atmosphere. But atoms do not collapse. Electrons do not fall into the nucleus. Thus, the attraction of oppositely charged particles may partly explain the motion of electrons about the nucleus of an atom, but we must look further for the rest of the explanation.

4.2 Electromagnetic radiation Visible light is one kind of electromagnetic radiation. Other kinds of electromagnetic radiation are X rays, ultraviolet and infrared light, and radio waves. Electromagnetic radiations are forms of energy that travel through space as waves. They move at the rate of 3.00×10^8 m/s, the speed of light in a vacuum. For any wave motion, the speed equals the product of the *frequency* (the number of waves passing a given point in one second) and the *wavelength*. For electromagnetic radiation

$$c = f\lambda$$

where c is the speed of light, f is the frequency, and λ (the Greek letter lambda) is the wavelength. Since c is the same for all electromagnetic radiation, the product $f\lambda$ is a constant, and λ is inversely proportional to f. See Figure 4-1.

Inverse proportion is explained in Section 1.22(2).

Direct proportion is explained in Section 1.22(1).

In addition to its wave characteristics, electromagnetic radiation has some properties of particles. Electromagnetic radiation is transferred to matter in units or *quanta* of energy called ***photons.*** The energy of a photon is proportional to the frequency of the radiation. Thus, the energy of a photon and the frequency of the radiation are related by

$$E = hf$$

Here, E is the energy of the photon, h is a proportionality constant called *Planck's constant,* and f is the frequency of the radiation. Planck's constant is the same for all types of electromagnetic radiation. Since f multiplied by Planck's constant equals E, f and E are directly proportional.

Energy is transferred to matter in photon units. Therefore, the absorption of a photon by an atom increases its energy by a definite quantity, hf. An atom that has absorbed energy in this way is called an *excited* atom. When excited atoms radiate energy, the radiation is given off in photon units as shown in Figure 4-2.

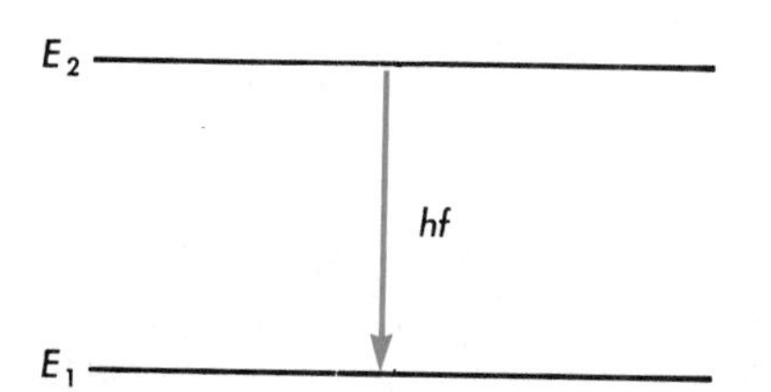

Figure 4-2. When an excited atom with energy E_2 returns to energy E_1, it gives off a photon having energy $E_2 - E_1 = E = hf$.

4.3 Spectra of atoms Atoms may be excited by heating them in a flame or in an electric arc. As these excited atoms return to their normal energy states, they give off light of a characteristic color. An example is the yellow-orange light given off by sodium atoms in a glass rod heated in a burner flame. Atoms of gases can be excited by passing high-voltage electricity through the gas contained inside a glass tube. The red light of neon advertising signs is a familiar example. If hydrogen gas is used in such a tube, it glows with a characteristic lavender color as seen in Figure 4-3.

Figure 4-3. Hydrogen atoms are excited when high-voltage electricity is passed through a glass tube containing hydrogen gas. The lavender glow is characteristic of hydrogen gas.

When observed through a spectroscope, the lavender-colored light of hydrogen gas reveals lines of particular colors, as shown in Figure 4-4. Such a spectrum is called a bright-line spectrum. It indicates that the light given off by excited atoms has only certain wavelengths. Light of a particular wavelength has a definite *frequency* ($c = f\lambda$) and a characteristic color. A definite frequency also means a definite energy ($E = hf$). Hence the bright-line hydrogen spectrum shows that excited hydrogen atoms emit (give off) photons having only certain energies. Now, recall from Section 4.2 that λ is inversely proportional to f, but that f is directly proportional to E. So, λ is inversely proportional to E. A long wavelength is thus associated with less energy and a short wavelength is associated with more energy.

Emitted photons have only certain energies. These energies represent differences between the energies of atoms before and after radiation. Therefore, these energies of atoms are *fixed and definite quantities.* And because each species of atom has its own characteristic spectrum, each atom must have its own characteristic energy possibilities. This evidence indicates that the energy changes that occur from time to time within an atom involve definite amounts of energy rather than a continuous flow of energy.

The continuous spectrum at the end of Chapter 24 shows the relationship between wavelength and color of light.

The energy changes of an excited atom returning to its normal energy state are actually changes in the energy of its electrons. Therefore, a diagram that shows the electron energy levels of the atom can be devised. Figure 4-5 includes lines in the ultraviolet and infrared regions of the hydrogen spectrum. Figure 4-6 gives the corresponding electron energy-level transitions in the hydrogen atom. For example, when electrons in excited hydrogen atoms move from the second energy level to the first

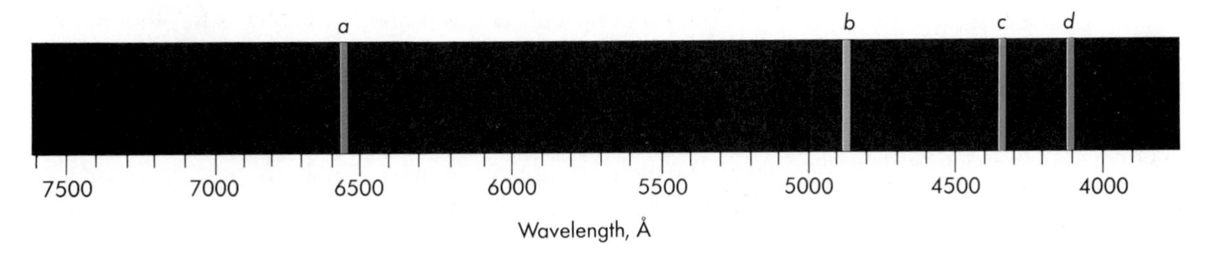

Figure 4-4. The visible bright-line spectrum of hydrogen seen through a spectroscope.

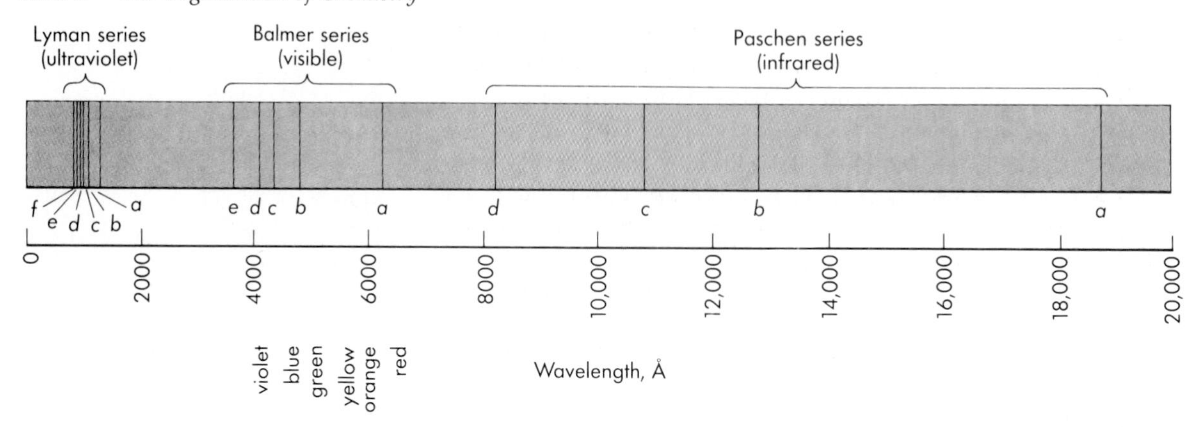

Figure 4-5. Representative lines in the hydrogen spectrum. The small letter below each line indicates which of the energy-level transitions in Figure 4-6 produces it.

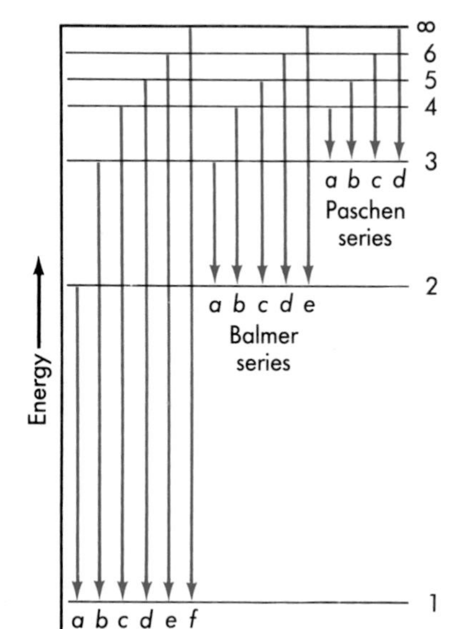

Figure 4-6. An electron energy-level diagram for hydrogen showing some of the transitions that are possible in this atom. Transitions that leave an electron in a particular final energy level belong to a particular spectral series. Some of these series are named for the people who discovered them.

energy level, the energy released produces ultraviolet line *a* in the Lyman series, wavelength 1215 Å. Similarly, electrons in excited hydrogen atoms that move from the fourth energy level to the second energy level release energy that produces the greenish-blue line *b* in the Balmer series, wavelength 4861 Å.

The idea of electron energy levels in the hydrogen atom was developed by Niels Bohr in 1913. The definite energy levels of the atom indicate two properties of the electron moving about the hydrogen nucleus. First, it can move at only certain distances from the nucleus. Second, it can move with only certain speeds. Bohr's theory states that electrons do not give off energy when they remain in given energy levels. They only give off energy when they change to lower energy levels. This helps explain why electrons in atoms do not lose energy, fall into the nucleus, and cause the atom to collapse.

This model works well in explaining the spectra of the one-electron hydrogen atom. But it does not satisfactorily explain the spectra of more complex atoms.

4.4 Wave-mechanics concept of an atom During the past half century, the work of theoretical physicists, including Heisenberg, de Broglie, and Schrödinger, helped develop a theory of atomic structure based on wave mechanics. The basic ideas of wave mechanics are beyond the scope of a high school chemistry course. It will be useful, however, for us to consider some of its conclusions.

The motion of an electron about an atom is not in a definite path like that of the earth about the sun. In fact, it is impossible to determine an electron's path without changing that path! So an electron's location can be given only in terms of probabilities. This location is described by an *orbital*. An orbital may be thought of as a highly probable location where an electron may be found. The motion of the single hydrogen electron creates a spherical *electron cloud* surrounding the nucleus. The electron cloud gives size and shape to an atom. It prevents two free atoms (or portions of free atoms) from occupying the same space.

In Section 4.1, we recognized that the attraction of the positively charged nucleus for the negatively charged electrons was only a partial explanation for the motion of electrons about the nucleus and consequently for the size of the electron cloud. Now we are ready to consider the explanation further. If the attraction of the positively charged nucleus for the negatively charged electrons were the only action operating, electrons would be drawn directly toward the nucleus. Somehow, moving electrons must produce an action that counteracts this attraction. An explanation can be found if the electron is considered not as a particle, but as a wave. We can do this when we visualize the motion of the electrons about an atom as producing an electron cloud. The electron cloud then becomes a diffuse region of both matter and electric charge. As electrons are drawn closer to the nucleus by the attraction of opposite charges, the electron cloud gets smaller. According to the theory of wave mechanics, the smaller the electron cloud, the shorter an electron's wavelength becomes. As an electron's wavelength becomes shorter, its energy increases. (λ is inversely proportional to E, Section 4.3.) Eventually a condition is reached in which the energy of an electron is great enough to counterbalance the attraction of the nucleus for it. Electrons do not tend to go closer to the nucleus or farther away from it. Thus, the size of the electron cloud is stabilized, and the radius and volume of the atom are determined.

4.5 Quantum numbers The mathematics of wave mechanics shows that the energy state of an electron in an atom may be described by a set of four numbers. These are called *quantum numbers*. Quantum numbers describe the orbital the electron occupies in terms of: (1) distance from the nucleus; (2) shape; (3) position with respect to the three axes in space; and (4) the direction of spin of the electron in the orbital. As you read on, refer to the corresponding columns in Table 4-1.

The *principal quantum number*, symbolized by n, indicates the most probable distance of the electron from the nucleus of the atom. It is a positive whole number, having values 1, 2, 3, etc. The principal quantum number is the main energy-level designation, or identifying number, of an orbital. The 1st energy level is closest to the nucleus with others at increasing distances. Electrons in the 1st energy level have the lowest energies. Electrons in higher energy levels have increasingly greater energies.

The *orbital quantum number* indicates the shape of the orbital the electron occupies. The letter designations for the first four orbital quantum numbers are *s*, *p*, *d*, and *f*. These are listed in order of ascending energies. For a particular energy level, the *s* orbital has the lowest energy. The *p* orbitals have higher energy

Table 4-1
QUANTUM NUMBER RELATIONSHIPS IN ATOMIC STRUCTURE

Principal quantum number (energy level) (n)	*Orbitals (n orbital shapes) (n sublevels)*	*Number of orbitals per sublevel*	*Number of orbitals per energy level (n^2)*	*Number of electrons per sublevel*	*Number of electrons per energy level ($2n^2$)*
1	*s*	1	1	2	2
2	*s* *p*	1 3	4	2 6	8
3	*s* *p* *d*	1 3 5	9	2 6 10	18
4	*s* *p* *d* *f*	1 3 5 7	16	2 6 10 14	32

than the *s* orbital, the *d* orbitals have higher energy than the *p* orbitals, and so on. Sometimes the *s* orbital is called the *s* sublevel, *p* orbitals the *p* sublevel, *d* orbitals the *d* sublevel, etc.

The number of possible orbital shapes in an energy level is equal to the value of the principal quantum number. In the 1st energy level, an orbital of only one shape is possible. So the first energy level has only an *s* orbital, or *s* sublevel. In the 2nd energy level, orbitals of two shapes are possible. The 2nd energy level has *s* and *p* orbitals. The 3rd energy level has three possible orbital shapes: *s*, *p*, and *d* orbitals. The 4th energy level has four possible orbital shapes; it has *s*, *p*, *d*, and *f* orbitals. In the *n*th energy level, orbitals of *n* shapes are possible.

The *magnetic quantum number* indicates the position of the orbital about the three axes in space. There is only one position in the space around the nucleus for an *s* orbital. There are three positions for a *p* orbital, five positions for a *d* orbital, and seven positions for an *f* orbital. See Figures 4-7 and 4-8.

In the 1st energy level, there is only one *s* orbital. In the 2nd energy level, there are one *s* orbital and three *p* orbitals, for a total of four orbitals. In the 3rd energy level, there are one *s* orbital, three *p* orbitals, and five *d* orbitals, for a total of nine orbitals. Do you see that the total number of orbitals per energy level is the square of the principal quantum number? In the *n*th energy level, there will be n^2 orbitals.

The *spin quantum number* indicates a property of the electron described by just two conditions. These conditions may be thought of as being like the right-handed or left-handed conditions of a glove. That is, they are like mirror images in being

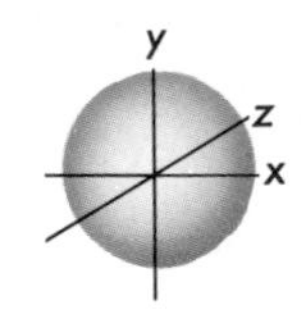

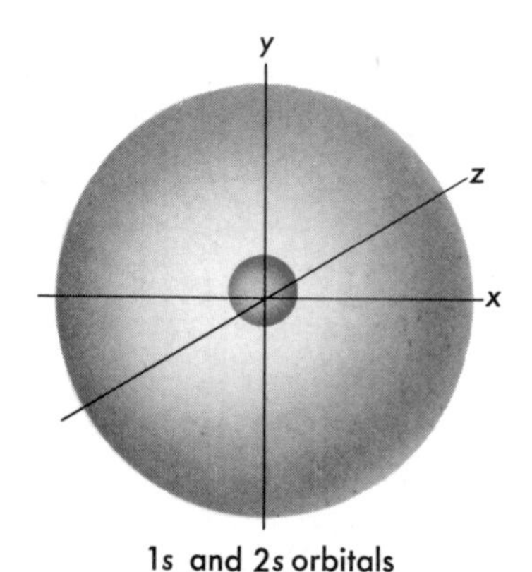

Figure 4-7. Position in the space about the nucleus of *s* orbitals. The *s* orbitals are spherical in shape.

"opposites" of each other. This property is called *electron spin.* Scientists often refer to the two possibilities for spin as clockwise and counterclockwise. Thus, each of the *positions of orbitals in the space around the nucleus described by the first three quantum numbers can be occupied by only two electrons having opposite spins.* No two electrons in an atom can have exactly the same set of four quantum numbers. This agrees with the observation that no two electrons in an atom have exactly the same energy.

If there are n^2 orbitals in the nth energy level, and each may be occupied by a maximum of two electrons, then the maximum number of electrons per energy level is $2n^2$.

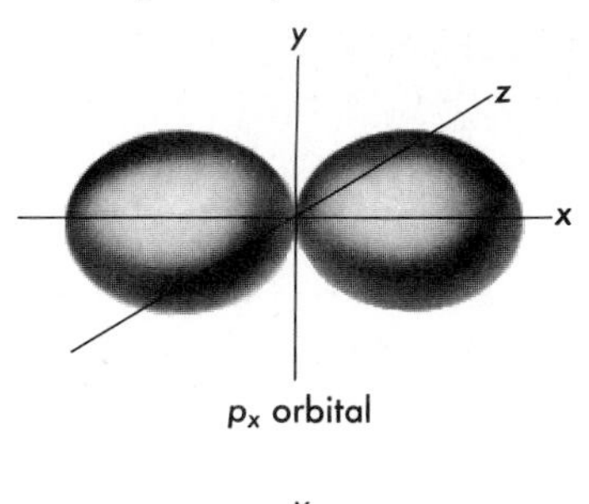

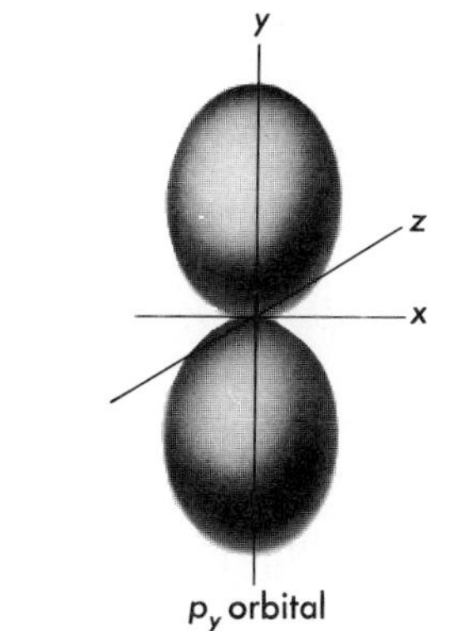

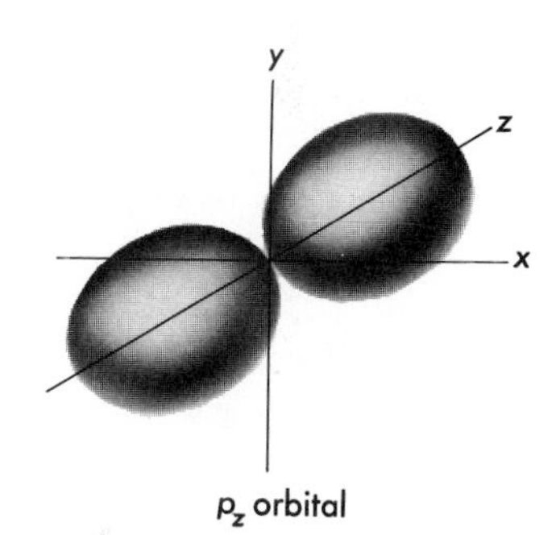

Figure 4-8. The three spatial positions of *p* orbitals. The *p* orbitals are shaped like a pair of ellipsoids tangent at points of greatest curvature at the nucleus. They are oriented along the three axes in space. The superposition of the three *p* orbitals produces a spherical electron cloud.

4.6 Electron configuration of atoms of first three series The quantum numbers that describe the arrangement of electrons about an atom are related to the energies of the electrons. The energies associated with the various orbitals as they become occupied by electrons are shown in Figure 4-9. The most stable state of an atom is called its *ground state.* In this condition, the electrons have the lowest possible energies. If the number of electrons in an atom is known, the arrangement of electrons of its ground state can be described. This is because electrons occupy the various orbitals in a reasonably definite order starting with those of lowest energy. The arrangement of electrons is called the *electron configuration.*

Hydrogen atoms have only one electron. In the ground state this electron moves in the 1*s* sublevel, the *s* sublevel of the 1st energy level. The two electrons of helium atoms both occupy the 1*s* sublevel. This may be shown in *orbital notation* as

$$\mathbf{H}\ \underset{1s}{\underline{\uparrow}} \qquad \mathbf{He}\ \underset{1s}{\underline{\downarrow\uparrow}}$$

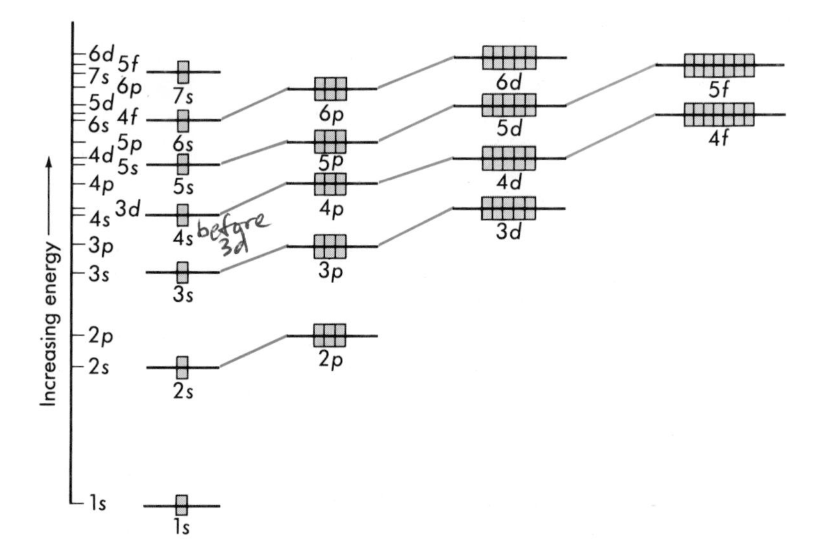

Figure 4-9. This chart shows the approximate relative energies of the atomic sublevels as they are occupied by electrons. Note how the sublevels of an energy level vary in energy and that the sublevels of higher energy levels overlap.

In orbital notation, one electron occupying an orbital is represented as $\underline{\uparrow}$. Two electrons occupying an orbital is represented as $\underline{\downarrow\uparrow}$. An unoccupied orbital is represented as __. The two helium electrons occupying the same orbital must have opposite spins. Two electrons of opposite spin in the same orbital are called an *electron pair.*

In *electron-configuration notation* hydrogen has the designation $1s^1$. This designation shows that hydrogen has one electron (represented by the superscript) in the *s* sublevel of the 1st energy level. Helium's electron structure is represented as $1s^2$. This means that helium has two electrons (represented by the superscript) in the *s* sublevel of the 1st energy level.

In *electron-dot notation,* hydrogen and helium are designated as follows:

H· **He:**

In this notation, the symbol represents the element, and the dots indicate the number of highest energy level electrons. The two dots written together, as shown in the helium notation, represent an *electron pair.*

The elements in the second series have electrons occupying the 1st and 2nd energy levels. Their ground-state electron arrangements may be represented by orbital notation, electron-configuration notation, and electron-dot notation, as shown in Table 4-2. Note that the orbital notations and electron-configuration notations show *all* of the electrons in the atom. The elec-

Table 4-2
ELECTRON NOTATIONS OF ATOMS IN THE SECOND SERIES

	Orbital notation			*Electron-configuration notation*	*Electron-dot notation*
	1s	**2s**	**2p**		
Li	↓↑	↑	_ _ _	$1s^2 2s^1$	Li·
Be	↓↑	↓↑	_ _ _	$1s^2 2s^2$	Be:
B	↓↑	↓↑	↑ _ _	$1s^2 2s^2 2p^1$	Ḃ:
C	↓↑	↓↑	↑ ↑ _	$1s^2 2s^2 2p^2$	·Ċ:
N	↓↑	↓↑	↑ ↑ ↑	$1s^2 2s^2 2p^3$	·N̈:
O	↓↑	↓↑	↓↑ ↑ ↑	$1s^2 2s^2 2p^4$	·Ö:
F	↓↑	↓↑	↓↑ ↓↑ ↑	$1s^2 2s^2 2p^5$	:F̈:
Ne	↓↑	↓↑	↓↑ ↓↑ ↓↑	$1s^2 2s^2 2p^6$	:N̈e:

tron-dot notations, however, show only the electrons in the *highest numbered* energy level. See Figure 4-10.

Observe that electrons do not pair up in *p* orbitals until each of the three *p* orbitals is occupied by a single electron. These single electrons have parallel spins. An atom such as that of neon has the *s* and *p* sublevels of its highest numbered energy level filled with eight electrons. Thus, it is said to have an *octet* in its highest numbered energy level.

The electron configurations of the elements of the third series are similar to those of the second series with successive electrons occupying the 3*s* and 3*p* orbitals. See Question 23.

6 3
4 Symbol 2
7 1
5 8

Figure 4-10. The order of placing electron dots, which represent highest-energy-level electrons. If there is only one dot in a position, it is centered. Dots 1 and 2 represent *s* electrons. Dots 3 through 8 represent *p* electrons. The *p* electrons do not form pairs until the three *p* orbitals have one electron each. An atom with electrons in all eight positions has an octet in its highest energy level.

4.7 Atoms of the fourth series The first two elements in the fourth series are potassium and calcium. Their atoms have the same electron configuration in the first three energy levels as the argon atom, $1s^22s^22p^63s^23p^6$. Their electron-dot symbols are

K· **Ca:**

These symbols show the presence of one and two electrons, respectively, in the 4*s* sublevel. The electron configuration of K is $1s^22s^22p^63s^23p^64s^1$ and of Ca, $1s^22s^22p^63s^23p^64s^2$. In the atoms of the next ten elements of this fourth series, the 3*d* sublevel is occupied in successive steps by the addition of electrons. The distribution of electrons in the ground state of these atoms is given in Table 4-3.

Table 4-3
STRUCTURE OF ATOMS IN THE FOURTH SERIES

Name	*Symbol*	*Atomic number*	*Number of electrons in sublevels*							
			1s	**2s**	**2p**	**3s**	**3p**	**3d**	**4s**	**4p**
potassium	K	19	2	2	6	2	6		1	
calcium	Ca	20	2	2	6	2	6		2	
scandium	Sc	21	2	2	6	2	6	1	2	
titanium	Ti	22	2	2	6	2	6	2	2	
vanadium	V	23	2	2	6	2	6	3	2	
chromium	Cr	24	2	2	6	2	6	5	1	
manganese	Mn	25	2	2	6	2	6	5	2	
iron	Fe	26	2	2	6	2	6	6	2	
cobalt	Co	27	2	2	6	2	6	7	2	
nickel	Ni	28	2	2	6	2	6	8	2	
copper	Cu	29	2	2	6	2	6	10	1	
zinc	Zn	30	2	2	6	2	6	10	2	
gallium	Ga	31	2	2	6	2	6	10	2	1
germanium	Ge	32	2	2	6	2	6	10	2	2
arsenic	As	33	2	2	6	2	6	10	2	3
selenium	Se	34	2	2	6	2	6	10	2	4
bromine	Br	35	2	2	6	2	6	10	2	5
krypton	Kr	36	2	2	6	2	6	10	2	6

Half-filled or completely filled sublevels have extra stability. The structures of both the chromium and copper atoms appear to be irregular. Chromium would be expected to have four 3*d* electrons and two 4*s* electrons. Instead, in the ground state, chromium atoms have five 3*d* electrons and one 4*s* electron. This $3d^5 4s^1$ structure must have higher stability and lower energy than the expected $3d^4 4s^2$ structure. (The $3d^4 4s^2$ configuration can occur in excited chromium atoms.) Apparently a half-filled 3*d* sublevel provides greater stability than a 3*d* sublevel with only four electrons. This greater stability occurs even though the electron-pair in the 4*s* sublevel is broken up. Copper would be expected to have nine 3*d* electrons and two 4*s* electrons. Instead, in the ground state, it has ten 3*d* electrons and one 4*s* electron. Here the filled 3*d* sublevel provides greater stability. (The $3d^9 4s^2$ configuration can occur in excited copper atoms.)

With the element zinc, the 3rd energy level is completely filled and there are two electrons in the 4th energy level. The remaining six elements in the fourth series are gallium, germanium, arsenic, selenium, bromine, and krypton. They all have completely filled 1st, 2nd, and 3rd energy levels. The electrons in the 4*s* and 4*p* sublevels are shown in these electron-dot symbols:

Ġa: **·Ġe:** **·Ȧs:** **·S̈e:** **:B̈r:** **:K̈r:**

Krypton is the last member of the fourth series. It is a gas that has an octet (two *s* and six *p* electrons) in its 4th energy level. Its electron-configuration notation is $1s^2 2s^2 2p^6 3s^2 3p^6 3d^{10} 4s^2 4p^6$. Observe that in this notation all the sublevels of an energy level are grouped together in *s, p, d,* and *f* order.

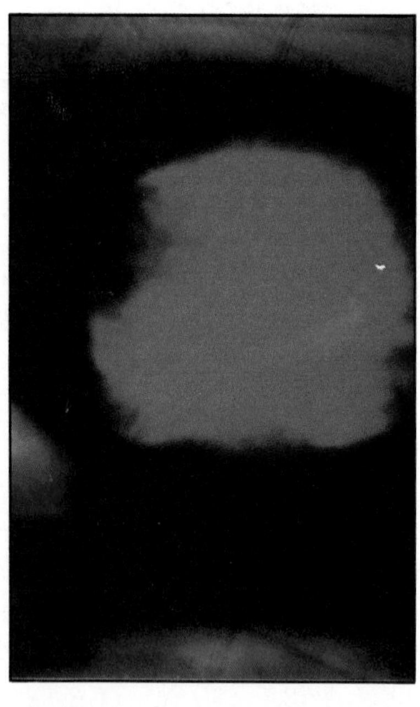

Figure 4-11. The compound einsteinium triiodide as seen through a microscope spectrophotometer.

4.8 Atoms of the fifth series The fifth series of elements, like the fourth, consists of 18 elements. The first two of these are rubidium and strontium. The lower energy levels of these elements resemble the krypton atom. The additional one and two electrons, respectively, occupy the 5*s* sublevel.

Rb· **Sr:**

In the atoms of the next ten elements, the five 4*d* sublevel orbitals become occupied by the successive addition of electrons.

The atoms of the element cadmium have completely filled 1st, 2nd, and 3rd energy levels. The 4*s*, 4*p*, and 4*d* sublevels are also filled. There are two electrons in the 5*s* sublevel. The atoms of the remaining six elements of the fifth series are indium, tin, antimony, tellurium, iodine, and xenon. The first four energy levels of these elements are similar to those of cadmium. Successive electrons occupy the 5*p* sublevel. See Table 4-4.

İn: **·Ṡn:** **·Ṡb:** **·T̈e:** **:Ï:** **:ẌE:**

Thus in the fifth series the addition of electrons to sublevels of two different energy levels proceeds as it did in the fourth series. Xenon, the last member of the series, has an octet in its 5th energy level and 4*s*, 4*p*, and 4*d* sublevels filled. See Table 4-4.

4.9 Atoms of the sixth series The sixth series of atoms is much longer than the others. It consists of 32 elements. The atoms of the first two, cesium and barium, have lower energy levels that resemble the xenon atom. Successive electrons occupy the 6*s* sublevel.

Cs· **Ba:**

At lanthanum, the lowest-energy *d* sublevel (here 5*d*) begins to fill, as 4*d* began to fill at yttrium and 3*d* began to fill at scandium. But with the very next element, cerium, something different happens. In the atoms of cerium and the next 12 elements of the sixth series, the 7 orbitals of the 4*f* sublevel are occupied by successive electrons. In atoms of the element ytterbium, the 4th energy level has all of its sublevels filled with 32 electrons.

The atoms of the next ten elements of the sixth series have successive electrons occupying the five orbitals of the 5*d* sublevel. These orbitals have the next higher energies.

The atoms of the remaining six elements of this series are thallium, lead, bismuth, polonium, astatine, and radon. They

Table 4-4
STRUCTURE OF ATOMS IN THE FIFTH SERIES

Name	*Symbol*	*Atomic number*	*Number of electrons in sublevels*										
			1*s*	2*s*	2*p*	3*s*	3*p*	3*d*	4*s*	4*p*	4*d*	5*s*	5*p*
rubidium	Rb	37	2	2	6	2	6	10	2	6		1	
strontium	Sr	38	2	2	6	2	6	10	2	6		2	
yttrium	Y	39	2	2	6	2	6	10	2	6	1	2	
zirconium	Zr	40	2	2	6	2	6	10	2	6	2	2	
niobium	Nb	41	2	2	6	2	6	10	2	6	4	1	
molybdenum	Mo	42	2	2	6	2	6	10	2	6	5	1	
technetium	Tc	43	2	2	6	2	6	10	2	6	5	2	
ruthenium	Ru	44	2	2	6	2	6	10	2	6	7	1	
rhodium	Rh	45	2	2	6	2	6	10	2	6	8	1	
palladium	Pd	46	2	2	6	2	6	10	2	6	10		
silver	Ag	47	2	2	6	2	6	10	2	6	10	1	
cadmium	Cd	48	2	2	6	2	6	10	2	6	10	2	
indium	In	49	2	2	6	2	6	10	2	6	10	2	1
tin	Sn	50	2	2	6	2	6	10	2	6	10	2	2
antimony	Sb	51	2	2	6	2	6	10	2	6	10	2	3
tellurium	Te	52	2	2	6	2	6	10	2	6	10	2	4
iodine	I	53	2	2	6	2	6	10	2	6	10	2	5
xenon	Xe	54	2	2	6	2	6	10	2	6	10	2	6

have the first four energy levels completely filled and filled 5*s*, 5*p*, and 5*d* sublevels. The 6*s* and 6*p* electrons are shown in these electron-dot symbols:

Tl: ·Pb: ·Bi: ·Po: :At: :Rn:

Radon, the last member of the sixth series, has an octet in its 6th energy level and 5*s*, 5*p*, and 5*d* sublevels filled. See Table 4-5.

4.10 Atoms of the seventh series The seventh series of elements is an incomplete series of which only 23 elements are known. Table 4-6 shows what is believed to be the arrangement of electrons in the 5th, 6th, and 7th energy levels. Appendix Table 7, Electronic Arrangement of the Elements, gives the complete electron configurations for all the elements.

Table 4-5
STRUCTURE OF ATOMS IN THE SIXTH SERIES

Name	*Symbol*	*Atomic number*		*Number of electrons in sublevels*						
				4d	*4f*	*5s*	*5p*	*5d*	*6s*	*6p*
cesium	Cs	55		10		2	6		1	
barium	Ba	56		10		2	6		2	
lanthanum	La	57		10		2	6	1	2	
cerium	Ce	58		10	2	2	6		2	
praseodymium	Pr	59		10	3	2	6		2	
neodymium	Nd	60		10	4	2	6		2	
promethium	Pm	61		10	5	2	6		2	
samarium	Sm	62		10	6	2	6		2	
europium	Eu	63		10	7	2	6		2	
gadolinium	Gd	64		10	7	2	6	1	2	
terbium	Tb	65		10	9	2	6		2	
dysprosium	Dy	66		10	10	2	6		2	
holmium	Ho	67		10	11	2	6		2	
erbium	Er	68	krypton	10	12	2	6		2	
thulium	Tm	69	structure	10	13	2	6		2	
ytterbium	Yb	70	plus	10	14	2	6		2	
lutetium	Lu	71		10	14	2	6	1	2	
hafnium	Hf	72		10	14	2	6	2	2	
tantalum	Ta	73		10	14	2	6	3	2	
tungsten	W	74		10	14	2	6	4	2	
rhenium	Re	75		10	14	2	6	5	2	
osmium	Os	76		10	14	2	6	6	2	
iridium	Ir	77		10	14	2	6	7	2	
platinum	Pt	78		10	14	2	6	9	1	
gold	Au	79		10	14	2	6	10	1	
mercury	Hg	80		10	14	2	6	10	2	
thallium	Tl	81		10	14	2	6	10	2	1
lead	Pb	82		10	14	2	6	10	2	2
bismuth	Bi	83		10	14	2	6	10	2	3
polonium	Po	84		10	14	2	6	10	2	4
astatine	At	85		10	14	2	6	10	2	5
radon	Rn	86		10	14	2	6	10	2	6

Table 4-6
STRUCTURE OF ATOMS IN THE SEVENTH SERIES

Name	*Symbol*	*Atomic number*		*Number of electrons in sublevels*						
				4f	*5d*	*5f*	*6s*	*6p*	*6d*	*7s*
francium	Fr	87		14	10		2	6		1
radium	Ra	88		14	10		2	6		2
actinium	Ac	89		14	10		2	6	1	2
thorium	Th	90		14	10		2	6	2	2
protactinium	Pa	91		14	10	2	2	6	1	2
uranium	U	92		14	10	3	2	6	1	2
neptunium	Np	93		14	10	4	2	6	1	2
plutonium	Pu	94	xenon	14	10	6	2	6		2
americium	Am	95	structure	14	10	7	2	6		2
curium	Cm	96	plus	14	10	7	2	6	1	2
berkelium	Bk	97		14	10	9	2	6		2
californium	Cf	98		14	10	10	2	6		2
einsteinium	Es	99		14	10	11	2	6		2
fermium	Fm	100		14	10	12	2	6		2
mendelevium	Md	101		14	10	13	2	6		2
nobelium	No	102		14	10	14	2	6		2
lawrencium	Lr	103		14	10	14	2	6	1	2
unnilquadium	Unq	104		14	10	14	2	6	2	2?
unnilpentium	Unp	105		14	10	14	2	6	3	2?
unnilhexium	Unh	106		14	10	14	2	6	4	2?
unnilseptium	Uns	107		14	10	14	2	6	5	2?
unniloctium	Uno	108		14	10	14	2	6	6	2?
unnilennium	Une	109		14	10	14	2	6	7	2?

SUMMARY

Electromagnetic radiations are forms of energy that travel through space as waves. The frequency of a wave and its wavelength are inversely proportional. Electromagnetic radiation also has some properties of particles. Electromagnetic radiation is transferred to matter in photon units. The energy of a photon is directly proportional to the frequency of the radiation. An atom that has absorbed energy is called an excited atom. When excited atoms radiate energy, the radiation is given off in photon units.

The bright-line spectra of atoms show that energy changes within atoms involve only fixed and definite quantities of energy. This is evidence that electrons occupy definite energy levels in atoms.

An orbital is a highly probable location where an electron may be found. The energy state of an electron in an atom may be described by four quantum numbers. The principal quantum number indicates the most probable distance of the electron from the nucleus of the atom. The orbital quantum number indicates the shape of the orbital the electron occupies. The magnetic quantum number indicates the position of the orbital about the three axes in space. The spin quantum number indicates the direction of spin of the electron. No two electrons in an atom can have exactly the same set of quantum numbers.

The most stable state of an atom is its ground state. If the number of electrons in an atom is known, then its electron configuration can be described because electrons occupy the various orbitals in a reasonably definite order. Electron configurations can be described by orbital notation, electron-configuration notation, and electron-dot notation. Two electrons of opposite spin in the same orbital are an electron pair. An atom with only the *s* and *p* orbitals of its highest numbered energy level filled with eight electrons is said to have an octet in its highest numbered energy level. Seven series of atoms have been identified. The final element in each of these series has an octet in its highest numbered energy level.

VOCABULARY

electromagnetic radiation
electron cloud
electron pair
frequency
ground state
magnetic quantum number
octet
orbital
orbital quantum number
photon
principal quantum number
quanta
spectrum
spin quantum number
wavelength

QUESTIONS

Group A

1. *(a)* What opposing actions operate on a satellite in orbit around the earth? *(b)* In what direction does each operate? *(c)* How do they compare when the satellite is in a stable orbit? *A3a*
2. *(a)* Describe electromagnetic radiations. *(b)* Name some common forms of electromagnetic radiation. *A1, A3b*
3. *(a)* When do electromagnetic radiations have the properties of waves? *(b)* When do they have the properties of particles? *A3b*
4. How is an atom in its ground state transformed into an excited atom? *A1, A2b*
5. *(a)* The red line in the visible spectrum of hydrogen (Figures 4-4 and 4-5) has a wavelength of 6563 Å. Use Figure 4-6 to identify the electron-energy-level transition that produces this line. *(b)* Which electron-energy-level transition produces the blue line having a wavelength of 4340 Å? *A1, A2a, A3c*
6. List four characteristics of electrons according to the Bohr model of the hydrogen atom. *A2c, A3d*
7. Define *(a) orbital. (b) electron cloud.* *A1, A2c*
8. What properties of an atom are believed to be caused by the electron cloud? *A2c, A3e*
9. *(a)* Name the four kinds of quantum numbers. *(b)* What does each quantum number indicate? *A1, A3f*
10. *(a)* What is the shape of an *s* orbital? *(b)* How many *s* orbitals can there be in an energy level? *(c)* How many electrons can occupy such an orbital? *(d)* What characteristic must these electrons have? *(e)* Which is the lowest energy level having an *s* orbital? *A2c, A3f, A3g, A3h*
11. *(a)* What is the shape of a *p* orbital? *(b)* How many *p* orbitals can there be in an energy level? *(c)* How are they arranged with respect to one another? *(d)* Which is the lowest energy level having *p* orbitals? *A2c, A3f, A3g, A3h*
12. *(a)* Do any two electrons in an atom have exactly the same energy? *(b)* How does this affect the quantum numbers assigned to the electrons in an atom? *A3f*
13. Under what conditions may two electrons occupy the same orbital in an atom? *A3f*
14. Define *(a) electron pair. (b) octet.* *A1*

Group B

15. Derive the relationship between λ and E for electromagnetic radiation. *A3b*
16. *(a)* What evidence is there that electron en-

ergy transitions within an atom occur in definite amounts rather than as a continuous flow? *(b)* What does this lead us to believe about the energies of electrons? *A2a, A2b*

17. *(a)* From Figures 4-5 and 4-6, determine the approximate wavelength of the radiation produced by an electron transition in a hydrogen atom from the 5th energy level to the 1st energy level; *(b)* from the 5th energy level to the 3rd energy level. *A1, A2a, A2b, A3c*

18. *(a)* Theoretically what would happen to an electron's wavelength as an electron cloud becomes smaller? *(b)* What effect does this have on the electron's energy? *(c)* How does this determine the ultimate size of an electron cloud? *A3e*

19. *(a)* How many *d* orbitals can there be in an energy level? *(b)* How many *d* electrons can there be in an energy level? *(c)* Which is the lowest energy level having *d* orbitals? *A2c, A3g*

20. *(a)* How many *f* orbitals can there be in an energy level? *(b)* How many *f* electrons can there be in an energy level? *(c)* Which is the lowest energy level having *f* orbitals? *A2c, A3g*

21. How many electron pairs are there in the highest energy level of each of the following atoms: *(a)* nitrogen; *(b)* neon; *(c)* selenium; *(d)* potassium; *(e)* bromine? *A2d, A3f*

22. Which of the atoms in Question 21 has an octet in the highest energy level? *A2d, A3f*

23. On a sheet of paper, copy and complete the table below for the atoms in the third series. *Do not write in this book.* *A2d, A3f, B1*

24. How many energy levels are partially or fully occupied in the californium atom? *A3f, A3g*

25. *(a)* How many elements comprise the fourth series? *(b)* the fifth series? *(c)* Why are there not 8 elements as in the second and third series? *A3g, B2*

26. *(a)* How many elements are there in the sixth series? *(b)* Explain. *A3g, B2*

27. *(a)* What types of orbitals can be found in the 4th energy level? *(b)* How many orbitals of each type are there? *(c)* How many electrons can occupy each of these types of orbitals? *(d)* How many electrons are needed to completely fill the 4th energy level? *A2c, A2d, A3f, A3g*

28. Which sublevels of the 3rd energy level are filled *(a)* in the element phosphorus; *(b)* in the element arsenic; *(c)* in the element antimony? *A2c, A3f, A3g, B2*

29. What is a probable electron configuration for the as yet undiscovered element 110 ununnilium? *A2d, A3i, B2*

30. What is a probable electron-dot notation for the undiscovered element 112? *A2d, B2*

Chemical symbol	*Orbital notation*	*Electron-configuration notation*	*Electron-dot notation*
Na			
Mg			
Al			
Si			
P			
S			
Cl			
Ar			

Glenn T. Seaborg (American), along with Edwin McMillan (American), was awarded the 1951 Nobel Prize in chemistry for discovering and studying transuranium elements, that is, elements beyond uranium on the periodic table. He bombarded uranium and the resulting products with neutrons and produced new artificial elements, such as plutonium, americium, and curium. The transuranium elements are radioactive and differ little from each other because they have similar electronic structures. These elements have been used as sources of energy for ships and power plants.

The Periodic Law

Goals

In this chapter, you will gain an understanding of: • the historical development of the periodic table • the periodic law • the modern periodic table • some periodic properties: atomic radius, ionization energy, and electron affinity

5.1 Atomic weights and the properties of elements The law of definite composition states that every compound has a definite composition by weight. By chemical analysis, it is possible to determine the weight of any element that combines with a fixed weight of a particular element. Such a weight is called a *combining weight*. Between 1807 and 1818, Berzelius determined the combining weights with oxygen of 43 elements: almost all the elements known at his time.

The law of definite composition was explained in Section 2.9.

Chemists also recognized that some elements had similar properties. Lithium, sodium, and potassium were found to be similar metals, as were calcium, strontium, and barium. Chlorine, bromine, and iodine were found to be similar nonmetals. So were sulfur, selenium, and tellurium. In 1829 Johann Wolfgang Döbereiner (1780–1849), a German chemist, pointed out that in such groups of three elements, the combining weight of the middle element was the average of the combining weights of the other two elements.

By 1860, it became possible to establish whether the combining weight of an element was indeed its atomic weight or a multiple or submultiple of the atomic weight. Stas' atomic weight determinations, begun in 1860, far surpassed the earlier work of Berzelius. Stas used techniques of both analysis and synthesis. His materials were very pure, his balance was a precision instrument, and he exercised the utmost care in his experimentation. The availability of Stas' atomic weights stimulated the search for a meaningful relationship between the properties of elements and their atomic weights.

Some of the ways this decision is made are described in Chapters 7 and 11.

Analysis and synthesis are defined in Section 2.8.

In 1865, John A. R. Newlands (1838–1898), an English chemist, arranged the elements in order of increasing atomic weight for the first time. He found that similar elements were 7 or a multiple of 7 elements apart. Lithium, the 2nd element, was similar in properties to sodium, the 9th element, and potassium, the 16th element. Fluorine, the 8th element, was similar to chlorine, the 15th element, and to bromine, the 29th element. Iodine was the 42nd element in atomic weight order, not the 43rd as Newlands' arrangement predicted. However, he felt that an even more accurate atomic weight determination would eventually clear up this discrepancy.

5.2 Mendeleev's periodic table In 1869, the Russian chemist Dmitri Mendeleev (men-deh-*lay*-eff) (1834–1907) published a table of elements based on both their properties and the order of their atomic weights. See Figure 5-1. The elements are arranged in vertical columns in atomic-weight order so that elements having similar chemical properties are in the same horizontal row. Mendeleev realized that all the elements were probably not yet discovered. He carefully studied the properties of the known elements and then left gaps in his table. Note the question marks in the table corresponding to atomic weights 45, 68, and 70. He predicted that new elements would be discovered to fill these gaps. He also predicted the properties of these new elements. When scandium, gallium, and germanium were discovered,

Figure 5-1. Even though Mendeleev's original periodic table appears quite different from the one we use today, there are similarities. How many elements do we now classify as belonging in the same groups? Which elements did Mendeleev predict would be discovered?

но въ ней, мнѣ кажется, уже ясно выражается примѣнимость выставляемаго мною начала ко всей совокупности элементовъ, пай которыхъ извѣстенъ съ достовѣрностію. На этотъ разъ я и желалъ преимущественно найдти общую систему элементовъ. Вотъ этотъ опытъ:

			Ti=50	Zr=90	?=180.
			V=51	Nb=94	Ta=182.
			Cr=52	Mo=96	W=186.
			Mn=55	Rh=104,4	Pt=197,4
			Fe=56	Ru=104,4	Ir=198.
			Ni=Co=59	Pl=106,6,	Os=199.
H=1			Cu=63,4	Ag=108	Hg=200.
	Be=9,4	Mg=24	Zn=65,2	Cd=112	
	B=11	Al=27,4	?=68	Ur=116	Au=197?
	C=12	Si=28	?=70	Sn=118	
	N=14	P=31	As=75	Sb=122	Bi=210
	O=16	S=32	Se=79,4	Te=128?	
	F=19	Cl=35,5	Br=80	I=127	
Li=7	Na=23	K=39	Rb=85,4	Cs=133	Tl=204
		Ca=40	Sr=87,6	Ba=137	Pb=207.
		?=45	Ce=92		
		?Er=56	La=94		
		?Yt=60	Di=95		
		?In=75,6	Th=118?		

а потому приходится въ разныхъ рядахъ имѣть различное измѣненіе разностей, чего нѣтъ въ главныхъ числахъ предлагаемой таблицы. Или же придется предполагать при составленіи системы очень много недостающихъ членовъ. То и другое мало выгодно. Мнѣ кажется притомъ, наиболѣе естественнымъ составить

Table 5-1
COMPARISON OF MENDELEEV'S PREDICTIONS WITH ACTUAL PROPERTIES OF GERMANIUM

Property	*Mendeleev's prediction*	*Actual value*
atomic weight	72	72.59
density, g/cm^3	5.5	5.32
atomic volume, cm^3	13	13.6
combining power	4	4
specific heat, cal/g C°	0.073	0.077
density of dioxide, g/cm^3	4.7	4.228
molecular volume of dioxide, cm^3	22	24.74
boiling point of tetrachloride, °C	<100	84
density of tetrachloride, g/cm^3	1.9	1.8443
molecular volume of tetrachloride, cm^3	113	116.26

their properties agreed very well with Mendeleev's predictions. See Table 5-1.

From Mendeleev's table of the elements, it was apparent that a relationship did exist between the properties of elements and their atomic weights. When elements are arranged in order of increasing atomic weights, their chemical properties follow a pattern. Similar chemical properties reoccur at definite intervals. Mendeleev concluded that "the properties of the elements are in periodic dependence on their atomic weights." Thus, Mendeleev's table, and others like it, are called periodic tables.

Lothar Meyer (1830–1895), a German chemist, independently arrived at this same conclusion at about the same time Mendeleev did.

5.3 Two discoveries alter the periodic table During the 1890s Sir William Ramsay (1852–1916), a British chemist, discovered the noble gases neon, argon, krypton, and xenon. Another noble gas, helium, had been discovered on the sun in 1868. It was not really accepted as an element until it was found on the earth in 1895. Radon, the noble gas of highest atomic weight, was discovered in 1900. These six elements added an entirely new row to Mendeleev's periodic table. This row of elements goes between the F, Cl, Br, I row and the Li, Na, K, Rb, Cs row.

During 1913–1914 the English scientist Henry Gwyn-Jeffreys Moseley (1887–1915) performed X-ray experiments that showed the manner in which the number of protons per nucleus varied progressively from element to element. Moseley used his experimental evidence to determine the atomic-number order of the elements.

The atomic number of an atom is defined in Section 3.7 as the number of protons in its nucleus.

X rays are electromagnetic radiations. Light is another form of electromagnetic radiation. But, unlike light, *X rays are not visible and are of higher frequency and shorter wavelength than light.* X rays are produced when high-speed electrons strike a metal target in an evacuated tube. Moseley found that the wavelengths of the X rays produced depend on the kind of metal used as the target. He used various metals ranging in atomic weight from aluminum to gold as the targets. He found that the wavelengths of X rays became shorter as elements with more protons in their nuclei were used. The higher the atomic number of an element, the shorter the wavelength of the X rays produced when that element is used as the target in an X-ray tube.

In three cases Moseley found an unusual variation in the wavelengths of X rays between two successive elements. The variation was twice as great as his calculations indicated. He concluded that in these cases an element was missing from the periodic table. The three elements, technetium, promethium, and rhenium, have since been discovered. They fill the gaps that Moseley indicated.

When the elements in a periodic table are placed in order of increasing atomic numbers instead of increasing atomic weights, some of the problems of arrangement disappear. Arranged according to increasing atomic weights, potassium, 39.1, precedes argon, 39.9. Yet, when arranged according to properties, potassium follows argon. This agrees with the atomic numbers: argon, 18, potassium, 19. A similar case is that of tellurium and iodine.

5.4 The periodic law As stated in Section 5.2, Mendeleev concluded that the properties of elements are related in a periodic way to their atomic weights. Today, evidence shows that atomic numbers are better standards for establishing the order of the elements. Mendeleev's conclusion is now restated as the ***periodic law:*** *The physical and chemical properties of the elements are periodic functions of their atomic numbers.* In other words, (1) the properties of elements go through a pattern of change; (2) elements of similar properties occur at certain intervals, provided the elements are arranged in a periodic table in order of increasing atomic number.

5.5 Arrangement of the modern periodic table The modern periodic table is shown on pages 100–101. As you study this section frequent reference to these pages will help you understand the periodic table and its importance in chemistry.

Each element is assigned a separate block in the table. See Figure 5-2. In the center of the block is the chemical symbol for the element. Below the symbol is the atomic number of the element. Above the symbol is the atomic weight. To the right of each symbol are numbers. These numbers indicate the distribution of electrons in the energy levels of the atoms of this element. A horizontal row of blocks on the table is called a ***period*** or ***series.*** A vertical column is called a ***group*** or ***family.***

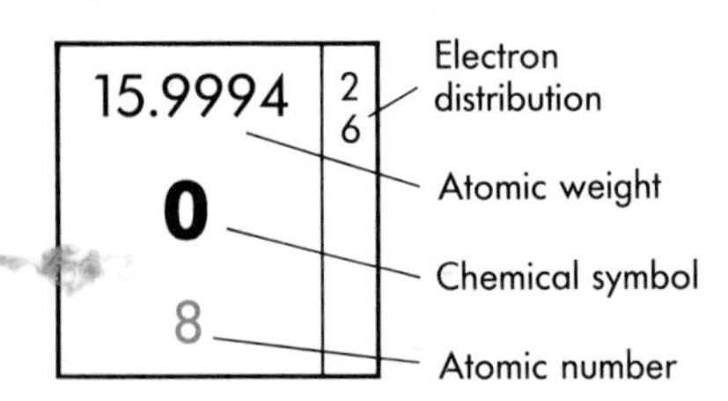

Figure 5-2. This figure shows the position of information found in each block of the periodic table.

Hydrogen, atomic number 1, is placed at the top of the table by itself because of its many unique properties. It is in the first column at the left of the table because its atoms have one electron in the highest energy level. Helium, atomic number 2, is at the top of the extreme right-hand column. Helium is classified as an inert gas because it does not react with other elements. It is the simplest member of the group of elements known as the *noble gases*. Note that helium atoms have two electrons in the 1st energy level, which makes the 1st energy level complete. Hydrogen and helium compose the first period of elements.

One unique property of hydrogen atoms is their single electron. If an electron is removed from a hydrogen atom, only the nucleus, one ten-thousandth of the diameter of the atom, remains. If an electron is removed from any other atom, enough other electrons remain to occupy the space around the nucleus and roughly maintain the diameter of the atom.

Figure 5-3 illustrates the relationship between the periodic table and the sublevels occupied by electrons in the structures of atoms of the elements. Refer to it also as you continue to study this section.

The second period consists of eight elements: (1) *lithium*, a soft, silvery, active metal, whose atoms have one electron in the 2nd energy level; (2) *beryllium*, a silvery metal, less active than

Figure 5-3. The periodic table consists essentially of blocks of elements whose structures add support to the modern atomic theory.

Sublevel Blocks of the Periodic Table

Period	Sublevels being filled	*s* sublevel block		TRANSITION ELEMENTS *d* sublevel block										*p* sublevel block					
		I	II											III	IV	V	VI	VII	VIII
1	1*s*	H 1	He 2																
2	2*s* 2*p*	Li 3	Be 4											B 5	C 6	N 7	O 8	F 9	Ne 10
3	3*s* 3*p*	Na 11	Mg 12											Al 13	Si 14	P 15	S 16	Cl 17	Ar 18
4	4*s* 3*d* 4*p*	K 19	Ca 20	Sc 21	Ti 22	V 23	Cr 24	Mn 25	Fe 26	Co 27	Ni 28	Cu 29	Zn 30	Ga 31	Ge 32	As 33	Se 34	Br 35	Kr 36
5	5*s* 4*d* 5*p*	Rb 37	Sr 38	Y 39	Zr 40	Nb 41	Mo 42	Tc 43	Ru 44	Rh 45	Pd 46	Ag 47	Cd 48	In 49	Sn 50	Sb 51	Te 52	I 53	Xe 54
6	6*s* 4*f* 5*d* 6*p*	Cs 55	Ba 56	Lu 71	Hf 72	Ta 73	W 74	Re 75	Os 76	Ir 77	Pt 78	Au 79	Hg 80	Tl 81	Pb 82	Bi 83	Po 84	At 85	Rn 86
7	7*s* 5*f* 6*d* 7*p*	Fr 87	Ra 88	Lr 103	Unq 104	Unp 105	Unh 106	Uns 107	Uno 108	Une 109									

RARE EARTH ELEMENTS
f sublevel block

Period	Sublevel	Series														
6	4*f*	Lanthanide series	La 57	Ce 58	Pr 59	Nd 60	Pm 61	Sm 62	Eu 63	Gd 64	Tb 65	Dy 66	Ho 67	Er 68	Tm 69	Yb 70
7	5*f*	Actinide series	Ac 89	Th 90	Pa 91	U 92	Np 93	Pu 94	Am 95	Cm 96	Bk 97	Cf 98	Es 99	Fm 100	Md 101	No 102

Periodic Table

METALS

Period	H
1	1.00794 **H** 1 1

Period	I	II
2	6.941 **Li** 3 2 1	9.01218 **Be** 4 2 2
3	22.98977 **Na** 11 2 8 1	24.305 **Mg** 12 2 8 2

TRANSITION ELEMENTS

Period	I	II							
4	39.0983 **K** 19 2 8 8 1	40.08 **Ca** 20 2 8 8 2	44.9559 **Sc** 21 2 8 9 2	47.88 **Ti** 22 2 8 10 2	50.9415 **V** 23 2 8 11 2	51.996 **Cr** 24 2 8 13 1	54.9380 **Mn** 25 2 8 13 2	55.847 **Fe** 26 2 8 14 2	58.9332 **Co** 27 2 8 15 2
5	85.4678 **Rb** 37 2 8 18 8 1	87.62 **Sr** 38 2 8 18 8 2	88.9059 **Y** 39 2 8 18 9 2	91.22 **Zr** 40 2 8 18 10 2	92.9064 **Nb** 41 2 8 18 12 1	95.94 **Mo** 42 2 8 18 13 1	[98] **Tc** 43 2 8 18 13 2	101.07 **Ru** 44 2 8 18 15 1	102.9055 **Rh** 45 2 8 18 16 1
6	132.9054 **Cs** 55 2 8 18 18 8 1	137.33 **Ba** 56 2 8 18 18 8 2	Lantha-nide Series 174.967 **Lu** 71 2 8 18 32 9 2	178.49 **Hf** 72 2 8 18 32 10 2	180.9479 **Ta** 73 2 8 18 32 11 2	183.85 **W** 74 2 8 18 32 12 2	186.207 **Re** 75 2 8 18 32 13 2	190.2 **Os** 76 2 8 18 32 14 2	192.22 **Ir** 77 2 8 18 32 15 2
7	[223] **Fr** 87 2 8 18 32 18 8 1	226.0254 **Ra** 88 2 8 18 32 18 8 2	Actinide Series [260] **Lr** 103 2 8 18 32 32 9 2	[261] **Unq** 104 2 8 18 32 32 10 2	[262] **Unp** 105 2 8 18 32 32 11 2	[263] **Unh** 106 2 8 18 32 32 12 2	[261?] **Uns** 107 2 8 18 32 32 13 2	[265] **Uno** 108 2 8 18 32 32 14 2	[266?] **Une** 109 2 8 18 32 32 15 2

Series							
Lanthanide Series	138.9055 **La** 57 2 8 18 18 9 2	140.12 **Ce** 58 2 8 18 20 8 2	140.9077 **Pr** 59 2 8 18 21 8 2	144.24 **Nd** 60 2 8 18 22 8 2	[145] **Pm** 61 2 8 18 23 8 2	150.36 **Sm** 62 2 8 18 24 8 2	151.96 **Eu** 63 2 8 18 25 8 2
Actinide Series	227.0278 **Ac** 89 2 8 18 32 18 9 2	232.0381 **Th** 90 2 8 18 32 18 10 2	231.0359 **Pa** 91 2 8 18 32 20 9 2	238.0289 **U** 92 2 8 18 32 21 9 2	237.0482 **Np** 93 2 8 18 32 22 9 2	[244] **Pu** 94 2 8 18 32 24 8 2	[243] **Am** 95 2 8 18 32 25 8 2

of the Elements

NONMETALS

			III	IV	V	VI	VII	Noble gases VIII
								4.00260 **He** 2 (2)
			10.81 **B** 5 (2, 3)	12.011 **C** 6 (2, 4)	14.0067 **N** 7 (2, 5)	15.9994 **O** 8 (2, 6)	18.998403 **F** 9 (2, 7)	20.179 **Ne** 10 (2, 8)
			26.98154 **Al** 13 (2, 8, 3)	28.0855 **Si** 14 (2, 8, 4)	30.97376 **P** 15 (2, 8, 5)	32.06 **S** 16 (2, 8, 6)	35.453 **Cl** 17 (2, 8, 7)	39.948 **Ar** 18 (2, 8, 8)
58.69 **Ni** 28 (2, 8, 16, 2)	63.546 **Cu** 29 (2, 8, 18, 1)	65.38 **Zn** 30 (2, 8, 18, 2)	69.72 **Ga** 31 (2, 8, 18, 3)	72.59 **Ge** 32 (2, 8, 18, 4)	74.9216 **As** 33 (2, 8, 18, 5)	78.96 **Se** 34 (2, 8, 18, 6)	79.904 **Br** 35 (2, 8, 18, 7)	83.80 **Kr** 36 (2, 8, 18, 8)
106.42 **Pd** 46 (2, 8, 18, 18, 0)	107.8682 **Ag** 47 (2, 8, 18, 18, 1)	112.41 **Cd** 48 (2, 8, 18, 18, 2)	114.82 **In** 49 (2, 8, 18, 18, 3)	118.69 **Sn** 50 (2, 8, 18, 18, 4)	121.75 **Sb** 51 (2, 8, 18, 18, 5)	127.60 **Te** 52 (2, 8, 18, 18, 6)	126.9045 **I** 53 (2, 8, 18, 18, 7)	131.29 **Xe** 54 (2, 8, 18, 18, 8)
195.08 **Pt** 78 (2, 8, 18, 32, 17, 1)	196.9665 **Au** 79 (2, 8, 18, 32, 18, 1)	200.59 **Hg** 80 (2, 8, 18, 32, 18, 2)	204.383 **Tl** 81 (2, 8, 18, 32, 18, 3)	207.2 **Pb** 82 (2, 8, 18, 32, 18, 4)	208.9804 **Bi** 83 (2, 8, 18, 32, 18, 5)	[209] **Po** 84 (2, 8, 18, 32, 18, 6)	[210] **At** 85 (2, 8, 18, 32, 18, 7)	[222] **Rn** 86 (2, 8, 18, 32, 18, 8)

RARE EARTH ELEMENTS

157.25 **Gd** 64 (2, 8, 18, 25, 9, 2)	158.9254 **Tb** 65 (2, 8, 18, 27, 8, 2)	162.50 **Dy** 66 (2, 8, 18, 28, 8, 2)	164.9304 **Ho** 67 (2, 8, 18, 29, 8, 2)	167.26 **Er** 68 (2, 8, 18, 30, 8, 2)	168.9342 **Tm** 69 (2, 8, 18, 31, 8, 2)	173.04 **Yb** 70 (2, 8, 18, 32, 8, 2)
[247] **Cm** 96 (2, 8, 18, 32, 25, 9, 2)	[247] **Bk** 97 (2, 8, 18, 32, 27, 8, 2)	[251] **Cf** 98 (2, 8, 18, 32, 28, 8, 2)	[252] **Es** 99 (2, 8, 18, 32, 29, 8, 2)	[257] **Fm** 100 (2, 8, 18, 32, 30, 8, 2)	[258] **Md** 101 (2, 8, 18, 32, 31, 8, 2)	[259] **No** 102 (2, 8, 18, 32, 32, 8, 2)

A value given in brackets denotes the mass number of the isotope of longest known half-life.

Figure 5-4(A). The first four elements of Period 3 of the periodic table.

lithium, whose atoms have two electrons in the 2nd energy level; (3) *boron*, a black solid with some nonmetallic properties, whose atoms have three electrons in the 2nd energy level; (4) *carbon*, a solid element with very distinctive chemical properties intermediate between those of metals and nonmetals: four electrons in the 2nd energy level; (5) *nitrogen*, a colorless gas with nonmetallic properties and five electrons in the 2nd energy level; (6) *oxygen*, a colorless gas having strong nonmetallic properties and six electrons in the 2nd energy level; (7) *fluorine*, a pale-yellow gas with very strong nonmetallic properties and seven electrons in the 2nd energy level; (8) *neon*, a colorless, inert (unreactive) gas having eight electrons in the 2nd energy level. See Appendix Table 7, Electron Arrangement of the Elements.

The elements of Period 2 range from an active metallic element (Li) through two metalloids (Be and B, whose properties are between those of the typical metal and nonmetal) to an active nonmetallic element (F). The last element in the period (Ne) is inert. This variation in properties from metallic through metalloidal to nonmetallic is accompanied by an increase in the number of 2nd energy electrons from 1 to 7. The inert element neon has 8 electrons, an octet, in the 2nd energy level.

The third period also consists of eight elements: (1) *sodium*, a soft, silvery, active metal similar to lithium, with one electron in its highest energy level, the 3rd energy level; (2) *magnesium*, a silvery metal similar in properties to beryllium, having two electrons in the 3rd energy level; (3) *aluminum*, a silvery metal with some nonmetallic properties and three electrons in the 3rd energy level; (4) *silicon*, a dark-colored, nonmetallic element with some properties resembling carbon, and with four electrons in the 3rd energy level; (5) *phosphorus*, a nonmetallic, solid element that forms compounds similar to those of nitrogen, with five electrons in the 3rd energy level; (6) *sulfur*, a yellow, nonmetallic solid element, six electrons in the 3rd energy level; (7) *chlorine*, a yellow-green gas with strong nonmetallic properties resembling those of fluorine and with seven electrons in the 3rd energy level; (8) *argon*, a colorless, inert gas that has eight electrons in the 3rd energy level. Figure 5-4 illustrates the variations in the properties of these third-period elements. Again the elements range from active metallic through metalloidal to active nonmetallic properties as the number of electrons in the 3rd energy level varies from 1 to 7. The element with an octet in its highest energy level is a noble gas.

Elements with similar properties have a similar arrangement of highest-energy-level electrons. They fall into the same group in the periodic table.

Group I of the periodic table is the *sodium family*, a group of six similar, very active, metallic elements. Their atoms all have only one *s* electron in the highest energy level. *Francium* is the most complex member of the sodium family. The position of francium in the periodic table indicates that it is probably the most active metal.

Group II also consists of six active metals whose chemical properties are very much alike. The atoms of each have two *s* electrons in the highest energy level. This is the *calcium family*. The most chemically active member of this family is *radium*.

The properties of elements in Group III vary from nonmetallic to metallic as the atoms become larger and heavier. The atoms of this group have three electrons, two *s* and one *p*, in the highest energy level.

The elements of Group IV vary in similar fashion. Their atoms have four electrons, two *s* and two *p*, in the highest energy level. Atoms of elements in both Group III and Group IV have very stable lower energy levels.

Group V is the *nitrogen family*. *Nitrogen* and *phosphorus*, the elements at the top of this family, are nonmetallic. The element *bismuth*, at the bottom of the table, is metallic. *Arsenic* and *antimony* exhibit both metallic and nonmetallic properties. The atoms of each of these elements have five electrons, two *s* and three *p*, in the highest energy level and very stable lower energy levels.

Group VI is the *oxygen family*. The properties of elements in this family vary from active nonmetallic to metallic as the atoms become larger and heavier. The atoms of each of these elements have six electrons, two *s* and four *p*, in the highest energy level and very stable lower energy levels.

The elements in Group VII, the *halogen family*, are very active nonmetals. Their atoms each have seven electrons, two *s* and five *p*, in the highest energy level and very stable lower energy levels. The most active member of the halogen family is its simplest element, *fluorine*.

Group VIII is the *noble-gas family*. With the exception of *helium* atoms, which have a pair of electrons in the highest energy level, atoms of these elements have an octet, two *s* and six *p* electrons, in the highest energy level. This is the greatest number of electrons found in the highest energy level. No compounds of helium, neon, and argon are known to exist. A few compounds of krypton, xenon, and radon have been prepared. The low-atomic-weight noble gases are not considered in a discussion of activity because they form no compounds. Thus, the activity of the elements ranges from the most active metal at the lower left corner of the periodic table to the most active nonmetal at the upper right corner.

The fourth period consists of 18 elements. It is the first long period. In addition to the eight elements in Groups I to VIII, there are also ten ***transition elements***. These are metallic elements whose atoms have one or two electrons in the highest energy level. Successive electrons usually occupy the group of five orbitals of the 3*d* sublevel.

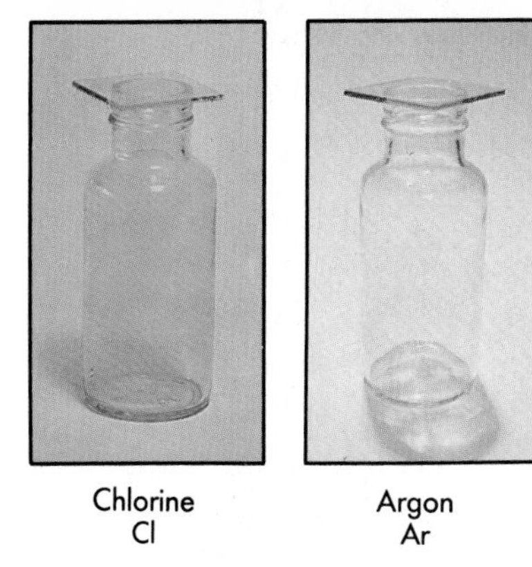

Figure 5-4(B). The last four elements of Period 3 of the periodic table.

The fifth period also consists of 18 elements. It includes ten transition elements, in which successive electrons occupy the group of five orbitals of the *4d* sublevel. The transition elements are all metals.

The sixth period consists of 32 elements. In addition to the elements in Groups I to VIII and the ten transition elements, there is a group of 14 ***rare earth elements.*** These elements have almost identical chemical properties. They compose the *lanthanide series.* The highest and next-to-highest energy levels of these atoms are almost the same. The highest energy level contains two electrons. The next-to-highest energy level usually contains eight electrons but in two of these elements contains nine. As the number of electrons in the 4th energy level increases from 18 to 32, successive electrons occupy the group of seven orbitals of the *4f* sublevel.

The seventh period of elements is incomplete at present. It is assumed to be similar to the sixth period. The rare earth elements in the period compose the *actinide series.* To date, 23 members of the seventh period are known or reported.

In the periodic table, the elements are roughly divided into metals, nonmetals, and noble gases. The line separating the metals from the nonmetals is zigzag. It runs diagonally down and to the right near the right side of the table. The elements that border this zigzag line are the *metalloids.* These elements show both metallic and nonmetallic properties under different conditions.

5.6 Size of atoms: a periodic property In Chapter 3 it was established that an atom consists of a central nucleus with electrons moving about it. The nucleus has a diameter about one ten-thousandth that of the atom. Most of the volume of an atom is occupied by moving electrons, whose complex motion makes the atoms appear as spheres, even though they are mostly empty space. The spherical electron cloud gives an atom its volume and excludes other atoms.

The volume of an atom is not a completely definite quantity because the boundary of an atom's electron cloud is not a distinct surface. Rather, the boundary is somewhat fuzzy and indefinite. An atom may be easily distorted when it combines with other atoms. But very great force must be used if it is to be compressed.

The reported "size" of an atom varies somewhat with the dimension measured and the method used to measure it. Scientists have measured the distance between adjacent nuclei in the crystalline forms of elements and in the molecules of gaseous elements. One-half of this distance, with slight correction, is used as the radius of one atom. The radius of an atom, and thus its volume, *do not increase regularly* with atomic number. Such an increase might be expected from the regular addition of an electron in successive elements. But atomic size varies in a periodic

Periodic Table of Atomic Radii

I	II											III	IV	V	VI	VII	VIII
0.32 **H** 1																	0.31 **He** 2
1.23 **Li** 3	0.89 **Be** 4											0.82 **B** 5	0.77 **C** 6	0.74 **N** 7	0.70 **O** 8	0.68 **F** 9	0.67 **Ne** 10
1.54 **Na** 11	1.36 **Mg** 12											1.18 **Al** 13	1.11 **Si** 14	1.06 **P** 15	1.02 **S** 16	0.99 **Cl** 17	0.98 **Ar** 18
2.03 **K** 19	1.74 **Ca** 20	1.44 **Sc** 21	1.32 **Ti** 22	1.22 **V** 23	1.18 **Cr** 24	1.17 **Mn** 25	1.17 **Fe** 26	1.16 **Co** 27	1.15 **Ni** 28	1.17 **Cu** 29	1.25 **Zn** 30	1.26 **Ga** 31	1.22 **Ge** 32	1.20 **As** 33	1.17 **Se** 34	1.14 **Br** 35	1.12 **Kr** 36
2.16 **Rb** 37	1.91 **Sr** 38	1.62 **Y** 39	1.45 **Zr** 40	1.34 **Nb** 41	1.30 **Mo** 42	1.27 **Tc** 43	1.25 **Ru** 44	1.25 **Rh** 45	1.28 **Pd** 46	1.34 **Ag** 47	1.48 **Cd** 48	1.44 **In** 49	1.40 **Sn** 50	1.40 **Sb** 51	1.36 **Te** 52	1.33 **I** 53	1.31 **Xe** 54
2.35 **Cs** 55	1.98 **Ba** 56	1.56 **Lu** 71	1.44 **Hf** 72	1.34 **Ta** 73	1.30 **W** 74	1.28 **Re** 75	1.26 **Os** 76	1.27 **Ir** 77	1.30 **Pt** 78	1.34 **Au** 79	1.49 **Hg** 80	1.48 **Tl** 81	1.47 **Pb** 82	1.46 **Bi** 83	1.46 **Po** 84	1.45 **At** 85	**Rn** 86
Fr 87	2.20 **Ra** 88	**Lr** 103	**Unq** 104	**Unp** 105	**Unh** 106	**Uns** 107	**Uno** 108	**Une** 109									

1.69 **La** 57	1.65 **Ce** 58	1.64 **Pr** 59	1.64 **Nd** 60	1.63 **Pm** 61	1.62 **Sm** 62	1.85 **Eu** 63	1.62 **Gd** 64	1.61 **Tb** 65	1.60 **Dy** 66	1.58 **Ho** 67	1.58 **Er** 68	1.58 **Tm** 69	1.70 **Yb** 70
2.0 **Ac** 89	1.65 **Th** 90	**Pa** 91	1.42 **U** 92	**Np** 93	**Pu** 94	**Am** 95	**Cm** 96	**Bk** 97	**Cf** 98	**Es** 99	**Fm** 100	**Md** 101	**No** 102

Atomic radii mostly from R.T. Sanderson, INORGANIC CHEMISTRY, Reinhold Publishing Corporation, New York, 1967, and subsequent private communication.

Figure 5-5. Periodic table showing radii of the atoms of the elements in angstrom units.

fashion as shown in Figure 5-5. This figure is a miniature periodic table with element symbols and atomic numbers in black. Atomic radii are shown in color. The radii of atoms of the elements are given in angstroms. The atomic radius is plotted as a function of the atomic number as shown in Figure 5-6. Two conclusions about the relationship between atomic radius and the periodic table are:

1. The atomic radius generally increases with atomic number in a particular group or family of elements. Each element in a group has one more energy level than the element above it. The increased nuclear charge decreases the radii of the energy levels by drawing them closer. But the addition of an energy level more than counteracts this effect.

2. From Group I to Group VIII in a period, there is a general decrease in the atomic radii of the elements. Each element has a greater positive nuclear charge than the one before it. This

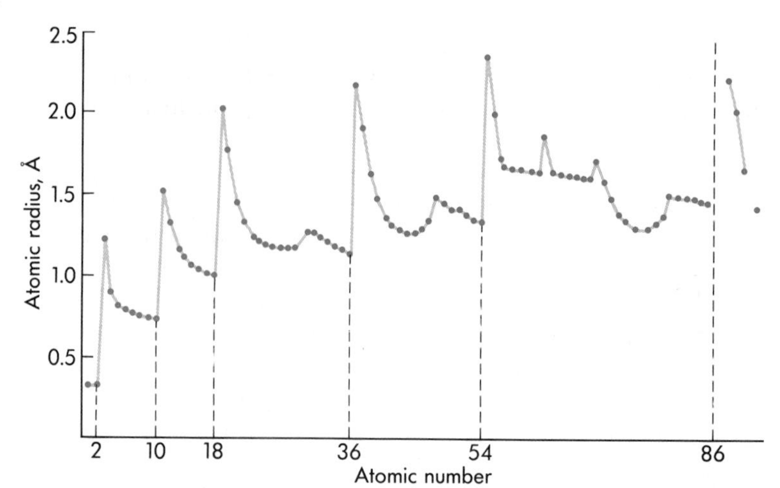

Figure 5-6. Graph showing atomic radius plotted as a function of atomic number.

greater charge results in a greater total force of attraction between the electrons and the nucleus. This greater force of attraction may explain why the electrons of the elements across a period are successively closer to the nucleus.

Because the number of protons in the nucleus of the elements increases from left to right in any one period, so does the number of electrons in the highest energy orbitals about the nucleus. As the number of electrons increases, the force of repulsion between them also increases. Irregularities in the atomic radius pattern may be caused by patterns in this force of repulsion.

5.7 Ionization energy A negatively charged electron is held in an atom by the attraction of the positively charged protons in the nucleus. By supplying energy, it is possible to remove an electron from an atom. Suppose that **A** is used as a symbol for an atom of any element and that → means "yields." Then this electron removal may be shown in equation form as

$$\mathbf{A} + \mathbf{energy} \rightarrow \mathbf{A}^{+} + \mathbf{e}^{-}$$

The particle $\mathbf{A}^{+}$ that remains after the removal of an electron, e^{-}, is an *ion* with a single positive charge. *An* ***ion*** *is an atom* (or sometimes a group of atoms) *that has a net positive or negative charge.* This net charge results from unequal numbers of positively charged protons and negatively charged electrons. *The energy required to remove an electron from an atom is its* ***ionization energy.*** The chart, Figure 5-7, shows the ionization energy required to remove one electron from an atom of each element. One unit in which ionization energy is expressed is kcal/mole. The first ionization energy of oxygen, for example, is 314 kcal/mole. This means that 314 kcal of energy must be supplied to remove the first, or least tightly held, highest-energy-level electron from each atom in one mole of oxygen atoms. (Recall that one mole of oxygen atoms is 6.02×10^{23} atoms.) Figure 5-8 is a graph showing first ionization energy plotted as a function of atomic number. From Figures 5-7 and 5-8, the following conclusions can be drawn:

1. *Low ionization energy* is characteristic of a *metal. High ionization energy* is characteristic of a *nonmetal.* The chemical inertness of the noble gases is strong evidence for the unusual stability, or resistance to change, of the highest-energy-level octet. As might be expected, the noble gases have unusually high ionization energies.

2. Within a group of nontransition elements, the ionization energy generally decreases with increasing atomic number. This is because increasing atomic number is accompanied by increasing atomic radius within such groups. The highest-energy-level electrons of the elements of higher atomic number within a group are farther from the nucleus. Thus, these electrons are attracted less by the nucleus. The ionization energy for removal of one highest-energy-level electron, therefore, is less as the

Figure 5-7. Periodic table showing first ionization energies of the elements in kilocalories per mole.

Periodic Table of First Ionization Energies

I	II											III	IV	V	VI	VII	VIII
314 H 1																	567 He 2
124 Li 3	215 Be 4											191 B 5	260 C 6	335 N 7	314 O 8	402 F 9	497 Ne 10
119 Na 11	176 Mg 12											138 Al 13	188 Si 14	242 P 15	239 S 16	299 Cl 17	363 Ar 18
100 K 19	141 Ca 20	151 Sc 21	157 Ti 22	155 V 23	156 Cr 24	171 Mn 25	181 Fe 26	181 Co 27	176 Ni 28	178 Cu 29	217 Zn 30	138 Ga 31	182 Ge 32	226 As 33	225 Se 34	272 Br 35	323 Kr 36
96 Rb 37	131 Sr 38	147 Y 39	158 Zr 40	159 Nb 41	164 Mo 42	168 Tc 43	170 Ru 44	172 Rh 45	192 Pd 46	175 Ag 47	207 Cd 48	133 In 49	169 Sn 50	199 Sb 51	208 Te 52	241 I 53	280 Xe 54
90 Cs 55	120 Ba 56	125 Lu 71	161 Hf 72	182 Ta 73	184 W 74	182 Re 75	201 Os 76	210 Ir 77	208 Pt 78	213 Au 79	241 Hg 80	141 Tl 81	171 Pb 82	168 Bi 83	196 Po 84	At 85	248 Rn 86
Fr 87	122 Ra 88	Lr 103	Unq 104	Unp 105	Unh 106	Uns 107	Uno 108	Une 109									

129 La 57	126 Ce 58	125 Pr 59	127 Nd 60	128 Pm 61	130 Sm 62	131 Eu 63	142 Gd 64	135 Tb 65	137 Dy 66	139 Ho 67	141 Er 68	143 Tm 69	144 Yb 70
159 Ac 89	Th 90	Pa 91	U 92	Np 93	134 Pu 94	138 Am 95	Cm 96	Bk 97	Cf 98	Es 99	Fm 100	Md 101	No 102

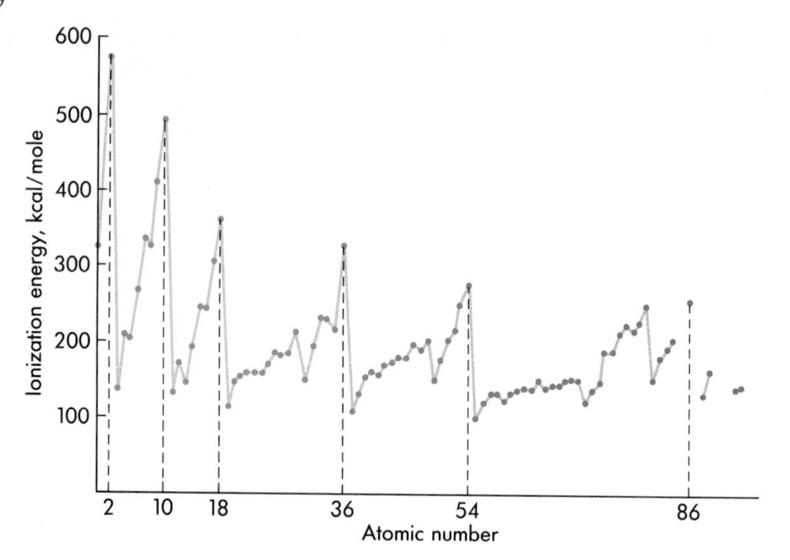

Figure 5-8. Graph showing first ionization energy as a function of atomic number.

atomic number of the atom is greater. This reasoning does not hold true for the transition elements.

3. Ionization energy does not vary uniformly from element to element within a series. Instead, it is a *periodic* property. In each series or period, the ionization energy increases from Group I to Group VIII. But the increase is not regular. There is a decrease in ionization energies between Groups II and III in Periods 2 and 3. This decrease occurs as the *s* sublevel is filled and the *p* sublevel is started. In these periods, there is also a decrease between Groups V and VI as the *p* sublevel becomes half-filled. In Periods 4, 5, and 6, there is a sharp decrease in the ionization energy between the last transition element and Group III. This decrease occurs as the *d* sublevel is filled and the *p* sublevel is started. These irregularities are apparently related to the extra stability of completed and half-completed sublevels.

5.8 Ionization energy to remove successive electrons It is possible to remove more than one electron from many-electron atoms.

$$\mathbf{Na + ionization\ energy\ 1st\ electron \rightarrow Na^{+} + e^{-}}$$

$$\mathbf{Na^{+} + ionization\ energy\ 2nd\ electron \rightarrow Na^{++} + e^{-}}$$

$$\mathbf{Na^{++} + ionization\ energy\ 3rd\ electron \rightarrow Na^{+++} + e^{-}}$$

Table 5-2 shows the electron configurations of sodium, magnesium, and aluminum atoms. It also gives the ionization energies required to remove successive electrons from atoms of these elements. Notice the increase in energy for each successive electron removed.

Table 5-2
ELECTRON CONFIGURATIONS AND IONIZATION ENERGIES OF SODIUM, MAGNESIUM, AND ALUMINUM

Elements	*Electron configuration*	*Ionization energy* (kcal/mole)			
		1st electron	*2nd electron*	*3nd electron*	*4th electron*
Na	$1s^2 2s^2 2p^6 3s^1$	119	$109\bar{0}$	1652	2281
Mg	$1s^2 2s^2 2p^6 3s^2$	176	347	1848	2519
Al	$1s^2 2s^2 2p^6 3s^2 3p^1$	138	434	656	2767

It is not surprising that the ionization energy increases with each electron removed from an atom. After all, each successive electron must be removed from a particle with an increasingly greater net positive charge. A closer examination of the variations in ionization energies is now needed.

For sodium atoms, there is a great increase between the first and second ionization energies. The first electron, a 3*s* electron, is rather easily removed. But to remove the second electron almost ten times as much energy is needed. This increase occurs because the second electron is a 2*p* electron in a much lower energy level. The lower the energy level, the greater the energy needed to remove an electron from the attraction of the nucleus. Figure 5-9 shows the energy-level transitions involved for the elements sodium, magnesium, and aluminum.

The removal of electrons from magnesium atoms requires little ionization energy for the first two electrons, since both are 3*s* electrons. More energy is required to remove the first 3*s* electron from magnesium atoms than the first 3*s* electron from sodium atoms. This increase occurs because magnesium atoms have a greater nuclear charge and the 3*s* electrons are closer to the nucleus. But to remove the third electron from magnesium atoms requires between five and six times as much energy as is needed to remove the second. This increase occurs because the third electron is a 2*p* electron. Since 2*p* electrons of magnesium are at a much lower energy level than 3*s* electrons, there is a great increase in ionization energy.

It is easy to remove the first three electrons from aluminum atoms. In fact, it is easier to remove the first electron from aluminum atoms than it is to remove the first electron from magnesium atoms. A look at the electron configuration indicates that the first aluminum electron is a 3*p* electron. It is in a slightly higher energy sublevel than the first magnesium electron, which is a 3*s* electron. There is a great increase in ionization energy between the third and fourth aluminum electrons. This increase is explained by the fact that the fourth electron is a 2*p* electron.

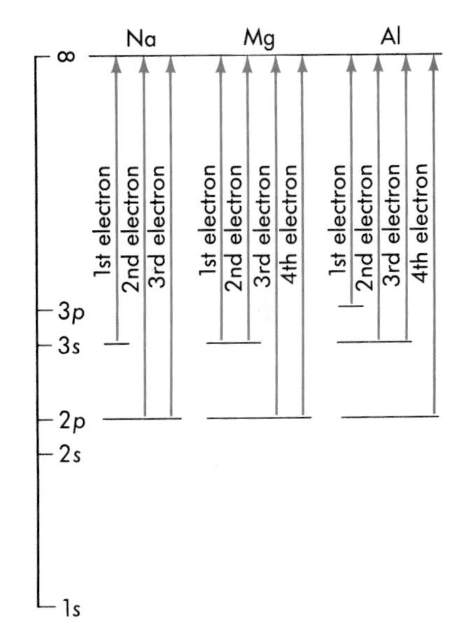

Figure 5-9. Energy-level transitions for the removal of successive electrons from sodium, magnesium, and aluminum atoms. The infinity sign (∞) on the energy-level scale indicates the removal of the electron from the atom.

This electron is in a much lower energy level than the first three electrons, all of which were 3rd energy-level electrons.

The order in which electrons are removed from atoms by ionization is not the same as the order in which electrons occupy orbitals in the structures described in Sections 4.6–4.10. This difference is caused by the shifting of the relative positions of energy sublevels as they are occupied by electrons. Electrons are removed from atoms by ionization in the reverse of the order given by the electron-configuration notation. The electron-configuration notation of iron is $1s^22s^22p^63s^23p^63d^64s^2$. The first electron removed from iron atoms is a 4s electron. So is the second. The third electron is a 3d electron, as are the fourth, fifth, sixth, and so on.

See Question 40 at the end of chapter.

5.9 Electron affinity Neutral atoms can acquire additional electrons. The measure of this tendency is *electron affinity*. ***Electron affinity*** *is the energy change that occurs when an electron is acquired by a neutral atom.* When an electron is added to a neutral atom, an ion with a single negative charge is formed. An amount of energy, the electron affinity, is either released or absorbed.

For most atoms the electron affinity is energy that is released. In equation form, the acquiring of an electron and the release of energy may be expressed as

$$A + e^- \rightarrow A^- + \text{energy} \qquad \textbf{(Equation 1)}$$

It may be helpful to review Sections 2.13 and 2.14.

Since energy is released, the change is exothermic. The negative ion will be more stable than the neutral atom. Because the energy is released, it is given a positive sign, as shown in Equation 1. Like ionization energy, electron affinity may be measured in kcal/mole. The electron affinity indicates how tightly an additional electron is bound to an atom. If the electron affinity is positive and low, the electron is weakly bound. If the electron affinity is positive and high, the electron is strongly bound.

For some atoms the electron affinity is energy that is absorbed. In equation form, this change may be expressed as

$$\mathbf{A + e^- + energy \rightarrow A^-}$$

Since energy is absorbed, the change is endothermic. A more useful form of this equation is similar to Equation 1, with the energy term to the right of the "yields" sign:

$$A + e^- \rightarrow A^- - \text{energy} \qquad \textbf{(Equation 2)}$$

In the case of an endothermic reaction, the energy change is given a negative sign. See Equation 2. If the electron affinity is negative, energy must be used in forming the negative ion. Thus, the negative ion will be less stable than the neutral atom.

Periodic Table of Electron Affinities

I	II											III	IV	V	VI	VII	VIII
17.4 **H** 1																	(−5.3) **He** 2
14.3 **Li** 3	(−18.0) **Be** 4											6.4 **B** 5	29.2 **C** 6	0.0 **N** 7	33.7 **O** 8	78.4 **F** 9	(−6.9) **Ne** 10
12.6 **Na** 11	−3.5 **Mg** 12											10.2 **Al** 13	31.9 **Si** 14	17.2 **P** 15	47.9 **S** 16	83.4 **Cl** 17	(−8.5) **Ar** 18
11.6 **K** 19	(−16.1) **Ca** 20	(11.5) **Sc** 21	1.8 **Ti** 22	12.1 **V** 23	15.4 **Cr** 24	(−15.0) **Mn** 25	3.8 **Fe** 26	15.3 **Co** 27	26.5 **Ni** 28	28.3 **Cu** 29	−11.5 **Zn** 30	(6.9) **Ga** 31	27.7 **Ge** 32	18.4 **As** 33	46.6 **Se** 34	77.6 **Br** 35	(−9.2) **Kr** 36
11.2 **Rb** 37	(−5.1) **Sr** 38	(16.1) **Y** 39	9.8 **Zr** 40	20.6 **Nb** 41	17.2 **Mo** 42	−13.4 **Tc** 43	15.9 **Ru** 44	26.2 **Rh** 45	12.9 **Pd** 46	30.0 **Ag** 47	−7.6 **Cd** 48	6.9 **In** 49	28.8 **Sn** 50	24.2 **Sb** 51	45.4 **Te** 52	70.6 **I** 53	(−9.7) **Xe** 54
10.9 **Cs** 55	(−12.5) **Ba** 56	11.5 **Lu** 71	5.1 **Hf** 72	7.4 **Ta** 73	18.8 **W** 74	3.5 **Re** 75	(25.8) **Os** 76	36.1 **Ir** 77	49.1 **Pt** 78	53.2 **Au** 79	−13.8 **Hg** 80	6.9 **Tl** 81	8.4 **Pb** 82	21.8 **Bi** 83	(43.4) **Po** 84	(64.6) **At** 85	(−10.4) **Rn** 86
(10.4) **Fr** 87	**Ra** 88	**Lr** 103	**Unq** 104	**Unp** 105	**Unh** 106	**Uns** 107	**Uno** 108	**Une** 109									

11.5 **La** 57	11.5 **Ce** 58	0 **Pr** 59	−6.9 **Nd** 60	−6.9 **Pm** 61	6.9 **Sm** 62	−6.9 **Eu** 63	11.5 **Gd** 64	11.5 **Tb** 65	−6.9 **Dy** 66	−6.9 **Ho** 67	−6.9 **Er** 68	6.9 **Tm** 69	−6.9 **Yb** 70
Ac 89	(0) **Th** 90	**Pa** 91	(25.4) **U** 92	**Np** 93	**Pu** 94	**Am** 95	**Cm** 96	**Bk** 97	**Cf** 98	**Es** 99	**Fm** 100	**Md** 101	**No** 102

Values in parentheses are estimated.
Data compiled by E.C.M. Chen and W.E. Wentworth, *J. Chem. Ed.*, August 1975 and subsequent private communications.

Figure 5-10. Periodic table showing the electron affinities of the elements in kilocalories per mole. The values in parentheses are estimated ones.

Figure 5-10 gives the known experimental values for electron affinities in kilocalories per mole of atoms. Estimated values are given for some of the other atoms. These data are graphed in Figure 5-11 on the next page. From these figures, several observations can be made.

The Group II elements, the elements of the zinc subfamily (Zn, Cd, Hg), and the noble gases have negative electron affinities. This is evidence that the atoms of these elements have filled highest *s* or *p* sublevels and stable lower energy levels. An electron can be added to such atoms only by supplying energy. The addition of an electron to these atoms produces a singly charged negative ion that is less stable than the atom.

If only the experimental values are considered, a general decrease in electron affinity in the numbered families from the third through the sixth periods is apparent. The added electron

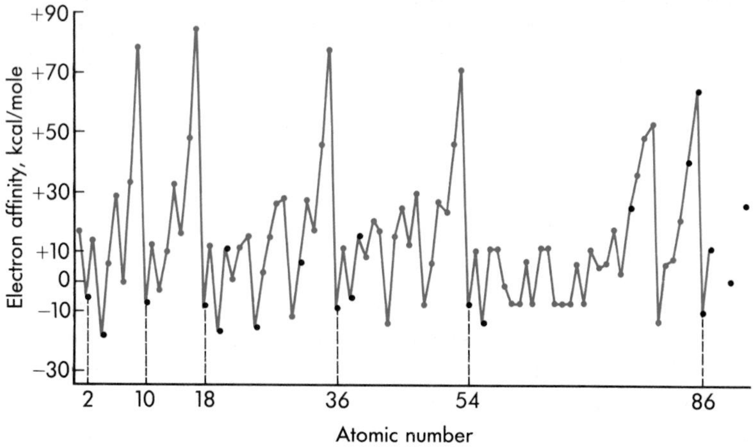

Figure 5-11. Graph showing electron affinity as a function of atomic number. Experimental data are shown by dots in color; estimated data are shown by black dots.

takes a position in the atom farther from the nucleus. Not as much energy is released; the electron is not as strongly held.

For the Group I elements there is a decrease in electron affinity between the second and third periods. But for Groups III, IV, V, VI, and VII, there is an increase between the second and third periods. The Period 2 elements are small in size and frequently show irregularities in properties when compared with other elements in their group.

There is an increase in electron affinity between Groups III and IV. The addition of an electron to a Group III atom results in an ion with two electrons in the outer *p* sublevel. The addition of an electron to a Group IV atom results in an ion with three electrons in the outer *p* sublevel. These three electrons constitute a half-filled *p* sublevel. The greater amount of energy released when an electron is added to a Group IV atom than to a Group III atom is evidence of the greater stability of a half-filled *p* sublevel than one with only two electrons.

In most cases there is a decrease in electron affinity between Groups IV and V. The Group V atoms have half-filled *p* sublevels, and there is little attraction for another electron.

There is a general increase in electron affinity among the atoms of Groups V, VI, and VII. The addition of electrons to Group V, VI, and VII atoms produces ions with increasingly stable highest energy levels. The addition of an electron to a Group VII atom produces an ion with an octet in the highest energy level. The quite high electron affinities of the halogens is further evidence for the stability of highest-energy-level octets.

Ionization energy and electron affinity are important concepts in understanding how chemical compounds are formed from atoms of metallic and nonmetallic elements.

5.10 Value of the periodic table In former years, the periodic table served as a check on atomic weight determinations. It also was used in the prediction of new elements. These uses are now outdated. Today the periodic table serves as a useful and systematic classification of elements according to their properties. The occurrence of properties such as atomic size, ionization energy, and electron affinity at regular intervals has already been described. This information is valuable in determining the types of compounds that certain elements form. Though not perfect, the periodic table makes the study of chemistry easier.

SUMMARY

About 1869, Mendeleev classified the elements according to physical and chemical properties by arranging them in order of increasing atomic weight. He concluded that the properties of elements are in periodic dependence on their atomic weights.

Moseley found that X rays could be used to determine the atomic number of an element. An increase in the number of protons in the nucleus of atoms of the metal used as the target in an X-ray tube shortens the wavelength of the X rays produced. When elements are arranged in order of their atomic numbers, some discrepancies in Mendeleev's arrangement disappear. As a result of this discovery, the periodic law is now stated as: The physical and chemical properties of the elements are periodic functions of their atomic numbers.

In the modern periodic table, the families of elements with similar properties, also known as groups, are in vertical columns. Each element in a family has a similar number of electrons in its highest energy level. At the left of the table, the most active elements are at the bottom. At the right of the table, they are at the top. A row of elements is called a period. In a given period, the properties of the elements gradually pass from strong metallic to strong nonmetallic nature. The last member of a period is a noble gas. Atomic size, ionization energy, and electron affinity vary from element to element in a periodic fashion.

VOCABULARY

actinide series
atomic radius
electron affinity
family
group
ion
ionization energy
lanthanide series
noble gas
period
periodic law
rare earth elements
series
transition elements
X rays

QUESTIONS

Group A

1. (*a*) What property of an element determined the order in which it was placed by Mendeleev in his periodic table? (*b*) What property determines the order in which an element is placed today? *A3c, A6*
2. How did Mendeleev use his knowledge of the physical and chemical properties of elements (*a*) in preparing his periodic table; (*b*) in predicting new elements? *A6*
3. What is the relationship between X-ray wavelengths and the target element used to produce them? *A3c, A6*
4. State the periodic law. *A3a*

5. (*a*) What information appears in each block of the periodic table? (*b*) How is this information arranged? *A3b*
6. (*a*) What is a group or family of elements? (*b*) What position does a family occupy in the periodic table? *A1*
7. (*a*) What is a series or period of elements? (*b*) What position does a period occupy in the periodic table? *A1*
8. (*a*) List the elements in the second period in order. (*b*) How does the number of electrons in the second energy level vary in these elements? (*c*) How do the properties of these elements vary? *A2b, A3g*
9. How do the electron configurations of elements that have similar properties compare? *A2a, A3g*
10. How do the elements in Group I vary in activity? *A2a, A3g*
11. How do the elements in Group VII vary in activity? *A2a, A3g*
12. (*a*) What term is applied to the elements bordering the zigzag line separating the metals from the nonmetals? (*b*) What characteristics do these elements have? *A1, A2a, A3k*
13. The naturally occurring isotopes of argon are Ar-36, Ar-38, and Ar-40. Those of potassium are K-39, K-40, and K-41. From the atomic weights of these elements, estimate which isotope of each element is most abundant. *A3b*
14. Why may measurements of the radius of an atom by different methods give different results? *A2c, A2d*
15. (*a*) Compare the atomic radii of the Group I elements with the radii of all the other elements in their period. (*b*) How does this comparison relate to the structure of the atoms of these elements? *A3i*
16. Compare the atomic radius of each Group VIII element with the atomic radii of the other *numbered group* elements in the same period. *A3i*
17. (*a*) What is the atomic radius of potassium; (*b*) of bromine? (*c*) What structural difference is related to the difference in radii? *A3i*
18. (*a*) What is the atomic radius of fluorine; (*b*) of iodine? (*c*) What structural differences are related to the difference in radii? *A3i*
19. Write an equation to represent (*a*) the removal of the single 3*s* electron from a sodium atom; (*b*) the addition of a 3*p* electron to a chlorine atom. (*c*) Describe the particles produced by these reactions. *A3j*
20. (*a*) Is the ionization energy of metals low or high? (*b*) What does this observation indicate about their atomic structure? *A3j*
21. (*a*) Is the ionization energy of nonmetals low or high? (*b*) What does this observation indicate about their atomic structure? *A3j*
22. Compare the first ionization energies of lithium and fluorine. What differences in structure are related to the difference in ionization energy? *A3j*
23. Compare the first ionization energies of magnesium and tellurium. What difference in structure is related to the difference in ionization energy? *A3j, A5*
24. Define: *electron affinity*. *A1*
25. Write an equation to represent the addition of an electron (*a*) to a potassium atom; (*b*) to a bromine atom; (*c*) to a krypton atom. (*d*) Compare the strength with which the electron is bound in the three atoms. *A3l*

Group B

26. It is not an unusual occurrence in science for two persons, Mendeleev and Meyer for example, to make a similar discovery at about the same time. Why do you believe this occurs?
27. (*a*) Which group of elements was missing from Mendeleev's periodic table? (*b*) Why? *A3d*
28. (*a*) What are X rays? (*b*) Describe how X rays are produced. *A3c*
29. (*a*) How did Mendeleev know where to leave gaps for undiscovered elements in his periodic table? (*b*) How did Moseley know where to leave gaps for undiscovered elements? *A6*
30. Explain why hydrogen is placed separately above Group I in the periodic table. *A3e, A3f*
31. Define the following terms: (*a*) *transition ele-*

ment; (*b*) *rare earth element.* (*c*) In which periods of elements do each appear? *A1, A4a*

32. (*a*) How does the atomic radius vary with atomic number within each *numbered group* of elements? (*b*) How does the theory of atomic structure explain this variation? *A3i*

33. (*a*) How does the atomic radius vary with atomic number within each period of elements? (*b*) How does the theory of atomic structure explain this variation? *A3i*

34. Two atoms are of the same size but the atomic number of one is much higher than that of the other. (*a*) How would you expect their ionization energies to compare? (*b*) Why? *A3j, A5*

35. (*a*) If energy must be supplied to remove an electron from the highest energy level of an atom, which is more stable, the atom or the resulting ion? (*b*) If energy is released during the addition of an electron to a neutral atom, which is more stable, the atom or the resulting ion? *A3j, A5*

36. (*a*) Which *numbered group* elements have negative electron affinities? (*b*) How is this property related to the structure of the atoms of these elements? *A3l*

37. What determines the number of elements in each period of the periodic table? *A2b*

38. How many 6th-energy-level orbitals would theoretically be filled in element 118? *A2b*

39. Why is the periodic table useful? *A3m*

40. On a separate sheet of paper, copy and complete the following table. *Do not write in this book.*

On the basis of the electron configuration, explain the variation in ionization energies for successive electrons. *A3k*

Chemical element	*Electron-configuration notation*	*Ionization energy* (kcal/mole)			
		1st electron	*2nd electron*	*3rd electron*	*4th electron*
K		100	729	1054	1405
Ca		141	274	1174	1547
Ga		138	473	708	1470

In 1954, the Nobel Prize in chemistry was given to Linus Pauling (American) for his studies into the nature of the chemical bond. He studied crystals and gas molecules to determine chemical bond lengths and atomic radii. He found that the current theories did not explain adequately the properties of some compounds, so he introduced the concepts of resonance, fractional bonds, and hydrogen bonds. Other research dealt with biochemical problems. He has also written extensively about the benefits of vitamin C. In 1962, Pauling received the Nobel Peace Prize.

Chemical Bonds

6

GOALS

In this chapter, you will gain an understanding of: • ionic and covalent bonding • empirical, molecular, electron-dot, ionic, and structural chemical formulas • oxidation and reduction • oxidation-number rules • monatomic and polyatomic ions • hybridization • electronegativity and its relationship to chemical bonding • resonance

6.1 Elements form compounds In Chapter 2 you learned about the process of chemical change. Forming compounds from elements is one kind of chemical change. In succeeding chapters the theory of the structure of the atoms of the elements was described. In this chapter, atomic theory will be used to explain *how* atoms combine to form the other particles of substances. You will learn what these particles are, how atoms are held together in these particles, and how these particles vary in size and shape.

Here is a series of chemical formulas for some well-known compounds:

HCl	$NaCl$
H_2O	$CaCl_2$
NH_3	$AlCl_3$
CH_4	CCl_4

Notice that each formula in the left column contains hydrogen. The first formula shows that one atom of hydrogen combines with one atom of chlorine. The second formula shows that two atoms of hydrogen combine with one atom of oxygen. The third and fourth formulas show that three atoms of hydrogen combine with one atom of nitrogen, but four atoms of hydrogen combine with one atom of carbon. In the four formulas, different numbers of hydrogen atoms have combined with one atom of other elements.

These formulas were originally established by chemical analysis. Methods of working out chemical formulas from the results of analyses are described in Chapters 7 and 11. Ways in which these compounds are formed are described in Chapter 8 and later chapters throughout this book. It is not necessary at this point to know how these compounds are formed or how these formulas have been determined.

Now look at the second column. Figure 6-1 shows the compounds represented by these formulas. Each formula contains chlorine. One chlorine atom combines with one sodium atom. Two chlorine atoms combine with one calcium atom. Three chlorine atoms combine with one aluminum atom, but four chlorine atoms combine with one carbon atom. Different numbers of

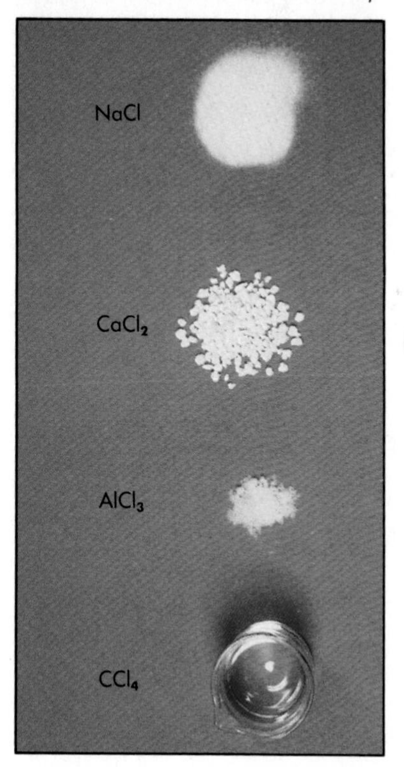

Figure 6-1. In these compounds, different numbers of chlorine atoms combine with single atoms of other elements.

chlorine atoms combine with one atom of other elements. Why is there a difference in the number of hydrogen and chlorine atoms that will combine with a single atom of another element? Is there any relation between the structure of an atom and the number of other atoms with which it will combine?

6.2 Valence electrons and chemical bonds The electrons in the highest energy level of an atom play an important part in the formation of compounds. The electrons in an *incomplete* highest energy level are called ***valence electrons***. The remainder of the atom, excluding the valence electrons, is called the ***kernel*** of the atom. In the formation of chemical compounds from elements, *valence electrons are usually either transferred from the highest energy level of one atom to the highest energy level of another atom or shared among the highest energy levels of the combining atoms.* This transfer or sharing of electrons produces ***chemical bonds***. (The formation of some chemical bonds by atoms of transition and rare earth elements involves not only electrons in the highest energy level but also those of an incomplete next-to-highest energy level.)

Electron transfer results in ***ionic bonding****, and electron sharing produces* ***covalent bonding***. When an atom of one element combines chemically with an atom of another element, both atoms usually attain a stable highest energy level having a noble-gas configuration. *This kind of electron structure has chemical stability.*

Electron-dot symbols for the noble gases (Sections 4.6 to 4.9) showing their highest-energy-level configurations are

He:, :N̈e:, :Är:, :K̈r:, :Ẍe:, :R̈n:.

Energy changes are always involved in the process of electron transfer or electron sharing. In *most* cases, energy is given off when compounds are formed. The process of electron transfer is *always* exothermic, and that of electron sharing is *usually* exothermic. In a *few* cases of compound formation by electron sharing, energy is absorbed, and the process is endothermic.

Recall from Section 2.14 that the products of exothermic reactions have less energy than the reactants. The products of endothermic reactions have more energy than the reactants.

Ionic Bonding

6.3 Ionic bonding In the formation of a compound by ionic bonding, electrons are actually transferred from the highest energy level of one atom to the highest energy level of a second atom. By this process both atoms usually attain highest energy levels with noble-gas configurations. For example, sodium reacts with chlorine to form sodium chloride. In this reaction the single 3*s* electron of a sodium atom is transferred to the singly occupied 3*p* orbital of a chlorine atom.

Some experimental evidence for electron transfer is described in Section 22.6.

Orbital notation is explained in text Section 4.6.

	1s	2s	2p	3s	3p
Na	↓↑	↓↑	↓↑ ↓↑ ↓↑	↑	— — —
Cl	↓↑	↓↑	↓↑ ↓↑ ↓↑	↓↑	↓↑ ↓↑ ↓

The sodium, now deficient in one electron, has the stable electron configuration of neon. The chlorine, now with one excess electron, has the stable electron configuration of argon. Only one atom of each element is required for the electron transfer that produces these stable electron configurations. Thus, the *formula* of the compound is **NaCl.** *A* ***chemical formula*** *is a shorthand method of representing the composition of a substance by using chemical symbols.*

The particles produced by this transfer of an electron are no longer electrically neutral atoms of sodium and chlorine. They are: (1) a *sodium ion* with a single excess positive charge, and (2) a *chloride ion* with a single excess negative charge.

	1s	2s	2p			3s	3p		
Na^+	↓↑	↓↑	↓↑	↓↑	↓↑	—	—	—	—
Cl^-	↓↑	↓↑	↓↑	↓↑	↓↑	↓↑	↓↑	↓↑	↓↑

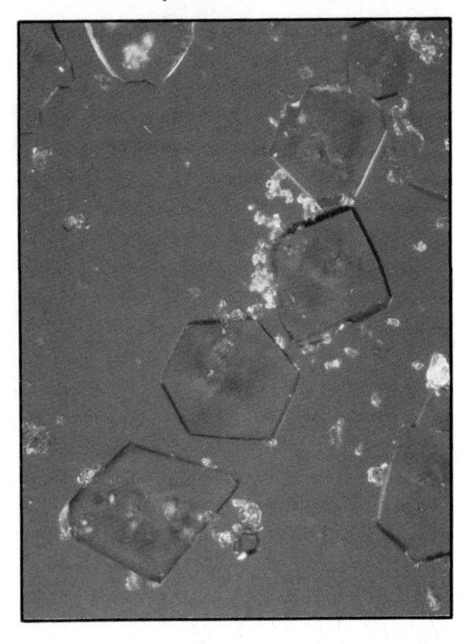

Figure 6-2. Magnified crystals of the ionic compound sodium fluoride. Notice their uniformity.

These ions are arranged systematically in crystals of sodium chloride in the ratio of 1 sodium ion to 1 chloride ion as shown in Figure 6-3.

The formula NaCl, which represents the composition of the compound sodium chloride, is an ***empirical formula.*** An empirical formula indicates (1) the kinds of atoms in the compound formed and (2) the simplest whole-number ratio of the atoms in the compound. The formula $Na_{17}Cl_{17}$ shows the kinds and ratio of atoms that make up the compound sodium chloride. But the empirical formula NaCl represents this information in the simplest way.

An ion is an atom or group of atoms that has a net positive or negative charge.

Table 6-1 shows the number of protons and electrons in the atoms and ions of sodium and chlorine and the charges that result. It also shows their symbols and each one's radius given in angstroms.

Using only the 3rd energy level electrons, the electron-dot symbol for an atom of sodium is

Na·

That for an atom of chlorine is

$\cdot\ddot{\underset{\cdot\cdot}{Cl}}:$

When these atoms react, sodium atoms become sodium ions. The electron-dot symbol for a sodium ion is

Na^+

The chlorine atoms become chloride ions, which have the electron-dot symbol

$:\ddot{\underset{\cdot\cdot}{Cl}}:^-$

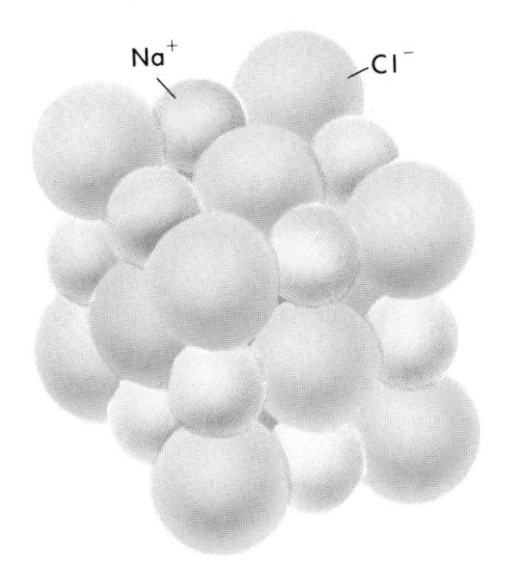

Figure 6-3. This diagram shows the arrangement of the sodium and chloride ions in a portion of a sodium chloride crystal.

Table 6-1
DATA ON ATOMS AND IONS OF SODIUM AND CHLORINE

	Sodium atom	*Sodium ion*	*Chlorine atom*	*Chloride ion*
number of protons	11	11	17	17
number of electrons	11	10	17	18
net charge	0	+1	0	−1
symbol	Na	Na^+	Cl	Cl^-
radius, Å	1.54	0.97	0.99	1.81

These ions form the compound sodium chloride. The electron-dot formula for sodium chloride may be written as

$$\mathbf{Na^+ \,{:}\underset{\cdot\cdot}{\overset{\cdot\cdot}{Cl}}{:}^-}$$

or as a simpler ionic formula, $\mathbf{Na^+Cl^-}$.

Note the symbols for electrons, ◦ and •, which are used here and in other electron-dot formulas in this chapter. These symbols are used only to show the origin of the electrons in the completed shells. They *do not mean* that electrons from different atoms are different from each other. All electrons, regardless of the atom from which they originate, are identical. However, when two electrons occupy the same orbital, the electrons must have opposite spins if the orbital is to be stably occupied.

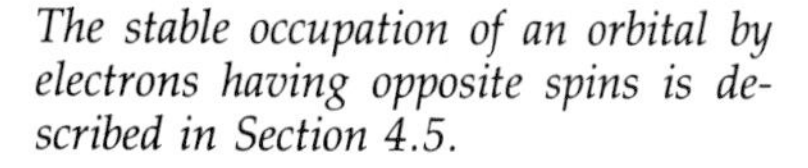
The stable occupation of an orbital by electrons having opposite spins is described in Section 4.5.

6.4 Energy change in ionic bonding Sodium ions and chloride ions in a common salt crystal like that shown in Figure 6-4 can be formed from widely separated sodium and chlorine

Figure 6-4. The regular shape of a naturally formed crystal of sodium chloride results from the regular arrangement of sodium ions and chloride ions in the crystal.

atoms. In order *to study the energy change involved,* this chemical change can be assumed to consist of three separate reactions. The first reaction, the removal of one electron from each sodium atom, is endothermic. The amount of energy required is the *ionization energy* of sodium. For one mole of sodium atoms,

Ionization energy is explained in text Section 5.7.

$$\textbf{1 mole Na + 119 kcal} \rightarrow \textbf{1 mole Na}^{+} \textbf{ + 1 mole e}^{-}$$

Only one electron can be readily removed from each sodium atom, because a great increase in ionization energy occurs between the first and second electrons. (See Section 5.8.)

The second reaction is the addition of one electron to each neutral chlorine atom. This reaction is exothermic. The energy given off is the *electron affinity* of chlorine. For one mole of chlorine atoms,

Electron affinity is explained in text Section 5.9.

$$\textbf{1 mole Cl + 1 mole e}^{-} \rightarrow \textbf{1 mole Cl}^{-} \textbf{ + 83 kcal}$$

In the third reaction, the oppositely charged sodium ions and chloride ions take up their positions in the sodium chloride crystal. This reaction is also exothermic.

$$\textbf{1 mole Na}^{+} \textbf{ + 1 mole Cl}^{-} \rightarrow \textbf{1 mole Na}^{+}\textbf{Cl}^{-} \textbf{ + 189 kcal}$$

The first reaction requires energy (119 kcal). This is less than the sum of the energies given off in the second and third reactions (83 kcal + 189 kcal). The overall effect is that energy is given off [(83 kcal + 189 kcal) − (119 kcal) = 153 kcal]. The summation of these three separate reactions is

$$\begin{array}{llll}
\textbf{1 mole Na} & \textbf{+ 119 kcal} & \rightarrow \textbf{1 mole Na}^{+} & \textbf{+ 1 mole e}^{-} \\
\textbf{1 mole Cl} & \textbf{+ 1 mole e}^{-} & \rightarrow \textbf{1 mole Cl}^{-} & \textbf{+ 83 kcal} \\
\textbf{1 mole Na}^{+} & \textbf{+ 1 mole Cl}^{-} & \rightarrow \textbf{1 mole Na}^{+}\textbf{Cl}^{-} & \textbf{+ 189 kcal} \\
\hline
\textbf{1 mole Na} & \textbf{+ 1 mole Cl} & \rightarrow \textbf{1 mole Na}^{+}\textbf{Cl}^{-} & \textbf{+ 153 kcal}
\end{array}$$

The net process of electron transfer is exothermic. All formations of ionic compounds from their elements are exothermic.

6.5 Oxidation and reduction Another aspect of the reactions by which sodium ions and chloride ions are produced should now be examined. The formation of a sodium ion from a sodium atom may be considered to involve the loss of an electron. The equation for this change might be written

$$\textbf{Na} - \textbf{e}^{-} \rightarrow \textbf{Na}^{+}$$

However, chemists prefer to show only the substances *entering* a reaction at the *left* of the "yields" sign. The substances *formed by* the reaction are written at the *right* of the "yields" sign. Consequently, the equation showing the formation of a sodium ion from a sodium atom should be written

$$\textbf{Na} \rightarrow \textbf{Na}^{+} + \textbf{e}^{-} \qquad \textbf{(Equation 1)}$$

The *oxidation state* of an element is represented by a signed number called an ***oxidation number.*** Oxidation numbers are assigned according to a set of seven rules. These rules will be introduced as needed in this chapter.

Rule 1: *The oxidation number of an atom of a free element is zero.*

Rule 2: *The oxidation number of a monatomic (one-atomed) ion is equal to its charge.*

It is apparent from these rules that: (1) the oxidation number of elemental sodium is zero, $\overset{0}{\mathbf{Na}}$, and (2) the oxidation number of the sodium ion is plus one, $\overset{+1}{\mathbf{Na}}{}^{+}$. Note that an oxidation number is written above a symbol, while an ion charge is written as a right superscript. Equation 1 can now be rewritten to include these oxidation numbers.

$$\overset{0}{\mathrm{Na}} \rightarrow \overset{+1}{\mathrm{Na}}{}^{+} + e^{-}$$

From this equation it is apparent that the oxidation state of sodium has changed from 0 to +1; the oxidation state has become *more positive. A chemical reaction in which an element attains a more positive oxidation state is called* **oxidation.** The particle whose oxidation state becomes *more positive* is said to be *oxidized.* In the preceding reaction, the sodium atom is oxidized to a sodium ion since its oxidation state has become more positive. The reaction is an oxidation.

The formation of a chloride ion from a chlorine atom involves the gain of an electron.

$$\mathrm{Cl} + e^{-} \rightarrow \mathrm{Cl}^{-} \qquad \textbf{(Equation 2)}$$

The oxidation number of elemental chlorine is zero: $\overset{0}{\mathbf{Cl}}$ (Rule 1). The oxidation number of the chloride ion is minus one: $\overset{-1}{\mathbf{Cl}}{}^{-}$ (Rule 2). Equation 2 may now be rewritten

$$\overset{0}{\mathrm{Cl}} + e^{-} \rightarrow \overset{-1}{\mathrm{Cl}}{}^{-}$$

This equation shows that the oxidation state of chlorine has changed from 0 to −1; the oxidation state has become *more negative. A chemical reaction in which an element attains a more negative oxidation state is called* **reduction.** The particle whose oxidation state becomes *more negative* is said to be *reduced.* The chlorine atom is reduced to the chloride ion since its oxidation state has become more negative. The above reaction is a reduction.

In the reaction

$$\overset{0}{\mathrm{Na}} + \overset{0}{\mathrm{Cl}} \rightarrow \overset{+1}{\mathrm{Na}}{}^{+}\,\overset{-1}{\mathrm{Cl}}{}^{-}$$

elemental sodium is oxidized and elemental chlorine is reduced.

The substance that is reduced is called the ***oxidizing agent.*** (In the reaction represented by the equation, chlorine is the oxidizing agent.) At the same time, *the substance that is oxidized* is called the ***reducing agent.*** (Sodium is the reducing agent in this reaction.) **NaCl** is the correct empirical formula for the compound sodium chloride. Note that the algebraic sum of the oxidation numbers written above this formula is zero.

Rule 3: *The algebraic sum of the oxidation numbers of all the atoms in the formula of a compound is zero.*

Figure 6-5. Magnesium and bromine react vigorously to form magnesium bromide.

6.6 Formation of magnesium bromide from its elements The reaction of bromine and magnesium in forming magnesium bromide is shown in Figure 6-5. In forming magnesium bromide, the two 3*s* electrons of the magnesium are transferred. *Both* 3*s* electrons must be transferred if magnesium is to acquire the stable electron configuration of the noble gas neon. Recall that the ionization energies of the two 3*s* electrons in magnesium are low. There is a great increase in ionization energy between the second and third electrons in a magnesium atom. Hence, only two electrons can be removed chemically. The 4th energy level of the bromine atom already contains seven electrons. (Eight, of course, is the number needed for a noble-gas configuration.) Thus, a single bromine atom has a place for only one of the two electrons that the magnesium atom transfers. So two bromine atoms are needed to react with one magnesium atom. Each bromine atom gains one electron. The diagram representing this electron transfer is shown below.

The empirical formula for the compound magnesium bromide is **$MgBr_2$**. The magnesium ions in this compound each have two excess positive charges. The bromide ions each have a single excess negative charge. These particles are arranged in orderly fashion in crystals of magnesium bromide. The ratio of the particles is 2 bromide ions to 1 magnesium ion. See Table 6-2 and also Figure 6-6. The subscript $_2$ following Br in the formula indicates that there are two bromide ions to each magnesium ion

Initially	1*s*	2*s*	2*p*	3*s*	3*p*	3*d*	4*s*	4*p*	Finally
:Br:	↓↑	↓↑	↓↑ ↓↑ ↓↑	↓↑	↓↑ ↓↑ ↓↑	↓↑ ↓↑ ↓↑ ↓↑ ↓↑	↓↑	↓↑ ↓↑ ↑	:Br:$^-$
Mg:	↓↑	↓↑	↓↑ ↓↑ ↓↑	↓↑	— — —	— — — — —	—	— — —	Mg^{++}
:Br:	↓↑	↓↑	↓↑ ↓↑ ↓↑	↓↑	↓↑ ↓↑ ↓↑	↓↑ ↓↑ ↓↑ ↓↑ ↓↑	↓↑	↓↑ ↓↑ ↓	:Br:$^-$

Chemical Change

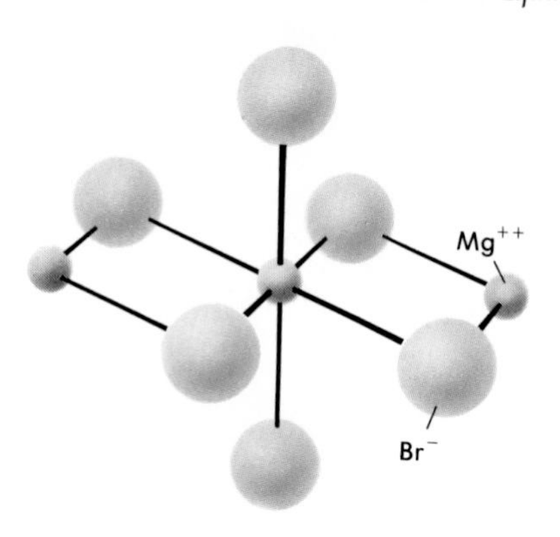

Figure 6-6. This diagram shows the arrangement of the magnesium and bromide ions in a portion of a magnesium bromide crystal.

Table 6-2
DATA ON ATOMS AND IONS OF MAGNESIUM AND BROMINE

	Magnesium atom	*Magnesium ion*	*Bromine atom*	*Bromide ion*
number of protons	12	12	35	35
number of electrons	12	10	35	36
net charge	0	+2	0	−1
symbol	Mg	Mg^{++}	Br	Br^-
radius, Å	1.36	0.66	1.14	1.96

in the compound. When no subscript is used, as with Mg, one atom or monatomic ion is understood.

The electron-dot symbols for atoms of magnesium and bromine are

$$\mathbf{Mg:} \qquad \mathbf{\cdot\ddot{\underset{..}{Br}}:}$$

and for the magnesium ion and bromide ion are

$$\mathbf{Mg^{++}} \qquad \mathbf{:\ddot{\underset{..}{Br}}:^-}$$

The electron-dot formula for magnesium bromide is

$$\mathbf{:\ddot{\underset{..}{Br}}:^-\ Mg^{++}\ :\ddot{\underset{..}{Br}}:^-}$$

and the ionic formula is $\mathbf{Mg^{++}Br^-{}_2}$. Note that in the formula $\mathbf{Mg^{++}Br^-{}_2}$, the subscript $_2$ applies to *both* the symbol **Br** and its charge $^-$. It shows that there are two $\mathbf{Br^-}$ ions with each $\mathbf{Mg^{++}}$ ion in the formula.

See Problem 6 at the end of the chapter to calculate the energy change for the reaction.

Energy is required to remove two electrons from one magnesium atom. But this energy is less than the electron affinity of two bromine atoms plus the energy released when one magnesium ion and two bromide ions take up their positions in a magnesium bromide crystal. Thus, the formation of the compound magnesium bromide from widely separated magnesium and bromine atoms is another example of the exothermic nature of electron transfer.

In forming the compound magnesium bromide, elemental magnesium atoms are *oxidized* from $\overset{0}{\mathbf{Mg}}$ to $\overset{+2}{\mathbf{Mg}}{}^{++}$. The oxidation state of magnesium becomes more positive as it changes from 0 to +2.

$$\overset{0}{\mathbf{Mg}} \rightarrow \overset{+2}{\mathbf{Mg}}{}^{++} + \mathbf{2e^-}$$

At the same time, for each magnesium atom oxidized, two elemental bromine atoms are *reduced* from $\overset{0}{\mathbf{Br}}$ to $\overset{-1}{\mathbf{Br}}{}^-$. The oxidation state of bromine becomes more negative as it changes from 0 to −1.

$$\mathbf{2\overset{0}{Br} + 2e^- \rightarrow 2\overset{-1}{Br}{}^-}$$

Bromine is the oxidizing agent; magnesium is the reducing agent.

In the formula $\overset{+2}{Mg^{++}}\ \overset{-1}{Br^{-}}{}_2$, the algebraic sum of +2 and 2(−1) equals zero. Note how the algebraic sum of the oxidation numbers of the atoms in a formula is determined. Each oxidation number must be multiplied by the number of atoms or monatomic ions having that oxidation number. The formulas of ionic compounds merely indicate the relative numbers of positive and negative ions that combine. Therefore, *all formulas for ionic compounds are empirical.*

Practice Problems

For the formation of compound $Li^{+}{}_2O^{--}$ by ionic bonding:

1. Draw the electron-transfer diagram using orbital notation.
2. Write the electron-dot formula for the compound.
3. Write an equation showing the oxidation.
4. Write an equation showing the reduction.
5. Identify the oxidizing agent and reducing agent.

6.7 Elements with several oxidation numbers Many elements exhibit more than one oxidation state. Some differences in the oxidation number of an element depend on the kind of bond that it forms with other elements. However, another factor is important. Transition elements, with from four to seven electron energy levels, can readily transfer the electrons in the highest energy level. Many of them can also transfer one or two electrons from the next-to-highest energy level with very little additional energy. The electrons in excess of an octet in the next-to-highest energy level are those available for transfer. Iron is such a transition element. In forming compounds, it can transfer two 4*s* electrons. In more energetic reactions, a 3*d* electron can also be transferred. Thus its oxidation number can be +2 or +3. This variable number of transferable electrons accounts for the variable oxidation state of many transition metals. See Figure 6-7. The table of Common Elements, inside the front cover and Appendix Table 5, lists common oxidation numbers.

Figure 6-7. Compounds of transition metals have different colors depending on their oxidation states. See Questions 39, 40, and 41 at the end of this chapter.

6.8 Relative sizes of atoms and ions Once again, look at Tables 6-1 and 6-2 accompanying Sections 6.3 and 6.6. These tables give the radii of the atoms and ions of sodium, magnesium, chlorine, and bromine. Notice the great difference in radius between an atom and the ion formed from it.

It is characteristic of metals to form positive ions. *Positive ions are called* ***cations.*** It is to be expected that metallic ions would be smaller than the corresponding metallic atoms since the highest-energy-level electrons are no longer present. As a result, the remaining electrons are drawn closer to the nucleus by its unbalanced positive charge.

Nonmetallic elements form negative ions. *Negative ions are called **anions**.* Nonmetallic ions are larger than the corresponding nonmetallic atoms. Electrons have been added, completing an octet in the highest energy level. Since the total positive charge of the nucleus remains the same, the average force of attraction for each electron decreases because there are more electrons.

Table 6-3 shows the sizes of representative atoms and ions. From these data and the generalizations above, it follows that:

1. Within a group or family of elements, the ion size increases with atomic number because of energy-level addition.

2. Within a period of elements, the Group I, II, and III cations show a sharp decrease in size. On the other hand, the Group VII anion is only slightly smaller than the Group VI anion. Group I and II cations have the same electron configurations, but the Group II cation's greater nuclear charge draws the electrons much closer. Group VI and VII anions have identical electron configurations too. The Group VII anion's greater nuclear charge draws the electrons somewhat closer.

Figures 6-3 and 6-6 show the arrangement of ions in crystals of the compounds sodium chloride and magnesium bromide. Such arrangements depend on the relative numbers and sizes of each kind of ion present. Crystals are described in more detail in Chapter 12.

Covalent Bonding

6.9 Covalent bonding: hydrogen molecules In covalent bonding electrons are not transferred from one atom to another. Instead, they are *shared* by the bonded atoms. In forming a single

Table 6-3
RADII OF REPRESENTATIVE ATOMS AND IONS, IN ANGSTROMS

	Group I		*Group II*		*Group III*		*Group VI*		*Group VII*	
Period 2	Li	1.23	Be	0.89	B	0.82	O	0.70	F	0.68
	Li^+	0.68	Be^{++}	0.35	B^{+++}	0.23	O^{--}	1.40	F^-	1.33
Period 3	Na	1.54	Mg	1.36	Al	1.18	S	1.02	Cl	0.99
	Na^+	0.97	Mg^{++}	0.66	Al^{+++}	0.51	S^{--}	1.84	Cl^-	1.81
Period 4	K	2.03	Ca	1.74	Ga	1.26	Se	1.17	Br	1.14
	K^+	1.33	Ca^{++}	0.99	Ga^{+++}	0.62	Se^{--}	1.98	Br^-	1.96
Period 5	Rb	2.16	Sr	1.91	In	1.44	Te	1.36	I	1.33
	Rb^+	1.47	Sr^{++}	1.12	In^{+++}	0.81	Te^{--}	2.21	I^-	2.20
Period 6	Cs	2.35	Ba	1.98	Tl	1.48				
	Cs^+	1.67	Ba^{++}	1.34	Tl^{+++}	0.95				

covalent bond, two atoms mutually share one of their electrons. These two shared electrons (with opposite spins) effectively fill an orbital in each atom. They make up a *covalent electron pair* that forms the bond.

The atoms of the common elemental gases, hydrogen, oxygen, nitrogen, fluorine, and chlorine, form stable diatomic (two-atomed) *molecules* by covalent bonding. *A* ***molecule*** *is the smallest chemical unit of a substance that is capable of stable independent existence.* For these elemental gases, the smallest chemical units capable of stable independent existence are diatomic units. Single atoms of the five gases listed above are chemically unstable under most conditions. Hence, these diatomic units make up molecules of these gases.

In diatomic hydrogen molecules each hydrogen atom shares its single 1*s* valence electron with the other hydrogen atom. These two electrons move about both nuclei so that each atom, in effect, has its 1*s* orbital filled. In essence, each hydrogen atom has the stable electron configuration of a helium atom.

H	$\frac{\uparrow}{1s}$
H	$\frac{\downarrow}{1s}$

The *electron-dot formula* for a molecule of hydrogen is **H:H.**

Frequently chemists indicate a shared pair of electrons by a dash (—) instead of the symbol (:). Thus the formula for a molecule of hydrogen may be written **H—H.** This type of formula is called a *structural formula.*

The *molecular formula* for hydrogen is H_2. The numerical subscript indicates the number of atoms per molecule. *A formula that indicates the actual composition of a molecule is called a* ***molecular formula.***

Hydrogen molecules are linear (straight-line), as shown in Figure 6-8. All diatomic molecules are linear because two points determine a straight line.

Hydrogen molecules have greater stability than separate hydrogen atoms. This statement is supported by the fact that energy is given off in the reaction

2 moles H atoms → 1 mole H_2 molecules + 104 kcal energy

One mole of H_2 molecules has lower energy and is more stable than two moles of uncombined H atoms. The reverse action is that of separating hydrogen molecules into the atoms of which they are made. Energy is required for this reaction.

1 mole H_2 + 104 kcal → 2 moles H

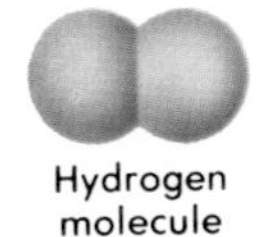

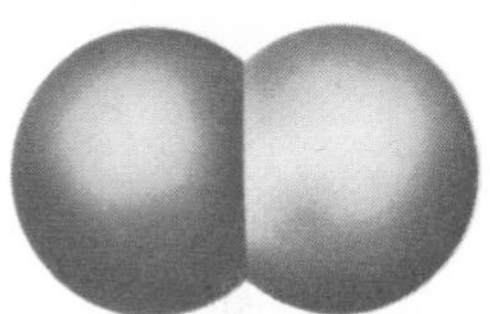

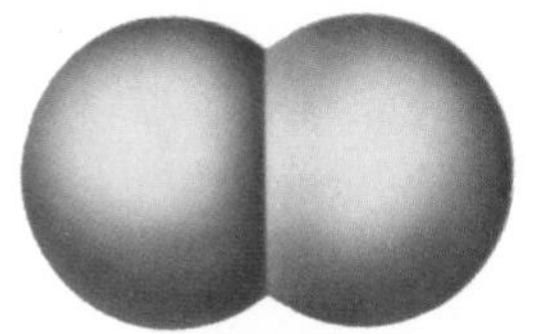

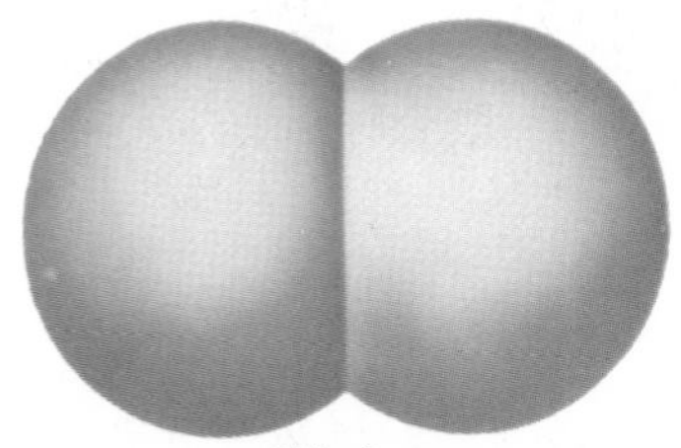

Figure 6-8. The molecules of the common gases, such as those of hydrogen, oxygen, nitrogen, and chlorine represented here, consist of two atoms joined by covalent bonding. These are all examples of linear molecules.

The energy required, 104 kcal/mole, is called the *bond energy* of the **H—H** bond. ***Bond energy** is the energy required to break chemical bonds and form neutral atoms.* It is usually expressed in kcal per mole of bonds broken.

Diatomic molecules of elements are considered to be free elements. Thus, each atom in the molecule may be assigned a zero oxidation number (Rule 1).

6.10 Chlorine molecules Diatomic chlorine molecules can be formed in the same manner as hydrogen molecules. Each atom shares one electron with the other and, in effect, fills an incomplete $3p$ orbital in each. This gives both atoms the stable electron arrangement of the noble gas argon. Both atoms essentially have an octet in the 3rd energy level.

	1s	2s	2p	2p	2p	3s	3p	3p	3p
Cl	↓↑	↓↑	↓↑	↓↑	↓↑	↓↑	↓↑	↓↑	↑
Cl	↓↑	↓↑	↓↑	↓↑	↓↑	↓↑	↓↑	↓↑	↓

The electron-dot formula for a molecule of chlorine is

$$:\ddot{\underset{\cdot\cdot}{Cl}}:\ddot{\underset{\cdot\cdot}{Cl}}:$$

Its structural formula is

Cl—Cl

and its molecular formula is Cl_2. Chlorine molecules are linear (straight-line) like those of hydrogen. See Figure 6-8.

Energy is required to separate chlorine molecules into chlorine atoms.

$$\textbf{1 mole } Cl_2 + \textbf{58 kcal} \rightarrow \textbf{2 moles Cl}$$

The bond energy of the **Cl—Cl** bond is 58 kcal/mole, and the chlorine molecules are more stable than separate chlorine atoms.

Diatomic molecules of the other halogens have similar formulas and shapes. Their bond energies are given in Table 6-4.

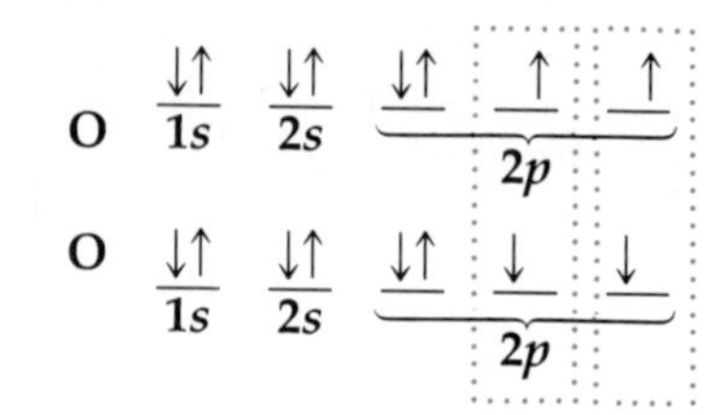

6.11 Oxygen molecules Oxygen also exists as diatomic molecules. But the bonding in an oxygen molecule is rather complex. It is not shown very satisfactorily by orbital and electron-dot formulas. A possible orbital notation for an oxygen molecule is shown in the left margin. The corresponding electron-dot formula for an oxygen molecule is

$$\dot{\underset{\cdot\cdot}{O}}::\dot{\underset{\cdot\cdot}{O}}$$

Its structural formula is

O=O

and its molecular formula is O_2.

Note that *two pairs* of electrons are shared in the oxygen molecule. This sharing of two electron pairs makes a *double covalent bond.* A double covalent bond is represented in a structural formula by a double dash. The oxygen molecule is linear. The bond energy of the **O═O** bond is 119 kcal/mole. Oxygen molecules are more stable than separate oxygen atoms.

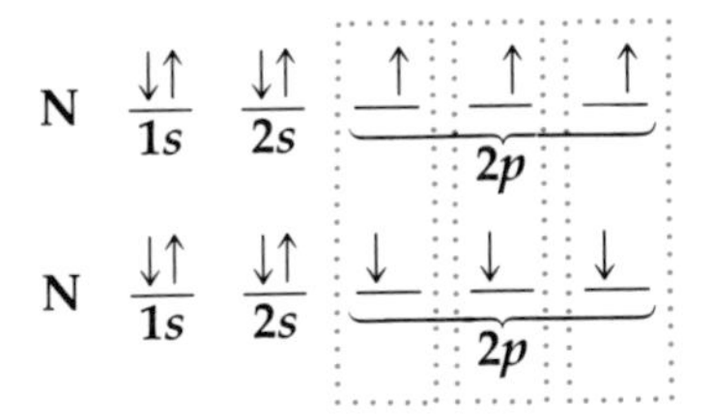

6.12 Nitrogen molecules The structure of the nitrogen molecule indicates the sharing of *three pairs* of electrons. Nitrogen molecules contain a *triple covalent bond,* as shown in the right margin. The electron-dot formula for the nitrogen molecule is

:N:::N:

Its structural formula is

N≡N

and its molecular formula is $\mathbf{N_2}$. Note that in a structural formula a triple covalent bond is represented by a triple dash. Nitrogen molecules are linear molecules. The bond energy of the **N≡N** bond is 226 kcal/mole. With such a high bond energy, nitrogen molecules are much more stable than individual nitrogen atoms.

6.13 Covalent bonding of unlike atoms: hydrogen chloride molecules Atoms of different elements may combine by covalent bonding. The reaction between hydrogen and chlorine shown in Figure 6-9 is such a reaction. A hydrogen atom and a chlorine atom combine by covalent bonding to form a hydrogen chloride molecule. In this molecule, the 1*s* hydrogen electron and a 3*p* chlorine electron complete an orbital, as shown below.

H										↑ (1*s*)
Cl	↓↑	↓↑	↓↑	↓↑	↓↑	↓↑	↓↑	↓↑	↓	
	1*s*	2*s*	2*p*			3*s*	3*p*			

The electron-dot formula for hydrogen chloride is

H:C̈l:

Its structural formula is

H—Cl

and its molecular formula is **HCl.** The HCl molecule is linear, as shown in Figure 6-10.

Energy is required to separate HCl molecules into H and Cl atoms.

1 mole HCl + 103 kcal → 1 mole H + 1 mole Cl

Hydrogen chloride molecules are more stable than separate hydrogen and chlorine atoms. The bond energy of the **H—Cl** bond is 103 kcal/mole.

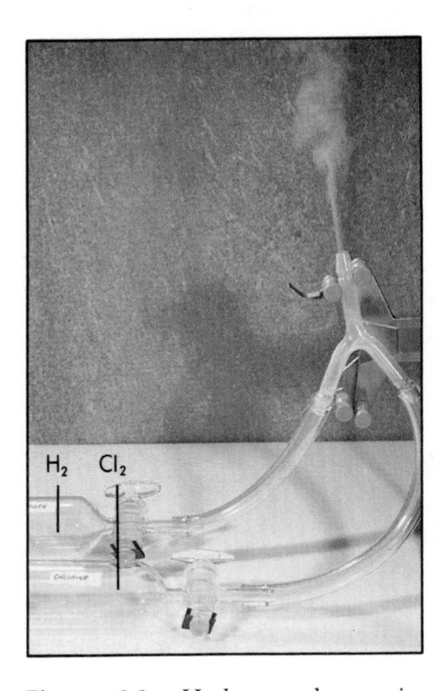

Figure 6-9. Hydrogen burns in chlorine to form hydrogen chloride, a colorless gas. Hydrogen chloride and moisture in the air form the fog visible in the photo.

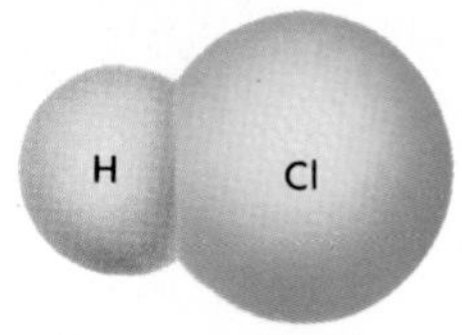

Hydrogen chloride molecule

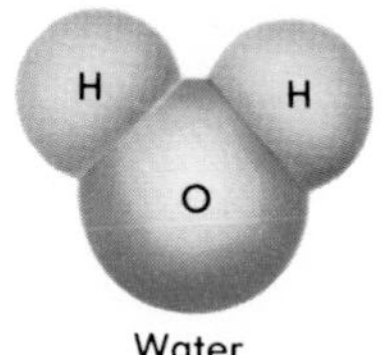

Water molecule

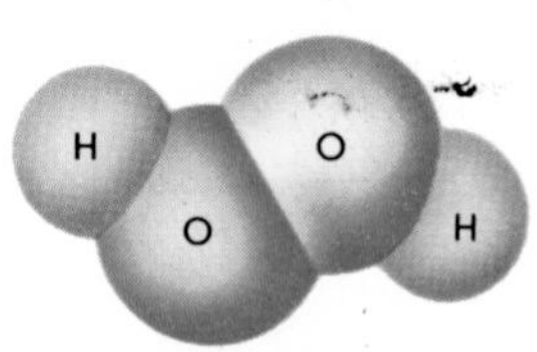

Hydrogen peroxide molecule

Figure 6-10. Hydrogen chloride, water, and hydrogen peroxide are compounds whose simplest particles are molecules composed of covalently bonded atoms. Hydrogen chloride molecules are linear; water molecules are bent molecules; hydrogen peroxide molecules are double-bent molecules.

Rule 4: *In compounds, the oxidation number of hydrogen is +1.*

By Rule 4 the oxidation number of hydrogen in hydrogen chloride is +1. By Rule 3 (Section 6.5) the oxidation number of chlorine must be −1.

Bond energies can be used to calculate the approximate energy of a chemical reaction. See Table 6-4. This method of calculation assumes that the energy change in a reaction is the net result of the breaking and making of chemical bonds.

The reaction for the formation of one mole of hydrogen chloride molecules can be used as an example. One-half mole of hydrogen molecules and one-half mole of chlorine molecules combine as follows:

$$\tfrac{1}{2}\text{ mole } H_2 + \tfrac{1}{2}\text{ mole } Cl_2 \rightarrow 1\text{ mole HCl}$$

For calculation purposes this reaction may be considered to occur in three steps. The first step is the breaking of bonds in one-half mole of hydrogen molecules.

$$\tfrac{1}{2}\text{ mole } H_2 + 52\text{ kcal} \rightarrow 1\text{ mole H}$$

(The equation is written for one-half mole of H_2 molecules. The energy that is needed is one-half the bond energy per mole, 104 kcal.)

The second step is the breaking of bonds in one-half mole of chlorine molecules.

$$\tfrac{1}{2}\text{ mole } Cl_2 + 29\text{ kcal} \rightarrow 1\text{ mole Cl}$$

(Again the energy involved is one-half the bond energy per mole, 58 kcal.)

The third reaction is the reverse of the **H—Cl** bond energy reaction written earlier in this section.

$$1\text{ mole H} + 1\text{ mole Cl} \rightarrow 1\text{ mole HCl} + 103\text{ kcal}$$

Combining these three equations and adding algebraically,

$$\begin{array}{lll}
\tfrac{1}{2}\text{ mole } H_2 & + 52\text{ kcal} & \rightarrow \cancel{1\text{ mole H}} \\
\tfrac{1}{2}\text{ mole } Cl_2 & + 29\text{ kcal} & \rightarrow \cancel{1\text{ mole Cl}} \\
\cancel{1\text{ mole H}} & + \cancel{1\text{ mole Cl}} & \rightarrow 1\text{ mole HCl} + 103\text{ kcal} \\
\hline
\tfrac{1}{2}\text{ mole } H_2 & + \tfrac{1}{2}\text{ mole } Cl_2 & \rightarrow 1\text{ mole HCl} + 22\text{ kcal}
\end{array}$$

Thus, the reaction is exothermic. The amount of energy released is 22 kcal.

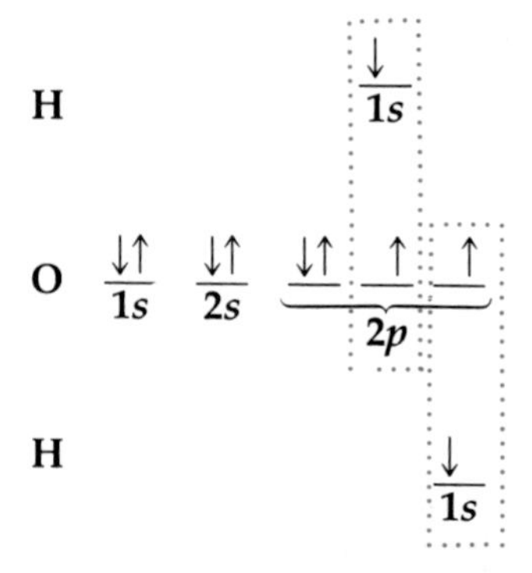

6.14 Water molecules The common compound water consists of molecules formed by covalent bonding of two hydrogen atoms with one oxygen atom. The orbital representation of this bonding is shown in the left margin. The electron-dot formula for water is

$$\begin{array}{c} H\!:\!\ddot{O}\!: \\ \ \ \ \ddot{H} \end{array}$$

Its structural formula is

$$\mathrm{H{-}O{-}H}\ \text{(bent)}$$

and its molecular formula is $\mathbf{H_2O}$.

A molecule containing three atoms can have one of only two shapes. Either (1) the atoms lie on the same straight line and form a linear molecule, or (2) the atoms do not lie on one straight line and form a bent molecule. Figure 6-10 shows that the hydrogen atoms and oxygen atom in a water molecule do not lie on the same straight line. Water molecules are bent molecules. Very careful measurements reveal that the distances between the oxygen atom and each of the two hydrogen atoms are the same. This evidence suggests that the two hydrogen atoms act the same way in the molecule.

Energy is required to break up water molecules into the atoms of which they are made.

$$\textbf{1 mole } \mathbf{H_2O} + \textbf{222 kcal} \rightarrow \textbf{2 moles H} + \textbf{1 mole O}$$

Water molecules have lower energy and are more stable than separate hydrogen and oxygen atoms. Two **H—O** bonds are broken per molecule of water decomposed. Therefore, the **H—O** bond energy is 111 kcal/mole, or one-half that shown in the equation.

Table 6-4
BOND ENERGIES

Bond	*Energy* (kcal/mole)
H—H	104
N—N	39
N≡N	226
O—O	35
O=O	119
F—F	38
Cl—Cl	58
Br—Br	46
I—I	36
C—H	99
N—H	93
O—H	111
F—H	135
Cl—H	103
Br—H	87
I—H	71
C—F	116
C—Cl	78
C—Br	66
C—I	57

6.15 Hydrogen peroxide molecules It is possible for more than one kind of molecule to be formed from the same kinds of atoms. A second compound that may be formed from hydrogen and oxygen is hydrogen peroxide, a well-known bleaching and oxidizing agent. The orbital notation of hydrogen peroxide is shown in the right margin. Its electron-dot formula is

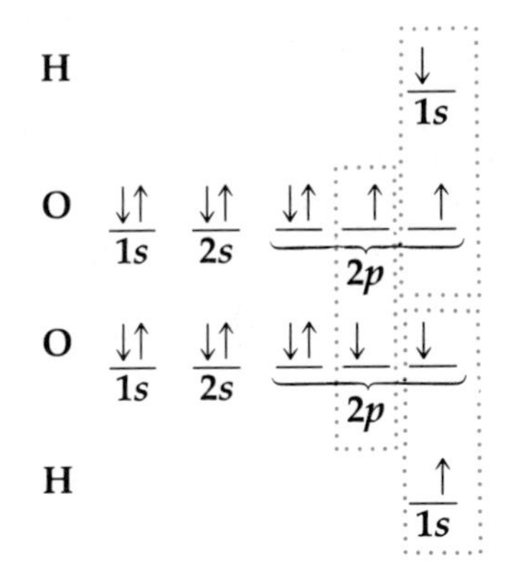

$$\mathrm{H{:}\ddot{O}{:}}$$
$$\mathrm{{:}\ddot{O}{:}H}$$

Its structural formula is

$$\begin{array}{l}\mathrm{H{-}O}\\ \quad\ |\\ \quad\ \mathrm{O{-}H}\end{array}$$

Its molecular formula is $\mathbf{H_2O_2}$. Experimental evidence indicates that the hydrogen peroxide molecule is a double-bent shaped molecule. (See Figure 6-10.)

Rule 5: *In compounds, the oxidation number of oxygen is −2.*

Since further exceptions to Rules 4 and 5 are discussed later in this chapter, use the statements of these rules in Section 6.25 for review purposes.

One exception to this rule is that in peroxides the oxidation number of oxygen is −1. In water the oxidation number of hydrogen is +1 and of oxygen is −2. On the other hand, in hydrogen peroxide the oxidation number of hydrogen is +1 and of oxygen is −1. (Use Rule 3 to check this result.)

6.16 Other molecules containing hydrogen Table 6-5 gives some examples of covalent bonding between hydrogen and nitrogen and between hydrogen and carbon. The oxidation number of hydrogen in each molecule is +1. But note that the oxidation numbers of nitrogen and carbon vary from molecule to molecule.

Ammonia molecules, NH_3, have three hydrogen atoms so spaced about a nitrogen atom that the molecule has the shape of a pyramid. Think of the hydrogen atoms as forming the base of the pyramid and the nitrogen atom the peak. In methane, CH_4, the hydrogen atoms are symmetrically spaced in three dimensions about the carbon atom. Methane molecules have a regular tetrahedral shape. A regular tetrahedron is a solid figure that has four sides, each side an equilateral triangle.

Consider the examples of covalent bonding illustrated thus far. Except for hydrazine (Table 6-5), all of the bonded atoms are more stable because they have filled orbitals. They have lower energy than the unbonded atoms. Thus, the reactions in which these molecules are formed from atoms are exothermic. Only the combination of nitrogen and hydrogen atoms to form hydrazine, N_2H_4, is endothermic.

The neutral particle that results from the covalent bonding of atoms is a molecule. Its composition is represented by a molecular formula.

Table 6-5
DATA ON SOME REPRESENTATIVE MOLECULES

	Ammonia	*Hydrazine*	*Methane*	*Ethane*
molecular formula	NH_3	N_2H_4	CH_4	C_2H_6
structural formula	H H \\ / N | H	H H \\ / N—N / \\ H H	H | H—C—H | H	H H | | H—C—C—H | | H H
electron-dot formula	H :N:H H	H H :N:N: H H	H H:C:H H	H H H:C:C:H H H
oxidation numbers	H = +1 N = −3	H = +1 N = −2	H = +1 C = −4	H = +1 C = −3
model of molecule				

Practice Problems

Calculate the energy change for the exothermic reactions:

1. $\frac{1}{2}$ mole H_2 + $\frac{1}{2}$ mole $F_2 \rightarrow$ 1 mole HF *ans.* 64 kcal
2. $\frac{1}{2}$ mole N_2 + $\frac{3}{2}$ mole $H_2 \rightarrow$ 1 mole NH_3 *ans.* $1\bar{0}$ kcal

6.17 Hybridization In a methane molecule, CH_4, one carbon atom is covalently bonded to four hydrogen atoms. The hydrogen atoms are symmetrically arranged as if at the vertices of a regular tetrahedron with the carbon atom at the center. The carbon-hydrogen bond angles in this molecule are all 109.5°. See Figure 6-11.

The orbital notation for carbon atoms (as shown in the right margin) indicates that the valence electrons should be two 2*s* and two 2*p* electrons. However, when carbon atoms combine, it is believed that one of the 2*s* electrons acquires enough energy to occupy a 2*p* orbital. Thus, the bonding electrons of a carbon atom are one 2*s* electron and three 2*p* electrons, as shown in the right margin. A carbon atom, therefore, can form four covalent bonds, as it does with hydrogen atoms in the methane molecule.

C ↓↑ (1*s*) ↓↑ (2*s*) ↑ ↑ — (2*p*)

C ↓↑ (1*s*) ↑ (2*s*) ↑ ↑ ↑ (2*p*)

Because one of the valence electrons of the carbon atom is a 2*s* electron and the other three are 2*p* electrons, one of the carbon-hydrogen bonds in methane might be expected to be different from the other three. Experimentally, however, this is not the case. All the bonds are equivalent. This difference can be explained by assuming that *hybridization* of the one 2*s* and three 2*p* orbitals occurs. Four equivalent orbitals are produced.

***Hybridization** is the combining of two or more orbitals of nearly the same energy into new orbitals of equal energy.* The hybrid orbitals of the carbon atom are called sp^3 (read "*sp*-three") orbitals. Hybrid orbitals of this type result from the combination of the one *s* orbital and three *p* orbitals. In orbital notation, the effect of hybridization can be shown as

C ↓↑ (1*s*) ↑ ↑ ↑ ↑ ($2sp^3$)

As mentioned, these four sp^3 orbitals point toward the corners of a regular tetrahedron from the carbon atom at their center. This arrangement permits the maximum separation of four orbitals grouped about a given point. This arrangement accounts for the regular tetrahedral shape of methane molecules.

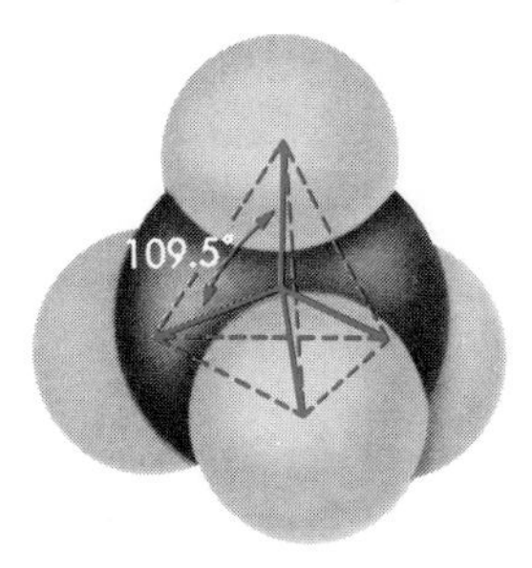

Figure 6-11. In the methane molecule, CH_4, a carbon atom is covalently and symmetrically bonded to four hydrogen atoms. The molecule's tetrahedral shape is the result of sp^3 hybridization of the carbon-atom orbitals.

The idea of hybrid tetrahedral orbitals can also account for the observed shape of the ammonia molecule. This is true even though ammonia has only three hydrogen atoms attached to a central nitrogen atom. The electron-dot symbol for a nitrogen atom is

$$\cdot\dot{\underset{\cdot}{N}}:$$

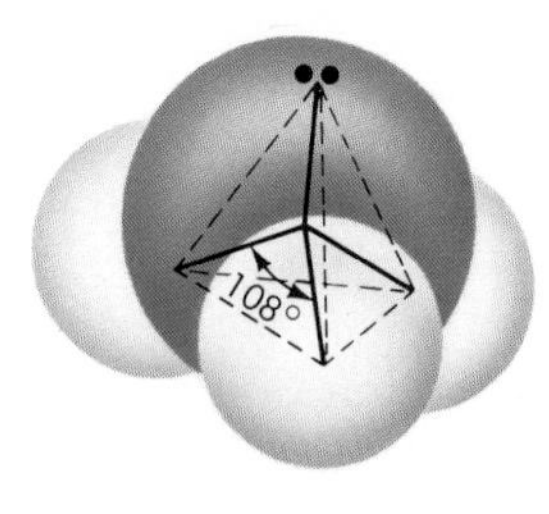

Figure 6-12. The pyramidal shape of the ammonia molecule, NH_3, may be explained by the sp^3 hybridization of nitrogen-atom orbitals.

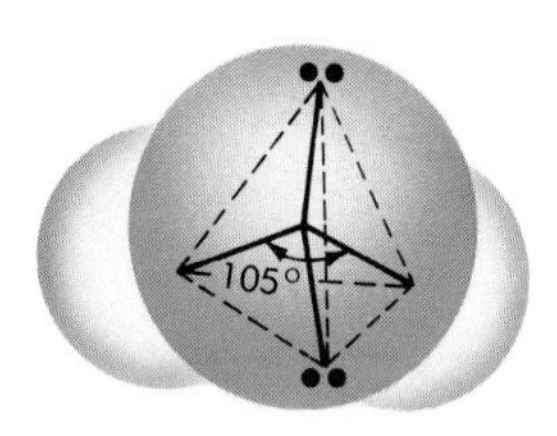

Figure 6-13. The bent shape of the water molecule, H_2O, may be explained by the considerable sp^3 hybridization of the oxygen-atom orbitals.

Imagine that the electrons in the four positions about the symbol are distributed among four equivalent tetrahedral orbitals. This arrangement gives one orbital with a pair of electrons and three orbitals with single electrons that can be shared with the electron from each of three hydrogen atoms. The result is a structure that resembles a pyramid, as shown in Figure 6-12. The three hydrogen atoms form the base of the pyramid. The unshared pair of nitrogen electrons forms the peak. The angles between the **N—H** bonds in ammonia are known to be 108°. This agrees well with the tetrahedral angle of 109.5°.

This idea can be extended to the structure of the water molecule. The electron-dot formula for this molecule is

H:Ö:
H

Here there are four tetrahedral hybrid orbitals. Two of these form the oxygen-hydrogen bonds. The other two are each a pair of unshared oxygen electrons. See Figure 6-13. The bond angle in the water molecule is only 105°. The unshared electron pairs on the oxygen atom tend to repel each other more than the shared electrons repel each other. These unequal opposing repulsions make the angle between the shared pairs decrease slightly in size.

Other types of hybridization are described in Chapter 18.

6.18 Size of molecules Molecules are very small. It is estimated that if a drop of water could be magnified to the size of the earth, the molecules composing it would be about one meter in diameter. But molecules vary greatly in size. The simple molecules of gases, consisting of one, two, or three atoms, have diameters of about 3×10^{-8} cm. Some virus protein molecules consist of about 7.5×10^5 atoms. These molecules have diameters of about 2.3×10^{-6} cm and have been photographed with an electron microscope. On the angstrom scale, the range of molecular diameters is from 3 Å to 230 Å.

ELECTRONEGATIVITY

6.19 Electronegativity In ionic bonding, electrons are *completely transferred* from the highest energy levels of metallic atoms to the highest energy levels of nonmetallic atoms. In covalent bonding, electrons are *shared* in the highest energy levels of the bonded atoms.

If two covalently bonded atoms are alike, their attractions for the shared electrons are equal. The electrons are most probably distributed equally about both atoms. Each atom remains electrically neutral, even though they are bonded together. *A covalent bond in which there is an equal attraction for the shared electrons and a resulting balanced distribution of charge is called a* ***pure,*** *or* ***non-***

polar, covalent bond. If two different atoms are covalently bonded, one atom may attract the shared electron pair more strongly than the other atom does. The electrons are not equally shared but are more closely held by the atom with stronger attraction. This atom is not electrically neutral. Instead it is slightly negative, though not as negative as a singly charged anion. The other atom is left slightly positive, though not as positive as a singly charged cation. *A covalent bond in which there is an unequal attraction for the shared electrons and a resulting unbalanced distribution of charge is called a* **polar covalent bond**. Polar covalent bonds are intermediate in nature between ionic bonds and pure covalent bonds. In ionic bonds, electron transfer is complete. In pure covalent bonds, electron sharing is equal. Polar covalent bonds are part covalent and part ionic in character.

Ionization energy is a measure of the strength with which a neutral atom holds a highest-energy-level electron. Electron affinity indicates the strength of the attraction between a neutral atom and an additional electron. Linus Pauling (b. 1901) and other chemists studied these two values, together with certain properties of molecules. They derived an arbitrary scale that indicates *the attraction of an atom for the shared electrons forming a bond between it and another atom. This property is called* ***electronegativity***. Atoms with high electronegativity have a strong attraction for electrons they share with another atom. Atoms with low electronegativity have a weak attraction for electrons they share with another atom. The relative electronegativities of two atoms indicate the type of bonding that can exist between atoms.

Figure 6-14 gives the electronegativity values of the elements. A study of the chart on the next page shows that:

1. Low electronegativity is characteristic of metals. The lower the electronegativity, the more active the metal. Thus, the lowest electronegativities are found at the lower left of the periodic table.

2. High electronegativity is characteristic of nonmetals. Thus, the highest electronegativities are found at the upper right of the periodic table. Fluorine is the most electronegative element. Oxygen is second.

3. Within the numbered groups or families, electronegativity generally decreases with increasing atomic number. In the transition element groups, there is usually only a slight variation.

4. Electronegativity increases within a period or series through the middle of the periodic table. It decreases slightly in the remaining metals and then increases usually to a maximum in Group VII or VIII.

6.20 Electronegativity and chemical bonding The electronegativities of two atoms bonded together can be used to determine the approximate percentage of ionic character of the bond between them. If element **A**, having the higher electronegativity

Periodic Table of Electronegativities

I	II											III	IV	V	VI	VII	VIII
2.1 **H** 1																	**He** 2
1.0 **Li** 3	1.5 **Be** 4											2.0 **B** 5	2.5 **C** 6	3.0 **N** 7	3.5 **O** 8	4.0 **F** 9	**Ne** 10
0.9 **Na** 11	1.2 **Mg** 12											1.5 **Al** 13	1.8 **Si** 14	2.1 **P** 15	2.5 **S** 16	3.0 **Cl** 17	**Ar** 18
0.8 **K** 19	1.0 **Ca** 20	1.3 **Sc** 21	1.5 **Ti** 22	1.6 **V** 23	1.6 **Cr** 24	1.5 **Mn** 25	1.8 **Fe** 26	1.8 **Co** 27	1.8 **Ni** 28	1.9 **Cu** 29	1.6 **Zn** 30	1.6 **Ga** 31	1.8 **Ge** 32	2.0 **As** 33	2.4 **Se** 34	2.8 **Br** 35	3.0 **Kr** 36
0.8 **Rb** 37	1.0 **Sr** 38	1.2 **Y** 39	1.4 **Zr** 40	1.6 **Nb** 41	1.8 **Mo** 42	1.9 **Tc** 43	2.2 **Ru** 44	2.2 **Rh** 45	2.2 **Pd** 46	1.9 **Ag** 47	1.7 **Cd** 48	1.7 **In** 49	1.8 **Sn** 50	1.9 **Sb** 51	2.1 **Te** 52	2.5 **I** 53	2.6 **Xe** 54
0.7 **Cs** 55	0.9 **Ba** 56	1.2 **Lu** 71	1.3 **Hf** 72	1.5 **Ta** 73	1.7 **W** 74	1.9 **Re** 75	2.2 **Os** 76	2.2 **Ir** 77	2.2 **Pt** 78	2.4 **Au** 79	1.9 **Hg** 80	1.8 **Tl** 81	1.8 **Pb** 82	1.9 **Bi** 83	2.0 **Po** 84	2.2 **At** 85	2.4 **Rn** 86
0.7 **Fr** 87	0.9 **Ra** 88	**Lr** 103	**Unq** 104	**Unp** 105	**Unh** 106	**Uns** 107	**Uno** 108	**Une** 109									

1.1 **La** 57	1.1 **Ce** 58	1.1 **Pr** 59	1.1 **Nd** 60	1.1 **Pm** 61	1.1 **Sm** 62	1.1 **Eu** 63	1.1 **Gd** 64	1.1 **Tb** 65	1.1 **Dy** 66	1.1 **Ho** 67	1.1 **Er** 68	1.1 **Tm** 69	1.1 **Yb** 70
1.1 **Ac** 89	1.3 **Th** 90	1.5 **Pa** 91	1.7 **U** 92	1.3 **Np** 93	1.3 **Pu** 94	1.3 **Am** 95	1.3 **Cm** 96	1.3 **Bk** 97	1.3 **Cf** 98	1.3 **Es** 99	1.3 **Fm** 100	1.3 **Md** 101	1.3 **No** 102

Figure 6-14. Periodic table showing the electronegativities of the elements on an arbitrary scale.

x_A, forms a bond with element **B**, having the lower electronegativity x_B, the percentage of ionic character of the bond **A—B** is given by:

$$\text{percentage of ionic character of bond A—B} = \frac{x_A - x_B}{x_A} \times 100\%$$

Let us first calculate the percentage of ionic character of bonds between metallic elements and distinctly nonmetallic elements. We will use sodium chloride, NaCl, as an example. From Figure 6-14, the electronegativity of sodium is 0.9; that of chlorine is 3.0. Substituting these values in the preceding equation,

$$\text{percentage of ionic character of bond Na—Cl} = \frac{3.0 - 0.9}{3.0} \times 100\% = 70\%$$

Practice Problems

1. Calculate the percentage of ionic character of a Ca—Br bond in $CaBr_2$. *ans.* 64%

2. Calculate the percentage of ionic character of a K—O bond in K_2O. *ans.* 77%

Bonds with more than 50% ionic character are considered to be ionic. These calculations confirm that the bonds between metallic elements and distinctly nonmetallic elements are ionic.

Next let us consider bonds between some nonmetallic elements. What is the percentage of ionic character of an H—O bond in the water molecule, H_2O?

$$\textbf{percentage of ionic character of bond H—O} = \frac{3.5 - 2.1}{3.5} \times 100\% = 40\%$$

Practice Problems

1. Calculate the percentage of ionic character of a C—Cl bond in CCl_4. *ans.* 17%
2. Calculate the percentage of ionic character of an H—S bond in H_2S. *ans.* 16%

Bonds with between 5% and 50% ionic character are considered to be polar covalent. Thus, H—O, C—Cl, and H—S bonds are all shown to be polar covalent. In the H—O bond, the hydrogen is somewhat positive and the oxygen somewhat negative; in the C—Cl bond, the carbon is somewhat positive and the chlorine is somewhat negative; in the H—S bond, the hydrogen is somewhat positive and the sulfur is somewhat negative.

Finally, let us consider bonds between two atoms of the same nonmetallic element. What is the percentage of ionic character of an O—O bond in an oxygen molecule, O_2?

$$\textbf{percentage of ionic character of bond O—O} = \frac{3.5 - 3.5}{3.5} \times 100\% = 0\%$$

Practice Problems

1. Calculate the percentage of ionic character of a Cl—Cl bond in Cl_2. *ans.* 0%
2. Calculate the percentage of ionic character of an I—I bond in I_2. *ans.* 0%

Bonds with below 5% ionic character are considered to be nonpolar covalent. The bonds between like atoms have no ionic character; they are nonpolar covalent bonds.

Chemical bonds have been classed as ionic bonds or covalent bonds. But it is apparent that these classifications are not clear and distinct. On the periodic table, the type of bonding gradually changes. Bonds are essentially ionic between the active metals of Groups I and II and oxygen or the halogens. But they are essentially covalent between metalloids and nonmetals as well as between two nonmetals. A third type of bonding, metallic bonding, occurs between atoms of metals. It will be described in Chapter 12.

The noble gases have electron configurations that are chemically very stable. It is possible, however, to produce compounds in which xenon is bonded to fluorine, chlorine, oxygen, or nitrogen. Compounds of krypton or radon with fluorine are also known.

The sixth rule for assigning oxidation numbers can now be introduced.

Rule 6: *In combinations involving nonmetals, the oxidation number of the less electronegative element is positive, and that of the more electronegative element is negative.*

Rule 6 leads to two further exceptions to the rules as already stated. First, hydrogen forms some ionic compounds with very active metals such as lithium and sodium. In these compounds, the hydrogen atom gains an electron from the metal and forms a *hydride* ion, $\mathbf{H^-}$. In hydrides the oxidation number of hydrogen is -1. Second, in compounds with fluorine, oxygen is the less electronegative element. In such compounds, therefore, oxygen must have a positive oxidation number, $+2$.

6.21 Molecular polarity If all the bonds in a molecule are nonpolar, the valence electrons are equally shared by the bonding atoms. Thus, there is a uniform distribution of electrons on the exterior of the molecule. This uniform distribution occurs regardless of the number of bonds and their direction in space. A molecule with such characteristics is a *nonpolar molecule.* Molecules such as H_2, Cl_2, O_2, and N_2 are nonpolar because the bonds in each molecule are nonpolar.

Diatomic molecules like HCl and HBr have only one bond, and it is a polar bond. These molecules have one somewhat more negative end and a somewhat less negative end. On the more negative end, the electron density (probability of finding electrons) is greater than on the less negative end. Molecules with such unbalanced electron distributions are *polar molecules.* Because polar molecules have two regions of different electric charge, they are called *dipolar molecules,* or simply *dipoles.*

If a molecule has more than one polar bond, the molecule as a whole may be nonpolar or polar depending on the arrangement in space of the bonds in the molecule. If the polar bonds in a molecule are all alike, the polarity of the molecule as a whole depends only on the arrangement in space of the bonds. Thus, water molecules are polar owing to a bent structure.

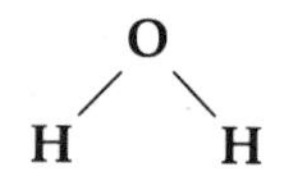

But carbon dioxide is nonpolar due to a linear structure.

O=C=O

Triangular boron trifluoride molecules, Figure 6-15,

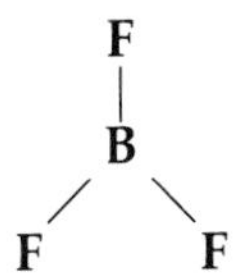

are nonpolar, but pyramidal ammonia molecules are polar.

N
H H H

Tetrahedral methane molecules are nonpolar.

H
H—C—H
H

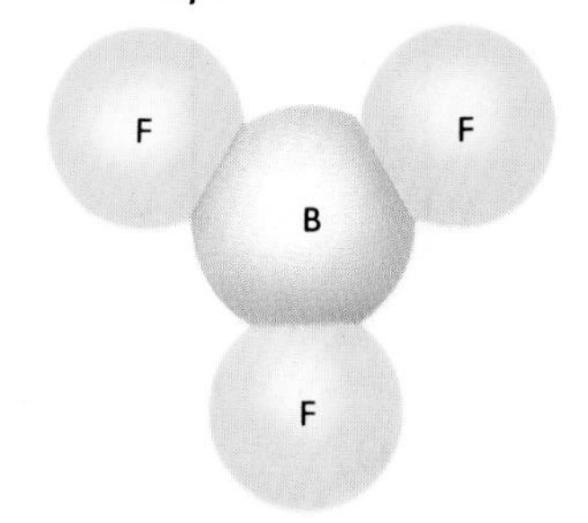

Figure 6-15. Even though the boron-fluorine bonds are polar bonds, their symmetrical arrangement in the molecule explains why boron trifluoride molecules are nonpolar.

A model of a carbon dioxide molecule is shown in Figure 17.11.

Resonance

6.22 Resonance If an attempt is made to draw an electron-dot formula for the molecular compound sulfur dioxide, the result is that two formulas may be written, each giving all three atoms in the molecule an octet. These are shown in the right margin.

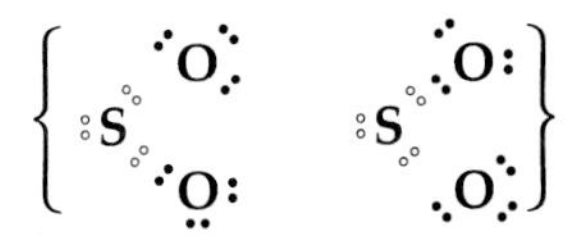

From these formulas, it might be suspected that the two sulfur-oxygen bonds in the molecule are different. Experimental evidence indicates, however, that these bonds are identical. Hence, neither of the formulas, taken alone, can be correct. Unfortunately, *bond identity cannot be satisfactorily represented by any single formula using the electron-dot notation system and keeping the octet rule*. A concept called ***resonance*** is used to describe such situations.

Sulfur dioxide molecules (see Figure 6-16) may be considered to have a structure intermediate between the two electron-dot structures given. This is signified by enclosing the formulas with braces. In this sense, the electron arrangement in the molecules is a *resonance hybrid* of the written structures. The term "resonance" is not very accurate in this case. It encourages the wrong idea that the structure of the molecule switches from one electron-dot formula to the other and that a pair of electrons is sometimes part of one bond and sometimes part of the other.

Sulfur dioxide molecules have only one real structure. The properties of a resonance hybrid do not switch from those of one electron-dot structure to those of the other. The properties are definite and are characteristic of the hybrid structure. The concept

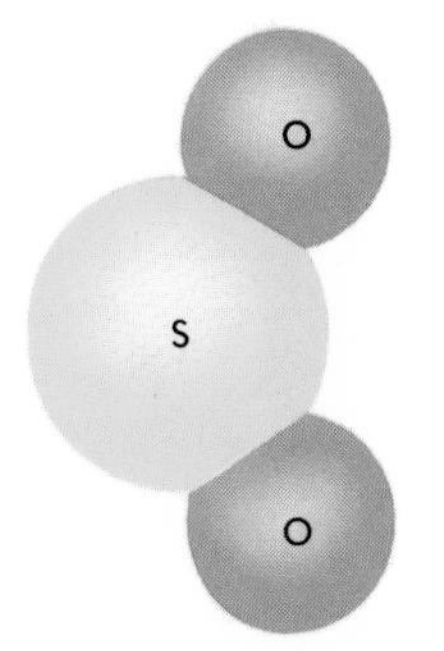

Figure 6-16. A model that shows the arrangement of atoms in sulfur dioxide molecules.

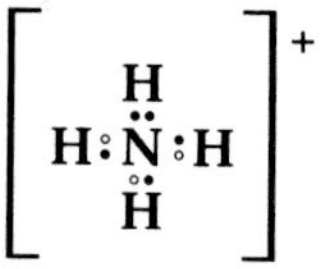

ammonium ion

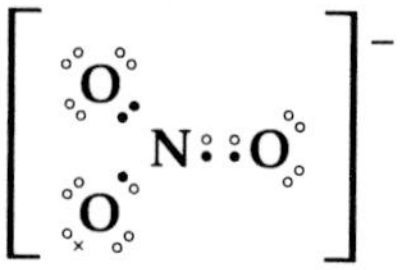

nitrate ion

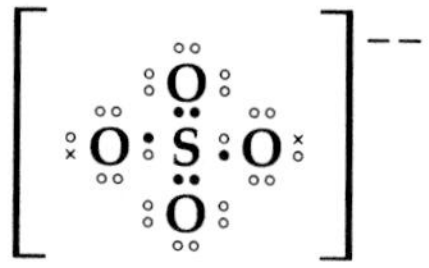

sulfate ion

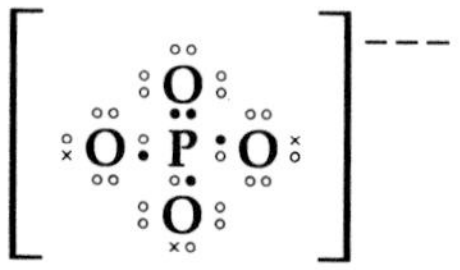

phosphate ion

of resonance is an attempt to make up for deficiencies in writing the electron-dot structures of certain molecules. Difficulties occur when the electron-pair and electron-octet rules are used to write the structures of these molecules. The difficulties lie in the method of writing formulas, not in the molecules represented.

6.23 Polyatomic ions Some covalently bonded groups of atoms act like single atoms in forming ions. These charged groups of covalently bonded atoms are called *polyatomic* (many-atomed) *ions*. Some common polyatomic ions are the sulfate ion, SO_4^{--}, the nitrate ion, NO_3^-, and the phosphate ion, PO_4^{---}. The bonds within these polyatomic ions are largely covalent. When combined, the groups of atoms have an excess of electrons and thus are negative ions. A common positive polyatomic ion, the ammonium ion, NH_4^+, is produced when a molecule of ammonia, NH_3, acquires a proton. Electron-dot representations of these polyatomic ions are shown in the left margin. (Models of these ions are shown in Figure 6-17.)

There are some covalent bonds in which both electrons that form the bond between two atoms come from only one of the bonded atoms. Each of the electron-dot formulas shown here has at least one such bond. In the ammonium ion, for example, three of the covalent nitrogen-hydrogen bonds consist of one shared electron from nitrogen and one shared electron from hydrogen. The other covalent nitrogen-hydrogen bond consists of two electrons supplied by a nitrogen atom only. A hydrogen ion (proton) can thus bond at this position and produce the single net positive charge of the NH_4^+ ion.

Looking at the structure another way, there are eleven protons in the ammonium ion (seven in the nitrogen nucleus and one in each of four hydrogen nuclei). There are only ten electrons (two 1*s* electrons of the nitrogen atom and eight valence electrons shown). With eleven protons and only ten electrons, an ammonium ion has a net charge of +1.

Rule 7: *The algebraic sum of the oxidation numbers of the atoms in the formula of a polyatomic ion is equal to its charge.*

In an ammonium ion, the oxidation number of hydrogen is +1 (Rules 4 and 6). But the algebraic sum of the oxidation numbers must equal +1 (Rule 7). Using x as the oxidation number of nitrogen:

$$\overset{x\ +1}{NH_4^+}$$

$$x + 4(+1) = +1 \text{ and } x = -3$$

The bonding, charges, and oxidation numbers of the elements in the other polyatomic ions can be examined in similar fashion. An electron represented by × is acquired by the ion from another element through electron transfer.

The electron-dot formula shown for the nitrate ion is only one of several possible formulas that can be written. Nitrate ions have a resonance-hybrid structure.

6.24 More about particles of matter Free and isolated atoms are rarely found in nature. Instead, atoms of most elements combine with one another at ordinary temperatures and form larger structural particles. Notable exceptions are the noble elements: helium, neon, argon, krypton, xenon, and radon. The atoms of these noble gases do not combine with each other and form larger particles. There is no distinction, therefore, between the atoms and molecules of these gases. A molecule of helium, **He**, is monatomic.

The atoms of some elements combine naturally and form pairs that exist as simple diatomic molecules. Atmospheric oxygen and nitrogen are two examples. Their molecules are represented as O_2 and N_2, respectively. Observe that O_2 means two atoms of oxygen are bonded together to form one oxygen molecule. On the other hand, **2O** means two separate unbonded oxygen atoms. The expression $3O_2$ means three molecules of oxygen, each of which consists of two oxygen atoms bonded together. The other elemental gases, hydrogen, fluorine, and chlorine, also exist as diatomic molecules, H_2, F_2, and Cl_2, respectively. The nonmetal bromine, a liquid at ordinary temperatures, exists as diatomic molecules, Br_2. Iodine forms molecular crystals in which each molecular particle is diatomic, I_2.

Other elements may form groups consisting of a larger number of atoms. Phosphorus may form molecules consisting of four atoms, written P_4. Sulfur molecules may be eight-atom particles, or S_8. The metallic elements generally exhibit crystalline structures. Their atoms are closely packed in regular patterns that show no simple molecular units. Each individual crystal is considered a single giant molecule.

Some compounds have distinct unit structures composed of simple molecules. Water is a familiar example. Water molecules consist of two hydrogen atoms and one oxygen atom, H_2O. The expression $2H_2O$ represents two molecules of water, each containing two atoms of hydrogen and one atom of oxygen. Similarly, $5H_2O$ signifies five molecules of water. When no other coefficient (the number before the formula) is used, it is understood that the coefficient is 1. Molecules of compounds range from a minimum of two atoms to large numbers of atoms.

Some substances show complex unit structures formed by groups of molecules or molecular aggregates. Still others have no molecular organization at all. Ordinary table salt, sodium chloride, consists of sodium and chloride ions distributed in a regular crystalline lattice pattern that is continuous to each face of the salt crystal. Simple molecules of sodium chloride do not exist except in the vapor state at high temperatures. In general,

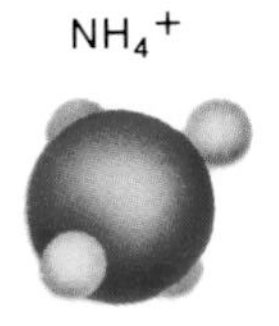

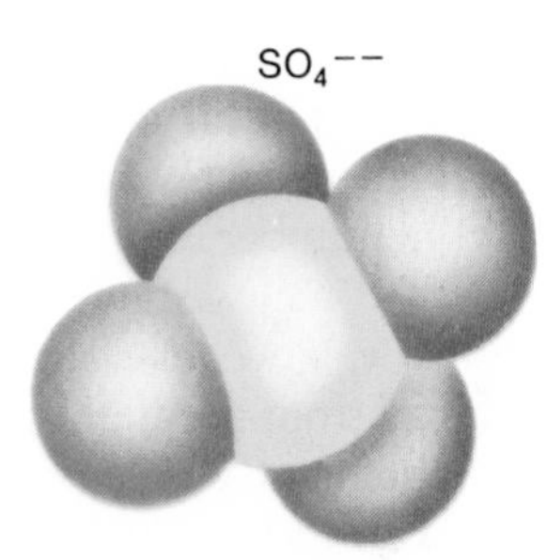

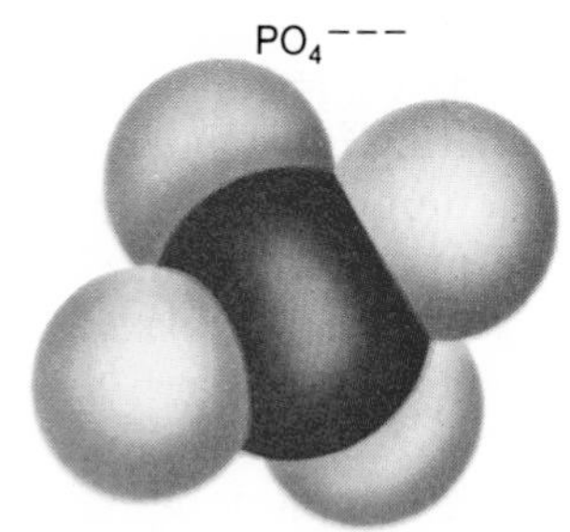

Figure 6-17. Models showing the arrangement of atoms in four common polyatomic ions.

a coefficient of a symbol or formula gives the number of particles whose composition is given by the symbol or formula; a subscript gives the number of atoms of a particular kind in the particle.

6.25 Summary of oxidation-number rules

1. The oxidation number of an atom of a free element is zero.
2. The oxidation number of a monatomic ion is equal to its charge.
3. The algebraic sum of the oxidation numbers of the atoms in the formula of a compound is zero.
4. In compounds, the oxidation number of hydrogen is +1, *except* in metallic hydrides, where its oxidation number is −1.
5. In compounds, the oxidation number of oxygen is −2, *except* in peroxides, where its oxidation number is −1. In compounds with fluorine, oxygen is the less electronegative element and has a positive oxidation number, +2.
6. In combinations involving nonmetals, the oxidation number of the less electronegative element is positive, and that of the more electronegative element is negative.
7. The algebraic sum of the oxidation numbers of the atoms in the formula of a polyatomic ion is equal to its charge.

SUMMARY

The electrons in an incomplete highest energy level of an atom, the valence electrons, usually fix the number of other atoms that may combine with it. In forming compounds from the elements, valence electrons are usually either transferred from the highest energy level of one atom to the highest energy level of another atom or shared among the highest energy levels of the combining atoms. This transfer or sharing of electrons produces chemical bonds. Electron transfer results in ionic bonding, and electron sharing produces covalent bonding. The process of electron transfer is always exothermic; electron sharing is usually exothermic. Bond energy is the energy required to break chemical bonds and form neutral atoms.

A chemical formula is a shorthand method of using chemical symbols to represent the composition of a substance. An empirical formula indicates the simplest whole-number ratio of the kinds of atoms in a compound. A molecule is the smallest chemical unit of a substance that is capable of stable independent existence. A molecule is also the neutral particle that results from the covalent bonding of atoms. A formula that indicates the actual composition of a molecule is called a molecular formula.

The oxidation state of an element is represented by a positive or negative number called an oxidation number. A chemical reaction in which an element attains a more positive oxidation state is called oxidation. The particle whose oxidation state becomes more positive is said to be oxidized. A chemical reaction in which an element attains a more negative oxidation state is called reduction. The particle whose oxidation state becomes more negative is said to be reduced. The substance that is oxidized is the reducing agent. The substance that is reduced is the oxidizing agent.

Positive ions are called cations; they are smaller than the atoms from which they are formed. Negative ions are called anions; they are larger than the atoms from which they are formed.

Hybridization is the combining of two or more orbitals of nearly the same energy into new orbitals of equal energy. Some molecular shapes can be explained by such hybridization.

A covalent bond in which there is an equal attraction for the shared electrons and a resulting balanced distribution of charge is called a pure, or nonpolar, covalent bond. A covalent bond in which there is an unequal attraction for the shared electrons and a resulting unbalanced distribution of charge is called a polar covalent bond. The attraction of an atom for the shared electrons forming a bond between it and another atom is called electronegativity. Electronegativity differences between atoms account for the different kinds of bonds.

Molecules with a uniform exterior electron distribution are nonpolar molecules. Molecules with unbalanced exterior electron distributions are polar molecules, or dipoles.

The concept of resonance is used to explain the structure of a molecule whose bond characteristics cannot be described adequately by a single electron-dot formula.

Some covalently bonded groups of atoms act like single atoms in forming ions. They are called polyatomic ions.

VOCABULARY

anion
bond energy
cation
chemical formula
covalent bond
diatomic
dipole
electron-dot formula
electronegativity
empirical formula
hybridization
ionic bond
kernel
molecular formula
molecule
monatomic
nonpolar covalent bond
oxidation
oxidation number
oxidizing agent
polar covalent bond
polyatomic ion
reducing agent
reduction
resonance
resonance hybrid
structural formula
valence electron

QUESTIONS

Group A

1. Define: (*a*) *valence electrons.* (*b*) *kernel.* *A1*
2. In what two ways do valence electrons form chemical bonds? *A2a, A3a*
3. (*a*) Name the two basic types of chemical bonding. (*b*) What kind of particle is formed by each type? *A2a, A3a*
4. (*a*) When one atom bonds to another atom, what type of highest-energy-level electron configuration does each usually attain? (*b*) Why is this electron configuration chemically stable? *A2b*
5. What is (*a*) an exothermic energy change? and (*b*) what is an endothermic energy change? (*c*) During which type(s) of bonding does each occur? *A3a*
6. Define: (*a*) *chemical formula.* (*b*) *empirical formula.* (*c*) *molecular formula.* *A1*
7. Identify the kind of formula used to represent the composition of (*a*) an ionic compound; (*b*) a covalent compound. *A3a*
8. Draw an electron-dot symbol for (*a*) a lithium atom; (*b*) a lithium ion. *A3b, A5a*
9. Draw an electron-dot symbol for (*a*) an oxygen atom; (*b*) an oxide ion. *A3b, A5a*
10. Using electron-dot symbols, represent a compound of lithium and oxygen. *A5a*
11. In writing a chemical equation, which substances usually appear (*a*) left of the "yields" sign? (*b*) right of the "yields" sign? *A2c*
12. Define: (*a*) *oxidation.* (*b*) *reduction.* *A1*
13. In a reaction involving oxidation and reduction, (*a*) why is the substance undergoing oxidation called the reducing agent? (*b*) why is the substance undergoing reduction called the oxidizing agent? *A6*
14. What is (*a*) a cation? (*b*) an anion? *A1*
15. Why is a bromine atom larger than a chlorine atom? *A2e*
16. Why is a strontium ion larger than a magnesium ion? *A2e*

17. For a fluorine molecule, draw: (*a*) its orbital notation; (*b*) its electron-dot formula; (*c*) its molecular formula. *A5a, A5b*

18. Define: (*a*) *atom*. (*b*) *molecule*. (*c*) For what group of elements are these particles considered identical? *A1, A4a*

19. Distinguish between a symbol and a chemical formula. *A1*

20. From a comparison of the bond energies of H_2, N_2, O_2, F_2, Cl_2, Br_2, and I_2, (*a*) which is the most stable molecule? (*b*) which is the least stable molecule? *A2f*

21. (*a*) What is hybridization? (*b*) How are the four tetrahedrally oriented sp^3 bonds of a carbon atom produced? *A7a*

22. Define: *electronegativity*. *A1*

23. What percentage of ionic character is characteristic of (*a*) ionic bonds? (*b*) pure covalent bonds? (*c*) polar covalent bonds? *A2j*

24. Assuming chemical union between the following pairs, indicate which element would have the positive oxidation number: (*a*) hydrogen-potassium; (*b*) chlorine-bromine; (*c*) fluorine-oxygen; (*d*) silicon-hydrogen; (*e*) hydrogen-sulfur. *A9, A10*

25. Assign oxidation numbers to each element in these compounds: (*a*) Fe_2O_3; (*b*) H_2SO_4; (*c*) $NaNO_2$; (*d*) $KClO_3$; (*e*) $Ca(OH)_2$. *A9*

26. Under what circumstances is the concept of resonance used? *A7b*

27. (*a*) Name the gaseous elements that have diatomic molecules. (*b*) By what symbol or formula is each represented? *A4b*

28. (*a*) Name the gaseous elements that have monatomic molecules. (*b*) By what symbol or formula is each represented? *A4a*

29. How many atoms of each element are represented by the following formulas: sugar, $C_{12}H_{22}O_{11}$; vinegar, $HC_2H_3O_2$; marble, $CaCO_3$; borax, $Na_2B_4O_7$; soap, $C_{17}H_{35}COONa$? *A3a*

30. What does each of the following represent? (*a*) Ne; (*b*) $3Cl_2$; (*c*) H_2F_2; (*d*) $5HNO_3$; (*e*) 4Sn; (*f*) $2Na^+I^-$; (*g*) Co; (*h*) CO; (*i*) $3O_2$; (*j*) $2O_3$. *A3a*

Group B

31. Draw the orbital notation for (*a*) a potassium atom; (*b*) a potassium ion. *A3b, A5b*

32. Draw the orbital notation for (*a*) a sulfur atom; (*b*) a sulfide ion. *A3b, A5b*

33. Using orbital notation, show how an ionic compound of potassium and sulfur is formed. *A5b*

34. Explain why a strontium atom is smaller than a rubidium atom. *A2e*

35. Explain why an oxide ion is larger than a fluoride ion. *A2e*

36. For a hydrogen bromide molecule, HBr, draw: (*a*) its orbital notation; (*b*) its electron-dot formula. *A5a, A5b*

37. Hydrogen and sulfur form a simple molecular compound. Represent this compound using (*a*) electron-dot notation; (*b*) orbital notation. (*c*) What is the probable molecular formula of this compound? (*d*) What is its probable shape? *A3e, A5a, A5b*

38. (*a*) Draw the electron-dot symbol for phosphorus. (*b*) What is the probable shape of a PH_3 molecule? *A5a*

39. What is the oxidation number of chromium in the following: (*a*) potassium dichromate, $K^+{}_2Cr_2O_7{}^{--}$; (*b*) chromium(III) chloride, $Cr^{+++}Cl^-{}_3$? (*c*) If chromium(III) chloride is one of the products of a reaction in which potassium dichromate was one of the reactants, what kind of change has chromium undergone? (*d*) What name is given to chromium in this change? *A2d, A6, A9*

40. What is the oxidation number of iron (*a*) in $Fe^{++}SO_4{}^{--}$; (*b*) in $Fe^{+++}Cl^-{}_3$? (*c*) What process occurs when Fe^{++} is converted to Fe^{+++}? *A2d, A6, A9*

41. What is the oxidation number of manganese (*a*) in $K^+MnO_4{}^-$; (*b*) in $Ba^{++}MnO_4{}^{--}$; (*c*) in MnO_2, a dioxide; (*d*) in $Mn^{++}SO_4{}^{--}$? *A2d, A6, A9*

42. The four oxygen acids of chlorine are hypochlorous acid, $HClO$; chlorous acid, $HClO_2$; chloric acid, $HClO_3$; and perchloric acid, $HClO_4$. What is the oxidation number of chlorine in each acid? *A2d, A9*

43. What is the oxidation number of each element in these polyatomic ions? (*a*) $SO_3{}^{--}$; (*b*) $SO_4{}^{--}$; (*c*) $NO_2{}^-$; (*d*) $NO_3{}^-$; (*e*) $CO_3{}^{--}$; (*f*) $HCO_3{}^-$. *A2d, A9*

44. Oxygen atoms form a triatomic molecule, O_3, called ozone. (*a*) Draw an electron-dot formula for this molecule that satisfies the octet rule. (*b*) Is this the only formula you can draw for this molecule that satisfies the octet rule? (*c*) What is the explanation of the actual structure of an ozone molecule? *A5a, A7b*

45. Draw three different possible electron-dot formulas for the nitrate ion. *A8*

PROBLEMS

Group A

1. Calculate the percentage of ionic character of each of these bonds and classify the bond as ionic, polar covalent, or pure covalent. (*a*)K—Cl; (*b*) C—O; (*c*) Na—O; (*d*) C—H; (*e*) Br—Br; (*f*) Ca—S; (*g*) Li—H; (*h*) Cu—S; (*i*) N—N; (*j*) Hg—O. *A10, B1*

2. From bond-energy data, calculate the energy change for this reaction. Is the change exothermic or endothermic? *A2c, B3*

$$\frac{1}{2} \text{ mole } H_2 + \frac{1}{2} \text{ mole } Br_2 \rightarrow 1 \text{ mole } HBr$$

3. Using bond-energy data, determine the amount and kind of energy change for the following reaction. *A2c, B3*

$$1 \text{ mole } H_2 + \frac{1}{2} \text{ mole } O_2 \rightarrow 1 \text{ mole } H_2O$$

4. From bond-energy data, calculate the amount and kind of energy change for the following reaction.

$$1 \text{ mole } CCl_4 + 2 \text{ moles } HF \rightarrow 1 \text{ mole } CCl_2F_2 + 2 \text{ moles } HCl$$

When this reaction occurs, imagine that for each CCl_4 molecule reacting, two C—Cl bonds in each molecule of CCl_4 break, two H—F bonds break, two C—F bonds in each molecule of CCl_2F_2 form, and two H—Cl bonds form. *A2c, B3*

Group B

5. Calculate the percentage of ionic character of the bonds and also consider the geometry of the structure. Then classify each of the following as ionic crystal, polar covalent molecule, or nonpolar covalent molecule. (*a*) $MgCl_2$; (*b*) CF_4, consisting of a central carbon atom and four symmetrically arranged fluorine atoms; (*c*) HBr; (*d*) SO_2, consisting of an oxygen atom, a sulfur atom, and a second oxygen atom arranged in a bent molecule; (*e*) P_4. *A3a, A3e, A10, B1*

6. Calculate the energy change for the following reaction. *A2c, B2*

$$1 \text{ mole } Mg + 2 \text{ moles } Br \rightarrow 1 \text{ mole } Mg^{++}Br^{-}{}_2$$

As in Section 6.6, this reaction may be assumed to consist of three separate reactions.

$$1 \text{ mole } Mg + \text{energy} \rightarrow 1 \text{ mole } Mg^{++} + 2 \text{ moles } e^-$$

The energy required is the sum of the ionization energies for removing the first and second electrons from a magnesium atom. Obtain energy data from Chapter 5.

$$2 \text{ moles } Br + 2 \text{ moles } e^- \rightarrow 2 \text{ moles } Br^- + \text{energy}$$

The energy released is the electron affinity of *two* moles of bromine atoms. Obtain energy data from Chapter 5.

$$1 \text{ mole } Mg^{++} + 2 \text{ moles } Br^- \rightarrow 1 \text{ mole } Mg^{++}Br^{-}{}_2 + 587 \text{ kcal}$$

Set up these three equations in suitable form and do the required calculations. Is the overall reaction exothermic or endothermic?

7. From bond-energy data, determine the kind and amount of energy change for the following reaction. *A2c, B3*

$$1 \text{ mole } H_2 + 1 \text{ mole } O_2 \rightarrow 1 \text{ mole } H_2O_2$$

Assume four reaction steps: (1) breaking bonds in H_2 molecules; (2) breaking bonds in O_2 molecules; (3) combining of H atoms and O atoms into H—O groups; (4) combining of H—O groups into H_2O_2 molecules by forming O—O bonds between them.

8. From bond-energy data, determine the kind and amount of energy change for the following reaction. *A2c, B3*

$$1 \text{ mole } N_2 + 2 \text{ moles } H_2 \rightarrow 1 \text{ mole } N_2H_4 \text{ (hydrazine)}$$

Jaroslav Heyrovský (Czechoslovakian) received the Nobel Prize in chemistry in 1959 for discovering and developing polarography, a method used for chemical analysis. The polarographic method involves applying a voltage to a cell having a dropping mercury electrode and then measuring the galvanometer deflections. This procedure provides much information about the composition of a solution and the processes taking place in it. This method is very sensitive and can detect a millionth of a mole in 0.1 mL of solution. Heyrovský became director of his country's Polarography Institute in 1950.

Chemical Composition

7

GOALS

In this chapter, you will gain an understanding of: • writing chemical formulas • naming compounds from their formulas • molecular and empirical formulas • formula and molecular weights • the calculation of percentage composition of a compound • the relationship between atomic theory, law of definite composition, law of multiple proportions, and percentage composition • the mole concept • the determination of empirical and molecular formulas

7.1 Common ions and their charges The formula of a compound is useful only if it correctly represents the composition of the substance. The composition of a compound is determined by chemical analysis. Its formula is then derived from the analysis data by applying concepts from atomic theory and chemical bonding. This procedure for determining the formulas of compounds is discussed in Section 7.12.

It is possible to write the formulas of many common compounds without composition data from chemical analyses. For ionic compounds, this is done by using the positively and negatively charged ions that constitute these compounds. A list of common ions and their charges is presented in Table 7-1. To become proficient in writing the formulas of compounds correctly, the ions and their charges listed in this table should be memorized.

Observe that, with the exception of ammonium ions and mercury(I) ions, all positively charged ions listed in Table 7-1 are monatomic. They have the same names as the elements from which they are derived. In some cases the positive or negative charge on a monatomic ion is directly related to the position of the element in the periodic table. The elements of Groups I, II, and VII are examples.

The formula of a compound represents the relative number of atoms of each element present.

Recall that atoms of the Group I metallic elements are characterized by the presence of a single, unpaired *s* electron in the highest energy level. First ionization energies in this group are low; second ionization energies are much higher. This suggests that the single *s* electron is removed with relative ease from a Group I atom, leaving a +1 monatomic ion with the stable electron configuration of the preceding noble gas.

Representative ionization energies are listed in Table 5-2.

Table 7-1
COMMON IONS AND THEIR CHARGES

+1	+2	+3
ammonium, NH_4^+	barium, Ba^{++}	aluminum, Al^{+++}
copper(I), Cu^+	calcium, Ca^{++}	chromium(III), Cr^{+++}
potassium, K^+	copper(II), Cu^{++}	iron(III), Fe^{+++}
silver, Ag^+	iron(II), Fe^{++}	
sodium, Na^+	lead(II), Pb^{++}	
	magnesium, Mg^{++}	
	mercury(I), Hg_2^{++}	
	mercury(II), Hg^{++}	
	nickel(II), Ni^{++}	
	zinc, Zn^{++}	
−1	**−2**	**−3**
acetate, $C_2H_3O_2^-$	carbonate, CO_3^{--}	phosphate, PO_4^{---}
bromide, Br^-	chromate, CrO_4^{--}	
chlorate, ClO_3^-	dichromate, $Cr_2O_7^{--}$	
chloride, Cl^-	oxide, O^{--}	
fluoride, F^-	peroxide, O_2^{--}	
hydrogen carbonate, HCO_3^-	sulfate, SO_4^{--}	
hydrogen sulfate, HSO_4^-	sulfide, S^{--}	
hydroxide, OH^-	sulfite, SO_3^{--}	
iodide, I^-		
nitrate, NO_3^-		
nitrite, NO_2^-		

Atoms of the Group II metallic elements have two paired *s* electrons in the highest energy level. Both first and second ionization energies are relatively low; third ionization energies are much higher. With the loss of its two *s* electrons, a Group II element forms a +2 monatomic ion with the stable electron configuration of the preceding noble gas.

Review electron affinity in Section 5.9.

Atoms of the Group VII nonmetallic elements have seven highest-energy-level electrons and a high positive electron affinity. The acquisition of a single electron by a Group VII atom results in a −1 monatomic ion with the stable configuration of the following noble gas.

The atoms of some metallic elements form ions in more than one oxidation state. For example, copper forms Cu^+ and Cu^{++} ions. The name of each copper ion includes its oxidation number as a Roman numeral in parentheses. The copper(I) ion is simply called the "copper-one" ion and is represented by Cu^+. The copper(II) ion is called the "copper-two" ion and is represented by Cu^{++}. This Roman numeral notation is not used with metals that form ions in only a single oxidation state.

An older system for distinguishing the metallic ions of an element formed in different oxidation states is still in limited use. By means of an appropriate suffix, the oxidation states are identified as a lower or higher state. The suffix *-ous* is used for the lower state and *-ic* for the higher state.

If the symbol for the element is derived from its Latin name, the Latin root may be used. For example, Fe, the symbol for iron, is derived from the Latin *ferrum*. The iron(II) ion, Fe^{++}, becomes the ferr*ous* ion. The iron(III) ion, Fe^{+++}, becomes the ferr*ic* ion. This older system for naming metallic ions is not used in this book. However, some common name equivalents are listed in Table 7-2 for comparison purposes.

Many of the common negative ions in Table 7-1 are polyatomic, as is the positive ammonium ion. They are covalently bonded groups of atoms that form ions just as single atoms do. It will be helpful to review polyatomic ions and their electron-dot configurations in Section 6.23. Refer also to the table inside the front cover or to Appendix Table 6 for a more extensive list of common ions and their charges.

7.2 Writing chemical formulas As stated above, the formulas for many *ionic* compounds can easily be derived from the table of ions and their charges. It is not necessary to become involved with the details of atomic structure and chemical bonding to do this. The first step in writing such a formula is to recognize the positive and negative ions named in the compound. Using this information, the ion notations (including their charges) are written in the order named. Then the number of each kind of ion is adjusted as required to provide *total* positive and negative ionic charges of equal magnitudes. This procedure is called the *ion-charge* method of writing formulas. Its use depends on a compound having been named systematically on the basis of its composition. The following examples illustrate the ion-charge procedure.

Table 7-2
METALLIC ION NAME EQUIVALENTS

Old system		*New system*	
chromic	Cr^{+++}	chromium(III)	Cr^{+++}
cobaltous	Co^{++}	cobalt(II)	Co^{++}
cobaltic	Co^{+++}	cobalt(III)	Co^{+++}
ferrous	Fe^{++}	iron(II)	Fe^{++}
ferric	Fe^{+++}	iron(III)	Fe^{+++}
cuprous	Cu^{+}	copper(I)	Cu^{+}
cupric	Cu^{++}	copper(II)	Cu^{++}
mercurous	Hg^{+}	mercury(I)	Hg_2^{++}
mercuric	Hg^{++}	mercury(II)	Hg^{++}
plumbous	Pb^{++}	lead(II)	Pb^{++}
plumbic	Pb^{++++}	lead(IV)	Pb^{++++}
stannous	Sn^{++}	tin(II)	Sn^{++}
stannic	Sn^{++++}	tin(IV)	Sn^{++++}

Sodium chloride consists of positive sodium ions and negative chloride ions. The sodium ion is represented by Na^+, and the chloride ion is represented by Cl^-. When writing formulas for ionic compounds, *the total charge of the first (positive) part of the compound must be equal and opposite to the total charge of the second (negative) part of the compound.* The total charge for an ion in the formula is the product of the charge of the ion and the number of that ion taken. For the sodium chloride example, the charge of one sodium ion is equal and opposite to the charge of one chloride ion. The formula for sodium chloride is simply *NaCl*, indicating that there is one of each kind of ion represented.

The formula for *calcium chloride* is derived in the same manner. The calcium ion is represented by Ca^{++}, and the chloride ion is Cl^-. The total charge of the negative part of the compound must be equal to that of the positive part. Thus, two chloride ions are needed with one calcium ion. Observe that one calcium ion has a charge of +2, one chloride ion has a charge of −1, and the total charge of two chloride ions is −2. The formula is written as $CaCl_2$. The subscript $_2$ indicates that the composition of calcium chloride is two chloride ions per calcium ion. No subscript is required with Ca since only one Ca^{++} is represented in the formula $CaCl_2$. When no subscript is written in the formula, the subscript $_1$ is understood.

The compound named *iron(III) oxide* consists of iron(III) ions, Fe^{+++}, and oxide ions, O^{--}. The total positive charge and total negative charge represented in the formula for iron(III) oxide must be equal. The lowest common multiple of 3 (Fe^{+++}) and 2 (O^{--}) is 6. The formula must indicate a total positive charge of +6 and a total negative charge of −6. Therefore, 2 Fe^{+++} ions and 3 O^{--} ions are required in the formula. Iron(III) oxide has the formula $Fe^{+++}{}_2O^{--}{}_3$, or simply Fe_2O_3. Conversely, if the oxide ion represents oxygen atoms in the −2 oxidation state, the formula Fe_2O_3 must signify that iron ions present are in the +3 oxidation state.

What is the formula for *aluminum bromide?* The aluminum ion is Al^{+++} and the bromide ion is Br^-. Three Br^- ions are needed to balance the positive charge of one Al^{+++} ion. The formula is $AlBr_3$.

Observe that this ion-charge method of writing formulas yields empirical formulas. An empirical formula shows the simplest whole-number ratio of atoms in a compound.

Figure 7-1. Jöns Jakob Berzelius introduced letter symbols for elements early in the nineteenth century. He conducted meticulous chemical analyses that resulted in the first systematic determination of atomic weights.

7.3 Writing the formulas for other compounds The formula for *lead(II) sulfate* is easy to write. The lead(II) ion is Pb^{++}, the sulfate ion is SO_4^{--}, and the charges are equal and opposite. Thus, the empirical formula for lead(II) sulfate has just one Pb^{++} ion and one SO_4^{--} ion, $PbSO_4$.

In the formula for *magnesium hydroxide*, the magnesium ion is Mg^{++} and the hydroxide ion, a polyatomic ion, is OH^-. Two

OH^- ions are required for the negative charge to be equal and opposite to the positive charge of one magnesium ion. More than one polyatomic ion must be represented in the formula. When writing the formula for magnesium hydroxide, parentheses are used to enclose the two-atomed hydroxide ion that is to be taken twice. Then the subscript $_2$ is written outside the parentheses, $(OH)_2$. This clearly shows that the *entire* OH^- ion is taken twice. The complete formula for magnesium hydroxide is $Mg(OH)_2$, not $MgOH_2$. This incorrect formula, $MgOH_2$, represents one oxygen atom and two hydrogen atoms. *Parentheses are not used when only one polyatomic ion is required in a formula.* For example, the formula for *potassium hydroxide*, KOH, represents one K^+ ion and one OH^- ion. The parentheses would serve no purpose in this formula.

Polyatomic ions are covalent structures with an electric charge. Review Section 6.23.

As another example, consider the formula for *lead(II) acetate.* The lead(II) ion is Pb^{++}, and the acetate ion is the polyatomic $C_2H_3O_2^-$. For each lead ion, two acetate ions are required in the formula. Following the scheme used in writing the formula for magnesium hydroxide, the formula for lead(II) acetate becomes $Pb(C_2H_3O_2)_2$. Observe that the *two* $C_2H_3O_2^-$ ions required are represented by $(C_2H_3O_2)_2$.

If a metallic atom forms more than one ion, a Roman numeral is used in the name of the ion to identify its oxidation state.

Ammonium sulfate has two polyatomic ions in its formula. The ammonium ion is NH_4^+, and the sulfate ion is SO_4^{--}. In order to achieve total equality of + and − charges, two NH_4^+ ions are required with one SO_4^{--} ion. Of course, two ammonium ions are represented in the formula by enclosing the NH_4^+ ion in parentheses with a subscript $_2$ outside. The formula is then written as $(NH_4)_2SO_4$.

Finally, consider the formula for *tin(IV) chromate.* (The metallic ions formed by atoms of tin are listed in the table inside the front cover and in Appendix Table 6.) The tin(IV) ion is Sn^{++++}, and the chromate ion is CrO_4^{--}. Again, the total positive and negative charges represented in the formula must be equal. Because the positive charge is supplied in units of 4 as Sn^{++++} and the negative charge is supplied in units of 2 as CrO_4^{--}, the equal-charge requirement is satisfied with a ratio of Sn^{++++} ions to CrO_4^{--} ions of 2 : 4. The formula can be written as $Sn_2(CrO_4)_4$. However, an empirical formula requires the smallest whole-number ratio of constituent units. Since the ratio 2 : 4 equals 1 : 2, the smallest whole-number ratio of Sn^{++++} ions to CrO_4^{--} ions in the tin(IV) chromate is 1 : 2. The formula is correctly written as $Sn(CrO_4)_2$.

It must be recognized that the ion-charge method has limitations. A formula can give no more information than that required to write it. It is possible to write the formula for a compound and then learn that such a compound does not exist! On the other hand, there are many covalent compounds that *do* exist but whose formulas cannot be written using the ion-charge method.

Table 7-3
GREEK NUMERICAL PREFIXES

Number	*Prefix*
1	mono-
2	di-
3	tri-
4	tetra-
5	penta- or pent-
6	hexa-
7	hepta-
8	octa-
9	*nona-
10	deca-

*(Latin)

Table 7-4
NITROGEN-OXYGEN SERIES OF BINARY COMPOUNDS

Formula	*Name*
N_2O	dinitrogen monoxide
NO	nitrogen monoxide
N_2O_3	dinitrogen trioxide
NO_2	nitrogen dioxide
N_2O_5	dinitrogen pentoxide

A Roman numeral used in the name of a metallic ion does not appear in the formula of a compound containing the ion.

The names of binary compounds have -ide *endings.*

7.4 Naming compounds from their formulas The names of many chemical compounds consist of two words: the name of the first part of the formula and the name of the second part. $BaSO_4$ is called barium sulfate (Ba^{++} represents the barium ion; SO_4^{--} represents the sulfate ion). $FeCl_3$ is the formula for iron(III) chloride. Notice that there are two possible oxidation states for iron. One is iron(II), with an oxidation number of +2; the other is iron(III), with an oxidation number of +3. There are 3 chloride ions associated with one iron ion in the formula $FeCl_3$. Therefore, the iron ion has a charge of +++, and the compound is *iron(III) chloride.* $FeCl_2$ is *iron(II) chloride.*

The use of Roman numerals to indicate oxidation states does not always provide simple and useful names for compounds. This system is commonly used with *ionic compounds* in which the metallic ion has different possible oxidation states. Another system, using Greek numeral prefixes, is used with certain *binary covalent* compounds. These prefixes are listed in Table 7-3. ***Binary compounds*** *are compounds that are composed of only two elements.* Some elements form more than one covalent compound with another element. For example, there are two covalent compounds of sulfur and oxygen, SO_2 and SO_3. Nitrogen forms a series of five different compounds with oxygen. The formulas and corresponding names of this series are given in Table 7-4. The names of these compounds must provide a way to distinguish them. Covalent binary compounds are named as follows:

1. The first word of the name is made up of (*a*) a prefix indicating the number of atoms of the first element appearing in the formula, if more than one; and (*b*) the name of the first element in the formula.

2. The second word of the name is made up of (*a*) a prefix indicating the number of atoms of the second element appearing in the formula, if more than one compound of these two elements exists; (*b*) the root of the name of the second element; and (*c*) the suffix *-ide*, which means that *only* the two elements named are present.

Carbon monoxide is written CO. Only one atom of the first element appears in the formula, so no prefix is used with the first word. It consists only of the name of the first element, *carbon*. The prefix *mon-* is used in the second word of the name because only one atom of oxygen appears in this formula, but more than one compound of carbon and oxygen exists. *Ox-* is the root of the name of the element oxygen. It is followed by the suffix *-ide*.

In a similar manner, CO_2 is *carbon dioxide*. The prefix *di-* indicates the presence of two oxygen atoms in the compound. The prefix indicating three is *tri-*; four is *tetra-*; and five is *pent-* or *penta-* (Table 7-3). These prefixes can be used with both the first and second words in the name. Some examples of the use of numerical prefixes are: $SbCl_3$, *antimony trichloride;* CCl_4, *carbon tetrachloride;* and As_2S_5, *diarsenic pentasulfide*. Refer to the table inside the front cover or to Appendix Table 5 for the usual oxidation states of the common elements.

A few negative polyatomic ions have names with an *-ide* suffix. The hydroxide ion, OH^-, and cyanide ion, CN^-, are examples. Compounds formed with either of these polyatomic ions obviously are not binary. Similarly, compounds formed with the positive ammonium ion, NH_4^+, are not binary. All binary compounds end in *-ide*, but not all compounds with *-ide* endings are binary. For example, NaOH, $Ca(CN)_2$, and NH_4Cl have names that end in *-ide*, but they are not binary compounds.

Practice Problems

1. Write the formulas for the following compounds: (*a*) sodium chromate, (*b*) magnesium bromide, (*c*) nickel(II) nitrate, (*d*) calcium phosphate, (*e*) mercury(I) sulfate, (*f*) ammonium carbonate.
   ans. (*a*) Na_2CrO_4, (*b*) $MgBr_2$, (*c*) $Ni(NO_3)_2$, (*d*) $Ca_3(PO_4)_2$, (*e*) Hg_2SO_4, (*f*) $(NH_4)_2CO_3$
2. Name the following compounds: (*a*) AgI, (*b*) $Cu(OH)_2$, (*c*) $Pb(C_2H_3O_2)_2$, (*d*) $(NH_4)_2S$, (*e*) FeO, (*f*) As_2O_3.
   ans. (*a*) silver iodide, (*b*) copper(II) hydroxide, (*c*) lead(II) acetate, (*d*) ammonium sulfide, (*e*) iron(II) oxide, (*f*) diarsenic trioxide

7.5 Significance of chemical formulas You have learned to write formulas for many chemical compounds by using your knowledge of the charges of the ions composing them. When it is known that a substance exists as simple molecules, its formula represents one molecule of the substance. Such a formula is known as a ***molecular formula.***

The molecular structure of some substances is not known. Other substances have no simple molecular structure. For these substances, the formula represents (1) the elements in the substances and (2) the simplest whole-number ratio of the atoms of these elements. In these cases, the formula is called an ***empirical,*** or simplest, ***formula.***

An empirical formula states the simplest whole-number ratio of atoms of the constituent elements in a compound.

In Section 6.20 you learned how the approximate percentage of ionic character of the chemical bonds between atoms is determined from the electronegativities of the bonded atoms. Furthermore, you learned that bonds with more than 50% ionic character are considered to be ionic, that bonds with 5% to 50% ionic character are polar covalent, and that bonds with below 5% ionic character are nonpolar covalent.

From these generalizations, one can reasonably assume that *metal-nonmetal* compounds are essentially *ionic* and their formulas are *empirical formulas.* Similarly, *nonmetal-nonmetal* compounds are *molecular* and their formulas are *molecular formulas.*

When no specific information on the structure of a compound is available, one can use these loose guidelines to decide whether a formula is an empirical formula or a molecular formula.

The chemical formula of a compound reveals important information about the compound. For example, hydrogen sulfide has a nonmetal-nonmetal composition. It has the molecular formula H_2S. The formula reveals that each molecule of hydrogen sulfide is composed of *two atoms of hydrogen and one atom of sulfur.*

Observe that the molecular formula H_2S is also its empirical formula.

The atomic weight of hydrogen is 1.0 and the atomic weight of sulfur is 32.1. Hence this molecular formula signifies that the *formula weight of hydrogen sulfide* is 34.1.

2 atoms H × atomic weight	**1.0 =**	**2.0**
1 atom S × atomic weight	**32.1 =**	**32.1**
formula weight H_2S	**=**	**34.1**

*The **formula weight** of any compound is the sum of the atomic weights of all the atoms represented in the formula.*

The compound sodium chloride has the empirical formula NaCl. It is a crystalline solid that has no molecular structure but is composed of an orderly arrangement of sodium and chloride ions. This empirical formula indicates the relative number of atoms (as ions) of each element present in the compound. It shows that there are equal numbers of Na^+ ions and Cl^- ions in any quantity of NaCl. The atomic weight of sodium is 23.0 and that of chlorine is 35.5. Hence the empirical formula denotes a formula weight for sodium chloride of 23.0 + 35.5, or 58.5.

7.6 Molecular weight You have learned that a molecular formula represents one molecule of a substance. The formula H_2S is a molecular formula. It represents one molecule of hydrogen sulfide. The formula weight, 34.1, is then the relative weight of *one molecule* of hydrogen sulfide. *The formula weight of a molecular substance is its **molecular weight.***

In the strictest sense, it is not correct to use the "molecular weight" for a nonmolecular substance. Sodium chloride, for example, is represented by an empirical formula. Empirical formulas have "formula weights." The term *formula weight* is a more general term than *molecular weight* and is preferred. In elementary chemical calculations the distinction is not important.

7.7 Formula weight of a compound To find the total weight of all the members of your chemistry class, you must add the weights of the individual members of the class. Similarly, to find the formula weight of any substance for which a formula is given, you must add the atomic weights of all the atoms represented in the formula. Consider the formula for cane sugar, $C_{12}H_{22}O_{11}$, as an example. Approximate atomic weights found inside the front cover and in Appendix Table 5 can be used.

Formula weight is the sum of the atomic weights of all the atoms represented in a formula.

Number of atoms	Atomic weight	Total weight
12 of C	12.0	12 × 12.0 = 144.0
22 of H	1.0	22 × 1.0 = 22.0
11 of O	16.0	11 × 16.0 = 176.0
formula weight (molecular weight) = 342.0		

Molecular weights are formula weights of molecular substances.

Calcium hydroxide's formula is $Ca(OH)_2$. The subscript $_2$ following the parentheses indicates that there are two polyatomic hydroxide ions per calcium ion. Thus, the formula $Ca(OH)_2$ includes atomic weights of one calcium atom, two oxygen atoms, and two hydrogen atoms.

Atomic weights apply to monatomic ions and atoms alike.

Number of atoms	Atomic weight	Total weight
1 of Ca	40.1	1 × 40.1 = 40.1
2 of O	16.0	2 × 16.0 = 32.0
2 of H	1.0	2 × 1.0 = 2.0
		formula weight = 74.1

The atomic weights of the elements are relative weights based on an atom of carbon-12 that has an assigned value of exactly 12. In the quantitative study of chemical reactions, the atomic weights and formula weights indicate the relative quantities of elements and compounds that combine or react. These quantities can be expressed in any desired units. Thus, atomic weights, formulas, and formula weights are important in chemical calculations.

Practice Problems

1. The formula for sodium hydrogen carbonate (baking soda) is $NaHCO_3$. What is its formula weight? *ans.* 84.0
2. (*a*) Write the formula for magnesium acetate. (*b*) Determine its formula weight. *ans.* (*a*) $Mg(C_2H_3O_2)_2$, (*b*) 142.3

7.8 Percentage composition of a compound Frequently it is important to know the composition of a compound in terms of the *mass percentage* of each element of which the compound is made. You may want to know the percentage of iron in the compound iron(III) oxide. Or, you may want to know the percentage of oxygen in potassium chlorate. This knowledge enables you to determine the amount of potassium chlorate needed to supply enough oxygen for a laboratory experiment.

The chemical formula for a compound can be used directly to determine its formula weight. This is done by adding the atomic weights of all the atoms represented. The formula weight represents *all*, or 100 percent, of the composition of the substance as

Figure 7-2. Students determining the composition of a compound in a laboratory experiment.

indicated by the formula. The part of the formula weight contributed by each element represented in the formula is the product of the atomic weight of that element × the number of atoms of that element. The fractional part due to each element is

$$\frac{\text{(atomic weight} \times \text{number of atoms) of the element}}{\text{formula weight of the compound}}$$

The percentage of each element present in the compound is therefore a fractional part of 100 percent of the compound.

$$\frac{\text{(atomic weight} \times \text{number of atoms) of the element}}{\text{formula weight of the compound}} \times 100\% \text{ of the compound}$$

The mass percentage of an element in a compound can be expressed as

$$\frac{\text{(atomic weight} \times \text{number of atoms) of the element}}{\text{formula weight of the compound}} \times 100\% \text{ of the compound} = \% \text{ element}$$

Consider the compound iron(III) oxide. The formula is Fe_2O_3. What is the percentage of each of the elements present?

1. *Formula weight of* Fe_2O_3

$$\begin{aligned} \text{total atomic weight Fe} &= 2 \times 55.8 = 111.6 \\ \text{total atomic weight O} &= 3 \times 16.0 = 48.0 \\ \hline \text{formula weight } Fe_2O_3 &= 159.6 \end{aligned}$$

2. *Percentage of Fe*

$$\frac{\text{total atomic weight Fe}}{\text{formula weight } \cancel{Fe_2O_3}} \times 100\% \cancel{Fe_2O_3} = \% \text{ Fe}$$

Using three significant figures,

$$\frac{112 \text{ Fe}}{16\bar{0} \text{ } Fe_2O_3} \times 100\% \text{ } Fe_2O_3 = 70.0\% \text{ Fe}$$

3. *Percentage of O*

$$\frac{\text{total atomic weight O}}{\text{formula weight } \cancel{Fe_2O_3}} \times 100\% \cancel{Fe_2O_3} = \% \text{ O}$$

$$\frac{48.0 \text{ O}}{16\bar{0} \text{ } Fe_2O_3} \times 100\% \text{ } Fe_2O_3 = 30.0\% \text{ O}$$

Of course, since there is no third element present, the percentage of oxygen is 100.0% − 70.0% = 30.0%.

Observe that the formulas Fe_2O_3 in step 2 cancel, leaving the symbol Fe in the result. In step 3, the formulas Fe_2O_3 cancel, leaving the symbol O in the result. The symbols remaining are the ones expected for each of these results. Many errors in the solutions to problems in chemistry can be avoided by consistently labeling each quantity properly and solving the expression for both the unit and the numerical magnitude of the result.

As a second example, the percentage composition of a crystallized (hydrated) form of sodium carbonate, $Na_2CO_3 \cdot 10H_2O$, will be determined. The raised dot followed by $10H_2O$ indicates that the crystals contain 10 molecules of water of crystallization (water of hydration) for each 2 Na^+ ions *or* for each CO_3^{--} present.

1. *Formula weight* $Na_2CO_3 \cdot 10H_2O$

	2 Na	2 × 23.0 =	46.0	
	1 C	1 × 12.0 =	12.0	
	3 O	3 × 16.0 =	48.0	
$10H_2O$	20 H	20 × 1.0 =	$2\bar{0}$	$18\bar{0}$
	10 O	10 × 16.0 =	$16\bar{0}$	
		formula weight =	286	

2. *Percentage of Na*

$$\frac{46.0 \text{ Na}}{286 \text{ Na}_2\text{CO}_3 \cdot 10\text{H}_2\text{O}} \times 100\% \text{ Na}_2\text{CO}_3 \cdot 10\text{H}_2\text{O} = 16.1\% \text{ Na}$$

3. *Percentage of C*

$$\frac{12.0 \text{ C}}{286 \text{ Na}_2\text{CO}_3 \cdot 10\text{H}_2\text{O}} \times 100\% \text{ Na}_2\text{CO}_3 \cdot 10\text{H}_2\text{O} = 4.2\% \text{ C}$$

4. *Percentage of O* (in CO_3^{--} ion)

$$\frac{48.0 \text{ O}}{286 \text{ Na}_2\text{CO}_3 \cdot 10\text{H}_2\text{O}} \times 100\% \text{ Na}_2\text{CO}_3 \cdot 10\text{H}_2\text{O} = 16.8\% \text{ O}$$

5. *Percentage of* H_2O

$$\frac{18\bar{0} \text{ H}_2\text{O}}{286 \text{ Na}_2\text{CO}_3 \cdot 10\text{H}_2\text{O}} \times 100\% \text{ Na}_2\text{CO}_3 \cdot 10\text{H}_2\text{O} = 62.9\% \text{ H}_2\text{O}$$

Approximate atomic weights usually have no more than two or three significant figures. The result is not improved by carrying out the computations beyond the precision of the data. Thus the sum of the mass percentages may only approximate 100%.

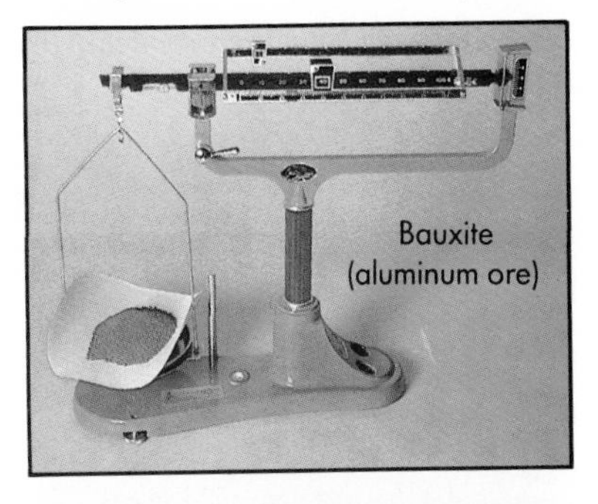

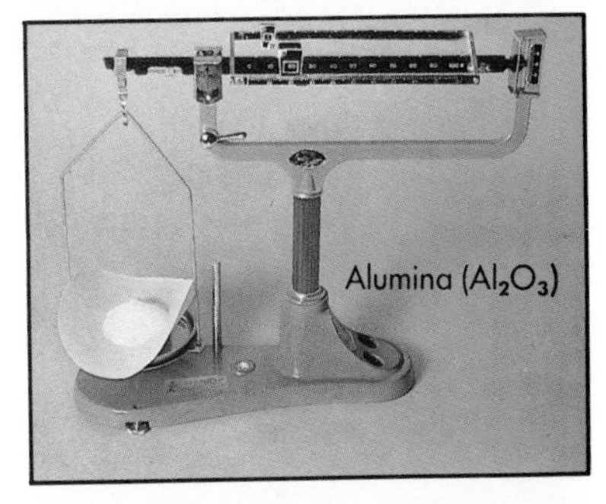

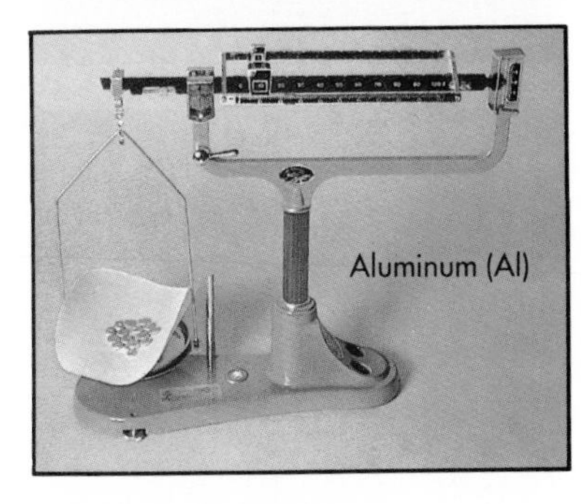

Figure 7-3. A supply of bauxite is 50.0% aluminum oxide (alumina), Al_2O_3. Assuming all the available Al is extracted in the recovery process, how many kg of metallic Al are produced by processing 1.00 metric ton (1.00×10^3 kg) of this bauxite?

7.9 Law of definite composition and the atomic theory The law of definite composition (Section 2.9) states, in effect, that the percentage composition of a chemical compound is constant. This characteristic of constant composition of chemical compounds depends on two facts: (1) the relative mass of the atoms of an element is constant; and (2) the proportion in which atoms are present in a compound is constant. The idea that, under most conditions, atomic weights are constant is basic to the atomic theory. For practical purposes, the atomic weight used for an element is the average relative mass of the naturally occurring mixture of its isotopes. (Review discussion in Section 3.14.) For example, the average relative mass (atomic weight) of the hydrogen atom is 1.0079. That of the chlorine atom is 35.453.

The law of definite composition in its present form may not hold rigorously for certain metallic sulfides and oxides under all reaction conditions.

Elements in compounds are combined in simple whole-number ratios.

The atomic theory explains how atoms combine in a constant proportion by mass. Individual atoms can lose, gain, or share only a definite number of electrons. Therefore, the mass of one element that can combine with a given mass of another element is limited. In forming hydrogen chloride, for example, only one hydrogen atom can combine with one chlorine atom. One hydrogen atom can share only 1 electron. One chlorine atom has room for only 1 electron to complete its octet of 3rd-energy-level electrons. So the ratio of these two atoms that combine can only be 1 : 1. It cannot be 1 : 2, or 3 : 2, or any other ratio. Consequently, a molecule of hydrogen chloride always consists of one atom of hydrogen, atomic weight 1.0079, and one atom of chlorine, atomic weight 35.453. The molecule has a molecular weight of 36.461. Furthermore, hydrogen chloride always contains $\frac{1.0079}{36.461}$ parts by mass of hydrogen, or 2.7643% hydrogen, and $\frac{35.453}{36.461}$ parts by mass of chlorine, or 97.235% chlorine.

Figure 7-4. Compounds formed by the same two elements combined in different proportions may have very different chemical properties. In (top photo), water, H_2O, is unreactive with the green felt. In (bottom photo), hydrogen peroxide, H_2O_2, bleaches the green dye in the felt.

7.10 Law of multiple proportions It is not uncommon for two or more compounds to be composed of the same two elements. Hydrogen and oxygen form a series of two compounds, water (H_2O) and hydrogen peroxide (H_2O_2). The constituent elements, hydrogen and oxygen, are present in each of these two compounds in unvarying but different *mass* proportions.

For water, the composition is about 2 parts hydrogen to 16 parts oxygen, or 1 part hydrogen to 8 parts oxygen. For hydrogen peroxide, the composition is 1 part hydrogen to 16 parts oxygen. Considering quantities of these two compounds *that have the same masses of hydrogen*, the relative masses of the second element (oxygen) are related *as a ratio of small whole numbers.*

Consider, for example,

H_2O	**1 g of H**	**and**	**8 g of O**
H_2O_2	**1 g of H**	**and**	**16 g of O**

The *masses of hydrogen* in this series, 1 g and 1 g, are in the constant ratio of 1 : 1. The *masses of oxygen,* 8 g and 16 g, are in the simple whole-number ratio of 1 : 2. Now observe that the ratio of *atoms of hydrogen* in the series of two compounds is 1 : 1, whereas the ratio of *atoms of oxygen* is 1 : 2.

Iron and chlorine also form a series of two compounds that exhibit similar simple whole-number relationships. Consider

$FeCl_2$	**56 g of Fe**	**and**	**71 g of Cl**
$FeCl_3$	**56 g of Fe**	**and**	**106.5 g of Cl**

The *masses of iron* in this series, 56 g and 56 g, are in the constant ratio of 1 : 1. The *masses of chlorine,* 71 g and 106.5 g, are in the simple whole-number ratio of 2 : 3. Again observe that the ratio of *atoms of iron* in the series is 1 : 1, whereas the ratio of *atoms of chlorine* is 2 : 3. The ***law of multiple proportions*** describes such series of compounds. *If two or more different compounds composed of the same two elements are analyzed, the masses of the second element combined with a fixed mass of the first element can be expressed as a ratio of small whole numbers.*

This law was first proposed by John Dalton in connection with his atomic theory. He recognized the possibility that two kinds of atoms could combine in more than one way and in more than one proportion. But only whole atoms could be involved in such combinations. Therefore, Dalton reasoned, the masses of the second atom combined with fixed masses of the first atom would have to be in the ratio of small whole numbers. This ratio is the same as the ratio of the actual numbers of atoms of the second element combined with a fixed number of atoms of the first element. The information summarized in the law of multiple proportions strongly supports the atomic theory.

Today scientists recognize two facts that support the law of multiple proportions. (1) Some elements exist in more than one oxidation state. (2) Some elements combine in more than one way with another element. Iron can exist in compounds as iron(II) ions with an oxidation number of +2 or as iron(III) ions with an oxidation number of +3. You have seen that iron and chlorine combine to form two different compounds: $FeCl_2$, iron(II) chloride, and $FeCl_3$, iron(III) chloride. The ratio of the numbers of atoms of chlorine combined with a single atom of iron is 2 : 3. As recognized by Dalton, the relative numbers of atoms that combine are proportional to the masses that combine. Therefore, the ratio of the masses of chlorine combined with a fixed mass of iron in these two compounds is also 2 : 3.

Analysis of the composition and molecular structure of water and hydrogen peroxide leads to these electron-dot formulas:

```
               H
H:Ö:         :Ö:Ö:
  H             H
water    hydrogen peroxide
```

These two formulas represent the two ways in which hydrogen and oxygen atoms combine. Each molecule of water contains 2 hydrogen atoms combined with only 1 oxygen atom. Each molecule of hydrogen peroxide contains 2 hydrogen atoms combined with 2 oxygen atoms. The numbers of atoms combined are proportional to the masses combined. So the ratio of the masses of oxygen combined with the same mass of hydrogen in these two compounds is 1 : 2, a ratio of small whole numbers.

Practice Problems

(Refer to Table 7-4.)

1. Based on the constant number of 2 atoms of nitrogen, express the ratios of oxygen atoms through the binary nitrogen-oxygen series. *ans.* N = 2:2:2:2:2; O = 1:2:3:4:5

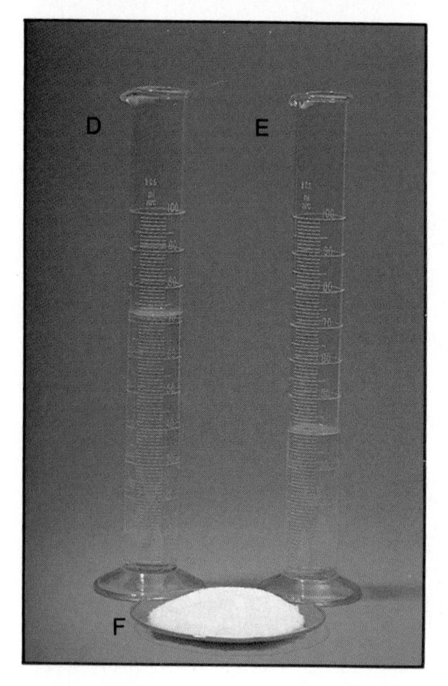

Figure 7-5. A comparison of mole quantities of several substances: (a) 1 mole (18.0 g) of H_2O; (b) 1 mole (27.0 g) of Al; (c) 1 mole (63.5 g) of Cu; (d) 4 moles (72.0 g) of H_2O; (e) 1 mole (32.0 g) of CH_3OH (methanol); and (f) 1 mole (58.5 g) of NaCl.

2. Recognizing that the oxidation state of oxygen through the series is −2, assign oxidation numbers to nitrogen in each compound of the series. *ans.* O: −2 −2 −2 −2 −2; N: +1 +2 +3 +4 +5
3. Research the properties of the nitrogen-oxygen series to determine which member is (*a*) called "laughing gas" and (*b*) a major air pollutant. *ans.* (*a*) compound with 63.6% N; (*b*) compound with 30.4% N

7.11 Mole concept Imagine that all the people on the earth were assigned the task of counting the molecules in a tablespoon of water. If each person counted at the rate of one molecule per second, it would take approximately 8×10^6 years to complete the project. The number of molecules involved staggers the imagination!

Fortunately, chemists are not ordinarily faced with the problem of weighing out a certain number of molecules of a compound or atoms of an element. They do frequently need to weigh out equal numbers of atoms or molecules of different substances. A knowledge of the atomic weights of the elements allows this to be done very simply.

In Sections 3.14 and 3.15, four important quantitative definitions were recognized.

1. The number of carbon-12 atoms in the defined quantity of 12 grams of this nuclide is the *Avogadro number*, 6.02×10^{23}.

2. The amount of substance containing the Avogadro number of any kind of chemical unit is a *mole* of that substance.

3. If the *atomic weight* of an element is expressed in gram units, the quantity is the *gram-atomic weight* of the element.

4. One *gram-atomic* weight of an element contains one *mole* of atoms of the element.

It is evident from these definitions that 12 g of carbon (one gram-atomic weight) consists of 1 mole of carbon atoms, the Avogadro number of atoms. Similarly, 1.0 g of hydrogen (one gram-atomic weight) consists of 1 mole of hydrogen atoms, the Avogadro number of atoms. These two quantities, 12 g of carbon and 1.0 g of hydrogen, contain the same number of atoms. Thus, any given masses of carbon and hydrogen that are in the ratio of 12 : 1 (the ratio of their atomic weights) must also have the same number of atoms.

To obtain 5.0 moles of oxygen atoms, 5.0×16 g $= 8\bar{0}$ g of oxygen must be measured. To obtain 5.0 moles of sulfur atoms, $5.0 \times 3\bar{2}$ g $= 160$ g of sulfur must be measured. These two quantities, $8\bar{0}$ g of oxygen and 160 g of sulfur, contain the same number of atoms. Observe that the ratio of these masses equals the ratio of the atomic weights of the elements

$$\frac{8\bar{0}\text{ g}}{160\text{ g}} = \frac{16}{32} = \frac{1}{2}$$

Using similar logic and the atomic weights of other elements, the following important generalization can be recognized: *If the mass quantities of two elements are in the same ratio as their atomic weights, they contain the same number of atoms.* Recall that quantities of substances are measured in gram (mass) units. These masses of substances are sometimes referred to as "weights" because "weighing" methods are used to determine them. However, you should remember that quantities measured in gram units are, in fact, mass quantities. Atomic weights that express the average relative masses of atoms of different elements are most useful when they are expressed in gram units. *The **gram-atomic weight** of an element is the mass of one mole of atoms of that element.* Thus, the mass of one mole of O atoms is 16 g of oxygen, of one mole of S atoms is 32 g of sulfur, and of one mole of Fe atoms is 56 g of iron.

The mole concept is extended to include the molecules of molecular substances. Oxygen, O_2, is one of several elemental gases composed of *diatomic molecules*. One mole of O_2 molecules consists of two moles of O atoms. Two moles of O atoms have a mass of 2×16 g = 32 g. Consequently, one mole of O_2 molecules must have this same mass of 32 g. This value is the molecular weight of O_2, 32, expressed in gram units. *The mass of a molecular substance expressed in grams equal to its molecular weight is its **gram-molecular weight**.* An extension of this logic leads to the observation that one mole of H_2 molecules has a mass of 2.0 g, and one mole of Cl_2 molecules has a mass of 71.0 g. Thus, *the gram-molecular weight of a diatomic molecular element is the mass of one mole of molecules of that element.* Observe that one mole of O atoms has a mass of 16 g, but one mole of O_2 molecules has a mass of 32 g. Also observe that 16 g of O atoms, 32 g of O_2 molecules, 1.0 g of H atoms, 2.0 g of H_2 molecules, 35.5 g of Cl atoms, and 71.0 g of

The gram-molecular weight is the gram-formula weight of a molecular substance.

Table 7-5
MASS-MOLE RELATIONSHIPS

1 gross = 144 objects

If you compare		The masses of the gross-quantity packages are different because the mass of each kind of object is different. However, the ratios between the masses of the packages, 1440 g: 720 g: 43200 g, are the same as the ratios between the masses of the individual objects, 10.0 g: 5.0 g: $30\overline{0}$ g, since each package contains the same number of objects.
144 erasers @ 10.0 g each	= 1440 g	
144 pencils @ 5.0 g each	= 720 g	
144 tablets @ $30\overline{0}$ g each	= 43200 g	

1 mole = 6.02×10^{23} particles

If you compare		The mass of the mole varies just as the mass of the gross varies. It depends upon what the particles are. However, the ratios between the masses of the moles are the same as the ratios between the masses of the individual particles since each mole contains the same number of particles.
6.02×10^{23} helium atoms	= 4.0 g	
6.02×10^{23} hydrogen molecules	= 2.0 g	
6.02×10^{23} water molecules	= 18.0 g	
6.02×10^{23} NaCl formula units	= 58.5 g	

Cl_2 molecules all consist of the same number of *particles*—atoms or molecules, as the case may be.

One mole of H_2O molecules contains two moles of H atoms and one mole of O atoms. Two moles of H atoms has a mass of 2.0 g; one mole of O atoms has a mass of 16 g. Thus, by addition, one mole of H_2O molecules has a mass of 18 g. This mass is the gram-molecular weight of water. One mole of methane, CH_4, contains one mole of C atoms and four moles of H atoms. One mole of C atoms has a mass of 12 g; four moles of H atoms has a mass of 4 g. One mole of CH_4 molecules has a mass of 16 g. This is its gram-molecular weight. From these examples, it is evident that the *gram-molecular weight of a molecular substance is the mass of one mole of molecules of the substance.* There is the same number of molecules in 32 g of O_2, 18 g of H_2O, and 16 g of CH_4. One-mole quantities of all molecular substances contain the same number of molecules, the Avogadro number, 6.02×10^{23} molecules. It follows that *any mass quantities* of oxygen, water, and methane that are in the ratio 32 : 18 : 16 *contain equal numbers of molecules.*

Gram-formula weights are formula weights expressed in grams.

The mole concept is also extended to include ionic substances that are expressed by empirical formulas. For example, the *gram-formula weight* (formula weight in grams) of sodium chloride is 23.0 g + 35.5 g = 58.5 g. One mole of sodium atoms (as Na^+ ions) has a mass of 23.0 g; one mole of chlorine atoms (as Cl^- ions) has a mass of 35.5 g. Thus, 58.5 g is the mass of one mole of sodium chloride. One mole of sodium chloride contains one mole of Na^+ ions *and* one mole of Cl^- ions.

The symbol for an element not only represents one atom of the element, but also the mass of one mole of atoms of that element. A formula for a diatomic molecule of an element represents one molecule of that element. The formula also represents the mass of one mole of these molecules. Similarly, the formula for a compound represents the mass of one mole of that compound as well as its composition. These mole relationships are summarized in Table 7-6.

7.12 Calculating empirical formulas The empirical formula of a compound consists of the symbols for the constituent elements in their smallest whole-number ratio. In Section 7.8 the percentage composition of a compound is determined from its formula. In this section the formula of a compound is determined from its percentage composition, obtained experimentally by chemical analysis. Most empirical formulas of compounds are originally determined by this analysis method. A compound is analyzed to identify the elements present and to determine their mass ratios or the percentage composition. The empirical formula is then calculated from these data. By knowing the mass per mole of atoms (gram-atomic weight) of each element, the mass ratios are reduced to atom ratios.

The simplest whole-number ratio of the atoms of different elements in a compound provides its empirical formula.

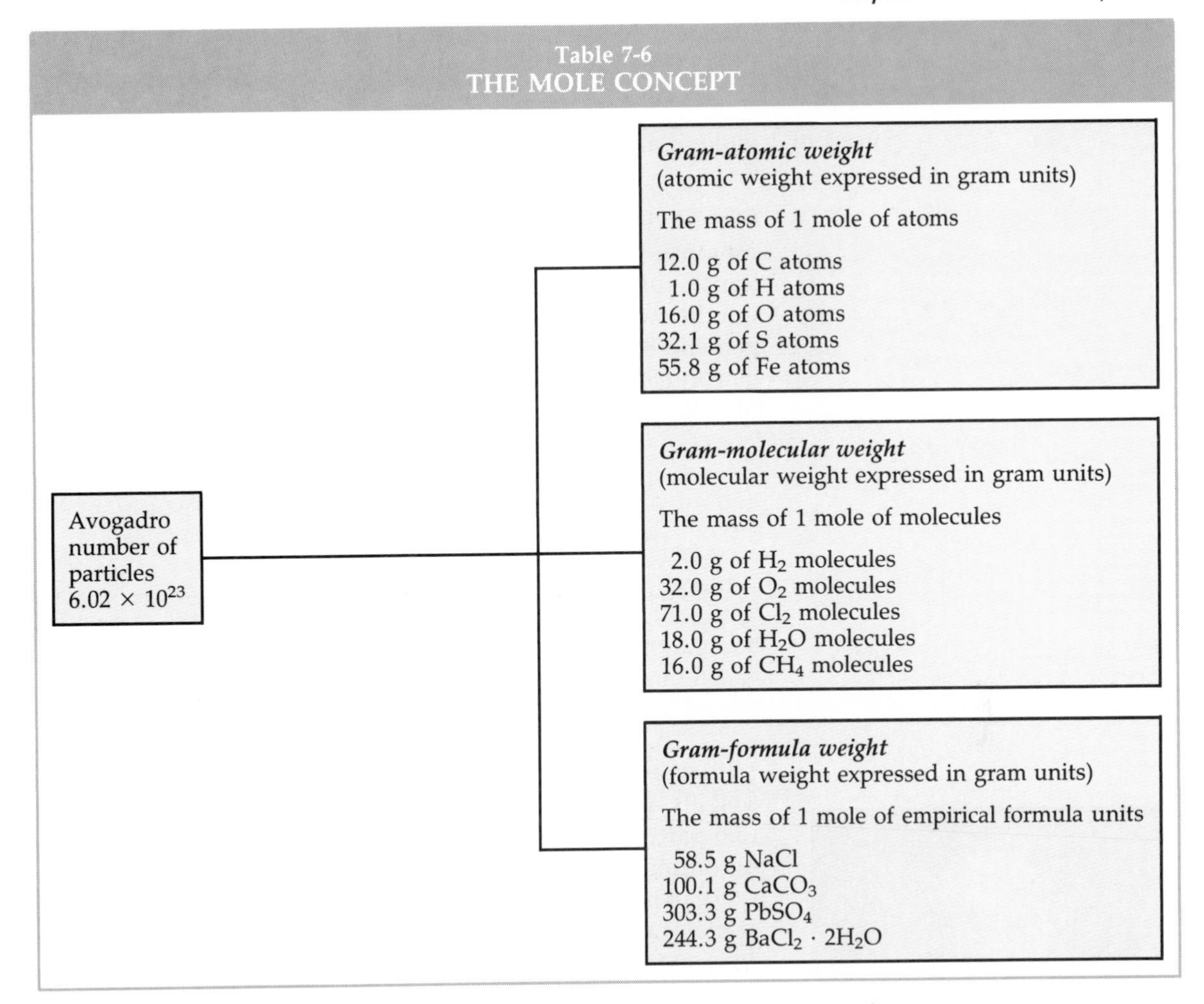

Edward W. Morley (1838–1923) of Western Reserve University found that 1.0000 part by mass of hydrogen combines with 7.9396 parts by mass of oxygen to form 8.9396 parts by mass of water vapor. Every 8.9396 parts of water formed requires 1.0000 part of hydrogen and 7.9396 parts of oxygen. Thus, water consists of

$$\frac{\textbf{1.0000 part H}}{\textbf{8.9396 parts water}} \times \textbf{100\% water} = \textbf{11.186\% hydrogen}$$

and

$$\frac{\textbf{7.9396 parts O}}{\textbf{8.9396 parts water}} \times \textbf{100\% water} = \textbf{88.814\% oxygen}$$

From these data the relative number of atoms of hydrogen and oxygen in water is easily determined. This is done by comparing the mass percentages of the elements or their actual masses by analysis to their respective gram-atomic weights (masses per mole of atoms).

$$\frac{\text{mass of the element}}{\text{mass of 1 mole of atoms of the element}} = \text{moles of atoms of an element}$$

1. *From percentage composition data.* The simplest way to think of percentage composition is in terms of *parts per hundred.* For example, Morley's experiments show that (rounding to 3 significant figures) 11.2% of water is hydrogen and 88.8% is oxygen. In other words, 100.0 parts of water consist of 11.2 parts hydrogen and 88.8 parts oxygen. Similarly, there are 11.2 g of hydrogen and 88.8 g of oxygen per 100.0 g of water. How many moles of hydrogen and oxygen atoms are present in 100.0 g of water?

$$\text{H:} \quad \frac{11.2 \not{g}\ \text{H}}{1.01 \not{g}/\text{mole}} = 11.1 \text{ moles H}$$

$$\text{O:} \quad \frac{88.8 \not{g}\ \text{O}}{16.0 \not{g}/\text{mole}} = 5.55 \text{ moles O}$$

Recall that 1 mole of atoms of one element is the same number of atoms as 1 mole of atoms of any other element (the Avogadro number). Therefore, the relative number of atoms is

$$\text{H:O} = 11.1{:}5.55$$

2. *From relative mass data.* Because approximate atomic weights are used, Morley's relative masses are rounded off to 1.00 part hydrogen and 7.94 parts oxygen in 8.94 parts water. Accordingly, each 8.94-g quantity of water produced requires 1.00 g of hydrogen and 7.94 g of oxygen. As before, the numbers of moles of hydrogen and oxygen atoms in 8.94 g of water are determined and then reduced to the relative number of atoms of each constituent element.

$$\text{H:} \quad \frac{1.00 \not{g}\ \text{H}}{1.01 \not{g}/\text{mole}} = 0.990 \text{ mole H}$$

$$\text{O:} \quad \frac{7.94 \not{g}\ \text{O}}{16.0 \not{g}/\text{mole}} = 0.496 \text{ mole O}$$

The relative number of atoms is

$$\text{H:O} = 0.990{:}0.496$$

From these calculations, the empirical formula of the compound water can be written as

$$H_{11.1}O_{5.55} \quad \text{or} \quad H_{0.990}O_{0.496}$$

Both formulas show the correct ratio of hydrogen atoms to oxygen atoms in water. However, according to the atomic theory, only whole atoms combine chemically. These atom ratios must be converted to their simplest whole-number values. This is accomplished by dividing each ratio by its smaller term, shown as

Figure 7-6. Edward W. Morley was a skillful chemical analyst. His very precise determinations of the densities of oxygen and hydrogen and of the mass ratio in which they combine, published in 1895, were the results of 12 years of research.

To determine an empirical formula:
1. Convert gram ratio to mole ratio.
2. Adjust mole ratio to simplest whole-number ratio.

$$\text{H} : \text{O}$$

$$\frac{11.1}{5.55} : \frac{5.55}{5.55}$$

$$2.00 : 1.00$$

and

$$\text{H} : \text{O}$$

$$\frac{0.990}{0.496} : \frac{0.496}{0.496}$$

$$2.00 : 1.00$$

The empirical formula for water is, therefore, H_2O.

Sometimes the operation just performed does not yield a simple whole-number ratio. In such instances the simplest whole-number ratio can be found by expressing the result as fractions and clearing. **CAUTION:** In some problems, dividing by the smallest term may result in such ratios as 1:2.01, 1:2.98, or 1:3.99. In such situations, do not attempt to clear the fractions; simply round off to 2, 3, or 4, respectively. The whole-number ratios are then 1:2, 1:3, and 1:4. These operations are illustrated in the Sample Problems 7-1, 7-2, 7-3.

SAMPLE PROBLEM 7-1

A compound is found by analysis to contain 75.0% carbon and 25.0% hydrogen. What is the empirical formula?

SOLUTION

Since the compound is 75.0% carbon (75.0 parts/100.0), 75.0 g per 100.0 g is carbon. Similarly 25.0 g per 100.0 g of the compound is hydrogen. The number of moles of atoms of each element in 100.0 g of the compound is determined by the following expression:

$$\frac{\textbf{mass of the element}}{\textbf{mass of 1 mole of atoms of the element}} = \textbf{moles of atoms of an element}$$

C: $$\frac{75.0\ \cancel{g}\ \text{C}}{12.0\ \cancel{g}/\text{mole}} = 6.25 \text{ moles C}$$

H: $$\frac{25.0\ \cancel{g}\ \text{H}}{1.01\ \cancel{g}/\text{mole}} = 24.8 \text{ moles H}$$

Relative number of atoms: $$\text{C:H} = 6.25:24.8$$

Smallest ratio of atoms: $$\frac{6.25}{6.25} : \frac{24.8}{6.25} = 1:4$$

$$\textbf{Empirical formula} = CH_4$$

SAMPLE PROBLEM 7-2

A compound contains carbon, 81.7%, and hydrogen, 18.3%. Find the empirical formula.

SOLUTION

Each 100.0-g quantity of the compound contains 81.7 g of carbon and 18.3 g of hydrogen, as shown by the percentage composition.

$$\frac{\text{mass of the element}}{\text{mass of 1 mole of atoms of the element}} = \text{moles of atoms of an element}$$

C: $$\frac{81.7\ \text{g C}}{12.0\ \text{g/mole}} = 6.81 \text{ moles C}$$

H: $$\frac{18.3\ \text{g H}}{1.01\ \text{g/mole}} = 18.1 \text{ moles H}$$

Relative number of atoms: $C:H = 6.81:18.1$

Smallest ratio of atoms: $$\frac{6.81}{6.81} : \frac{18.1}{6.81} = 1:2.66$$

Simplest whole-number ratio: $1:2.66 = 1:2\frac{2}{3} = \frac{3}{3}:\frac{8}{3} = 3:8$

Empirical formula $= C_3H_8$

SAMPLE PROBLEM 7-3

The decomposition of 11.47 g of a compound of copper and oxygen yields 9.16 g of copper. What is the empirical formula for the compound?

SOLUTION

Since the compound is composed of only copper and oxygen, the mass of oxygen removed in the decomposition process must be

$$11.47\ \text{g} - 9.16\ \text{g} = 2.31\ \text{g oxygen}$$

$$\frac{\text{mass of the element}}{\text{mass of 1 mole of atoms of the element}} = \text{moles of atoms of an element}$$

Cu: $$\frac{9.16\ \text{g Cu}}{63.5\ \text{g/mole}} = 0.144 \text{ mole Cu}$$

O: $$\frac{2.31\ \text{g O}}{16.0\ \text{g/mole}} = 0.144 \text{ mole O}$$

Relative number of atoms: $Cu:O = 0.144:0.144$

Smallest ratio of atoms: $1:1$

Empirical formula $= CuO$

Practice Problems

1. A $10\overline{0}$-g sample of a compound is found by analysis to contain 39.3 g of sodium and 60.7 g of chlorine. Find its empirical formula. *ans.* NaCl
2. Analysis: hydrogen, 2.77%; chlorine, 97.3%. What is the empirical formula? *ans.* HCl
3. Analysis: hydrogen, 5.92%; sulfur, 94.1%. What is the empirical formula? *ans.* H_2S
4. Analysis: calcium, 20.0%; bromine, 80.0%. What is the empirical formula? *ans.* $CaBr_2$
5. Analysis: nitrogen, 9.66%; hydrogen, 2.79%; iodine, 87.6%. Find the empirical formula. *ans.* NH_4I

7.13 Calculating molecular formulas The analysis of a substance enables one to determine its empirical formula. This simplest formula of a molecular substance may or may not be its molecular formula. The empirical formula for the gas methane, a molecular compound, is found to be CH_4 in Sample Problem 7-1. Any multiple of CH_4, such as C_2H_8, C_3H_{12}, or C_nH_{4n}, represents the same ratio of carbon and hydrogen atoms. How then can the correct molecular formula be determined?

It is not possible to decide which formula is the molecular formula unless the molecular weight of the substance has been determined. Some substances lend themselves to known methods of determining molecular weights, and some do not. The methods will be discussed in Chapters 11 and 13. However, if the molecular weight is known, it is a simple matter to decide which multiple of the empirical formula is the molecular formula.

The correct multiple of the empirical formula can be represented by the subscript $_x$. Then

$$\textbf{(empirical formula)}_x = \textbf{molecular formula}$$

Molecular weight must be known before a molecular formula can be determined.

and (empirical formula weight)$_x$ can be equated to the known molecular weight.

$$\textbf{(empirical formula weight)}_x = \textbf{molecular weight}$$

In the case of methane, the molecular weight is known to be 16.0. The equation is

Molecular weight is a whole-number multiple of the empirical formula weight.

$$(\mathbf{CH_4\ weight})_x = \mathbf{16.0}$$

$$[\mathbf{12.0 + 4(1.0)}]_x = \mathbf{16.0}$$

$$\mathbf{x = 1}$$

$$\textbf{molecular formula} = (\mathbf{CH_4})_1 \textbf{ or } \mathbf{CH_4}$$

Hence the empirical formula for methane is also the molecular formula. You have seen that this is also true in the case of water. For another example, see Sample Problem 7-4.

SAMPLE PROBLEM 7-4

Hydrogen peroxide is found by analysis to consist of 5.9% hydrogen and 94.1% oxygen. Its molecular weight is determined to be 34.0. What is the correct formula?

SOLUTION

1. The empirical formula, determined from the analysis by the method described in Section 7.12, is

$$\mathbf{HO}$$

2. The molecular formula determined from the empirical formula and molecular weight is

$$(\text{HO weight})_x = 34.0$$

$$(1.0 + 16.0)_x = 34.0$$

$$x = 2$$

$$\text{molecular formula} = (HO)_2 \text{ or } H_2O_2$$

Practice Problems

1. A certain hydrocarbon compound is found by analysis to have the empirical formula CH_3. Its molecular weight is 30.0. What is the molecular formula? *ans.* C_2H_6
2. A compound is found by analysis to consist of 40.1% sulfur and 59.9% oxygen. Its molecular weight is 80.1. (*a*) What is its empirical formula? (*b*) What is its molecular formula? (*c*) What is the name of the compound? *ans.* (*a*) SO_3, (*b*) SO_3, (*c*) sulfur trioxide
3. A compound of nitrogen and sulfur is found by analysis to have the empirical formula NS_2. Its molecular weight is 156.4. (*a*) What is its molecular formula? (*b*) What is the name of the compound? *ans.* (*a*) N_2S_4 (*b*) dinitrogen tetrasulfide

SUMMARY

The formulas for common ionic compounds can be written using the ion-charge method. The ion notations are written in the order used in the name of the compound. The number of each kind of ion is then adjusted to make the total positive and negative ionic charges equal.

The names of ionic compounds include the names of the constituent ions. If the metallic ion present has different possible oxidation states, Roman numerals are added to the name of the ion to indicate the oxidation state. A system of prefixes is used in the names of covalent compounds when an element forms more than one compound with another element. For binary compounds, the *-ide* ending means that only two elements are present.

The gram-atomic weights of elements and the gram-formula weights of compounds represent the mass of a mole of these substances.

The percentage composition of a compound can be calculated if the formula is known. Conversely, the empirical formula can be calculated if the composition of a compound is known. Furthermore, if the compound is molecular and its molecular weight is known, the molecular formula can be determined.

VOCABULARY

Avogadro number
binary compound
empirical formula
formula weight
gram-atomic weight
gram-formula weight
gram-molecular weight
law of multiple proportions
mole
molecular weight
percentage composition

QUESTIONS

Group A

1. To what characteristics of the Group I metallic elements can the ease with which their atoms form +1 monatomic ions be attributed? *A4a*
2. How can the ease with which the atoms of Group II elements form +2 monatomic ions be explained? *A4a*
3. What information can be derived from the molecular formula NH_3? *A4c*
4. Write the name and charge of each of the following monatomic ions: (*a*) Ag^+; (*b*) Fe^{+++}; (*c*) Ca^{++}; (*d*) Na^+; (*e*) Cu^+; (*f*) Cu^{++}; (*g*) O^{--}; (*h*) Hg^{++}; (*i*) S^{--}; (*j*) Br^-; (*k*) Fe^{++}; (*l*) I^-. *A2b*
5. Write the name and charge of each of the following polyatomic ions: (*a*) $C_2H_3O_2^-$; (*b*) SO_3^{--}; (*c*) Hg_2^{++}; (*d*) $Cr_2O_7^{--}$; (*e*) SO_4^{--}; (*f*) CrO_4^{--}; (*g*) O_2^{--}; (*h*) HCO_3^-; (*i*) NO_2^-; (*j*) ClO_3^-; (*k*) OH^-; (*l*) HSO_4^-. *A2b, A3a*
6. Write the symbol or formula and charge of each of the following: (*a*) ammonium ion; (*b*) carbonate ion; (*c*) nickel(II) ion; (*d*) aluminum ion; (*e*) lead(II) ion; (*f*) potassium ion; (*g*) chloride ion; (*h*) fluoride ion; (*i*) nitrate ion; (*j*) magnesium ion; (*k*) chromium(III) ion; (*l*) phosphate ion; (*m*) barium ion; (*n*) zinc ion; (*o*) lead(IV) ion. *A2b, A3a*
7. Distinguish between the terms *formula weight* and *molecular weight*. *A1*
8. Write the formula for each of the following compounds: (*a*) lead(II) acetate; (*b*) iron(III) hydroxide; (*c*) calcium nitrate; (*d*) sodium hydrogen carbonate; (*e*) sodium peroxide. *A2c, B1*
9. Name the compound represented by each of the following formulas: (*a*) $Mg_3(PO_4)_2$; (*b*) Al_2S_3; (*c*) CrF_3; (*d*) $HgCl_2$; (*e*) $KClO_3$. *B2*
10. Write the formula for each of the following compounds: (*a*) barium sulfate; (*b*) zinc oxide; (*c*) silver iodide; (*d*) nickel(II) carbonate; (*e*) potassium sulfite. *B1*
11. Name the compound represented by each formula: (*a*) $(NH_4)_2Cr_2O_7$; (*b*) Cu_2SO_4; (*c*) $Cu(OH)_2$; (*d*) $Hg_2(C_2H_3O_2)_2$; (*e*) AgBr. *B2*
12. Write the formula for each of these compounds: (*a*) iron(II) bromide; (*b*) ammonium hydrogen sulfate; (*c*) chromium(III) sulfate; (*d*) nickel(II) oxide; (*e*) aluminum oxide. *B1*
13. Name the following compounds: (*a*) $PbCr_2O_7$; (*b*) $ZnCrO_4$; (*c*) Cu_2CO_3; (*d*) $Ca_3(PO_4)_2$; (*e*) $Mg(OH)_2$. *B2*
14. Write formulas for these compounds: (*a*) iron(II) iodide; (*b*) mercury(II) nitrate; (*c*) nickel(II) fluoride; (*d*) copper(II) nitrate; (*e*) potassium hydrogen carbonate. *B1*
15. Name the following compounds: (*a*) Hg_2Br_2; (*b*) Ag_2S; (*c*) BaO_2; (*d*) $KHSO_4$; (*e*) $CaSO_3$. *B2*

QUESTIONS

Group B

NOTE: For Questions 16 through 18, apply the nomenclature system for binary covalent compounds. Refer to the table of common elements inside the front cover or Appendix Table 5, and to Table 7-3.

16. Name each of the following binary compounds: (*a*) SO_2; (*b*)SO_3; (*c*) CO; (*d*) CO_2; (*e*) CS_2; (*f*) CCl_4; (*g*) $SbCl_3$; (*h*) As_2S_3; (*i*) As_2S_5; (*j*) $BiBr_3$. *A2e*

17. Write the chemical formulas for the following binary compounds: (*a*) bismuth trioxide; (*b*) dinitrogen monoxide; (*c*) dinitrogen trioxide; (*d*) dinitrogen pentoxide; (*e*) dinitrogen pentasulfide; (*f*) oxygen difluoride; (*g*) diphosphorus pentoxide (empirical) and tetraphosphorus decaoxide (molecular); (*h*) silicon tetrabromide; (*i*) silicon dioxide; (*j*) silicon disulfide. *A2e*

18. Name the following binary compounds: (*a*) SF_4; (*b*) SF_6; (*c*) PCl_3; (*d*) SCl_4; (*e*) PCl_5; (*f*) $SiCl_4$; (*g*) CBr_4; (*h*) HI; (*i*) HCl; (*j*) HBr. *A2e*

19. How are the atomic theory, the law of definite composition, and the principle of percentage composition related? *A1, A4b*

20. How is the law of multiple proportions related to the atomic theory? *A1, A4b*

PROBLEMS

Group A

1. What is the formula weight of potassium chlorate, $KClO_3$? *B3a*

2. Determine the formula weight of acetic acid, $HC_2H_3O_2$. *B3a*

3. Sucrose (cane sugar) has the formula $C_{12}H_{22}O_{11}$. Determine its formula weight. *B3a*

4. Determine the formula weight of glycerol, $C_3H_8O_3$. *B3a*

5. Potassium dichromate has the formula $K_2Cr_2O_7$. Determine its formula weight.

6. Crystalline magnesium sulfate (Epsom salts) has the formula $MgSO_4 \cdot 7H_2O$. What is its formula weight? *B3a*

7. Determine the formula weight of each of the following compounds: (*a*) K_2CO_3: (*b*) N_2H_4; (*c*) HgO; (*d*) $CuSO_4 \cdot 5H_2O$; (*e*) H_2SO_4; (*f*) $MgBr_2$; (*g*) Al_2S_3; (*h*) $Ca(NO_3)_2$; (*i*) $Fe_2(Cr_2O_7)_3$; (*j*) $KMnO_4$. *B3a*

8. Baking powders contain sodium hydrogen carbonate, $NaHCO_3$. Calculate its percentage composition. *B3b*

9. What is the percentage composition of a soap having the formula $C_{17}H_{35}COONa$? *B3b*

10. Vinegar contains acetic acid, $HC_2H_3O_2$. Find its percentage composition. *B3b*

11. What is the percentage composition of each of these compounds: (*a*) SO_2; (*b*) $Ca(OH)_2$; (*c*) $Ca(H_2PO_4)_2 \cdot H_2O$; (*d*) $MgSO_4 \cdot 7H_2O$?

12. Which of these compounds contains the highest percentage of nitrogen: (*a*) $Ca(NO_3)_2$; (*b*) $AgNO_3$; (*c*) $(NH_4)_2SO_4$? *B3b*

13. A strip of pure copper, mass 7.536 g, is heated with oxygen to form a compound of copper and oxygen, mass 9.433 g. What is the percentage composition of the compound? *B3b*

14. You are given 25.0 g of (*a*) CaO; (*b*) $Na_2CO_3 \cdot 10H_2O$; (*c*) $BaCl_2 \cdot 2H_2O$; (*d*) $(NH_4)_2SO_4$; (*e*) $Fe(NO_3)_3 \cdot 6H_2O$; (*f*) $Al_2(SO_4)_3 \cdot 18H_2O$; (*g*) K_2CrO_4. How many moles of each compound do you have? *B3c*

15. Calculate the mass of (*a*) 1.00 mole of chlorine atoms; (*b*) 5.00 moles of nitrogen atoms; (*c*) 3.00 moles of bromine molecules; (*d*) 6.00 moles of hydrogen chloride; (*e*) 10.0 moles of magnesium sulfate; (*f*) 2.50 moles of potassium iodide; (*g*) 0.500 mole of silver nitrate; (*h*) 0.100 mole of sodium chloride. *B3c*

16. How many moles of iron can be recovered from 2.500 metric tons (10^3 kg/metric ton) of Fe_3O_4? *B3c*

17. Cinnabar, an ore of mercury, has the formula HgS. Calculate the number of moles of mercury recovered from 5.00 kg of cinnabar. *B3c*
18. Calculate the percentage of CaO in $CaCO_3$. *B3b*
19. Calculate the percentage of H_2O in $CuSO_4 \cdot 5H_2O$. *B3b*
20. Calculate the percentage of copper in each of these minerals: (*a*) cuprite, Cu_2O; (*b*) malachite, $CuCO_3 \cdot Cu(OH)_2$; and (*c*) cubanite, $CuFe_2S_4$. *B3b*

PROBLEMS

Group B

21. What is the percentage composition of the drug chloramphenicol, $C_{11}H_{12}N_2O_5Cl_2$? *B3b*
22. One compound of platinum and chlorine is known to consist of 42.1% chlorine. Another consists of 26.7% chlorine. Determine the two empirical formulas and write the name of the compound represented by each formula. *B2, B3e*
23. What is the empirical formula of silver fluoride? It is 85% silver. *B3e*
24. Analysis: 39.3% sodium and 60.7% chlorine. What are the empirical formula and the name of the compound? *B3e, B2*
25. Analysis: phosphorus, 43.67%; oxygen, 56.33%. What are the empirical formula and the name of the compound? *B3e*
26. Analysis: potassium, 24.58%; manganese, 34.81%; oxygen, 40.50%. Determine the empirical formula. *B3e*
27. Calculate the empirical formula of a compound having 37.70% sodium, 22.95% silicon, and 39.34% oxygen. *B3e*
28. Analysis: aluminum, 15.8%; sulfur, 28.1%; oxygen 56.1%. What are the empirical formula and the name of this compound? *B3e, B2*
29. A compound has the composition: Na, 28.05%; C, 29.26%; H, 3.66%; O, 39.02%. What are the empirical formula and the name of the compound? *B3e, B2*
30. The analysis of a chemical compound shows the following: nitrogen, 21.20%; hydrogen, 6.06%; sulfur, 24.30%; oxygen, 48.45%. Find the empirical formula and name the compound. *B3e, B2*
31. A compound has this composition: potassium, 44.82%; sulfur, 18.39%; oxygen, 36.79%. Determine the empirical formula and name the compound. *B3e, B2*
32. Analysis: iron, 48.2%; carbon, 10.4%; oxygen, 41.4%. Determine the empirical formula and name the compound. *B3e, B2*
33. Analysis: calcium, 24.7%; hydrogen, 1.2%; carbon, 14.8%; oxygen, 59.3%. Determine its empirical formula and name the compound. *B3e, B2*
34. An oxide of iron has the composition: Fe, 72.4%; O, 27.6%. What is the empirical formula of this compound? *B3e*
35. By analysis, the composition of a gas is: C, 92.3%; H, 7.7%. Its molecular weight is 26.0. What is its molecular formula? *B3e, B3f*
36. Analysis: C, 80.0%; H, 20.0%. The molecular weight is 30.0. What is the molecular formula of this compound? *B3e, B3f*
37. Analysis: iodine, 76.0%; oxygen, 24.0%. The molecular weight is 334. (*a*) Determine the molecular formula. (*b*) What is the oxidation number of the iodine in this compound? *B3e, B3f*
38. The mass percentages of carbon in its two oxides are 42.8% and 27.3%. Use these data to show the law of multiple proportions. *A3d*

Odd Hassel (Norwegian), along with Derek H. R. Barton (British), was awarded the 1969 Nobel Prize in chemistry for determining the actual three-dimensional shape of many molecules. He determined crystal and molecular structures by using X-ray and electron diffraction and by measuring electric dipole moments. His studies of cyclohexane and its derivatives indicated that the molecule exists in two forms. This led to the development of conformational analysis, which helped revolutionize the concepts of molecular structure. Hassel later studied structures of compounds containing halogens.

Equations and Mass Relationships

8

GOALS

In this chapter, you will gain an understanding of: • the difference between the information provided by word equations and formula equations • the significance of balanced chemical equations • the procedure in writing balanced chemical equations • the four general types of chemical reactions • the activity series of the elements and its use in predicting reactions and completing equations • stoichiometry and the solution of mass-mass problems by the mole method

CHEMICAL EQUATIONS

8.1 Writing equations The most useful way of representing the chemical changes that occur during chemical reactions is by writing *chemical equations. A* ***balanced chemical equation*** *is a concise, symbolic expression for a chemical reaction.* It is a *quantitative* statement which indicates the proportionate number of moles of each chemical species that enters into the reaction and is formed by the reaction.

The first step in writing chemical equations involves describing the basic facts of the reaction systems in *word equations.*

1. *Word equations.* This verbal equation is a brief statement that gives the names of the chemical species involved in the reaction. In a reaction system, the chemical species that react are called the *reactants,* and the species that are formed are called the *products.* The word equation does not give the *quantities* of reactants used or products formed. Thus, a word equation has only *qualitative,* or descriptive, significance.

Experiments show that the combustion of hydrogen, whether in air or in an atmosphere of pure oxygen, produces water as the only combustion product. The word equation for this combustion reaction is

hydrogen + oxygen → water

The arrow (→) in this expression is the "yields" sign. It conveys the information that a chemical reaction occurs. It is read, *"react to yield,"* or just *"yields."* The word equation is ordinarily read, *"hydrogen and oxygen react to yield water,"* or simply, *"hydrogen and oxygen yield water."* The word equation signifies that hydrogen and oxygen react chemically to form water. Thus, it briefly states an experimental fact. It does not specify the conditions under which the reaction occurs, the relative quantities of hydrogen and oxygen used, or the relative quantity of water formed.

2. *Formula equations.* Suppose you replace the names of the reactants, hydrogen and oxygen, and the name of the product, water, with their respective formulas. The equation becomes

$$H_2 + O_2 \rightarrow H_2O \quad \text{(not balanced)}$$

The law of conservation of matter and energy is fundamental in science. From this law a very useful generalization results: *In ordinary chemical reactions, the total mass of the reacting substances is equal to the total mass of the products formed.* This statement may be referred to as ***the law of conservation of atoms.***

By adusting for the relative quantities of reactants and products as found by experiment, the equation can be written as a *balanced formula equation.* This equation agrees with the law of conservation of atoms.

The balanced formula equation is commonly called the chemical equation.

$$2H_2 + O_2 \rightarrow 2H_2O$$

This agreement is verified by comparing the total number of atoms of hydrogen and oxygen on the left side of the reaction sign (→) to their respective totals on the right. Two molecules of hydrogen contains 4 atoms of hydrogen; two molecules of water also contains 4 hydrogen atoms. One molecule of oxygen contains 2 atoms of oxygen; two molecules of water also contains 2 oxygen atoms. Thus a chemical equation is similar to an algebraic equation. They both express equalities. *Until it is balanced, a chemical equation cannot express a chemical equality and is not a valid equation.* The yields sign (→) has the meaning of an equals sign (=). In addition, the yields sign indicates the direction in which the reaction proceeds.

3. *Significance of equations.* This chemical equation now signifies much more than a word equation. It tells you

(*a*) the relative proportions of the reactants, hydrogen and oxygen, and the product, water, and

(*b*) that **2 molecules** of hydrogen react with **1 molecule** of oxygen to form **2 molecules** of water.

Most importantly, because one mole of a molecular substance contains the Avogadro number of molecules, the equation tells you

(*c*) that **2 moles** of hydrogen molecules reacts with **1 mole** of oxygen molecules to form **2 moles** of water molecules.

Because the mass of one mole of a molecular substance is its gram-molecular weight, the equation tells you

(*d*) that **4 g** of hydrogen reacts with **32 g** of oxygen to form **36 g** of water.

Furthermore, since these quantities are proportionate, the equation tells you

$2H_2$	+	O_2	→	$2H_2O$
2 molecules hydrogen	+	1 molecule oxygen	→	2 molecules water
2 moles hydrogen	+	1 mole oxygen	→	2 moles water
4 grams hydrogen	+	32 grams oxygen	→	36 grams water
any mass quantity of hydrogen	+	mass of oxygen in ratio 8:1 with mass of hydrogen	→	mass of water in ratio 9:1 with mass of hydrogen

Figure 8-1. The chemical equation is a symbolic expression of a chemical reaction. It indicates the proportionate number of quantity units of each chemical species in the reaction system.

(e) that any masses of hydrogen and oxygen that are in the ratio of **1 : 8,** respectively, and that react to form water yield a mass of water that is related to the masses of hydrogen and oxygen as **1 : 8 : 9.**

Finally, in any equation the equality exists in both directions. If $x + y = z$, then $z = x + y$. Therefore, the equation tells you

(f) that the decomposition of **2 moles** of water yields **2 moles** of hydrogen and **1 mole** of oxygen.

From these six statements it is clear that balanced formula equations have quantitative significance. *Quantitative information relates to the measured quantities of reactants and products in a reaction system.* The chemical equations represent experimentally established facts about reactions. They indicate the identities and relative quantities of reactants and products in a chemical reaction system. However, equations reveal nothing about the mechanisms by which the reactants become restructured into the final products.

It is possible to write an equation for a reaction that does not occur. For example, gold and oxygen *do not* react directly to form gold(III) oxide, Au_2O_3. Yet an equation for the nonreaction can be written and balanced to conform with the law of conservation of atoms. Even though the equation is properly written and balanced, it is invalid because it is contrary to known facts.

Figure 8-2. A mixture of zinc and iodine reacts chemically under favorable conditions to form zinc iodide. This reaction occurs (bottom dish) when water is added dropwise to the zinc-iodine mixture. Heat generated by the reaction vaporizes some iodine, as shown by the violet vapor.

8.2 Factors in equation writing A chemical equation has no value unless it is accurate in every detail. Three factors must be considered in writing a balanced equation.

1. *The equation must represent the facts.* If you are to write the equation for a reaction, you need to know the facts concerning the reaction. You need to know all the reactants and all the products. The chemist relies upon analysis for these facts.

2. *The equation must include the symbols and formulas for all elements and compounds that are used as reactants and formed as products.* You must know these symbols and formulas and be sure that they are correctly written. The elements that exist as diatomic molecules are oxygen, hydrogen, nitrogen, fluorine, chlorine, bromine, and iodine. These elements are listed in Table 8-1 for review purposes. Other elements are usually considered to be monatomic (one-atomed) structures. Your knowledge of the oxidation states of the elements and the ion-charge method of writing correct formulas will enable you to satisfy this requirement for correct symbols and formulas.

3. *The law of conservation of atoms must be satisfied.* There must be the same number of atoms of each species represented on each side of the equation. A new species of atom cannot be represented on the product side and no species of atom can disappear from the reactant side. These are the *balancing requirements.*

Table 8-1
ELEMENTS WITH DIATOMIC MOLECULES

Element	*Symbol*	*Atomic number*	*Molecular formula*	*Structural formula*	*Phase at room temperature*
hydrogen	H	1	H_2	H—H	gas
nitrogen	N	7	N_2	N≡N	gas
oxygen	O	8	O_2	O=O	gas
fluorine	F	9	F_2	F—F	gas
chlorine	Cl	17	Cl_2	Cl—Cl	gas
bromine	Br	35	Br_2	Br—Br	liquid
iodine	I	53	I_2	I—I	solid

Balanced equations have
1. the chemical facts;
2. correct formulas;
3. atoms conserved.

They are met by adjusting the *coefficients* of the formulas of reactants and products. You must adjust these coefficients to the *smallest possible whole numbers* that satisfy the law of conservation of atoms.

8.3 Procedure in writing equations To write the chemical equation that represents a chemical reaction, you must proceed in steps that satisfy the three factors in equation writing identified in Section 8.2.

1. *Represent the facts.*
2. *Write correct formulas for compounds balanced as to oxidation number or ion charge.* (Formulas for elemental substances with diatomic molecules must also be correctly written.)
3. *Balance the equation according to the law of conservation of atoms.*

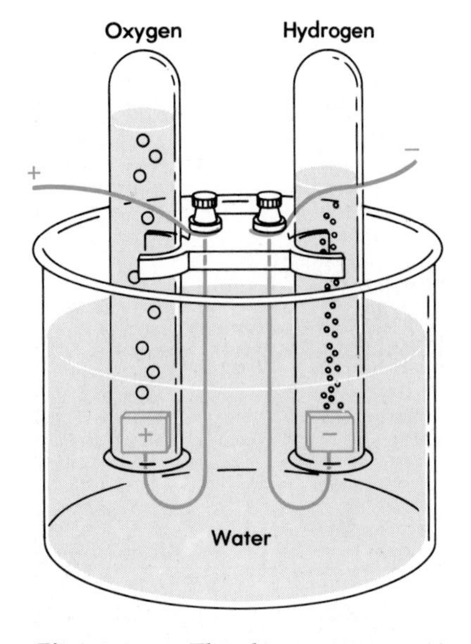

Figure 8-3. The decomposition of water to hydrogen and oxygen by means of an electric current. Hydrogen is collected at the negative electrode, and oxygen is collected at the positive electrode. Visually estimate the volume ratio of the products of this reaction, $V_{H_2} : V_{O_2}$.

Consider the chemical reactions that follow and write the chemical equation for each reaction described. When an electric current is passed through water made slightly conductive, the elements hydrogen and oxygen are formed as decomposition products. First represent the facts by the word equation. Then write and balance the formula equation for this reaction.

Step 1: *What are the facts?* The only reactant is water. The only products are hydrogen and oxygen. The word equation is

$$\textbf{water} \rightarrow \textbf{hydrogen} + \textbf{oxygen}$$

Now substitute the formulas for these substances.

$$H_2O \rightarrow H_2 + O_2 \quad \textbf{(not balanced)}$$

Step 2: *Are the formulas correctly written?* The oxidation number of hydrogen is +1 and of oxygen −2. Thus the formula for water is correctly written as H_2O. Both hydrogen and oxygen exist in the free state as diatomic molecules; their molecular formulas are correctly written as H_2 and O_2.

Step 3: *Is the equation balanced as to atoms?* On the left, 1 molecule of water is represented. It consists of 2 hydrogen atoms and 1 oxygen atom. On the right of the yields sign (→), 1 molecule of hydrogen consisting of 2 atoms and 1 molecule of oxygen consisting of 2 atoms are represented. However, *there is only 1 atom of oxygen on the left.* How can this difference be reconciled? A subscript $_2$ may *not* be added to the oxygen of the water formula since this subscript would change a formula that is now correctly written. *Once the formulas of the substances in the equation are written correctly, their subscripts must not be changed.* This rule applies because the number of atoms in a molecule is an established experimental fact. However, the number of molecules or moles of a substance in an equation can be changed by changing the coefficient.

Change coefficients, not subscripts, to balance atoms in an equation.

Suppose the number of water molecules is increased to 2. This change can be made by placing the coefficient 2 ahead of the formula H_2O, making it $2H_2O$. Now the equation shows 2 molecules of water, each having 1 oxygen atom. This change gives the necessary 2 atoms of oxygen on the left.

$$2H_2O \rightarrow H_2 + O_2 \quad \textbf{(not balanced)}$$

Two molecules of water have a total of 4 atoms of hydrogen. The right side of the equation must now be adjusted to represent 4 atoms of hydrogen. This is accomplished by placing the coefficient 2 ahead of the hydrogen molecule, making it $2H_2$. There is now a total of 4 atoms of hydrogen represented on the right. The equation reads

$$2H_2O \rightarrow 2H_2 + O_2$$

The same number of atoms of each element is represented on both sides of the equation. Furthermore, the equation has the lowest possible whole-number ratio of coefficients. Thus the equation is balanced.

To indicate the physical phases of the various species in the reaction (the reactant, water, to the left of the yields sign and the products, hydrogen and oxygen, to the right), the balanced equation is written

$$2H_2O(l) \rightarrow 2H_2(g) + O_2(g)$$

The abbreviations commonly used in this way are (s) solid, (l) liquid, (g) gas, and (aq) water solution. In this text, the abbreviations are used when they contribute to a better understanding of the reaction represented by the equation.

When sulfur burns in oxygen (or in air) it combines with the oxygen gas to form sulfur dioxide gas. These reaction facts can be written as

$$\textbf{sulfur + oxygen} \rightarrow \textbf{sulfur dioxide}$$

or

$$S(s) + O_2(g) \rightarrow SO_2(g)$$

Figure 8-4. Sulfur burns in oxygen to form sulfur dioxide gas, SO_2.

Molecular oxygen is diatomic and is represented as O_2. The binary name "sulfur dioxide" indicates that its formula is SO_2. All formulas are correctly written. The number of atoms of sulfur and of oxygen are the same on both sides of the equation. No further adjustments are required; the equation is balanced.

Oxygen can be prepared in the laboratory by heating mercury(II) oxide. The facts are: when solid mercury(II) oxide is heated, it yields liquid metallic mercury and oxygen gas.

mercury(II) oxide → mercury + oxygen

By substituting the proper symbols and formulas, the word equation becomes

$$HgO \rightarrow Hg + O_2 \quad \text{(not balanced)}$$

Phase symbols in an equation:
(s) = solid
(l) = liquid
(g) = gas
(aq) = water solution

Mercury(II) has an oxidation number of +2 and oxygen has an oxidation number of −2. The formula of mercury(II) oxide is correct as written. However, the equation is not balanced. Two molecules of HgO must decompose to yield the 2 atoms of the diatomic oxygen molecule. After making this adjustment, 2 additional atoms of mercury must also appear on the right. The balanced equation, including the phase of each species, is

$$2HgO(s) \rightarrow 2Hg(l) + O_2(g)$$

Zinc reacts with hydrochloric acid (a water solution of hydrogen chloride gas). The reaction produces hydrogen gas and a solution of zinc chloride. These facts can be represented by

zinc + hydrochloric acid → zinc chloride + hydrogen

Dispense with the word-equation step when you no longer need it as the initial expression of the facts.

Or the facts can be represented directly by the initial formula equation

$$Zn + HCl \rightarrow ZnCl_2 + H_2 \quad \text{(not balanced)}$$

CAUTION: *All formulas in the equation must be correctly written before the atom-balancing step is undertaken.*

This equation shows 1 hydrogen atom on the left and 2 on the right. Hydrogen atoms are balanced by representing 2 molecules of hydrogen chloride as 2HCl. Note that this change also gives 2 chlorine atoms on the left, which balance the 2 chlorine atoms on the right. The balanced equation is

$$Zn(s) + 2HCl(aq) \rightarrow ZnCl_2(aq) + H_2(g)$$

Do not become discouraged if equations cause you some initial difficulty. Common mistakes made by students while learning the equation-balancing technique are (1) neglecting to verify the correctness of each formula in the "facts" equation and (2) altering a correctly written formula in an attempt to establish the required number of atoms for a balanced equation. Keep in mind that once a formula is correctly written, *subscripts cannot be added, deleted, or changed*. Formulas must be written correctly *before* proceeding with the final atom-balancing step. As you continue your study of chemistry and gain experience in the laboratory, the equations that now seem difficult will become easy.

This next reaction occurs in one step of a water-purification process. Aluminum sulfate and calcium hydroxide are added to water containing unwanted suspended matter. The two substances react in water to produce two insoluble products, aluminum hydroxide and calcium sulfate. These products settle out, taking the suspended matter with them.

The two *reactants* are aluminum sulfate and calcium hydroxide. Their respective formulas (by ion-charge balancing) are $Al_2(SO_4)_3$ and $Ca(OH)_2$. The two *products* of the reaction are aluminum hydroxide, $Al(OH)_3$, and calcium sulfate, $CaSO_4$. Having determined the correct formula for each species involved in the reaction, the word equation can be omitted and the initial "facts" equation can be written directly.

$$\mathbf{Al_2(SO_4)_3 + Ca(OH)_2 \rightarrow Al(OH)_3 + CaSO_4} \quad \textbf{(not balanced)}$$

The next step is to balance the equation. On the left, the formula $Al_2(SO_4)_3$ represents 2 Al atoms. On the right, 2 Al atoms are realized by placing the coefficient 2 in front of $Al(OH)_3$. The equation now reads

CAUTION: *Do not attempt to balance atoms until after all formulas have been written correctly.*

$$\mathbf{Al_2(SO_4)_3 + Ca(OH)_2 \rightarrow 2Al(OH)_3 + CaSO_4} \quad \textbf{(not balanced)}$$

Three polyatomic SO_4^{--} ions are indicated on the left. Therefore the coefficient 3 is placed in front of $CaSO_4$. The equation now reads

$$\mathbf{Al_2(SO_4)_3 + Ca(OH)_2 \rightarrow 2Al(OH)_3 + 3CaSO_4} \quad \textbf{(not balanced)}$$

Next observe that there must be 3 Ca atoms on the left to balance the 3 Ca atoms now on the right. The coefficient 3 is placed in front of $Ca(OH)_2$. This coefficent also gives 6 OH^- ions on the left which balance the 6 OH^- ions on the right. The equation is now balanced.

$$\mathbf{Al_2(SO_4)_3 + 3Ca(OH)_2 \rightarrow 2Al(OH)_3(s) + 3CaSO_4(s)}$$

The (s) notations indicate that both products are insoluble. They leave the reaction environment as *precipitates*.

Precipitate: *(1) A substance that separates from a solution as a result of a physical or chemical change. (2) The separation of an insoluble substance from a solution.*

Hence, to write chemical equations correctly you need to:

1. know the symbols of the common elements;
2. know the usual oxidation numbers or ionic charges of the common elements and polyatomic ions;
3. know the facts relating to the reaction for which an equation is to be written;
4. assure that all formulas are correctly written before attempting to balance the equation;
5. balance the equation for atoms of all elements represented so that the lowest possible ratio of whole-number coefficients results.

Practice Problems

Write the balanced equations for the following reactions:

1. zinc + chlorine → zinc chloride *ans.* $Zn + Cl_2 \rightarrow ZnCl_2$
2. sodium bromide + chlorine → sodium chloride + bromine *ans.* $2NaBr + Cl_2 \rightarrow 2NaCl + Br_2$
3. sodium hydrogen carbonate → sodium carbonate + water + carbon dioxide *ans.* $2NaHCO_3 \rightarrow Na_2CO_3 + H_2O + CO_2$
4. ammonium chloride + calcium hydroxide → calcium chloride + ammonia (NH_3) + water *ans.* $2NH_4Cl + Ca(OH)_2 \rightarrow CaCl_2 + 2NH_3 + 2H_2O$
5. calcium hydroxide + carbon dioxide → calcium carbonate + water *ans.* $Ca(OH)_2 + CO_2 \rightarrow CaCO_3 + H_2O$
6. potassium hydroxide + carbon dioxide → potassium hydrogen carbonate *ans.* $KOH + CO_2 \rightarrow KHCO_3$
7. hydrogen sulfide + lead(II) chloride → hydrogen chloride + lead(II) sulfide *ans.* $H_2S + PbCl_2 \rightarrow 2HCl + PbS$
8. ammonia (NH_3) + hydrogen sulfide → ammonium sulfide *ans.* $2NH_3 + H_2S \rightarrow (NH_4)_2S$

8.4 General types of chemical reactions There are several different ways of classifying chemical reactions. No single scheme is entirely satisfactory. In elementary chemistry, it is helpful to recognize reactions that fall into the four categories given below. Later you will learn other ways in which chemical reactions can be classified. The four main types of reactions are:

Composition reactions are also called synthesis reactions.

Reactants in composition reactions aren't always elements.

1. *Composition reactions,* in which two or more substances combine to form a more complex substance. Composition reactions have the general form

$$A + X \rightarrow AX$$

Examples:

Iron and sulfur combine to form iron(II) sulfide.

$$Fe + S \rightarrow FeS$$

Magnesium burns in air (combines with oxygen) to form magnesium oxide. This combustion is shown in Figure 8-5.

$$2Mg + O_2 \rightarrow 2MgO$$

Water and sulfur trioxide combine to form hydrogen sulfate (sulfuric acid).

$$H_2O + SO_3 \rightarrow H_2SO_4$$

Figure 8-5. A composition reaction. Magnesium metal burns in air to form magnesium oxide, MgO.

2. *Decomposition reactions,* which are the reverse of the first type. Here one substance breaks down to form two or more simpler substances. Decomposition reactions have the general form

$$AX \rightarrow A + X$$

Examples:
Water is decomposed, yielding hydrogen and oxygen.

$$2H_2O \rightarrow 2H_2(g) + O_2(g)$$

When heated, potassium chlorate decomposes, yielding potassium chloride and oxygen.

$$2KClO_3 \rightarrow 2KCl + 3O_2(g)$$

Mercury(II) oxide decomposes when heated. The products are metallic mercury and oxygen. See Figure 8-6.

Products of decomposition reactions aren't always elements.

$$2HgO \rightarrow 2Hg(l) + O_2(g)$$

3. *Replacement reactions,* in which one substance is replaced in its compound by another substance. Replacement reactions have the general form

$$A + BX \rightarrow AX + B$$

or

$$Y + BX \rightarrow BY + X$$

Examples:
Iron replaces copper in a solution of copper(II) sulfate, yielding iron(II) sulfate and copper.

$$Fe + CuSO_4 \rightarrow FeSO_4 + Cu(s)$$

The Fe atom loses 2 electrons to the Cu^{++} ion and replaces it in the solution as an Fe^{++} ion. The Cu^{++} ion gains 2 electrons and leaves the solution as a Cu atom.

Copper replaces silver in a solution of silver nitrate, and copper(II) nitrate and silver are the products.

$$Cu + 2AgNO_3 \rightarrow Cu(NO_3)_2 + 2Ag(s)$$

The Cu atom loses 2 electrons, 1 to each of the 2 Ag^+ ions, and replaces them in the solution as a Cu^{++} ion. The 2 Ag^+ ions gain 1 electron each and leave the solution as Ag atoms.

This reaction is shown in Figure 8-7. Silver nitrate solutions are colorless; copper(II) nitrate solutions are blue. Observe the color of the solution and the spongy deposit of silver on the copper strip in the beaker on the right.

Chlorine replaces iodine in a solution of potassium iodide, yielding potassium chloride and iodine.

$$Cl_2 + 2KI \rightarrow 2KCl + I_2$$

In this replacement reaction, the iodide ion, I^-, loses an electron to a chlorine atom. The I^- ion becomes an I atom; iodine remains in solution as I_2 molecules. The Cl atom becomes a chloride ion, Cl^-. Potassium ions, K^+, do not participate in the replacement reaction.

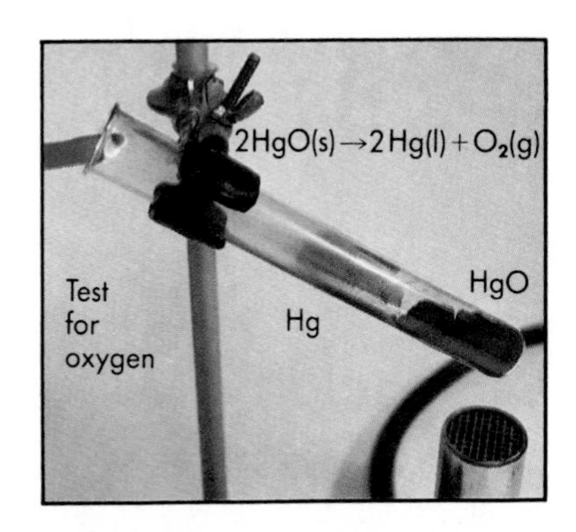

Figure 8-6. A decomposition reaction. With mild heating, mercury(II) oxide decomposes to metallic mercury and oxygen gas.

Figure 8-7. A replacement reaction. Copper atoms replace silver ions from the colorless silver nitrate solution, forming the blue copper nitrate solution and the spongy white deposit of silver on the copper strip.

In these first three types of reactions (composition, decomposition, and replacement), some change in the sharing of electrons occurs, or there is a transfer of electrons from one atom to another. Usually, but not always, there are changes in oxidation states.

4. *Ionic reactions,* which involve no transfer of electrons. Instead, ions in solution combine to form a product that leaves the reaction environment. Ionic reactions may have the general form

$$A^{+}(aq) + B^{-}(aq) \rightarrow AB$$

Examples:

A solution of sodium chloride, containing sodium ions and chloride ions, is added to a solution of silver nitrate, containing silver ions and nitrate ions. A reaction occurs between the silver ions and chloride ions. A white precipitate of silver chloride is formed.

$$Ag^{+}(aq) + Cl^{-}(aq) \rightarrow Ag^{+}Cl^{-}(s)$$

Aqueous ions that do not participate in an ionic reaction are called spectator ions.

The sodium ions and nitrate ions do not participate in the reaction but remain in solution as aqueous ions. By evaporating the water, they can be recovered as ionic crystals of sodium nitrate. The sodium ions and silver ions can be regarded as having exchanged places. For this reason, ionic reactions are sometimes referred to as *exchange* reactions.

Formula equations can be written for ionic reactions to identify the substances used as sources of the participating ions. These equations have the general form

$$AX + BY \rightarrow AY + BX$$

Figure 8-8. An ionic reaction. When aqueous potassium iodide is added to aqueous lead(II) nitrate, insoluble lead(II) iodide separates as a yellow precipitate.

In this sense, the ionic reaction just described is written

$$NaCl + AgNO_3 \rightarrow NaNO_3 + AgCl(s)$$

When a solution of potassium iodide is added to a solution of lead(II) nitrate, insoluble lead(II) iodide separates as a yellow precipitate. See Figure 8-8. The potassium ions and nitrate ions remain in solution and do not participate in the ionic reaction. The net equation is

$$Pb^{++}(aq) + 2I^{-}(aq) \rightarrow PbI_2(s)$$

In order for an ionic reaction to occur, a product must be formed that separates ions from the reaction environment (the solution). The product may be a solid precipitate, an insoluble gas, or a new molecular species (water).

Use the solubility table, Appendix Table 12, to help determine whether a proposed ionic reaction occurs.

8.5 Six kinds of decomposition reactions You have learned that the first step in writing a balanced equation is to represent the facts about the reaction. You must know what substances react and what products are formed. In this connection, it can be helpful to recognize general reaction patterns among chemically similar substances.

For example, many metallic carbonates have a similar reaction pattern when heated. They decompose into the corresponding metallic oxide and carbon dioxide gas. When this reaction pattern is known, equations can be written for the decomposition of several different metallic carbonates. Other patterns are evident in decomposition and replacement reactions.

Decomposition reactions are promoted by heat or electricity.

The kinds of decomposition reactions generally recognized are:

1. *Metallic carbonates, when heated, form metallic oxides and carbon dioxide.* When heated, calcium carbonate, $CaCO_3$, forms calcium oxide, CaO. Carbon dioxide, CO_2, is given off as a gas.

$$CaCO_3 \rightarrow CaO + CO_2(g)$$

Ammonium carbonate, $(NH_4)_2CO_3$, because of the nonmetallic nature of the ammonium ion, decomposes in a special manner. The equation for this reaction is

$$(NH_4)_2CO_3 \rightarrow 2NH_3(g) + H_2O(g) + CO_2(g)$$

Ammonia, steam, and carbon dioxide are produced.

2. *Many metallic hydroxides, when heated, decompose into metallic oxides and water.* If calcium hydroxide, $Ca(OH)_2$, is strongly heated, steam is given off and calcium oxide, CaO, remains.

$$Ca(OH)_2 \rightarrow CaO + H_2O(g)$$

Sodium hydroxide and potassium hydroxide are common exceptions to this rule.

3. *Metallic chlorates, when heated, decompose into metallic chlorides and oxygen.* This is the type of reaction used to prepare oxygen from potassium chlorate.

$$2KClO_3 \rightarrow 2KCl + 3O_2(g)$$

Knowing these general decomposition reaction patterns helps you predict the decomposition products of other similar reactions.

4. *Some acids, when heated, decompose into nonmetallic oxides and water.* Acids may be formed by the composition reaction of certain nonmetallic oxides and water. The reactions described below involve the reverse process—the decomposition of the acid. Carbonic acid yields water and carbon dioxide gas.

$$H_2CO_3 \rightarrow H_2O + CO_2(g)$$

Sulfurous acid yields water and sulfur dioxide gas.

$$H_2SO_3 \rightarrow H_2O + SO_2(g)$$

The two reactions above take place quite readily at room temperature. The following reaction occurs at higher temperatures.

$$H_2SO_4 \rightarrow H_2O + SO_3(g)$$

5. *Some oxides, when heated, decompose.* Most oxides are very stable compounds. There are only a few that decompose on heating. Two such reactions are

$$2HgO \rightarrow 2Hg + O_2(g)$$

$$2PbO_2 \rightarrow 2PbO + O_2(g)$$

6. *Some decomposition reactions are produced by an electric current.* The following reactions are typical.

$$2H_2O \xrightarrow{\text{(electricity)}} 2H_2(g) + O_2(g)$$

$$2NaCl \xrightarrow{\text{(electricity)}} 2Na + Cl_2(g)$$

Important reaction conditions are sometimes shown above or below the yields sign in a chemical equation.

The separation of a compound into simpler substances by an electric current is called ***electrolysis.***

8.6 Four kinds of replacement reactions The quantities of energy involved in replacement reactions are generally smaller than those in composition and decomposition reactions. The possibility of reaction depends on the relative activities of the elements involved. An experimentally derived *activity series,* as in Table 8-2 (Section 8.8), is generally useful for writing replacement reactions. Four kinds of replacement reactions are considered here.

1. *Replacement of a metal in a compound by a more active metal.* One reaction of this type involving zinc and a solution of copper(II) sulfate, $CuSO_4$, is demonstrated in Figure 8-9. Zinc replaces the copper in the solution. This reaction indicates that zinc is a more active metal than copper.

$$Zn + CuSO_4 \rightarrow ZnSO_4 + Cu(s)$$

2. *Replacement of hydrogen in water by metals.* The very active metals, such as potassium, calcium, and sodium, react vigorously with water. They replace half the hydrogen to form metallic hydroxides. The reaction represented by the following equation is typical.

$$Ca + 2H_2O \rightarrow Ca(OH)_2 + H_2(g)$$

At elevated temperatures less active metals, such as magnesium, zinc, and iron, react with water (steam) to replace hydrogen. Because of the high temperature involved, oxides rather

Figure 8-9. Zinc is a more active metal than copper and replaces copper in the aqueous copper(II) sulfate solution, forming the colorless zinc sulfate solution.

than hydroxides are formed. Metals less active than iron do not react measurably with water.

Refer to the activity series when determining whether a proposed replacement reaction occurs.

3. *Replacement of hydrogen in acids by metals.* Many metals react with certain acids, such as hydrochloric acid and dilute sulfuric acid. These metals replace hydrogen in the acids to form the corresponding metallic compounds. This method is commonly used for the laboratory preparation of hydrogen, in which sulfuric acid reacts with zinc.

$$Zn + H_2SO_4 \rightarrow ZnSO_4 + H_2(g)$$

4. *Replacement of halogens.* The halogens are the elements in Group VII of the periodic table. These elements are fluorine, chlorine, bromine, and iodine. Experiments show that fluorine is the most active halogen; it replaces the other three halogens in their compounds. Chlorine replaces bromine and iodine in their compounds. Bromine replaces only iodine. The replacement by chlorine of bromide ions in a potassium bromide solution is an example of the halogen replacement reaction.

$$Cl_2 + 2KBr \rightarrow 2KCl + Br_2$$

Chlorine replaces iodide ions, forming the corresponding chloride, NaCl.

$$Cl_2 + 2NaI \rightarrow 2NaCl + I_2$$

Bromine replaces iodide ions, but not as vigorously as chlorine does.

$$Br_2 + 2KI \rightarrow 2KBr + I_2$$

8.7 Many reactions are reversible Frequently, the products of a chemical reaction can react to produce substantial amounts of the original reactants. Hydrogen can be used as a reducing agent to separate certain metals from their oxides. If dry hydrogen gas is passed over hot magnetic iron oxides, iron and steam are produced.

$$4H_2 + Fe_3O_4 \rightarrow 3Fe + 4H_2O(g)$$

If the procedure is reversed and steam is passed over hot iron, magnetic iron oxide and hydrogen are formed.

$$3Fe + 4H_2O \rightarrow Fe_3O_4 + 4H_2(g)$$

Such reactions are said to be reversible. Reversible reactions are identified by two yields signs that are pointing in opposite directions ($\leftrightarrows$).

$$3Fe + 4H_2O \leftrightarrows Fe_3O_4 + 4H_2$$

Conditions may be such that both reactions occur at the same time. That is, if none of the products leaves the field of action, they may react to form the original reactants. Under such circumstances, an equilibrium (state of balance) may develop with the two opposing reactions proceeding at equal rates. After equilibrium is reached, the quantities of all the reactants remain constant. The subject of equilibrium reactions is discussed in Chapter 21.

8.8 The activity series of the elements In general, the ease with which the atoms of a metal lose electrons determines the ease with which the metal forms compounds. In the replacement reaction

$$\mathbf{A + BX \rightarrow AX + B}$$

metal **A** gives up electrons to metal **B** and replaces it. Thus metal **A** is shown to be more active than metal **B.** If metal **B** is immersed in a solution of the compound **AX,** metal **B** does not replace metal **A.** Atoms of a more active metal lose electrons to positively charged ions of a less active metal under proper reaction conditions. Similarly, atoms of more active nonmetals acquire electrons from negatively charged ions of less active nonmetals.

Zinc, being a more active metal than copper, replaces copper from a copper(II) sulfate solution. This replacement reaction is described in Section 8.6 and is shown in Figure 8-9.

$$\mathbf{Zn + Cu^{++}SO_4^{--} \rightarrow Zn^{++}SO_4^{--} + Cu}$$

Each Zn atom loses two electrons to a Cu^{++} ion to form a Zn^{++} ion.

$$\mathbf{Zn \rightarrow Zn^{++} + 2e^-}$$

Each Cu^{++} ion that gains two electrons forms a Cu atom.

$$\mathbf{Cu^{++} + 2e^- \rightarrow Cu}$$

Observe that SO_4^{--} ions do not participate in this replacement reaction. The net equation is

$$\mathbf{Zn(s) + Cu^{++}(aq) \rightarrow Zn^{++}(aq) + Cu(s)}$$

When copper is placed in a solution of zinc sulfate, no reaction is observed. Copper is a less active metal than zinc and does not replace zinc from a solution of a zinc compound.

Many spontaneous chemical reactions can be carried out in such a way that electric energy is given off. From a study of such reactions, chemists are able to devise an activity series of elements that helps predict the course of replacement reactions. Some composition and decomposition reactions can likewise be predicted with the aid of an activity series. The series presented in Table 8-2 lists important common elements in descending order of their metallic and nonmetallic activities.

Table 8-2
ACTIVITY SERIES OF THE ELEMENTS

Metals	*Nonmetals*
lithium	fluorine
potassium	chlorine
calcium	bromine
sodium	iodine
magnesium	
aluminum	
zinc	
chromium	
iron	Decreasing activity ↓
nickel	
tin	
lead	
HYDROGEN	
copper	
mercury	
silver	
platinum	
gold	

120.5

Figure 8-10. The very active metal, sodium, vigorously displaces hydrogen from water and forms the metallic hydroxide, NaOH.

The relative positions of the elements in the activity series support the application of some of the folowing generalizations to composition, decomposition, and replacement reactions.

1. Each element in the list displaces from a compound any of the elements below it. The larger the interval between elements in the series, the more vigorous the replacement reaction.

2. All metals above hydrogen displace hydrogen from hydrochloric acid or dilute sulfuric acid.

3. Metals above magnesium vigorously displace hydrogen from water. Magnesium displaces hydrogen from steam.

4. Metals above silver combine directly with oxygen; those near the top do so rapidly.

5. Metals below mercury form oxides only indirectly.

6. Oxides of metals below mercury decompose with mild heating.

7. Oxides of metals below chromium easily undergo reduction to metals by heating with hydrogen.

8. Oxides of metals above iron resist reduction by heating with hydrogen.

9. Elements near the top of the series are never found free in nature.

10. Elements near the bottom of the series are often found free in nature.

Mass Relationships

8.9 Stoichiometry The determination of empirical formulas for compounds is always the result of experimentation. Empirical formulas are derived from the relative numbers of moles of atoms or ions of the elements present in compounds. Therefore, the empirical formulas indicate the relative numbers of atoms or ions present. An empirical formula reveals nothing about the nature of the association of these particles or whether a substance actually exists in simple molecular units. Nevertheless, empirical formulas are very useful in calculations involving the combining and reacting relationships among substances.

The branch of chemistry that deals with the numerical relationships of elements and compounds and the mathematical proportions of reactants and products in chemical reactions is known as ***stoichiometry*** (stoy-key-*om*-eh-tree). The determinations of percentage composition of compounds and of empirical formulas are discussed in Chapter 7. The calculations for these determinations are based upon *stoichiometric relationships.* You are now ready to learn to solve stoichiometric problems involving the mass relations of reactants and products in chemical reactions. To do this, you will need an understanding of the mole concept and some skill in writing and balancing chemical equations. The concept of mole volumes of gases is introduced in Chapter 11. This concept will enable you to use simple computations to solve a great number of problems involving mass and volume relationships of gaseous reactants and products. These simple mass and volume relationships exist only for gases.

8.10 Mole relationships of reactants and products When carbon burns in the oxygen of the air, carbon dioxide, a covalent molecular gas, is produced.

$$C + O_2 \rightarrow CO_2(g)$$

The balanced equation indicates the mole proportions of the reactants and products. It also gives the composition of each substance in terms of the kinds of elements and the relative number of each kind of atom present. Thus the equation signifies that 1 mole of carbon atoms combines with 1 mole of oxygen molecules to yield 1 mole of carbon dioxide molecules. This may be indicated as follows:

C	+	O_2	$\rightarrow$	$CO_2(g)$
1 mole		**1 mole**		**1 mole**
= 12.0 g		**= 32.0 g**		**= 44.0 g**

The mole proportions of reacting substances and products convert readily to equivalent mass quantities, as shown above. An equation is used in this way to determine the mass of one substance that reacts with, or is produced from, a definite mass of another. This is one of the common problems chemists are called upon to solve.

8.11 Mole method of solving mass-mass problems The mole concept is both important and practical in chemistry. Here is an example of its use in solving a typical mass-mass problem.

Suppose you want to determine the mass of calcium oxide produced by heating 50.0 g of calcium carbonate. Observe that the mass of the reactant is given and the mass of a product is required. From the data given in the problem and the facts known concerning this reaction, you can *set up the problem*. This can be done in four steps.

The correct solution to a problem requires a balanced equation.

Step 1: *Write the balanced equation.*

Step 2: *Show the problem specifications: what is given and what is required.* To do this, write the mass of calcium carbonate, 50.0 g, above the formula $CaCO_3$. Let X represent the unknown mass of calcium oxide produced, and place it above the formula CaO.

Step 3: *Show the mole proportions established by the balanced equation.* This is done by writing under each substance in the problem the number of *moles* indicated by the equation.

Step 4: *Determine the mass of 1 mole of each substance involved in the problem.* These masses are written below the equation setup. The problem is now ready to be solved.

Step 2:	50.0 g	X	
Step 1:	$\mathbf{CaCO_3}$ →	**CaO** +	$\mathbf{CO_2(g)}$
Step 3:	1 mole	1 mole	

Step 4: 1 mole $CaCO_3$ = 100.1 g
1 mole CaO = 56.1 g

The number of moles of $CaCO_3$ *given in the problem* is found by multiplying the given mass of $CaCO_3$, 50.0 g, by the fraction $\frac{\text{mole}}{100.1 \text{ g}}$, or

$$\mathbf{50.0 \not{g}\ CaCO_3 \times \frac{mole}{100.1 \not{g}} = moles\ CaCO_3}$$

The balanced equation indicates that for each mole of $CaCO_3$ decomposed, 1 mole of CaO is produced. So from the given mass of $CaCO_3$ you can produce

$$\mathbf{50.0 \not{g}\ \cancel{CaCO_3} \times \frac{\cancel{mole}}{100.1 \not{g}} \times \frac{1\ mole\ CaO}{1\ \cancel{mole}\ \cancel{CaCO_3}} = moles\ CaO}$$

To determine the mass of CaO produced, multiply by the mass of 1 mole of CaO.

$$\mathbf{50.0 \not{g}\ \cancel{CaCO_3} \times \frac{\cancel{mole}}{100.1 \not{g}} \times \frac{1\ \cancel{mole}\ CaO}{1\ \cancel{mole}\ \cancel{CaCO_3}} \times \frac{56.1\ g}{\cancel{mole}} = g\ CaO}$$

It is always prudent to *estimate* the answer to a problem before starting the computations. A simple units-operation check of the solution setup will reveal the answer unit and the correctness of the setup. For this problem, the unit operations indicate a result in g CaO. Thus the problem solution is set up correctly. A numerical estimate indicates an answer of approximately 30 ($\frac{1}{2} \times 1 \times$ about 60). The problem solution can now be expected to yield a reliable answer of approximately 30 g CaO.

$$\mathbf{X = 50.0\ \cancel{g}\ \cancel{CaCO_3} \times \frac{\cancel{mole}}{100.1\ \cancel{g}} \times \frac{1\ \cancel{mole}\ CaO}{1\ \cancel{mole}\ \cancel{CaCO_3}} \times \frac{56.1\ g}{\cancel{mole}}}$$

$$\mathbf{X = 28.0\ g\ CaO}$$

The calculated result, 28.0 g CaO, is in good agreement with the estimated value of approximately 30 g CaO. This mass-mass problem-solving technique is further illustrated in Sample Problems 8-1 and 8-2.

SAMPLE PROBLEM 8-1

How many grams of potassium chlorate must be decomposed to yield 30.0 g of oxygen?

SOLUTION

First, set up the problem by writing the balanced equation. *Second,* write the specifications of the problem above the equation. *Third,* write the number of moles of each specified substance under its formula. And *fourth,* calculate the mass/mole of each of the specified substances. Let **X** represent the mass of potassium chlorate decomposed.

Step 2: X 30.0 g
Step 1: $\mathbf{2KClO_3 \rightarrow 2KCl + 3O_2(g)}$
Step 3: 2 moles 3 moles

Step 4: $\mathbf{1\ mole\ KClO_3 = 122.6\ g}$
$\mathbf{1\ mole\ O_2 = 32.0\ g}$

$$\mathbf{X = \underbrace{30.0\ \cancel{g}\ \cancel{O_2} \times \frac{\cancel{mole}}{32.0\ \cancel{g}}} \times \underbrace{\frac{2\ \cancel{moles}\ KClO_3}{3\ \cancel{moles}\ \cancel{O_2}}} \times \underbrace{\frac{122.6\ g}{\cancel{mole}}} = \underbrace{76.6\ g\ KClO_3}}$$

Operation 1 gives expression for moles O_2 given

Operation 2 gives expression for moles $KClO_3$ required

Operation 3 gives expression for grams $KClO_3$ required

Operation 4 involves unit check, estimate of answer, and arithmetic operations

SAMPLE PROBLEM 8-2

(*a*) How many grams of oxygen are required to oxidize $14\bar{0}$ g of iron to iron(III) oxide? (*b*) How many moles of iron(III) oxide are produced?

SOLUTION

The problem setup for Parts (*a*) and (*b*) is

$$
\begin{array}{llcccc}
\textbf{Step 2:} & 14\bar{0}\ \text{g} & & X & & Y \\
\textbf{Step 1:} & 4\text{Fe} & + & 3O_2(g) & \rightarrow & 2Fe_2O_3 \\
\textbf{Step 3:} & 4\ \text{moles} & & 3\ \text{moles} & & 2\ \text{moles}
\end{array}
$$

$$
\begin{array}{ll}
\textbf{Step 4:} & 1\ \text{mole Fe} = 55.8\ \text{g} \\
 & 1\ \text{mole}\ O_2 = 32.0\ \text{g}
\end{array}
$$

The solution setup for Part (*a*) is

$$X = 14\bar{0}\ \cancel{\text{g Fe}} \times \frac{\cancel{\text{mole}}}{55.8\ \cancel{\text{g}}} \times \frac{3\ \cancel{\text{moles}}\ O_2}{4\ \cancel{\text{moles Fe}}} \times \frac{32.0\ \text{g}}{\cancel{\text{mole}}} = 60.2\ \text{g}\ O_2$$

The solution setup for Part (*b*) is

$$Y = 14\bar{0}\ \cancel{\text{g Fe}} \times \frac{\cancel{\text{mole}}}{55.8\ \cancel{\text{g}}} \times \frac{2\ \text{moles}\ Fe_2O_3}{4\ \cancel{\text{moles Fe}}} = 1.25\ \text{moles}\ Fe_2O_3$$

Practice Problems

Use factor-label techniques and maintain the proper numbers of significant figures in all calculations.

1. An excess of chlorine gas is bubbled into an aqueous solution containing 4.65 g of potassium bromide. (*a*) What mole quantity of molecular bromine is produced? (*b*) What is the mass of this mole quantity of bromine?
   ans. (*a*) 0.0195 mole Br_2 (*b*) 3.12 g Br_2
2. When magnesium burns in air, it combines with oxygen to form magnesium oxide. (*a*) What mass of oxygen combines with 10.0 g of magnesium in this reaction? (*b*) What mass of magnesium oxide is produced?
   ans. (*a*) 6.58 g O_2 (*b*) 16.6 g MgO

SUMMARY

A chemical equation is a concise, symbolized statement of a chemical reaction. Useful quantitative information about a reaction is provided by a balanced formula equation.

Three conditions must be met in a balanced equation. (1) The equation must represent the facts of the reaction, the substances that react and the products formed. (2) The symbols or formulas for all reactants and products must be written correctly. (3) The law of conservation of atoms must be satisfied.

In some chemical reactions, changes occur in the sharing of electrons, or electrons are transferred from one reactant to another. In other reactions, aqueous ions combine to form a product that separates from the reaction environment. Chemical reactions may be separated into four main types: (1) composition; (2) decomposition; (3) replacement; and (4) ionic.

In chemical reactions, the numerical relationships of elements and compounds and the proportions of reactants and products are called stoichiometry. Problems dealing with mass relationships of the reactants and products in chemical reactions are stoichiometric problems. Solutions of stoichiometric problems are based on the fact that balanced equations establish the quantitative relationships among reactants and products. The quantities in these equations are expressed in moles.

VOCABULARY

activity series
chemical equation
composition reaction
decomposition reaction
electrolysis
ionic reaction
law of conservation of atoms
mass-mass problem
mole
precipitate
product
qualitative
quantitative
reactant
replacement reaction
reversible reaction
spectator ion
stoichiometry
synthesis
word equation
yield sign

EQUATIONS

Group A

Write balanced chemical equations for these reactions and identify the type of reaction. *Do not write in this book.* *A2–A5*

1. zinc + sulfur → zinc sulfide.
2. potassium chloride + silver nitrate → silver chloride(s) + potassium nitrate.
3. calcium oxide + water → calcium hydroxide.
4. sodium hydroxide + hydrochloric acid (HCl) → sodium chloride + water.
5. magnesium bromide + chlorine → magnesium chloride + bromine.
6. sodium chloride + sulfuric acid (H_2SO_4) → sodium sulfate + hydrogen chloride(g).
7. aluminum + iron(III) oxide → aluminum oxide + iron.
8. ammonium nitrite → nitrogen(g) + water.
9. silver nitrate + nickel → nickel(II) nitrate + silver(s).
10. hydrogen + bromine → hydrogen bromide(g).

Complete the word equation and write the balanced chemical equation. Give a reason for the product(s) in each case. Consult the activity series in Table 8-2, and solubilities in Appendix Table 12, as necessary. *A2–A5*

Composition reactions:

11. sodium + iodine →
12. calcium + oxygen →
13. hydrogen + chlorine →

Decomposition reactions:

14. nickel(II) chlorate →
15. barium carbonate →
16. zinc hydroxide →

Replacement reactions:

17. aluminum + sulfuric acid →
18. potassium iodide + chlorine →
19. iron + copper(II) nitrate → iron(II) nitrate +

Ionic reactions:

20. silver nitrate + zinc chloride →
21. copper(II) hydroxide + acetic acid ($HC_2H_3O_2$) →
22. iron(II) sulfate + ammonium sulfide →

Group B

If the word equation is complete, write and balance the chemical equation. If the word equation is incomplete, complete it. Write and balance the formula equation. Tell the type of reaction. Give a reason for the product(s). *A2–A5*

23. barium chloride + sodium sulfate →
24. calcium + hydrochloric acid →
25. iron(II) sulfide + hydrochloric acid → hydrogen sulfide(g) +
26. zinc chloride + ammonium sulfide →
27. ammonia (NH_3) + oxygen → nitric acid (HNO_3) + water.
28. magnesium + nitric acid →

29. potassium + water →
30. sodium iodide + bromine →
31. silver + sulfur →
32. sodium chlorate →
33. carbon + steam (H_2O) → carbon monoxide(g) + hydrogen(g).
34. zinc + lead(II) acetate →
35. iron(III) hydroxide →
36. iron(III) oxide + carbon monoxide → iron + carbon dioxide(g).
37. lead(II) acetate + hydrogen sulfide →
38. aluminum bromide + chlorine →
39. magnesium carbonate →
40. iron(III) chloride + sodium hydroxide →
41. calcium oxide + diphosphorus pentoxide → calcium phosphate.
42. chromium + oxygen →
43. sodium + water →
44. calcium carbonate + hydrochloric acid →
45. calcium hydroxide + phosphoric acid (H_3PO_4) →
46. sodium carbonate + nitric acid →
47. aluminum hydroxide + sulfuric acid →
48. sodium sulfite + sulfuric acid →
49. copper + sulfuric acid → copper(II) sulfate + water + sulfur dioxide(g).
50. calcium hydroxide + ammonium sulfate → calcium sulfate + water + ammonia(g).

PROBLEMS

Group A

1. When a mass of 12.7 g of magnesium is burned in air, it combines with oxygen to form magnesium oxide. (*a*) Write the chemical equation for this reaction. (*b*) What mole ratios of reactants and products are indicated by the balanced equation? (*c*) Determine the mass of 1 mole of each reactant and product. (*d*) How many moles of magnesium are burned? (*e*) How many moles of oxygen are required? (*f*) How many moles of magnesium oxide are formed? (*g*) How many grams of oxygen are required? (*h*) How many grams of magnesium oxide are formed? *A2–A5, B1*

2. When a mass of 32.5 g of nitrogen reacts with hydrogen under appropriate conditions, ammonia (NH_3) is formed as the only product. (*a*) Write the chemical equation for this reaction. (*b*) What mole ratios of reactants and products are indicated by the balanced equation? (*c*) Determine the mass of 1 mole of each reactant and product. (*d*) How many moles of nitrogen react? (*e*) How many moles of hydrogen are required? (*f*) How many moles of ammonia are formed? (*g*) What mass of hydrogen is required? (*h*) What mass of ammonia is formed? *A2–A5, B1*

3. A mass of 37.5 g of mercury(II) oxide is decomposed by heating. The reaction is $2HgO \rightarrow 2Hg + O_2$. (*a*) How many moles of mercury(II) oxide are decomposed? (*b*) How many moles of oxygen are prepared? (*c*) How many grams of oxygen are prepared? *A2–A5, B1*

4. A mass of 50.0 g of potassium chlorate is decomposed by heating. (*a*) How many moles of potassium chlorate are decomposed? (*b*) How many moles of oxygen are prepared? (*c*) What is the mass of oxygen prepared? *A2–A5, B1*

5. A quantity of zinc reacts with sulfuric acid to produce 0.10 g of hydrogen. (*a*) How many moles of hydrogen are produced? (*b*) How many moles of zinc are required? (*c*) How many grams of zinc are required? *A2–A5*

6. Sodium chloride reacts with 15.0 g of silver nitrate in water solution. (*a*) How many moles of silver nitrate react? (*b*) How many moles of sodium chloride are required? (*c*) How many grams of sodium chloride are required? *A2–A5, B1*

7. (*a*) How many moles of silver chloride are precipitated in the reaction of Problem 6? (*b*) How many grams of silver chloride is this? *A2–A5, B1*

8. In a reaction between sulfur and oxygen, the only product is 80.0 g of sulfur dioxide. How many grams of sulfur were burned? *A2–A5, B1*

9. How many grams of hydrogen are required to completely convert 35 g of hot magnetic iron oxide (Fe_3O_4) to elemental iron? Steam is the other product of the reaction. *A2–A5, B1*
10. How many grams of copper(II) oxide can be formed by oxidizing 1.00 kg of copper? *A2–A5, B1*
11. What mass of silver is precipitated when a mass of 50.0 g of copper reacts with an excess of silver nitrate in solution? *A2–A5, B1*
12. Suppose a mass of 10.0 g of iron(II) sulfide reacts with an excess of hydrochloric acid. How many grams of hydrogen sulfide gas are produced? *A2–A5, B1*

Group B

13. An excess of sulfuric acid reacts with 165 g of barium peroxide. (*a*) How many moles of hydrogen peroxide are produced? (*b*) How many moles of barium sulfate? *A2–A5, B1*
14. A mass of 130 g of zinc was added to a solution containing $10\bar{0}$ g of HCl. After the reaction ceased, 41 g of zinc remained. How many moles of hydrogen were produced? *A2–A5, B1*
15. A mixture of 10.0 g of powdered iron and 10.0 g of sulfur is heated to its reaction temperature in an open crucible. (*a*) How many grams of iron(II) sulfide are formed? (*b*) The reactant in excess is oxidized. How many grams of its oxide are formed? *A2–A5*
16. What mass in grams of calcium hydroxide can be produced from 1.20 kg of limestone, calcium carbonate? (Decomposition of calcium carbonate with heat yields calcium oxide and carbon dioxide. Calcium hydroxide is formed by the reaction of calcium oxide and water.) *A2–A5, B1*
17. How many grams of air are required to complete the combustion of 93 g of phosphorus to diphosphorus pentoxide, assuming the air to be 23% oxygen by mass? *A2–A5, B1*
18. How many metric tons of carbon dioxide can be produced from the combustion of 5.00 metric tons of coke that is 85.5% carbon? *A2–A5, B1*
19. (*a*) What mass of H_2SO_4 in grams is required in a reaction with an excess of aluminum to produce 0.50 mole of aluminum sulfate? (*b*) How many moles of hydrogen are also produced? *A2–A5, B1*
20. A certain rocket uses butane, C_4H_{10}, as fuel. How many kilograms of liquid oxygen should be carried for the complete combustion of each 1.00 kg of butane to carbon dioxide and water vapor? *A2–A5, B1*
21. When a mass of 85 g of ethane gas, C_2H_6, is burned completely in air, carbon dioxide and water vapor are formed. (*a*) How many moles of carbon dioxide are produced? (*b*) How many moles of water are produced? *A2–A5, B1*
22. (*a*) How many grams of sodium sulfate are produced in the reaction between $15\bar{0}$ g of sulfuric acid and an excess of sodium chloride? (*b*) How many grams of sodium chloride are used? (*c*) How many grams of hydrogen chloride are also produced? *A2–A5, B1*

Harold Clayton Urey (American) received the 1934 Nobel Prize in chemistry for his discovery of heavy hydrogen, which he named deuterium. He proposed the existence of deuterium after comparing the experimental atomic weight of hydrogen with that of oxygen-16. He was able to concentrate deuterium by distilling liquid hydrogen near its triple point. Its presence was indicated by the appearance of new lines in the spectrum of hydrogen. Deuterium and heavy water (water whose molecular structure contains deuterium) are now industrially produced by electrolysis.

Two Important Gases: Oxygen and Hydrogen

9

GOALS

In this chapter you will gain an understanding of: • the occurrence and discovery of oxygen • the preparation, properties, and uses of oxygen • allotropy: ozone • the occurrence and discovery of hydrogen • the preparation, properties, and uses of hydrogen • isotopes: protium and deuterium

9.1 Introduction Thus far your study of chemistry has been concerned mostly with the theory of atomic structure and its explanation of the nature of elements and compounds. You have learned about common types of chemical reactions. The writing of chemical formulas and equations has been explained. Some calculations involving formulas and equations have been described, and you have had problems for practice. Now a more detailed description of gases, liquids, and solids than that given in Chapter 1 will be presented. But before gases in general are considered, two important gaseous elements, oxygen and hydrogen, will be studied.

Gases, liquids, and solids are described in Section 1.8.

OXYGEN

9.2 Occurrence of oxygen Oxygen, atomic number 8, has several characteristics that make it a very important element. It is the most abundant element in the earth's crust. In fact, it is estimated that, considered together, the entire earth, the waters on the earth, and the atmosphere surrounding the earth contain more oxygen atoms than atoms of any other single element. After hydrogen and helium, oxygen atoms rank third in abundance in the known universe. Oxygen is necessary for the support of plant and animal life. Oxygen combines with all other elements except the lower-atomic-weight noble gases.

You may wish to review the explanation of the earth's composition in text Section 2.7.

About one-fifth of the earth's atmosphere by volume is oxygen. Animals living in water get their oxygen from the small amount that is dissolved in water. Oxygen in the air and oxygen that is dissolved in water are examples of *free oxygen*. As the free element, oxygen consists of diatomic covalent molecules, $\mathbf{O_2}$.

Oxygen that has united with other elements in compounds is called *combined oxygen*. Combined oxygen is much more plentiful than free oxygen. Water contains almost 89% oxygen by mass in

Figure 9-1. Free oxygen comprises about one-fifth of the earth's atmosphere by volume. Combined oxygen is abundant in water and in the rocks of the earth's crust.

combination with hydrogen. Such minerals as clay, sand, and limestone contain a large percentage of combined oxygen. Oxygen is one of the elements present in most of the rocks and minerals of the earth's crust. See Figure 9-1.

9.3 Discovery of oxygen The discovery of oxygen is related to the work of three eighteenth-century scientists. Karl Wilhelm Scheele (*shay*-luh) (1742–1786), a Swedish pharmacist, prepared oxygen in 1771 by heating manganese dioxide with concentrated sulfuric acid.

In this reaction, for each $\overset{+4}{Mn}$ that is reduced to $\overset{+2}{Mn}$, an $\overset{-2}{O}$ is oxidized to $\overset{0}{O}$.

$$2MnO_2 + 2H_2SO_4 \rightarrow 2MnSO_4 + 2H_2O + O_2(g)$$

Scheele carried out experiments with oxygen which proved to him that oxygen was different from other known gases. He also identified oxygen as that part of the air involved in combustion and oxidation of metals. However, it was not until 1776 that the announcement of Scheele's discovery was published.

Meanwhile, Joseph Priestley (1733–1804), an English clergyman and scientist, prepared oxygen in 1774 by using a lens to focus the sun's rays on mercury(II) oxide. When this powdery red oxide is heated strongly, it decomposes. Oxygen gas is evolved, and liquid mercury remains.

The decomposition of mercury(II) oxide is shown in Figure 8-6.

$$2HgO \rightarrow 2Hg + O_2(g)$$

Priestley was delighted to find that a candle would flare up and continue to burn brightly in oxygen. Upon inhaling some of the gas, Priestley observed that he "felt peculiarly light and easy for some time afterwards." News of Priestley's work was published in 1775. Some of his laboratory apparatus is shown on the next page.

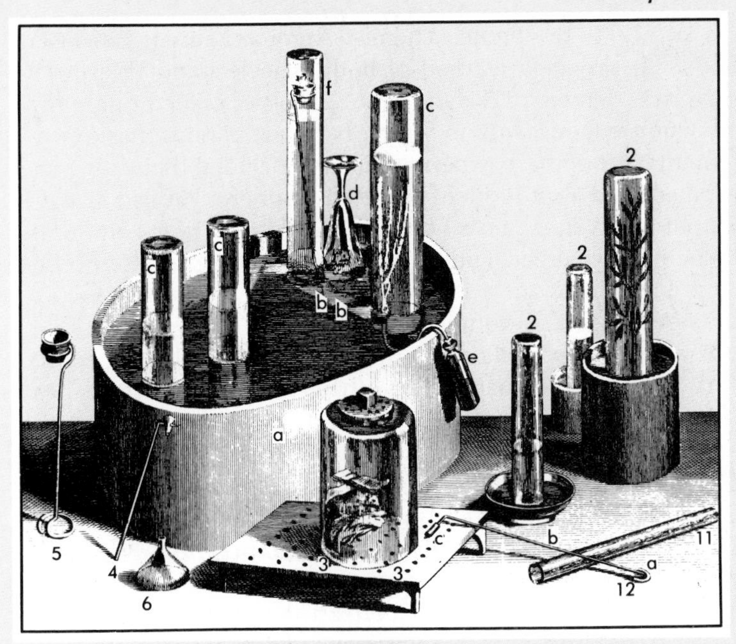

Karl Wilhelm Scheele, a Swedish pharmacist pictured above, prepared oxygen and studied some of its properties in 1771.

Some of the laboratory apparatus used by Priestley in his experiments on the properties of gases is shown top-left. The pneumatic trough a *is fitted with a shelf* bb. *The trough is filled with water. Two collecting jars* c *and* c *stand in the water. A gas is being produced by a chemical reaction in the generator bottle* e *and delivered through tubing into a jar* c *on the shelf of the trough. An inverted glass* d *contains a mouse breathing a confined gas. Jar* f *contains a cup mounted on a wire stand, also shown as* 5. *A growing plant is surrounded by a gas in large jar* 2. *Compare this equipment with that which you use for collecting and testing gases.*

Lavoisier's great contributions to chemistry involved extending and interpreting the experiments of others, such as Scheele and Priestley (codiscoverers of oxygen) and Cavendish (discoverer of hydrogen). In the engraving pictured at left, Lavoisier is shown carrying out an experiment in which he is using an electric spark to ignite a mixture of oxygen and hydrogen in an attempt to determine the composition of water.

In late 1774, the French chemist Antoine-Laurent Lavoisier (1743–1794) privately learned of both Scheele's and Priestley's discoveries. Between 1772 and 1786, Lavoisier carried out many combustion experiments in which he made careful mass measurements. From these experiments he concluded that oxygen is an element and a component of the atmosphere. He also discovered that when substances burn in air or oxygen they gain mass because of the oxygen with which they combine. These observations brought about a reevaluation of the phlogiston theory and its eventual abandonment.

Oxygen was given its name by Lavoisier. Scheele and Priestley are considered codiscoverers of this element. See page 199.

9.4 Preparation of oxygen 1. *By the decomposition of hydrogen peroxide.* This is a safe and convenient laboratory method of preparing oxygen. A dilute (6%) solution of hydrogen peroxide in water decomposes very slowly at room temperature. But manganese dioxide may be used as a catalyst to increase the rate of decomposition. If hydrogen peroxide solution is allowed to react drop by drop with manganese dioxide powder at room temperature, the hydrogen peroxide decomposes rapidly and smoothly. See Figure 9-2. Oxygen gas is given off, and water is formed.

A catalyst is a substance or combination of substances that alters the rate of a chemical reaction without itself being permanently changed. The purpose and use of catalysts are described in text Section 2.13(5).

$$2H_2O_2 \rightarrow 2H_2O(l) + O_2(g)$$

The oxygen gas produced in the flask passes through the delivery tube into a water-filled inverted bottle. As the oxygen rises in the bottle, it displaces the water. This method of collecting gases is known as *water displacement*. It is used for gases that are not very soluble in water.

2. *By adding water to sodium peroxide.* Sodium peroxide, Na_2O_2, is prepared by burning sodium in air. If water is allowed to drop onto sodium peroxide in a generator like that shown in Figure 9-2, oxygen is liberated.

$$2Na_2O_2 + 2H_2O \rightarrow 4NaOH + O_2(g)$$

This is another convenient laboratory method for preparing small quantities of oxygen since it does not require heat. **CAUTION:** *The sodium hydroxide solution that is formed must be disposed of carefully. It can burn the skin and destroy certain textile fibers.*

Decomposition of metallic chlorates, as a type of decomposition reaction, is explained in Section 8.5(3).

3. *By heating potassium chlorate.* Potassium chlorate is a white, crystalline solid. When this compound is heated, it decomposes. Oxygen is given off, and potassium chloride is left as a residue.

$$2KClO_3 \rightarrow 2KCl + 3O_2(g)$$

Manganese dioxide is usually mixed with the potassium chlorate in this laboratory preparation. It acts as a catalyst by lowering the decomposition temperature of potassium chlorate. **CAUTION:** *The decomposition of potassium chlorate is dangerous. It should*

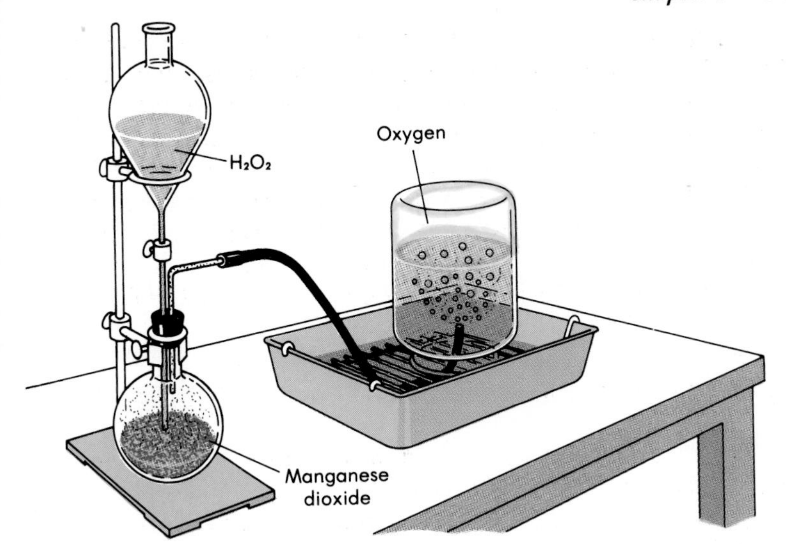

Figure 9-2. Oxygen may be prepared in the laboratory by the catalytic decomposition of hydrogen peroxide. Manganese dioxide is used as the catalyst. Oxygen is collected by water displacement. This same generator setup may be used for preparing oxygen by the action of water (placed in the dropping funnel) on sodium peroxide (placed in the flask).

be carried out with only small quantities of reagent-grade chemicals. Carefully assembled, inspected, and well-shielded equipment should be used under close supervision. Potassium chlorate is extremely dangerous when mixed with combustible materials.

4. *By the electrolysis of water.* Figure 9-3 represents a laboratory apparatus in which water may be decomposed by an electric current. During the electrolysis of water, a direct current is passed through the water. Oxygen gas collects at the positive terminal and hydrogen gas at the negative terminal. Sulfuric acid is added to make the water a better conductor of electricity.

Electrolysis is the production of a chemical change by electricity. Decomposition by electrolysis is explained in Section 8.5(6).

$$2H_2O \rightarrow 2H_2(g) + O_2(g)$$

Large quantities of electric energy are needed to decompose the water. Industrially, this method yields oxygen of the highest purity. The hydrogen, which is produced along with it, is sold as a by-product.

5. *From liquid air.* This is the common industrial method for preparing oxygen. Air condenses to a liquid if it is greatly compressed and at the same time is cooled to a very low temperature (−190 °C). Liquid air consists largely of oxygen and nitrogen. Liquid oxygen may be separated from liquid nitrogen on the basis of boiling-point difference. Liquid nitrogen boils at −195.8 °C, or about thirteen degrees lower than liquid oxygen at −183.0 °C. By a process that simultaneously involves both evaporation and condensation, liquid air can be separated into liquid oxygen that is at least 99.5% pure and nitrogen gas about 98% pure.

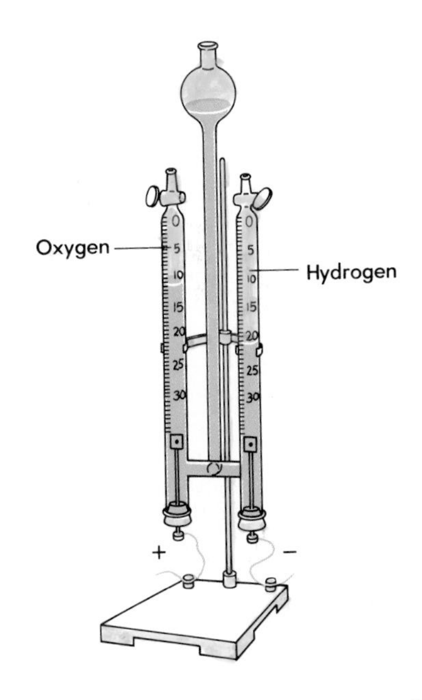

Figure 9-3. Water is decomposed by electrolysis into its constituent elements, oxygen and hydrogen.

9.5 Physical properties of oxygen Pure oxygen is a colorless, odorless, and tasteless gas. It is slightly denser than air. At 0 °C and one atmosphere pressure, one liter of oxygen has a mass of 1.43 g. Under the same conditions, one liter of air has a mass of

Table 9-1 PROPERTIES OF OXYGEN	
atomic number	8
atomic weight	15.9994
electron configuration	2, 6
oxidation numbers	−2, −1, +2
melting point	−218.4 °C
boiling point	−183.0 °C
density, 0 °C, 1 atm	1.429 g/L
atom radius	0.70 Å
ion radius, O^{--}	1.40 Å

1.29 g. Oxygen is slightly soluble in water. The colder the water is, the more oxygen it can dissolve. About five liters of oxygen can be dissolved in 100 liters of water at 0 °C, but only three liters of oxygen can be dissolved at 20 °C.

Any gas can be converted into a liquid if it is simultaneously compressed under a great enough pressure and cooled sufficiently. Liquid oxygen is pale-blue in color and boils at −183.0 °C. It is attracted by a magnet, which leads chemists to believe that there are unpaired electrons in an oxygen molecule. Thus the electron-dot formula for an oxygen molecule is sometimes written as

:Ö:Ö:

rather than

:O::O:

In 1983 the commercial production of oxygen in the United States was over thirteen million metric tons. Oxygen ranked fourth, after sulfuric acid, nitrogen, and lime in tonnage produced.

Further cooling of liquid oxygen results in its freezing to a pale-blue crystalline solid at −218.4 °C. Properties of oxygen are listed in Table 9-1.

9.6 Chemical properties of oxygen Oxygen is one of the most active elements. It combines with other elements to form compounds called *oxides. An **oxide** is a compound consisting of oxygen and usually one other element; in oxides, the oxidation number of oxygen is −2.* When oxides are formed by direct combination of the elements, the reaction is exothermic. Generally, oxides are very stable compounds. Pure oxygen is much more active than air.

Oxygen reacts with the metals of Groups I and II. High temperatures are required to start reactions with the lower-atomic-weight metals lithium, sodium, potassium, magnesium, and calcium. The higher-atomic-weight metals rubidium, cesium, strontium, and barium, on the other hand, react spontaneously at room temperature. The electronegativity difference between oxygen and each of these metals is large. The oxygen compounds formed are ionic compounds.

The composition reaction of magnesium burning in the oxygen of the air is shown in Figure 8-5.

Lithium reacts with oxygen to form lithium oxide.

$$4Li + O_2 \rightarrow 2Li^+{}_2O^{--}(s)$$

Depending on conditions, sodium reacts with oxygen to form sodium oxide or sodium peroxide. Heating sodium with dry oxygen at 180 °C yields sodium oxide.

$$4Na + O_2 \xrightarrow{180\ °C} 2Na^+{}_2O^{--}(s)$$

Combustion of sodium in oxygen yields sodium peroxide.

$$2Na + O_2 \rightarrow Na^+{}_2O_2{}^{--}(s)$$

The combustion of potassium, rubidium, and cesium in oxygen yields the peroxides.

The Group II metals react with oxygen under the temperature conditions just stated to yield oxides. The reaction with barium is typical.

$$2Ba + O_2 \rightarrow 2Ba^{++}O^{--}(s)$$

The reactions of nonmetals and oxygen generally occur at the high temperatures of combustion. The electronegativity difference between oxygen and the nonmetals is small. The resulting oxides contain covalent bonds and exist as molecules. The following reactions are examples of this type of oxide formation.

$$S + O_2 \rightarrow SO_2(g)$$
$$C + O_2 \rightarrow CO_2(g)$$
$$2H_2 + O_2 \rightarrow 2H_2O(l)$$

Figure 9-4. Iron burns brilliantly in pure oxygen to form the iron oxide Fe_3O_4.

Reactions between oxygen and metals other than those of Groups I and II may occur slowly at room temperature. They will occur, sometimes quite rapidly, if the temperature is raised. The electronegativity difference between oxygen and these metals is of intermediate value, from about 0.8 to 1.8.

Iron at room temperature and in the presence of moisture unites slowly with oxygen. The resulting product is iron(III) oxide, commonly called rust.

$$4Fe + 3O_2 \rightarrow 2Fe_2O_3(s)$$

A strand of steel picture wire or a small bundle of steel wool, heated red hot and plunged into pure oxygen, burns brilliantly and gives off bright sparks, as shown in Figure 9-4. Molten drops of Fe_3O_4, another oxide of iron, are formed in this reaction but quickly solidify as they cool.

The detailed explanation of the structures of solids is given in Chapter 12.

$$3Fe + 2O_2 \rightarrow Fe_3O_4(s)$$

Such metals as tin, lead, copper, and zinc unite with oxygen to form oxides. These reactions occur slowly when the elements are cold, but more rapidly when they are heated. Oxides of metals such as gold and platinum are formed only at high temperatures. The oxides of metals of intermediate electronegativity difference usually have structures that are more complex than those of ionic compounds or simple molecular compounds.

Figure 9-5. During steel making in a basic-oxygen furnace, impurities are burned off by directing a stream of oxygen from above into the molten metal.

9.7 The test for oxygen A blazing splint continues to burn in air, but it burns more vigorously in pure oxygen. A glowing splint, lowered into a bottle of pure oxygen, bursts into flame immediately. If a glowing splint, lowered into a bottle of colorless, odorless gas, bursts into flame, the gas *is* oxygen. This test is commonly used to identify oxygen.

9.8 Uses of oxygen 1. *As an essential for life.* Almost all living things require oxygen in order to live. The exceptions are certain

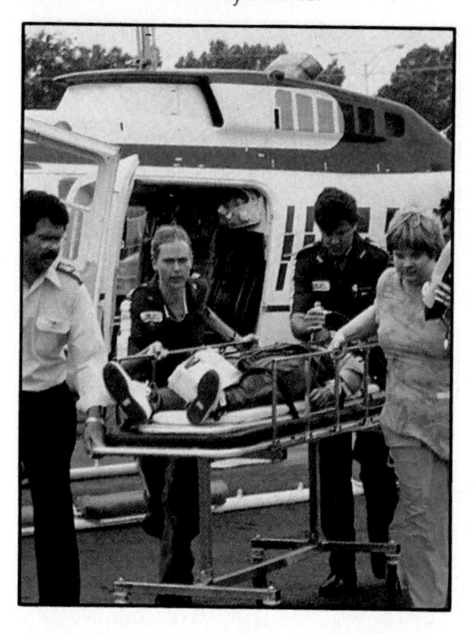

Figure 9-6. Portable oxygen equipment is being used for emergency treatment of patients whose breathing may have been impaired.

Figure 9-7. The liquid effluent from a sewage-processing plant is treated with oxygen by sprinkling it into the air over rocks.

microscopic organisms that can live with or without oxygen, and others that cannot live in the presence of oxygen.

2. *In the iron and steel industry.* This is the most important industrial use of oxygen. Oxygen is used in large quantities in the basic-oxygen method of making steel shown in Figure 9-5. It is also used for removing surface impurities from steel slabs and for welding and cutting steel. Welding and cutting iron and steel are done using oxygen-acetylene torches.

3. *For producing chemical compounds.* Depending on the proportion of oxygen used, the temperature, and the selection of catalyst, different products are formed when pure oxygen reacts with coal, natural gas, or liquid fuels. Sometimes a mixture of hydrogen and carbon monoxide results. This mixture can be used to make synthetic gasoline, methanol, or ammonia. Under other conditions, partially oxygenated carbon and hydrogen compounds are formed. They can be used to make detergents, synthetic rubber, or compounds used as antifreeze.

4. *For medical treatment.* Chronic lung diseases from smoking and occupational exposure to dust reduce the surface area of the lungs and lower the rate at which oxygen enters the blood. Pure oxygen or air to which more oxygen has been added is sometimes given to persons suffering from pneumonia, asthma, and other lung diseases. Oxygen concentrating machines provide a continuous, stationary supply of oxygen. Small portable oxygen tanks are used when mobility is needed. Oxygen is also administered to persons who may be too weak to inhale a normal quantity of air because of heart attacks. The increased concentration of oxygen in the blood helps limit the amount of heart muscle that dies because of the heart attack.

In cases of asphyxiation from inhaling smoke or suffocating gases, from apparent drowning, or from electric shock, oxygen may be administered by using a resuscitator. See Figure 9-6.

5. *For water and sewage treatment.* Fountains, cascades, and other aerating devices are used in water treatment. The oxygen thus introduced into the water helps remove odors of decomposing animal and vegetable matter, of growing microorganisms, and of dissolved gases. Aeration makes the water more tasty for drinking purposes.

The microorganisms used for purification in sewage treatment plants require large amounts of oxygen. Air or oxygen is bubbled into the waste water to provide the necessary oxygen. See Figure 9-7.

6. *For rocket propulsion.* Liquid oxygen is used in many rockets and missiles. The fuel may be a kerosene-like liquid or liquid hydrogen. For the rapid burning needed to propel rockets, oxygen must be supplied in large quantities. Oxygen is needed whether the rocket is moving through the earth's atmosphere or in interplanetary space. See Figure 9-8.

Ozone

9.9 Occurrence of ozone A peculiar odor in the air is often noticed after lightning has flashed nearby. This odor is due to the presence of ozone, O_3, another form of elemental oxygen. Electric discharges through the air, such as sparks from electric machinery or lightning flashes, convert some of the oxygen of the air into ozone.

The ultraviolet rays from the sun change some of the oxygen in the upper atmosphere into ozone. In the stratosphere, between 15 km and 50 km above the earth, ozone occurs to the extent of 5 to 10 parts per million. The greatest concentration of ozone is found at an altitude of about 25 km.

The absorption of ultraviolet rays by this upper-atmosphere layer of ozone protects the earth's surface from most of the sun's ultraviolet radiation. The sun's ultraviolet radiation is one cause of human skin cancer. Too much ultraviolet radiation can cause increased sunburning, skin aging, and eye damage. It can also damage crops, cause cancer in livestock, produce changes in the climate, and bring about environmental changes that would upset the natural habitats of plants and animals.

Figure 9-8. Liquid oxygen is used for the very rapid fuel combustion needed during a space shuttle lift-off.

9.10 Preparation of ozone Ozone is commonly produced by passing oxygen through an ozone generator like that shown in Figure 9-9. This device consists of glass tubes that are partially covered with layers of metal foil. The inner layers of metal foil are connected to one terminal of an induction coil or static electricity machine. The outer layers are connected to the other terminal. The discharge of electricity from one layer to the other provides the energy that converts some of the oxygen passing through the generator to ozone.

When oxygen is converted to ozone, energy is absorbed. The preparation of ozone from oxygen is an endothermic reaction. Ozone, therefore, has a higher energy content than oxygen. Ozone is less stable and more active than oxygen. Three molecules of oxygen form two molecules of ozone. This is because oxygen, O_2, has 2 atoms per molecule, while ozone, O_3, has 3 atoms per molecule.

$$3O_2 + \text{energy} \rightarrow 2O_3(g)$$

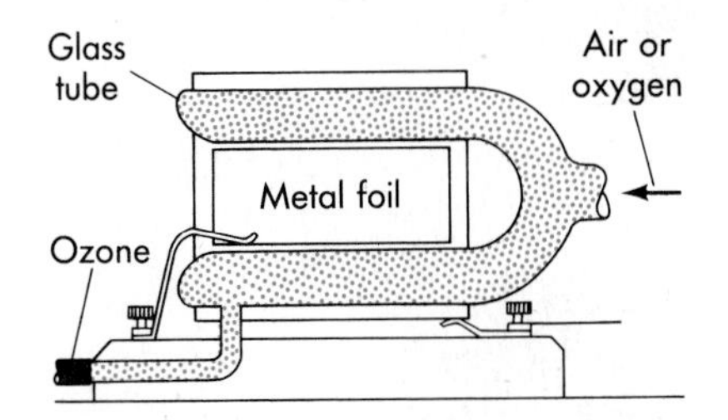

Figure 9-9. An ozone generator. Some of the oxygen passing through the generator is converted to ozone by the electric discharge from one metal-foil layer to the other.

9.11 The difference between oxygen and ozone Several chemical elements exist in two or more different forms. Oxygen is one of these elements. Other elements such as carbon, sulfur, and phosphorus also occur in different forms. *The existence of an element in two or more forms in the same physical phase is known as **allotropy**.* The different forms of such elements are, therefore, called *allotropes*.

Another reason for allotropy is described in Section 17.3.

Ordinary oxygen and ozone are allotropes of the element oxygen. Allotropes are generally given different names: in this case, oxygen and ozone. One reason for allotropy is that an element has two or more kinds of molecules, each with different numbers of atoms. Thus, oxygen's allotropy is due to the existence of oxygen molecules, O_2, and ozone molecules, O_3. See Figure 9-10.

9.12 The structure of ozone molecules Ozone molecules are bent molecules with electron-dot formulas similar to those of sulfur dioxide. The angle between the two oxygen-oxygen bonds is about 117°.

$$\left\{ \begin{array}{ccc} & & \ddot{O}: \\ :O & & \\ & & \ddot{O}: \end{array} \qquad \begin{array}{ccc} & & \ddot{O}: \\ :O & & \\ & & \ddot{O}: \end{array} \right\}$$

Experimental evidence shows that there is no difference between the two oxygen-oxygen bonds in the ozone molecule. Each bond has a length of 1.28 Å. Thus, the concept of resonance must be used to describe ozone's molecular structure. The electron arrangement in the ozone molecule is a resonance hybrid of the electron-dot formulas written above. Each oxygen-oxygen bond in the molecule is intermediate in properties between single and double covalent bonds.

You may wish to review how resonance explains the structure of sulfur dioxide molecules, Section 6.22.

The oxygen-oxygen bond energy in the ozone molecule is 72 kcal/mole; it is 118 kcal/mole in the oxygen molecule. This bond energy difference means that the oxygen-oxygen bonds in the ozone molecule are weaker than the oxygen-oxygen bond in the oxygen molecule. This bond energy difference is not unexpected. Even though the bonded atoms are the same kind, the bonds in the ozone molecule are longer, 1.28 Å, than those in the oxygen molecule, 1.21Å.

Bond energy is the energy required to break chemical bonds. See Sections 6.9 and 6.11.

Oxygen molecule

Ozone molecule

Figure 9-10. Oxygen molecules consist of two atoms of oxygen. Ozone molecules consist of three atoms of oxygen. Oxygen and ozone are allotropes.

9.13 The properties of ozone Ozone is a poisonous, blue gas with an irritating and pungent odor. It is denser than oxygen and much more soluble in water. Some physical properties of oxygen and ozone are compared in Table 9-2. Because ozone molecules have more energy than oxygen molecules, ozone is one of the most vigorous oxidizing agents known. It destroys bacteria and causes many colors to fade rapidly. Ozone is very explosive.

The presence of ozone in amounts over 0.25 part per million in the cabin atmosphere of jet airplanes that fly long distances in the stratosphere can cause chest pain, coughing, headache, and eye irritation. The amount of ozone in airplanes can be reduced by special charcoal filters.

Table 9-2
PHYSICAL PROPERTIES OF OXYGEN AND OZONE

Allotrope	*Molecular formula*	*Color*	*Odor*	*Density, 0 °C, 1 atm* (g/L)	*Melting point* (° C)	*Boiling point* (° C)	*Solubility in water, 0 °C* ($cm^3/100\ cm^3\ H_2O$)
oxygen	O_2	colorless	odorless	1.429	−218.4	−183.0	4.89
ozone	O_3	blue	pungent, irritating	2.144	−192.7	−111.9	49

9.14 Several important uses of ozone The uses of ozone are applications of its oxidizing properties. Ozone destroys bacteria, fungi, and algae and has been used for many years as a water-purifying agent. In concentrations of less than one part per million, it completely sterilizes and deodorizes water; it also removes certain objectionable impurities, such as iron and manganese compounds. See Figure 9-11. Ozone may be used to disinfect sewage. It is also used in preparing some complex compounds containing carbon, hydrogen, and oxygen. These compounds are lubricants, resin softeners, and pharmaceuticals.

HYDROGEN

9.15 Occurrence of hydrogen Hydrogen is a gas at ordinary temperatures. It ranks ninth in abundance by mass among the chemical elements in the earth's surface environment. In the known universe, however, there are more atoms of hydrogen than of any other chemical element. On the earth, hydrogen is usually combined with other elements in a variety of compounds. There are more compounds of hydrogen than of any other element. Free, or elemental, hydrogen exists as covalent diatomic molecules, $\mathbf{H_2}$. Free hydrogen is not common on earth because of its reactivity.

Very small traces of elemental hydrogen, probably derived from volcanoes and coal mines, do exist in the air. However, these amounts are so small that hydrogen is not listed as one of the important gases of the atmosphere. Tremendous quantities of elemental hydrogen occur in the sun and the stars. In our solar system, the planets Jupiter and Saturn consist mostly of liquid hydrogen. One-ninth of water by mass is combined hydrogen, and all acids contain this element. Hydrogen is a constituent of nearly all plant and animal tissues. Fuels such as natural gas, wood, coal, and oil contain hydrogen compounds.

Figure 9-11. Ozone generators produce ozone used to purify and decolorize drinking water.

9.16 Early history of hydrogen In the seventeenth century, scientists observed that a combustible gas was produced when sulfuric acid reacted with iron. The English scientist Henry Cavendish (1731–1810) is usually credited with the discovery of hydrogen because in 1766 he prepared a quantity of the gas and observed its properties. Cavendish observed that hydrogen burns. He called the gas "inflammable air." In 1783, he showed that water is the only product formed when hydrogen burns in air. That same year Lavoisier suggested the name *hydrogen,* a word derived from two Greek words meaning "water producer."

9.17 Preparation of hydrogen *1. From acids by replacement.* This is the usual laboratory method of preparing hydrogen. All acids contain hydrogen, which may be displaced by reaction with metals that are above hydrogen in the activity series. Several different acids and metals can be used. For example, iron, zinc, or magnesium will react with either hydrochloric acid or sulfuric acid to produce hydrogen. The rate at which hydrogen is given off in such reactions depends upon several factors: the amount of metal surface exposed to the acid; the temperature; the strength and kind of acid used; the kind of metal used; and the purity of the metal.

Replacement reactions of this type are explained in Section 8.6(3). The activity series appears in Table 8-2.

Figure 9-12 shows one type of apparatus commonly used for the laboratory preparation of hydrogen. Zinc is put into the generator bottle and either dilute sulfuric or hydrochloric acid is then added through the funnel tube. The hydrogen is collected by water displacement. The equations for the chemical reactions of zinc with sulfuric acid, H_2SO_4, and hydrochloric acid, HCl, are

$$\mathbf{Zn + H_2SO_4 \rightarrow ZnSO_4 + H_2(g)}$$
$$\mathbf{Zn + 2HCl \rightarrow ZnCl_2 + H_2(g)}$$

If the formulas of the acids are compared with the formulas of the zinc compounds produced, it can easily be seen that an atom

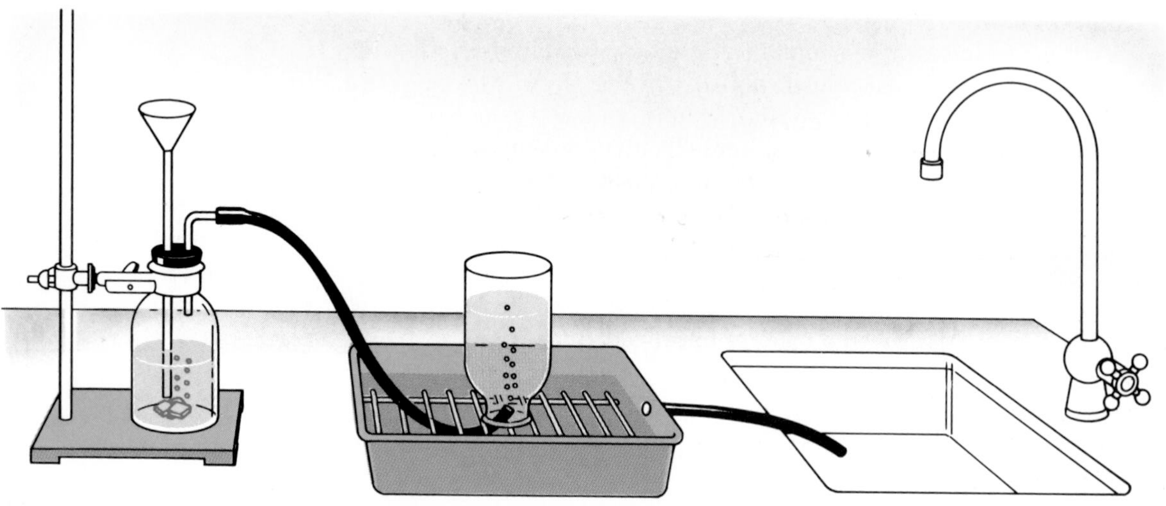

Figure 9-12. Laboratory setup for the preparation of hydrogen. Any metal more active than hydrogen can be used to replace the hydrogen from any one of many acids. The metal is put in the wide-mouth generator bottle, and the acid is added through the funnel tube. Hydrogen is collected by water displacement.

of zinc has replaced two atoms of hydrogen. The zinc sulfate or zinc chloride products are dissolved in the excess water in the generator. Either can be recovered as a white solid by evaporating the water.

2. *From water by replacement.* Sodium is a silvery metal, soft enough to be easily cut with a knife and of low enough density to float on water. It reacts vigorously with water and displaces hydrogen from it. Each sodium atom replaces one of the hydrogen atoms in a molecule of water. To show this reaction more clearly, the formula for water is written as HOH, instead of the usual H_2O.

Replacement of hydrogen in water by metals is described in Section 8.6(2).

$$2Na + 2HOH \rightarrow 2NaOH + H_2(g)$$

not good example

This equation shows that each water molecule has had *one* of its hydrogen atoms replaced by a sodium atom. The sodium hydroxide produced can be recovered as a white, crystalline solid if the excess water is evaporated.

Potassium is a metal similar to sodium. It is below sodium in Group I in the periodic table and is more active. It displaces hydrogen from water so vigorously that the heat of the reaction ignites the hydrogen.

Figure 24-2 is a photo showing the reaction of potassium and water.

$$2K + 2HOH \rightarrow 2KOH + H_2(g)$$

Magnesium reacts slowly with *boiling* water. Calcium, in the same family as magnesium but more reactive, replaces hydrogen from cold water.

$$Ca + 2HOH \rightarrow Ca(OH)_2 + H_2(g)$$

At a high temperature, iron displaces hydrogen from steam. This iron-steam reaction is the only one of the water-replacement methods of preparing hydrogen described that is used commercially. All the other methods are laboratory methods.

3. *From water by electrolysis.* In the electrolysis of water, hydrogen as well as oxygen is produced. Commercially, if oxygen is the main product, hydrogen becomes a useful by-product. This method is used in the United States for producing pure hydrogen wherever cheap electricity is available. It is also a laboratory method of preparing hydrogen.

4. *From hydrocarbons.* Hydrocarbons are compounds of hydrogen and carbon and are commonly derived from petroleum or natural gas. If a hydrocarbon such as methane, CH_4, reacts with steam in the presence of a nickel catalyst at a temperature of about 850 °C, hydrogen and carbon monoxide are produced.

$$CH_4 + H_2O \rightarrow CO(g) + 3H_2(g)$$

When this mixture of gaseous products is cooled and compressed, the carbon monoxide liquefies and can be separated from the hydrogen gas. This is the principal commercial method of producing hydrogen.

Table 9-3
PROPERTIES OF HYDROGEN

atomic number	1
atomic weight	1.00794
electron configuration	1
oxidation numbers	+1, −1
melting point	−259.1 °C
boiling point	−252.9 °C
density, 0 °C, 1 atm	0.0899 g/L
atom radius	0.32 Å
ion radius, H^-	1.53 Å

5. *From water by hot carbon.* This is another commercial method of producing hydrogen. When steam is passed over red-hot coal or coke, a mixture of gases called *water gas* is formed. It consists mainly of hydrogen and carbon monoxide. The equation for the reaction is

$$C + H_2O \rightarrow CO(g) + H_2(g)$$

Frequently, the carbon monoxide from either the hydrocarbon-steam process or the water-gas process is converted to carbon dioxide by adding steam to the carbon monoxide-hydrogen mixture and passing the gases over a catalyst such as iron oxide, at a temperature below 500 °C.

$$CO + H_2O \rightarrow CO_2(g) + H_2(g)$$

More hydrogen is produced from the steam, and the resulting carbon dioxide is separated by dissolving it in water under moderate pressure.

9.18 Physical properties of hydrogen Hydrogen gas is colorless, odorless, and tasteless. It is the gas of lowest density. Its density is only one-fourteenth that of air. One liter of hydrogen at 0 °C under one atmosphere pressure has a mass of 0.0899 gram. It is less soluble than oxygen in water. Because of its low solubility, hydrogen is collected in the laboratory by water displacement. Properties of hydrogen are listed in Table 9-3.

In 1898, the Scotch chemist and physicist James Dewar (1842–1923) succeeded in converting hydrogen into a liquid by simultaneously cooling the gas to a very low temperature and applying very high pressure. Liquid hydrogen is clear and colorless and only one-fourteenth as dense as water. Thus, liquid hydrogen is the liquid of lowest density, with a mass of only about 70 grams per liter. Under atmospheric pressure, liquid hydrogen boils at −252.9 °C. When a part of the liquid is evaporated, the remainder freezes to an icelike solid. Its melting point is −259.1 °C. Solid hydrogen is also the solid of lowest density, having a mass of about 80 grams per liter.

Do not confuse adsorption *with* absorption. *Absorption is the taking up of one material through the entire mass of another.*

Hydrogen is *adsorbed* on certain metals such as platinum and palladium. ***Adsorption*** *is an acquisition of one substance by the surface of another.* Heat is evolved when adsorption occurs. During adsorption, widely separated gas molecules are brought much closer together. This process causes the molecules to give up energy, which appears as heat. Finely divided platinum can adsorb hydrogen so rapidly that the heat given off may ignite the hydrogen.

Hydrogen, as a gas, a liquid, or a solid, is a nonconductor of electricity under usual conditions. The elements, other than hydrogen, that have a single valence electron are metals. Metals have the general property of being conductors of electricity. Sci-

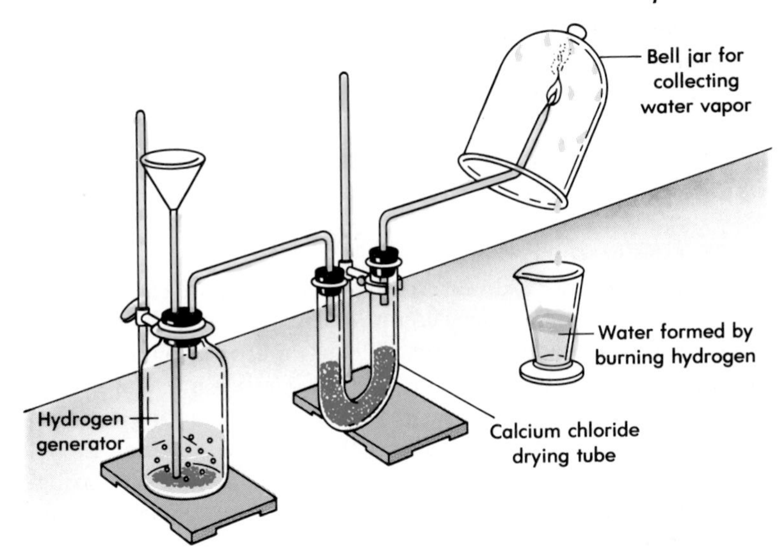

Figure 9-13. Hydrogen is produced by the action of a metal and dilute acid in the generator. The gas then passes through a drying tube filled with calcium chloride, $CaCl_2$. Here the water vapor carried over with the hydrogen from the generator is removed. The dry hydrogen burns at the jet tip in the air-filled bell jar. The water vapor formed during combustion condenses on the cool walls of the bell jar and drops into the collecting vessel. When hydrogen is burned in air, water is the only product.

entists believe that it is possible, under very high pressure, to convert hydrogen into a liquid or a solid having metallic properties. Much of the inner structure of the planet Jupiter is believed to consist of liquid metallic hydrogen.

9.19 Chemical properties of hydrogen 1. *Reactions with nonmetals*. The electronegativity of hydrogen, 2.1, is equal to or less than the electronegativity of other nonmetals. Consequently, hydrogen reacts with nonmetals to form molecular compounds. The polarity of the covalent bonds in these compounds depends on the electronegativity of the nonmetal. The bonds range in polarity from the nonpolar H—P bond to the highly polar H—F bond.

Electronegativity data are given in Figure 6-14 and Sections 6.19 and 6.20.

Hydrogen burns in air or oxygen with a very hot, pale blue, and nearly invisible flame. Water is the only product. See the diagram in Figure 9-13.

$$2H_2 + O_2 \rightarrow 2H_2O(g)$$

Hydrogen does not support combustion. If a bottle of hydrogen is held mouth downward while a blazing splint is thrust slowly upward into the bottle, the hydrogen ignites and burns at the mouth of the bottle. But the splint does not burn in the atmosphere of hydrogen inside the bottle. See Figure 9-14.

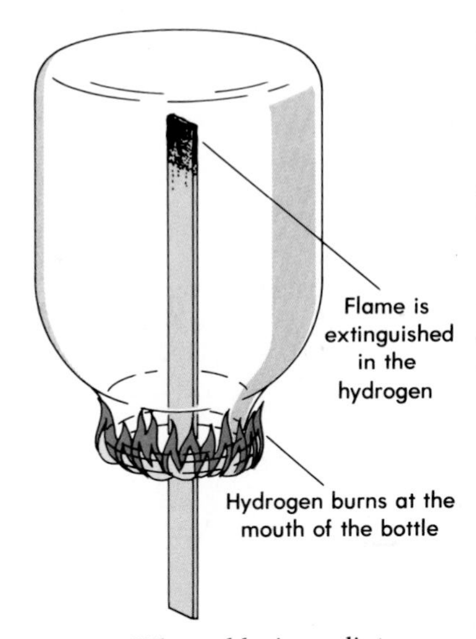

Figure 9-14. When a blazing splint is thrust upward into a bottle of hydrogen, the hydrogen is ignited and burns at the mouth of the bottle. The splint does not burn inside the bottle in an atmosphere of hydrogen. Hydrogen does not support combustion.

Hydrogen is not a very active element at ordinary temperatures. A mixture of hydrogen and oxygen must be heated to 800 °C or ignited at a lower temperature by an electric spark to make the gases combine. Then they combine explosively to form water.

Hydrogen and chlorine do not combine when they are mixed in the dark. But in the presence of direct sunlight, they unite explosively to form hydrogen chloride. A jet of hydrogen will burn in chlorine. The same equation for these chemical changes is written as

$$H_2 + Cl_2 \rightarrow 2HCl(g)$$

Under suitable conditions, hydrogen can unite with nitrogen to form ammonia, NH_3, a very important compound.

$$3H_2 + N_2 \rightarrow 2NH_3(g)$$

2. *Reactions with metals.* Hydrogen reacts with many nontransition metals to form binary compounds called *hydrides.* In these compounds hydrogen is the more electronegative element. The hydrides of Group I metals and of the Group II metals calcium, strontium, and barium are ionic compounds. At room temperature they are white crystalline solids. In these compounds, hydrogen is present as the H^-, hydride ion, and has an oxidation number of -1. Metallic hydrides can be produced by heating the metal in an atmosphere of hydrogen.

$$2Na + H_2 \rightarrow 2Na^+H^-(s)$$
$$Ba + H_2 \rightarrow Ba^{++}H^-{}_2(s)$$

Some complex metal hydrides such as $LiAlH_4$ are used as reducing agents in the production of pharmaceuticals, flavors, fragrances, and fine chemicals. The formation of iron-titanium hydride is a safe way to store large quantities of hydrogen in a small volume at ordinary temperatures.

9.20 The test for hydrogen A colorless, odorless, and tasteless gas that burns in air or oxygen with a nearly colorless flame and yields water as the *only* product *is* hydrogen. This is the test that is commonly used to identify hydrogen.

9.21 Uses of hydrogen *1. For making hydrogen compounds.* Ninety percent of the elemental hydrogen produced today is used for preparing two useful hydrogen compounds, ammonia and methanol. Ammonia, NH_3, is made by direct union of nitrogen and hydrogen. Ammonia is then used as the starting point for making fertilizers, explosives, dyestuffs, and many other important compounds. Methanol, CH_3OH, which is used as a solvent and for preparing other compounds, is made from hydrogen and carbon monoxide.

Large quantities of hydrogen chloride, HCl, are prepared by direct combination of hydrogen and chlorine. Hydrogen chloride is then dissolved in water to produce hydrochloric acid.

2. *In petroleum refining*. Hydrogen is used in refining thick, sulfur-containing crude oil. The addition of hydrogen makes the oil more liquid and removes the sulfur. These changes improve the oil's quality as a fuel. Sulfur is also removed from low-boiling petroleum products by reaction with hydrogen.

3. *For solidifying oils*. Millions of pounds of cottonseed oil are

changed each year from liquid oil to solid or semisolid fat by hydrogenation. Finely divided nickel is the catalyst used in this reaction. Some of the molecules in the liquid oil combine with hydrogen atoms to produce a substance that is solid at room temperature but still a liquid at body temperature. Most vegetable shortenings found on the market today are examples of such hydrogenated oil.

Peanut, corn, soybean, and coconut oil are also hardened by hydrogenation to make margarine. Some fish oils lose their offensive odor when hydrogenated and thus become suitable for making soap. Lard is sometimes hydrogenated to produce a whiter, firmer product.

Hydrogenation: chemical addition of hydrogen to a material.

4. *As a reducing agent.* Hydrogen can remove oxygen from metallic oxides such as CuO, SnO, PbO, FeO, CoO, and Bi_2O_3.

$$CuO + H_2 \rightarrow Cu + H_2O(g)$$

Hydrogen could be used as a reducing agent to separate these metals from their oxide ores. More often, however, carbon in the form of coke is used as the reducing agent because it is usually less expensive and more convenient.

Some metals must be worked in an atmosphere free of oxygen, called a *reducing atmosphere.* A reducing atmosphere prevents the unwanted reaction of the metal with oxygen at the high temperatures needed for the processes. Tungsten, a metal that is used for making the filaments of electric lamps, is worked in a reducing atmosphere. Surrounding tungsten with hydrogen in a closed furnace prevents the oxidation of the metal.

5. *As a fuel.* Nearly all fuels contain hydrogen, either free or combined with other elements. Coal gas and oil gas contain hydrogen in quantity. Methane, CH_4, is a major component of natural gas. Hydrogen is used as a fuel for the oxyhydrogen torch. Pure hydrogen makes an excellent fuel but is more expensive than other available gaseous fuels.

STOP O

X **9.22 Deuterium** Deuterium (described in Section 3.6) was discovered and named in 1931 by scientists at Columbia University headed by Harold C. Urey (1893–1981). In addition to its occurrence on the earth, deuterium exists in the atmosphere of Jupiter, in our sun, in distant stars and nebulae, and in interstellar space. The discoveries of deuterium in space were made by ultraviolet spectroscopes and radio telescopes from the earth's surface or from aboard earth satellites. By studying the relative amounts of protium (hydrogen-1) atoms and deuterium atoms in our galaxy, scientists hope to learn more about the origin and nature of the universe.

Protium atoms consist of a proton nucleus and an orbital electron. Deuterium atoms have a proton and a neutron in the nucleus, and an orbital electron. Consequently, deuterium atoms

Models showing the structures of protium and deuterium are shown in Figure 3-4.

Table 9-4
PHYSICAL PROPERTIES OF PROTIUM AND DEUTERIUM

Isotope	Symbol	Atomic mass	Molecular formula	Melting point (° C)	Boiling point (° C)	Density of liquid (g/L)
protium	H	1.007825	H_2	−259.1	−252.9	0.0709 (−252.7 °C)
deuterium	D	2.01410	D_2	−254.6	−249.7	0.169 (−250.9 °C)

have a mass about double that of protium atoms. This mass difference between hydrogen isotopes is proportionally larger than between isotopes of any other element. The mass difference is related to differences in physical properties. See Table 9-4.

Differences in chemical properties of most isotopes are so slight that they are not usually considered. But because of the significant mass difference between its isotopes, this is not true for hydrogen. Protium and deuterium react the same way and form similar compounds. But the rates of their reactions are different. Reactions of deuterium and deuterium compounds are generally much slower than those of protium and protium compounds. This rate difference makes possible the electrolytic separation of deuterium oxide from water.

The separation of deuterium oxide from water by electrolysis is described in Section 12.20.

SUMMARY

Oxygen, the earth's most abundant element, is found in the air, in water, and in the earth's crust. Scheele and Priestley were codiscoverers of oxygen. Oxygen was shown by Lavoisier to be an element and was named by him. Oxygen can be prepared in the laboratory by decomposing hydrogen peroxide, adding water to sodium peroxide, heating potassium chlorate, or electrolyzing water. Commercially, oxygen is produced from liquid air or by electrolyzing water.

Oxygen is a colorless, odorless, tasteless gas which is slightly denser than air and slightly soluble in water. Liquid oxygen is slightly attracted by a magnet. Oxygen is an active element that combines with other elements to form oxides. An oxide is a compound consisting of oxygen and usually one other element; in oxides the oxidation number of oxygen is −2. Oxygen does not burn, but supports combustion.

Oxygen is necessary for life. It is used extensively in the iron and steel industry and for producing many chemical compounds. Patients who have difficulty breathing are sometimes administered oxygen. Oxygen aids in the purification of water and sewage and supports the combustion of fuels in rockets and missiles.

Ozone is a more active form of oxygen produced by electric discharge. It exists naturally in the stratosphere, where it shields the earth from much of the sun's ultraviolet radiation.

The existence of an element in two or more forms in the same physical phase is known as allotropy. The different forms of such elements are allotropes. Oxygen, O_2, and ozone, O_3, are allotropes of the element oxygen. The structure of the ozone molecule is a resonance hybrid of two different structures. Ozone is a vigorous oxidizing agent. It is used in water and sewage purification and in preparing chemicals.

Hydrogen is the most abundant element in the universe. It is found in water, acids, living tissues, and many fuels. It was discovered by Cavendish and named by Lavoisier. In the laboratory, hydrogen is prepared from acids by replacement, from water by replacement, and from water by electrolysis. Industrially, hydrogen is prepared from hydrocarbons, from water by hot carbon, and from water by electrolysis.

Hydrogen is a colorless, odorless, tasteless gas that has the lowest density of any material known. It is very slightly soluble in water. Platinum and palladium adsorb hydrogen. Adsorption is an acquisition of one substance by the surface of another. It is believed that at very low temperatures and extremely high pressures, hydrogen can acquire metallic properties.

Hydrogen burns with a very hot flame that is nearly invisible. Hydrogen does not support combustion. Hydrogen is not active at ordinary temperatures. A mixture of hydrogen and oxygen, when ignited, produces an explosion, and water vapor is the only product. Hydrogen and chlorine react to form hydrogen chloride. Hydrogen and nitrogen can be made to react; ammonia is the product. Hydrogen reacts with many nontransition metals to form hydrides.

Hydrogen is used for making hydrogen compounds, principally ammonia and methanol. It is used in petroleum refining, for solidifying oils, as a reducing agent, and also as a source of fuel.

Deuterium, the hydrogen isotope with mass number 2, has different physical properties from protium, hydrogen-1. Chemically it reacts more slowly than protium. This is the result of the proportionally great mass difference between the two isotopes.

VOCABULARY

absorption
adsorption
allotrope
allotropy
combined oxygen
deuterium
free oxygen
hydride
hydrogenation
liquid air
oxide
ozone
reducing atmosphere
water displacement
water gas

QUESTIONS

Group A

1. *(a)* What is *free* oxygen? *(b)* Where is free oxygen found in greatest abundance on the earth? *A1, A2a*
2. *(a)* What is *combined* oxygen? *(b)* Name some common materials that are examples of combined oxygen. *A1, A2a*
3. Scheele and Priestley used different chemical reactions in their discoveries of oxygen. Write a balanced formula equation for the reaction each used. *A4a*
4. *(a)* Define: *catalyst*. *(b)* What is the purpose of the catalyst in the preparation of oxygen from hydrogen peroxide? *(c)* What is the purpose of the catalyst in the preparation of oxygen from potassium chlorate? *A1, A3a*
5. In the laboratory preparation of oxygen, the gas is collected by water displacement. Why? *A1, A3a*
6. *(a)* Name the two most abundant substances in liquid air. *(b)* When liquid air is warmed, the gas that evaporates first is 93% N_2 and 7% O_2. When the last portion is evaporating, the gas is 55% N_2 and 45% O_2. Explain. *A3a, A3b*
7. Give five physical properties of oxygen gas. *A3b*
8. Distinguish: *oxide, dioxide, peroxide.* *A3b*
9. Describe the test for oxygen. *A3c*
10. List three ways in which oxygen is used in the iron and steel industry. *A3d*
11. Why is oxygen useful in treating water and sewage? *A3b, A3d*
12. Representative compounds of oxygen and the other Period Two elements have the formulas: Li_2O, BeO, B_2O_3, CO_2, N_2O_3, and OF_2. Assign oxidation numbers to each element in these formulas.
13. How and where is ozone produced naturally in the atmosphere? *A2a*
14. What is the effect of ozone on the sun's ultraviolet radiation? *A3e*

15. Define: *(a) allotropy. (b) allotrope.* *A1*
16. What is the reason oxygen exhibits allotropy? *A4b*
17. Ozone is used to destroy bacteria, fungi, and algae. What property of ozone is responsible for this effect? *A3b*
18. Why is there so much more combined hydrogen than free hydrogen on the earth? *A3b*
19. Which is more abundant in the explorable universe, oxygen or hydrogen? *A2a*
20. From Appendix Table 2 obtain the mass numbers of the known hydrogen isotopes. *(a)* What particles make up the nucleus of each isotope? *(b)* What is the electron configuration of each isotope?
21. In the laboratory preparation of hydrogen, *(a)* what are the usual reactants? *(b)* the usual products? *A3a*
22. Give five physical properties of hydrogen gas. *A3b*
23. Distinguish: *adsorption, absorption.* *A1*
24. *(a)* Does hydrogen burn? *(b)* Does hydrogen support combustion? *(c)* Describe a simple demonstration that supports your answers. *A3b*
25. Describe the test for hydrogen. *A3c*
26. What is the most important commercial use for elemental hydrogen? *A3d*
27. *(a)* What is a reducing atmosphere? *(b)* Describe a specific use of a reducing atmosphere. *A3b, A3d*
28. Scientists are actively studying many possible ways to decompose water using sunlight as the energy source. Their goal is to produce hydrogen inexpensively. Why would this be environmentally desirable?

Group B

29. Lavoisier did not discover any new substances. He did not design any new laboratory apparatus. He did not develop any better preparation methods. Yet Lavoisier is recognized as one of the greatest eighteenth-century chemists. Why? *A5*
30. *(a)* Write balanced formula equations for each of the four laboratory methods of preparing oxygen. *(b)* Assign oxidation numbers to each oxygen atom in these equations. *(c)* For each equation, determine whether the oxygen was oxidized, reduced, or both. *A4c*
31. What property of oxygen leads chemists to believe that the oxygen molecule contains unpaired electrons? *A3b, A4b*
32. Write balanced formula equations for the reaction between: *(a)* potassium and oxygen to yield potassium peroxide. *(b)* strontium and oxygen. *(c)* selenium and oxygen, in which the product is a solid. *(d)* phosphorus and oxygen to yield $P_4O_{10}(s)$. *(e)* zinc and oxygen. *A4d(1)*
33. Is OF_2 an oxide or a fluoride? Explain. *A3b*
34. How does air enriched with oxygen help persons with heart or lung diseases? *A3b, A3d*
35. How do oxygen molecules and ozone molecules compare *(a)* in energy content? *(b)* in activity? *A3b*
36. Show how the bonding in the ozone molecule is an example of resonance. *A4b*
37. Even though hydrogen had been known much earlier, Cavendish is usually credited with its discovery. Explain. *A2b*
38. *(a)* From Figures 4-5 and 4-6, determine the approximate wavelength of the radiation produced by an electron transition in a hydrogen atom from the 4th energy level to the 1st energy level; *(b)* from the 4th energy level to the 3rd energy level.
39. Write balanced formula equations for the reaction between: *(a)* potassium and hydrogen. *(b)* calcium and hydrogen. *(c)* hydrogen and lead(II) oxide. *(d)* hydrogen and bismuth(III) oxide. *(e)* deuterium and oxygen. *A4d(2)*
40. List as many factors as you can that affect the rate at which hydrogen is evolved from hydrochloric acid by reaction with coarse iron filings. *A3a*
41. *(a)* From its position in the periodic table, would you expect rubidium to react with water more or less vigorously than sodium? *(b)* Why? *A3a*
42. Write the balanced formula equation for the reaction by which hydrogen is prepared from methane. *A4c*

43. In the water-gas reaction, $C + H_2O \rightarrow CO(g) + H_2(g)$, identify *(a)* the oxidizing agent; *(b)* the reducing agent; *(c)* the substance oxidized; *(d)* the substance reduced.

44. What change in the physical properties of hydrogen would identify the transition from nonmetallic hydrogen to metallic hydrogen? *A3b*

45. *(a)* What product is formed when hydrogen combines with bromine? *(b)* with one atom of carbon? *(c)* What type of bond does hydrogen form with these two elements? *(d)* Of what type of particle do these compounds consist? *A3b*

46. *(a)* What product is formed when hydrogen combines with lithium? *(b)* with cesium? *(c)* What type of bond does hydrogen form with these two elements? *(d)* Of what type of particle do these compounds consist? *A3b*

47. Iron(II) oxide reacts with hydrogen to form iron and water vapor. Which substance is *(a)* the oxidizing agent; *(b)* the reducing agent; *(c)* oxidized; *(d)* reduced? *A4d(2)*

48. *(a)* Protium atoms and deuterium atoms are both classed as hydrogen atoms, yet their physical properties differ. Explain. *(b)* How do their chemical properties differ? *A3b*

PROBLEMS

Group A

1. What volume does the gram-molecular weight of oxygen occupy at 0 °C and 1 atm pressure? Obtain density data from Table 9-1 on page 202.

2. Calculate the percentage of hydrogen in *(a)* NaH; *(b)* BaH_2; *(c)* $LiAlH_4$; *(d)* $NaBH_4$; *(e)* Mg_2NiH_4.

3. A compound consists of 74.2% sodium and 25.8% oxygen. Find the empirical formula.

4. Binary compounds of hydrogen and the Period Two elements have the formulas LiH, BeH_2, BH_3, CH_4, NH_3, H_2O, and HF. *(a)* Calculate the percentage of ionic character of the bonds in each compound. *(b)* How does the bonding in these compounds vary with the increasing atomic number of the Period Two element?

5. From Table 3-2 obtain the percentages of O–16, O–17, and O–18 in naturally occurring oxygen. Calculate the number of atoms of each isotope in one mole of atoms of naturally occurring oxygen.

Group B

6. An oxide of tin contains 78.77% tin. What is the empirical formula?

7. Many metallic ores are oxygen compounds, as listed below. Calculate the percentage of: *(a)* holmium in Ho_2O_3; *(b)* molybdenum in $PbMoO_4$; *(c)* boron in $Na_2B_4O_7 \cdot 4H_2O$; *(d)* beryllium in $3BeO \cdot Al_2O_3 \cdot 6SiO_2$; *(e)* cobalt in $CoO \cdot 2Co_2O_3 \cdot 6H_2O$.

8. For an experiment, it is necessary to prepare 1.00 L of oxygen at 0 °C and 1 atm pressure. To produce the oxygen, calculate the mass of *(a)* mercury(II) oxide that must be decomposed. *(b)* hydrogen peroxide that must be decomposed catalytically using manganese dioxide. *(c)* sodium peroxide that must be reacted with water. *(d)* potassium chlorate that must be decomposed catalytically using manganese dioxide. *(e)* water that must be electrolyzed.

9. As lightning passes through air, nitrogen monoxide is formed: $\frac{1}{2}N_2 + \frac{1}{2}O_2 \rightarrow NO$. The bond energy of N=O is 151 kcal/mole. What is the amount and kind of energy change in this reaction? Obtain other data from Table 6-4 on page 131.

10. A student in the laboratory must prepare enough hydrogen by reacting a metal with excess hydrochloric acid to fill five 250-mL bottles at 0 °C and 1 atm pressure. The student decides to use enough metal to actually prepare 1.50 times the required amount of hydrogen so as to compensate for losses that might occur during the preparation. Calculate the mass of each of the following metals that is necessary to produce the hydrogen: *(a)* zinc. *(b)* magnesium. *(c)* iron. *(d)* aluminum. *(e)* nickel. Based on the activity series, which reaction should be most vigorous? least vigorous?

The 1936 Nobel Prize in chemistry was awarded to Peter J. W. Debye (Dutch) for his studies of dipole moments and the diffraction of X rays and electrons in gases. The dipole moment studies showed how the centers of positive and negative charge in a molecule are aligned, thus giving new information on which molecules are polar, how atoms are arranged in molecules, and the distances between atoms. He also extended the Arrhenius theory of ionization (see Chapter 14) by helping to develop the Debye-Hückel theory, which states that ions in solution are associated with solvent molecules.

The Gas Laws

10

GOALS

In this chapter you will gain an understanding of: • the three basic assumptions of the kinetic theory • the characteristic properties of gases • the kinetic theory description of a gas; an ideal gas • the attractive forces between gas molecules • the dependence of gas volume on temperature and pressure • the Kelvin temperature scale • Boyle's law and Charles' law • solving gas-law problems that involve changes in volume, pressure, temperature, and nature of the liquid displaced

10.1 Kinetic theory In Chapter 1, matter was described as existing in three physical phases—gas, liquid, and solid. In later chapters, it was explained that the particles making up various substances are atoms, molecules, or ions. A physical model will now be used to help you understand the properties of gases, liquids, and solids. This model explains the properties in terms of (1) the forces between the particles of matter and (2) the energy these particles possess. This useful model is the *kinetic theory*.

Most of the data in support of the kinetic theory comes from indirect observation. It is almost impossible to observe the behavior of individual particles of matter. However, scientists can observe the behavior of large groups of particles. From the results of these observations, they can describe the average behavior of the particles under study.

The three basic assumptions of the *kinetic theory* are:

1. *Matter is composed of very tiny particles.* The chemical properties of the particles of matter depend on their composition. Their physical properties depend on the forces they exert on each other and the distance separating them.

2. *The particles of matter are in continual motion.* Their average kinetic energy (energy of motion) depends on temperature.

3. *The total kinetic energy of colliding particles remains constant.* When individual particles collide, some lose energy while others gain energy. But there is no overall energy loss. Collisions of this type are said to be *elastic* collisions.

10.2 Observed properties of gases Past and current studies of gases have revealed four basic characteristic properties:

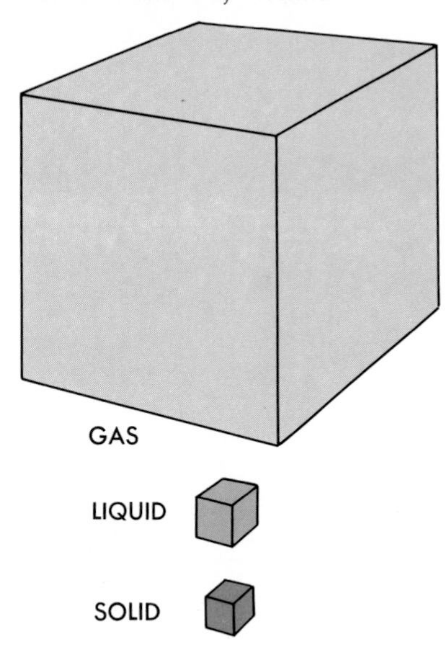

Figure 10-1. The density of many substances in the gaseous phase is about 1/1000 of their density in the liquid and solid phases. This diagram shows the relative volumes of 1.5 g of oxygen as a gas, a liquid, and a solid. The oxygen gas occupies 1050 mL, whereas liquid oxygen occupies only 1.3 mL and solid oxygen occupies only 1.0 mL.

1. *Expansion.* A gas does not have a definite shape or a definite volume. It completely fills any container in which it is confined. For example, the shape and volume of the air in balloons of various sizes and shapes, in automobile tires, and in air mattresses are determined by the container.

2. *Pressure.* When a toy balloon is inflated, it becomes larger because the pressure on its inside surface is increased. If the air is allowed to escape, the balloon becomes smaller because the pressure is decreased. Raising the temperature of the air in the balloon by warming it causes the balloon to become larger. This observation indicates that pressure increases with an increase in temperature. As the balloon cools, its size decreases because the pressure decreases.

3. *Low density.* The density of a gas is about $\frac{1}{1000}$ of the density of the same substance in the liquid or solid phase. The densities of gaseous, liquid, and solid oxygen and hydrogen are typical data. Oxygen gas has a density of 1.429 g/L (0.001429 g/mL) at 0 °C and 1 atmosphere pressure, whereas liquid oxygen has a density of 1.149 g/mL at −183 °C, and solid oxygen has a density of 1.426 g/mL at −252.5 °C. Figure 10-1 shows the relative volumes of oxygen as a gas, a liquid, and a solid. Hydrogen gas has a density of 0.0899 g/L (0.0000899 g/mL) at 0 °C and 1 atmosphere pressure. Liquid hydrogen has a density of 0.0708 g/mL at −253 °C. Solid hydrogen has a density of 0.0706 g/mL at −262 °C.

4. *Diffusion.* If the stopper is removed from a container of ammonia, the irritating effects of this gas on the eyes, nose, and throat soon become evident throughout the room. When the chemistry class makes the foul-odored hydrogen sulfide gas in the laboratory, objections may come from other students and teachers in all parts of the building. This process of spreading out spontaneously (without additional help) to occupy a space uniformly is characteristic of all gases. It is known as *diffusion.* Figure 10-2 shows the diffusion of bromine vapor.

10.3 Kinetic-theory description of a gas According to the kinetic theory, a gas consists of very small independent particles that move at random in space and experience elastic collisions. This theoretical description is of an imaginary gas called the ***ideal gas.***

The particles of substances that are gases at room temperature are molecules. Some of these molecules consist of a single atom (He, Ne, Ar). Many consist of two atoms (O_2, H_2, HCl, etc.). Others consist of several atoms (NH_3, CH_4, C_2H_2, etc.). Matter in the gaseous phase occupies a volume of the order of 1000 times that which it occupies in the liquid or solid phase. Thus, molecules of gases are much farther apart than those of liquids or solids. This difference accounts for the much lower density of gases as compared to solids or liquids. Even so, 1 mL

Figure 10-2. The rate of diffusion of bromine vapor in air. Diffusion has occurred for 2 minutes in the left cylinder, whereas diffusion has occurred for 20 minutes in the right cylinder.

of a gas at 0 °C and 1 atmosphere pressure will contain about 3×10^{19} molecules. Many ordinary molecules have diameters of the order of 4Å, or 4×10^{-10} m. In gases, these molecules are widely separated. They are, on an average, about 4×10^{-19} m (or about 10 diameters) apart at 0 °C and 1 atmosphere pressure. The kinetic energy of the molecules of a gas (except near its condensing temperature) overcomes the attractive forces between them. The molecules of a gas are essentially independent particles that travel in random directions at high speed as suggested by Figure 10-3. This speed is of the order of 10^3 m/s at 0 °C and is independent of the pressure. At this speed, molecules of a gas under 1 atmosphere pressure travel about 10^{-7} m before colliding with other gas molecules or with the walls of the container. Because gas molecules collide and exchange energy, their speeds will vary. They undergo about 5×10^9 collisions per second.

The expansion and diffusion of gases are both explained by the fact that gas molecules are essentially independent particles. They move through space until they strike other gas molecules or the walls of the container. A gas moves very rapidly into an evacuated container. Gaseous diffusion is slowed down, but not prevented, by the presence of other gases. The rate of diffusion of one gas through another depends on three properties of the intermingling gas molecules: their speeds, their diameters, and the attractive forces between them.

Hydrogen diffuses rapidly because hydrogen molecules are smaller and move about with greater speed than the larger, heavier molecules of other gases at the same temperature. The diffusion of hydrogen can be demonstrated by placing a bottle filled with hydrogen above another bottle filled with air, as

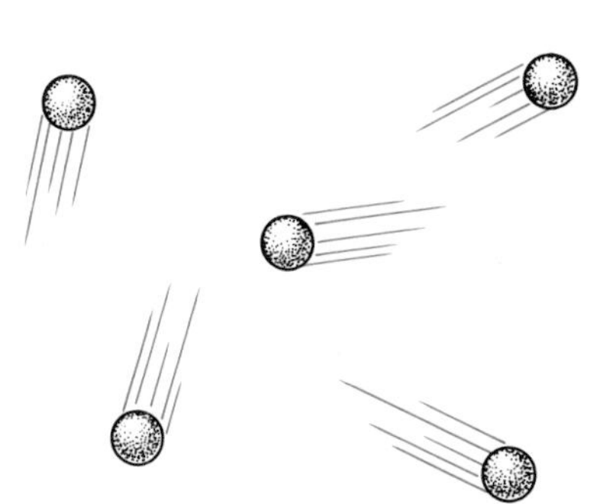

Figure 10-3. Molecules of a gas are widely separated and move very rapidly.

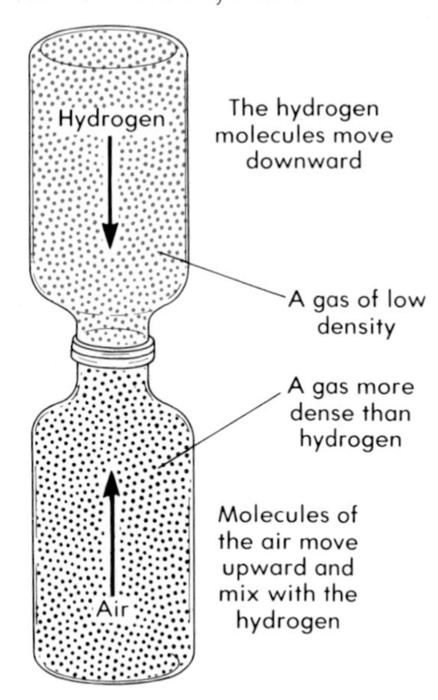

Figure 10-4. Even though hydrogen is less dense than air, hydrogen diffuses downward and the air upward.

shown in Figure 10-4. After a few minutes, the mouth of each bottle is held in the flame of a laboratory burner. The resulting explosions show that the hydrogen molecules moved so that there were some in both bottles.

Even if two gases are separated by a porous barrier, such as a membrane or an unglazed porcelain cup, diffusion takes place through the pores. See Figure 10-5. An unglazed porcelain cup is closed by a rubber stopper through which a piece of glass tubing has been inserted. When a large beaker filled with hydrogen is placed over the porcelain cup, hydrogen molecules diffuse into the cup faster than the molecules of the gases in air diffuse out of the cup and into the beaker. This creates a pressure in the cup that forces gas out the end of the tube.

Gas pressure results from many billions of moving molecules continuously hitting the walls of the container. If the number of molecules within the container is increased, the number that strike any area of the inside surface increases. Therefore, the pressure on the inside surface increases. If the temperature of the gas is raised, the molecules on an average have more kinetic energy. They move more rapidly and collide more energetically with the walls of the container. These more frequent and more energetic collisions with the container walls increase the pressure. The pressure drops when the number of molecules is decreased or the temperature is lowered.

Figure 10-6 shows the distribution of molecular speeds in a gas at two different temperatures. From these graphs two conclusions can be drawn:

1. All molecules of a gas do not have the same speed.
2. An increase in temperature increases the average rate at which gas molecules move.

The kinetic energy of a molecule is related to its speed by the equation

$$E_k = \frac{1}{2} mv^2$$

where E_k is the kinetic energy of the molecule, m is its mass, and v is its speed. Since the molecules of a gas do not all have the same speed, they will not all have the same kinetic energy. Since the average molecular speed varies with the temperature, the average kinetic energy of the molecules of a gas varies with the temperature. Thus, the temperature of a gas provides an indication of the average kinetic energy of the molecules. The higher the temperature, the higher the average kinetic energy. The lower the temperature, the lower the average kinetic energy.

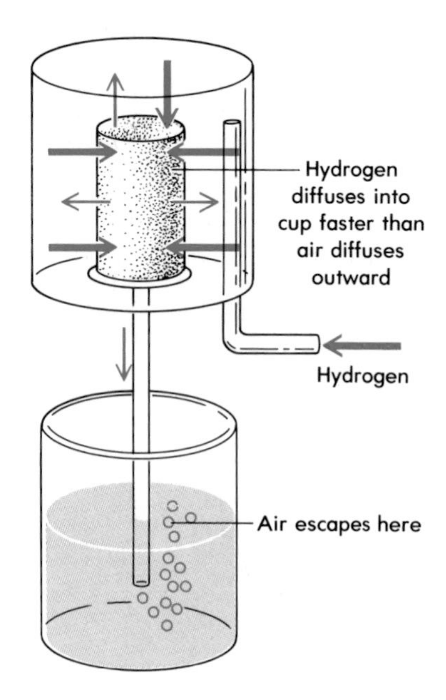

Figure 10-5. Because hydrogen diffuses into the cup faster than air diffuses outward, the net pressure in the cup increases, forcing gas bubbles out of the tube.

10.4 Attractive forces between gas molecules The lowest temperature at which a substance can exist as a gas at atmospheric pressure is the *condensation temperature* of the gas. At this temperature, the kinetic energy of the gas particles is not sufficient to overcome the forces of attraction between the particles.

The gas condenses to a liquid. The kinetic theory states that the temperature of a substance is a measure of the kinetic energy of its particles. So a study of condensation temperatures of various substances should provide an idea of the magnitude (size) of the forces of attraction between particles of matter. Table 10-1 gives selected condensation temperatures.

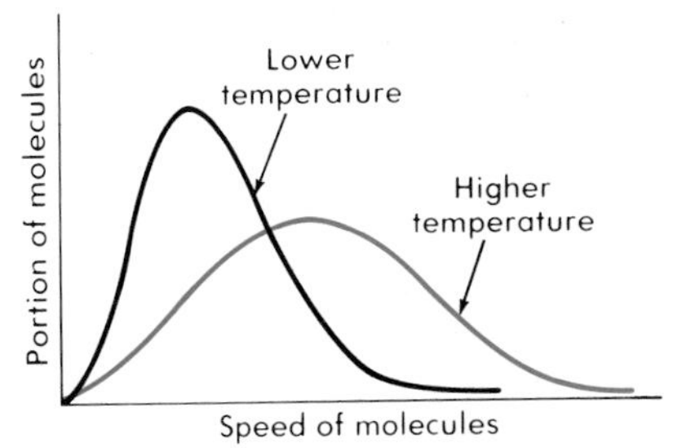

Figure 10-6. Molecular speed distribution in a gas measured at different temperatures.

Substances such as H_2, O_2, and CH_4 (methane) consist of low-molecular-weight nonpolar covalent molecules. They can exist as gases at very low temperatures. Evidently the attractive forces between such molecules in the gaseous phase are very small. More complex substances such as CCl_4 and C_6H_6 (benzene) have condensation temperatures somewhat above room temperature. They have higher-molecular-weight nonpolar molecules. The forces of attraction between such molecules must be greater than those between similar but less complex molecules.

Ammonia (NH_3) and H_2O consist of polar covalent molecules. Notice that their molecular weights are low and their molecular structures are simple. Even so, their condensation temperatures are considerably above those of nonpolar molecules of the same molecular weight.

The condensation temperatures of ionic compounds such as sodium chloride, covalent network substances such as diamond, and metals are all very high. Evidently the attractive forces between particles of such substances are very strong. The nature of these forces will be described in Chapter 12.

The attractive forces between molecules are called *van der Waals forces*. These forces are important only when molecules are very close together. Therefore, van der Waals forces are not important in gases unless the gas molecules are under very high pressure or are at a temperature near their condensation temperature. Van der Waals forces are of two types. One type, called *dispersion interaction*, exists between all molecules. The other type, called *dipole-dipole attraction*, exists between polar molecules only.

The strength of dispersion interaction depends on the number of electrons in a molecule and the tightness with which the electrons are held. The greater the number of electrons and the less tightly they are bound, the more powerful is the attractive force of dispersion interaction. This effect can be observed most easily between nonpolar molecules, where dispersion interaction is the only type of attractive force. Thus, for nonpolar molecules in general, the higher the molecular weight, the higher the condensation temperature. This generalization can be observed in Table 10-1. Compare the condensation temperatures of oxygen, hydrogen, and methane with those of carbon tetrachloride and benzene. The energy associated with dispersion interaction is only a few tenths of a kilocalorie per mole.

Dipole-dipole attraction is the attraction between the oppositely charged portions of neighboring polar molecules. Recall

Table 10-1
CONDENSATION TEMPERATURES OF REPRESENTATIVE SUBSTANCES

Type of substance	*Substance*	*Condensation temperature* (1 atm, °C)
nonpolar covalent molecular	H_2	−253
	O_2	−183
	CH_4	−164
	CCl_4	77
	C_6H_6	80
polar covalent molecular	NH_3	−33
	H_2O	100
ionic	NaCl	1413
	MgF_2	2239
covalent network	$(SiO_2)_x$	2230
	C_x (diamond)	4200
metallic	Hg	357
	Cu	2567
	Fe	2750
	W	5660

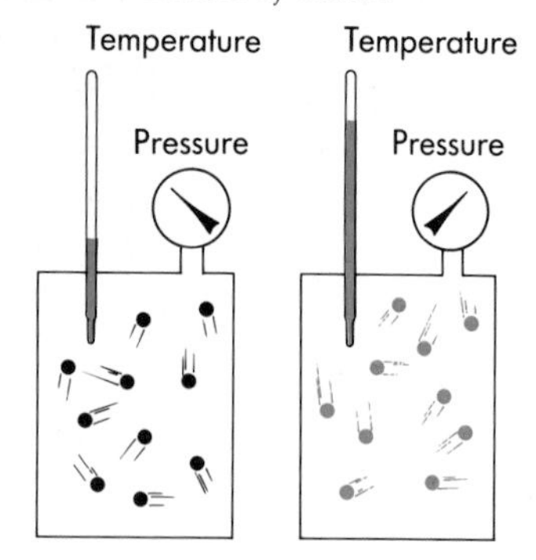

Figure 10-7. At a constant volume, the pressure a gas exerts increases as the temperature of the gas increases.

from Section 6.21 that polar molecules are sometimes called dipoles. Dipole-dipole attractive forces as well as dispersion interaction forces act between polar molecules. This combination of forces accounts for the much higher condensation temperatures of polar molecules as compared to nonpolar molecules of similar complexity. For example, methane, ammonia, and water have comparable molecular weights. Yet the condensation temperature of ammonia is about 130 C° higher than that of methane. The condensation temperature of water is over 130 C° higher than that of ammonia. The energy associated with dipole-dipole attraction can be as high as 6 kilocalories per mole.

10.5 Dependence of gas volume on temperature and pressure

A given number of molecules can occupy widely different volumes. The expression "a liter of air" means little unless the temperature and pressure at which it is measured are known. A liter of air can be compressed to a few milliliters in volume. It can also expand to fill an auditorium. Steel cylinders containing oxygen and hydrogen are widely used in industry. They have an internal volume of about 55 liters. When such cylinders are returned "empty," they still contain about 55 liters of gas, although when they were delivered "full" they may have had 100 times as many molecules of the gas compressed within the cylinder.

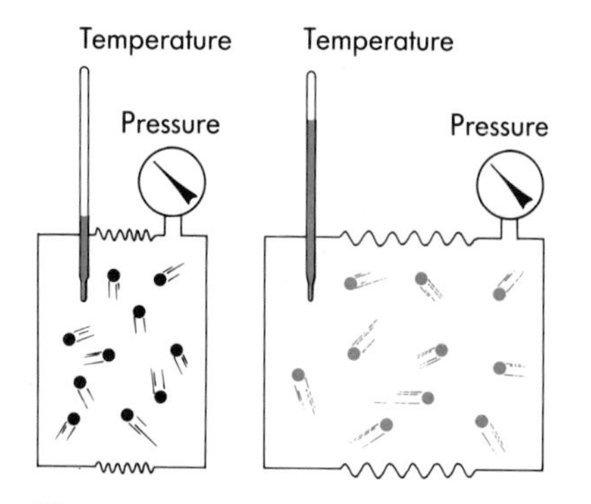

Figure 10-8. At a constant pressure, the volume a gas occupies increases as the temperature of the gas increases.

You have learned that the temperature of a gas is an indication of the average kinetic energy of its molecules. The higher the temperature of a gas, the more kinetic energy its molecules have, and the more rapidly they move about. The pressure that a gas exerts on the walls of its container is the result of the collisions of gas molecules with the walls. Gas molecules that have more kinetic energy (because they are at a higher temperature) undergo more frequent and more energetic collisions with the walls. If the volume of one mole of gas molecules remains constant, the pressure exerted by the gas increases as its temperature is raised. This relationship is shown in Figure 10-7. It follows that the pressure exerted by one mole of gas molecules decreases as the temperature is lowered.

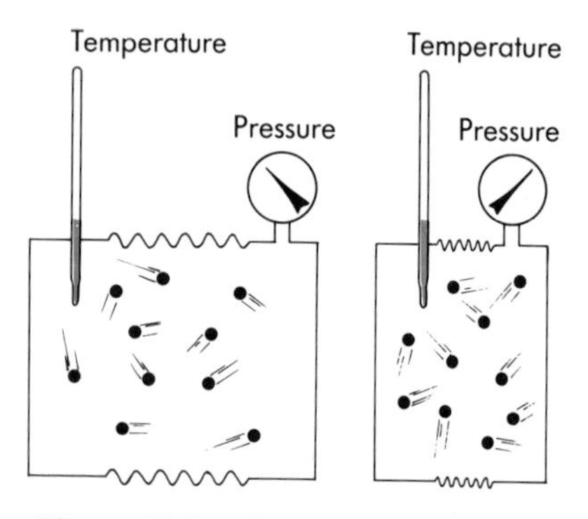

Figure 10-9. At a constant temperature, the pressure that a gas exerts increases as the volume of the gas decreases.

Furthermore, if the pressure exerted by one mole of gas molecules is to remain the same as the temperature increases, the volume that the gas occupies must increase. Since the molecules move faster at higher temperatures, they strike the walls of the container more frequently and with more force. Suppose the area of the wall is increased. Then the force of collisions on a unit area of the wall can remain the same as at the lower temperature, and the pressure will remain the same. The area that the molecules strike can be enlarged by enlarging the volume of the container as shown in Figure 10-8. On the other hand, if the pressure remains constant and the temperature decreases, the volume that one mole of gas molecules occupies must decrease.

Finally, suppose the temperature of one mole of gas molecules remains constant. Then the pressure exerted by the gas becomes greater as the volume that the gas occupies becomes smaller. See Figure 10-9. And similarly, the pressure exerted by one mole of gas molecules is less if the volume available to the gas is larger. In light of these consequences, the kinetic theory satisfactorily explains how gas volumes are related to the temperature and pressure of the gas. Accordingly, both temperature and pressure must be considered when measuring the volume of a gas.

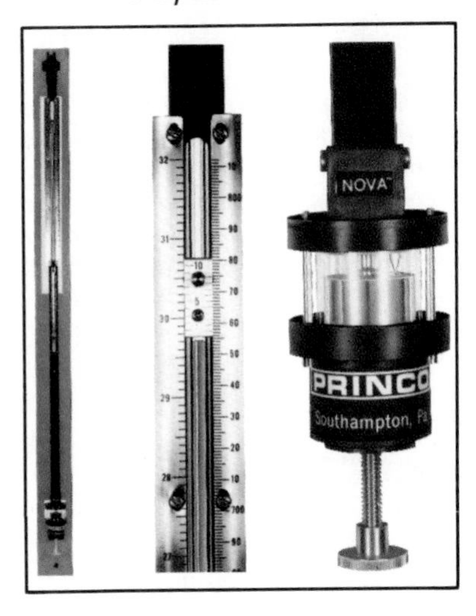

Figure 10-10. A barometer is used to measure atmospheric pressure. Left, a laboratory mercurial barometer. Center, a close-up of the top of the mercury column showing the height-measuring scales. Right, the adjustable reservoir with indicator pin for setting the height of the mercury level that is exposed to the atmosphere.

10.6 Standard temperature and pressure As explained in Section 10.5, the volume of a gas depends greatly on temperature and pressure. For this reason, it is very helpful to have a standard temperature and a standard pressure for use in measuring or comparing gas volumes. ***Standard temperature*** *is defined as exactly zero degrees Celsius.* It is the temperature of melting ice. This temperature was selected because pure water is widely available and the melting temperature of ice is not significantly affected by pressure changes. ***Standard pressure*** *is defined as the pressure exerted by a column of mercury exactly 760 millimeters high.* This value is used as the standard pressure because it is the average atmospheric pressure at sea level. Temperatures are easily measured with an accurate thermometer. The pressure of a gas in simple gas experiments can be determined from a properly corrected barometer reading. A mercurial barometer is shown in Figure 10-10. Standard temperature and pressure are commonly abbreviated as STP.

10.7 Variation of gas volume with pressure: Boyle's law If a rubber ball filled with air is squeezed, the volume of the gas inside is decreased. But the gas expands again when the pressure is released. The English chemist and physicist Robert Boyle (1627–1691) was the first scientist to make careful measurements showing the relationship between pressures and volumes of gases. The results of these and later experiments by other scientists established that the pressure on a gas and its volume are inversely proportional. Doubling the pressure on a gas reduces its volume by one-half. Reducing the pressure on a gas to one-third enables it to expand and occupy a volume three times as great. A graph of data like these is shown in Figure 10-11. This relationship has been formulated into a law that bears Boyle's name. ***Boyle's law*** is stated: *The volume of a definite quantity of dry gas is inversely proportional to the pressure, provided the temperature remains constant.*

This law is expressed mathematically as

$$\frac{V}{V'} = \frac{p'}{p}$$

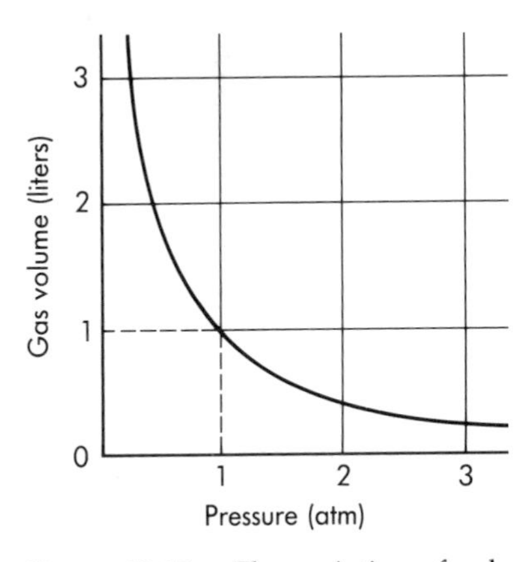

Figure 10-11. The variation of volume with pressure at constant temperature of 1 liter of an ideal gas measured at 1 atmosphere pressure. $pV = constant$.

Robert Boyle is considered to be the founder of modern chemistry because (1) he recognized that chemistry is a worthwhile field of learning and not just a branch of medicine or alchemy; (2) he introduced careful experimentation into chemistry; and (3) he defined an element as something that could be combined with other elements to form compounds but could not itself be decomposed—a definition that held for over 200 years.

V is the original volume, V' the new volume, p the original pressure, and p' the new pressure. Solving the expression for V', Boyle's law takes the form:

$$V' = V\frac{p}{p'}$$

10.8 Using Boyle's law Suppose 30.0 mL of hydrogen gas, collected when the barometer reading is $74\bar{0}$ mm, stands until the barometric pressure has risen to $75\bar{0}$ mm. If the temperature is unchanged, the volume of gas becomes smaller because of the increased pressure. The new gas volume measures 29.6 mL.

Using Boyle's law, the new volume resulting from the pressure change can be calculated without actually experiencing the pressure change. The new volume V' is p/p', or $74\bar{0}$ mm/$75\bar{0}$ mm of the original volume V, 30.0 mL. Substituting this value in the Boyle's law formula, the new volume is calculated as

If you need to review inverse proportion, see Section 1.22.

$$V' = Vp/p'$$
$$V' = 30.0 \text{ mL} \times 74\bar{0} \text{ mm}/75\bar{0} \text{ mm}$$
$$V' = 29.6 \text{ mL}$$

Later, if the pressure falls to $72\bar{0}$ mm, the new gas volume V' is calculated as

$$V' = 30.0 \text{ mL} \times 74\bar{0} \text{ mm}/72\bar{0} \text{ mm}$$
$$V' = 30.8 \text{ mL}$$

Sample Problem 10-1 gives another example of Boyle's law. Although Sample Problems 10-1 and 10-2 and the Practice Problems that follow them can be partially solved by "cancellation," you will find it much easier to do the arithmetic of gas-law problems if you use a calculator.

SAMPLE PROBLEM 10-1

A $19\bar{0}$-mL sample of hydrogen is collected when the pressure is $72\bar{0}$ mm of mercury. What volume will the gas occupy at $76\bar{0}$ mm pressure?

SOLUTION

$$V' = Vp/p'$$
$$V'_{76\bar{0}\text{ mm}} = 19\bar{0} \text{ mL} \times 72\bar{0} \text{ mm}/76\bar{0} \text{ mm}$$
$$V'_{76\bar{0}\text{ mm}} = 18\bar{0} \text{ mL}$$

Practice Problem

Oxygen, 37.5 mL, is collected at a pressure of $77\bar{0}$ mm. What is its volume at $75\bar{0}$ mm pressure? *ans.* $V'_{75\bar{0}\text{ mm}} = 38.5$ mL

10.9 Variation of gas volume with temperature Bread dough rises when put in a hot oven. The increase in temperature causes the bubbles of carbon dioxide gas within the dough to expand. The rather large increase in the volume of the dough during baking shows that the gas must expand considerably as the temperature increases. In fact, gases expand many times more per degree rise in temperature than do liquids and solids.

Jacques Charles (1746–1823), a French scientist, was the first to make careful measurements of the changes in volume of gases with changes in temperature. His experiments revealed that:

1. All gases expand or contract at the same rate with changes in temperature, provided the pressure is unchanged.
2. The change in volume amounts to $\frac{1}{273}$ of the original volume at 0 °C for each Celsius degree the temperature is changed.

You may start with a definite volume of a gas at 0 °C and experiment by heating it. Just as the whole of anything may be considered as made up of two halves, $\frac{2}{2}$, or three thirds, $\frac{3}{3}$, so this volume can be considered as $\frac{273}{273}$. If the gas is warmed one Celsius degree, it expands $\frac{1}{273}$ of its original volume. Its new volume is $\frac{274}{273}$. In the same manner, the gas expands $\frac{100}{273}$ when it is heated 100 C°. Such expansion, added to the original volume, makes the new volume $\frac{373}{273}$. Any gas warmed 273 Celsius degrees expands $\frac{273}{273}$. Its volume is doubled, as represented by the fraction $\frac{546}{273}$.

A gas whose volume is measured at 0 °C contracts by $\frac{1}{273}$ of this volume if it is cooled 1 C°. Its new volume is $\frac{272}{273}$ of its original volume. Cooling the gas to −100 °C reduces the volume by $\frac{100}{273}$. In other words, the gas shrinks to $\frac{173}{273}$ of its original volume. At this rate, if the gas were cooled to −273 °C, it would lose $\frac{273}{273}$ of its volume. Its new volume would be zero. Such a situation cannot occur, however, because all gases become liquids before such a low temperature is reached. This rate of contraction with cooling applies only to gases.

10.10 Kelvin temperature scale In Section 1.15 the Celsius temperature scale was described. This scale is based on the *triple point* of water. *The* ***triple point*** *of pure water is that single temperature and pressure condition at which water exists in all three phases at equilibrium.* On the Celsius scale, the triple point of water has a temperature of 0.01 °C.

You may wish to review Section 1.15.

From measuring the variation of gas volume with temperature at low pressures, scientists believe that −273.15 °C is the lowest possible temperature. At this temperature a body would have lost all the heat that it is possible for it to lose. Scientists have come very close to this lowest possible temperature, but theoretically it is impossible to reach. The interval between the lowest possible temperature, −273.15 °C, and the triple-point temperature of water, 0.01 °C, is 273.16 C°.

The physicist Sir William Thomson (1824–1907), better known by his title Lord Kelvin, invented the Kelvin temperature

Table 10-2
COMPARISON OF TEMPERATURES ON THE KELVIN AND CELSIUS SCALES

Celsius scale (°C)	*Kelvin scale* (K)
100°	373
50°	323
20°	293
5°	278
4°	277
3°	276
2°	275
1°	274
0°	273
−100°	173
−273°	0

scale. A Kelvin degree is the same temperature interval as a Celsius degree. But 0 K is the lowest possible temperature, −273.15 °C. The temperature of the triple point of water is 273.16 K. The lowest possible temperature, 0 K, is frequently called *absolute zero.* Temperatures measured on the Kelvin scale are often called *absolute temperatures.*

The normal freezing point of water is 0.01 C° lower than the triple-point temperature of water. On the Celsius scale the normal freezing point of water is 0.00 °C. Recall that a Kelvin degree is the same temperature interval as a Celsius degree. So the normal freezing point of water is also 0.01 K lower than the triple-point temperature of water, 273.16 K, or 273.15 K. Thus, 0.00 °C = 273.15 K. In many calculations, 273.15 K is rounded off to 273 K. Refer to the temperature scales in Table 10-2.

Kelvin temperature = Celsius temperature + 273

In Section 10.3, you learned that the temperature of a gas provides an indication of the average kinetic energy of the molecules. Since the Kelvin scale starts at what is believed to be the lowest possible temperature, the average kinetic energy of the molecules of a gas is directly proportional to the Kelvin temperature of the gas.

If you need to review direct proportion, see Section 1.22.

10.11 Charles' law Thermometers are not graduated (marked) to give Kelvin-scale readings. But use of the Kelvin scale does give results that correspond with actual volume changes observed in gases as shown in Figure 10-12. In problems dealing with changes in gas volumes as temperatures vary, the Kelvin scale eliminates the use of zero and of negative numbers. Using the Kelvin temperature scale, ***Charles' law*** can be stated: *The volume of a definite quantity of dry gas varies directly with the Kelvin temperature, provided the pressure remains constant.*

Charles' law may be expressed mathematically as

$$\frac{V}{V'} = \frac{T}{T'}$$

where V is the original volume, V' the new volume, T the original Kelvin temperature, and T' the new Kelvin temperature. Solving the expression for V',

$$V' = V\frac{T'}{T}$$

Sample Problem 10-2 illustrates the use of this formula.

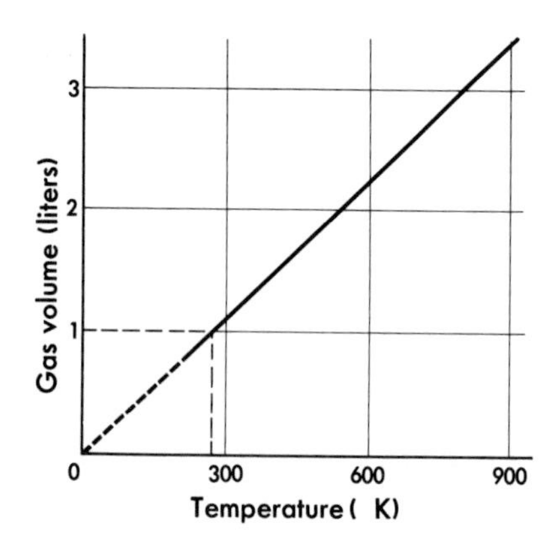

Figure 10-12. The variation in volume with Kelvin temperature at constant pressure of 1 liter of an ideal gas measured at 273 K. V/T = *constant.*

SAMPLE PROBLEM 10-2

A 14.9 mL volume of gas is measured at 25 °C. If the pressure remains unchanged, what will be the volume of the gas at 0 °C?

SOLUTION

Change the Celsius temperatures to Kelvin temperatures:

$$25\ °C + 273 = 298\ K;\ \bar{0}\ °C + 273 = 273\ K$$
$$V' = VT'/T$$
$$V'_{\bar{0}\ °C} = 14.9\ mL \times 273\ K/298\ K$$
$$V'_{\bar{0}\ °C} = 13.6\ mL$$

Practice Problems

1. A 14.9-mL volume of gas is measured at 25 °C. If the pressure remains unchanged, what will be the volume of the gas at $4\bar{0}$ °C? *ans.* $V'_{4\bar{0}\ °C} = 15.6$ mL
2. A gas volume, $15\bar{0}$ mL, is measured at 27 °C. What is the gas volume at $\bar{0}$ °C if the pressure does not change? *ans.* $V'_{\bar{0}\ °C} = 136$ mL

10.12 Use of Boyle's and Charles' laws combined Calculation of the new volume of a gas when both temperature and pressure are changed involves no new principles. The new volume is the same whether the changes in temperature and pressure are done together or in either order. The original volume is first multiplied by a ratio of the pressures. In this way the new volume corrected for pressure alone is determined. Then this answer is multiplied by a ratio of the Kelvin temperatures. This gives the new volume corrected for both pressure and temperature. Expressed mathematically,

$$V' = V \times \frac{p}{p'} \times \frac{T'}{T}$$

Sample Problem 10-3 illustrates the use of this formula.

SAMPLE PROBLEM 10-3

A gas measures 27.5 mL at 22 °C and $74\bar{0}$ mm pressure. What will be its volume at 15 °C and 755 mm pressure?

SOLUTION

$$22\ °C = 295\ K;\ 15\ °C = 288\ K$$
$$V' = V \times p/p' \times T'/T$$
$$V'_{15\ °C,\ 755\ mm} = 27.5\ mL \times 74\bar{0}\ mm/755\ mm \times 288\ K/295\ K$$
$$V'_{15\ °C,\ 755\ mm} = 26.3\ mL$$

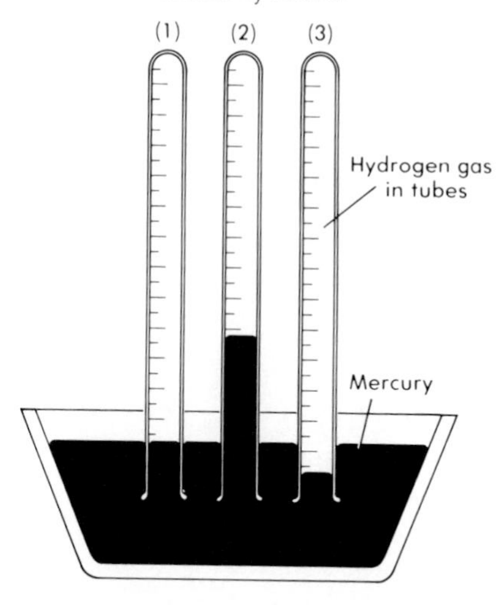

Figure 10-13. In (1) the pressure of the hydrogen is the same as that of the atmosphere. In (2) the pressure of the hydrogen is less than that of the atmosphere. In (3) the pressure of the hydrogen is greater than that of the atmosphere.

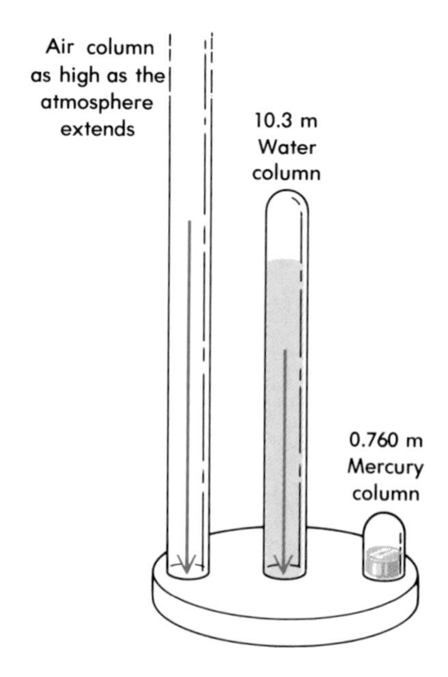

Figure 10-14. The pressure of the atmosphere supports a column of water 13.5 times as high as the column of mercury it supports.

Practice Problem

A gas occupies 42.3 mL at 24 °C and 725 mm pressure. What will be its volume at 12 °C and $75\bar{0}$ mm pressure?
ans. $V'_{12\ °C,\ 75\bar{0}\ mm} = 39.2$ mL

10.13 Pressure of a gas collected by displacement of mercury You have learned how gas volume varies with pressure and temperature changes. But no practical laboratory methods of making the necessary measurements have as yet been considered. These laboratory operations will now be explained. In the laboratory, a gas may be collected and its volume measured by using a long graduated tube closed at one end. This tube is called a *eudiometer* (you-dee-*om*-eh-ter), Figure 10-13.

Suppose some hydrogen is delivered into a eudiometer that was previously filled with mercury. As hydrogen enters the tube, it bubbles to the top and pushes the mercury down. Suppose enough hydrogen is added to make the level of the mercury inside the tube the same as the level of the mercury in the bowl, as in (1), Figure 10-13. *When these two levels are the same, the pressure of the hydrogen is the same as that of the atmosphere.* This pressure can be found by reading a barometer. The volume of hydrogen is read from the graduations (markings) on the eudiometer.

Suppose, however, that not enough hydrogen is delivered into the eudiometer to make the mercury levels the same even when the eudiometer rests on the bottom of the bowl of mercury. Then the mercury level inside the tube is above the level outside the tube, as in (2), Figure 10-13. The pressure of the gas inside the tube is less than the pressure of the air outside. Otherwise, the enclosed gas would push the mercury down to the same level as that outside the tube. To determine the pressure of the hydrogen, the difference between the levels of the mercury inside and outside the tube must be *subtracted* from the barometer reading. The gas volume is read, as before, from the eudiometer graduations.

Suppose enough hydrogen is delivered into the eudiometer to force the mercury level inside the tube below the outside level. Then the pressure of the gas inside the tube is greater than the pressure of the air outside, as in (3), Figure 10-13. To determine the pressure of the gas in this case, the difference between the levels of the mercury inside and outside the tube must be *added* to the barometer reading. To make this measurement and that of the gas volume directly would not be practical, since mercury is not transparent. It is easier to raise the eudiometer in the bowl of mercury until the mercury levels inside and outside the tube are the same. Then the gas pressure inside will be the same as that read on the barometer. The volume of hydrogen can now be read from the graduations on the eudiometer. See Sample Problems 10-4 and 10-5.

SAMPLE PROBLEM 10-4

What is the pressure of the gas in a eudiometer (gas-measuring tube) when the mercury level in the tube is 17 mm higher than that outside? The barometer reads 735 mm.

SOLUTION

Since the mercury level inside is higher than that outside, the pressure of the gas in the eudiometer must be less than atmospheric pressure. Accordingly, the difference in levels is subtracted from the barometric pressure to obtain the pressure of the gas,

735 mm − 17 mm = 718 mm, the pressure of the gas

SAMPLE PROBLEM 10-5

The volume of oxygen in a eudiometer is 37.0 mL. The mercury level inside the tube is 22.0 mm higher than that outside. The barometer reading is 742.0 mm. The temperature is 18 °C. What will be the volume of the oxygen at STP?

SOLUTION

Note: When STP conditions are involved, 0 °C and 760 mm are considered exact quantities. No bars are required over the zeros. These terms have no effect on the number of significant figures in the calculated result.

1. Correction for difference in levels:

742.0 mm − 22.0 mm = 720.0 mm

2. Conversion of Celsius temperatures to Kelvin temperatures:

18 °C + 273 = 291 K; 0 °C + 273 = 273 K

3. Correction for change in pressure and temperature:

$$V' = V \times p/p' \times T'/T$$
$$V'_{STP} = 37.0 \text{ mL} \times 720.0 \text{ mm}/760 \text{ mm} \times 273 \text{ K}/291 \text{ K}$$
$$V'_{STP} = 32.9 \text{ mL}$$

10.14 Pressure of a gas collected by water displacement In elementary work, gases are usually collected by water displacement rather than by mercury displacement. (Mercury is very expensive and poisonous.) Water is $\frac{1}{13.5}$ as dense as mercury. Therefore, a given gas pressure will support a column of water 13.5 times as high as an equivalent column of mercury. (Refer to Figure 10-14 on the left-hand page). When a gas is collected by

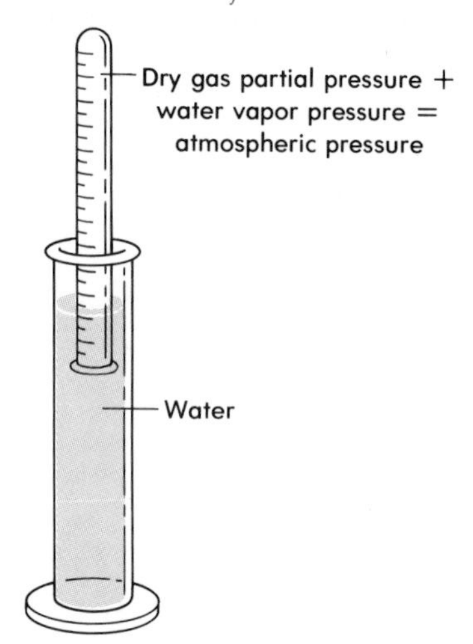

Figure 10-15. Because the liquid levels inside and outside this eudiometer are the same, the sum of the partial pressures of the confined gas and water vapor equals atmospheric pressure. To determine the dry gas pressure, the water vapor pressure must be subtracted from the atmospheric pressure.

water displacement, pressure corrections are made just as with mercury displacement. *But a difference in water levels must first be divided by 13.5 to convert it to its equivalent height in terms of a column of mercury.*

In advanced work, gases are often collected by mercury displacement. The advantage of this method is that mercury does not evaporate measurably at room temperatures. When a gas is bubbled through water, however, the collected gas always has some water vapor mixed with it. Water vapor, like other gases, exerts pressure. Since the gas pressure is the result of the collision of the various gas molecules with the walls of the container, *the total pressure of the mixture of gases* (the collected gas and the water vapor) *is the sum of their partial pressures.* This is a statement of ***Dalton's law of partial pressures.*** The partial pressure is the pressure each gas would exert if it alone were present. The partial pressure of the water vapor, called *water vapor pressure,* depends only on the temperature of the water. Appendix Table 8 gives the pressure of water vapor at different water temperatures. *To determine the partial pressure of the dry gas* (unmixed with water vapor), *the vapor pressure of water at the given temperature is subtracted from the total pressure of the gas in the tube.* Refer to the diagram in Figure 10-15 when solving Sample Problems 10-6 and 10-7.

SAMPLE PROBLEM 10-6

Oxygen is collected in a eudiometer by water displacement. The water level inside the tube is 27.0 mm higher than that outside. The temperature is 25.0 °C. The barometric pressure is 731.0 mm. What is the partial pressure of the dry oxygen in the eudiometer?

SOLUTION

To convert the difference in water levels to an equivalent difference in mercury levels, the difference in water levels is divided by 13.5.

27.0 mm ÷ 13.5 = 2.0 mm, the equivalent difference in mercury levels

Since the level inside the tube is higher than that outside, the difference in levels must be subtracted from the barometric pressure.

731.0 mm − 2.0 mm = 729.0 mm

Appendix Table 8 indicates that the water vapor pressure at 25.0 °C is 23.8 mm. To correct for the water vapor pressure, this pressure must be subtracted from the pressure corrected for difference in levels.

729.0 mm − 23.8 mm = 705.2 mm, the partial pressure of the dry oxygen

SAMPLE PROBLEM 10-7

A gas-measuring tube contains 38.4 mL of air collected by water displacement at a temperature of 20.0 °C. The water level inside the eudiometer is 139 mm higher than that outside. The barometer reading is 740.0 mm. Calculate the volume of dry air at STP. (See Figure 10-16.)

SOLUTION

Note that in this problem the barometric pressure is measured to the nearest 0.1 mm, while the water level difference is measured to the nearest unit millimeter.

1. Correction for difference in levels:

$$139 \text{ mm} \div 13.5 = 10.3 \text{ mm}$$

Since the water level inside is higher than that outside, the correction must be subtracted from the barometric pressure.

$$740.0 \text{ mm} - 10.3 \text{ mm} = 729.7 \text{ mm}$$

2. Correction for water vapor pressure: Appendix Table 8 indicates that the water vapor pressure at 20.0 °C is 17.5 mm. This correction is subtracted:

$$729.7 \text{ mm} - 17.5 \text{ mm} = 712.2 \text{ mm}$$

3. Correction for pressure and temperature changes:

$$V' = V \times p/p' \times T'/T$$
$$V'_{STP} = 38.4 \text{ mL} \times 712.2 \text{ mm}/760 \text{ mm} \times 273 \text{ K}/293 \text{ K}$$
$$V'_{STP} = 33.5 \text{ mL}$$

Practice Problem

A volume of oxygen, 32.8 mL, is collected by water displacement at 23.0 °C. The water level inside the eudiometer tube is 29.6 mm higher than that outside. The barometer reading is 737.0 mm. What is the volume of dry oxygen at STP?
ans. $p = 713.7$ mm; $V'_{STP} = 28.4$ mL

10.15 Behavior of real gases Boyle's and Charles' laws describe the behavior of the *ideal gas.* Real gases consist of molecules of finite size that do exert forces on each other. These forces affect the behavior of the molecules. But at temperatures near room temperature and at pressures of less than a few atmospheres, real gases conform closely to the behavior of an ideal gas. Under these conditions of temperature and pressure, the spaces separating the molecules are large enough so that the size of the molecules and forces between them have little effect.

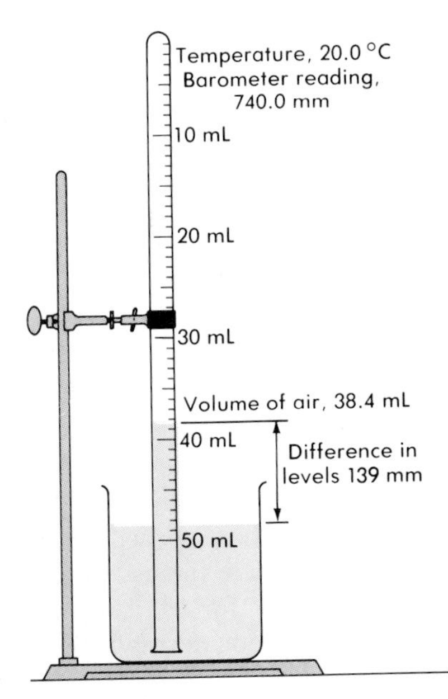

Figure 10-16. Laboratory setup for a gas-volume conversion problem.

Figure 10-17. Jacques Charles and M. N. Robert's first ascent in a hydrogen balloon at the Tuilleries, Paris, Dec. 1, 1783.

Boyle's law applies to real gases with a fairly high degree of accuracy. But it does not apply to gases under very high pressure. Under such pressures the molecules are close enough together to attract each other, and the gas is almost at its condensation point.

Charles' law holds for real gases with considerable accuracy, except at low temperature. Under this condition, gas molecules move more slowly and molecular attraction exerts a greater influence. Thus, Charles' law does not apply at temperatures near the point at which a gas condenses into a liquid.

SUMMARY

The kinetic theory helps explain the properties of gases, liquids, and solids in terms of the forces between the particles of matter and the energy these particles possess. The three basic assumptions of the kinetic theory are: (1) matter is composed of very tiny particles; (2) the particles of matter are in continual motion; and (3) the total kinetic energy of colliding particles remains constant.

Gases have four characteristic properties: expansion, pressure, low density, and diffusion. These may be explained by describing the motion of the widely spaced particles in a gas.

The attractive forces between molecules are van der Waals forces. These are of two types: dispersion interaction, which exists between all molecules; and dipole-dipole attraction, which exists between polar molecules only.

If the volume that a certain number of gas molecules occupies remains constant, the pressure exerted by the gas increases as its temperature increases. If the pressure exerted by this number of gas molecules remains the same as the temperature is increased, the volume that the gas occupies increases. At a constant temperature, the gas exerts increasing pressure as its volume becomes smaller. The volume of a gas is related to its temperature and pressure.

Standard temperature is zero degrees Celsius. Standard pressure is the pressure exerted by a column of mercury exactly 760 millimeters high. The temperature of a gas is measured by means of a thermometer, and its pressure is measured by using a barometer. When a gas is collected over water, some water vapor becomes mixed with the gas. The amount of pressure that results from water vapor varies according to the temperature.

Boyle's law: The volume of a definite quantity of dry gas is inversely proportional to the pressure, provided the temperature remains constant. Jacques Charles found that all gases expand or contract at the same rate. Gases expand $\frac{1}{273}$ of the volume at 0 °C for each Celsius degree the temperature is raised. The lowest possible temperature is −273 °C. The Kelvin temperature scale has its 0 K reading at −273 °C. The readings on the Kelvin scale are 273 degrees higher than on the Celsius scale. Charles' law: The volume of a definite quantity of dry gas varies directly with the Kelvin temperature, provided the pressure remains constant. Boyle's law and Charles' law may be combined in the formula $V' = V \times p/p' \times T'/T$.

VOCABULARY

absolute zero
barometer
Boyle's law
Charles' law
condensation temperature
Dalton's law of partial pressures
diffusion
dipole-dipole attraction
dispersion interaction
elastic collision
eudiometer
expansion
ideal gas
Kelvin scale
kinetic theory
pressure
real gas
STP
triple point
van der Waals forces
water vapor pressure

QUESTIONS

Group A

1. List the three basic assumptions of the kinetic theory. *A3b*
2. What is an ideal gas? *A1, A3r*
3. A gas sample is held in a closed container under constant temperature conditions. Explain why the gas pressure remains constant. *A3l, A3p*
4. In a sample of oxygen, all the molecules have exactly the same speed at the same temperature—true or false? Explain. *A3k*
5. How are the temperature of a gas and the kinetic energy of its molecules related? *A3j, A3k*
6. Define: *condensation temperature*. *A1*
7. Describe the two types of attractive forces that may exist between the molecules of molecular substances. *A3o*
8. Imagine that on two successive days you obtained 1.00-L samples of the air in your classroom. Will the two samples contain the same number of molecules? Explain. *A3p*
9. Define: *standard pressure; standard temperature*. *A1, A2*
10. State Boyle's law *(a)* in words; *(b)* mathematically. *A4*
11. *(a)* If you are given a Celsius temperature, how do you calculate the corresponding Kelvin temperature? *(b)* If you are given a Kelvin temperature, how do you calculate the corresponding Celsius temperature? *A4*
12. What two conclusions about the expansion and contraction of gases with changes in temperature were drawn from Jacques Charles' experiments? *A5*
13. State Charles' law *(a)* in words; *(b)* mathematically. *A4*
14. In the laboratory, some nitrogen gas is collected over mercury in a eudiometer. How may the pressure of the nitrogen compare with the air pressure in the laboratory? *B5*

Group B

15. If it is assumed that the molecules of a solid or a liquid are in contact with each other but those of a gas are about 10 diameters apart,

why is the volume occupied by a gas about 1000 times that of the solid or liquid? *A3c*

16. Explain these properties of a gas in terms of the kinetic theory: *(a)* expansion; *(b)* pressure; *(c)* low density; *(d)* diffusion. *A3c, A3g*
17. Suppose the porous cup of Figure 10–5 is surrounded by carbon dioxide molecules. Carbon dioxide molecules are larger and heavier than the molecules of oxygen and nitrogen in the air. *(a)* Will air bubble from the glass tube? *(b)* Will water rise into the glass tube? *(c)* Explain. *A3i*
18. In terms of the kinetic theory explain what happens to a gas at its condensation temperature. *A3m, A3n*
19. Compare the strength of the attractive forces between the particles of nonpolar covalent molecular substances and polar covalent molecular substances as indicated by their condensation temperatures. *A3m, A3n, A3o*
20. Compare the strength of the attractive forces between the particles of molecular substances and the particles of ionic, covalent network, and metallic substances, as indicated by their condensation temperatures. *A3m, A3n, A3o*
21. At constant volume, how is the pressure exerted by a gas related to the Kelvin temperature? *A3p*
22. At constant pressure, how is the volume occupied by a gas related to the Kelvin temperature? *A3p*
23. At constant temperature, how is the volume occupied by a gas related to its pressure? *A3p*
24. A 50.0-mL sample of hydrogen is collected over mercury in one eudiometer, and a second 50.0-mL sample of hydrogen is collected over water in another eudiometer. The liquid levels inside and outside each eudiometer are the same. *(a)* How do the hydrogen pressures compare? *(b)* Why? *B5, B6*
25. How is the barometer reading corrected to give the pressure of the dry gas when *(a)* the gas is collected by displacement of mercury, level inside the eudiometer the same as that outside; *(b)* the gas is collected by displacement of mercury, level inside the eudiometer higher than that outside; *(c)* the gas is collected by displacement of water, level inside the eudiometer the same as that outside; *(d)* the gas is collected by displacement of water, level inside the eudiometer higher than that outside? *A2, B5, B6*
26. Boyle's and Charles' laws describe the behavior of the ideal gas. Under what conditions do they describe the behavior of real gases? *A3r*

PROBLEMS

GROUP A

Use cancellation if possible.

1. Some oxygen occupies $15\bar{0}$ mL when its pressure is $72\bar{0}$ mm. How many milliliters will it occupy when its pressure is $75\bar{0}$ mm? *B1*
2. A nitrogen sample collected when the pressure is $80\bar{0}$ mm has a volume of $19\bar{0}$ mL. What volume, in milliliters, will the nitrogen occupy at standard pressure? *B1*
3. Some hydrogen has a volume of $30\bar{0}$ mL when the pressure is $74\bar{0}$ mm. How many milliliters will the hydrogen occupy at $70\bar{0}$ mm pressure? *B1*
4. A sample of helium has a volume of 200.0 mL at 73.0 cm pressure. What pressure, in centimeters of mercury, is needed to reduce the volume to 50.0 mL? *B1*
5. Convert the following temperatures to Kelvin scale: *(a)* 25 °C; *(b)* $7\bar{0}$ °C; *(c)* −18 °C; *(d)* −253 °C. *B2*
6. Given $10\bar{0}$ mL of hydrogen gas collected when the temperature is 27 °C, how many milliliters will the hydrogen occupy at 42 °C? *B3*
7. A sample of argon has a volume of 155 mL when its temperature is 37 °C. At what Celsius temperature will the volume of the ar-

gon be 125 mL? *B3*

8. An oxygen sample occupies 80.0 mL at a temperature of −33 °C. What will be its volume in milliliters at 33 °C? *B3*
9. Some methane gas occupies 58.0 L at 17 °C. What will be the volume of the gas in liters at standard temperature? *B3*
10. Convert to standard conditions: $304\bar{0}$ mL of hydrogen measured at 27 °C and $70\bar{0}$ mm pressure. *B4*
11. Convert to standard conditions: $38\bar{0}$ mL of nitrogen at $3\bar{0}$ °C and 808 mm pressure. *B4*
12. Convert to standard conditions: $19\bar{0}$ mL of oxygen at −23 °C and $75\bar{0}$ mm pressure. *B4*
13. A sample of ethane collected when the temperature is 27 °C and the pressure is 80.0 cm measures $30\bar{0}$ mL. Calculate the volume in milliliters at −3 °C and 75.0 cm pressure. *B4*
14. Given $12\bar{0}$ mL of oxygen measured at 17 °C and $38\bar{0}$ mm pressure, what volume, in milliliters, will the gas occupy at 307 °C and $60\bar{0}$ mm pressure? *B4*

Group B

These problems will be much easier to solve if you use a calculator.

15. Hydrogen, 38.5 mL, was collected in a eudiometer by displacement of mercury. The mercury level inside the eudiometer was 28 mm higher than that outside. The temperature was 23 °C and the barometric pressure was 733 mm. Convert the volume of hydrogen to STP. *B7*
16. Carbon dioxide collected by displacement of mercury in an inverted graduated cylinder occupies 65 mL. The mercury level inside the cylinder is 15 mm higher than that outside; temperature, 18 °C; barometer reading, 754 mm. Convert the volume of carbon dioxide to STP. *B7*
17. Argon is collected by water displacement in a eudiometer. Gas volume, 29.2 mL; liquid levels inside and outside the eudiometer are the same; temperature, 21 °C; barometer reading, 742.0 mm. Convert the volume to that of dry argon at STP. *B7*
18. Some nitrogen is collected by displacement of water in a gas-measuring tube. Gas volume, 42.3 mL; liquid levels inside and outside the tube are the same; temperature, 25 °C; barometer reading, 737.7 mm. Convert the volume to that of dry nitrogen at STP. *B7*
19. A volume of 34.7 mL of oxygen is collected by water displacement. The water level inside the eudiometer is 43 mm higher than that outside. Temperature, 22.5 °C; barometer reading, 724.5 mm. Convert the volume to that of dry oxygen at STP. *B7*
20. At 18.5 °C and 740.2 mm barometric pressure, 16.3 mL of hydrogen is collected by water displacement. The liquid level inside the gas-measuring tube is 237 mm higher than that outside. Convert the volume to that of dry hydrogen at STP. *B7*
21. The density of oxygen at STP is 1.43 g/L. What is the mass of exactly one liter of oxygen, if the pressure increases by $4\bar{0}$ mm of mercury but the temperature is unchanged? *B1*
22. The density of nitrogen is 1.25 g/L at STP. Find the mass of exactly one liter of nitrogen at a temperature of 39 °C, if the pressure remains unchanged. *B3*
23. The density of carbon dioxide at STP is 1.98 g/L. Find the mass of exactly one liter of carbon dioxide at a temperature of 27 °C and at a pressure of $90\bar{0}$ mm of mercury. *B4*
24. At $2\bar{0}$ °C and 735 mm pressure, $45\bar{0}$ mL of argon has a mass of 0.722 g. What is the density of argon in g/L at STP? *B4*
25. At a temperature of 25 °C and a pressure of $80\bar{0}$ mm, a helium sample occupies $40\bar{0}$ mL. To what Celsius temperature must the helium be cooled if its volume is to be reduced to $36\bar{0}$ mL when the pressure is $75\bar{0}$ mm? *B4*

Sir William Ramsay (British) received the 1904 Nobel Prize in chemistry for discovering the noble gases and determining their positions in the periodic table. Earlier studies had shown a discrepancy in the specific gravities of nitrogen samples obtained from different sources. Ramsay found the difference to be a result of the presence of a new element, argon. Then he also isolated the elements neon, krypton, xenon, and helium. These new elements formed a family and allowed the development of Bohr's atomic theory. The methods Ramsay used to separate the gases were later applied in industry.

Molecular Composition of Gases

11

GOALS

In this chapter you will gain an understanding of: • Gay-Lussac's law of combining volumes of gases • the relationship between Gay-Lussac's law, Avogadro's principle, and the molecular composition of gases • the molar volume of a gas • solving chemical problems involving gases • the gas constant • the distinction between real gases and the ideal gas

11.1 Law of combining volumes of gases In Chapter 2, you learned that the observations made by Proust led to the law of definite composition. Applying this law to his investigation of the masses of combining substances, Dalton developed the atomic theory. Meanwhile, the Swedish chemist Berzelius was improving methods of chemical analysis. And at this same time, the French chemist Joseph Louis Gay-Lussac (1778–1850) was experimenting with gaseous substances. He was particularly interested in measuring and comparing the volumes of gases that react with each other. He also measured and compared the volumes of any gaseous products.

Gay-Lussac investigated the reaction between hydrogen and oxygen. He observed that when the volumes of reactants and products were measured at the same temperature and pressure, 2 liters of hydrogen reacted with 1 liter of oxygen to form 2 liters of water vapor.

2 liters hydrogen + 1 liter oxygen → 2 liters water vapor

This equation can be written in a more general form to show the simple relationship between the volumes of the reactants and the volume of the product.

2 volumes hydrogen + 1 volume oxygen → 2 volumes water vapor

Gay-Lussac also found that 1 liter of hydrogen combined with 1 liter of chlorine to form 2 liters of hydrogen chloride gas.

1 volume hydrogen + 1 volume chlorine → 2 volumes hydrogen chloride

From a third experiment, Gay-Lussac discovered that 1 liter

of hydrogen chloride combined with 1 liter of ammonia to produce a white powder. No hydrogen chloride or ammonia remained.

1 volume hydrogen chloride + 1 volume ammonia → ammonium chloride(s)

Another French chemist, Claude Louis Berthollet (1748–1822), recognized a similar relationship in experiments with hydrogen and nitrogen. He found that 3 liters of hydrogen always combined with 1 liter of nitrogen to form 2 liters of ammonia.

3 volumes hydrogen + 1 volume nitrogen → 2 volumes ammonia

In 1808, Gay-Lussac summarized the results of these experiments and stated the principle that bears his name. ***Gay-Lussac's law of combining volumes of gases*** *states: Under the same conditions of temperature and pressure, the volumes of reacting gases and of their gaseous products are expressed in ratios of small whole numbers.*

Proust had demonstrated the definite proportion of elements in a compound. Dalton's atomic theory had explained this regularity in the composition of substances. However, Dalton had pictured an atom of one element combining with an atom of another element to form a single particle of the product. He could not explain why *one* volume of hydrogen reacted with *one* volume of chlorine to form *two* volumes of hydrogen chloride gas. To explain this volume relationship in terms of Dalton's theory would require that atoms be subdivided. Dalton had described the atoms of elements as "ultimate particles" not capable of subdivision. Here was a disagreement between Dalton's theory and Gay-Lussac's observations. Was there an explanation to resolve the difficulty?

11.2 Avogadro's principle In 1811, Avogadro proposed a possible explanation for Gay-Lussac's simple ratios of combining gases. This explanation was to become one of the basic principles of chemistry, although its importance was not understood until after Avogadro's death.

Avogadro's explanation was that *equal volumes of all gases, under the same conditions of temperature and pressure, contain the same number of molecules.* Avogadro decided upon this believable explanation after studying the behavior of gases. He immediately saw its application to Gay-Lussac's volume ratios.

Avogadro further reasoned that the numbers of molecules of all gases, as reactants and products, *must be in the same ratio as their respective gas volumes.* Thus the composition of water vapor

2 volumes hydrogen + 1 volume oxygen → 2 volumes water vapor

could be represented as 2 molecules of hydrogen combining with 1 molecule of oxygen to produce 2 molecules of water vapor.

2 molecules hydrogen + 1 molecule oxygen → 2 molecules water vapor (Equation 1)

Avogadro reasoned that the oxygen molecules must somehow be equally divided between the two molecules of water vapor formed. *Thus, each molecule of oxygen must consist of* ***at least*** *two identical parts (atoms).* Avogadro did not reject the atoms of Dalton. He merely stated that they did not exist as independent, basic particles. Instead, they were grouped into molecules that consisted of two identical parts. The simplest such molecule would, of course, contain two atoms.

Avogadro's reasoning applied equally well to the combining volumes in the composition of hydrogen chloride gas.

1 volume hydrogen + 1 volume chlorine → 2 volumes hydrogen chloride

1 molecule hydrogen + 1 molecule chlorine → 2 molecules hydrogen chloride (Equation 2)

Each molecule of hydrogen must consist of two identical parts. After the reaction one of these parts is found in each of the two molecules of hydrogen chloride. Likewise, each chlorine molecule must consist of two identical parts. One of these parts is found in each of the two molecules of hydrogen chloride.

By Avogadro's explanation, the simplest molecules of hydrogen, oxygen, and chlorine each contain two atoms. The simplest possible molecule of water contains two atoms of hydrogen and one atom of oxygen. The simplest molecule of hydrogen chloride contains one atom of hydrogen and one atom of chlorine.

The correctness of Avogadro's explanation is so well recognized today that it has become known as ***Avogadro's principle.*** It is supported by the kinetic theory of gases, and chemists use it widely when determining molecular weights and molecular formulas.

Figure 11-1. Gay-Lussac and Biot making their balloon ascension for scientific observations in 1804.

11.3 Molecules of active gaseous elements are diatomic By applying Avogadro's principle to Gay-Lussac's law of combining volumes of gases, the simplest possible makeup of elemental gases can be determined. But whether the simplest structure of an elemental gas is the correct one must be known. To do this, the molecular formula for the product of each composition reaction described so far must be determined. In Sections 7.12 and 7.13, empirical and molecular formulas were discussed. You will recall that an empirical formula can be determined from percentage composition data obtained by chemical analysis. The molecular formula can then be calculated if the molecular weight is known. The molecular weights of gases can be determined from their densities. The molecular weights of solids and liquids that exist as molecules can be found by measuring the freezing or boiling temperatures of their solutions.

Chemists have analyzed hydrogen chloride gas and, with the aid of atomic weights, have determined its empirical formula to be HCl. The molecular formula could be the same as the empirical formula, or it could be any multiple of the empirical formula such as H_2Cl_2, H_3Cl_3, etc. To decide which formula is correct, chemists must experimentally determine the molecular weight. By measuring the density of hydrogen chloride, they have found its molecular weight to be 36.5. So the molecular formula must be HCl. The weight of one atom of hydrogen is 1.0, and the weight of one atom of chlorine is 35.5, giving a calculated molecular weight of 36.5. Any other multiple of the empirical formula gives a calculated molecular weight that is too high compared to the molecular weight determined experimentally. Thus, a molecule of hydrogen chloride contains *only one* atom of hydrogen and *only one* atom of chlorine.

Equation 2 in Section 11.2 indicates that two molecules of hydrogen chloride (2HCl) are formed from one molecule of hydrogen and one molecule of chlorine. But 2HCl contains two atoms of hydrogen. These two hydrogen atoms must have been provided by the one molecule of hydrogen reactant. Therefore this molecule must be a two-atomed, or *diatomic,* molecule, H_2. Similarly, 2HCl contains two atoms of chlorine which must have been provided by the one molecule of chlorine reactant. This chlorine molecule must therefore be a diatomic molecule, Cl_2. The equation can be correctly written as

$$\mathbf{H_2 + Cl_2 \rightarrow 2HCl}$$

A model of this equation is shown in Figure 11-2.

Similarly, oxygen molecules are proved to be diatomic because H_2O is the known molecular formula of water vapor.

$$\mathbf{2H_2 + O_2 \rightarrow 2H_2O}$$

This equation is shown in model form in Figure 11-3.

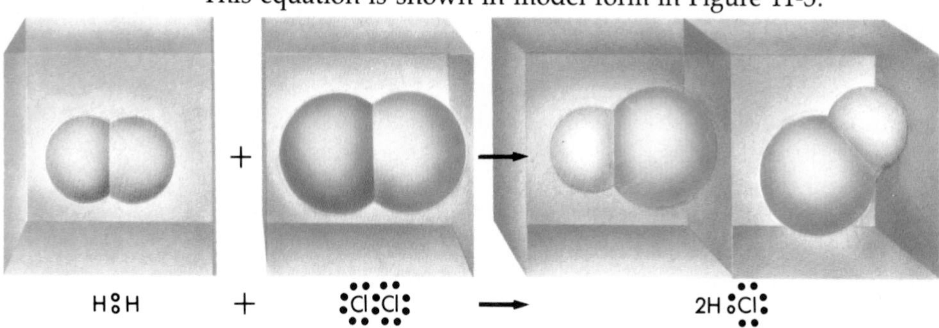

Figure 11-2. As HCl is known to be the correct molecular formula for hydrogen chloride gas, the molecules of hydrogen and chlorine are proved to be diatomic.

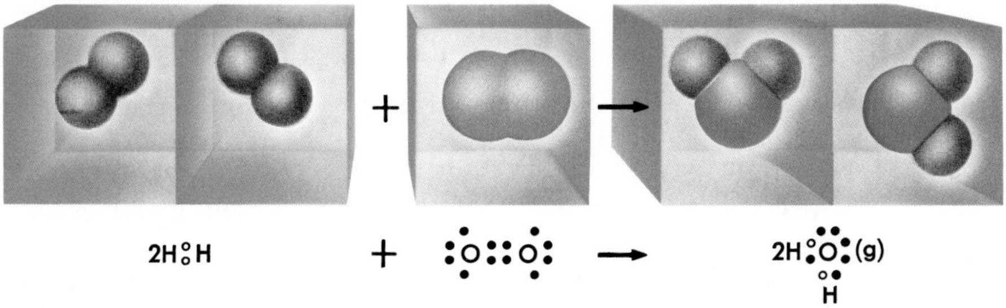

Figure 11-3. The correct molecular formula for water vapor is H_2O. Applying Avogadro's principle to the gas volumes observed by Gay-Lussac shows that hydrogen and oxygen molecules are diatomic.

One additional gaseous reaction deserves examination. Berthollet found that *3 volumes* of hydrogen combined with *1 volume* of nitrogen to form *2 volumes* of ammonia. By Avogadro's principle, *3 molecules* of hydrogen combine with *1 molecule* of nitrogen to form *2 molecules* of ammonia. Analysis of ammonia reveals that it is composed of 82% nitrogen and 18% hydrogen. The atomic weights of nitrogen and hydrogen are 14 and 1.0, respectively. The molecular weight of ammonia is known from gas density measurements to be 17.0. Therefore, the molecular formula is determined as follows:

At this point, you may wish to review the calculations explained in Sections 7.12 and 7.13.

$$\text{N:} \quad \frac{\mathbf{82\ g\ N}}{\mathbf{14\ g/mole}} = \mathbf{5.9\ moles\ N}$$

$$\text{H:} \quad \frac{\mathbf{18\ g\ H}}{\mathbf{1.0\ g/mole}} = \mathbf{18\ moles\ H}$$

$$\text{N:H} = \frac{\mathbf{5.9}}{\mathbf{5.9}} : \frac{\mathbf{18}}{\mathbf{5.9}} = \mathbf{1.0{:}3.0}$$

Empirical formula = $\mathbf{NH_3}$
Molecular formula = $\mathbf{(NH_3)_x}$
and $\mathbf{(NH_3\ weight)_x}$ = **17.0**
thus **x** = **1**
Molecular formula = $\mathbf{NH_3}$

Since each molecule of NH_3 contains 1 nitrogen atom, the 2 molecules of NH_3 produced must contain a total of 2 nitrogen atoms. Therefore, 2 nitrogen atoms must come from the 1 molecule of nitrogen reactant. Again the 2 molecules of NH_3 produced must contain a total of 6 hydrogen atoms. Therefore,

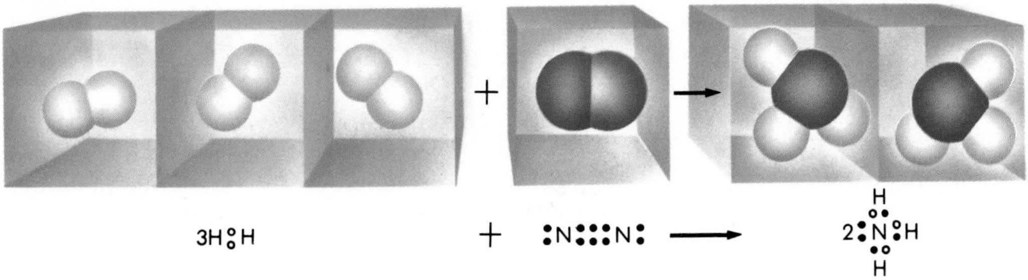

Figure 11-4. Application of Avogadro's principle to Gay-Lussac's combining volumes shows that molecules of elemental gas reactants are diatomic.

these 6 atoms must come from the 3 molecules of hydrogen reactant. These relations can be summarized as:

3 volumes hydrogen + 1 volume nitrogen → 2 volumes ammonia

3 molecules hydrogen + 1 molecule nitrogen → 2 molecules ammonia

$$3H_2 + N_2 \rightarrow 2NH_3$$

Both nitrogen and hydrogen molecules are diatomic. A model of this reaction is shown in Figure 11-4.

11.4 Molecules of the noble gases are monatomic By using the methods described in the preceding sections, it can be shown that the molecules of all *ordinary* gaseous elements contain *two* atoms. Other methods have shown that the *noble* gaseous elements, such as helium and neon, have only *one* atom to each molecule. None of these methods applies to solids and may not even apply to the vapors of certain elements that are liquid or solid at room temperature. For example, at high temperatures the molecules of mercury and iodine vapors are known to consist of only one atom each.

Particles of matter are described in Section 6.24.

11.5 Molar volume of a gas Oxygen gas consists of diatomic molecules. One mole of $\mathbf{O_2}$ contains the Avogadro number of molecules (6.02×10^{23}) and has a mass of 31.9988 g. Likewise, one mole of $\mathbf{H_2}$ contains the same number of molecules but has a mass of 2.016 g. Helium is a monatomic gas. One mole of **He** contains the Avogadro number of monatomic molecules and has a mass of 4.003 g. One-mole quantities of all molecular substances contain the Avogadro number of molecules.

The Avogadro number and the mole are explained in Section 3.15.

The volume occupied by one mole (or by one gram-molecular weight) of a gas at STP is called its ***molar volume.*** Since moles of gases all have the same numbers of molecules, Avogadro's principle indicates that they must occupy equal volumes under the same conditions of temperature and pressure. *The molar volumes of all gases are equal under the same conditions.* This has great practical significance in chemistry. You will now see how the molar volume of gases can be determined.

Recall from text Section 10.6 that STP means standard temperature and pressure.

The densities of gases represent the masses of *equal numbers* of their respective molecules measured at STP. Since different gases have different densities, the mass of an *individual molecule* of one gas must be different from the mass of an individual molecule of a different gas.

Density is defined in Section 1.7 as mass per unit volume.

The density of hydrogen is 0.0899 g/L, measured at STP. A mole of hydrogen, or 1 g-mol wt of hydrogen, has a mass of 2.016 g. (G-mol wt is an abbreviation for gram-molecular weight.) Since 0.0899 g of H_2 occupies 1 L volume at STP, what volume will 2.016 g of H_2 occupy under the same conditions? The molar volume expressed in liters bears the same relation to 1 L as 2.016 g bears to 0.0899 g. This proportionality can be expressed as follows:

$$\frac{\text{molar volume of } H_2}{1\text{ L}} = \frac{2.016\text{ g}}{0.0899\text{ g}}$$

Solving for molar volume,

$$\text{molar volume of } H_2 = \frac{2.016\text{ g} \times 1\text{ L}}{0.0899\text{ g}}$$

$$\text{molar volume of } H_2 = 22.4\text{ L}$$

The density of oxygen is 1.43 g/L. A mole of oxygen has a mass of 31.9988 g. Following the reasoning in the case of hydrogen, the molar volume of O_2 is

$$\frac{\text{molar volume of } O_2}{1\text{ L}} = \frac{32.0\text{ g}}{1.43\text{ g}}$$

$$\text{molar volume of } O_2 = \frac{32.0\text{ g} \times 1\text{ L}}{1.43\text{ g}}$$

$$\text{molar volume of } O_2 = 22.4\text{ L}$$

Computations with data on other gases would yield similar results. However, it is clear from Avogadro's principle that this is unnecessary. The proportion used above can be generalized to read as follows:

$$\frac{1\text{ molar volume}}{1\text{ L}} = \frac{\text{g-mol wt}}{\text{mass of 1 L}}$$

Transposing terms,

$$\frac{\text{mass of 1 L}}{1\text{ L}} = \frac{\text{g-mol wt}}{1\text{ molar volume}}$$

But

$$\frac{\text{mass of 1 L}}{1\text{ L}} = \text{density } (D) \text{ of a gas}$$

and

$$1\text{ molar volume} = 22.4\text{ L}$$

So

$$D\text{ (of a gas)} = \frac{\text{g-mol wt}}{22.4\text{ L}}$$

and

$$\text{g-mol wt} = D \times 22.4\text{ L}$$

Figure 11-5. Hot air balloons rise because the air inside the bag is warmer and, therefore, less dense than the surrounding air. The heat for the "hot-air" balloon comes from a propane burner, which produces a flame that reaches up into the bag to heat the air (bottom photo).

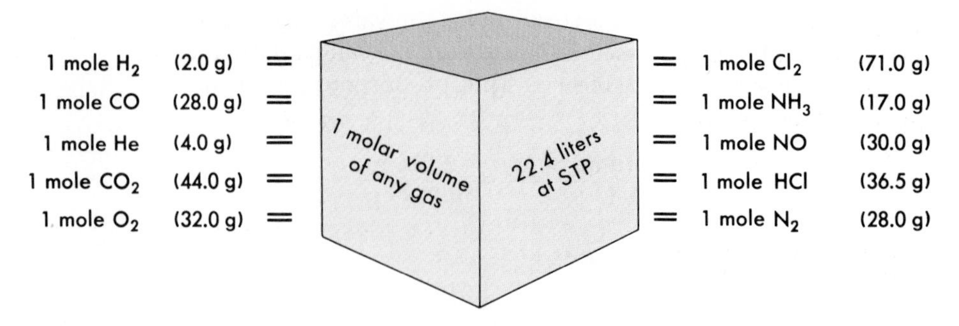

Figure 11-6. At STP, 22.4 liters of all gases have the same number of molecules, and the mass of each volume in grams is numerically equal to its molecular weight.

Thus, as shown in Figure 11–6, the *gram-molecular weight* (mass in grams of one mole) *of a gaseous substance is the mass, in grams, of 22.4 liters of the gas measured at STP*. In other words, it is simply the density of the gas multiplied by the constant 22.4 L. Similarly, the density of a gas can be found by dividing its gram-molecular weight by the constant, 22.4 L.

If the molecular formula for a gas is known, its density can be determined directly from the formula. For example, the mass of one mole of SO_2 is 64.1 g. Thus

$$D_{SO_2} = \frac{64.1\ \text{g}}{22.4\ \text{L}} = 2.86\ \text{g/L at STP}$$

Practice Problems

1. What is the density of CO_2 at STP?
   ans. $D_{CO_2} = 1.96$ g/L at STP
2. What is the mass of $35\bar{0}$ mL CO_2 at STP?
   ans. $D = m/V$, $m = 0.686$ g
3. What is the volume of 25.0 g CO_2 at STP?
   ans. $V = 12.8$ L
4. What is the molecular weight of a gas that has a density of 1.25 g/L at STP? *ans.* mol wt = 28.0

11.6 Molecular weights of gases determined experimentally

The molecular weights of gases, or of the vapors of substances that vaporize without decomposition, can be determined by the *molar-volume method*.

Conversion of a gas volume at given conditions to STP is explained in Section 10.14.

It is generally impractical in the laboratory to directly weigh a molar volume (22.4 L) of a gas or vapor at STP. Indeed, some substances otherwise suitable for this method are liquids or even solids at STP conditions. Fortunately, it is not necessary to weigh a full molar volume of a gas or vapor to find its molecular weight. Any quantity of a gas or vapor that can be weighed to determine its mass precisely can be used. Its volume can be measured under any suitable conditions of temperature and pressure. This volume is then converted to STP, and the mass of 22.4 L is calculated. Sample Problem 11-1 shows how this experimental method is used to determine molecular weight.

SAMPLE PROBLEM 11-1

A gas sample, mass 0.467 g, is collected by water displacement at $2\bar{0}$ °C and $75\bar{0}$ mm pressure. Its volume is $20\bar{0}$ mL. What is the molecular weight of the gas?

SOLUTION

Correction for water vapor pressure: From Appendix Table 8, the water vapor pressure at $2\bar{0}$ °C is 17.5 mm. This water vapor pressure is subtracted from the barometric pressure to obtain the pressure of the dry gas.

$$75\bar{0} \text{ mm} - 17.5 \text{ mm} = 732 \text{ mm}$$

Conversion of Celsius temperatures to Kelvin temperatures:

$$2\bar{0}\ °\text{C} + 273 = 293 \text{ K};\ \bar{0}\ °\text{C} + 273 = 273 \text{ K}$$

Correction for pressure and temperature changes:

$$V' = V \times p/p' \times T'/T$$

$$V'_{\text{STP}} = 20\bar{0} \text{ mL} \times \frac{732 \text{ mm}}{76\bar{0} \text{ mm}} \times \frac{273 \text{ K}}{293 \text{ K}}$$

$$V'_{\text{STP}} = 179 \text{ mL, or } 0.179 \text{ L}$$

The quantity of 0.467 g per 0.179 L at STP is an expression of the density of the gas and can be substituted for it. Thus

$$\text{g-mol wt} = D \times 22.4 \text{ L}$$

$$\text{g-mol wt} = \frac{0.467 \text{ g}}{0.179 \text{ L}} \times 22.4 \text{ L}$$

$$\text{g-mol wt} = 58.4 \text{ g}$$

$$\text{mol wt} = 58.4$$

The volume occupied by a known mass of a gas under any conditions of temperature and pressure can also be calculated if the molecular formula of the gas is known. The molecular formula provides the mass of 1 mole of the gas. One mole of any gas occupies the molar volume, 22.4 L, at STP. By proportion, the volume of the known mass of the gas at STP can be determined. Then, by applying the gas laws, the volume at any temperature and pressure can be computed. This type of calculation is shown in Sample Problem 11-2.

SAMPLE PROBLEM 11-2

What is the volume of 12 g of carbon dioxide gas, CO_2, at $2\bar{0}$ °C and at 740 mm pressure?

SOLUTION

The formula CO_2 indicates that the molecular weight is 44. Now 44 g (1 mole) of CO_2 occupies 22.4 L (1 molar volume) at STP. The volume occupied by 12 g will be represented as X.

$$\frac{44 \text{ g}}{22.4 \text{ L}} = \frac{12 \text{ g}}{\text{X}}$$

Solving for X,

$$\text{X} = 12 \text{ g} \times \frac{22.4 \text{ L}}{44 \text{ g}} = 6.1 \text{ L, at STP}$$

Correction for pressure and temperature changes:

$$V' = V \times p/p' \times T'/T$$

$$V'_{2\bar{0}\text{ °C, 740 mm}} = 6.1 \text{ L} \times \frac{760 \text{ mm}}{740 \text{ mm}} \times \frac{293 \text{ °K}}{273 \text{ °K}}$$

$$V'_{2\bar{0}\text{ °C, 740 mm}} = 6.7 \text{ L}$$

11.7 Chemical problems involving gases As shown in text Section 8.1, the equation for a chemical reaction expresses quantities of reactants and products in *moles*. The numerical coefficients in the balanced equation indicate the number of moles of each substance.

Frequently, a reactant or a product of a reaction is a gas. Indeed, in certain reactions, *all* reactants and products may be gaseous. Avogadro's principle indicates that single moles of all such gases have the same volume under the same conditions of temperature and pressure. At STP, a mole of any gas occupies 1 molar volume, 22.4 L. *Consequently, the mole relationships in the equation are also the volume relationships of gases.*

Remember that the behavior of real gases is not described exactly by the ideal gas laws. Calculations that involve the molar volume as 22.4 L can give only approximately correct answers.

Ideal-gas laws and real-gas behavior are described in Section 10.15.

There are two general types of problems that involve chemical equations and the volumes of gases.

1. ***Gas volume–gas volume problems.*** In these problems, a certain *volume of a gas* reactant or product is given. The *volume of another gas* reactant or product is required.

2. ***Mass–gas volume problems.*** Here a certain *mass* of a reactant or product is given and the *volume of a gas* reactant or product is required, or vice versa.

Figure 11-7. The recycling of gases requires knowing the chemical principles of gases. At a sewage plant in Zurich, Switzerland, methane gas is produced during the fermentation process of sludge. The gas is then used for further processing of sludge. Excess methane gas is piped to an incinerator plant where it is used in the production of electricity.

11.8 Gas volume–gas volume problems The *volume* of one *gaseous substance* is given in gas volume–gas volume problems. In this type of problem, you are asked to determine the *volume* of another *gaseous substance* involved in the chemical action. Recall that single moles of all gases at the same temperature and pressure occupy the same volume. Thus, in a balanced equation, the volumes of gases are proportional to the number of moles shown by the numerical coefficients. To illustrate,

$2CO(g)$	+	$O_2(g)$	→	$2CO_2(g)$
2 moles		**1 mole**		**2 moles**
2 volumes		**1 volume**		**2 volumes**

The balanced equation signifies that 2 moles of CO reacts with 1 mole of O_2 to produce 2 moles of CO_2. From Avogadro's principle, 2 volumes (liters, milliliters, etc.) of carbon monoxide reacts with 1 volume (liter, milliliter, etc.) of oxygen to produce 2 volumes (liters, milliliters, etc.) of carbon dioxide. (In this relationship it is assumed, of course, that the temperature and pressure of all three gases are the same.) Thus, 10 liters of CO would require 5 liters of O_2 for complete combustion and would produce 10 liters of CO_2. The volume relationship is 2:1:2 under the same conditions of temperature and pressure.

Since the above reaction is exothermic, the gas that is produced expands because of the rise in temperature. The volume relations apply only after the temperature of the gaseous product has been lowered to that of the reactants at the beginning of the reaction.

The conditions of temperature and pressure must be known in order to determine which substances exist as gases. Whenever the conditions are not stated, they are assumed to be standard. Consider the complete combustion of methane.

$CH_4(g)$	+	$2O_2(g)$	→	$CO_2(g)$	+	$2H_2O(l)$
1 mole		**2 moles**		**1 mole**		**2 moles**
1 volume		**2 volumes**		**1 volume**		

The formula for water is followed by the symbol (l) because water is a liquid at temperatures under $10\bar{0}$ °C. Assume that the volumes of the gaseous reactants are measured under ordinary room conditions. If so, water cannot be included in the volume ratio. But the reactants, methane and oxygen, and the product, carbon dioxide, are gases. Their volume relationship is seen to be 1:2:1.

Gas volume–gas volume problems are very simple to solve. The problem setup is similar to that of mass–mass problems (see Section 8.11). However, it is not necessary to use atomic weights to convert moles of the specified gases to their respective masses as represented in the equation. Once set up, most gas volume–gas volume problems can be solved by inspection. The following example shows how these problems are commonly solved.

Suppose you wish to know the volume of hydrogen that combines with 5.0 L of nitrogen to form ammonia gas. The problem is set up this way:

$$\begin{array}{ccccc} \mathbf{X} & & \mathbf{5.0\ L} & & \mathbf{Y} \\ \mathbf{3H_2(g)} & + & \mathbf{N_2(g)} & \rightarrow & \mathbf{2NH_3(g)} \\ \mathbf{3\ moles} & & \mathbf{1\ mole} & & \mathbf{2\ moles} \end{array}$$

The equation shows that H_2 and N_2 combine in the ratio of 3 moles to 1 mole. From Avogadro's principle, these gases must combine in the ratio of 3 volumes to 1 volume. Thus, 5.0 liters of nitrogen requires 15 liters of hydrogen for complete reaction.

Mathematically, the problem can be set up

$$X = 5.0\ L\ \cancel{N_2} \times \frac{3\ \cancel{moles}\ H_2}{1\ \cancel{mole}\ \cancel{N_2}}$$

Solving,

$$X = 15\ L\ H_2$$

Since the equation shows that 2 moles of NH_3 is produced, it is apparent by inspection that $1\bar{0}$ L of NH_3 is formed from 5.0 L of N_2.

$$Y = 5.0\ L\ \cancel{N_2} \times \frac{2\ \cancel{moles}\ NH_3}{1\ \cancel{mole}\ \cancel{N_2}}$$

$$Y = 1\bar{0}\ L\ NH_3$$

See Sample Problem 11-3.

SAMPLE PROBLEM 11-3

Assuming air to be 20.9% oxygen by volume: (*a*) How many liters of air must enter a carburetor to complete the combustion of 30.0 L of octane vapor? (*b*) How many liters of carbon dioxide are formed? (All gases are measured at the same temperature and pressure.)

SOLUTION

Octane has the formula C_8H_{18}. Its complete combustion produces carbon dioxide and water. Note that it is only the *oxygen* of the air that reacts with octane. Therefore it is necessary first to determine the volume of oxygen required. Let X be this volume and Y the volume of CO_2 formed. The problem setup is

$$\begin{array}{cccc} 30.0\ \text{L} & \text{X} & \text{Y} & \\ 2C_8H_{18}(g) + & 25O_2(g) \rightarrow & 16CO_2(g) + & 18H_2O(l) \\ 2\ \text{moles} & 25\ \text{moles} & 16\ \text{moles} & \end{array}$$

(*a*) Solving

$$X = 30.0\ \text{L}\ \cancel{C_8H_{18}} \times \frac{25\ \cancel{\text{moles}}\ O_2}{2\ \cancel{\text{moles}\ C_8H_{18}}}$$

$$X = 375\ \text{L}\ O_2$$

Now 375 L of O_2 is 20.9% of the air required. So

$$\text{air required} = 375\ \text{L} \times \frac{100\%}{20.9\%} = 1790\ \text{L}$$

(*b*) Solving as before,

$$Y = 30.0\ \text{L}\ \cancel{C_8H_{18}} \times \frac{16\ \cancel{\text{moles}}\ CO_2}{2\ \cancel{\text{moles}\ C_8H_{18}}}$$

$$Y = 24\bar{0}\ \text{L}\ CO_2$$

Reminder: Because this is a gas volume–gas volume problem, the volumes of air and CO_2 computed are those that would be measured at the temperature and pressure of the octane vapor prior to its combustion. Under such conditions, the water product would be a liquid. Thus its quantity is not considered in solving the problem.

11.9 Mass–gas volume problems This type of problem involves the relation between the *volume of gas* and the *mass* of another substance in a reaction. In some cases, the mass of the substance is given, and the volume of the gas is required. In others, the volume of the gas is given, and the mass of the substance is required.

As an illustration, suppose you wish to determine how many grams of calcium carbonate, $CaCO_3$, must be decomposed to produce 5.00 L of carbon dioxide, CO_2, at STP. The problem setup is

$$\begin{array}{ccc} \text{X} & & 5.00\ \text{L} \\ CaCO_3(s) \rightarrow & CaO(s) + & CO_2(g) \\ 1\ \text{mole} & & 1\ \text{mole} \end{array}$$

$$1\ \text{mole}\ CaCO_3 = 10\bar{0}\ \text{g}$$

$$1\ \text{mole}\ CO_2 = 22.4\ \text{L}$$

The four operations in solving mass–mass problems are explained in Section 8.11.

1. *Determine the number of moles of the substance given.*
2. *Determine the number of moles of the substance whose mass is required.*
3. *Determine the mass of the substance required.*
4. *Check the units; estimate the answer. Perform the arithmetic operations and compare the calculated result with the estimated one.*

Observe that the molar volume (22.4 L) is used in place of the mass/mole (44 g/mole) of CO_2. This is possible because each mole of gas occupies 22.4 L at STP (*and only at STP*). The problem can now be solved by the method used for mass–mass problems.

The volume of CO_2 multiplied by the fraction $\frac{\text{mole}}{22.4\ \text{L}}$ will indicate the number of moles of CO_2 given (operation 1):

$$\mathbf{5.00\ L\ CO_2 \times \frac{mole}{22.4\ L} = number\ of\ moles\ CO_2}$$

Then, after operations 2, 3, and 4 of the solution of a mass–mass problem,

$$\mathbf{X = 5.00\ \cancel{L}\ \cancel{CO_2} \times \frac{\cancel{mole}}{22.4\ \cancel{L}} \times \frac{1\ \cancel{mole}\ CaCO_3}{1\ \cancel{mole}\ \cancel{CO_2}} \times \frac{10\bar{0}\ g}{\cancel{mole}} = 22.3\ g\ CaCO_3}$$

Practice Problems

1. How many liters of CO_2 at STP can be produced by the decomposition of 20.0 g of $CaCO_3$?

ans. $X = 20.0\ \cancel{g}\ \cancel{CaCO_3} \times \frac{\cancel{mole}}{10\bar{0}\ \cancel{g}} \times \frac{1\ \cancel{mole}\ CO_2}{1\ \cancel{mole}\ \cancel{CaCO_3}} \times \frac{22.4\ L}{\cancel{mole}} = 4.48\ L\ CO_2$

2. How many liters of H_2 at STP can be produced by the reaction of 4.60 g Na and excess H_2O? *ans.* X = 2.24 L H_2
3. How many grams of Na are needed to react with H_2O and liberate 40$\bar{0}$ mL H_2 at STP? *ans.* X = 0.821 g Na

11.10 Gases not measured at STP Gases are seldom measured under standard conditions of temperature and pressure. But *only* gas volumes under standard conditions can be placed in a proportion with the molar volume of 22.4 L. Therefore, *gas reactants measured under conditions other than STP must first be corrected to STP before proceeding with mass–gas volume calculations.* These corrections are performed in agreement with the gas laws in Chapter 10.

For example, suppose that the gas in question is a *product* whose volume is to be measured under conditions other than STP. *You must first calculate the volume at STP from the chemical equation.* Then convert this volume at STP to the required conditions of temperature and pressure by proper application of the gas laws.

Gas volume–gas volume calculations do not require STP corrections since volumes of gases are related to moles rather than to molar volumes. Thus, in gas volume—gas volume problems, it is necessary only that all measurements of gas volumes be made at the same temperature and pressure.

11.11 Gases collected by water displacement The volume of any gas in a mass–gas volume problem is calculated under STP conditions and must be corrected for any other specified conditions of temperature and pressure. If this gas is collected by water displacement, the vapor pressure of the water must be taken into account. The partial pressure of the gas is the difference between the measured pressure and the partial pressure exerted by water vapor at the specified temperature.

Suppose that a gaseous product is collected by water displacement. The volume of this product is to be determined at 23 °C and 752 mm pressure by a mass–gas volume calculation. The volume at STP is computed from an appropriate chemical equation. The vapor pressure of water at 23 °C is found in Appendix Table 8 to be 21.1 mm. Thus

This type of calculation is explained in Section 10.14.

$$V'_{23\,°C,\ 752\ mm} = V_{STP} \times \frac{76\bar{0}\ mm}{(752 - 21)\ mm} \times \frac{296\ K}{273\ K}$$

Sample Problem 11-4 shows the complete solution to this type of problem.

SAMPLE PROBLEM 11-4

What volume of oxygen, collected by water displacement at $2\bar{0}$ °C and 750.0 mm pressure, can be obtained by the decomposition of 55.0 g of potassium chlorate?

SOLUTION

A mass is given and a gas volume is required. The volume of the gas at STP is first found from the chemical equation. The problem setup is as follows:

$$\begin{array}{ccc} 55.0\ g & & X \\ 2KClO_3(s) \rightarrow & 2KCl(s) + & 3O_2(g) \\ 2\ moles & & 3\ moles \end{array}$$

This type of problem setup is described in Section 11.9.

$$1\ mole\ KClO_3 = 122.6\ g$$
$$1\ mole\ O_2 = 22.4\ L$$

Solution by moles

$$X = 55.0\ g\ \cancel{KClO_3} \times \frac{\cancel{mole}}{122.6\ g} \times \frac{3\ \cancel{moles}\ O_2}{2\ \cancel{moles\ KClO_3}} \times \frac{22.4\ L}{\cancel{mole}}$$

$$X = 15.1\ L\ O_2\ at\ STP$$

The vapor pressure of water at $2\bar{0}$ °C is found to be 17.5 mm.

Correction for change in pressure and temperature:

$$V' = V \times p/p' \times T'/T$$

$$V'_{2\bar{0}\ °C,\ 750.0\ mm} = 15.1\ L \times \frac{76\bar{0}\ mm}{(750.0 - 17.5)\ mm} \times \frac{293\ °K}{273\ °K}$$

$$V'_{2\bar{0}\ °C,\ 750.0\ mm} = 16.8\ L$$

11.12 The gas constant From Avogadro's principle you know that the volume of a mole is the same for all gases under the same conditions of temperature and pressure. The volume of any gas is directly proportional to the number of moles (n) of the gas, if pressure and temperature are constant. The $\propto$ means "varies directly as."

Direct and inverse proportions are explained in Section 1.22.

$$V \propto n \text{ (} p \text{ and } T \text{ constant)}$$

From Boyle's law you know that the volume of a gas is inversely proportional to the pressure applied to it if the quantity of gas (number of moles of gas) and temperature are constant. The inverse relationship is shown by placing the pressure in the denominator.

$$V \propto \frac{1}{p} \text{ (} n \text{ and } T \text{ constant)}$$

Similarly, from Charles' law you know that the volume of a gas is directly proportional to the Kelvin temperature if the pressure and quantity of gas remain constant.

$$V \propto T \text{ (} p \text{ and } n \text{ constant)}$$

Thus,

$$V \propto n \times \frac{1}{p} \times T$$

By inserting a proportionality constant R of suitable dimensions, this proportion can be restated as an equation.

$$V = nR\left(\frac{1}{p}\right)T$$

or

$$pV = nRT$$

The proportionality constant R is known as the *gas constant.* When the quantity of gas is expressed in moles, R has the same value for all gases. Usually the gas volume V is expressed in *liters,* the quantity n in *moles,* the temperature T in *degrees Kelvin,* and the pressure p in *atmospheres* (abbreviated atm). Of course, the standard pressure of $76\bar{0}$ mm is *1 atmosphere.* So

$$\frac{\text{pressure in mm Hg}}{76\bar{0}\text{ mm Hg/atm}} = \text{pressure in atm}$$

The dimensional units of the gas constant R can be determined from the ideal-gas equation as follows:

$$pV = nRT$$

Solving for R,

$$R = \frac{pV}{nT}$$

Using the units stated,

$$R = \frac{\text{atm} \times \text{L}}{\text{moles} \times \text{K}}$$

Thus R must have the dimensions *L atm/mole K.*

Careful measurements of the density of oxygen at low pressures yield the molar volume of 22.414 L. This is accepted as the accurate molar volume of an ideal gas. It represents the volume occupied by exactly 1 mole of ideal gas under conditions of exactly 1 atm and 273.15 K. Substituting in the ideal-gas equation and solving the expression for R,

$$R = \frac{pV}{nT} = \frac{1\ \text{atm} \times 22.414\ \text{L}}{1\ \text{mole} \times 273.15\ \text{K}}$$

$$R = 0.082057\ \text{L atm/mole K}$$

Suppose the properties of an unknown gas are examined at a temperature of 28 °C and 74$\bar{0}$ mm pressure. It is found that the mass of 1 L is 5.16 g under these conditions. You wish to determine the molecular weight of the gas. This calculation can be made directly by using the gas constant R, 0.0821 L atm/mole K, and the ideal-gas equation, $pV = nRT$. The use of R in the ideal-gas equation enables moles per liter of gas to be computed.

$$T = 273° + 28\ °\text{C} = 301\ \text{K}$$

$$V = 1\ \text{L}$$

$$p = \frac{74\bar{0}\ \text{mm}}{76\bar{0}\ \text{mm/atm}} = 0.974\ \text{atm}$$

$$pV = nRT$$

Solving for n moles,

$$n = \frac{pV}{RT}$$

$$n = \frac{0.974\ \cancel{\text{atm}} \times 1\ \cancel{\text{L}}}{\frac{0.0821\ \cancel{\text{L}}\ \cancel{\text{atm}}}{\text{mole}\ \cancel{\text{K}}} \times 301\ \cancel{\text{K}}}$$

$$n = \frac{0.974\ \text{mole}}{0.0821 \times 301} = 0.0394\ \text{mole}$$

This calculation shows that the experimental mass of 1 L of the gas, 5.16 g, is 0.0394 mole.

Since 0.0394 mole has a mass of 5.16 g, the mass of 1 mole is

$$1\ \cancel{\text{mole}} \times \frac{5.16\ \text{g}}{0.0394\ \cancel{\text{mole}}} = 131\ \text{g}$$

Therefore, the molecular weight of the gas is 131.

The ideal-gas equation simplifies the solution of mass–gas volume problems under nonstandard conditions. The number of moles of gas required is calculated by the mole method. This result can then be substituted in the ideal-gas equation, together with the pressure and temperature conditions. The gas volume is then calculated directly. Conversely, if the volume of a gas under nonstandard temperature and pressure conditions is known, the ideal-gas equation can be used to calculate the *number of moles* of the gas. This quantity can then be used to complete the solution of a gas volume–mass problem by the mole method.

Practice Problems

1. What volume does 0.0250 mole H_2 occupy at 0.821 atm pressure and $30\bar{0}$ °K? *ans.* $V = 0.750$ L
2. Some oxygen, $25\bar{0}$ mL, is measured at 25 °C and $72\bar{0}$ mm pressure. How many moles is this? *ans.* $n = 0.00968$ mole

11.13 Real gases and the ideal gas Precise experiments involving molar volumes of gases show that all gases vary slightly from the ideal-gas characteristics assigned to them by the gas laws and Avogadro's principle. This does not mean that the laws are only approximately true. Rather, it indicates that real gases do not behave as ideal gases over wide ranges of temperature and pressure. They act most like ideal gases at low pressures and high temperatures.

The behavior of real gases is described earlier in Section 10.15.

Two factors contribute to the difference between the behavior of real gases and the equations for an ideal gas:

1. Compression of a gas is *limited* by the fact that the molecules themselves occupy space, even though the volume of the molecules is extremely small.

2. Compression of a gas is *aided* by the fact that van der Waals (attractive) forces, however weak, do exist between the molecules of a gas.

Only when these two opposing tendencies within the gas balance exactly will the gas respond as an ideal gas.

When expressed to five significant figures, the molar volume of an ideal gas is 22.414 L. Gases such as ammonia and chlorine, which at ordinary temperatures are not far above their condensation points, show rather marked deviations from 22.4 L as the molar volume. The molar volumes of ammonia and chlorine, measured at standard conditions, are 22.09 L and 22.06 L, respectively. Gases such as oxygen and nitrogen, which have low condensation points, behave more nearly as ideal gases under ordinary conditions. The molar volumes of both oxygen and nitrogen at standard conditions are 22.39 L. For most gases, deviations from ideal-gas performance through ordinary ranges of temperature and pressure do not exceed two percent.

SUMMARY

According to Gay-Lussac's law, under the same conditions of temperature and pressure, the volumes of reacting gases and their gaseous products are expressed in ratios of small whole numbers.

Avogadro suggested that equal volumes of all gases, under similar conditions of temperature and pressure, contain the same number of molecules. He applied this idea to Gay-Lussac's volume ratios and obtained the ratios of the numbers of reacting molecules. Avogadro's principle, Gay-Lussac's law, and molecular-weight data, can be used to show that molecules of the active elemental gases are diatomic.

The volume occupied by one mole of a gas at STP is called the molar volume of the gas. Molar volumes of all gases are equal and are found experimentally to be 22.4 liters at STP. This relationship provides a simple experimental method for determining the molecular weights of gases and of other molecular substances that can be vaporized without undergoing decomposition.

There are two general types of problems involving gas volumes: gas volume–gas volume problems and mass–gas volume problems. Both types are based on an application of Avogadro's principle.

When gas volumes are measured under conditions other than STP, they must be adjusted to STP before being placed in a calculation with the molar volume, 22.4 liters. Calculations from chemical equations yield gas volumes at standard temperature and pressure. If gas volumes at other conditions are required, conversion is made using the gas laws.

The gas laws can be developed into a useful equation involving a gas constant. This equation enables certain calculations with gas measurements to be made easily.

An ideal gas conforms exactly to the gas laws and Avogadro's principle. Real gases deviate slightly from the behavior of the ideal gas except when the responsible molecular characteristics balance out.

VOCABULARY

Avogadro's principle
gas constant
ideal gas equation
law of combining volumes of gases
molar volume
real gas

PROBLEMS

In the absence of stated conditions of temperature and pressure, assume gases at STP.

Group A

1. Calculate the density of chlorine gas, Cl_2, at STP to three significant figures. *A1, A2, A3d, B1*
2. Determine the density of hydrogen bromide gas, HBr, at STP, to three significant figures. *A3d, B1*

3. What is the mass in grams of 1.00 L of ethane gas, C_2H_6, at STP? *A3d, B1*
4. The mass of 1.00 L of a gas at STP is 1.52 g. What is the molecular weight of the gas? *A3d, B2*
5. Obtain the density of fluorine at STP from Appendix Table 16. (*a*) From this density, calculate the molecular weight of fluorine. (*b*) Using the calculated molecular weight, determine the number of atoms in a molecule of fluorine. *A3a, A3d, B2*
6. Hydrogen is the gas of lowest density. What is the mass in grams of $35\bar{0}$ mL of hydrogen at STP? *A3d, B1*
7. At standard conditions, $25\bar{0}$ mL of a gas has a mass of 0.7924 g. What is the molecular weight of the gas? *A3d, B2*
8. Calculate the mass in grams of $65\bar{0}$ mL of CO_2 at STP. *A3d, B1*
9. If the mass of $75\bar{0}$ mL of methane is 0.537 g at STP, what is the molecular weight of methane? *A3d, B2*
10. The compounds NH_3, PH_3, and AsH_3 are all gases at STP. (*a*) Determine their molecular weights to three significant figures. (*b*) What is the density of each at STP? *A3d, B1*
11. Calculate the mass in grams of 7.00 L of N_2, NH_3, and C_2H_2. *A3d, B1*
12. How many liters of hydrogen and nitrogen are required to produce $3\bar{0}$ L of ammonia gas? *B5*
13. Carbon monoxide burns in oxygen and forms carbon dioxide. (*a*) How many liters of carbon dioxide are produced when 25 L of carbon monoxide burns? (*b*) How many liters of oxygen are required? *B5*
14. Acetylene gas, C_2H_2, burns in oxygen and forms carbon dioxide and water vapor. (*a*) How many liters of oxygen are needed to burn 15.0 L of acetylene? (*b*) How many liters of carbon dioxide are formed? (*c*) How many moles of water are produced? *B5*
15. Ethane gas, C_2H_6, burns in air and produces carbon dioxide and water vapor. (*a*) How many liters of carbon dioxide are formed when 20.0 L of ethane is burned? (*b*) How many moles of water are formed? *B5*
16. How many liters of air are required to furnish the oxygen to complete the reaction in Problem 15? (Assume the air to be 20.9% oxygen by volume.) *B5*
17. If $25\bar{0}$ mL of hydrogen and $20\bar{0}$ mL of oxygen are mixed and ignited, (*a*) what volume of oxygen remains uncombined? (*b*) What volume of water vapor is formed if all these gases are measured at $15\bar{0}$ °C? *B5*
18. How many grams of sodium are needed to release 1.5 L of hydrogen from water? *B6*
19. (*a*) How many liters of hydrogen are required to convert 22.5 g of hot copper (II) oxide to metallic copper? (*b*) How many moles of water are formed? *B6*
20. When 125 g of zinc reacts with 125 g HCl, how many liters of hydrogen are formed? (Note: First determine which reactant is in excess.) *B6*
21. (*a*) How many liters of oxygen can be produced by the electrolysis of 75.0 g of water? (*b*) How many liters of hydrogen are produced simultaneously? *B6*
22. (*a*) How many grams of copper will be produced when hydrogen reacts with 50.0 g of hot copper (II) oxide? (*b*) How many liters of hydrogen are required? (*c*) How many moles of water are formed? *B6*
23. How many liters of hydrogen will be produced by the action of 15.0 g of calcium metal and excess hydrochloric acid? Calcium chloride is the other product. *B6*
24. Zinc reacts with dilute sulfuric acid to form hydrogen and zinc sulfate in solution. If 8.00 g of zinc reacts with excess sulfuric acid, (*a*) what volume of hydrogen is liberated? (*b*) What mass of zinc sulfate is formed? *B6*

Group B

25. A gaseous compound has the analysis: nitrogen, 63.64%; oxygen, 36.36%. Its density is 1.977 g/L. Determine (*a*) its empirical formula; (*b*) its molecular weight; (*c*) its molecular formula. *A3d, B2*
26. The mass of 1.00 L of a certain gaseous element, collected at a pressure of $73\bar{0}$ mm of mercury and a temperature of 27 °C, has a mass of 1.25 g. (*a*) What is the molecular weight of the gas? (*b*) What is its identity? *A3a, A3d, B3*

27. It is found that 1.00 L of nitrogen combines with 1.00 L of oxygen in an electric arc and forms 2.00 L of a gas. By analysis, this gas contains 46.7% nitrogen and 53.3% oxygen. Its density is determined to be 1.34 g/L. (*a*) Find the empirical formula of the product. (*b*) What is the molecular formula? (*c*) Using the information of this problem and the arguments of Avogadro, determine the number of atoms per molecule of nitrogen and oxygen. *A3a, A4, B2*

28. What will be the volume in liters of 2.0 g of CS_2 vapor at 726 mm pressure and $7\bar{0}$ °C? *B4*

29. A 5.00-L flask is filled with a gas at STP. (*a*) How many gas molecules are in the flask? (*b*) The flask is then attached to a high-vacuum pump and evacuated until the pressure is only 1.00×10^{-4} mm. Assuming no temperature change, how many molecules remain in the flask? *A2*

30. A gas sample, mass 0.427 g, occupies 125 mL at $2\bar{0}$ °C and 745 mm. Calculate the molecular weight of the gas. *B3*

31. (*a*) When 14.5 g of sulfur burns, how many liters of sulfur dioxide at STP are produced? (*b*) What volume will this gas occupy at 25 °C and 755 mm pressure? *B6*

32. If 40.0 mL of hydrogen measured at 22 °C and 738 mm pressure is to be prepared from magnesium and hydrochloric acid, how many grams of magnesium must be used? *B6*

33. Oxygen, 15.0 L, is collected by water displacement at 21 °C and 745.0 mm pressure. Calculate the mass of the oxygen in grams. *B4*

34. How many liters of hydrogen, collected by water displacement at 25 °C and 755.0 mm pressure, can be obtained from 5.40 g of zinc and an excess of sulfuric acid? *B6*

35. Aluminum, 5.20 g, reacts with excess sulfuric acid to form hydrogen and aluminum sulfate. What volume of hydrogen, collected over water at $2\bar{0}$ °C and $77\bar{0}$ mm pressure, is produced? *B6*

36. Carbon disulfide, CS_2, burns in the oxygen of the air to form carbon dioxide, CO_2, and sulfur dioxide, SO_2. How many liters of dry air, 20.9% oxygen by volume, measured at 29 °C and 744 mm pressure are needed for the complete combustion of 2.50 moles of CS_2? *B6*

37. What is the volume of the mixture of CO_2 and SO_2 produced in the reaction of Problem 36 if measured under the same conditions as the air used in the reaction? *B6*

38. Chlorine gas may be generated in the laboratory by a reaction between manganese dioxide and hydrogen chloride. The equation is

$$\mathbf{MnO_2(s) + 4HCl(aq) \rightarrow MnCl_2(aq) + 2H_2O(l) + Cl_2(g)}$$

(*a*) How many grams of MnO_2 are required to produce 5.00 L of Cl_2 gas at STP? (*b*) How many grams of HCl are required? *B6*

39. In Problem 38, the HCl is available as a water solution that is 37.4% hydrogen chloride by mass. The solution has a density of 1.189 g/mL. What volume of HCl solution (hydrochloric acid) must be supplied to the reaction?

40. How many grams of charcoal, 90.0% carbon, must be burned to produce $10\bar{0}$ L of CO_2 measured at 25 °C and $74\bar{0}$ mm pressure? *B6*

41. How many grams of chlorine gas are contained in a 5.00-L flask at 27 °C and $72\bar{0}$ mm pressure? *B4*

42. What temperature must be maintained to insure that a 1.00-L flask containing 0.0400 mole of oxygen will show a continuous pressure of 745 mm? *A3e, B7*

43. At 12.0 °C and $74\bar{0}$ mm, 1.07 L of a gas has a mass of 1.78 g. Calculate the molecular weight of the gas from the ideal-gas equation. *B7*

44. What is the molecular weight of a gas if 11.6 g of the gas occupies 10.4 L at $75\bar{0}$ mm pressure and 29.0 °C? *B7*

45. From the ideal-gas equation, $pV = nRT$, and the density of a gas defined as the mass per unit volume, $D = m/V$, prove that the density of a gas at STP is directly proportional to its molecular weight. *B7*

Irving Langmuir (American) was awarded the Nobel Prize in chemistry in 1932 for his studies in the surface chemistry of liquids and solids. He studied how oil is adsorbed on the surface of water and determined what affects film formation. He also investigated the surfaces of solid substances that act as catalysts and also catalytic poisons. In other work, Langmuir extended the life of the light bulb, developed a vacuum pump and high-vacuum tubes for radio, and attempted to induce rainfall by seeding clouds.

Liquids–Solids–Water

12

GOALS

In this chapter you will gain an understanding of: • properties and kinetic-theory description of liquids • physical equilibria and equilibrium vapor pressure • LeChatelier's principle • properties and kinetic-theory description of solids • nature of crystals • molecular structure and properties of water • preparation, properties, and uses of hydrogen peroxide

LIQUIDS

12.1 Properties of liquids All liquids have several easily observed properties in common.

Phases of matter are first described in Section 1.8.

1. *Definite volume.* A liquid occupies a definite volume; it does not expand and completely fill its container as does a gas. A liquid can have one free surface; its other surfaces must be supported by the container walls.

2. *Fluidity.* A liquid can be made to flow or can be poured from one container to another. A liquid takes the shape of its container.

3. *Noncompressibility.* If water at 20 °C is subjected to a pressure of 1000 atmospheres, its volume decreases by only 4%. This behavior of water is typical of all liquids. Even under very high pressure, liquids are compressed only slightly.

4. *Diffusion.* Suppose you slowly pour some ethanol (ethyl alcohol) down the side of a graduated cylinder already half full of water. If you pour carefully, the alcohol can be made to float on the water. At first, a fairly definite boundary exists between the liquids. However, if you let this system stand, the boundary becomes less and less distinct. Some of the water diffuses into the alcohol, and at the same time some of the alcohol diffuses into the water. If the cylinder stands undisturbed for some time, the alcohol and water completely mix.

5. *Evaporation.* If a liquid is left in an open container, it may gradually disappear. Spontaneously, many liquids slowly change into vapors at room temperature.

12.2 Kinetic-theory description of a liquid Liquids are denser than gases. This fact indicates that the particles of liquids are much closer together than those of gases. Further, liquids are

The basic assumptions of the kinetic theory are listed in Section 10.1.

Densities of gases and liquids are compared in Section 10.2(3).

Table 12-1
LIQUID-PHASE TEMPERATURE RANGES OF REPRESENTATIVE SUBSTANCES

Type of substance	*Substance*	*Temperature range of liquid phase* (1 atm, °C)
nonpolar covalent molecular	H_2	−259– −253
	O_2	−218– −183
	CH_4	−182– −164
	CCl_4	−23– 77
	C_6H_6	6– 80
polar covalent molecular	NH_3	−78– −33
	H_2O	0– 100
ionic	NaCl	801– 1413
	MgF_2	1266– 2239
covalent network	$(SiO_2)_x$	1610– 2230
	C_x (diamond)	3700– 4200
metallic	Hg	−39– 357
	Cu	1083– 2567
	Fe	1535– 2750
	W	3410– 5660

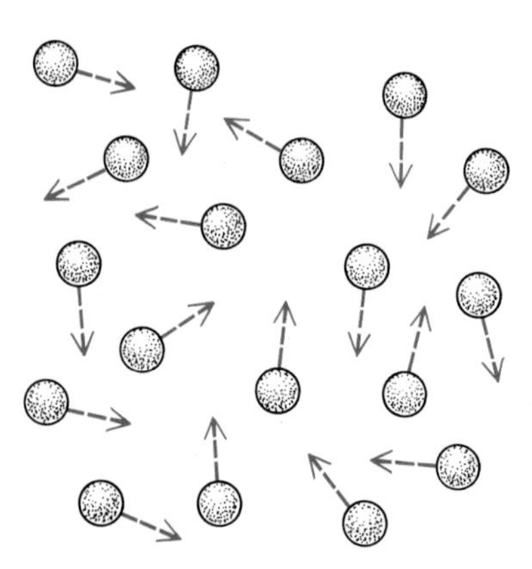

Figure 12-1. The attractive forces between the particles of a liquid are strong enough so that a liquid has a definite volume but weak enough so that the particles can move with respect to one another.

practically noncompressible. Therefore, the particles of a liquid must be almost as close together as it is possible for them to be. A liquid has a definite volume and can have one free surface. This observation shows that a body of liquid holds together. Consequently, the attractive forces among the particles of a liquid must be much stronger than those among the particles of a gas.

Liquids are fluid and take the shape of their containers. These observations indicate that single particles or groups of particles in a liquid move with respect to one another as suggested by Figure 12-1. The kinetic energy of the liquid particles must be large enough to make this motion possible despite the attractive forces among them.

Substances composed of low-molecular-weight nonpolar molecules are liquids only at temperatures below room temperature. Higher-molecular-weight nonpolar molecular substances can be liquids at room temperature. Hence, the forces of attraction among nonpolar molecules are the relatively weak dispersion interaction forces, which were described in Section 10.4.

Substances composed of low-molecular-weight polar molecules may be liquids at room temperature. Thus, the attractive force among polar molecules is the stronger combination of dispersion interaction and dipole-dipole attraction forces.

Most metals, ionic compounds, and covalent network substances are liquids only at temperatures well above room temperature. Therefore, it is believed that the attractive forces among the particles of these substances are much stronger than van der Waals forces. Table 12-1 gives the range of temperatures over which these different substances can exist as liquids.

There is abundant evidence that the particles of liquids are in motion. Very small particles of a solid suspended in water or some other liquid can be viewed through a microscope. They are observed to move about in a random manner. The motion is increased by using smaller particles and higher temperatures. Thus, the observed random motion is believed to be caused by collisions with molecules of the liquid. See Figure 12-2.

Liquid molecules intermingle because of their motion. This movement results in diffusion. The diffusion of liquids is slower than that of gases because the molecules of liquids move more slowly and are closer together. Their movement in a given direction is thereby hindered.

Water and some other liquids, such as perfume, evaporate fairly rapidly. Evaporation occurs when some molecules acquire enough kinetic energy to escape from the surface into the vapor phase as shown in Figure 12-3. The vapor molecules of liquids in closed containers exert pressure, as do the molecules of all confined gases. This vapor pressure reaches some maximum value depending on the temperature and nature of the substance.

The molecules of the vapors of liquids and solids have properties similar to those of gases. On cooling, gases and vapors may condense and become liquids. With further cooling, they may ultimately become solids.

12.3 Dynamic equilibrium Suppose a cover is placed over a container partially filled with a liquid. It appears that evaporation of the liquid continues for a while and then ceases. This apparent situation can be examined in terms of the kinetic theory. The temperature of the liquid is proportional to the average kinetic energy of all the molecules of the liquid. Most of these molecules have energies very close to the average. However, as a result of collisions with other molecules, some have very high energies and a few have very low energies at any given time. The motions of all are random.

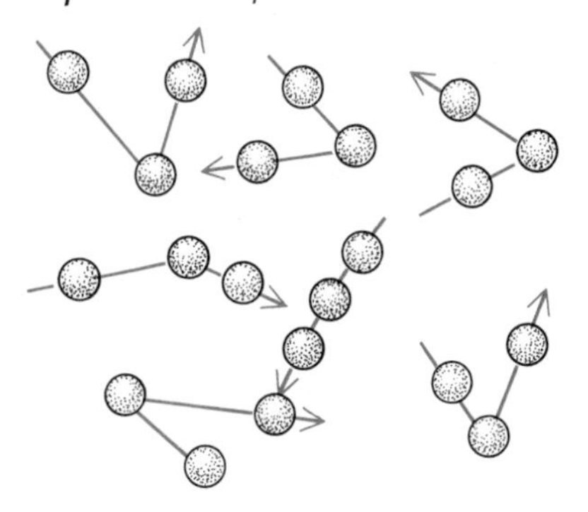

Figure 12-2. This enlarged diagram shows the movement of particles of paint as they are bombarded by invisible molecules of the liquid in which they are suspended.

Some high-energy molecules near the surface are moving toward the surface. These molecules may overcome the attractive forces of the surface molecules completely and escape, or evaporate. But some of them then collide with molecules of gases in the air or other vapor molecules and rebound into the liquid. See Figure 12-4(A). As the evaporation continues, the concentration of vapor molecules continues to increase. This causes the chance of collisions of escaping molecules with vapor molecules to increase. Consequently, the number of vapor molecules rebounding into the liquid increases.

The definitions of and the distinction between gas and vapor are given in Section 1.8.

Eventually, the number of vapor molecules returning to the liquid equals the number of liquid molecules evaporating. Beyond this point, there will be no *net* increase in the *concentration* of vapor molecules. That is, there will be no change in the number of vapor molecules per unit volume of air above the liquid. But the motions of the molecules do not cease. The two actions, *evaporation* and *condensation, do not cease* either. They merely *continue at equal rates.* An *equilibrium* is reached between the rate of liquid molecules evaporating and the rate of vapor molecules condensing. See Figure 12-4(B). This *equilibrium* is a dynamic condition in which *opposing changes* occur at *equal rates.* Since this dynamic equilibrium involves only physical changes, it is referred to as ***physical equilibrium:*** *a dynamic state in which two opposing physical changes occur at equal rates in the same system.*

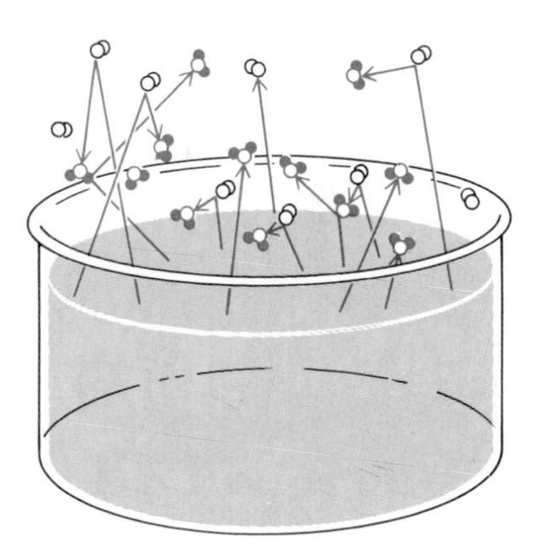

Figure 12-3. Water evaporates because some molecules acquire sufficient kinetic energy to escape from the surface into the vapor phase. Some molecules rebound into the surface after colliding with molecules of gases in the air or with other water vapor molecules.

The evaporation process can be represented as

$$\textbf{liquid + energy} \rightarrow \textbf{vapor}$$

Then the condensation process will be

$$\textbf{vapor} \rightarrow \textbf{liquid + energy}$$

The state of dynamic equilibrium occurring in a confined space is

$$\textbf{liquid + energy} \rightleftarrows \textbf{vapor}$$

Because of the double "yields" sign in an equation representing an equilibrium reaction, the equation is read in either direction.

The *forward* reaction is represented when the equation is read from *left to right:* liquid + energy → vapor. The *reverse* reaction is represented when the equation is read from *right to left:* vapor → liquid + energy.

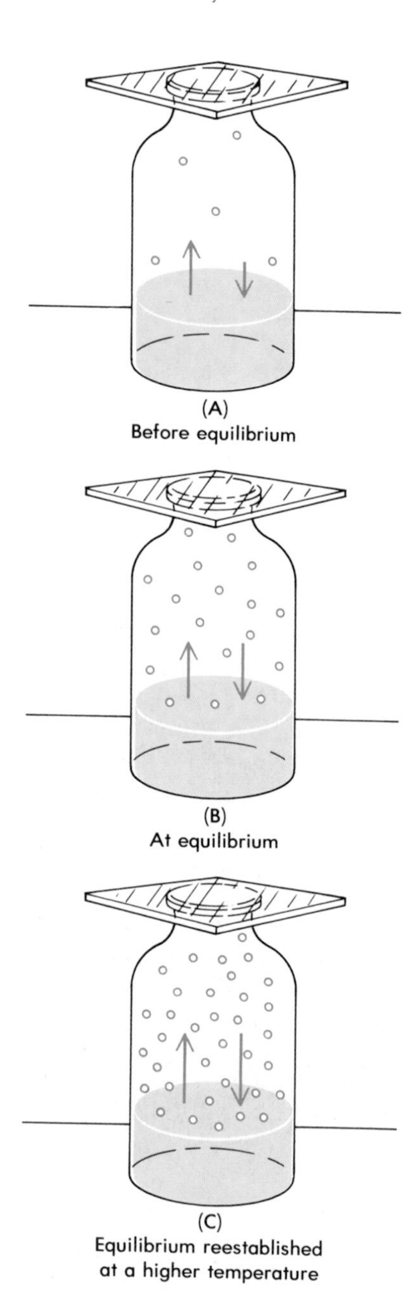

Figure 12-4. An example of physical equilibrium and the influence of temperature. The different lengths of the arrows indicate the relative rates of evaporation (up) and condensation (down). The pressure exerted by the vapor molecules at equilibrium is known as the equilibrium vapor pressure of that particular liquid.

12.4 Equilibrium vapor pressure The vapor molecules of liquids in closed containers exert pressure, as do the molecules of all confined gases. But when equilibrium is reached, there is no further net change in the system. The concentration of vapor molecules in the space above the liquid surface remains constant. Thus, at equilibrium, there is a vapor pressure characteristic of the liquid present in the system. It is known as the *equilibrium vapor pressure* of the liquid. ***Equilibrium vapor pressure** is the pressure exerted by a vapor in equilibrium with its liquid.*

What is the effect on a liquid-vapor equilibrium system if the temperature of the liquid is raised? Again, this situation can be examined in terms of the kinetic theory. The rise in temperature means that the average kinetic energy of the liquid molecules has been increased. A relatively larger number of liquid molecules now possess enough energy to escape through the liquid surface. Thus, the rate of evaporation is increased. In this way, the liquid-vapor equilibrium is *disturbed.* The concentration of vapor molecules above the liquid surface is increased. More vapor molecules, in turn, increase the chances of collision with escaping molecules and cause an increase in the rate of condensation. Soon the equilibrium is reestablished but at a *higher equilibrium vapor pressure.* See Figure 12-4(C).

All liquids have characteristic forces of attraction between their molecules. If the attractive forces are strong, there is less tendency for the liquid to evaporate. The equilibrium vapor pressure of such a liquid is correspondingly low. Glycerol is an example of a liquid with a low equilibrium vapor pressure at room temperature. Conversely, attractive forces between liquid molecules may be relatively weak. Then the liquid tends to evaporate readily, with a resulting high equilibrium vapor pressure. Ether is such a liquid. The equilibrium vapor pressure of a liquid depends *only* on the *nature of the liquid and its temperature.*

12.5 Le Chatelier's principle In systems that have attained equilibrium, opposing actions occur at equal rates. Any change that alters the rate of either the forward or the reverse action disturbs the equilibrium. It is often possible to displace an equilibrium in a desired direction by changing the equilibrium conditions of the system.

In 1888, the French chemist Henri Louis Le Chatelier (luh-*shah*-teh-lee-ay) (1850–1936) set forth an important principle. This principle is the basis for much of our knowledge of equilibrium. ***Le Chatelier's principle*** may be stated as follows: *If a system* *at equilibrium is subjected to a stress, the equilibrium will be displaced in such direction as to relieve the stress.* This principle is a general law that applies to all kinds of dynamic equilibria. It can be applied to the

$$\textbf{liquid + energy} \rightleftarrows \textbf{vapor}$$

system under discussion.

You have already learned something about how the kinetic theory applies to this system. You know that the system may reach equilibrium at a given temperature. If the temperature is then raised, equilibrium can be reestablished, but at a greater vapor concentration. *The rise in temperature is a stress on the system.* According to Le Chatelier's principle, the equilibrium is displaced in the direction that relieves this stress. The equilibrium is displaced in the direction that counteracts the temperature rise; it is displaced in the direction that absorbs energy. In this case, the forward (left-to-right) reaction is endothermic. The forward reaction absorbs heat energy and displaces the equilibrium to the right. This displacement means that the vapor concentration is higher when equilibrium is reestablished. Similarly, Le Chatelier's principle can be applied to a lowering of the temperature of the system at equilibrium. Here, the reverse (right-to-left) reaction is favored. Equilibrium is reestablished at the lower temperature with a reduced vapor concentration. The equilibrium is displaced in the direction that counteracts the temperature decrease; it is displaced in the direction that releases energy. Thus, Le Chatelier's principle is used to predict the dependence of equilibrium vapor pressure on temperature.

Suppose the temperature of the system is kept constant but the volume that the system occupies is altered. This constant-temperature condition means that the vapor pressure at equilibrium remains constant. It also means that the concentration of vapor molecules (density of the vapor) at equilibrium remains constant.

First, the volume that the system occupies is increased. The volume of the liquid cannot change measurably, but the volume of the vapor can. When the volume of the vapor increases at constant temperature, its pressure must drop (Boyle's law). In order to restore the vapor to its equilibrium concentration, more vapor molecules must be produced. Le Chatelier's principle indicates that the equilibrium must shift to the right and more liquid must evaporate. Equilibrium is then restored with the same vapor pressure and the same concentration of vapor molecules as before. But with a larger volume of vapor, the actual number of vapor molecules must be greater. The number of liquid molecules is, therefore, necessarily reduced. Energy is required for this evaporation. Since the temperature of the system is kept constant, the system must absorb this needed energy from its surroundings.

Figure 12-5. Henri Louis Le Chatelier, a French mining engineer and chemist, was a brilliant scientist, teacher, writer, and editor. He did work in the science and technology of metals, high-temperature measurement, microscopy, ceramics, cements, chemical mechanics, and the theory of combustion gases. His important contribution to the understanding of the direction in which a physical or chemical change will occur is now known as Le Chatelier's principle.

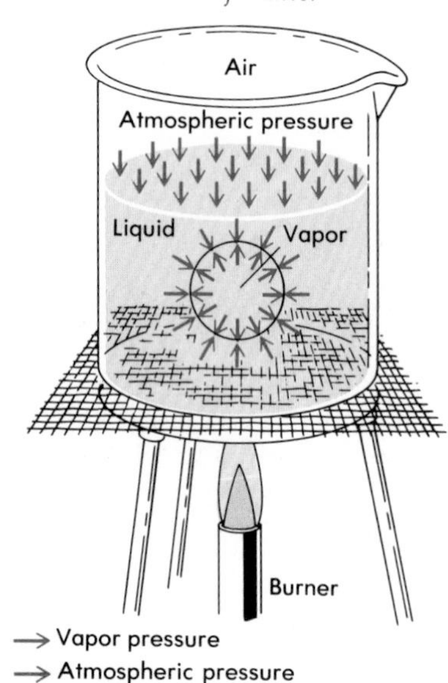

Figure 12-6. A liquid boils when its equilibrium vapor pressure becomes equal to the prevailing atmospheric pressure.

If the volume that the system occupies is reduced, the pressure of the vapor increases. To restore equilibrium, this increased pressure must be reduced to the equilibrium vapor pressure at the same temperature. Le Chatelier's principle indicates that some vapor molecules must condense to liquid molecules. The reverse reaction, condensation, is favored. Equilibrium is once again established. This time there are fewer vapor molecules and a greater number of liquid molecules. But the same concentration of vapor molecules exists as before. The energy given off during condensation is transferred by the sytem to the surroundings in order to maintain the system's constant temperature.

12.6 Boiling of liquids You now have some understanding of equilibrium and how equilibrium vapor pressures arise. You can now apply this knowledge to the phenomenon of *boiling*.

Pressure exerted uniformly on the surface of a confined liquid is transmitted undiminished in every direction throughout the liquid (Pascal's law). Consider a beaker of water heated over a Bunsen flame, as in Figure 12-6. Vapor bubbles first appear at the bottom of the beaker, where the water is hottest. They diminish in size and disappear completely as they rise into cooler water. According to Pascal's law, atmospheric pressure presses from all directions perpendicular to the surface of the vapor bubble, collapsing it. Only when the equilibrium vapor pressure exerted by the vapor molecules on the liquid at the surface of the bubbles is equal to the atmospheric pressure can the vapor bubble be maintained as it rises through the liquid.

As the temperature of the water increases, the vapor pressure also increases. *Ultimately a temperature is reached at which the equilibrium vapor pressure is equal to the pressure of the atmosphere acting on the surface of the liquid.* At this temperature, the vapor bubbles maintain themselves in the liquid. They present to the liquid a greatly increased liquid-vapor surface. This allows evaporation (a surface phenomenon) to occur at a greatly increased rate. The liquid is said to *boil*. *The* ***boiling point*** *of a liquid is the temperature at which the equilibrium vapor pressure of the liquid is equal to the prevailing atmospheric pressure.* Vapor pressure curves for ether, alcohol, water, and glycerol are shown in Figure 12-7. If the pressure on the surface of a liquid is increased, the boiling point of the liquid is raised. If the pressure is decreased, the boiling point of the liquid is lowered.

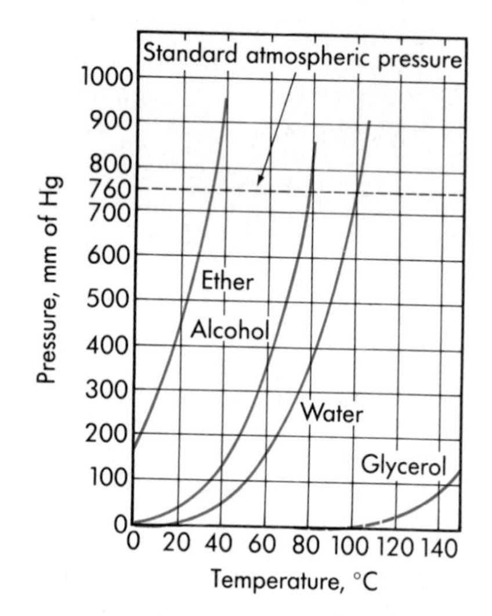

Figure 12-7. The equilibrium vapor pressures of some common liquids as a function of temperature.

The boiling point of water is exactly 100 °C at *standard atmospheric pressure*. This temperature is known as the *standard* (or *normal*) *boiling point of water*. When the boiling points of liquids are given, standard pressure conditions are understood. Ether, which has a high equilibrium vapor pressure, boils at 34.6 °C. The boiling point of glycerol, mentioned earlier (Section 12.4) for its low equilibrium vapor pressure, is 290 °C.

During boiling, the temperature of a liquid remains constant. The temperature of the vapor is the same as that of the liquid. Hence, the kinetic energies of linear motion of liquid and vapor molecules must be the same. But energy must be supplied for boiling to continue. This energy is absorbed by the liquid as it becomes a vapor. The energy separates the molecules in the liquid to the much wider spacing of molecules in the gas. It effects this separation by overcoming the attractive forces between the molecules. Thus, this supplied energy increases the potential energy of the molecules. Refer to Figure 2-13.

The heat energy required to vaporize one mole of liquid at its standard boiling point is its *standard molar heat of vaporization*. Its magnitude is a measure of the degree of attraction between the liquid molecules.

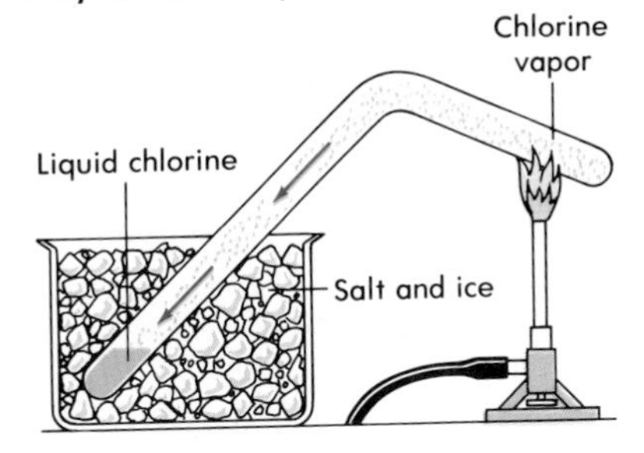

Figure 12-8. By using a tube like the one shown above, Faraday succeeded in liquefying chlorine, sulfur dioxide, and several other gases having high critical temperatures.

12.7 Liquefying gases Michael Faraday (1791–1867) discovered that it is possible to liquefy certain gases by cooling and compressing them at the same time. He used a thick-walled sealed tube of the type shown in Figure 12-8. With this apparatus, Faraday liquefied chlorine, sulfur dioxide, and some other gases. One end of the glass tube containing the chlorine gas was strongly heated. That caused the gas in the heated end of the tube to expand. The expanding gas exerted pressure on the gas in the other end of the tube, which was cooled in a freezing mixture. Cooling and compression in this manner converted the gaseous chlorine into liquid chlorine.

The modern method of liquefying a gas is more complicated. The first step involves compressing the gas and then removing the heat of compression. Compressing a gas always raises its temperature since energy is acquired by the molecules of a gas when work is done to push them closer together. In liquefying gases, the heat of compression is absorbed by a suitable coolant. The gas molecules thereby lose the energy acquired during compression. The compressed gas is cooled to the same temperature it had before compression. Now the molecules possess the same kinetic energy as they had before compression but are now closer.

The second step in liquefying a gas is to permit the cool compressed gas to expand without absorbing external energy. When a compressed gas expands, the molecules lose energy as they do work in spreading apart against the force of molecular attraction. This energy loss by the molecules is observed as a decrease in the temperature of the gas. Remember that the temperature of the compressed gas was that which it had *before* compression. So the *expanded* gas is now at a much lower temperature than originally. By repeating this compression, cooling, and expansion cycle, the temperature of the gas is lowered still further.

Liquefying a gas is the result of a combination of lowered temperature and increased pressure. The increased pressure

Figure 12-9. Teenage campers at winter site in Fairlee, Vermont using portable propane grill. Propane, a gaseous fuel, is often used at trailer parks and at campsites for cooking, water heating, space heating, and refrigeration. The supply of propane is contained as a liquid under pressure in metal bottles.

Table 12-2 CRITICAL TEMPERATURES AND PRESSURES

Gas	*Critical temperature* (°C)	*Critical pressure* (atm)
water	374.2	218.3
sulfur dioxide	157.5	77.9
chlorine	144.0	76.1
carbon dioxide	31.0	72.8
oxygen	−118.4	50.1
nitrogen	−146.9	33.5
hydrogen	−239.9	12.8

crowds the gas molecules together. The lowered temperature slows their movement. Ultimately, they are slowed down greatly and crowded together very closely. At this point, the attractive forces between the molecules cause them to condense to a liquid. See Figure 12-8.

Scientists have found that above a certain characteristic temperature it is impossible to liquefy a gas by pressure alone. Above this temperature, the kinetic energy of the molecules is great enough to overcome the attractive forces between them. Thus the gas does not liquefy no matter how great the pressure applied. *The highest temperature at which it is possible to liquefy a gas with any amount of pressure is called its* ***critical temperature.*** *The pressure required to liquefy a gas at its critical temperature is called its* ***critical pressure.*** *The volume occupied by one mole of a gas under these conditions is called its* ***critical volume.*** The critical temperature and critical pressure of several common gases are in Table 12-2.

From these data, it is apparent that two conditions are necessary to liquefy a gas:

1. Its temperature must be lowered to or below its critical temperature.

2. Its pressure must be raised to or above the vapor pressure of the liquified gas at this temperature.

These two conditions enable the attractive forces between the molecules to effect condensation. All gases can be liquefied.

In Section 1.8, a vapor was initially defined as the gaseous phase of a substance that is a solid or liquid at normal temperatures. Now that the meaning of critical temperature has been explained, a vapor can be defined more precisely. *A* ***vapor*** *is a gas at a temperature below its critical temperature.*

12.8 Critical temperature and molecular attraction The critical temperature of a gas has been defined as the temperature above which it cannot be liquefied no matter how great the pressure. Thus, the magnitude of the critical temperature of a gas serves as a measure of the attractive forces between its molecules. The higher the critical temperature of a gas, the larger is the attractive force between its molecules. The lower the critical temperature of a gas, the smaller is the attractive force between its molecules.

The high critical temperature of water is shown in Table 12-2. This high critical temperature indicates that the forces of attraction between polar water molecules are very great. In fact, they can cause water vapor to liquefy even at 374 °C. The critical temperature of sulfur dioxide is less than that of water. Thus, the attractive forces between sulfur dioxide molecules must be less than those between water molecules. This difference is expected because sulfur dioxide molecules are less polar than water molecules. Consequently, sulfur dioxide can be condensed to a liquid only below 157 °C.

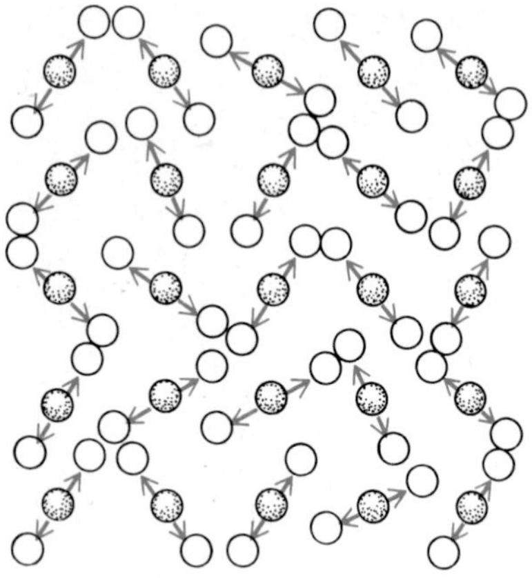

Figure 12-10. The particles of a solid vibrate about fixed equilibrium positions.

Attractive forces also exist between nonpolar covalent molecules such as chlorine, carbon dioxide, oxygen, nitrogen, and hydrogen. These attractive forces are dispersion interaction forces. Such forces generally increase with an increase in the complexity of a nonpolar molecule. Thus, the higher the molecular weight of such a molecule, the higher its critical temperature. This phenomenon is clearly illustrated in Table 12-2. Notice the descending order of the critical temperatures of chlorine, carbon dioxide, oxygen, nitrogen, and hydrogen.

Solids

12.9 Properties of solids Some of the general properties of solids that can easily be seen are

1. *Definite shape.* A solid maintains its shape. Unlike liquids and gases, it does not flow under ordinary circumstances. The shape of a solid is independent of its container.
2. *Definite volume.* All the surfaces of a solid are free surfaces. Thus, the volume of a solid is also independent of its container.
3. *Noncompressibility.* The pressures required to decrease the volumes of solids are even greater than pressures required for liquids. For all practical purposes, solids are noncompressible. Solids such as wood, cork, sponge, etc., may *seem* to be compressible. But remember that these materials are very porous. Compression does not reduce the volume of the solid portion of such substances significantly. It merely reduces the volume of the pores of the solid.
4. *Very slow diffusion.* Suppose a lead plate and a gold plate are placed in close contact. After several months, particles of gold can be detected in the lead and vice versa. This observation is evidence that diffusion occurs even in solids, although at a *very slow rate.*
5. *Crystal formation.* Solids may be described as either *crystalline* or *amorphous.* Crystalline solids have a regular arrangement of particles, whereas amorphous solids have a random particle arrangement.

12.10 Kinetic-theory description of a solid Particles of solids are held close together in fixed positions by forces that are stronger than those between particles of liquids. This is how scientists explain the definite shape and volume of a solid as well as its noncompressibility. Whether the particle arrangement is orderly or not determines whether the solid is crystalline or amorphous. The particles of a solid vibrate back and forth weakly about fixed equilibrium positions as suggested by Figure 12-10. Their kinetic energy is related to the extent of this vibratory motion. Their kinetic energy is proportional to the temperature of the solid. At low temperatures, the kinetic energy is small. At higher temper-

Figure 12-11. Frost, the solid phase of water, makes artistic patterns when lacing a leaf, tree branches, and a spider's web.

atures, it is larger. Crystal particles vibrate extensively at high temperatures. But there is relatively little diffusion in any solid because the vibration is about fixed positions.

As you study this section, it will be helpful to refer again to Figure 2-13.

12.11 Changes of phase involving solids The physical change of a liquid to a solid is called *freezing*, and involves a loss of energy by the liquid.

$$\textbf{liquid} \rightarrow \textbf{solid + energy}$$

Since this change occurs at constant temperature, the liquid and solid particles must have the same kinetic energy. The energy loss is a loss of *potential* energy. The particles lose potential energy as the forces of attraction do work on them.

The reverse physical change, *melting*, also occurs at constant temperature.

$$\textbf{solid + energy} \rightarrow \textbf{liquid}$$

It involves a gain of potential energy by the particles of the solid as they do work against the attractive forces in becoming liquid particles.

For pure crystalline solids, the temperatures at which these two processes occur coincide. That is, the freezing point and the melting point are the same. For pure water, both processes occur at 0 °C. Ice melts at 0 °C and forms liquid water; water freezes at 0 °C and forms ice. The heat energy required to melt one mole of solid at its melting point is its *molar heat of fusion*.

If ice gradually disappears in a mixture of ice and water, the melting process clearly is proceeding faster than the freezing process. Suppose, however, that the relative amounts of ice and water remain unchanged in the mixture. Then both processes must be proceeding at equal rates and a state of physical equilibrium is indicated.

$$\textbf{solid + energy} \rightleftarrows \textbf{liquid}$$

Not all particles of a solid have the same energy. A surface particle of a solid may acquire sufficient energy to overcome the attractive forces holding it to the body of the solid. Such a particle may escape from the solid and become a vapor particle.

$$\textbf{solid + energy} \rightarrow \textbf{vapor}$$

If a solid is placed in a closed container, vapor particles cannot escape from the system. Eventually they come in contact with the solid and are held by its attractive forces.

$$\textbf{vapor} \rightarrow \textbf{solid + energy}$$

Thus a solid in contact with its vapor can reach a state of equilibrium.

$$\textbf{solid + energy} \rightleftarrows \textbf{vapor}$$

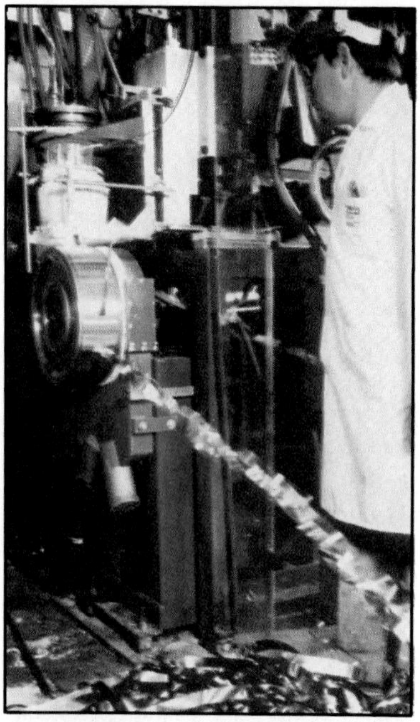

Figure 12-12. A molten iron alloy made into an amorphous metallic glass. The iron alloy is melted and is squirted under pressure through a small opening onto the surface of the rapidly rotating, very cold copper cylinder. A very thin ribbon of metallic glass is formed and peels off the cylinder.

In such a case, the solid exhibits a characteristic equilibrium vapor pressure. Like that of a liquid, the equilibrium vapor pressure of a solid depends only on the temperature and the substance involved. Some solids like camphor and naphthalene (moth crystals) have fairly high equilibrium vapor pressures. They evaporate noticeably when exposed to air. Solids like carbon dioxide (Dry Ice) and iodine have equilibrium vapor pressures that rise very rapidly as the temperature is raised. The equilibrium vapor pressures of these solids equal atmospheric pressure before the solids melt. In such cases the solids vaporize directly, without passing through the liquid phase. The change of phase from a solid to a vapor is known as *sublimation.*

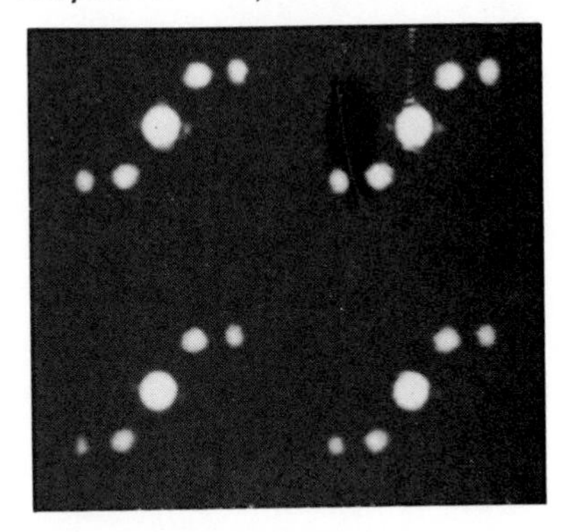

Figure 12-13. Images of atoms in a crystal. By using X rays, a computer, and laser light, an image showing the location and relative sizes of atoms in a crystal can be produced. The one shown in the photograph is of a complex crystal of magnesium bromide tetrahydrofuran, a compound containing magnesium, bromine, carbon, hydrogen, and oxygen. The magnesium atoms are the largest spots. The two next largest dots on opposite sides of the magnesium atoms are oxygen atoms, and the smallest outermost dots are carbon atoms. This image is evidence of the regularity that is characteristic of the structure of crystals.

12.12 Amorphous solids The term *amorphous solids* refers to those that appear to have random particle arrangement. Many solids that scientists once thought were amorphous have been found to have a partially crystalline structure. Charcoal is such a solid.

Materials like glass and paraffin are considered amorphous. These materials have the properties of solids. That is, they have definite shape and volume and diffuse slowly. But they do not have the orderly arrangement of particles characteristic of crystals. They also lack sharply defined melting points. In many respects, they resemble liquids that flow very slowly at room temperature.

Amorphous forms of metal alloys have recently been prepared. They are called *metallic glasses.* They are made by *very rapid* cooling of a thin film of melted metal as shown in Figure 12-12. The cooling occurs so rapidly that the metal atoms do not have time to arrange themselves in a crystal pattern. Metallic glasses may find use in applications based on their unusual nondirectional magnetic properties. Some of these applications are in electric transformer cores, in the heads of magnetic tape recorders and players, and in magnetic bubble lattice memories in computer systems.

12.13 Nature of crystals Most substances exist as solids in some characteristic crystalline form. *A* ***crystal*** *is a homogeneous portion of a substance bounded by plane surfaces making definite angles with each other, giving a regular geometric form.* See Figure 12-13.

Scientists determine the arrangement of particles composing a crystal by mathematical analysis of the crystal's diffraction patterns. These patterns appear clearly on photographs produced when the crystal is illuminated by X rays. An example of an X-ray diffraction photograph is shown in Figure 12-14. Every crystal structure shows a pattern of points that describes the arrangement of its particles. This pattern of points is known as the *crystal lattice.* The smallest portion of the crystal lattice that exhibits the pattern of the lattice structure is called the *unit cell.*

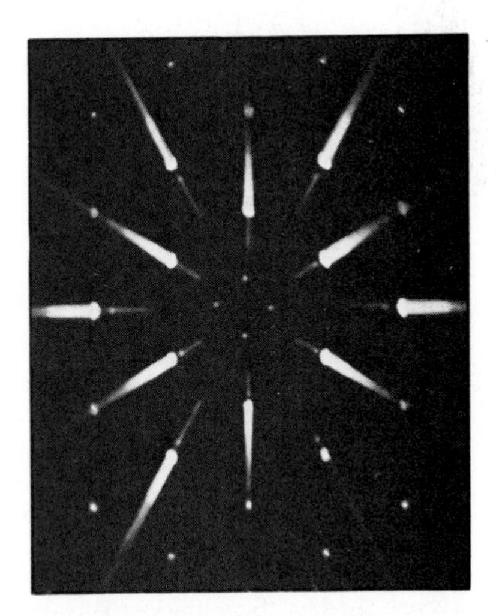

Figure 12-14. An X-ray diffraction photograph of ice. Chemists use X-ray diffraction in their study of crystal structure.

The unit cell defines the kind of symmetry to be found throughout a crystalline substance.

The classification of crystals by shape is a part of the science of *crystallography*. Shape classification helps chemists to identify crystals. Any crystal can be placed in one of the seven crystalline systems shown in Figure 12-15.

1. *Cubic.* The three axes are at right angles, as in a cube, and are of equal length.

2. *Tetragonal.* The three axes are at right angles to each other, but only the two lateral axes are of equal length.

3. *Hexagonal.* Three equilateral axes intersect at angles of 60°. A vertical axis of variable length is at right angles to the equilateral axes.

4. *Trigonal.* There are three equal axes and three equal oblique (not right-angle) intersections.

5. *Orthorhombic.* Three unequal axes are at right angles to each other.

6. *Monoclinic.* There are three unequal axes, with one oblique intersection.

7. *Triclinic.* There are three unequal axes and three unequal oblique intersections.

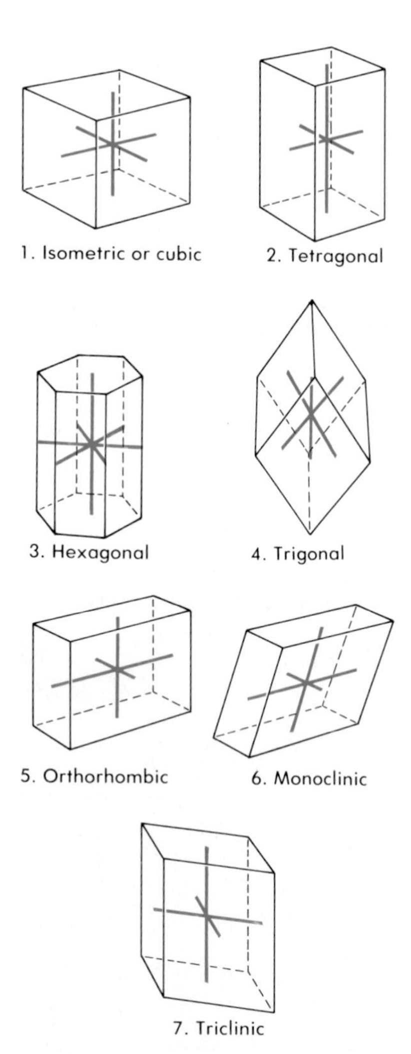

Figure 12-15. Schematic diagram of the seven basic crystal systems. Photos of crystals representing each of the seven basic crystal systems are shown on page 273.

Crystals of common salt, NaCl, are cubic. This cubic nature can be seen by sprinkling a little table salt on a black surface and examining the crystals with a magnifying lens. Andradite, $Ca_3Fe(SiO_4)_3$, and alum crystals, $K_2SO_4 \cdot Al_2(SO_4)_3 \cdot 24H_2O$, are also cubic, but the alum crystals are formed as *octahedrons* (regular eight-sided solids). An example of a tetragonal crystal is wulfenite, $PbMoO_4$. Hexagonal crystals are formed by beryl, $Be_3Al_2(SiO_3)_6$, while elbaite, $H_8Na_2Li_3Al_3B_6Al_{12}Si_{12}O_{62}$, forms trigonal crystals. Stibnite, Sb_2S_3, forms orthorhombic crystals. The compound azurite, $Cu_3(CO_3)_2(OH)_2$, forms monoclinic crystals. Chalcanthite or copper(II) sulfate pentahydrate, $CuSO_4 \cdot 5H_2O$, forms blue triclinic crystals. See photographs on page 273.

Crystals of many chemical compounds are formed when their solutions evaporate or when their hot, saturated solutions cool. Crystals also form when certain substances change from the liquid to the solid phase and when others change from the gaseous to the solid phase. Most of you are familiar with snowflake crystals, formed when water vapor changes directly to the solid phase. Melted sugar, sulfur, and iron form crystals in a similar manner when they change from the liquid to the solid phase. In some cases, crystals grow from the solid phase.

12.14 Binding forces in crystals The regularity of crystal structures is their most fascinating feature. When possible, ions or atoms or molecules arrange themselves in positions of least energy. The more opportunity there is for particles to do this during the crystal formation, the more symmetrical and regular

ANDRADITE (CUBIC)
$Ca_3Fe(SiO_4)_3$

AZURITE (MONOCLINIC)
$Cu_3(CO_3)_2(OH)_2$

BERYL (HEXAGONAL)
$Be_3Al_2(SiO_3)_6$

STIBNITE (ORTHORHOMBIC)
Sb_2S_3

WULFENITE (TETRAGONAL)
$PbMoO_4$

CHALCANTHITE (TRICLINIC) $CuSO_4 \cdot 5H_2O$

ELBAITE (TRIGONAL) $H_8Na_2Li_3Al_3B_6Al_{12}Si_{12}O_{62}$

Table 12-3
MELTING POINTS AND BOILING POINTS OF REPRESENTATIVE SUBSTANCES

Type of substance	*Substance*	*Melting point* (°C)	*Boiling point* (1 atm, °C)
nonpolar covalent molecular	H_2	−259	−253
	O_2	−218	−183
	CH_4	−182	−164
	CCl_4	−23	77
	C_6H_6	6	80
polar covalent molecular	NH_3	−78	−33
	H_2O	0	100
ionic	NaCl	801	1413
	MgF_2	1266	2239
covalent network	$(SiO_2)_x$	1610	2230
	C_x (diamond)	3700	4200
metallic	Hg	−39	357
	Cu	1083	2567
	Fe	1535	2750
	W	3410	5660

the crystals will be. Thus, the more slowly crystals form, the more closely their shapes approach perfect regularity.

The seven crystalline systems are described in terms related to symmetry. But it is frequently more useful to classify crystals according to the types of lattice structure. Is the crystal lattice *ionic, covalent network, metallic,* or *covalent molecular?* Ionic and covalent molecular crystal lattices represent the two extremes of bonding. Covalent network and metallic crystal lattices are intermediate types. See Table 12-3.

1. *Ionic crystals.* The ionic crystal lattice consists of positive and negative ions arranged in a characteristic regular pattern. No molecular units are evident within the crystal. The strong binding forces result from the attraction of positive and negative charges. Consequently, ionic crystals are hard and brittle, have rather high melting points, and are good insulators. Generally, ionic crystals result from Group I or Group II metals combining with Group VI or Group VII nonmetals, or the nonmetallic polyatomic ions.

2. *Covalent network crystals.* The covalent network crystal lattice consists of an array of atoms. Each of the atoms shares electrons with neighboring atoms. The binding forces are strong covalent bonds that extend in fixed directions. The resulting crystals are compact, interlocking, covalent network structures. They may be considered to be giant molecules. They are very

hard and brittle, have rather high melting points, and are nonconductors. Diamond, silicon carbide, silicon dioxide, and oxides of transition metals are of this type. See Figure 12-16.

3. *Metallic crystals.* The metallic crystal lattice consists of positive ions surrounded by a cloud of valence electrons. This cloud is commonly referred to as the *electron "gas."* The binding force is the attraction between the positive ions of the metal and the electron "gas." The valence electrons are donated by the atoms of the metal and belong to the crystal as a whole. These electrons are free to migrate throughout the crystal lattice. This electron mobility explains the high electric conductivity associated with metals. The hardness (resistance to wear) and melting points of metallic crystals vary greatly for different metals. Sodium, iron, tungsten, copper, and silver are typical examples of metallic crystals that have good electric conductivity. But they are quite different in terms of other characteristics, such as hardness and melting point.

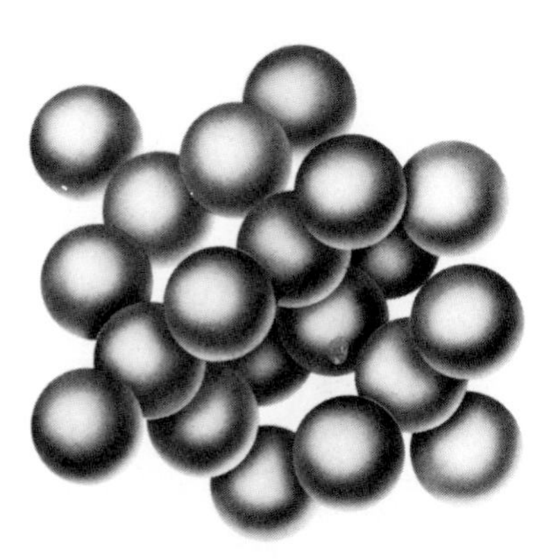

Figure 12-16. Diamond forms covalent network crystals that may be considered to be giant molecules.

4. *Covalent molecular crystals.* The covalent molecular crystal lattice consists of an orderly arrangement of individually distinct molecules. If the molecules are nonpolar, the binding force is the relatively weak dispersion interaction force. If the molecules are polar, both types of van der Waals attractions make up the binding force. The covalent chemical bonds that bind the atoms within the molecules are much stronger than the forces that form the crystal lattice. Thus,molecular crystals have low melting points, are relatively soft and volatile (easily vaporized), and are good insulators. Iodine, carbon dioxide, water, and hydrogen form crystals of this type. See Figure 12-17. Ice crystals are described in Section 12.17.

WATER

12.15 Occurrence of water On the earth, water is the most abundant and essential liquid. The oceans, rivers, and lakes cover about 75% of the surface of the earth. Significant quantities of water are frozen in glaciers, which cover almost 10% of the land surface. Water vapor is always present in the air, and there are large quantities of underground water.

Water is essential to human, animal, and vegetable life. From 70% to 90% of the weight of living things is water. The chemical reactions of life processes take place in water; water is frequently also a reactant or product of such reactions.

The Viking mission produced evidence that leads scientists to believe that the planet Mars has an underground layer of ice. The polar regions of Mars are covered with a layer consisting mostly of ice. Studies have revealed that water vapor is present in Jupiter's atmosphere. Photographs and other data obtained by the two Voyager spacecraft showed the presence of large amounts of water ice on the surfaces of Europa and Ganymede,

Figure 12-17. Solid iodine forms covalent molecular crystals.

two of the moons of Jupiter, and on the surface of the particles making up the main rings of Saturn. About 60% to 70% of the mass of most of Saturn's moons is believed to be ice.

12.16 Physical properties of water Pure water is a transparent, odorless, tasteless, and almost colorless liquid. The faint blue-green color of water is apparent only in deep layers.

Any odor or taste in water is caused by impurities such as dissolved mineral matter, dissolved liquids, or dissolved gases. The strong odor and taste of water from some mineral springs are caused by the presence of such substances in detectable quantity.

The expansion that occurs when water freezes is unusual. When most liquids freeze, they contract.

Depending on temperature and pressure, water may exist as a vapor, liquid, or solid. Liquid water changes to ice at 0 °C under standard pressure, 760 mm of mercury. As water solidifies, it gives off heat and *expands* one-ninth in volume. Consequently, ice has a density of about 0.9 g/cm^3. The density of ice increases slightly as ice is cooled below 0 °C. The molar heat of fusion of ice at 0 °C is 1.436 kcal.

When water at 0 °C is warmed, it contracts until its temperature reaches 4 °C. Then water gradually expands as its temperature is raised further. Thus water occupies its minimum volume and has its maximum density at 4 °C. *At its temperature of maximum density, 4 °C, one milliliter of water has a mass of one gram.*

When the pressure on the surface of water is *one atmosphere* (760 mm of mercury), water boils at a temperature of 100 °C. The molar heat of vaporization of water at 100 °C is 9.70 kcal. The steam that is formed by boiling water occupies a much greater volume than the water from which it was formed. When one liter of water at 100 °C is completely boiled away, the steam occupies about 1700 liters at 100 °C and 1 atmosphere pressure.

When water is heated in a closed vessel so that the steam cannot escape, the pressure on the water's surface increases. As a result, the boiling temperature of the water is raised above 100 °C. But suppose that the air and water vapor above the liquid in a closed vessel are partially removed by means of a vacuum pump. Then the pressure is decreased and the water boils at a lower temperature than 100 °C. Pressure cookers are used for cooking food because the higher temperature of the water cooks the food in a shorter time. Vacuum evaporators are used to concentrate milk and sugar solutions. Under reduced pressure, the water boils away at a temperature low enough so that the sugar or milk is not scorched.

Figure 12-18. On Sept. 25, 1977, the Viking 2 lander's camera took this picture of the Martian landscape showing the surface whitened by frozen water.

12.17 Structure and properties of water molecules Water molecules are composed of two atoms of hydrogen and one atom of oxygen joined by polar covalent bonds. Studies of the crystal structure of ice indicate that these atoms are not joined in a straight line. Instead, the molecule is bent, with a structure that can be represented as

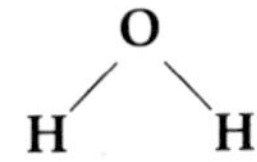

The angle between the two hydrogen-oxygen bonds is about 105°. This was explained in Section 6.17 as evidence of considerable sp^3 hybridization of the oxygen-atom orbitals.

Since oxygen is more strongly electronegative than hydrogen, the bonds in a water molecule are polar. The electronegativity values of hydrogen and oxygen indicate that H—O bonds have about 40% ionic character. Thus, the electrons are not distributed perfectly uniformly about the molecule. On the average, they are clustered slightly about the oxygen nucleus. This gives the oxygen part of the molecule a partial negative charge and leaves the hydrogen parts with a partial positive charge. Since the polar covalent bonds in this molecule do not lie on the same straight line, the molecule as a whole is polar. Water molecules, being polar, are sometimes called water dipoles.

The polarity of water molecules enables them to be attracted to one another. This mutual attraction causes water molecules to *associate,* or join together into groups of molecules. One slightly positive hydrogen atom of a water molecule may weakly, but effectively, attract the slightly negative oxygen atom of a second water molecule. In this way, one hydrogen serves as a link between the oxygen atoms of two water molecules by sharing electrons with them. This situation is an example of a *hydrogen bond. A* ***hydrogen bond*** *is a weak chemical bond between a hydrogen atom in one polar molecule and a very electronegative atom in a second polar molecule.* A hydrogen of the second water molecule may be attracted to the oxygen of a third water molecule, and so on. In this way, a group of molecules is formed. The number of molecules in such a group decreases with an increase in temperature. But there are usually from four to eight molecules per group in liquid water. The formation of molecular groups by hydrogen bonding causes water to be liquid at room temperature. Other substances, such as methane, CH_4, have nonpolar molecules similar in size and mass to water molecules. But these substances do not undergo hydrogen bonding and are gases at room temperature. Hydrogen bonding seems to occur only between hydrogen and highly electronegative, small, nonmetallic atoms such as oxygen, fluorine, and nitrogen.

Ice consists of H_2O molecules arranged in a definite hexagonal structure. They are held together by hydrogen bonds in a rather open hexagonal pattern, as Figure 12-19 illustrates. As heat is applied to ice, the increased energy of the atoms and molecules causes them to vibrate more vigorously. This stretches the hydrogen bonds, and the ice expands as it is heated.

The structure of ice is so porous that small molecules such as methane, CH_4, can be trapped in the cavities and be held there by dispersion interaction. Such nonstoichiometric crystalline solids are called inclusion compounds. Their formation in a cold pipeline carrying moist natural gas can plug the pipeline.

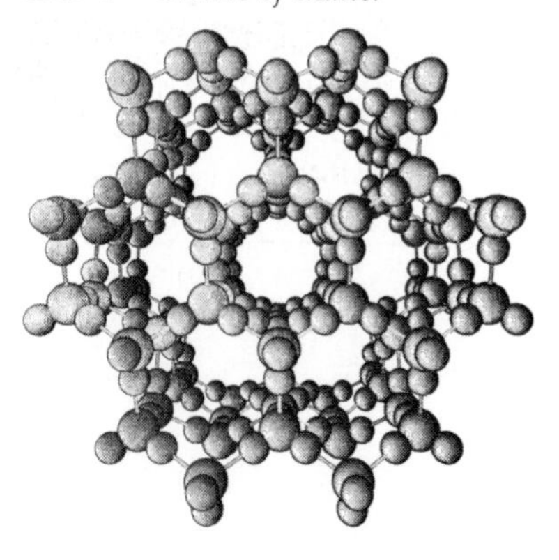

Figure 12-19. A computer-generated drawing of a portion of an ice crystal. The larger spheres are oxygen atoms. The smaller spheres are hydrogen atoms. The "sticks" represent hydrogen bonds.

When the melting point of ice is reached, the energy of the atoms and molecules is so great that the rigid open lattice structure of the ice crystals breaks down. The ice turns into water. The hydrogen bonds in water at 0 °C are longer than those in ice. But they are more flexible. Thus, the groups of liquid molecules can crowd together more compactly than can the groups of ice molecules. As a result, H_2O molecules occupy less volume as water than they do as ice. Water is denser than ice.

As water is warmed from 0 °C, two phenomena having opposite effects occur:

1. The breaking down of some hydrogen bonds enables water molecules to crowd closer together.

2. The increased energy of the water molecules causes them to overcome molecular attractions more effectively and spread apart.

Up to 4 °C, the first effect predominates and water increases in density. Above 4 °C, the first phenomenon continues to occur, but the effect of the second becomes much greater. Thus, above 4 °C, the density of water decreases. Table 12-4 shows the variation of the density of water with temperature.

Groups of water molecules must absorb enough energy to break up into single molecules before water boils. This energy requirement makes the boiling point of water relatively high. It also makes it necessary to use a large amount of heat to vaporize water at its normal boiling point.

12.18 Chemical behavior of water

1. *Stability of water.* Water is a very ***stable compound.*** *A stable compound is one that does not break up, or decompose, easily.* Mercury(II) oxide, on the other hand, is a rather ***unstable compound.*** *This means that it does not require much energy to decompose it into its elements.* Water is so stable that it does not decompose a measurable amount until its temperature reaches about 2700 °C. The stability of water is evidence of the strength of the covalent bonds between the oxygen and hydrogen atoms.

The type of replacement reaction, replacement of hydrogen in water by metals, is described in Section 8.6(2).

2. *Behavior with metals.* Very active metals such as sodium and potassium react with cold water, setting hydrogen free and forming metallic hydroxide solutions.

$$2Na(s) + 2HOH(l) \rightarrow 2NaOH(aq) + H_2(g)$$

Magnesium reacts with boiling water to form magnesium hydroxide and hydrogen. When heated red hot, iron reacts with steam to form iron oxide and hydrogen. Aluminum and zinc also react with water at high temperatures.

The reactions of water with metallic oxides are composition reactions. See Section 8.4(1).

3. *Behavior with metallic oxides.* The oxides of many metals are insoluble, and water has little or no effect on them. But water does react with the ionic oxides of the very active metals. The oxides of sodium, potassium, calcium, and barium unite with

water to form soluble hydroxides. Soluble metallic hydroxides are compounds whose water solutions have *basic* properties. Some basic properties of a solution are a slippery feeling, a bitter taste, and the ability to change red litmus to blue. Calcium hydroxide is formed when water is added to calcium oxide, CaO.

$$\mathbf{CaO(s) + H_2O(l) \rightarrow Ca(OH)_2(s)}$$

Compounds such as calcium oxide, CaO, are known as *anhydrides.* The word anhydride means "without water." Since it forms a solution with basic properties when water is added to it, calcium oxide is called a *basic anhydride. A* ***basic anhydride*** *is the oxide of a metal that unites with water to form a solution having basic properties.*

4. *Behavior with oxides of nonmetals.* The oxides of such nonmetals as carbon, sulfur, and phosphorus are molecular compounds with polar covalent bonds. They unite with water to form a solution with *acidic* properties. Two acidic properties are a sour taste and the ability to turn blue litmus to red. For example, water unites with carbon dioxide to form carbonic acid, H_2CO_3.

$$\mathbf{CO_2(g) + H_2O(l) \rightarrow H_2CO_3(aq)}$$

Carbon dioxide is an anhydride that, with water, forms a solution having acidic properties. Therefore, it is called an *acid anhydride. An* ***acid anhydride*** *is the oxide of a nonmetal that unites with water to form a solution having acidic properties.*

5. *Water of crystallization.* In some crystals, many positive ions and a few negative ions are surrounded by a definite number of water molecules. These crystals are formed by evaporating the water from their solutions. This water is called *water of crystallization* or *water of hydration. A crystallized substance that contains water of crystallization is a* ***hydrate.*** Each hydrate holds a definite proportion of water that is necessary for the formation of its crystal structure. For example, blue crystals of copper(II) sulfate consist of copper(II) ions, sulfate ions, and water molecules. Each copper(II) ion is surrounded by four water molecules. Each sulfate ion is associated with one water molecule. The formula of this substance is written as

$$\mathbf{CuSO_4 \cdot 5H_2O}$$

This formula is empirical in two ways:

1. The empirical formula $CuSO_4$ shows the composition of the crystal with respect to copper(II) and sulfate ions.

2. The water of crystallization is shown as the *total* per formula unit of copper(II) sulfate. This total is expressed as $\cdot\ 5H_2O$. If $CuSO_4 \cdot 5H_2O$ is heated to above 150 °C, the water of crystallization is driven off.

$$\mathbf{CuSO_4 \cdot 5H_2O(s) \rightarrow CuSO_4(s) + 5H_2O(g)}$$

Table 12-4
DENSITY OF WATER

Temperature (°C)	*Density* (g/mL)
0	0.99987
1	0.99993
2	0.99997
3	0.99999
4	1.00000
5	0.99999
6	0.99997
7	0.99993
8	0.99988
9	0.99981
10	0.99973
15	0.99913
20	0.99823
25	0.99707
30	0.99567
40	0.99224
50	0.98807
60	0.98324
70	0.97781
80	0.97183
90	0.96534
100	0.95838

The reactions of water with the oxides of nonmetals are also composition reactions, Section 8.4(1).

Figure 12-20. Water acts to promote some reactions, as shown when effervescent alkalizing tablets are added to water.

The substance that remains is called an ***anhydrous compound.*** Anhydrous copper(II) sulfate, $CuSO_4$, is a white powder. It can be prepared by gently heating the blue $CuSO_4 \cdot 5H_2O$ crystals in a test tube. The fact that water turns anhydrous copper(II) sulfate blue can be used as a *test for water.* Other examples of hydrates are the compounds $ZnSO_4 \cdot 7H_2O$, $CoCl_2 \cdot 6H_2O$, and $Na_2CO_3 \cdot 10H_2O$. Some hydrates have two or more forms that are stable over different temperature ranges.

Many other compounds form crystals that do not require water of crystallization. Examples are NaCl, KNO_3, and $KClO_3$.

When certain hydrates are exposed to the air, they lose some or all of their molecules of water of crystallization. Sodium carbonate decahydrate crystals, $Na_2CO_3 \cdot 10H_2O$, are an example.

$$Na_2CO_3 \cdot 10H_2O(s) \rightarrow Na_2CO_3 \cdot H_2O(s) + 9H_2O(g)$$

The loss of some or all molecules of water of crystallization when a hydrate is exposed to the air is called ***efflorescence.*** Efflorescence occurs when the water vapor pressure of the hydrate is greater than the partial pressure of the water vapor in the air surrounding it. The water vapor pressure of hydrates varies greatly. $Na_2CO_3 \cdot 10H_2O$ has a high water vapor pressure and effloresces quite rapidly. $CuSO_4 \cdot 5H_2O$, on the other hand, has a low water vapor pressure. It can be exposed to the air at room temperature without efflorescence taking place.

6. *Water promotes many chemical changes.* One good example of these changes is the reaction of an effervescent alkalizing tablet, which is a dry mixture. As long as the tablet is kept dry, no chemical action occurs. When the tablet is dropped in water, the substances in the mixture dissolve and react immediately. Bubbles of gas are given off, as shown in Figure 12-20. Mixtures of many other dry substances do not react until water is added. The way water promotes chemical changes will be more fully explained in Unit 5.

12.19 Deliquescence Suppose some calcium chloride granules are placed on a watch glass and exposed to moist air for half an hour. The granules become wet and may even form a solution with water from the air. *Certain substances take up water from the air and form solutions. This property is called* ***deliquescence.*** Deliquescent substances are very soluble in water. Their concentrated solutions have water vapor pressures lower than the normal range of partial pressures of water vapor in the air. Consequently, such substances and their solutions absorb water vapor from the air more rapidly than they give it off. The absorption of water results in a gradual dilution of the solution involved. Absorption and dilution continue until the water vapor pressure of the solution and the partial pressure of water vapor in the surrounding air are equal.

Many insoluble materials such as silk, wool, hair, and tobacco take up water vapor from the air. The water molecules may be held in pores and imperfections of the solid. All such materials, along with deliquescent substances, are classed as *hygroscopic.*

12.20 Deuterium oxide Most water molecules are composed of hydrogen atoms with mass number 1 and oxygen atoms with mass number 16. But there are other possible types of water molecules. There are two natural isotopes of hydrogen with mass numbers 1 and 2. There are three natural isotopes of oxygen with mass numbers 16, 17, and 18. The possible combinations of these five nuclides give nine types of water molecules. In liquid water, these molecules are associated most commonly in chains of from four to eight units. In water, there is also a very small proportion of hydronium (H_3O^+) ions, hydroxide ions, and oxide ions. These ions are formed from the various isotopes of hydrogen and oxygen. Thus water is a complex mixture of many kinds of molecules and ions.

Particles other than ordinary water molecules exist in only small traces in a water sample. But one such type of water molecule has been studied rather extensively. This is the *deuterium oxide* molecule, D_2O. The symbol D is used to represent an atom of the isotope of hydrogen with mass number 2. Electrolysis separates D_2O from H_2O. The D_2O molecules are not as readily decomposed by the passage of electric current as are H_2O molecules. Thus, the concentration of D_2O molecules increases as H_2O molecules are decomposed. From 2400 L of water, 83 mL of D_2O that is 99% pure can be obtained.

Deuterium oxide is about 10% denser than ordinary water. It boils at 101.42 °C, freezes at 3.82 °C, and has its maximum density at 11.6 °C. Delicate tests have been devised for detecting deuterium oxide. It has been used as a "tracer" in research work on living organisms. By tracing the course of deuterium oxide molecules through such organisms, scientists have gained new information about certain life processes. Deuterium oxide usually produces harmful effects on living things, particularly when present in high concentrations. The most important use of deuterium oxide is in nuclear reactors. You will learn more about this use of deuterium oxide in Chapter 31.

12.21 Hydrogen peroxide In addition to water, H_2O, the elements hydrogen and oxygen form another compound, hydrogen peroxide, H_2O_2. The bonding, structure, and oxidation states of the elements in hydrogen peroxide molecules were described in Section 6.15. More details of the bond lengths and bond angles in the hydrogen peroxide molecule are shown in Figure 12-21.

A nearly pure water solution of hydrogen peroxide can be prepared in the laboratory by reacting equivalent quantities of a

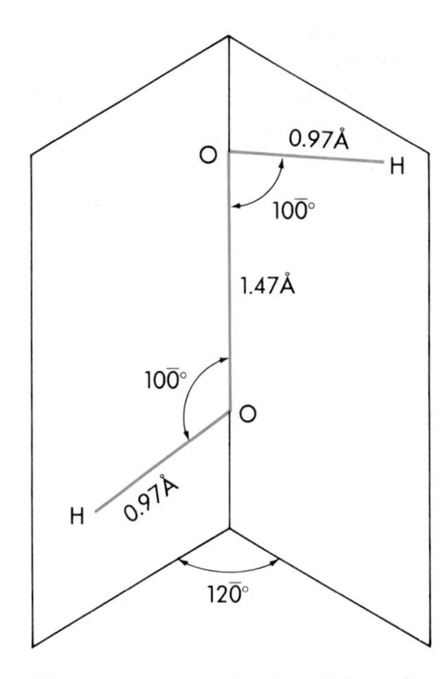

Figure 12-21. The bond lengths and bond angles of a free H_2O_2 molecule in its lowest energy state.

solution of barium peroxide, BaO_2, and ice-cold dilute sulfuric acid, H_2SO_4.

$$BaO_2(aq) + H_2SO_4(aq) \rightarrow H_2O_2(aq) + BaSO_4(s)$$

The $BaSO_4$ is insoluble and can be separated from the H_2O_2 solution by filtration.

The name of this complex carbon-hydrogen-oxygen compound is 2-ethyl-9, 10-dihydroxyanthracene. You will be able to understand this name and the structure of this compound after studying Chapters 18 and 19.

The commercial preparation involves the indirect combination of hydrogen and oxygen to form hydrogen peroxide by means of a very complex carbon-hydrogen-oxygen compound, $C_{16}H_{14}O_2$, and a palladium catalyst.

Recall from Sections 10.4 and 12.2 that condensation temperatures and boiling temperatures are related to molecular complexity.

Some physical properties of water and hydrogen peroxide are given in Table 12-5. Pure hydrogen peroxide is a colorless liquid. Hydrogen peroxide solutions have a faint odor similar to that of ozone, but are practically tasteless. Hydrogen peroxide vapor consists of widely separated H_2O_2 molecules, but in the liquid phase the H_2O_2 molecules associate by hydrogen bonding. This association together with a molecular weight nearly double that of water accounts for the high boiling point of hydrogen peroxide. Solid H_2O_2 is a compact crystalline substance in which the molecules are held together by hydrogen bonding. Unlike water, when solid H_2O_2 melts it expands (as most substances do). Water is denser than ice, but solid H_2O_2 is denser than liquid H_2O_2.

Hydrogen peroxide decomposes into water and oxygen.

$$2H_2\overset{-1}{O}_2 \rightarrow 2H_2\overset{-2}{O} + \overset{0}{O}_2(g)$$

Observe that during this decomposition, for each oxygen atom that is oxidized from $\overset{-1}{O}$ to $\overset{0}{O}$, another oxygen atom is reduced from $\overset{-1}{O}$ to $\overset{-2}{O}$. Thus the decomposition of hydrogen peroxide involves the oxidation and reduction of atoms having the same oxidation state. The rate of this decomposition reaction is increased by an increase in temperature, an increase in the concentration of the H_2O_2, and the presence of catalysts.

Manganese dioxide was used as a catalyst to increase the rate of decomposition of hydrogen peroxide in the laboratory preparation of oxygen, Section 9.4(1).

Hydrogen peroxide is used to bleach natural and artificial fibers and to purify gas and liquid waste materials produced by

Table 12-5
PHYSICAL PROPERTIES OF WATER AND HYDROGEN PEROXIDE

Property	*Water*	*Hydrogen peroxide*
melting point, °C	0.00	−0.41
boiling point, °C	100.0	150.2
density of solid, −20 °C	0.919	1.71
density of liquid, 20 °C	0.998	1.450
heat of fusion, kcal/mole	1.436	2.987

industrial plants. It is also used in uranium mining and metal extraction processes and as a starting material for preparing other compounds in which oxygen has an oxidation number of −1. The 3% solution in water may be used as a mild antiseptic.

SUMMARY

All liquids have the properties of definite volume, fluidity, noncompressibility, diffusion, and evaporation. These properties can be explained by the spacing, motion, and attractive forces of molecules of liquids.

Physical equilibrium is a dynamic state in which two opposing physical changes occur at equal rates in the same system. Equilibrium vapor pressure is the pressure exerted by a vapor in equilibrium with its liquid. The equilibrium vapor pressure of a liquid depends on only the nature of the liquid and its temperature. The principle of Le Chatelier is: If a system at equilibrium is subjected to a stress, the equilibrium will be displaced in such a direction as to relieve the stress. Stresses on a physical equilibrium system can be changes in pressure, temperature, and volume. The boiling point of a liquid is the temperature at which the equilibrium vapor pressure of the liquid equals the prevailing atmospheric pressure.

Liquefying a gas requires two steps: (1) compressing the gas and then removing the heat of compression; (2) permitting the cool compressed gas to expand without absorbing external energy. The highest temperature at which it is possible to liquefy a gas with any amount of pressure is called its critical temperature. The pressure required to liquefy a gas at its critical temperature is called its critical pressure. The volume of one mole of gas under these conditions is called its critical volume. Critical temperature is related to the strength of the attractive forces between molecules. A vapor is a gas at a temperature below its critical temperature.

General properties of solids are definite shape, definite volume, noncompressibility, very slow diffusion, and crystal formation. These properties can also be explained by the spacing, motion, and attractive forces of particles of solids.

Freezing is the physical change of a liquid to a solid. The reverse physical change is melting. The change of phase from a solid to a vapor is known as sublimation.

Amorphous solids are those that appear to have a random particle arrangement. A crystal is a homogeneous portion of a substance bounded by plane surfaces making definite angles with each other, giving a regular geometric form. There are seven crystal systems. Crystalline materials can have ionic, covalent-network, metallic, or covalent-molecular lattice structures.

Water is the most abundant and essential liquid on earth, and water molecules are widespread in the known universe. Pure water is transparent, odorless, tasteless, and almost colorless. Impurities may affect its odor and taste. Water freezes at 0 °C and boils at 100 °C under standard pressure. Water molecules are polar molecules. They are linked by hydrogen bonds into groups of from four to eight molecules in liquid water. A hydrogen bond is a weak chemical bond between a hydrogen atom in one polar molecule and a very electronegative atom in a second polar molecule.

Water is a very stable compound. It reacts with active metals such as sodium and potassium. Steam reacts with red-hot iron. The ionic oxides of very active metals are basic anhydrides, and the covalent oxides of nonmetals are acid anhydrides. Many crystals contain water of crystallization which can be driven off by heat, forming anhydrous compounds. Water promotes many chemical changes.

The loss of some or all molecules of water of crystallization when a hydrate is exposed to the air is called efflorescence. Certain substances that take up water from the air and form solutions are deliquescent. Insoluble materials that take up water vapor from the air and deliquescent substances are hygroscopic.

Deuterium oxide is used in research work on living organisms and in nuclear reactors.

Hydrogen peroxide is prepared in the laboratory by reacting barium peroxide and sulfuric acid. Commercially it is prepared by the catalytic combination of hydrogen and oxygen. Hydrogen peroxide is a colorless liquid with a faint odor. Hydrogen peroxide molecules associate by hydrogen bonding. The decomposition of hydrogen peroxide involves oxidation and reduction of oxygen atoms in the −1 oxidation state. Because of its oxidizing properties, hydrogen peroxide is used for bleaching and purification processes.

VOCABULARY

acid anhydride
amorphous solid
anhydrous
basic anhydride
boiling
boiling point
critical pressure
critical temperature
critical volume
crystal
deliquescence
efflorescence
equilibrium vapor pressure
forward reaction
freezing
hydrate
hydrogen bond
Le Chatelier's principle
melting
molar heat of fusion
molar heat of vaporization
physical equilibrium
reverse reaction
stable compound
sublimation
vapor

QUESTIONS

Group A

1. *(a)* List five general properties of liquids. *(b)* Explain each of these properties in terms of the kinetic theory. *A3a*
2. How do scientists know that the particles of a liquid are in constant motion? *A3a*
3. The equation representing an evaporation-condensation equilibrium is: **liquid + energy ⇄ vapor.** (a) What is meant by the "forward" reaction? *(b)* The "reverse" reaction? *(c)* At equilibrium, how do these reactions compare? *A3c*
4. An open wide-mouth bottle half-full of water is on the laboratory table. Will an equilibrium vapor pressure be attained in the space above the water? Explain. *A3c*
5. A stoppered 250-mL Erlenmeyer flask contains $10\bar{0}$ mL of alcohol. The flask is maintained at $2\bar{0}$ °C until the liquid-vapor equilibrium is established. If the flask is then placed in a large beaker of cold water at $1\bar{0}$ °C, explain how the liquid-vapor equilibrium is disturbed. *A3c, A3d*
6. State Le Chatelier's principle. *A3c, A7*
7. *(a)* If the pressure on a water surface is increased, how is the boiling temperature of the water changed? *(b)* If the pressure on a water surface is decreased, how is the boiling temperature of the water changed? *A3d, A3e*
8. What is meant by *(a)* critical temperature? *(b)* critical pressure? *(c)* critical volume? *A1*
9. What two conditions are needed to liquefy a gas? *A3f*
10. Define *vapor*. *A1*
11. *(a)* List five general properties of solids. *(b)* Explain each of these properties in terms of the kinetic theory. *A3g*
12. What is the distinction between the terms *evaporation* and *sublimation?* *A1*
13. How do these crystal systems compare? *(a)* Cubic and trigonal. *(b)* Tetragonal and orthorhombic. *(c)* Orthorhombic, monoclinic, and triclinic. *A1, A3i*
14. List the general properties of solids composed of *(a)* atoms in a covalent network structure; *(b)* ions; *(c)* metal ions in an electron "gas"; *(d)* molecules. *A3j*
15. *(a)* Where does water occur on the earth? *(b)* Where does water occur in other parts of the solar system? *A3k*

16. List seven physical properties of water. *A3l*
17. *(a)* If 1.00 mL of water freezes, what volume of ice is formed? *(b)* If 1.00 mL of water boils, what volume of steam is formed? *A3l*
18. *(a)* Write a formula equation for the melting of one mole of ice at 0 °C to water at 0 °C, including the quantity of energy involved. *(b)* Similarly, write an equation for the boiling of one mole of water at 100 °C to steam at 100 °C. *A3d, A3l*
19. *(a)* Describe the structure of a water molecule. *(b)* Explain why water molecules are polar. *A3m*
20. What is a hydrogen bond? *A1, A3m*
21. *(a)* Compare the molecular weights of methane, CH_4, and water, H_2O. *(b)* Compare their boiling points and explain why they are so different. *A3d*
22. What does the term *nonstoichiometric* mean when describing inclusion compounds?
23. *(a)* Define: *stable compound, unstable compound.* *(b)* Give an example of each. *A1*
24. *(a)* From Table 8-2, Activity Series of the Elements, select five metals that react with water. *(b)* Describe the reaction conditions. *A3o*
25. *(a)* Define: *acid anhydride, basic anhydride.* *(b)* Give an example of each. *A1*
26. What is the significance of the raised dot in $NiCl_2 \cdot 6H_2O$? *A3o*
27. A package of washing soda, $Na_2CO_3 \cdot 10H_2O$, labeled "one pound" was found to weigh only 14 ounces. Was the packer necessarily dishonest? Explain. *A3o*
28. Why are effervescent alkalizing tablets individually sealed in metal foil? *A4*
29. Why is anhydrous calcium chloride used to lower the humidity in enclosed areas? *A3o, A4*
30. How is D_2O separated from ordinary water? *A3p*
31. What are the uses of D_2O? *A3p*
32. *(a)* Describe the structure of the H_2O_2 molecule. *(b)* Is the H_2O_2 molecule polar? *(c)* Why? *A3q*
33. Assign oxidation numbers to each atom in the *(a)* water molecule; *(b)* hydrogen peroxide molecule.
34. *(a)* Write a balanced formula equation for the laboratory preparation of hydrogen peroxide. *(b)* What type of reaction is this? *(c)* Why does it go to completion? *A6c*
35. List three uses of hydrogen peroxide. *A3q*

Group B

36. What kinds of particles compose substances that are liquids *(a)* well below room temperature; *(b)* at room temperature; *(c)* well above room temperature? *A3b*
37. The system **alcohol(l) + energy ⇄ alcohol(g)** is at equilibrium in a closed container at 25 °C. What will be the effect of each of the following stresses on the equilibrium? *(a)* Temperature is raised to 45 °C. *(b)* Volume of container is doubled. *(c)* Barometer falls from $76\bar{0}$ mm to 735 mm. *A3c*
38. *(a)* Use Figure 12-7 to determine the temperature at which water will boil when the pressure on the water surface is $5\bar{0}0$ mm. *(b)* What is the boiling temperature of alcohol at this pressure? *(c)* Of ether? *A3d, A3e*
39. In the late thirteenth century Marco Polo traveled through the mountainous Pamir region of Asia, which is now the border of the Soviet Union with Afghanistan and China. Here there are many peaks over 6,000 meters high. In his *Travels,* Polo writes of this area, "And I assure you that because of this great cold, fire is not so bright here nor of the same color as elsewhere, and food does not cook well." What chemical explanations can you give for his observations? *A3e*
40. *(a)* What is the effect of compression on the temperature of a gas? *(b)* Why? *A3f*
41. *(a)* What is the effect of expansion on the temperature of a gas? *(b)* Why? *A3f*
42. *(a)* Is it possible to liquefy carbon dioxide at $10\bar{0}$ °C? *(b)* Is it possible to liquefy chlorine at $10\bar{0}$ °C? *(c)* Explain. *A3d, A3f*
43. *(a)* The molar heat of vaporization is added to one mole of a liquid at its standard boiling point. How is the energy of the particles of liquid affected? *(b)* The molar heat of fusion is withdrawn from one mole of a liquid at its standard freezing point. How is the energy of the particles of liquid affected? *A3d*

44. *(a)* What is a metallic glass? *(b)* How are metallic glasses made? *(c)* Give some uses of metallic glasses. *A4*
45. A bottle of alum crystals was erroneously labeled "sodium chloride." How could the error be detected at once by an alert chemistry student? *A3i, A4*
46. Natural camphor, $C_{10}H_{16}O$, is a crystalline compound, density 0.990 g/cm^3. The rhombohedral crystals are soft, have a fragrant, penetrating odor, and melt at 179 °C. Liquid camphor boils at 209 °C. What type of lattice structure does camphor have? *A3j*
47. Why is water essential on earth? *A3k*
48. The fifth century B.C. Greek philosopher, Empedocles, believed that all matter was composed of four "elements," earth, air, fire, and water. These "elements" were joined or separated by the force of attraction or the force of repulsion. Why do you think Empedocles selected water as one of the four "elements?"
49. Why does ice occupy a larger volume than the water from which it is formed? *A2d*
50. The system **ice + energy** $\rightleftarrows$ **water** is at equilibrium at 0 °C in an open vessel. What will be the effect on the temperature of the system if *(a)* a small amount of warm water is added; *(b)* a small amount of ice is added; *(c)* the pressure on the system is reduced? *A3c, A3l*
51. Why does water have a point of maximum density at 4 °C? *A3n*
52. Water is such a stable compound that it does not decompose measurably until it is heated to about 2700 °C. What does this fact indicate about the strength of the bonds in the water molecule? *A3m*
53. What particles other than H_2O molecules are present in pure water? *A3l, A3m*
54. *(a)* What is the molecular weight of O_2? From Appendix Table 16, obtain its boiling point. *(b)* What is the molecular weight of H_2O_2? From Table 12-5, obtain its boiling point. *(c)* How do the molecular weights compare? *(d)* How do the boiling points compare? *(e)* Explain. *A3b, A3q*
55. *(a)* Write a balanced formula equation for the decomposition of hydrogen peroxide. *(b)* Assign oxidation numbers to each atom. *(c)* Identify the oxidation reaction. *(d)* Identify the reduction reaction. *A6c*

PROBLEMS

GROUP A

1. What is the mass of $25\overline{0}$ mL of water at 25 °C? *A3n*
2. *(a)* Convert 50.0 g of H_2O to moles. *(b)* How many H_2O molecules is this?
3. A mixture of 50.0 mL of hydrogen and 35.0 mL of oxygen is ignited by an electric spark. What gas remains? What is its volume in milliliters?
4. A mixture of 40.0 mL of oxygen and 100.0 mL of hydrogen is ignited. What gas remains and how many milliliters does it occupy?
5. $SO_2(g)$ is an acid anhydride and $MgO(s)$ is a basic anhydride. *(a)* Assign oxidation numbers to each element. *(b)* Calculate the percent ionic character of the bonds in each compound and indicate the type of substance. *(c)* Write balanced formula equations for the reactions of these compounds with water. *(d)* Identify any elements that are oxidized or reduced in these reactions. *A6a*
6. Calculate the mass of hydrogen and the mass of oxygen required to produce $30\overline{0}$ g of water.
7. What is the empirical formula of hydrated crystals composed of 55.16% $CoSO_4$ and 44.84% water?
8. *(a)* How many milliliters of hydrogen are needed for complete reaction with 27.5 mL of oxygen? *(b)* How many moles of water are formed?
9. A mixture of hydrogen and oxygen has a volume of $15\overline{0}$ mL. When the mixture is ignited, the gases react completely. How many millimoles of water are formed?

10. *(a)* How many grams of BaO_2 are required for reaction with H_2SO_4 to produce 75.0 g of H_2O_2? *(b)* How many grams of $BaSO_4$ are also formed? *A6c*

11. Calculate the percentage of zinc, fluorine, and water in $ZnF_2 \cdot 4H_2O$.

GROUP B

12. A very rare type of water molecule consists of two atoms of deuterium and one atom of oxygen-18. *(a)* What is the molecular weight of this molecule to five significant figures? Obtain data from Section 3.14. *(b)* What is the percentage composition of this molecule? *A3m*

13. *(a)* Calculate the percentage of water in $Na_2CO_3 \cdot 10H_2O$. *(b)* How many grams of anhydrous sodium carbonate can be obtained by heating $10\bar{0}$ g of $Na_2CO_3 \cdot 10H_2O$? *A6b*

14. If 124.8 g of copper(II) sulfate crystals is heated to drive off the water of crystallization, the loss of mass is 45.0 g. What is the percentage of water in hydrated copper(II) sulfate? *A6b*

15. The anhydrous copper(II) sulfate in problem 14 was found to contain copper, 31.8 g; sulfur, 16.0 g; and oxygen, 32.0 g. What is the empirical formula of hydrated copper(II) sulfate crystals?

16. How many milliliters of hydrogen, collected over water at $2\bar{0}$ °C and 730.0 mm pressure, can be prepared by the electrolysis of 0.325 g of water?

17. A compound of iron, sulfur, and oxygen forms a hydrate upon crystallization from a water solution. The percentage composition of the hydrate is iron, 20.1%; sulfur, 11.6%; oxygen, 63.3%; and hydrogen, 5.0%. What is the empirical formula of the hydrate?

18. Hydrogen peroxide is catalytically decomposed to form oxygen and water. If 0.500 g of hydrogen peroxide is decomposed, how many milliliters of oxygen are produced? The oxygen is collected over water at 25 °C and 740.0 mm pressure. The water level inside the gas collecting tube is 92 mm higher than that outside. *A6c*

19. *(a)* The volume of a water molecule is 15 Å^3. From this information, calculate the volume in milliliters that one mole of water should occupy. *(b)* From the gram-molecular weight of water and its density at 4 °C, calculate the volume in milliliters actually occupied by one mole of water. *(c)* What is the meaning of the difference between these two results? *A2c*

20. Using data from Table 6-4, calculate the energy change in the reaction **1 mole $H_2O_2 \rightarrow$ 1 mole H_2O + $\frac{1}{2}$ mole O_2**. Assume four reaction steps: (1) breaking O—O bonds in H_2O_2 molecules; (2) breaking O—H bonds; (3) forming H_2O molecules from H and from O—H; (4) forming O_2 molecules.

21. The density of carbon tetrachloride at O °C is 1.600 g/mL. *(a)* Calculate the volume in milliliters occupied by a molecule of CCl_4. *(b)* Assuming the molecule to be spherical, calculate its approximate diameter in Å.

22. Obtain from Figure 5-5 the radius of an atom of mercury. *(a)* Assuming this atom to be spherical, what is its volume in Å^3? *(b)* If the liquid mercury atoms are packed in a cubic array with six nearest neighbors, what is the volume in milliliters of 1.00 mole of liquid mercury? *A3i*

23. It is estimated that the effective worldwide water supply is about 9×10^3 km^3 per year and that a person requires for all needs about 4×10^2 m^3 per year. *(a)* How large a population could theoretically be supported? *(b)* If the earth's population is only 4 billion people, why is supplying sufficient water such a problem in so many areas?

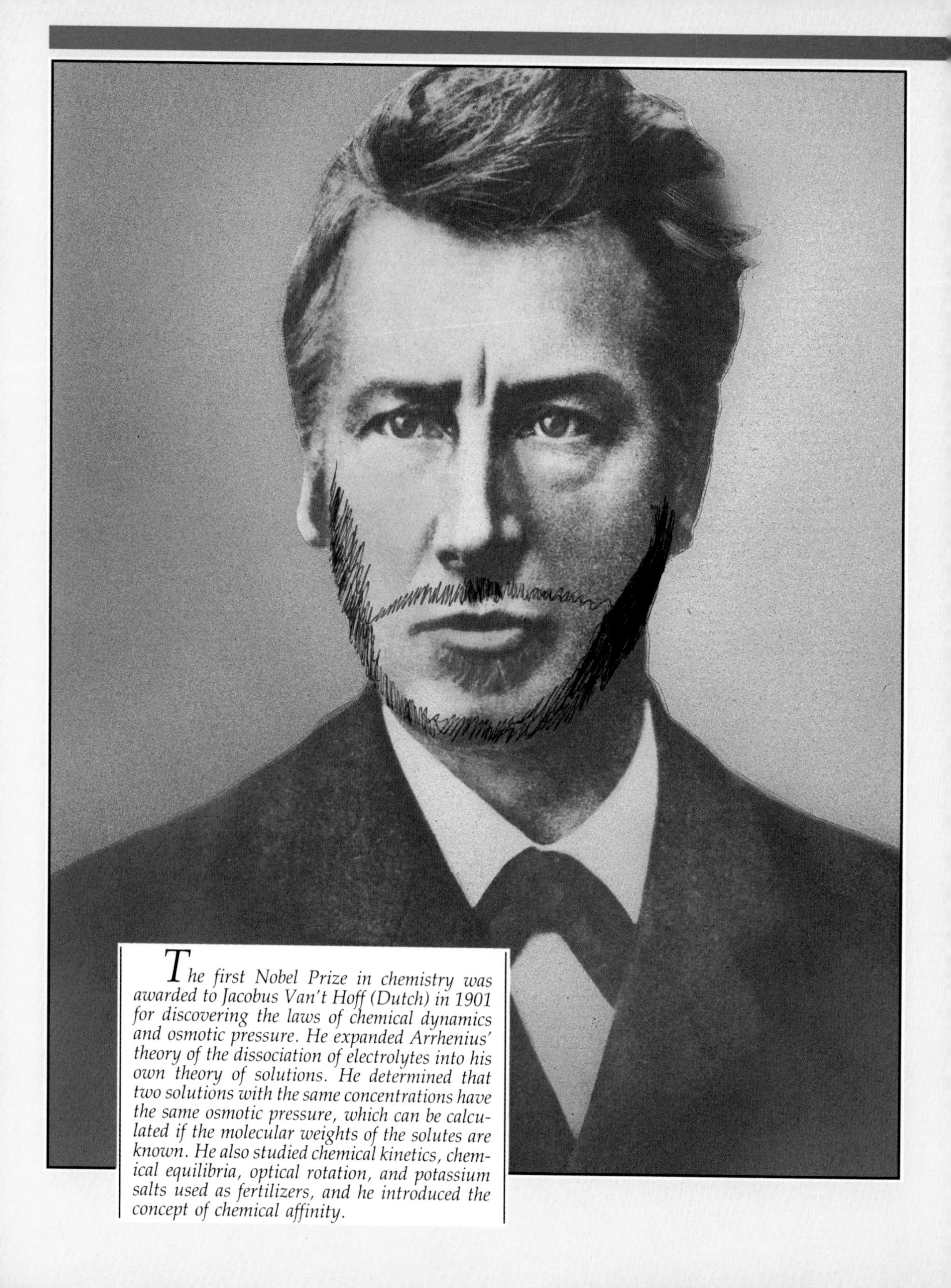

The first Nobel Prize in chemistry was awarded to Jacobus Van't Hoff (Dutch) in 1901 for discovering the laws of chemical dynamics and osmotic pressure. He expanded Arrhenius' theory of the dissociation of electrolytes into his own theory of solutions. He determined that two solutions with the same concentrations have the same osmotic pressure, which can be calculated if the molecular weights of the solutes are known. He also studied chemical kinetics, chemical equilibria, optical rotation, and potassium salts used as fertilizers, and he introduced the concept of chemical affinity.

Solutions

13

Goals

In this chapter you will gain an understanding of: • suspensions and solutions, their properties and composition • the selective nature of solvents and how it is related to solvent structure • solution equilibrium • effects of temperature and pressure on solubility • the dissolving mechanism and heat of solution • solution concentration in terms of molality • colligative properties of solutions

The Solution Process

13.1 Solutions and suspensions Suppose a lump of sugar is dropped into a beaker of water. The sugar gradually disappears and is said to *dissolve* in the water. Careful examination of a drop of this sugar-water mixture under a microscope does not reveal the dissolved sugar. If more sugar is added to the water in this "thought experiment," it also dissolves. However, if this process of adding sugar is continued, a point is reached at which no additional sugar dissolves.

We taste sugar-water mixtures commonly when we drink sweetened tea, coffee, or soft drinks. The sweet taste of these liquids indicates that sugar is present in the water. In our *thought experiment*, molecules of sugar become uniformly distributed among the molecules of water; the properties of the sugar are present in all parts of the liquid. *The mixture of sugar in water is homogeneous throughout.* It is an example of a *solution*.

A ***solution*** *is a homogeneous mixture of two or more substances.* The dissolving medium is called the *solvent*; the substance dissolved is called the *solute*. The simplest solution consists of molecules of a single solute distributed throughout a single solvent. In the example of the sugar-water solution, sugar is the solute and water is the solvent.

Water is the most common solvent.

Not all substances form true solutions in water. If clay is mixed with water, for example, very little clay actually dissolves. Particles of clay are huge compared to molecules of water. The result is a muddy, heterogeneous mixture called a *suspension*. Because the components of the mixture have different densities, they readily separate into two distinct phases. However, some very small particles, still much larger than water molecules, are

Figure 13-1. A beam of light used to distinguish a colloidal suspension from a true solution. The jar at the left contains a water solution of sodium chloride. The jar at the right contains a colloidal suspension of gelatin in water. The path of light is clearly visible in the suspension, but is not visible in the solution.

kept permanently suspended. They are bombarded from all sides by water molecules, and this bombardment keeps them from settling out. Such mixtures may appear to be homogeneous, but careful examination shows that they are not true solutions. Mixtures of this type are called *colloidal suspensions*. One simple way of distinguishing between a true solution and a colloidal suspension is shown in Figure 13-1.

The term *colloid* was originally applied to sticky substances such as starch and glue. *Colloids* are substances that, when mixed with water, do not pass through parchment (semipermeable) membranes. In contrast, substances such as sugar and salt form *true solutions* with water and do pass through parchment membranes. These substances were originally called *crystalloids* by early chemists who believed that certain substances formed suspensions and others formed solutions. The distinction between a colloidal suspension and a solution is shown in Figure 13-2.

Today, chemists recognize that under certain conditions some substances are nondiffusing and colloidal. Under different conditions they are soluble. *The state of subdivision* of a substance, rather than its chemical nature, determines whether it forms a suspension or a true solution when dispersed in a second medium. For example, a colloidal suspension can be formed if sodium ions and chloride ions are brought together in a medium in which sodium chloride is not soluble. Colloidal-size crystals, each consisting of many sodium ions and chloride ions, are formed in the medium.

A true solution is formed when a solute, as molecules or ions, is dispersed throughout a solvent to form a homogeneous mixture. A true solution consists of a *single phase*. The solute is said to be *soluble* in the *solvent*. A colloidal suspension, on the other hand, is a *two-phase system*. It has *dispersed particles*, rather than a solute, and a *dispersing medium*, rather than a solvent. The dispersed substance, *internal phase*, is not soluble in the dispersing medium, *external phase*. The system consists of finely divided particles that remain suspended in the medium. Colloidal suspensions are *heterogeneous mixtures*.

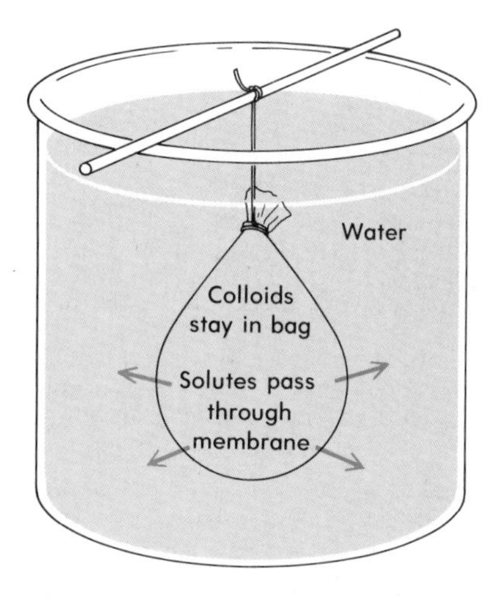

Figure 13-2. Dialysis: An ultrafiltration process for separating ions and molecules of solutions from dispersed particles of colloidal suspensions. The process is based on the difference in their rates of diffusion through a semipermeable membrane.

An ionic solid, such as sodium chloride, does not have a simple molecular structure. Instead, it has a crystal lattice composed of ions bound together by the strong electrostatic force of attraction between positive and negative charges. Soluble ionic substances form aqueous solutions that conduct an electric current. Such substances are called ***electrolytes***.

Generally, covalent substances that dissolve in water form molecular solutions that do not conduct an electric current. Such substances are called ***nonelectrolytes***. Many covalent acids are exceptions. When undissolved, their structures are molecular. However, their aqueous solutions do conduct an electric current. These acids are *electrolytes*.

Solutions of electrolytes have physical properties that are different from solutions of nonelectrolytes. Electrolytes will be considered in detail in Chapters 14 and 15. The remainder of this discussion of the properties of solutions will deal primarily with solutions of nonelectrolytes.

The colloidal state has been called the world of neglected dimensions. It lies between true solutions and coarse suspensions that separate on standing. Colloidal size ranges between ordinary molecular size and a size great enough to be seen through a microscope. Colloidal particles have dimensions ranging from approximately 10 Å to 10,000 Å. Ordinary simple molecules are only a few angstroms in diameter.

13.2 Types of solutions Matter can exist as a solid, liquid, or gas, depending on temperature and pressure. Therefore, nine different types of solutions are possible. These types of solutions are listed in Table 13-1.

All mixtures of gases are solutions because they consist of homogeneous systems of different kinds of molecules. Solutions of solids in liquids are very common. Because water is a liquid at ordinary temperatures, water vapor in air can be thought of as a liquid-in-gas solution. Solutions of gases in solids are rare. An example is the *condensation* of hydrogen on the surfaces of palladium and platinum. This phenomenon, called *adsorption*, approaches the nature of a solution.

Adsorption is a surface phenomenon. See Section 9.18.

Perhaps a more familiar example of adsorption is illustrated in the use of a charcoal gas mask to remove toxic gases from air before it is inhaled. A porous wafer of charcoal presents an unusually large surface area of carbon atoms. These surface carbon atoms attract gas molecules, especially of the polar type. When a mixture of toxic gases and air is passed over the surface of the charcoal, the toxic gas molecules are selectively adsorbed, and air passes through for respiration. Toxic gases are usually complex polar molecules. The air gases, mainly nitrogen and oxygen, are simple nonpolar molecules. Layers of other materials can be added to the charcoal wafer to remove smokes, dusts, mists, carbon monoxide, etc.

Substances that have similarities in their chemical makeup, such as silver and gold or alcohol and water, are apt to form solutions. Two liquids that are mutually soluble in each other in

Figure 13-3. Emulsions are colloidal dispersions of two immiscible liquids. (Left) Vinegar is a water solution of 4–5% acetic acid. When shaken with a salad oil, as in the preparation of a salad dressing, a temporary oil-in-vinegar emulsion is formed. It quickly separates into a layer of oil and a layer of vinegar. (Right) Mayonnaise is also an oil-in-vinegar emulsion. However, egg yolks, a colloidal material, act as an emulsifying agent that stabilizes the dispersed oil droplets with a protective film that prevents the coalescence of the oil droplets.

Table 13-1
TYPES OF SOLUTION

Solute	*Solvent*	*Example*
gas	gas	air
gas	liquid	soda water
gas	solid	hydrogen on platinum
liquid	gas	water vapor in air
liquid	liquid	alcohol in water
liquid	solid	mercury in copper
solid	gas	sulfur vapor in air
solid	liquid	sugar in water
solid	solid	copper in nickel

all proportions are said to be ***miscible***. Ethanol (ethyl alcohol) and water are miscible in all proportions. Similarly, ether and ethanol are completely miscible. Ether and water, on the other hand, are only slightly miscible. Acetone is completely miscible with water, alcohol, and ether. Solutions of solid metals, such as copper in nickel or stainless steel, are called *alloys*.

13.3 Solvents are selective High solubility occurs when solutes and solvents are structurally alike. In a very general sense, the *possibility* of solvent action is increased by a similarity in the composition and structure of substances. Chemists believe that the distribution of electronic forces helps to explain why solvents are *selective*. That is, it may explain why solvents dissolve some substances readily and others only to an insignificant extent.

The water molecule is a polar structure with a distinct negative region (the oxygen atom) and a distinct positive region (the hydrogen atoms). It is frequently referred to as the *water dipole*. The two polar covalent O—H bonds in water form an angle of about 105°. Thus the molecule as a whole is polar and behaves as a dipole (has a negative region and a positive region).

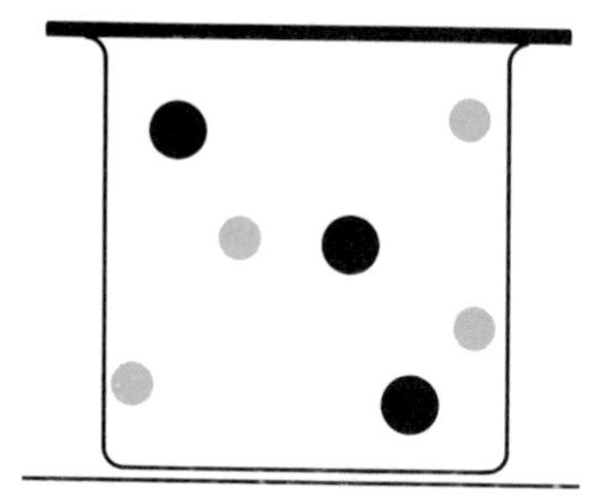

Gaseous solution

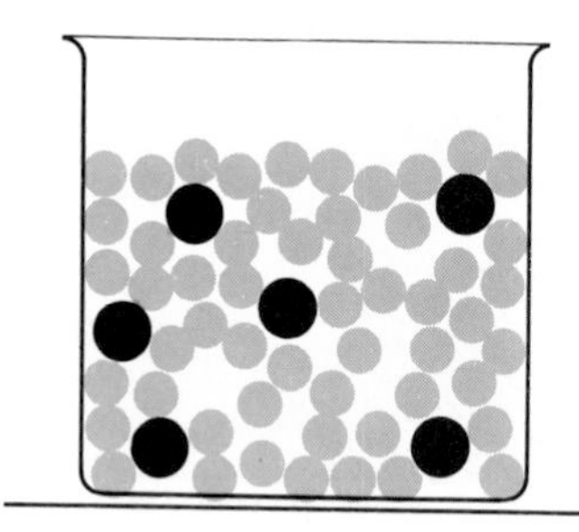

Liquid solution

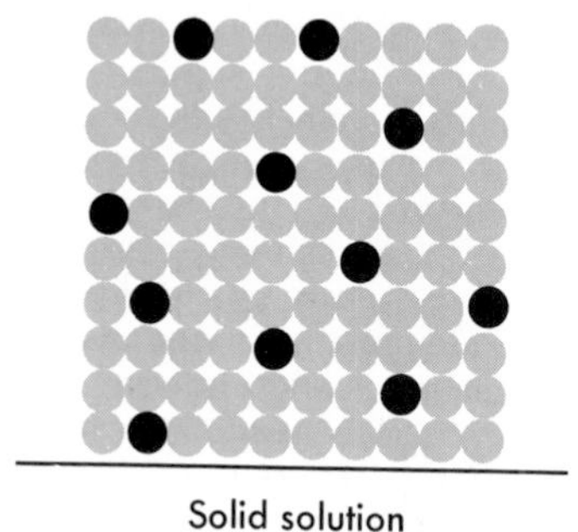

Solid solution

Figure 13-4. Models of solutions.

The carbon tetrachloride molecule, CCl_4, contains four polar covalent bonds. Each C—Cl bond is formed by electron sharing between an sp^3 hybrid orbital of the carbon atom and a p orbital of a chlorine atom. The set of four sp^3 orbitals of carbon (see Section 6.17) leads to the regular tetrahedral shape of the CCl_4 molecules, as shown in Figure 13-5. Because of the symmetrical arrangement of the four polar bonds, the molecule is nonpolar. Gasoline-type compounds, although unsymmetrical in bond arrangement, are practically nonpolar because the electronegativity difference between the hydrogen and carbon atoms of which they are composed is small.

If the rough rule that *like dissolves like* is applied to these solvents, water is expected to dissolve polar substances and carbon tetrachloride to dissolve nonpolar substances. Solute crystals composed of polar molecules or ions are held together by strong attractive forces. They are more likely to be attracted away from their solid structures by polar water molecules than by nonpolar solvents. Thus many crystalline solids like sodium chloride and molecular solids like sugar readily dissolve in water. Compounds that are insoluble in water, such as oils and greases, readily dissolve in nonpolar carbon tetrachloride.

Ethanol, C_2H_5OH, is typical of a group of solvents that dissolves both polar and nonpolar substances.

$$\begin{array}{ccccc} & H & H & & \\ H: & C: & C: & O: & H \\ & H & H & & \end{array}$$

There are five essentially nonpolar carbon-hydrogen bonds. There is also one carbon-carbon bond that is completely nonpolar. The carbon-oxygen bond and the hydrogen-oxygen bond are polar. As in water, the oxygen region of an ethanol molecule is more negative than the other regions. Thus an ethanol molecule has some polar character. This may account for the fact that ethanol is a good solvent for some polar and some nonpolar substances. As a solvent, ethanol is intermediate between the strongly polar water molecule and the nonpolar carbon tetrachloride molecule.

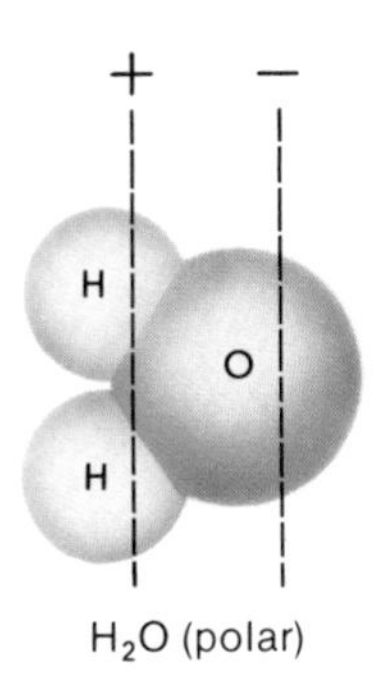

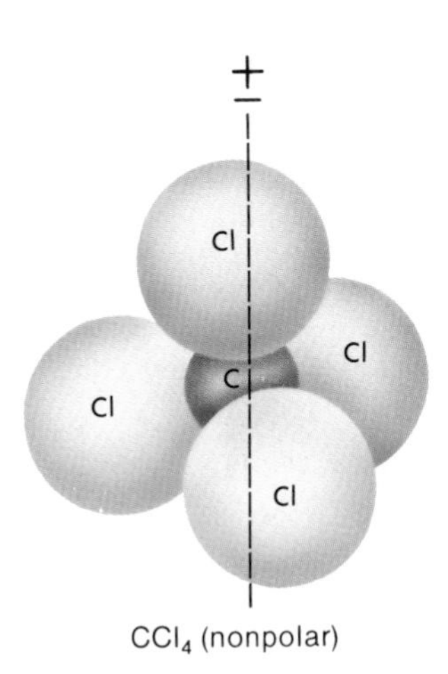

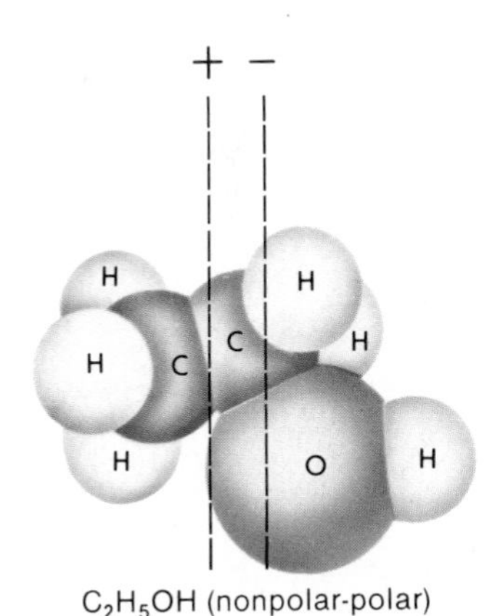

Figure 13-5. Molecular models of three common solvents. Differences in molecular structure may help to explain their selectivity.

13.4 Hydrogen bonds and properties of solvents *Electronegativity* is the measure of the tendency of an atom in a molecule to attract shared electrons (Section 6.19). Hydrogen atoms form distinctly polar covalent bonds with atoms of such highly electronegative elements as fluorine, oxygen, and nitrogen. The hydrogen end of these polar bonds is unique; it can be thought of essentially as an exposed proton. Shared electrons are more strongly attracted by the highly electronegative atom than by hydrogen. The effect of this unequal sharing is to leave the hydrogen nucleus (a proton) without an electron shield to isolate its positive charge.

Because of the unique character of the hydrogen end of the bond, the polar molecules formed may experience relatively strong intermolecular forces. The positive hydrogen end of one molecule may attract the highly electronegative atom of another molecule strongly enough to form a loose chemical bond. The bond between such molecules is called the *hydrogen bond* (Section 12.17). Although hydrogen bonds between hydrogen and fluorine are stronger, those with oxygen are by far the most common. See Figure 13-6.

It is the uniqueness of the polar bonds between atoms of hydrogen and those of highly electronegative elements that accounts for hydrogen bonding between their molecules. All elements that share electrons with highly electronegative elements differ from hydrogen in one significant way. Their atoms have lower energy-level electrons that continue to isolate or shield their nuclear charges when the polar bonds are formed. The positive ends of their bonds do not have the unique character of hydrogen in such polar bonds. Their molecules, however polar, do not exhibit hydrogen-bond behavior. The abnormally high boiling and melting points of water may be attributed in part to hydrogen bonds among water molecules. Also, the formation of hydrogen bonds between a solvent and a solute increases the solubility of the solute. Hydrogen bond formation between water and ethanol molecules may partially explain the complete miscibility of these two substances.

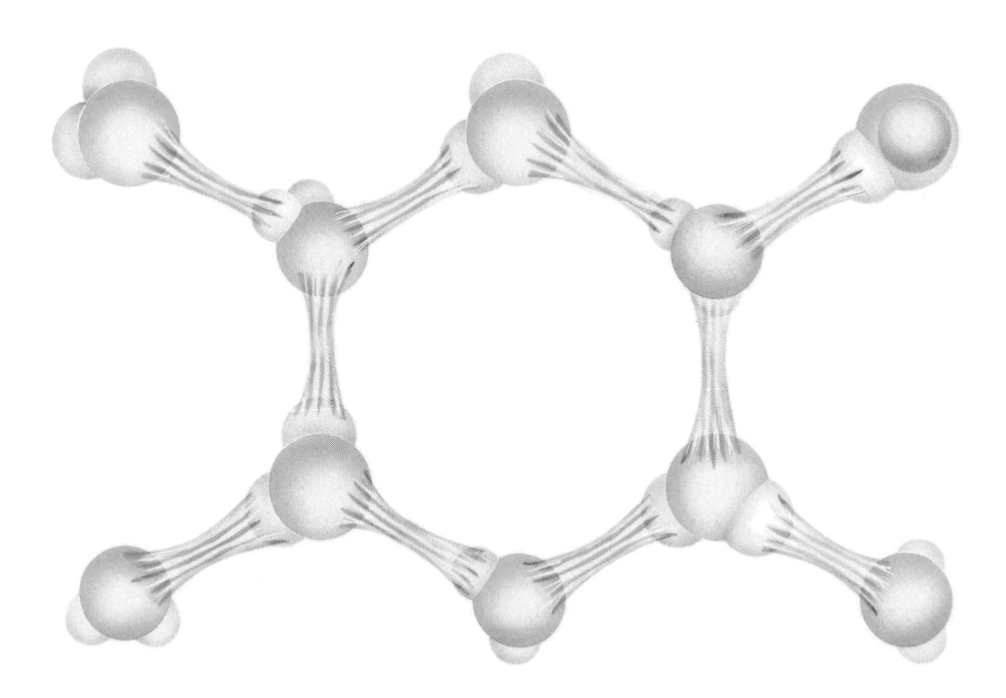

Figure 13-6. Hydrogen bond formation in an ice crystal.

Figure 13-7. A saturated solution contains the equilibrium concentration of solute under existing conditions, as evidenced in this bottle.

13.5 Solution equilibrium The solution process can be thought of as *reversible*. Again consider the lump of sugar dropped into a beaker of water. The sugar molecules that break away from the crystals and enter the water have completely random motions. Some of these molecules come in contact with the undissolved sugar. Here they are attracted by the sugar molecules in the crystal and become part of the crystal structure once more. In this way, the solution process includes the processes of dissolving and crystallizing.

At first, there are no sugar molecules in solution. The solution process occurs only in the direction of dissolving. Molecules leave the crystal structure and diffuse throughout the water. *As the solution concentration (number of sugar molecules per unit volume of solution) increases, the reverse process begins.* The rate at which the sugar crystals rebuild increases as the concentration of the solution increases. Eventually, if undissolved sugar remains, sugar crystals rebuild as fast as they dissolve.

At this point, the concentration of the solution is the maximum possible under existing conditions. Such a solution is said to be *saturated*. At saturation, solute crystallizes at the same rate as solute dissolves. These opposing processes proceeding at equal rates describe a solution in a state of dynamic *equilibrium*. ***Solution equilibrium*** *is the physical state in which the opposing processes of dissolving and crystallizing of a solute occur at equal rates. A* ***saturated solution*** *is one that contains the maximum proportion of dissolved solute to solvent under existing equilibrium conditions.* What visible evidence is there that the solution shown in Figure 13-7 is saturated?

If more water is added to the sugar solution, it is no longer saturated. It is now an *unsaturated* solution. The concentration of solute molecules is decreased and, consequently, the rate of the crystallization process is lowered. In terms of Le Chatelier's principle, this decrease in concentration of solute particles acts as a stress on the equilibrium. The stress is lessened by a decrease in the rate of crystallization. Fewer sugar molecules separate from the solution than enter it; the *same equilibrium concentration* of solute molecules is restored. Solution equilibrium exists when no more solute can be contained in a given quantity of solvent under existing conditions. *The **solubility** of a substance is defined as the maximum amount of that substance that can dissolve in a given amount of a solvent under specified conditions.*

See Section 12.5.

13.6 Influence of pressure on solubility All mixtures of gases are homogeneous, and the "solubility" of one gas in another is independent of pressure. The gas laws describe the behavior of mixtures of gases (gas-in-gas solution) just as they do of individual gases.

Ordinary changes in pressure affect the solubilities of liquids or solids in liquid solvents so slightly that they may be neglected. The solubility of gases in liquid solvents, however, always increases significantly with increasing pressure.

Carbonated beverages *effervesce* when poured into an open glass tumbler. At the bottling plant, carbon dioxide gas is forced into solution in the flavored water under a pressure of from 5 to 10 atmospheres, and the gas-in-liquid solution is sealed in the

Figure 13-8. A comparison of the masses of three common solutes that can be dissolved in 10$\bar{0}$ g of water at 0 °C. Convert these quantities to moles of solute per 10$\bar{0}$ g of water and compare them.

The "fizz" you hear when opening a bottle of carbonated beverage is an example of effervescence.

bottles. When the cap is removed, the pressure is reduced to 1 atmosphere, and some of the carbon dioxide escapes from solution as gas bubbles. *This rapid escape of a gas from a liquid in which it is dissolved is known as* ***effervescence.***

Solutions of gases in liquids reach equilibrium in about the same way that solids in liquids do. The attractive forces between gas molecules are relatively insignificant, and thus their motions are relatively free. If a gas is in contact with the surface of a liquid, gas molecules can easily enter the liquid surface. As the concentration of dissolved gas molecules increases, some molecules begin to escape from the liquid and reenter the gas phase above the liquid surface. An equilibrium is eventually established between the rates at which gas molecules enter the liquid phase and escape from it. While this equilibrium persists, the concentration of the gas-in-liquid solution remains constant. It is a dynamic equilibrium in which the entering and escaping rates of gas molecules are equal. Thus the solubility of the gas is limited to its equilibrium concentration in the solvent under existing conditions.

Equilibrium is a dynamic state: two opposing actions proceed at equal rates.

If the pressure of the gas above the liquid is increased, gas molecules enter the liquid surface at a higher rate. This tends to relieve the pressure and increase the concentration of dissolved gas molecules. In turn, the increasing solute concentration causes the rate at which gas molecules escape from the liquid surface to increase until the rates are again equal and equilibrium is restored at a higher gas solute concentration.

Again, in terms of Le Chatelier's principle, the increase in gas pressure acts as a stress on the initial equilibrium system. The stress is lessened by a decrease in pressure. Thus the solubility of the gas in the liquid is increased. *The solubility of a gas in a liquid is directly proportional to the pressure of the gas above the liquid.* This is a statement of ***Henry's law,*** named after the English chemist William Henry (1775–1836).

Henry's law relates gas solubility to pressure.

Gases that react chemically with their liquid solvents are generally more soluble than those that do not form compounds with the solvent molecules. Oxygen, hydrogen, and nitrogen are only slightly soluble in water. Ammonia, carbon dioxide, and sulfur dioxide are more soluble probably because they form the weak monohydrates $NH_3 \cdot H_2O$, $CO_2 \cdot H_2O$, and $SO_2 \cdot H_2O$ with the water solvent. Such gases deviate from Henry's law.

If a mixture of different gases is confined in a space of constant volume and temperature, *each gas exerts the same pressure it would if it alone occupied the space.* The pressure of the mixture as a whole is the *total* of the individual, or *partial,* pressures of the gases composing the mixture. You will recognize this statement as *Dalton's law of partial pressures,* discussed in Section 10.14. The partial pressure of each gas is proportional to the number of molecules of that gas in the mixture.

When a mixture of gases is in contact with a liquid, the solubility of each gas is proportional to its partial pressure. Assuming that the gases in the mixture do not react in any way when in solution, each gas dissolves to the same extent that it would if the other gases were not present.

Air is about 20% oxygen. When air is bubbled through water, only about 20% as much oxygen dissolves as would dissolve if pure oxygen were used at the same pressure. Oxygen remains dissolved in the water because it is in equilibrium with the oxygen in the air above the water. If the oxygen were removed from the air above the water, this equilibrium would be disturbed. By Le Chatelier's principle, the dissolved oxygen would eventually escape from the water. This is an important fact, considering the abundance of life that exists in water.

Chemistry students should know Le Chatelier's principle.

13.7 Temperature and solubility 1. *Gases in liquids*. A glass of water drawn from the hot water tap often appears milky. Tiny bubbles of air suspended throughout the water cause this cloudiness. The suspended air was originally *dissolved* in cold water. It was driven out of solution as the water was heated.

Raising the temperature of a gas-in-water solution increases the average speed of its molecules. Molecules of dissolved gas leave the solvent at a faster rate than gas molecules enter the solvent. This lowers the equilibrium concentration of the solute. Thus the solubility of a gas decreases as the temperature of the solvent increases. Appendix Table 11 shows how the solubility of gases varies with the kind of gas and the temperature.

2. *Solids in liquids*. An excess of sugar added to water results in an equilibrium between the sugar solute and the undissolved sugar crystals. This equilibrium is characteristic of a saturated solution.

If the solution is warmed, the equilibrium is disturbed, and solid sugar dissolves as the temperature of the solution rises. It is

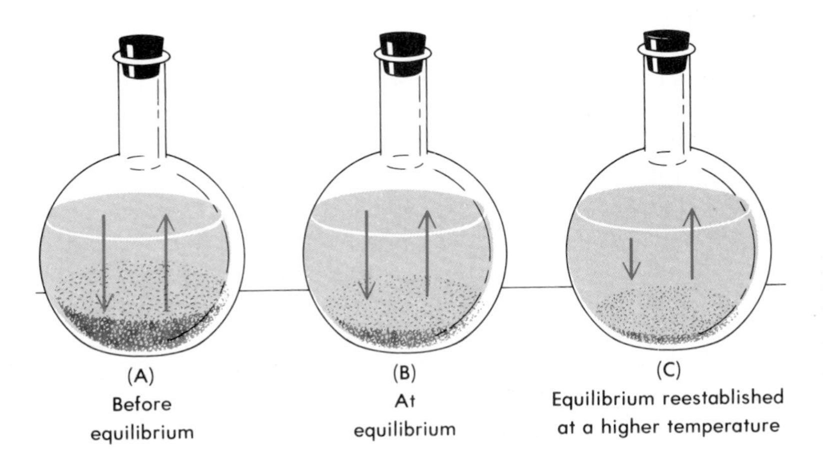

Figure 13-9. An example of solution equilibrium in which the solubility of solute increases with temperature. The different lengths of arrows indicate the relative rates of dissolving (up) and crystallizing (down).

Figure 13-10. Crystals of sugar grown on a string from a supersaturated solution of sugar in water.

Solubility of a solute depends on the temperature of the solvent.

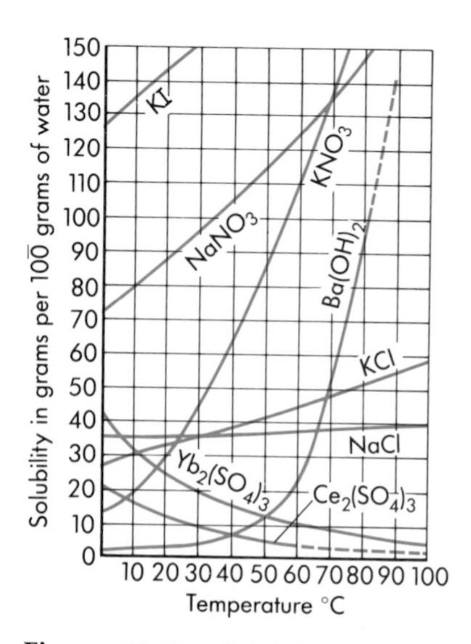

Figure 13-11. Solubility curves. The solubility of a solute is expressed in grams per 100 grams of water at a stated temperature.

evident that the solubility of the sugar in water has increased with the rise in temperature. A new solution equilibrium is eventually reached at the higher solution temperature, where it has a higher equilibrium concentration of solute sugar.

Cooling the solution causes dissolved sugar to separate as crystals. The separation of solute from the solution indicates that solubility diminishes as the temperature falls. No more than the equilibrium concentration of the solute can normally remain in solution. Thus lowering the temperature disturbs the equilibrium, and sugar crystallizes from solution faster than solid crystals dissolve.

When a hot saturated solution of sugar is allowed to cool, the excess solute usually, but not always, separates as expected. If, on cooling, crystallization of the excess solute does not occur, the solution is said to be *supersaturated*. The solution contains a higher concentration of solute than does the saturated solution at the lower temperature. Supersaturation can easily be demonstrated by using a solute that is much more soluble in hot water than in cold water (sodium thiosulfate or sodium acetate, for example). The hot saturated solution is filtered and left undisturbed to cool slowly.

A supersaturated solution is unstable and usually has a strong tendency to reestablish normal equilibrium. A sudden shock or disturbance may cause the excess solute to separate. A small crystal fragment of the solute dropped into the solution (called seeding) starts the crystallizing process instantly.

Many pure liquids can be cooled below their normal freezing points with similar treatment. They are called *supercooled* liquids. As expected, this condition is unstable, and the solutes usually show a strong tendency to solidify.

Increasing the temperature usually increases the solubility of solids in liquids. Sometimes, however, the reverse effect is observed. A certain rise in temperature may result in a large increase in solubility in one case, a slight increase in another case, and a definite decrease in still another. For example, the solubility of potassium nitrate in 100 g of water at 0 °C is about 14 g. Solubility increases to nearly 170 g when the temperature is raised to 80 °C. Under similar circumstances, the solubility of sodium chloride increases only about 2 g. The solubility of cerium sulfate *decreases* nearly 14 g. Typical solubility curves are shown in Figures 13-11 and 13-12. If the solubility curve for cane sugar in water were included in Figure 13-11, the graph would have to be extended considerably. At 0 °C, 179 g of sugar dissolves in 100 g of water. The solubility increases to 487 g at 100 °C. The solubility of solids depends upon the nature of the solid, the nature of the solvent, and the temperature. Solubility data for various substances in water are given in Table 13-2.

When a solid dissolves in a liquid, the solid may be thought

Table 13-2
SOLUBILITY OF SOLUTES AS A FUNCTION OF TEMPERATURE
(Grams of solute per 100 grams of H_2O)

Substance	Temperature, °C 0°	20°	40°	60°	80°	100°
$AgNO_3$	122	216	311	440	585	733
$Ba(OH)_2$	1.67	3.89	8.22	20.9	101	—
$C_{12}H_{22}O_{11}$	179	204	238	287	362	487
$Ca(OH)_2$	0.189	0.173	0.141	0.121	0.094	0.076
$Ce_2(SO_4)_3$	20.8	10.1	—	3.87	—	—
KCl	28.0	34.2	40.1	45.8	51.3	56.3
KI	128	144	162	176	192	206
KNO_3	13.9	31.6	61.3	106	167	245
Li_2CO_3	1.54	1.33	1.17	1.01	0.85	0.72
NaCl	35.7	35.9	36.4	37.1	38.0	39.2
$NaNO_3$	73	87.6	102	122	148	180
$Yb_2(SO_4)_3$	44.2	$37.5^{10°}$	$17.2^{30°}$	10.4	6.4	4.7
CO_2(gas at SP)	0.335	0.169	0.0973	0.058	—	—
O_2(gas at SP)	0.00694	0.00537	0.00308	0.00227	0.00138	0.0000

of as changing in phase to something resembling a liquid. Such a change is endothermic; heat is absorbed. Thus the temperature of the solution should be lowered as the solid dissolves. The solubility of the solid should increase as the temperature is raised. Deviations from this normal pattern may indicate some kind of chemical reaction between solute and solvent.

The most common solutions are solids dissolved in liquids.

3. *Liquids in liquids.* Similar logic applies to solutions of liquids in liquids. As no change in phase occurs when such solutions are prepared, small changes in temperature can be expected.

A large change in temperature, as in the case of sulfuric acid in water, suggests some type of chemical reaction between solute and solvent. With water as the solvent, this reaction may involve *hydration*, a clustering of water dipoles about the solute particles.

13.8 Increasing the rate of dissolving The rate at which a solid dissolves in a liquid depends on the solid and liquid involved. As a rule, the more nearly the solute and solvent are alike in structure, the more readily solution occurs. However, the rate of solution of a solid in a liquid can be increased in three ways.

1. *By stirring.* The diffusion of solute molecules throughout the solvent occurs rather slowly. Stirring or shaking the mixture aids in the dispersion of the solute particles by bringing fresh portions of the solvent in contact with the undissolved solid.

2. *By powdering the solid.* Solution action occurs only at the surface of the solid. By grinding the solid into a fine powder, the

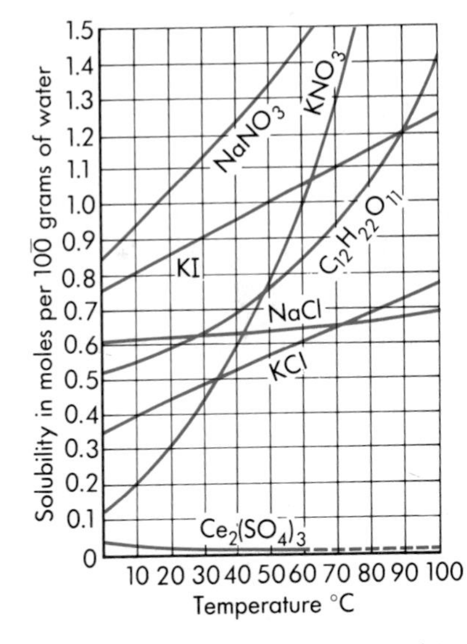

Figure 13-12. Solubility plotted in moles of solute per 100 grams of water as a function of temperature. Compare the relative positions of the curves with those shown in Figure 13-11.

surface area is greatly increased. Hence, finely powdered solids dissolve much more rapidly than large lumps or crystals of the same substance.

3. *By heating the solvent.* The rate of dissolving increases with temperature. If heat is applied to a solvent, the molecular activity increases. As a result, the dissolving action is speeded up. At the same time, the solubility of the substance increases if the dissolving process is endothermic.

The first two actions influence the rate of dissolving by increasing the effective contact area between solid and liquid. The third does so by producing a more favorable energy distribution among the particles of the solid. As the temperature is raised, the average kinetic energy of the solute particles is also raised. A larger portion of the particles has enough kinetic energy to overcome the binding forces and leave the surface of the solid.

13.9 Dissolving mechanisms The solution process is spontaneous, or self-acting. Even though the precise manner in which substances enter into solution is not fully understood, some probable mechanisms by which a solid dissolves in a liquid can be examined. At least three energetic actions are assumed to occur in the dissolving process.

1. Solute particles are separated from the solid mass (as a solid changing phase to a liquid). *This action absorbs energy.*

2. Solvent particles are moved apart to allow solute particles to enter the liquid environment. *This action also absorbs energy.*

3. Solute particles are attracted to solvent particles. *This action releases energy.*

If dissolving is endothermic, solubility increases with rising temperature.

If dissolving is exothermic, solubility decreases with rising temperature.

The first two of these actions are endothermic and the last one is exothermic. If the exothermic action is less than the combined effect of the two endothermic actions, the net change is endothermic. Consequently, the temperature of the solution *decreases* as the solid dissolves. This is the usual pattern for solid-in-liquid solutions. In such cases, heating the solution results in an increase in the solubility of the solid. If the net change is exothermic, the temperature of the solution *increases* as the solid dissolves. Heating such solutions results in a decrease in the solubility of the solid. Refer to Figure 13-11 for examples of dissolving processes that are endothermic and those that are exothermic. The reasons for these effects are discussed in the text, Section 13.10.

A possible model of the solution process is illustrated in Figure 13-13. Dissolving is aided by the attraction between solute and solvent particles. Solvent molecules move at random; some are attracted to surface molecules of undissolved crystals. As they cluster about these surface molecules, enough energy may be given up to enable solvent molecules to carry off solute molecules. The attraction between unlike molecules of solute and

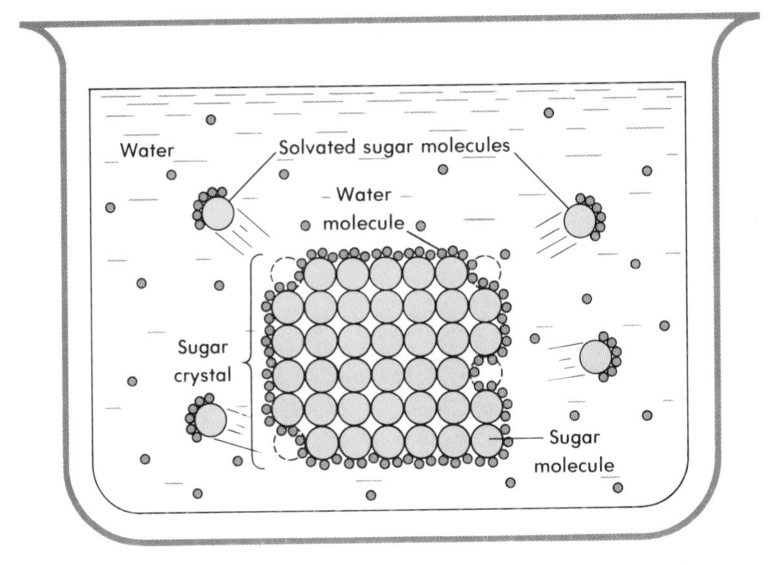

Figure 13-13. A model illustrating a possible mechanism for the dissolving process.

solvent is responsible for this process, which is known as *solvation*. The solute molecules that leave the crystal along with their cluster of solvent molecules are said to be *solvated*. If water is the solvent, the solvation process is known more specifically as *hydration*, and the solute molecules are said to be *hydrated*.

Solvation aids dissolving.

Hydration is solvation in which water is the solvent.

Natural processes generally lead to lower energy states (Section 2.14). Thus an endothermic energy change cannot account for the fact that the dissolving process occurs spontaneously. Perhaps the entropy change can.

The mixture of solute and solvent particles in a solution represents a more disordered state (higher entropy) than that of the unmixed solid and liquid. Because mixture, or higher entropy, is more probable than the unmixed states, the change toward increased entropy that accompanies the dissolving process may cause solution to occur even though the energy change is not toward a lower energy state. In such cases the tendency toward increased entropy overrides the tendency toward lower energy states.

How well is this logic sustained when it is applied to solutions of gases in liquids? The solute is in a more random state as a gas than as a dissolved solute in a liquid. Thus the higher entropy of the gaseous state opposes the dissolving process. For dissolving to occur, the energy change must be favorable (toward a lower state). It also must be great enough to overcome the unfavorable entropy change.

Accordingly, the dissolving process for a gas-in-liquid solution should be exothermic. If the temperature of the solution is raised, the solubility of the gas should be lowered. Experiments show this to be the case. Heat is evolved when a gas dissolves in water, and the solubility of the gas decreases as the temperature of the solution increases.

Table 13-3
HEATS OF SOLUTION
(kcal/mole solute in 200 moles H_2O)
[(s) = solid, (l) = liquid, (g) = gas at SP]

Substance	*Heat of solution*
$AgNO_3(s)$	+5.44
$CO_2(g)$	−4.76
$CuSO_4(s)$	−16.20
$CuSO_4 \cdot 5H_2O(s)$	+2.75
$HC_2H_3O_2(l)$	−0.38
$HCl(g)$	−17.74
$HI(g)$	−7.02
$H_2SO_4(l)$	−17.75
$KCl(s)$	+4.20
$KClO_3(s)$	+10.04
$KI(s)$	+5.11
$KNO_3(s)$	+8.52
$KOH(s)$	−13.04
$LiCl(s)$	−8.37
$Li_2CO_3(s)$	−3.06
$MgSO_4 \cdot 7H_2O(s)$	+3.80
$NaCl(s)$	+1.02
$NaNO_3(s)$	+5.03
$NaOH(s)$	−9.94
$Na_2SO_4 \cdot 10H_2O(s)$	+18.76
$NH_3(g)$	−8.28
$NH_4Cl(s)$	+3.88
$NH_4NO_3(s)$	+6.08

Negative heat of solution: The solute solubility decreases with rising solution temperature.

Positive heat of solution: The solute solubility increases with rising solution temperature.

13.10 Heat of solution From the foregoing discussion, it is clear that no single rule can be stated for changes of solubility with increasing temperature. Gases become less soluble in water as temperature is raised. Table 13-2 shows that some solids become more soluble in water as temperature is raised, others become less soluble, and still others experience little change in solubility.

As a solute dissolves in a solvent, heat is either given off or absorbed. Thus the total heat content of a solution is not the same as that of its separate components. *The difference between the heat content of a solution and the heat contents of its separate components is called the* ***heat of solution.*** The heat of solution is a measure of the heat energy absorbed or released as a solution is formed.

solute + solvent → solution + heat (exothermic)

or

solute + solvent + heat → solution (endothermic)

When the dissolving process is exothermic, the total heat content of the solution is lower than that of its separate components. The solution warms as dissolving proceeds. The heat of solution is said to be *negative.* When the process is endothermic, the heat content of the solution is higher than that of its components. The solution cools as dissolving proceeds and the heat of solution is said to be *positive.*

The dilution of a concentrated solution may cause the release or absorption of heat. Thus the heat of solution of any system depends upon the concentration of the final solution. For this reason, *heat of solution is measured in kilocalories per mole of solute dissolved in a specific number of moles of solvent.* Heats of solution for some common substances are given in Table 13-3. How do these data relate to the solubility information in Table 13-2?

Observe that the change in solubility of a substance with temperature is closely related to its heat of solution. The heat of solution of sodium chloride, for example, is nearly zero. The solubility change of sodium chloride with temperature is also very small.

In a saturated solution with undissolved solute, an equilibrium exists between the dissolving and crystallizing processes. Consider such a solution of KCl. The symbols (s) and (l) indicate solid and liquid, respectively.

$$\mathbf{KCl(s) + H_2O(l) + heat \rightleftarrows solution}$$

The heat of solution is +4.20 kcal/mole. The dissolving process is endothermic. The crystallizing process must then be exothermic. At equilibrium, the tendency toward lower energy (release of heat) as solute crystallizes just balances the tendency toward higher entropy (greater disorder) as crystals dissolve. Consequently, there is no net driving force in the system.

Suppose the solution is now heated. The increased heat content, as indicated by the rise in temperature, produces a stress on the equilibrium. According to Le Chatelier's principle, the stress is relieved by an increase in the rate of the *endothermic* process by which heat is absorbed. Thus dissolving proceeds faster than crystallizing until the concentration of KCl in solution is increased and the stress is relieved. Equilibrium is again established at a higher solute concentration. Solutes with positive heats of solution ordinarily become more soluble as the temperature of their solutions is raised.

The effect of temperature on saturated solutions of solutes with negative heats of solution is the reverse of the process just described. Here, the dissolving process is exothermic and the crystallizing process is endothermic. A rise in solution temperature disturbs the equilibrium, and the rate of the endothermic (crystallizing) process increases. Solute separates from the solution. Solutes with negative heats of solution ordinarily become less soluble as the temperature of their solutions is raised.

Figure 13-14. In the top photo, the two beakers of water are at the same temperature. In the bottom photo, a salt is added to the left beaker and the temperature of the aqueous solution rises. Is the dissolving process for this solute endothermic or exothermic? Does the solute have a positive or negative heat of solution?

13.11 Concentration of solutions The *concentration* of a solution depends upon the relative proportions of solute and solvent. The more solute that is dissolved in a solvent, the more *concentrated* the solution becomes. On the other hand, the more solvent that is added, the more *dilute* the solution becomes.

The terms *dilute* and *concentrated* are qualitative. In order to be of value, the concentrations of solutions must be known quantitatively. Chemists have developed several methods for expressing solution concentrations quantitatively. Three of these methods are introduced in this book. One method expresses solution concentration in terms of the *ratio of solute to solvent*. The other two methods express solution concentration in terms of the *ratio of solute to solution*.

When it is important to know the ratio of solute molecules to solvent molecules in a solution, the concentration is expressed in terms of *molality*. ***Molality*** *is the concentration of a solution expressed in moles of solute per kilogram of solvent.* The symbol for molality is the small letter *m*.

A *one-molal* (1-*m*) solution contains *one mole of solute per kilogram of solvent*. You will recognize that 0.5 mole of solute dissolved in 0.5 kg of solvent, or 0.25 mole of solute in 0.25 kg of solvent, also gives a 1-*m* solution. A *half-molal* (0.5-*m*) solution contains *one-half mole of solute per kilogram of solvent*. A *two-molal* (2-*m*) solution has *two moles of solute per kilogram of solvent*. Several exercises for expressing the concentration of solutions in terms of molality are given in Table 13-4.

Table 13-4
CONCENTRATION OF SOLUTIONS IN MOLALITY

Quantity of solute	*Quantity of solvent*	*Mass solute per mole*	*Conversion to moles solute*	*Conversion to kg solvent*	*Moles solute per kg solvent*	*Molality*
18.2 g HCl	$25\bar{0}$ g H_2O	$\frac{36.5\ g}{mole}$	$18.2\ g\ HCl \times \frac{mole}{36.5\ g}$	$25\bar{0}\ g\ H_2O \times \frac{kg}{10^3\ g}$	$\frac{0.499\ mole\ HCl}{0.250\ kg\ H_2O}$	= 2.00 *m*
2.50 g NH_3	175 g H_2O	$\frac{17.0\ g}{mole}$	$2.50\ g\ NH_3 \times \frac{mole}{17.0\ g}$	$175\ g\ H_2O \times \frac{kg}{10^3\ g}$	$\frac{0.147\ mole\ NH_3}{0.175\ kg\ H_2O}$	= 0.840 *m*
15.6 g NaCl	$50\bar{0}$ g H_2O	$\frac{58.5\ g}{mole}$	$15.6\ g\ NaCl \times \frac{mole}{58.5\ g}$	$50\bar{0}\ g\ H_2O \times \frac{kg}{10^3\ g}$	$\frac{0.267\ mole\ NaCl}{0.500\ kg\ H_2O}$	= 0.533 *m*
12.2 g I_2	$10\bar{0}$ g CCl_4	$\frac{254\ g}{mole}$	$12.2\ g\ I_2 \times \frac{mole}{254\ g}$	$10\bar{0}\ g\ CCl_4 \times \frac{kg}{10^3\ g}$	$\frac{0.0480\ mole\ I_2}{0.100\ kg\ CCl_4}$	= 0.480 *m*

A 0.1-molal solution contains 0.1 mole of solute per kilogram of solvent.

Molal solutions are important to chemists because, for a given solvent, *two solutions of equal molality have the same ratio of solute to solvent molecules.* (A kilogram, 1000 g, of solvent can be expressed in terms of moles since 1000 g ÷ number of grams/mole of solvent = number of moles of solvent.) Molality is preferred for expressing the concentration of solutions in procedures in which temperature changes may occur. Sample Problems 13-1 and 13-2 illustrate calculations involving solution concentrations in molalities.

SAMPLE PROBLEM 13-1

How many grams of $AgNO_3$ are needed to prepare a 0.125-*m* solution in $25\bar{0}$ mL of water?

SOLUTION

Molality expresses solution concentration in moles of solute per kilogram of solvent.

$$25\bar{0}\ mL\ H_2O = 25\bar{0}\ g\ H_2O$$

The gram-formula weight of $AgNO_3$ = $17\bar{0}$ g = mass of 1 mole $AgNO_3$. The 0.125-*m* $AgNO_3$ solution = 0.125 mole $AgNO_3$/kg H_2O. The mass of $AgNO_3$ required for $25\bar{0}$ g of H_2O to give the 0.125-*m* concentration must be determined. Moles of $AgNO_3$ must be converted to grams and grams of H_2O to kilograms. This is done by unit cancellations as follows:

$$\frac{0.125\ \cancel{mole}\ AgNO_3}{\cancel{kg\ H_2O}} \times \frac{17\bar{0}\ g}{\cancel{mole}} \times 25\bar{0}\ \cancel{g\ H_2O} \times \frac{\cancel{kg}}{10^3\ \cancel{g}} = 5.31\ g\ AgNO_3$$

SAMPLE PROBLEM 13-2

A solution contains 17.1 g of sucrose, $C_{12}H_{22}O_{11}$, dissolved in 125 g of water. Determine the molal concentration.

SOLUTION

The formula weight $C_{12}H_{22}O_{11}$ = 342. Thus 1 mole has a mass of 342 g. The concentration is 17.1 g sucrose/125 g H_2O. To express in terms of molality, grams of sucrose must be converted to moles, and grams of H_2O to kilograms, giving moles of sucrose per kilogram of water. This is accomplished by unit cancellations as follows:

$$\frac{17.1\ \cancel{g}\ C_{12}H_{22}O_{11}}{125\ \cancel{g}\ H_2O} \times \frac{\text{mole}}{342\ \cancel{g}} \times \frac{10^3\ \cancel{g}}{\text{kg}} = \frac{0.400\ \text{mole}\ C_{12}H_{22}O_{11}}{\text{kg}\ H_2O} = 0.400\ m$$

Practice Problems

1. What quantity of methanol, CH_3OH, is required to prepare a 0.244-*m* solution in $40\bar{0}$ g of water? *ans.* 3.12 g CH_3OH
2. What is the molality of a solution composed of 2.65 g of acetone, $(CH_3)_2CO$, dissolved in $20\bar{0}$ g of water? *ans.* 0.228 *m*

COLLIGATIVE PROPERTIES OF SOLUTIONS

13.12 Freezing-point depression of solvents In the preceding sections the nature of the solution process has been examined in some detail. It should not be surprising to find that the addition of a solute affects certain properties of the solvent.

Experiments involving vapor pressure show that *at any temperature the vapor pressure of a pure solvent is higher than that of the same solvent containing a nonvolatile solute.* The vapor pressure of a liquid is a measure of the *escaping tendency* of the liquid molecules. Thus the presence of solute particles in solution lowers the escaping tendency of the solvent molecules. This effect is reasonable if the solute particles are thought of as *decreasing the concentration* of solvent molecules.

A solute always lowers the vapor pressure of a liquid solvent.

See Section 12.4.

Figure 13-15 shows plots of vapor pressure of a solvent as a function of temperature. Curves are shown for the *pure solvent,* for the solvent in a *dilute solution* of molal concentration X, and for the solvent in a *dilute solution* of molal concentration 2X. Observe that at any given temperature *the vapor pressure of the solvent* in the dilute solutions decreases in proportion to the concentration of *solute particles.* This decrease in solvent vapor pressure has the effect of extending the liquid range of the solution. That is, the solution can exist in the liquid phase at both higher

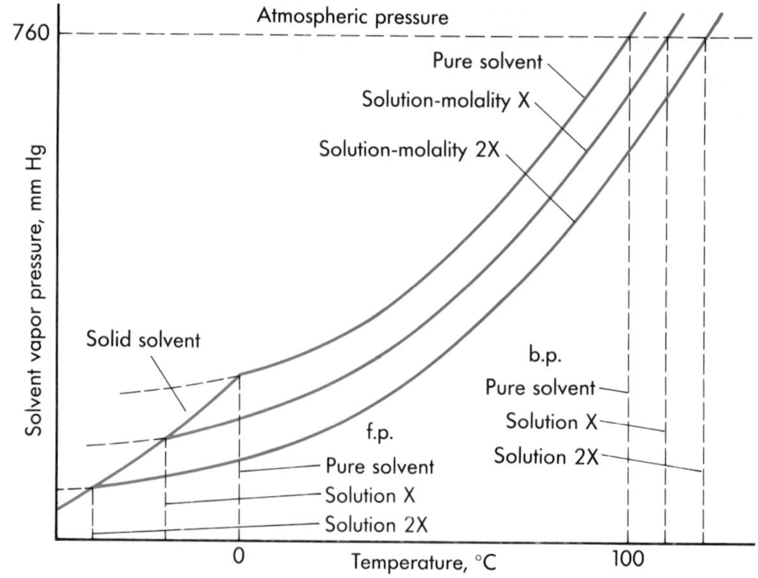

Figure 13-15. Vapor pressure of solvent as a function of temperature plotted for a pure solvent and dilute solutions of molalities X and 2X. Freezing-point depressions and boiling-point elevations are proportional to the molal concentrations of the solutions.

and lower temperatures than can the pure solvent. From this observation, what can you conclude about the boiling and freezing points of the solvent in dilute solutions?

Solutes depress the freezing points of solvents.

Salt water freezes at a lower temperature than fresh water. Sea water is a dilute solution of common salt, NaCl, and many other minerals. Suppose a sample of sea water is cooled enough for freezing to occur. The crystals produced are those of the *pure solvent* itself (in this case, water) and not of the solution. This phenomenon occurs when any dilute solution is cooled enough for solvent freezing to occur. *Solutes lower the freezing point of the solvent in which they are dissolved.* This fact is the basis for the common practice of adding ethylene glycol (permanent antifreeze) to the water in an automobile radiator.

Experiments with many dilute solutions of nonelectrolytes in water show that the *freezing-point depression* of the water is determined by the *number* of solute particles in a given quantity of water, not by the chemical identity of the particles. Dilute solutions of *equal molality*, but which contain different molecular solutes, yield the same freezing-point depression for water. Dilute solutions of different molality give different freezing-point depressions. These facts support the evidence in Figure 13-15 that *the freezing-point depression of the solvent is directly proportional to the molecular concentration of the solute.*

Properties, such as vapor pressure and freezing point of a solvent, that are influenced by the number but not the chemical identity of particles present are called *colligative properties. A* ***colligative property*** *of a system is a property that is determined by the number of particles present in the system but is independent of the properties of the particles themselves.*

Freezing-point depressions are observed for solvents containing volatile solutes as well as nonvolatile solutes.

Freezing-point depressions of solvents are represented by the symbol ΔT_f and are expressed in Celsius degrees, C°. The Greek letter Δ (delta) signifies "change in." Suppose the freezing point of a quantity of water is lowered from 0.00 °C to −0.36 °C by the addition of a small amount of some molecular solute. The freezing-point depression of the water is 0.36 C°.

$$\Delta T_f = 0.36\ \text{C}^\circ$$

The depression of the freezing point of water has been calculated for a 1-molal solution of any molecular solute in water. It has the constant value of 1.86 C°. *This freezing-point depression for a 1-molal water solution is called the* ***molal freezing-point constant, K_f,*** *for water.* It is 1.86 C°/molal. Since

$$\text{molality} = \frac{\text{moles solute}}{\text{kg solvent}}$$

K_f has the dimensions $\frac{\text{C}^\circ}{\text{mole solute/kg solvent}}$. The molal freezing-point constant for water is

$$K_f = \frac{1.86\ \text{C}^\circ}{\text{molal}} = \frac{1.86\ \text{C}^\circ}{\text{mole solute/kg H}_2\text{O}}$$

According to this value of K_f, a molecular solute added to 1 kg of water should lower the freezing point 1.86 C° per mole of solute dissolved. However, this relation holds experimentally only for dilute solutions. Even a 1-molal solution is concentrated enough so that the freezing-point depression is somewhat less than 1.86 C°. All solvents have their own characteristic molal freezing-point constants. The values of K_f for some common solvents are given in Table 13-5.

The depression of the freezing point of the solvent in a dilute solution of a molecular solute is *directly proportional* to the molal

Table 13-5
MOLAL FREEZING-POINT AND BOILING-POINT CONSTANTS

Solvent	*Normal f.p.* (°C)	*Molal f.p. constant, K_f* (C°/molal)	*Normal b.p.* (°C)	*Molal b.p. constant, K_b* (C°/molal)
acetic acid	16.6	3.90	118.5	3.07
acetone	−94.8	—	56.00	1.71
aniline	−6.1	5.87	184.4	3.22
benzene	5.48	5.12	80.15	2.53
carbon disulfide	−111.5	3.80	46.3	2.34
carbon tetrachloride	−22.96	—	76.50	5.03
ethanol	−114.5	—	78.26	1.22
ether	−116.3	1.79	34.42	2.02
naphthalene	80.2	6.9	218.0	5.65
phenol	40.9	7.27	181.8	3.56
water	0.00	1.86	100.0	0.51

The molal f.p. constant, K_f, for water is 1.86 C°.

concentration of the solution. The freezing-point depression, ΔT_f, *is equal to the product of the molal freezing-point constant,* K_f, *of the solvent and the molality, m, of the solution.*

$$\Delta T_f \propto m$$

$$\Delta T_f = K_f m$$

When K_f is expressed in C°/(mole solute/kg solvent) and m is expressed in moles solute/kg solvent, ΔT_f is the freezing-point depression in C°. Sample Problems 13-3 and 13-4 illustrate the use of this expression.

SAMPLE PROBLEM 13-3

A solution is prepared in which 17.1 g of sucrose, $C_{12}H_{22}O_{11}$, is dissolved in $20\bar{0}$ g of water. What is the freezing-point depression of the solvent?

SOLUTION

The gram-molecular weight of sucrose, $C_{12}H_{22}O_{11}$ = 342 g. Thus 342 g of sucrose = 1 mole. The mass, 17.1 g of sucrose, must be converted to moles. The mass, $20\bar{0}$ g of water, must be converted to kilograms. These conversions are accomplished by familiar unit-cancellation methods.

$$K_f \text{ for water} = \frac{1.86\ \text{C°}}{\text{molal}} = \frac{1.86\ \text{C°}}{\text{mole solute/kg } H_2O}$$

Solving for ΔT_f:

$$\Delta T_f = K_f m$$

$$\Delta T_f = \frac{1.86\ \text{C°}}{\text{mole } C_{12}H_{22}O_{11}/\text{kg } H_2O} \times \frac{17.1\ \text{g } C_{12}H_{22}O_{11}}{20\bar{0}\ \text{g } H_2O} \times \frac{\text{mole}}{342\ \text{g}} \times \frac{10^3\ \text{g}}{\text{kg}}$$

$$\Delta T_f = 0.465\ \text{C°}$$

Observe that the unit-cancellation operations yield the answer unit C°.

$$\Delta T_f = \frac{\text{C°}}{\cancel{\text{mole}}\ \cancel{C_{12}H_{22}O_{11}}/\cancel{\text{kg}}\ \cancel{H_2O}} \times \frac{\cancel{\text{g}}\ \cancel{C_{12}H_{22}O_{11}}}{\cancel{\text{g}}\ \cancel{H_2O}} \times \frac{\cancel{\text{mole}}}{\cancel{\text{g}}} \times \frac{\cancel{\text{g}}}{\cancel{\text{kg}}}$$

$$\Delta T_f = \text{C°}$$

SAMPLE PROBLEM 13-4

A water solution of a nonelectrolyte is found to have a freezing point of −0.23 °C. What is the molal concentration of the solution?

SOLUTION

The normal freezing point of water is 0.00 °C, so the freezing-point depression, ΔT_f, = 0.23 C°.

$$K_f \text{ (water)} = 1.86 \text{ C°/molal}$$

$$\Delta T_f = K_f m$$

Solving for *m*:

$$m = \frac{\Delta T_f}{K_f} = \frac{0.23 \text{ C°}}{1.86 \text{ C°/molal}} = 0.12 \text{ molal}$$

Practice Problems

1. A solution consists of 10.3 g of glucose, $C_6H_{12}O_6$, dissolved in 25$\bar{0}$ g of water. What is the freezing-point depression of the solvent? *ans.* 0.426 C°.
2. In a laboratory experiment the freezing point of a water solution of ethanol, C_2H_5OH, is found to be −0.325 °C. What is the molal concentration of this solution? *ans.* 0.175 *m*
3. A solution is prepared by dissolving 2.69 g of phenol, C_6H_5OH, in 20$\bar{0}$ g of water. *(a)* Determine the freezing-point depression of the solvent. *(b)* What is the freezing point of the solvent? *(c)* What is the molal concentration of the solution? *ans. (a)* 0.266 C°, *(b)* −0.266 °C, *(c)* 0.143 *m*

13.13 Boiling-point elevation of solvents The boiling point of a liquid is defined in Section 12.6 as the temperature at which the vapor pressure of the liquid is equal to the prevailing atmospheric pressure. A change in either atmospheric pressure or vapor pressure of the liquid will cause a corresponding change in its boiling point. If a nonelectrolyte is dissolved in the liquid, the vapor pressure of the liquid (now the solvent of the solution) is lowered, provided the solute is a nonvolatile substance. From Figure 13-15 it is evident that the boiling point of the solvent is higher than that of the pure liquid alone. Experiments with dilute solutions of nonvolatile solutes have shown *that the boiling-point elevation, ΔT_b, of the solvent is directly proportional to the molecular concentration of the solute.*

Nonvolatile solutes elevate the boiling points of solvents.

When the boiling-point elevation of a 1-molal water solution of a molecular solute is calculated from data derived from dilute solutions at standard pressure, its constant value is 0.51 C°. That is, the 1-*m* solution should boil at 100.51 °C at 760 mm pressure. *This boiling-point elevation of 0.51 C° for a 1-molal water solution is the **molal boiling-point constant, K_b,** for water.* It is 0.51 C°/molal.

Alcohol and ammonia are examples of volatile solutes in water solutions.

Thus the molal boiling-point constant for water is

$$K_b = \frac{0.51\ C°}{\text{molal}} = \frac{0.51\ C°}{\text{mole solute/kg } H_2O}$$

Sugar is a nonvolatile solute in water solution.

The actual elevation of the boiling point in concentrated solutions deviates somewhat from that indicated by K_b. Thus its validity as a proportionality constant is limited to dilute solutions. All solvents have their own characteristic molal boiling-point constants. The values of K_b for some common solvents are given in Table 13-5.

Since the elevation of the boiling point of the solvent in a dilute solution of a nonvolatile molecular solute is *directly proportional* to the molal concentration of the solution, the boiling-point elevation, ΔT_b, *is equal to the product of the molal boiling-point constant,* K_b, *of the solvent times the molality,* m, *of the solution.*

$$\Delta T_b \propto m$$

$$\Delta T_b = K_b m$$

The molal b.p. constant, K_b, for water is 0.51 C°.

When K_b is expressed in C°/(mole solute/kg solvent) and m is expressed in moles solute/kg solvent, T_b is the boiling-point elevation in C°.

The freezing points and boiling points of solutions of electrolytes are also depressed and elevated. However, they do not follow the simple relationships just described. Such solutes are not molecular, and the generalizations about molecular substances do not hold for them. The behavior of electrolytes in water solution is the subject of Chapter 14.

13.14 Molecular weights of solutes The application of the Avogadro principle discussed in Section 11.6 provides a way of determining the molecular weights of gases and volatile liquids. However, molecular weights of substances that decompose instead of vaporizing when heated cannot be determined by this molar-volume method.

The extent to which the freezing point and boiling point of a solution of a molecular solute differ from those of the pure solvent is related to the number of solute molecules in solution and not to their identity. Consequently, the experimental determination of the freezing-point depression or the boiling-point elevation of a solvent enables the molal concentration of that solution to be calculated.

This colligative property of solutions provides a method of determining the molecular weight of a substance that is soluble in water or some other solvent, and that does not react with its solvent. If the empirical formula of the solute has been determined from data, the molecular formula can also be expressed.

The molal freezing-point constants, K_f, and molal boiling-point constants, K_b, are known for many common solvents. If the concentration of a molecular solution in terms of the mass of solute and mass of solvent (for which K_f or K_b is known) is known, the freezing-point depression or boiling-point elevation can be determined. The molecular weight can then be calculated. The freezing-point method is favored in these molecular-weight determinations.

From Section 13.12, recall that

$$\Delta T_f = K_f m$$

$$\textbf{Molality } m = \frac{\textbf{moles solute}}{\textbf{kg solvent}} = \frac{\textbf{g solute/g-mol wt}}{\textbf{kg solvent}}$$

Thus,

$$\Delta T_f = K_f \times \frac{\textbf{g solute/g-mol wt}}{\textbf{kg solvent}}$$

Solving for gram-molecular weight:

$$\textbf{g-mol wt} = \frac{K_f \times \textbf{g solute}}{T_f \times \textbf{kg solvent}}$$

Sample Problem 13-5 illustrates this method of determining the molecular weight of a solute.

SAMPLE PROBLEM 13-5

It is found experimentally that 1.8 g of sulfur dissolved in $10\bar{0}$ g of naphthalene, $C_{10}H_8$, decreases the freezing point of the solvent 0.48 C°. What is the molecular weight of the solute?

SOLUTION

The molal freezing-point constant for naphthalene is 6.9 C°/molal (Table 13-5). Molality is expressed in terms of kilograms of solvent. Therefore, the quantity of naphthalene used must be converted from grams to kilograms, using the factor 10^3 g/kg.

$$\Delta T_f = K_f m$$

$$\textbf{where } m = \frac{\textbf{moles solute}}{\textbf{kg solvent}} = \frac{\textbf{g solute/g-mol wt}}{\textbf{kg solvent}}$$

$$\Delta T_f = K_f \times \frac{\textbf{g solute/g-mol wt}}{\textbf{kg solvent}}$$

$$\textbf{g-mol wt} = \frac{K_f \times \textbf{g solute}}{\Delta T_f \times \textbf{kg solvent}}$$

$$\text{g-mol wt} = \frac{6.9\ \cancel{\text{C}^\circ} \times 1.8\ \text{g}\ \cancel{\text{S}}}{\text{mole}\ \cancel{\text{S}}/\cancel{\text{kg}\ \text{C}_{10}\text{H}_8} \times 0.48\ \cancel{\text{C}^\circ} \times 10\bar{0}\ \cancel{\text{g}\ \text{C}_{10}\text{H}_8} \times \cancel{\text{kg}}/10^3\ \cancel{\text{g}}}$$

$$\text{g-mol wt} = 260\ \text{g/mole}$$

$$\text{mol wt} = 260$$

What does this result suggest about the composition of the sulfur molecule?

Practice Problems

1. When 1.56 g of methanol is dissolved in $20\bar{0}$ g of water, the freezing-point depression of the solvent is 0.453 C°. Determine the molecular weight of the solute. *ans.* 32.0
2. If 1.84 g of a molecular solute is dissolved in $15\bar{0}$ g of water, the freezing point of the solvent is −0.248 °C. What is the molecular weight of the solute? *ans.* 92.0

SUMMARY

A solution is a homogeneous mixture made up of two parts, a solute that dissolves and a solvent in which the solute dissolves. Solutes whose water solutions conduct electricity are called electrolytes. Solutes whose water solutions do not conduct are called nonelectrolytes. Solutions are single-phase systems.

Substances whose particles are too large to form solutions but small enough to form permanent suspensions are called colloids. A colloidal suspension is a two-phase system. In colloids, dispersed particles correspond to the solute of a solution, and a dispersing medium corresponds to the solvent of a solution.

The possibility of solubility is increased when substances are alike in composition and structure. The solubility of a solute is increased by hydrogen bonds between solute and solvent particles.

When a solution is in equilibrium, the opposing processes of dissolving and crystallizing occur at equal rates. Solution equilibrium limits the quantity of solute that can dissolve in a given quantity of solvent. Equilibrium is affected by temperature. The solubility of a solute is determined by the equilibrium concentration of the solute.

The solubility of a gas in a liquid is affected by pressure in accordance with Henry's law. Gases that react chemically with their liquid solvents are generally more soluble than those that do not. The solubility of a gas decreases as temperature increases. The solubility of most, but not all, solids in liquids increases with a rising temperature.

When the dissolving process is endothermic, the solution cools as dissolving proceeds, and the heat of solution is positive. When the dissolving process is exothermic, the solution warms, and the heat of solution is negative. Heats of solution are expressed in kilocalories per mole of solute dissolved in 200 moles of water.

The concentration of a solution can be expressed in molality as moles of solute per kilogram of solvent. Nonvolatile molecular solutes lower the freezing points and raise the boiling points of their solvents by characteristic amounts determined by the kinds of solvents and the solution concentrations. Each solvent has a specific molal freezing-point depression and molal boiling-point elevation. Specific molal freezing-point depression and molal boiling-point elevation are used to determine molecular weights of soluble substances that cannot be vaporized.

VOCABULARY

adsorption
alloy
boiling-point elevation
colligative property
colloid
colloidal suspension
concentrated
concentration
condensation
crystalloid
dilute
dispersed particles
dispersing medium
dissolve
effervescence
electrolyte
endothermic process
entropy
exothermic process
freezing-point depression
heat of solution
Henry's law
heterogeneous
homogeneous
hydration
hydrogen bond
immiscible
miscible
molal boiling-point constant
molal freezing-point constant
molality
nonelectrolyte
saturated
solubility
soluble
solute
solution
solution equilibrium
solvation
solvent
supercooled
supersaturated
suspension
unsaturated

QUESTIONS

Group A

1. Distinguish between a *solution* and a *colloidal suspension*. *A1, A3a, A4b*
2. Identify and define the components of a solution. *A1, A3b*
3. *(a)* What different types of solutions are possible? *(b)* Which of these solutions is the most common? *A4a*
4. Why are the terms *dilute* and *concentrated* not entirely satisfactory for expressing the concentration of solutions? *A1*
5. *(a)* Define *effervescence*. *(b)* Why does carbonated water effervesce when drawn from a soda fountain or when a bottle of "soda" water is opened? *A1, A2f*
6. What limits the quantity of a solute that can remain dissolved in a given quantity of solvent under fixed conditions? *A2c*
7. Describe the influence of pressure on the solubility of: *(a)* a solid in a liquid; *(b)* a gas in a liquid. *A2d, A2f*
8. Describe the influence of temperature on the solubility of: *(a)* a solid in a liquid; *(b)* a gas in a liquid. *A2g*
9. Distinguish between the terms *dissolve* and *melt*. *A1*
10. Two liquids are described as being *miscible*. *(a)* What information is conveyed by this statement? *(b)* Give an example of completely miscible liquids. *(c)* Give an example of slightly miscible liquids. *A1, A4d, A4f*
11. Describe the distinguishing characteristics of *polar* molecules. *A1, A2b*
12. A solution is said to be *saturated*. *(a)* What kind of visible evidence could indicate saturation? *(b)* Describe the *saturated solution* in terms of solution equilibrium. *A2c*
13. What methods can be used to increase the *rate* of solution of a solid in a liquid? *A2k*
14. Referring to Figure 13-11, determine the solubility (in g solute/100 g H_2O) of: *(a)* sodium chloride at 0 °C and 100 °C; *(b)* sodium nitrate at 10 °C and 75 °C; *(c)* cerium sulfate at 0 °C and 50 °C; *(d)* potassium nitrate at 10 °C and 60 °C. *A2i*
15. Suggest a procedure for preparing a supersaturated solution of aqueous potassium chloride. *A2h, A4g*
16. Ethanol (ethyl alcohol) is completely miscible with both water and ether. Explain the miscibility. *A2a, A4f*

QUESTIONS

GROUP B

17. A pond of water in a forest is at constant ambient temperature and its surface is in contact with the atmosphere above the pond. *(a)* What determines the quantity of oxygen dissolved in the pond water? *(b)* What occurs if the oxygen content of the air is seriously depleted during a forest fire? *(c)* Appraise the effect on aquatic animals living in the pond. *A2d, A2f, A2g*
18. Explain why ice cubes made with cold water may be cloudy whereas ice cubes made with hot water may be clear. *A2g*
19. Alcohol is a nonelectrolyte and is completely soluble in water. However, a molal solution of alcohol in water does not yield the molal boiling-point elevation of water. Explain. *A2p*
20. A saturated solution of copper(II) sulfate in water is required. The solute available is the crystalline hydrate $CuSO_4 \cdot 5H_2O$. *(a)* In what three ways can the dissolving process to saturation be hastened? *(b)* What final evidence indicates that the solution is saturated at room temperature? *A2g, A2h, A3d*
21. A saturated solution of calcium hydroxide in water is required. *(a)* To insure that the solution is saturated at room temperature, is it advisable to bring the solution to saturation with a solvent temperature initially *above* or *below* room temperature? *(b)* Justify your answer given in *(a)*. *A2g, A2h, A3d*
22. The carbon tetrachloride molecule, CCl_4, is nonpolar. However, each of its four single covalent bonds is polar. Refer to the space model of the CCl_4 molecule in Figure 13-5 and suggest a reason for the nonpolar character of the molecule. *A1, A2a, A3c*
23. Given: an aqueous solution of sodium nitrate which has a heat of solution (Table 13-3) of +5.03 kcal/mole. *(a)* Of the two solution processes, *dissolving* and *crystallizing*, which is *endothermic* and which is *exothermic? (b)* As dissolving proceeds and solution concentration increases, what change occurs in the solution temperature? *(c)* When the rates for dissolved solute to crystallize and crystals to dissolve are equal, what two solution conditions prevail? *(d)* The solution is now heated and its temperature rises. Apply the Le Chatelier principle and determine which rate, dissolving or crystallizing, is increased by the rise in temperature. *(e)* How is the solubility of a solute with a positive heat of solution ordinarily influenced by a rise in temperature of its solution? *A2c, A2g, A2n*
24. Given: an aqueous solution of lithium chloride which has a heat of solution (Table 13-3) of −8.37 kcal/mole. *(a)* Of the two solution processes, *dissolving* and *crystallizing*, which is *endothermic* and which is *exothermic? (b)* As dissolving proceeds and solution concentration increases, what change occurs in the solution temperature? *(c)* When the rates for dissolved solute to crystallize and crystals to dissolve are equal, what two solution conditions prevail? *(d)* The solution is now heated and its temperature rises. Apply the Le Chatelier principle and determine which rate, dissolving or crystallizing, is increased by the rise in temperature. *(e)* How is the solubility of a solute with a negative heat of solution influenced by a rise in temperature of its solution? *A2c, A2g, A2n*
25. Liquid methanol, CH_3OH, and water are miscible in all proportions. When 1 mole of CH_3OH (solute) is mixed with 10 moles of H_2O (solvent), the heat of solution is found to be −1.43 kcal. *(a)* Is the formation of solution accompanied by an increase or decrease in entropy? (State the argument upon which your answer is based.) *(b)* Does the change in entropy favor the separate components or the solution? *(c)* Is the dissolving process endothermic or exothermic? Justify your answer. *(d)* Does the energy change as indicated by the sign of the heat of solution favor the separate components or the solution? *(e)* Are the previous answers consistent with the fact that methanol and water are freely miscible? Explain. *A2n, A4d*

PROBLEMS

Group A

1. What mass in grams of sucrose, $C_{12}H_{22}O_{11}$, must be dissolved in $200\bar{0}$ g of water to yield a 0.100-*m* solution? *A20, B1*
2. What quantity of ethanol, C_2H_5OH, is required to prepare a 0.225-*m* solution in $25\bar{0}$ g of water? *A20, B1*
3. Determine the molality of a solution containing 42.0 g of glycerol, $C_3H_5(OH)_3$, in $75\bar{0}$ g of water. *A20, B1*
4. A solution of glucose, $C_6H_{12}O_6$, is prepared by dissolving 8.50 g of the glucose in $40\bar{0}$ g of water. What is the molality? *A20, B1*
5. A solution contains 85.0 g of methanol, CH_3OH, in $300\bar{0}$ g of water. Calculate the molality of the solution. *A20, B1*
6. Determine the molality of a solution containing 1.50 g of I_2 (solute) in $45\bar{0}$ g of CCl_4 (solvent). *A20, B1*
7. How many grams of water must be added to 45.0 g of glucose, $C_6H_{12}O_6$, to form a 0.250-*m* solution? *A20, B1*
8. A 0.400-*m* solution of naphthalene, $C_{10}H_8$, in benzene, C_6H_6, is required. If 30.0 g of naphthalene is available, how many grams of benzene must be used? *A20, B1*
9. What is the molality of a solution that contains 25.0 g of ethylene glycol, $C_2H_4(OH)_2$, in $10\bar{0}$ g of water? *A20, B1*
10. What is the freezing point of 250.0 g of water containing 12.5 g of a nonelectrolyte that has a molecular weight of 180? *B2*
11. A solution consists of 15 g sucrose, $C_{12}H_{22}O_{11}$, in $25\bar{0}$ g of water. What is the freezing point of the water? *B2*
12. What is the boiling point of the solution described in Problem 11? *B2*
13. A solution of iodine in benzene is found to have a freezing point of 4.1 °C. What is the molality of the solution? *A20, B1, B2*
14. A sucrose-in-water solution raises the boiling point of the solvent to 100.13 °C at standard pressure. Determine the molality of the solution. *A20, B1, B2*

PROBLEMS

Group B

15. A compound contains: carbon, 40.00%; hydrogen, 6.67%; oxygen, 53.33%. Tests show that 18.0 g of the compound dissolved in 1.00 kg of water raises the boiling point of the water 0.051 C°. *(a)* Determine the empirical formula. *(b)* Determine its molecular weight. *(c)* What is the molecular formula? *B2, B3*
16. The analysis of a compound yields: carbon, 32.0%; hydrogen, 4.0%; oxygen, 64.0%. It is found that 12.0 g of the compound added to $80\bar{0}$ g of water lowers the freezing point of the water 0.186 C°. *(a)* Determine the empirical formula. *(b)* Determine the molecular weight. *(c)* What is the molecular formula? *B2, B3*
17. By analysis, a compound consists of: carbon, 30.3%; hydrogen, 1.7%; bromine, 68%. The substance is soluble in benzene and 15.0 g of it lowers the freezing point of 150.0 g of benzene 2.1 C°. *(a)* What is the empirical formula of the solute? *(b)* Determine its molecular weight. *(c)* What is the molecular formula? *B2, B3*
18. Calculate the freezing point of an aqueous solution that boils at 100.46 °C at SP. *B1, B2, B3*

Svante August Arrhenius (Swedish) received the 1903 Nobel Prize in chemistry for his theory of electrolytic dissociation. He found that an electric current is conducted between two metal bars placed in water containing an electrolyte or in a substance whose molecules dissociate into charged particles called ions. These ions carry the electric current. As the concentration of the electrolyte changes, so does the conductivity. Arrhenius had many interests and later developed a fundamental theory of chemical reactions and a theory of cosmic physics.

Ionization

GOALS

In this chapter you will gain an understanding of: • the conductivity of solutions in terms of electrolytes and nonelectrolytes • the development of the modern theory of ionization • the dissociation process that occurs when ionic compounds dissolve in water • writing balanced ionic and net ionic equations • the ionization process that occurs when polar covalent compounds dissolve in water • the hydronium ion and how it is formed • the distinction between strong and weak electrolytes in terms of ionization • the effect of electrolytes on freezing and boiling points of solvents

14.1 Conductivity of solutions In Section 13.1 water-soluble substances are distinguished as either *electrolytes* or *nonelectrolytes,* depending on the electric conductivity of their aqueous solutions. Electrolytes form ionic solutions that conduct an electric current. Nonelectrolytes are covalent substances that form molecular solutions that do not conduct an electric current. Exceptions are covalent acids whose aqueous solutions are ionic conductors.

For an aqueous solution to be conductive, its solute must be capable of transporting an electric charge. The neutral molecules of a covalent solute are not charge carriers, and the solution does not conduct. The ions of an ionic solute are charge carriers, and the magnitude of current the solution conducts is related to the solute ion concentration.

The conductivity of a solution can be tested as shown in Figure 14-1. A lamp is connected in the test circuit with an ammeter, a battery (the source of current), a switch, and a pair of electrodes. The electrodes are conductors that make electric contact with the test solution. Observe that the lamp filament is connected in series with the test solution. For a current to be in the lamp filament (and in the meter), the test solution must provide a conducting path between the two electrodes. A nonconducting solution is, in effect, an open switch between the electrodes; thus there can be no current in the circuit.

If the liquid tested is a *good conductor* of electric current, the lamp filament glows brightly when the switch is closed and the meter registers a substantial current in the circuit. For a liquid

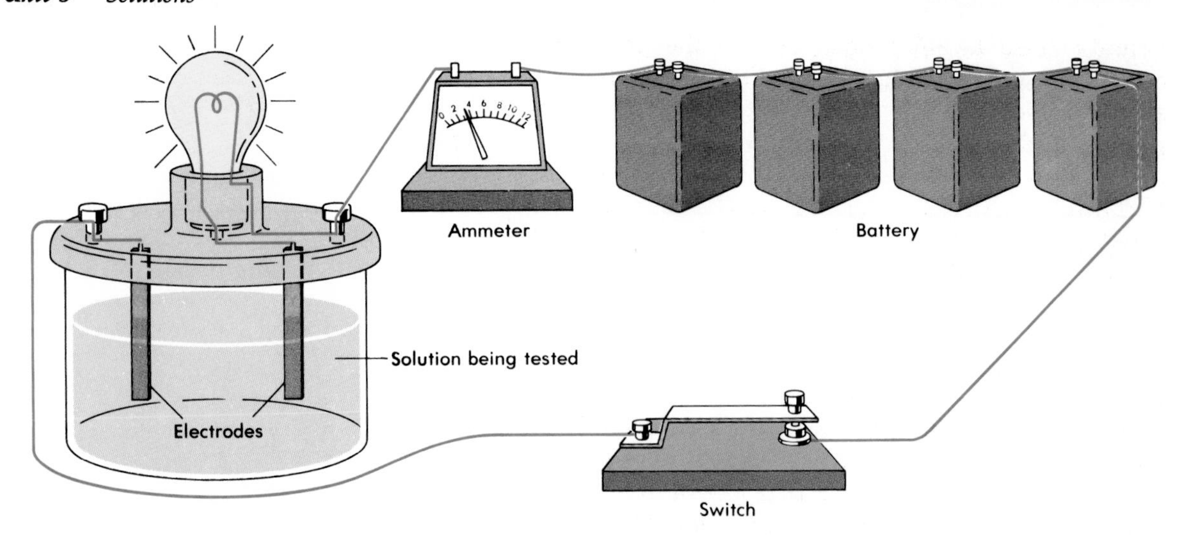

Figure 14-1. An apparatus used when testing the conductivity of solutions.

The term "pure water" *refers to water that contains no other kind of matter.*

Oxidation: loss of electrons. Reduction: gain of electrons.

that is only a *moderate conductor,* the lamp is not as bright and the meter registers a smaller current. If a liquid is a *poor conductor,* the lamp does not glow at all, and the meter registers only a feeble current.

When pure water is tested, the lamp does not glow, and the meter (depending on its sensitivity) may not register a current. Pure water is such a poor conductor that, except in special situations, it may be regarded as a *nonconductor.* Similarly, water solutions of such covalent substances as sugar, alcohol, and glycerol (glycerin) are nonconductors of electricity. These solutes are *nonelectrolytes.*

In general, all solutes whose water solutions conduct an electric current are *electrolytes.* Solutions of electrovalent substances such as sodium chloride, copper(II) sulfate, and potassium nitrate are conductors. These solutes are electrolytes. Hydrogen chloride is one of several covalent compounds whose water solution conducts an electric current. Therefore, these compounds are also electrolytes.

For an electric current to be conducted through a solution, there must be charged particles (ions) in the solution. These ions must be free to move or migrate through the solution. Chemical reactions by which electric charges enter and leave the solution environment must take place at the two electrodes. Consequently, an oxidation reaction must occur at one electrode, and a reduction reaction must occur simultaneously at the other electrode. (Review Section 6.5.) These electrode reactions will be treated in greater detail in Chapter 22.

Ionic compounds can act as conductors of electricity in another way. Any effect that overcomes the attraction between the ions allows them the mobility to conduct an electric current. In Section 14.5 you will see how water as a solvent provides ion mobility. Heating produces the same effect. If an ionic compound is heated until it melts, or *fuses*, the ions become mobile and can conduct an electric current through the melted substance. Some solid ionic compounds, such as silver nitrate and potassium chlorate, melt at fairly low temperatures. The electric conductivity of such fused salts can be demonstrated easily in the laboratory. Other ionic compounds, such as sodium chloride and potassium fluoride, must be heated to relatively high temperatures before they melt. When melted, they too conduct an electric current.

14.2 Electrolytes as solutes The molal freezing-point constant for water, 1.86 C°, is related to the influence of nonelectrolytes in water solution (Section 13.12). *Solutes that are electrolytes have a greater influence on the freezing point of their solvents than solutes that are not electroytes.* A 0.1-*m* solution of sodium chloride in water lowers the freezing point *nearly twice* as much as a 0.1-*m* solution of sugar. A 0.1-*m* solution of potassium sulfate or calcium chloride lowers the freezing point *nearly three times* as much as a 0.1-*m* solution of sugar. *Electrolytes in water solutions lower the freezing point nearly two, or three, or more times as much as nonelectrolytes in water solutions of the same molality.*

The molal boiling-point constant for water is 0.51 C°. Nonvolatile electrolytes in solution have a greater effect on the boiling point of the solvent than do nonelectrolytes. Dilute sodium chloride solutions have boiling-point elevations *almost twice* those of sugar solutions of equal molality. A 0.1-*m* solution of potassium sulfate shows *almost three times* the rise in boiling point as a 0.1-*m* solution of sugar. *Nonvolatile electrolytes in water solutions raise the boiling point nearly two, or three, or more times as much as nonelectrolytes in water solutions of the same molality.* The observations that freezing-point depressions and boiling-point elevations of electrolytes in aqueous solutions are not exactly two, three, or more times those of nonelectrolytes of the same molalities lead to the development of an important theory of interionic attractions in ionic solutions. This concept is discussed briefly in Section 14.10.

14.3 Behavior of electrolytes explained Michael Faraday first used the terms *electrolyte* and *nonelectrolyte* in describing his experiments on the conductivity of solutions. He assumed that ions were formed from molecules in solution by the electric potential difference (voltage) between the electrodes. He concluded that the formation of these ions was what enabled the solution to conduct an electric current. Later experiments showed that electrolytic solutions contained ions regardless of the presence of the charged electrodes.

In 1887, the Swedish chemist Svante Arrhenius (1859–1927) published a report on his study of solutions of electrolytes. In this report (written in 1883), he introduced the original *theory of*

Figure 14-2. Michael Faraday first used the terms *electrolyte* and *nonelectrolyte* to describe his experiments on the conductivity of solutions.

ionization. The assumption Arrhenius began with regarding the origin of ions in electrolytic solutions was different from Faraday's. Arrhenius assumed that charged electrodes were not necessary for ionization and he described many properties of electrolytes to show that this assumption was correct. He concluded that ions were produced by the *ionization* of molecules of electrolytes in water solution. He considered the ions to be electrically charged. When molecules ionized, they produced both positive ions and negative ions. The solution as a whole contained equal numbers of positive and negative charges. He also considered the ionization to be complete only in very dilute solutions. In more concentrated solutions, the ions were in equilibrium with *un-ionized* molecules of the solute.

These assumptions formed the basis of the theory of solutions of electrolytes. As chemists gained a better understanding of crystals and water molecules, some of the original concepts were modified or replaced. It is a great tribute to Arrhenius that his original theory served for so long. Present knowledge of the crystalline structure of ionic compounds was not available to him when, at the age of 24, he wrote his thesis on ionization.

According to the modern theory of ionization, the solvent plays an important part in the solution process. Water is by far the most important solvent. Knowledge of the polar nature of water molecules helps in understanding the solution process. The theory of ionization assumes

1. *that electrolytes in solution exist in the form of ions;*
2. *that an ion is an atom or a group of atoms that carries an electric charge;*
3. *that in the water solution of an electrolyte, the total positive ionic charge equals the total negative ionic charge.*

A review of Section 6.3 will be helpful at this point.

14.4 Structure of ionic compounds Ionic compounds result from the transfer of electrons from one kind of atom to another. Thus these compounds are not composed of neutral atoms, but consist of atoms that have lost or gained electrons. Atoms that *gain* electrons in forming the compound become ions with a *negative charge.* Those that *lose* electrons become ions with a *positive charge.* Atoms lose electric neutrality by forming ions and gain chemical stability by associating with other ions of opposite charge.

The properties of ions are very different from those of the atoms from which they are formed. Such differences accompany changes in electronic structure and the acquisition of charge resulting from the formation of ions. For example, a neutral sodium atom with a single 3*s* electron in the third energy level is different from a sodium ion. The sodium ion does not have the 3*s* electron and thus has one excess positive charge. There is an octet in the second energy level consisting of two 2*s* and six 2*p* electrons.

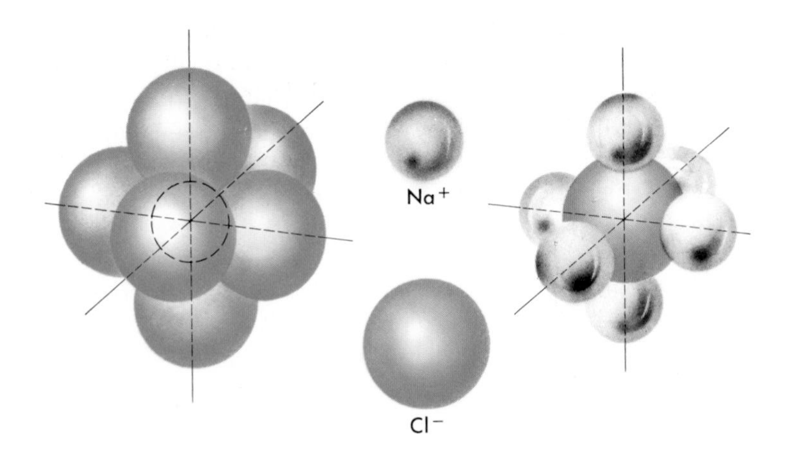

Figure 14-3. The packing of Na^+ and Cl^- ions in the NaCl crystal. In the model on the left, the six Cl^- ions are clustered about one Na^+ ion that cannot be seen here, but whose location is indicated by the dashed outline.

It is useful to remember that chemical properties are determined chiefly by the highest-energy-level electron configuration of an atom or an ion. If the outer electron structure changes, the properties change. The loss of the 3*s* electron gives sodium the stable electron configuration of neon. *The charge of a monatomic ion and its oxidation number are the same.* In fact, the charge of the ion determines its oxidation state.

Ionic compounds usually exist as crystals consisting of positive ions and negative ions arranged in close-packed, orderly structures. These structural configurations are determined by X-ray analysis.

The packing of Na^+ ions and Cl^- ions in the NaCl crystal is shown in Figure 14-3. Each Na^+ ion is surrounded by six Cl^- ions. Similarly, each Cl^- ion is surrounded by six Na^+ ions. The unit cell of the crystal lattice, Figure 14-4, reveals a face-centered cubic structure. Recall from Section 12.13 that the symmetry throughout a crystalline substance is defined by the unit cell that describes its lattice structure. Magnified crystal fragments of common table salt (sodium chloride) are shown in Figure 14-5. Observe the cubic configuration of the salt granules.

Numerous ionic compounds consisting of positive and negative ions in a 1 : 1 ratio have the face-centered cubic lattice configuration of sodium chloride. For example, most of the Group I metal halides have the sodium chloride lattice structure. All ionic compounds have characteristic lattice structures that are determined by the relative size and charge of their ions.

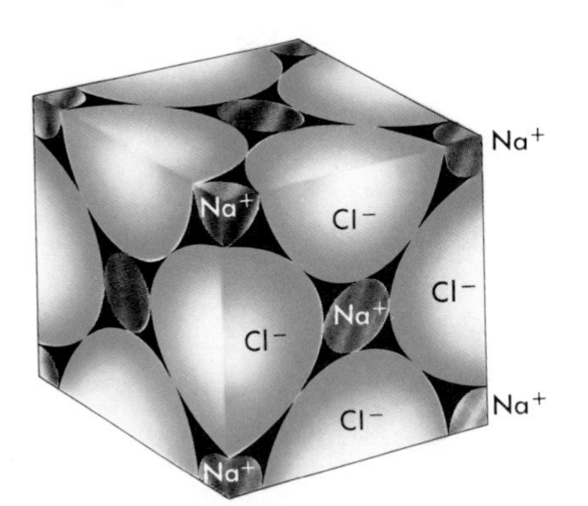

Figure 14-4. The sodium chloride unit cell has the face-centered cubic lattice configuration of positive and negative ions that characterize the crystals.

14.5 Hydration of ions Suppose a few crystals of sodium chloride are dropped into a beaker of water. The water dipoles exert an attractive force on the ions forming the surfaces of the crystals. The negative oxygen ends of several water dipoles exert an attractive force on a positive sodium ion. Similarly, the positive hydrogen ends of other water dipoles exert an attractive force on a negative chloride ion. These forces weaken the bond

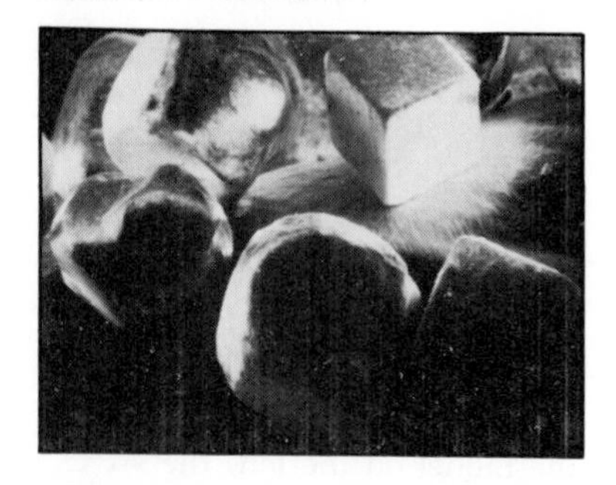

Figure 14-5. Crystals of common table salt (sodium chloride) magnified approximately 320 times.

by which the sodium and chloride ions are held together in the crystal lattice. The sodium and chloride ions then break away from the crystal lattice. They diffuse throughout the solution, loosely bonded to the solvent water molecules. See Figure 14-6.

Other sodium and chloride ions are similarly attracted by solvent molecules and diffuse in the solution. In this way, the salt crystal is gradually dissolved, and the ions are dispersed throughout the solution. *The separation of ions from the crystals of ionic compounds during the solution process is called* ***dissociation.*** The dissociation of sodium chloride crystals in water can be represented by use of an ionic equation:

$$Na^+Cl^- \text{ (solid)} \rightarrow Na^+ \text{ (in water)} + Cl^- \text{ (in water)}$$

Sodium chloride is said to *dissociate* when it is dissolved in water.

Solutions in water are commonly referred to as "aqueous solutions." In this sense, the symbol (aq) means "in water." This symbol, (aq), is used in a chemical equation to signify "in aqueous solution" just as the familiar symbols (s), (l), and (g) are used to signify "solid," "liquid," and "gas." Thus, the dissociation of an ionic compound in water solution is written as

$$Na^+Cl^-(s) \rightarrow Na^+(aq) + Cl^-(aq)$$

The number of water dipoles that attach themselves to the ions of the crystal depends largely upon the size and charge of the ion. *This attachment of water molecules to ions of the solute is called* ***hydration.*** The ions are said to be *hydrated.* The degree of hydration of these ions is somewhat indefinite. The water molecules are interchanged continuously from ion to ion and from ion to solvent.

Figure 14-6. When sodium chloride crystals are dissolved in water, the polar water molecules exert attracting forces that weaken the ionic bonds of the crystal. The + end of H_2O dipoles attracts Cl^- ions, and the − end of other H_2O dipoles attracts the Na^+ ions. The process of dissolving occurs as the sodium and chloride ions become hydrated.

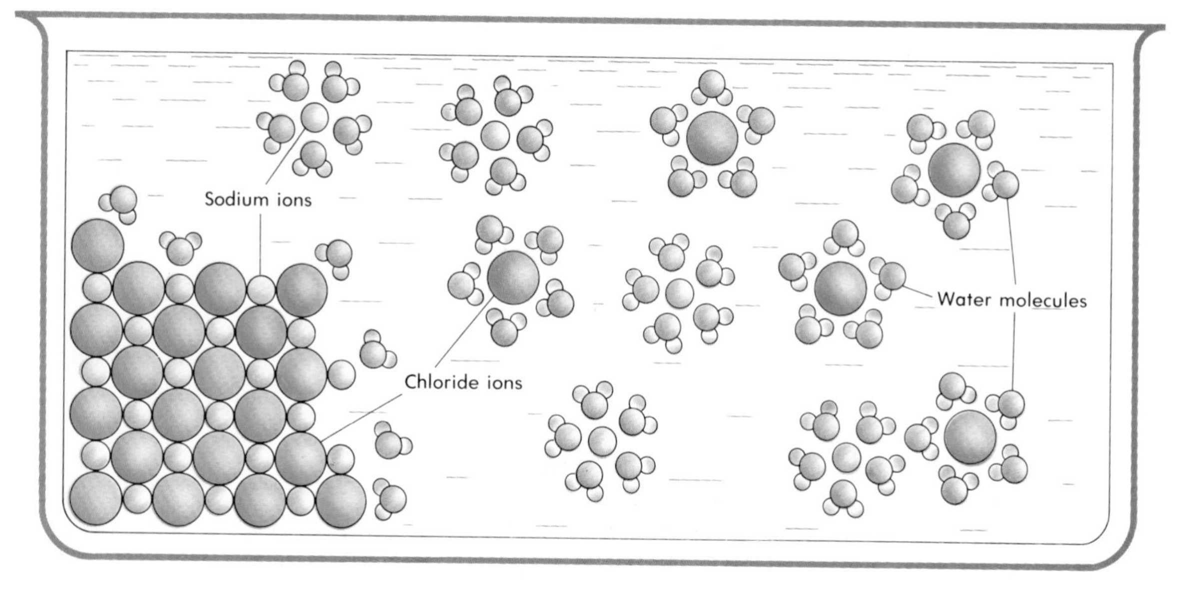

In certain cases, the water dipoles are not involved in reforming the crystal structure during the evaporation of the solvent. This situation occurs with sodium chloride, whose crystals do not contain water of crystallization. On the other hand, some ions retain a characteristic number of water molecules in reforming the crystal lattice of their salt in the hydrated form. An example is crystalline copper(II) sulfate, $CuSO_4 \cdot 5H_2O$. Four of the five water molecules are associated with the Cu^{++} ion. The fifth water molecule is associated with the SO_4^{--} ion.

Extensive hydration of the solute ions ties up a large portion of the solvent molecules. This reduces the number of *free* water molecules in the spaces separating hydrated ions of opposite charge. Attraction between ions then becomes stronger, and the crystal begins to form again. Eventually, the tendency for ions to re-form the crystal lattice reaches an *equilibrium* with the tendency for ions to be hydrated. At this point the practical limit of solubility is reached. The solution is saturated under existing conditions.

$$\mathbf{Na^+Cl^-(s) \rightleftarrows Na^+(aq) + Cl^-(aq)}$$

Here the tendency toward minimum energy (crystallizing) equals the tendency toward maximum entropy (dissolving).

Many ionic compounds that exist as crystalline solids are very soluble in water. They dissolve and produce solutions with high concentrations of hydrated ions. Typical examples are:

$$\mathbf{Ca^{++}Cl^-{}_2(s) \rightarrow Ca^{++}(aq) + 2Cl^-(aq)}$$
$$\mathbf{K^+Cl^-(s) \rightarrow K^+(aq) + Cl^-(aq)}$$
$$\mathbf{K^+ClO_3^-(s) \rightarrow K^+(aq) + ClO_3^-(aq)}$$
$$\mathbf{Ag^+NO_3^-(s) \rightarrow Ag^+(aq) + NO_3^-(aq)}$$

When a formula of the type $CaCl_2$ is written ionically, the form $Ca^{++}Cl^-{}_2$ is used. Observe that the subscript $_2$ is spaced out to mean two Cl^- ions, as in $Ca^{++}(Cl^-)_2$.

Even ionic compounds of very slight solubility in water show measurable dissociation tendencies. Low concentrations of ions are present in water solutions of slightly soluble ionic compounds. Silver chloride, AgCl, is so slightly soluble that it is ordinarily described as being insoluble. However, such "insoluble" compounds precipitate from solutions saturated with respect to their ions, regardless of how low their concentration may be. The dissociation equation for AgCl in aqueous solution is written as

Solubility of KCl in water at 20 °C is 4.5 moles/liter.

$$\mathbf{Ag^+Cl^-(s) \rightarrow Ag^+(aq) + Cl^-(aq)}$$

Both KCl and $AgNO_3$ have been described as very soluble ionic compounds. Their aqueous solutions contain hydrated ions which can be present in very high concentrations. These hydrated particles are K^+(aq) and Cl^-(aq) ions for the KCl solute. They are Ag^+(aq) and NO_3^-(aq) ions for the $AgNO_3$ solute. On the other hand, AgCl is only very slightly soluble in water. Its solubility at 20 °C is 1×10^{-5} mole/liter. This number means that

Solubility of $AgNO_3$ in water at 20 °C is 13 moles/liter.

only very low concentrations of both Ag^+ ions and Cl^- ions can be present in the same water solution.

Suppose fairly concentrated solutions of KCl and $AgNO_3$ are mixed. In a single-solution environment there are the four ionic species: $K^+(aq)$, $Cl^-(aq)$, $Ag^+(aq)$, and $NO_3^-(aq)$. However, the concentrations of Ag^+ and Cl^- ions greatly exceed the solubility of AgCl. Excess Ag^+ and Cl^- ions separate from the solution as a *precipitate* of solid AgCl. *The separation of a solid from a solution is called* ***precipitation.***

The empirical equation for this reaction can be written as

$$KCl + AgNO_3 \rightarrow KNO_3 + AgCl(s)$$

KCl, $AgNO_3$, and KNO_3 are all very soluble in water. Only their aqueous ions are present in the solution environment. A more useful representation is the *ionic equation,*

$$K^+(aq) + Cl^-(aq) + Ag^+(aq) + NO_3^-(aq) \rightarrow K^+(aq) + NO_3^-(aq) + Ag^+Cl^-(s)$$

In this form the equation clearly shows that the $K^+(aq)$ and $NO_3^-(aq)$ ions take no part in the reaction. Ions that take no part in a chemical reaction are referred to as ***"spectator ions."*** Suppose these spectator ions are ignored and only the reacting species are retained. The chemical reaction is then shown most effectively by the following *net ionic equation:*

$$Ag^+(aq) + Cl^-(aq) \rightarrow Ag^+Cl^-(s)$$

Sometimes there is no reason to write the complete empirical equation or the complete ionic equation for such a reaction. The net ionic equation, which includes only the participating chemical species, may be the most useful way of representing the reaction.

Practice Problems

For each of the following reactions, write the ionic equation and identify the spectator ions and the insoluble product. From this information, write the net ionic equation. As required, refer to Appendix Table 12 for solubility information.

1. Silver nitrate + ammonium chloride →
   ans. $Ag^+(aq) + Cl^-(aq) \rightarrow Ag^+Cl^-(s)$
2. Iron(III) sulfate + sodium hydroxide →
   ans. $2Fe^{+++}(aq) + 6OH^-(aq) \rightarrow 2Fe^{+++}(OH^-)_3(s)$
3. Zinc nitrate + ammonium sulfide →
   ans. $Zn^{++}(aq) + S^{--}(aq) \rightarrow Zn^{++}S^{--}(s)$
4. Mercury(I) nitrate + iron(II) iodide →
   ans. $Hg_2^{++}(aq) + 2I^-(aq) \rightarrow Hg_2^{++}I^-_2(s)$
5. Lead(II) chloride + potassium chromate →
   ans. $Pb^{++}(aq) + CrO_4^{--}(aq) \rightarrow Pb^{++}CrO_4^{--}(s)$

14.6 Some covalent compounds ionize Covalent bonds are formed when two atoms share electrons. The shared electrons move about the nuclei of both atoms joined by the covalent bond. If the two atoms differ in electronegativity, the shared electrons are drawn toward the more electronegative atom. The covalent bond is polar to some degree. The molecule formed by these two atoms is polar covalent with a negative region and a positive region.

When a polar molecule is dissolved in water, the water dipoles exert attractive forces on the oppositely charged regions of the solute molecule. These attractive forces oppose the bonding force within the molecule. If they exceed the bonding force, the covalent bond breaks, and the molecule is separated into simpler (charged) fragments. The charged fragments disperse in the solvent as hydrated ions. *The formation of ions from solute molecules by the action of the solvent is called* ***ionization.***

The smaller an ion and the higher its charge, the stronger the hydration tendency is likely to be.

Factors that favor the ionization of solute molecules are the weakness of their covalent bonds and the ability of the ions formed to associate with the solvent. The tendency for solute molecules to ionize in a suitable solvent is determined by the relative stabilities of the molecules and the separated ions in the solution.

The extent to which covalent solutes ionize varies over a wide range. Some substances are completely ionized in aqueous solutions. Others may be only slightly ionized. This difference among covalent solutes is related to their relative bond strengths and their entropy changes on ionization.

Consider an example of ionization. Hydrogen chloride is one of a series of compounds of hydrogen and a member of the Halogen Family of elements. These elements are fluorine, chlorine, bromine, and iodine. (See Section 5.5.) The compounds, collectively called *hydrogen halides,* are molecular and have single covalent bonds. All are gases and are very soluble in water. Hydrogen chloride, HCl, is the most important hydrogen halide.

As a pure liquid, hydrogen chloride *does not* conduct an electric current. When dissolved in water, the hydrogen chloride solution *does* conduct. Ions are present in the aqueous solution. A solution of hydrogen chloride in a nonpolar solvent such as benzene *does not* conduct. In conclusion, the ions in the aqueous solution result from a chemical reaction between hydrogen chloride molecules and water molecules. The solvent water dipoles are able to overcome the H—Cl bonds thereby separating the solute molecules into hydrogen ions, H^+, and chloride ions, Cl^-.

Arrhenius believed the ionization of HCl simply involved the dissociation of solute molecules into H^+ ions and Cl^- ions upon entering the solution. Today chemists recognize that single H^+ ions do not exist as free particles in water solutions. They do, however, show a strong tendency to become hydrated.

The H^+ is a bare hydrogen nucleus that can approach very close to a water molecule. This fact accounts for the strong hydrating effect of water molecules on H^+ ions, which increases the tendency for ionization. Most covalent compounds are nonelectrolytes and do not ionize in solution. Those that do ionize are most often hydrogen-containing compounds that can form H^+ ions.

The ionization of HCl in aqueous solution is illustrated with models in Figure 14-7. The ionization reaction is given by the following equation:

$$\mathbf{HCl(g) + H_2O(l) \rightarrow H_3O^+(aq) + Cl^-(aq)}$$

Omitting the phase notations for reactants and products, the equation can be written more simply as

$$\mathbf{HCl + H_2O \rightarrow H_3O^+ + Cl^-}$$

"Proton" *as used here refers to the* H^+ *ion, not to a proton from the nucleus of any of the atoms.*

The H_3O^+ *ion is a hydrated proton* ($H^+ \cdot H_2O$) *and is known as the* ***hydronium ion.*** A model of this ion is shown in Figure 14-8. Because of ionization, a solution of hydrogen chloride in water has distinctly different properties from those of the hydrogen chloride gas. The aqueous solution is given the name *hydrochloric acid.* The properties of acids will be studied in greater detail in Chapter 15.

The aluminum halides, binary compounds of aluminum with a halogen, are molecular. The energy required to remove the two 3*s* and one 3*p* electrons from each aluminum atom exceeds the energy available when the elements combine. The bonds are covalent with a partial ionic character.

Aluminum chloride has the structure Al_2Cl_6 in both liquid and vapor phases. A model of this molecule is shown in Figure 14-9. The structure of the solid is less certain but is thought to consist of $AlCl_3$ units. Thus the empirical formula $AlCl_3$ is generally used for this halide.

Aluminum chloride in the liquid phase is a very poor conductor of electricity. In water solution, however, it is a good conductor. This change indicates that ionization occurs during

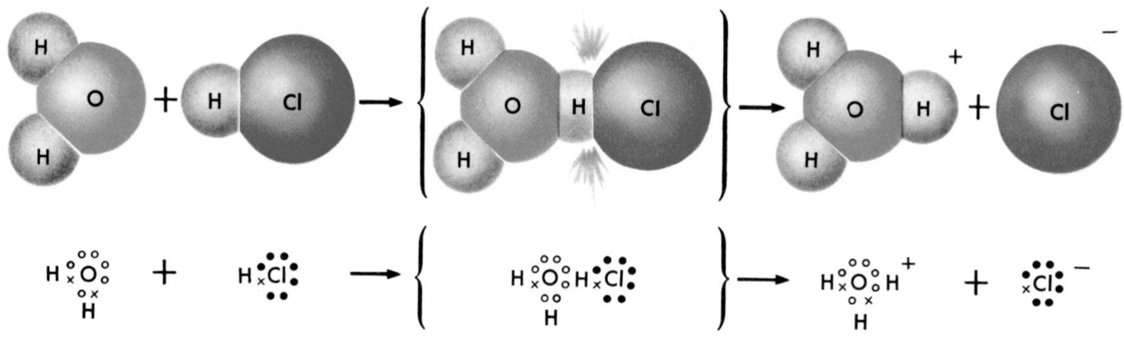

Figure 14-7. An ionization process. Water dipoles react with hydrogen chloride molecules to form hydronium ions and chloride ions.

the solution process. The aluminum ion has a strong tendency for hydration. The hydration process provides the energy needed to complete the transfer of three electrons from each aluminum atom to three chlorine atoms. The ionization can be represented as follows:

$$Al_2Cl_6 + 12H_2O \rightarrow 2Al(H_2O)_6^{+++} + 6Cl^-$$

or, more simply, using the empirical formula

$$AlCl_3 + 6H_2O \rightarrow Al(H_2O)_6^{+++} + 3Cl^-$$

Other hydrated aluminum ions probably form at the same time.

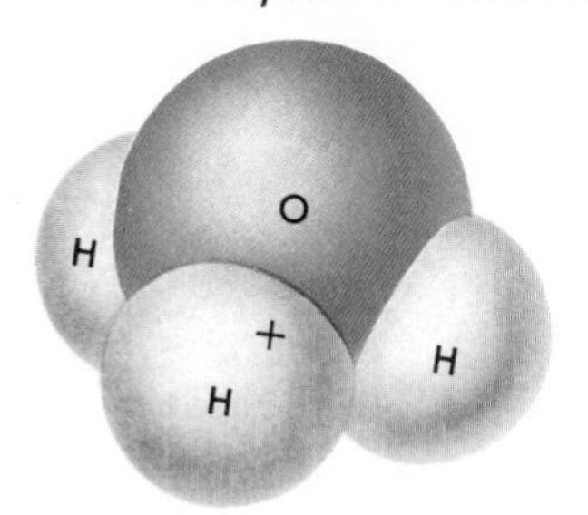

Figure 14-8. A model of the hydronium ion, H_3O^+.

14.7 Strength of electrolytes The strength of an electrolyte is determined by the concentration of its ions in solution. Ionic compounds in the solid phase are crystals composed of ions. Their solutions contain the solute only in the form of dispersed hydrated ions. The solutions are said to be completely ionized. Ionic substances, if very soluble in water, can form solutions with high concentrations of ions. Their water solutions are good conductors of electricity. These ionic substances are called *strong electrolytes*.

The single covalent bonds of the hydrogen halide molecules range from the very strong and very polar H—F bond, through the H—Cl and H—Br bonds, to the very weak and slightly polar H—I bond. Even moderately concentrated solutions of HCl, HBr, and HI are essentially completely ionized. They are strong electrolytes. Of these halides, HI has the weakest bonds and is the strongest electrolyte.

Aqueous solutions of hydrogen fluoride, HF, are only slightly ionized. The very strong H—F bond prevents an extensive ionization reaction with the solvent molecules. The concentrations of H_3O^+ ions and F^- ions remain low, and the concentration of HF molecules remains high. Hydrogen fluoride is a *weak electrolyte*. Solutions of weak electrolytes will contain low concentrations of ionic species and high concentrations of molecular species.

Acetic acid, a water solution of hydrogen acetate, $HC_2H_3O_2$, is a poor conductor. Its relative merit as a conductor is shown in Figure 14-10. The fact that the solution does conduct slightly indicates that some ionization has occurred. This ionization is shown in the reversible reaction

$$HC_2H_3O_2 + H_2O \rightleftarrows H_3O^+ + C_2H_3O_2^-$$

It must be assumed that the ion concentration is low. A 0.1-*m* solution of $HC_2H_3O_2$ is approximately 1% ionized; a 0.001-*m* solution is approximately 15% ionized. Of course, if 1% of the $HC_2H_3O_2$ molecules in solution is present in the form of ions,

Figure 14-9. A model of the aluminum chloride molecule, Al_2Cl_6, in both the liquid and vapor phases. The solid phase is thought to consist of $AlCl_3$ units. Consequently, the empirical formula $AlCl_3$ is used for this halide.

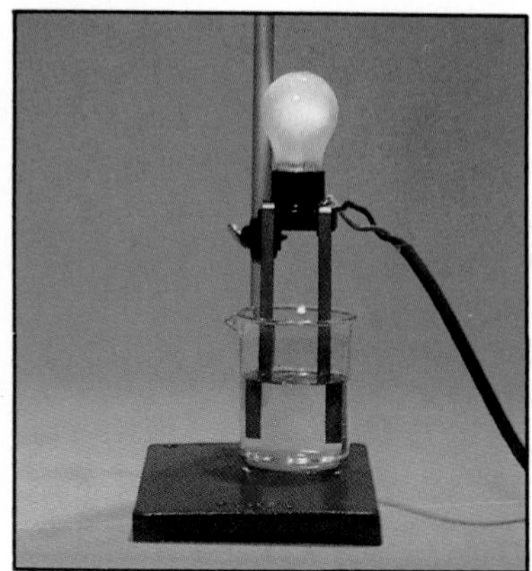

Figure 14-10. The brightness of the lamp filament indicates the conductivity of the solution tested. (Left) Distilled water is a nonconductor. (Center) Dilute acetic acid, a solution of a weak electrolyte, is a poor conductor. (Right) Dilute hydrochloric acid, a solution of a strong electrolyte, is a good conductor of electricity.

then 99% is present in the form of covalent molecules. Hydrogen acetate is a weak electrolyte.

Ammonia, NH_3, is a covalent compound and a gas at ordinary temperatures. It is extremely soluble in water and its water solution is a poor conductor. Ammonia ionizes slightly in water solution to give a low concentration of ammonium ions, NH_4^+, and hydroxide ions, OH^-. It is a weak electrolyte. The reaction with water is

$$NH_3(g) + H_2O(l) \rightleftarrows NH_4^+(aq) + OH^-(aq)$$

Ammonia-water solutions are sometimes called *ammonium hydroxide.* This is a traditional name, and it may be justified by the fact that many useful properties of aqueous ammonia are those of the few NH_4^+ and OH^- ions present. (A more plausible reason that it is still used might be that all the reagent bottles are now labeled "**AMMONIUM HYDROXIDE,**" and no one wants to change the labels.)

The terms *strong* and *weak* must not be confused with the terms *dilute* and *concentrated. Strong* and *weak* refer to the *degree of ionization. Dilute* and *concentrated* refer to the *amount of solute dissolved in a solvent.* The terms *strong* and *weak* are not precise descriptions of electrolytes. There are many degrees of strength and weakness, and the dividing lines are not clear-cut. The terms *dilute* and *concentrated* can be qualified in a similar way.

14.8 Ionization of water The ordinary electric-conductivity test of solutions described in Section 14.1 shows water to be a nonconductor. More sensitive electric-conductivity tests show that pure water has a slight, but measurable, conductivity. *At ordinary temperatures, water ionizes to the extent of approximately two molecules in a billion.* Thus, pure water ordinarily contains about two H_3O^+ ions and two OH^- ions per 10^9 H_2O molecules. Even though this concentration of ions is very low, it is important. The slight ionization of water is neglected when dealing with solutes

Figure 14-11. The formation of a hydrogen bond between two water dipoles may be an intermediate step in the ionization of water.

such as hydrogen chloride, which may be completely ionized in aqueous solutions.

The ionization of water probably begins with the formation of a hydrogen bond between two water molecules, as shown in Figure 14-11. The ionization products are hydronium ions and hydroxide ions. They are formed by the following reaction:

$$H_2O + H_2O \rightleftharpoons H_3O^+ + OH^-$$

In any reaction involving the hydronium ion, the water of hydration is always left behind. This water of hydration is often of little significance in the reaction process, and the ion can be written more simply as $\mathbf{H^+}$ or $\mathbf{H^+(aq)}$. In any case, it is understood that *all ions are hydrated in aqueous solutions.*

Chemists do not have conclusive evidence that H_3O^+ ions exist in water solution in precisely this form. Some evidence suggests a more aqueous structure such as $H_9O_4^+$ because of hydrogen bonding. When the ion is written H_3O^+ it is with the understanding that additional water molecules might be associated with it.

14.9 Electrolytes and the freezing and boiling points of solvents Molal solutions have a definite ratio of solute particles to solvent molecules (Section 13.11). The depression of the freezing point of a solvent by a solute is directly proportional to the number of particles of solute present. The same reasoning applies to the elevation of the boiling point of a solvent by a nonvolatile solute. Why, then, does one mole of hydrogen chloride dissolved in 1 kg of water lower the freezing point more than one mole of sugar does? Recall that a colligative property of

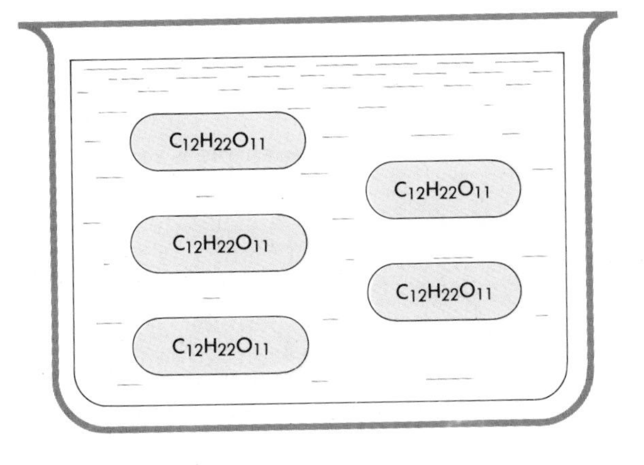

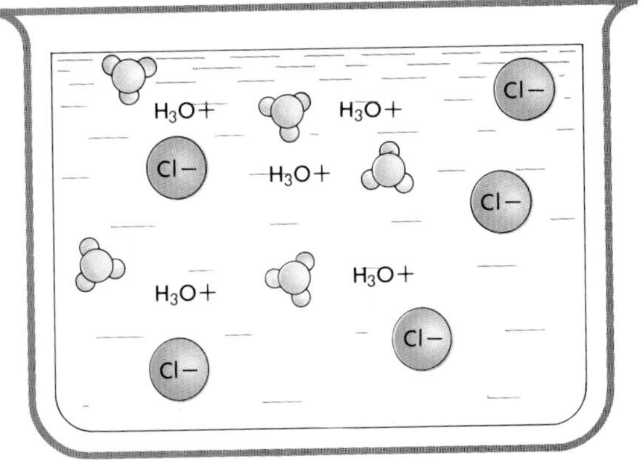

Figure 14-12. Five sugar molecules provide only five solute particles in solution. Five hydrogen chloride molecules, on the other hand, provide ten solute particles when dissolved in water.

solutions relates the freezing-point depression of a solvent to the number of solute particles present, not to their identity. The greater freezing-point depression is caused by the separation of each molecule of hydrogen chloride that ionizes into *two* particles. This comparison is illustrated in Figure 14-12.

Suppose that, in a concentrated solution, 90 out of every 100 molecules ionize. Then, for every 100 molecules in solution, 190 particles are formed (180 ions and 10 un-ionized molecules). Such a solute lowers the freezing point of its solvent 190/100 = 1.9 times as much as would a solute that does not ionize.

Suppose 100% of the hydrogen chloride molecules were ionized, as in a more dilute solution. The lowering of the freezing point should be double that caused by the solute in a solution of a nonelectrolyte having the same molality.

The following equation assumes the complete ionization of hydrogen sulfate in very dilute solutions.

$$H_2SO_4 + 2H_2O \rightarrow 2H_3O^+ + SO_4^{--}$$

Each molecule of hydrogen sulfate that ionizes completely forms *three ions.* Two are hydronium ions, each with one positive charge. One is a sulfate ion with two negative charges. A very dilute solution of a given molality should, if completely ionized, lower the freezing point of its solvent *three times as much* as a solution of a nonelectrolyte having the same molality. Nonvolatile electrolytes in solution raise the boiling points of solvents in a similar way because of their ionization.

Actually, hydrogen sulfate ionizes in aqueous solutions in two steps. The two hydrogen atoms in each molecule are ionized one at a time. In the first step, H_3O^+ ions and HSO_4^- ions are formed. In the second step, the HSO_4^- ions are further ionized to H_3O^+ ions and SO_4^{--} ions. Except in *very* dilute solutions, the second step may be far from complete. The ionization of H_2SO_4 is discussed further in Section 15.3.

Now consider a solution of an ionic substance such as calcium chloride in water. The dissociation equation is

$$Ca^{++}Cl^-{}_2(s) \rightarrow Ca^{++}(aq) + 2Cl^-(aq)$$

One mole of $CaCl_2$ dissociates and provides three times the Avogadro number of solute particles in solution. Recall that one mole of nonelectrolyte provides one mole of solute particles when completely dissolved in water. An example is one mole of sugar, which provides the Avogadro number of particles. You might expect a $CaCl_2$ solution of a given molality to lower the freezing point of its solvent three times as much as a nonelectrolyte having the same molality. However, experiments do not confirm this effect except for very dilute solutions. At ordinary concentrations, the freezing-point depressions of ionic solutes are less than those expected from ideal solutions. Table 14-1

Table 14-1
FREEZING-POINT DEPRESSIONS FOR AQUEOUS SOLUTIONS OF IONIC SOLUTES

Solute	*Concentration* (in moles solute/kg H_2O)				
	0.1	0.05	0.01	0.005	0.001
Values are observed molal freezing-point depressions in C° for concentrations indicated.					
$2K_f = 2 \times 1.86$ C° = 3.72 C°					
$AgNO_3$	3.32	3.42	3.60	—	—
KCl	3.45	3.50	3.61	3.65	3.66
KNO_3	3.31	3.43	3.59	3.64	—
LiCl	3.52	3.55	3.60	3.61	—
$MgSO_4$	2.25	2.42	2.85	3.02	3.38
NH_4Cl	3.40	3.49	3.58	3.62	—
NH_4NO_3	3.40	3.47	3.57	—	—
$3K_f = 3 \times 1.86$ C° = 5.58 C°					
$BaCl_2$	4.70	4.80	5.03	5.12	5.30
$Ba(NO_3)_2$	4.25	—	5.01	—	5.39
$CaCl_2$	4.83	4.89	5.11	—	—
$Cd(NO_3)_2$	5.08	—	5.20	5.28	—
$CoCl_2$	4.88	4.92	5.11	5.21	—
K_2SO_4	4.32	4.60	5.01	5.15	5.28
$ZnCl_2$	4.94	—	5.15	5.28	—
$4K_f = 4 \times 1.86$ C° = 7.44 C°					
$K_3Fe(CN)_6$	5.30	5.60	6.26	6.53	7.10

gives observed values of freezing-point depressions in Celsius degrees for several ionic solutes at different dilutions of their aqueous solutions. Observe the trend *toward* a whole number multiple of K_f (1.86 C° for water) as the concentrations decrease. This problem is examined in Section 14.10.

14.10 Apparent degree of ionization The larger the number of ions in a given volume of a solution, the better it conducts electricity. This fact suggests one way to determine the concentration of ions in a solution. Another way is to measure the lowering of the freezing point caused by an ionized solute in a measured amount of solvent. Suppose a 0.1-*m* solution of sodium chloride were to freeze at −0.372 °C [0° − (2 × 0.186 C°)]. You could assume the solute to be 100% ionized.

Actual measurements, however, give only an *apparent degree of ionization*. You have seen that electrovalent compounds, by the nature of their structure, must be 100% ionic in solution. Experimental results give an apparent degree of ionization somewhat less than 100%. For example, consider the 0.1-*m* solution of sodium chloride referred to above. This solution actually freezes at −0.346 °C, yielding a freezing-point depression of 0.346 C°

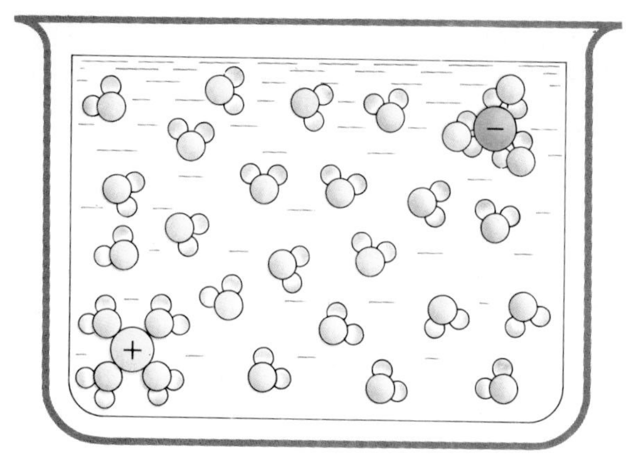

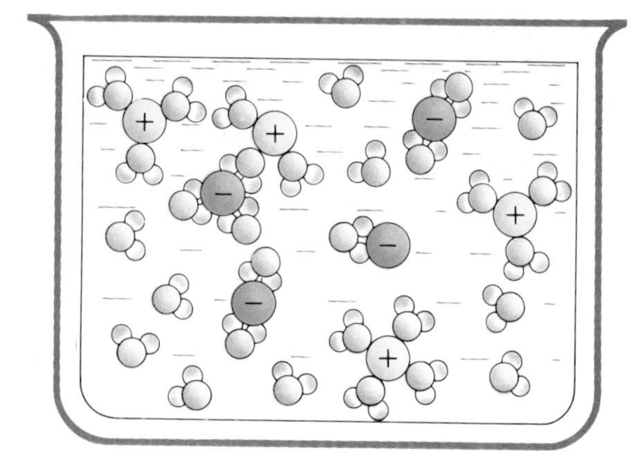

Figure 14-13. The ions in the dilute solution on the left are far apart and act independently. The activity of the ions in the solution on the right is somewhat restricted because of the concentration. Thus the apparent number of ions present may be less than the actual number.

instead of the predicted value of 0.372 C°. For many years, such differences prevented chemists from deciding whether a compound was completely ionized in water solution.

Today it is recognized that attractive forces exist between ions in aqueous solutions. These forces are small compared with those in the crystalline salt. However they do interfere with the movements of the aqueous ions, even in dilute solutions. Only in very dilute solutions is the average distance between ions great enough, and the attraction between ions small enough, for the aqueous solute ions to move about freely. Therefore the more dilute the sodium chloride solution, the more nearly the freezing-point depression approaches twice the value for a molecular solute.

These observations are consistent with the Debye-Hückel theory of interionic attraction. This theory accounts quantitatively for the attraction between dissociated ions of ionic solids in dilute water solutions. According to the Debye-Hückel theory, each ion is surrounded on average by more ions of opposite charge than of like charge. This clustering effect is illustrated in Figure 14-13. The effect is to hinder the movements of an ion. Thus the "ion activity" is less than that expected on the basis of

Table 14-2
INFLUENCE OF CONCENTRATION ON FREEZING POINT OF AQUEOUS SOLUTIONS OF NaCl

Concentration of NaCl (*m*)	*Freezing-point depression* (C°)	*Freezing-point depression/mole NaCl* (C°)
0.100	0.346	3.46
0.0100	0.0361	3.61
0.00100	0.00366	3.66
0.000100	0.000372	3.72

the number of ions known to be present. Table 14-2 gives the observed freezing-point depressions of aqueous solutions of sodium chloride at various concentrations. The table also shows freezing-point depressions per mole of NaCl calculated from the observed values.

In concentrated solutions, another effect on the freezing-point depression may arise from a shortage of solvent molecules. There may not be enough water molecules to hydrate completely all solute ions. In such a case, clusters of ions may act as a single solute unit.

SUMMARY

Ions that have mobility and are free to move through a liquid medium are able to conduct electricity. Two electrodes connected to a source of electric current must be in contact with the liquid for conduction to occur through it.

In crystal form, ionic compounds are composed of ions. When fused, their liquid phase consists of ions. They are soluble only in highly polar solvents like water and form solutions that contain ions. Ionic compounds conduct an electric current only when fused or dissolved in water. Their ions have mobility in either situation.

The process by which ionic substances are dissolved in water is called dissociation. In dissociation, ions that compose the compounds are torn apart by the hydrating action of the solvent, and they dissolve as hydrated ions.

Some covalent compounds form solutions that contain ions. As pure liquids, covalent compounds do not conduct. The ions are produced by chemical reaction between the solute molecules and the solvent (called ionization). Water is the best, but not the only, ionizing agent.

Substances whose water solutions conduct electricity are called electrolytes. Those whose water solutions do not conduct electricity are nonelectrolytes. Electrolytes are strong or weak depending on how well their solutions conduct an electric current. Solutions of strong electrolytes have high concentrations of ions. Those of weak electrolytes have low concentrations of ions. Water ionizes to a very small extent, forming hydronium ions and hydroxide ions in very low concentrations.

Electrolytes affect the freezing and boiling points of solvents to a greater extent than nonelectrolytes. Freezing-point depressions and boiling-point elevations of solvents vary with the number of solute particles in solution. The apparent degree of ionization increases with dilution of an electrolyte in solution. Ionic solutes are composed of ions and consequently are completely dissociated in solution. An apparent degree of ionization less than the actual ion concentration can be explained on the basis of the theory of interionic attraction.

VOCABULARY

aqueous solution
Debye-Hückel theory
dissociation
electrolyte
hydrated proton
hydration
hydronium ion
ionic bond
ionic equation
ionization
net ionic equation
nonpolar bond
nonpolar molecule
polar bond
polar molecule
precipitation
spectator ion
strong electrolyte
theory of ionization
water dipole
weak electrolyte

QUESTIONS

GROUP A

1. Define: *(a)* ion; *(b)* electrolyte; *(c)* nonelectrolyte; *(d)* dissociation; *(e)* ionization; *(f)* hydration. *A1*
2. State the important assumptions of the theory of ionization. *A4*
3. Distinguish between substances classified as electrolytes and nonelectrolytes. *A3a*
4. The theory of ionization is useful in explaining the behavior of what kind of solutes? *A3a*
5. Compare the effects on the freezing and boiling points of a solvent resulting from the addition of an electrolytic solute versus the addition of a nonelectrolytic solute. *A2h*
6. The water molecule is described as a polar molecule. Explain. *A1*
7. Write the equation for the ionization of water. *A2l, A6b*
8. Distinguish between an atom and an ion. *A2i*
9. What accounts for the stability of an ion? *A2j*
10. Describe the crystal structure of an ionic compound. *A2k*
11. Potassium chlorate has a melting point of 368 °C. Explain the fact that it conducts an electric current when in the liquid phase. *A2g*
12. *(a)* What is the role of solvent water molecules in the dissociation of an ionic compound? *(b)* How is the process reversed? *A3c*
13. What evidence indicates that the dissociation of ionic compounds is complete? *A3c*
14. Given: 0.01-*m* aqueous solutions of the following substances: *(a)* $NaBr$; *(b)* CH_3OH; *(c)* $Al_2(SO_4)_3$. Predict the approximate freezing-point depression of the solvent water for each solution. *A2h*

QUESTIONS

GROUP B

15. Distinguish between dissociation and ionization. *A3c*
16. *(a)* Describe the ionization of hydrogen chloride in water solution. *(b)* Write the equation for this reaction. *A2c*
17. Given: an aqueous solution of an electrolyte. Explain the abnormal freezing-point depression and boiling-point elevation of the solvent water. *A2h*
18. Describe solution equilibrium in a saturated solution of potassium nitrate containing an excess of the solute crystals. *A2k*
19. The carbon tetrachloride molecule has four polar covalent C—Cl bonds. However, CCl_4 is practically insoluble in water and the molecule does not ionize. Suggest a reason for these "nonpolar" characteristics of the CCl_4 molecule. A model of the CCl_4 molecule is shown in Figure 13-5. *A2o*
20. *(a)* Write the equation for the dissociation of copper(II) chloride in aqueous solution. *(b)* What is the theoretical freezing point of a one-molal aqueous solution of copper(II) chloride? *A6a*
21. In what two ways can the apparent degree of ionization be measured? *A2a, A2m*
22. Explain the discrepancy between measurements of the apparent degree of ionization and evidence that ionic compounds are 100% dissociated in aqueous solution. *A4*
23. *(a)* Describe the difference between a dilute aqueous solution of a strong electrolyte and a concentrated aqueous solution of a weak electrolyte. *(b)* Give an example of each kind of aqueous solution referred to in *(a)*. *A5c*
24. A 0.10-*m* solution is prepared by dissolving

0.10 mole of a substance in 1.0 kg of water. The freezing point of the solvent water is determined to be −0.36 °C. *(a)* What can you predict regarding the nature of the solution? *(b)* What can you conclude about the oxidation numbers of the solute particles? *A2m*

25. When ammonium chloride is dissolved in water, the dissolving process is endothermic. *(a)* What temperature change does the solution undergo? *(b)* Which is greater, the hydration energy or the lattice energy? *(c)* What do you consider the driving force that causes the dissolving process to proceed?
26. *(a)* What is the effect on the solubility of ammonium chloride when the solution of Question 25 is warmed? *(b)* Apply the principle of Le Chatelier to account for this change in solubility.
27. Write the equation for the ionization of hydrogen bromide gas in aqueous solution. *A6a*
28. Write the equation for the ionization of ammonia gas in aqueous solution. *A6a*

Questions 29–32: Each reactant is in aqueous solution. For each reaction *(a)* write the complete ionic equation; *(b)* underscore the spectator ions; and *(c)* write the net ionic equation. *A6b*

29. Barium chloride + magnesium sulfate →
30. Silver nitrate + potassium iodide →
31. Manganese(II) sulfate + ammonium sulfide →
32. Zinc bromide + calcium hydroxide →

PROBLEMS

Group B

1. It was determined by experiment that 278 drops of water were required to provide a volume of 15.0 mL. *(a)* Determine the number of molecules of water in each drop. *(b)* Determine the number of hydronium ions in each drop. *(c)* How many hydroxide ions are in each drop? *A21*
2. Determine the mass in grams of copper(II) sulfate pentahydrate that must be dissolved in $25\bar{0}$ g of water to prepare a 0.0155-*m* solution. *B*
3. A student collected 20.0 mL of dry HCl gas at 21.0 °C and 748 mm pressure and then dissolved the gas in 1.00 kg of water. Assuming complete ionization and no interionic attraction, calculate the freezing-point depression of the solvent water. *B*
4. Calculate the freezing point of $50\bar{0}$ g of water to which 10.5 g of ethyl alcohol, C_2H_5OH, has been added.
5. The composition of a substance was determined by analysis to be 10.1% carbon, 0.846% hydrogen, and 89.1% chlorine. It was found to be soluble in benzene, and when 2.50 g of the substance was dissolved in $10\bar{0}$ g of benzene, the freezing point of the solvent benzene was 4.41 °C. Determine the molecular formula of the substance. *B*

The 1964 Nobel Prize in chemistry was given to Dorothy Crowfoot Hodgkin (British) for determining the structures of biochemical compounds, such as penicillin and vitamin B_{12}. Fascinated with crystals from the age of ten, she became interested in the use of X-ray crystallography to determine the structures of complex macromolecules. During World War II, she studied crystalline samples of the salts and degradation products of penicillin and eventually determined its structure. Later, in determining the atomic arrangement in vitamin B_{12}, she was among the first to use electronic computers.

Acids, Bases, and Salts

15

GOALS

In this chapter you will gain an understanding of: • the importance of the four major industrial acids • the modern definitions of acids • the properties of aqueous acids • naming binary acids and oxyacids • the modern definitions of bases • the properties of hydroxides, protolysis • acid anhydrides and basic anhydrides • conjugate acid-base pairs • the nature and preparation of salts • naming salts

ACIDS

15.1 The nature of acids Substances whose aqueous solutions conduct an electric current are called *electrolytes*. Their solutions contain ions. Historically, such substances are classified as *acids, bases,* or *salts*. In this sense, ***acids*** *are substances that react with water to form hydronium ions,* H_3O^+. Acids defined in this way are limited to aqueous solutions. ***Bases*** *are electrolytes that provide hydroxide ions,* OH^-, *in aqueous solutions.* ***Salts*** *are ionic compounds of metal-nonmetal composition.* As solids, salts have ionic crystalline structures; in aqueous solutions, they are dissociated as separate metallic and nonmetallic hydrated ions. Bases and salts are treated in detail beginning with Sections 15.8 and 15.13 respectively.

When acids are ionized in water, their molecules contribute one or more hydrogen ions to the aqueous solution as the hydrated proton, $H^+ \cdot H_2O$ or H_3O^+, called the *hydronium ion*. See Section 14.6 and Figure 14-8. As a group, acids share a common set of properties to varying degrees.

1. All acids have a sour taste. **CAUTION:** *Never use the "taste test" to identify an acid in the chemistry lab.* You know the sour taste of lemon juice (citric acid) and cider vinegar (acetic acid).
2. Some acids are very corrosive, and some are poisonous.
3. Acids contain hydrogen, and some acids liberate hydrogen in reactions with certain metals.
4. Acids influence the color of acid/base indicators. The effect of acids on the color of indicators and their neutralization of bases are the concerns of Chapter 16.

Review Section 8.8 and Table 8-2.

Indicators are organic dyes that have one color in acidic solutions and another color in basic solutions.

Nearly all fruits contain acids, as do many other common foods. Lemons, oranges, and grapefruits contain citric acid. Apples contain malic acid. The souring of milk produces lactic acid. Rancid butter contains butyric acid. The fermentation of "hard" cider forms the acetic acid of vinegar. These acids, because of their origin and structure, are called *organic* acids.

The carboxyl group in organic acids:

```
     O
    //
 —C
    \
     O—H
```

Organic acids are covalent (molecular) structures that contain a specific atom configuration called the *carboxyl group* (−COOH). They are weak acids, being slightly ionized in water solution. To the extent that ionization occurs, the hydrogen atom of the carboxyl group becomes hydrated as an H_3O^+ ion. Organic acids are discussed in Chapter 19.

Acids manufactured from minerals are known as *inorganic* acids, or simply as *mineral* acids. These substances have been known for centuries because their properties are very distinctive. When the term "acid" is used in a very general sense, it usually refers to these compounds. They are the substances historically known as acids; their water solutions are called *aqueous acids.* Modern definitions of acids give the name "acid" to many substances that are not acids in this traditional sense.

Mineral acids occupy major roles in the chemical industry. Four mineral acids are among the largest-volume industrial chemicals produced in the United States.

15.2 Industrial acids If a manufacturing chemist were asked to name the most important acid, *sulfuric acid* would undoubtedly be the one mentioned. It is a very versatile industrial product. Because sulfuric acid is involved in so many technical and manufacturing processes, its consumption is an index to a country's industrialization and prosperity. More sulfuric acid is manufactured in the United States than any other chemical—approximately 35 million metric tons each year.

The second ranking industrial acid is *phosphoric acid* and the third is *nitric acid.* The dominant use of these three acids is in agriculture. The production of fertilizers consumes 70% of the sulfuric acid, 85% of the phosphoric acid, and 70% of the nitric acid produced in the United States.

Ranking fourth in importance is *hydrochloric acid.* The gastric secretion in the human stomach is about 0.4% hydrochloric acid. Its industrial uses have little in common with the three acids just mentioned. Hydrochloric acid is a volatile acid and is more expensive than sulfuric acid. It is an important laboratory reagent.

Reagents are substances, usually in solution, used to react with other substances or mixtures of substances to dissolve, precipitate, oxidize, or reduce them for analysis or use.

Phosphoric acid, H_3PO_4, is a moderately strong acid. It is made by reacting sulfuric acid with phosphate rock and by burning phosphorus and reacting the oxide with water. Its major uses are in making fertilizers and detergents. It is not as commonly used in the chemistry laboratory as are sulfuric, nitric, and hydrochloric acids. A brief description of each of these three common laboratory acids follows.

1. *Sulfuric acid.* H_2SO_4 is a dense, oily liquid with a high boiling point. *Concentrated* sulfuric acid contains 95% to 98% sulfuric

acid by mass (the balance is water). Its density is 1.84 g/mL. Common dilute sulfuric acid is made by adding 1 volume of concentrated acid to 6 volumes of water. Its major industrial uses are in producing phosphoric acid (fertilizers), cellulose products and other chemicals, and in refining petroleum.

CAUTION: *Add sulfuric acid to water slowly while stirring. Never add water to acid.*

2. *Nitric acid.* HNO_3 is a volatile liquid. Pure nitric acid is too unstable for commercial use. Concentrated nitric acid, a 70% solution in water, is fairly stable. It has a density of 1.42 g/mL. Common dilute nitric acid is made by adding 1 volume of concentrated acid to 5 volumes of water. A pure solution of nitric acid is colorless. However, it gradually becomes yellow on standing because of slight decomposition that forms brown nitrogen dioxide gas.

$$4HNO_3 \rightarrow 4NO_2 + O_2 + 2H_2O$$

Its major industrial uses are in producing fertilizers, explosives, and other chemicals.

3. *Hydrochloric acid.* HCl, or gaseous hydrogen chloride, is extremely soluble in water. It forms the colorless solution known as hydrochloric acid. Concentrated hydrochloric acid is a water solution containing about 37% hydrogen chloride. Its density is 1.19 g/mL. Common dilute hydrochloric acid is prepared by adding 1 volume of concentrated acid to 4 volumes of water. Such a solution contains approximately 7% hydrogen chloride. Its industrial uses include pickling steel and recovering magnesium from seawater.

Pickling: Immersion of iron or steel in an acid solution in order to remove surface oxides.

15.3 Aqueous acids Arrhenius provided a clue to the chemical nature of acids in his *Theory of Ionization.* He believed that acids ionize in water solutions to form hydrogen ions.

The acids just described are essentially covalently bonded structures. They have one element in common, *hydrogen.* Concentrated sulfuric and nitric acids are very poor conductors of electricity because they are only very slightly ionized. Liquid hydrogen chloride is a nonconductor. In dilute water solution, however, each of these substances is highly ionized because of the hydrating action of the water dipoles. Their ionization in water solutions can be represented by these equations:

$$H_2SO_4 + H_2O \rightarrow H_3O^+ + HSO_4^-$$
$$HNO_3 + H_2O \rightarrow H_3O^+ + NO_3^-$$
$$HCl + H_2O \rightarrow H_3O^+ + Cl^-$$

Hydronium ions, H_3O^+, are present in all of these solutions. It can be assumed that the acid properties of the solutions are due to these H_3O^+ ions.

When a sulfuric acid solution is diluted enough, the ionization of HSO_4^- ions may become appreciable.

$$HSO_4^- + H_2O \rightarrow H_3O^+ + SO_4^{--}$$

Table 15-1
COMMON AQUEOUS ACIDS

Strong acids	
$HClO_4$	$\rightleftharpoons H^+ + ClO_4^-$
HCl	$\rightleftharpoons H^+ + Cl^-$
HNO_3	$\rightleftharpoons H^+ + NO_3^-$
H_2SO_4	$\rightleftharpoons H^+ + HSO_4^-$
Weak acids	
HSO_4^-	$\rightleftharpoons H^+ + SO_4^{--}$
H_3PO_4	$\rightleftharpoons H^+ + H_2PO_4^-$
HF	$\rightleftharpoons H^+ + F^-$
$HC_2H_3O_2$	$\rightleftharpoons H^+ + C_2H_3O_2^-$
H_2CO_3	$\rightleftharpoons H^+ + HCO_3^-$
H_2S	$\rightleftharpoons H^+ + HS^-$
HCN	$\rightleftharpoons H^+ + CN^-$
HCO_3^-	$\rightleftharpoons H^+ + CO_3^{--}$

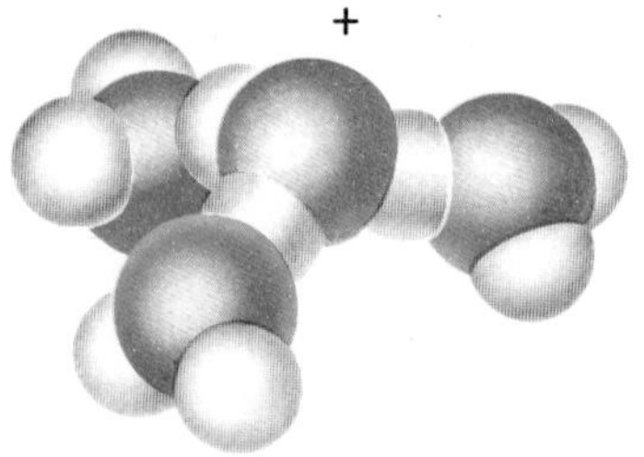

The ionization of HSO_4^- ions may be complete in very dilute solutions of sulfuric acid. If so, the equation is written

$$H_2SO_4 + 2H_2O \rightarrow 2H_3O^+ + SO_4^{--}$$

Note how this equation summarizes the two partial ionizations that occur with increasing dilution of the sulfuric acid solution.

$$\begin{array}{ll} \textbf{(1st stage)} & H_2SO_4 + H_2O \rightarrow H_3O^+ + HSO_4^- \\ \textbf{(2nd stage)} & HSO_4^- + H_2O \rightarrow H_3O^+ + SO_4^{--} \\ \hline \textbf{(summary)} & H_2SO_4 + 2H_2O \rightarrow 2H_3O^+ + SO_4^{--} \end{array}$$

Acids that ionize completely, or nearly so, in aqueous solution provide high concentrations of hydronium ions. Such concentrations characterize strong acids. Sulfuric, nitric, and hydrochloric acids are strong mineral acids. Substances that produce few hydronium ions in aqueous solution, such as acetic acid and carbonic acid, are weak acids. They ionize slightly in aqueous solution.

15.4 Modern definitions of acids The hydronium ion is a hydrated proton. In aqueous solution, it is considered to be in the hydrated form, $H^+ \cdot H_2O$, and is usually represented as H_3O^+. The proton is vigorously hydrated in water because of its small size and high positive charge density.

Chemists have found conclusive evidence that H_3O^+ ions exist in hydrated crystals of perchloric acid (a very strong acid) and in concentrated solutions of strong acids. In dilute aqueous solutions of acids, however, the proton hydration may be more extensive. Physical evidence, such as electric and thermal conductivities, suggests the formula $H^+ \cdot 4H_2O$. This formula corresponds to the ionic species $H_9O_4^+$. A model of the $H_9O_4^+$ ion having the spatial structure of a triagonal pyramid is shown in Figure 15-1.

Figure 15-1. A possible model of the $H^+ \cdot 4H_2O$ or $H_9O_4^+$ ion that suggests an H_3O^+ ion with three H_2O molecules attached through hydrogen bonds.

Other species of the hydrated proton have been suggested for dilute aqueous acid solutions. These species are $H^+ \cdot 2H_2O$ and $H^+ \cdot 3H_2O$, corresponding, respectively, to the species $H_5O_2^+$ and $H_7O_3^+$. Chemists write formulas of this kind only when the degree of hydration is itself the subject of discussion. Otherwise, for simplicity, the hydrated proton in aqueous solutions is written as $H^+(aq)$ or as the hydronium ion, H_3O^+.

When hydrogen chloride is dissolved in a nonpolar solvent, the solution remains a nonconductor. *This means that protons (hydrogen ions) are not released by molecules such as HCl unless there are molecules or ions present that can accept them.*

Hydrogen chloride dissolved in ammonia transfers protons to the solvent much as it does in water.

$$HCl + H_2O \rightarrow H_3O^+ + Cl^-$$
$$HCl + NH_3 \rightarrow NH_4^+ + Cl^-$$

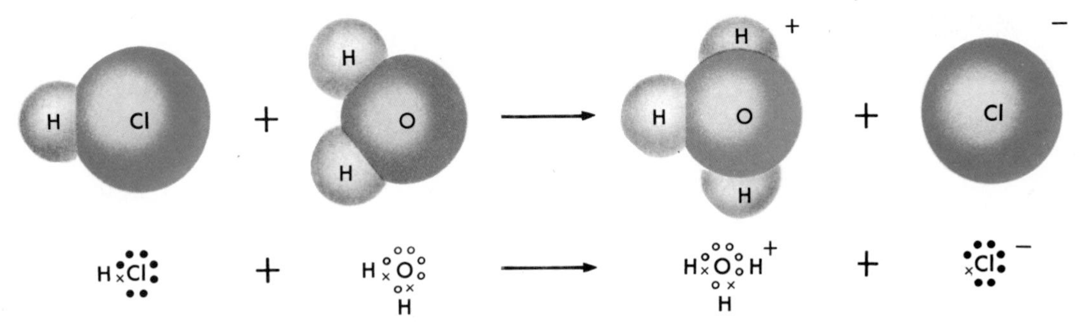

The similarity of these reactions is very clear when electron-dot formulas are used.

$$\mathrm{H{:}\ddot{\underset{..}{Cl}}{:} + H{:}\overset{..}{\underset{\underset{H}{..}}{O}}{:} \rightarrow H{:}\overset{..}{\underset{\underset{H}{..}}{O}}{:}H^{+} + {:}\ddot{\underset{..}{Cl}}{:}^{-}}$$

$$\mathrm{H{:}\ddot{\underset{..}{Cl}}{:} + H{:}\overset{..}{\underset{\underset{H}{..}}{N}}{:}H \rightarrow H{:}\overset{\overset{H^{+}}{..}}{\underset{\underset{H}{..}}{N}}{:}H + {:}\ddot{\underset{..}{Cl}}{:}^{-}}$$

Figure 15-2. When HCl is dissolved in water, a proton is donated to an H_2O molecule by a polar HCl molecule to form the hydronium ion, H_3O^+, and the chloride ion, Cl^-.

A proton is transferred to the ammonia molecule, forming the *ammonium ion,* just as one is transferred to the water molecule, forming the hydronium ion. See Figures 15-2 and 15-3. In each case, the proton is given up by the hydrogen chloride molecule. This molecule is said to be a *proton donor.* In 1923, the Danish chemist J. N. Brønsted defined all proton donors as acids. According to Brønsted's definition, *an **acid** is any species (molecule or ion) that gives up protons to another substance.* For example, hydrogen chloride is an acid in this sense even though it does not contain hydrogen ions as a pure substance.

Johannes Nicolaus Brønsted (1879–1947) spent most of his life as a professor of physical chemistry at the University of Copenhagen. He is recognized for his work on relationships between the forms of energy involved in physical and chemical changes, a method of separating isotopes, the catalytic properties and strengths of acids and bases, and the development of the acid-base theory explained here.

Water is an acid when ammonia is dissolved in it because water molecules donate protons to ammonia molecules.

$$\mathrm{H{:}\overset{..}{\underset{\underset{H}{..}}{N}}{:}H + H{:}\ddot{\underset{\underset{H}{..}}{O}}{:} \rightleftarrows H{:}\overset{\overset{H^{+}}{..}}{\underset{\underset{H}{..}}{N}}{:}H + {:}\ddot{\underset{\underset{H}{..}}{O}}{:}^{-}}$$

$$NH_3 + H_2O \rightleftarrows NH_4^+ + OH^-$$

This reaction is illustrated with space-filling models in Figure 15-4. Furthermore, in water solutions of the strong mineral acids described in Section 15.3, the hydronium ion, H_3O^+, is an acid.

Figure 15-3. When HCl is dissolved in ammonia, a proton is donated to an NH_3 molecule by a polar HCl molecule to form the ammonium ion, NH_4^+, and the chloride ion, Cl^-.

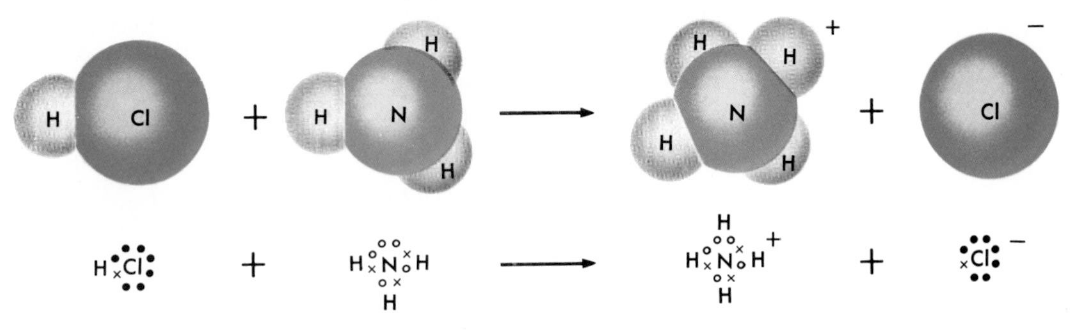

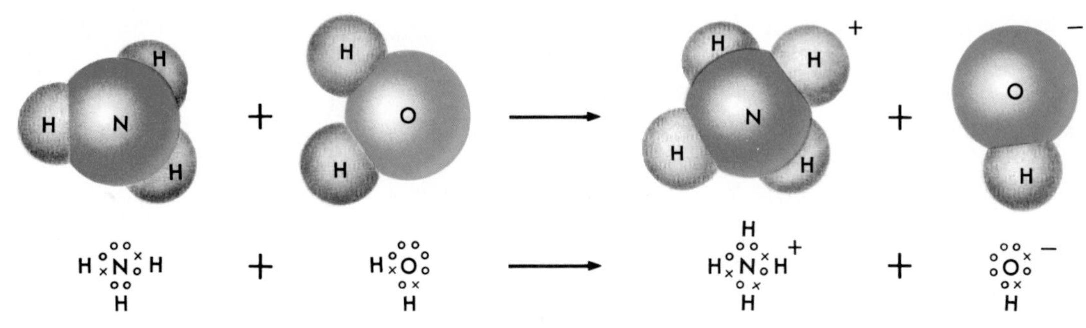

Figure 15-4. When ammonia is dissolved in water, water molecules are the proton donors, and ammonia molecules are the proton acceptors in this reaction.

The ion is the actual proton donor in reactions involving these solutions. This modern definition of acids can be applied to any molecule or ion capable of donating a proton to another molecule or ion.

15.5 Properties of aqueous acids 1. *Acids donate protons when they react with bases.* The many common properties of acids depend on this characteristic behavior. Acids that donate only one proton per molecule are called *monoprotic acids*. Examples are HCl, HNO_3, and $HC_2H_3O_2$. Sulfuric acid, H_2SO_4, and carbonic acid, H_2CO_3, are *diprotic* because they are capable of donating two protons per molecule. Phosphoric acid, H_3PO_4, is *tripotic.*

Relative strengths of aqueous acids are listed in Table 15-5.

2. *Acids contain ionizable hydrogen in covalent combination with a nonmetallic element or polyatomic species.* The strength of an acid depends upon the *degree* of ionization in water solution, not upon the *amount* of hydrogen in the molecule. Perchloric acid, $HClO_4$, hydrochloric acid, HCl, and nitric acid, HNO_3, are strong acids. They are almost completely ionized in water solutions. Each donates one proton per molecule. Acetic acid, $HC_2H_3O_2$, is a weak acid. It ionizes slightly in water to yield one proton and one acetate ion, $C_2H_3O_2^-$, per ionized molecule.

In Section 15.3, you learned that sulfuric acid ionizes in two stages depending on the amount of dilution:

$$H_2SO_4 + H_2O \rightleftarrows H_3O^+ + HSO_4^-$$
$$HSO_4^- + H_2O \rightleftarrows H_3O^+ + SO_4^{--}$$

The first stage is completed in fairly concentrated solutions. The second stage may be appreciable in more dilute solutions. All stages in the ionization of diprotic and triprotic acids occur in the same solution. Therefore, solutions of H_2SO_4 may contain H_3O^+, HSO_4^-, and SO_4^{--} ions. The concentration of ions formed in the first stage is very much greater than the concentration of ions formed in the second stage.

The triprotic phosphoric acid ionizes in three stages.

$$H_3PO_4 + H_2O \rightleftarrows H_3O^+ + H_2PO_4^-$$
$$H_2PO_4^- + H_2O \rightleftarrows H_3O^+ + HPO_4^{--}$$
$$HPO_4^{--} + H_2O \rightleftarrows H_3O^+ + PO_4^{---}$$

Thus a solution of H_3PO_4 may contain each of the following species: H_3PO_4, H_3O^+, $H_2PO_4^-$, HPO_4^{--}, and PO_4^{---}. As with diprotic acids, the concentration of ions formed in the first stage is much greater than the concentration of ions formed in the second stage. Similarly, the concentration of ions formed in the second stage is much greater than the concentration of ions formed in the third stage.

3. *Acids have a sour taste.* We describe lemons, grapefruits, and limes as "sour." These fruits contain weak acids in solution. A solid acid tastes sour as it dissolves in the saliva to form a water solution. Most laboratory acids are very corrosive (they destroy the skin) and are powerful poisons. They must never be tasted or allowed to come in contact with skin or clothing.

Figure 15-5. Blue litmus indicator changes to red in an aqueous acid.

CAUTION: *Never use the "taste test" in the laboratory.*

4. *Acids affect indicators.* If a test strip of blue *litmus* is placed in contact with an acid solution, the blue color changes to red. See Figure 15-5. Litmus is a dye extracted from certain lichens. Other substances are also used as indicators. *Phenolphthalein* (feen-ull-*thall*-een) is colorless in the presence of acids and red in the presence of bases. *Methyl orange* turns red in acid solutions and yellow in basic solutions.

5. *Acids neutralize hydroxides.* Solutions of an acid and a metallic hydroxide can be mixed in chemically equivalent quantities, in which case each cancels the properties of the other. This process is called *neutralization* and is an example of an ionic reaction. The products of neutralization are a salt and water. The salt is recovered in crystalline form by evaporating the water. The acid and the hydroxide neutralize each other.

Suppose a solution containing 1 mole of NaOH is added to a dilute solution containing 1 mole of HCl. The reaction is represented empirically by the following equation:

$$HCl + NaOH \rightarrow NaCl + H_2O$$

Molecular HCl is ionized in dilute solution. The reactants present are H_3O^+ and Cl^- ions.

$$HCl + H_2O \rightarrow H_3O^+ + Cl^-$$

Ionic NaOH is dissociated in water solution. The reactants in this solution are Na^+ and OH^- ions.

$$Na^+OH^- \rightarrow Na^+ + OH^-$$

The reactants in the separate solutions are present as aqueous ions. The original empirical equation can now be written as an ionic equation.

$$H_3O^+ + Cl^- + Na^+ + OH^- \rightarrow Na^+ + Cl^- + 2H_2O$$

The Na^+ and Cl^- ions remain dissociated in the combined solution. They are spectator ions. The ionic species that actually participate in the neutralization reaction are H_3O^+ and OH^- ions.

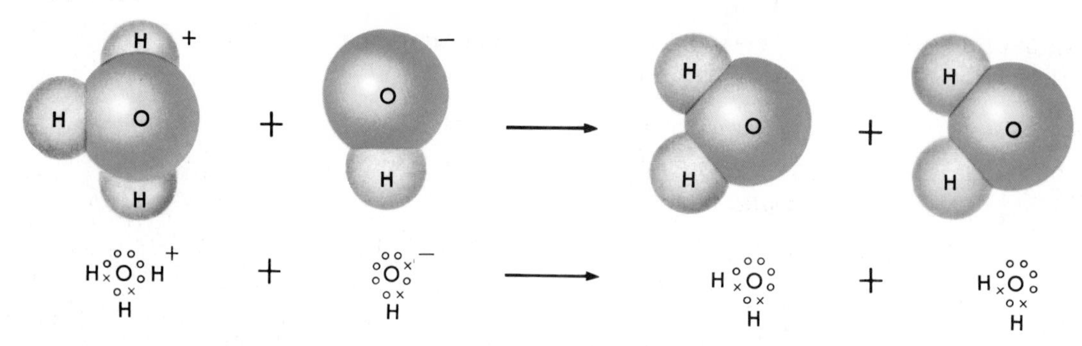

Figure 15-6. In neutralization reactions, hydronium ions and hydroxide ions will combine to form essentially un-ionized water molecules.

They form essentially un-ionized water. The net ionic equation is written as

$$H_3O^+ + OH^- \rightarrow 2H_2O$$

The neutralization reaction between an acid and a base is illustrated with space-filling models in Figure 15-6.

The neutralization reaction is entirely between hydronium ions from the acid and hydroxide ions from the soluble metallic hydroxide. In all neutralizations of very soluble hydroxides by strong acids, the reaction is the same. The nonmetallic ions of the acid and metallic ions of the hydroxide undergo no chemical change. However, you still may prefer to write the complete equation because it shows what pure substances were the original reactants and what salt can be recovered by evaporating the water solvent.

6. *Acids react with many metals.* Solutions of *nonoxidizing* acids, such as hydrochloric acid and dilute sulfuric acid, react with metals above hydrogen in the replacement series (Table 8-2) to liberate hydrogen gas and form a salt. The replacement reaction between zinc and dilute sulfuric acid is typical.

$$Zn(s) + H_2SO_4(aq) \rightarrow ZnSO_4(aq) + H_2(g)$$

Written ionically,

$$Zn(s) + 2H^+(aq) + SO_4^{--}(aq) \rightarrow Zn^{++}(aq) + SO_4^{--}(aq) + H_2(g)$$

Observe that $SO_4^{--}(aq)$ is a spectator ion. Only the species participating in the reaction are included in the net ionic equation.

$$Zn(s) + 2H^+(aq) \rightarrow Zn^{++}(aq) + H_2(g)$$

Upon evaporation of the water solvent, Zn^{++} ions and SO_4^{--} ions separate from solution as crystals of zinc sulfate.

Solutions of *oxidizing* acids, such as nitric acid and hot, concentrated sulfuric acid, react with metals both above and below hydrogen in Table 8-2. These reactions are more complex than the simple replacement example given above. Hot, concentrated sulfuric acid reacts with zinc to form zinc sulfate, hydrogen sulfide gas, and water.

Nitric acid and hot, concentrated sulfuric acid are vigorous oxidizing agents. See Section 6.5 for a review of oxidizing and reducing agents.

$$4Zn(s) + 5H_2SO_4(aq) \rightarrow 4ZnSO_4(aq) + H_2S(g) + 4H_2O(l)$$

Dilute nitric acid reacts with copper to form copper(II) nitrate, nitrogen monoxide, and water.

$$3Cu(s) + 8HNO_3(aq) \rightarrow 3Cu(NO_3)_2(aq) + 2NO(g) + 4H_2O(l)$$

With concentrated nitric acid, copper forms copper(II) nitrate, nitrogen dioxide, and water.

$$Cu(s) + 4HNO_3(aq) \rightarrow Cu(NO_3)_2(aq) + 2NO_2(g) + 2H_2O(l)$$

7. *Acids react with oxides of metals.* Salts and water are the products of these reactions. As an example, consider the reaction of copper(II) oxide and sulfuric acid.

$$CuO + H_2SO_4 \rightarrow CuSO_4 + H_2O$$

Written ionically,

$$CuO + 2H^+(aq) + SO_4^{--} \rightarrow Cu^{++} + SO_4^{--} + H_2O$$

The net reaction is

$$CuO + 2H^+(aq) \rightarrow Cu^{++} + H_2O$$

Crystalline $CuSO_4$ is recovered only by evaporating the solvent water.

8. *Acids react with carbonates.* These reactions produce a salt, water, and carbon dioxide, which is given off.

$$CaCO_3 + 2HCl \rightarrow CaCl_2 + H_2O + CO_2(g)$$

The salt $CaCl_2$ is recovered as a product of the reaction upon evaporation of the solvent water.

This reaction of an acid and a carbonate provides a simple test for carbonates, as with calcium carbonate. Being insoluble in water, calcium carbonate is found in such forms as limestone, marble, and sea shells. A sea shell placed in an acid solution gives the "fizzy" production of CO_2 gas.

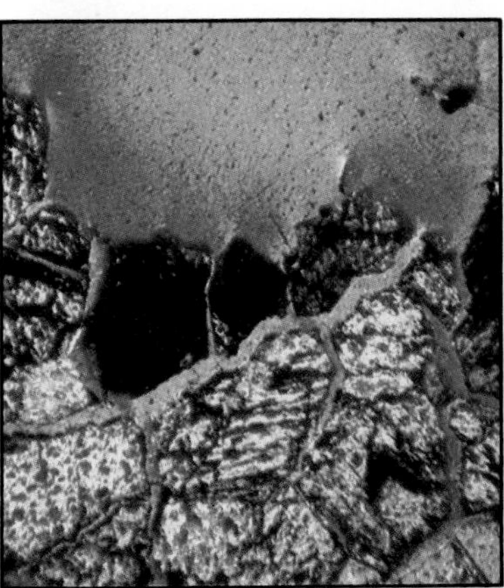

Figure 15-7. A marble statue, the victim of "acid" rain. When fossil fuels are burned, oxides of nitrogen and (usually) sulfur are released into the atmosphere, where they dissolve in moisture to form acidic solutions. These acids react with marble, a form of calcium carbonate, and slowly corrode it. The bottom photo is a photomicrograph of the corrosion.

15.6 Naming aqueous acids Some acids are *binary* compounds. They contain only *two* elements, hydrogen and another nonmetal. Other acids contain oxygen as a third element. They are often referred to as *oxyacids*.

1. *Binary acids.* The name of an acid having only two elements begins with the prefix *hydro-*. The root of the name of the nonmetal in combination with hydrogen follows this prefix. The name ends with the suffix *-ic*. This scheme is illustrated by the examples given in Table 15-2. Water solutions of the binary compounds listed are acids known by the names given in the right column.

2. *Oxyacids.* These acids contain hydrogen, oxygen, and a third element. The formulas and names of the series of oxyacids of chlorine illustrate the general method of naming such acids. The name for the water solution of each oxychlorine acid is shown at the right in the following series.

Table 15-2
NAMES OF BINARY ACIDS

Formula	*Name of pure substance*	*Name of acid*
HF	hydrogen fluoride	hydrofluoric acid
HCl	hydrogen chloride	hydrochloric acid
HBr	hydrogen bromide	hydrobromic acid
HI	hydrogen iodide	hydriodic acid
H_2S	hydrogen sulfide	hydrosulfuric acid

HClO	**hypo-chlor-ous acid**
$HClO_2$	**chlor-ous acid**
$HClO_3$	**chlor-ic acid**
$HClO_4$	**per-chlor-ic acid**

In all these oxyacids, chlorine is the central element. For this reason the root *-chlor-* is used in each case. $HClO_3$ is named *chlor-ic acid*. It contains the chlorate group, and no prefix is used. The oxyacid of chlorine that contains *more* oxygen per molecule than chloric acid is called *per-chlor-ic acid*. The prefix *per-* is a contraction of *hyper*, which means *above*. The acid containing *less* oxygen per molecule than chloric acid is called *chlor-ous acid*. The oxyacid of chlorine that contains *still less* oxygen than chlorous acid has the prefix *hypo-*, the root *-chlor-*, and the suffix *-ous*. The prefix *hypo-* means *below*.

To use this scheme, it is only necessary to know the formula and name of one oxyacid in any series. The formula and name of the member of each series that you should memorize follows.

$HClO_3$	**chloric acid**
HNO_3	**nitric acid**
$HBrO_3$	**bromic acid**
H_2SO_4	**sulfuric acid**
H_3PO_4	**phosphoric acid**

The names of common oxyacids are given in Table 15-3.

15.7 Acid anhydrides Only fluorine is more electronegative than oxygen. Its compound with oxygen, OF_2, is a fluoride rather than an oxide. Other elements form oxides with oxygen. The oxidation number of oxygen in these compounds is -2. Oxides range structurally from ionic to covalent. The more ionic oxides involve the highly electropositive metals on the left side of the periodic table. The oxides formed with nonmetals on the right side of the periodic table are covalent molecules.

Many of the molecular (nonmetallic) oxides are gases at ordinary temperatures. Examples are carbon monoxide, CO, carbon dioxide, CO_2, and sulfur dioxide, SO_2. Diphosphorus pentoxide, on the other hand, is a solid. The name "diphosphorus

Table 15-3
NAMES OF OXYACIDS

Formula	Name of pure substance	Name of acid
$HC_2H_3O_2$	hydrogen acetate	acetic acid
H_2CO_3	hydrogen carbonate	carbonic acid
$HClO$	hydrogen hypochlorite	hypochlorous acid
$HClO_2$	hydrogen chlorite	chlorous acid
$HClO_3$	hydrogen chlorate	chloric acid
$HClO_4$	hydrogen perchlorate	perchloric acid
HNO_2	hydrogen nitrite	nitrous acid
HNO_3	hydrogen nitrate	nitric acid
H_3PO_3	hydrogen phosphite	phosphorous acid
H_3PO_4	hydrogen phosphate	phorphoric acid
H_2SO_3	hydrogen sulfite	sulfurous acid
H_2SO_4	hydrogen sulfate	sulfuric acid

pentoxide" is that of the empirical formula P_2O_5. (The molecular formula is now known to be P_4O_{10}.) Most nonmetallic oxides react with water to form *oxyacids.*

All oxyacids contain one or more oxygen-hydrogen (O—H) groups in the covalent structure. They are called *hydroxyl* groups. They are not to be confused with O—H groups existing as OH^- ions whose compounds are called *hydroxides.* The hydroxyl group is arranged in the molecule in such a manner that it may donate a proton. This arrangement gives the molecule its acid character.

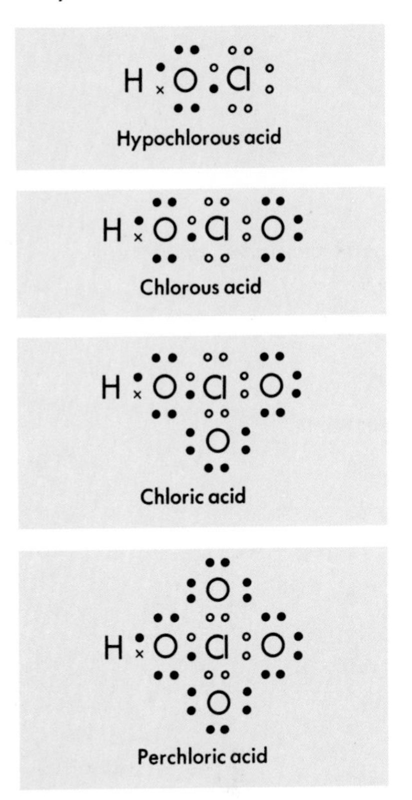

Figure 15-8. Electron-dot formulas for the four oxyacids of chlorine.

Figure 15-8 shows the electron-dot formulas of the four oxyacids of chlorine. Each formula contains the O—H group, not as an ion but as a group covalently bonded to the central Cl atom. Aqueous solutions of these solutes are acidic because the O—H bond is broken during ionization. The O—H group provides the proton donated by the oxychlorine acid molecule.

When carbon dioxide dissolves in water, a very small amount of it reacts chemically with the solvent to form carbonic acid.

$$CO_2 + H_2O \rightleftarrows H_2CO_3$$

These H_2CO_3 molecules can ionize and give the aqueous solution a very low concentration of H_3O^+ ions.

$$H_2CO_3 + H_2O \rightleftarrows H_3O^+ + HCO_3^-$$

The net equation that shows the acid-producing behavior of CO_2 in aqueous solution is

$$CO_2 + 2H_2O \rightleftarrows H_3O^+ + HCO_3^-$$

Carbonic acid and carbon dioxide differ in composition by just a molecule of water. Thus, carbon dioxide is called the *acid anhydride* of carbonic acid. The name *anhydride* means "without water." *Oxides that react with water to form acids, or that are formed by the removal of water from acids, are known as* ***acid anhydrides.***

Acid anhydride: See Section 12.18(4).

Binary acids do not contain oxygen and do not have an anhydride form. Thus the reaction between an acid anhydride and water cannot be used to prepare such acids. However, it is an important method of preparing some oxyacids.

Sulfur dioxide is the acid anhydride of sulfurous acid.

$$SO_2 + H_2O \rightleftarrows H_2SO_3$$

Acid anhydrides form oxyacids.

Sulfur trioxide is the acid anhydride of sulfuric acid.

$$SO_3 + H_2O \rightleftarrows H_2SO_4$$

These anhydrides are important in the manufacture of sulfuric acid. Sulfuric acid, because it is cheap and has a high boiling point, can be used in the laboratory to produce several other acids. Although most hydrochloric acid is now produced commercially by other methods, the reaction for producing hydrogen chloride gas is an example.

$$H_2SO_4 + 2NaCl \rightarrow Na_2SO_4 + 2HCl(g)$$

Nitric acid can be produced by the reaction of sulfuric acid with a nitrate. However, this process is not used commercially for preparing HNO_3 because it is more expensive than other processes.

Bases

15.8 The nature of bases Several substances long known as bases are commonly found in homes. Household ammonia, an ammonia-water solution, is a familiar cleaning agent. Lye is a commercial grade of sodium hydroxide, NaOH, used for cleaning clogged sink drains. Limewater is a solution of calcium hydroxide, $Ca(OH)_2$. Milk of magnesia is a suspension of magnesium hydroxide, $Mg(OH)_2$, in water; it is used as an antacid, a laxative, and an antidote for strong acids. All of these hydroxides are *aqueous bases.*

Arrhenius considered a base to be any soluble hydroxide that neutralized an acid when their solutions were mixed. You now know that the only reaction occurring in such a neutralization is between hydronium ions and the hydroxide ions. The nonmetal of the acid and the metal of the hydroxide remain in solution as hydrated ions. Hydronium ions combine with hydroxide ions to form water.

$$H_3O^+(aq) + OH^-(aq) \rightarrow 2H_2O(l)$$

Ammonia-water solutions and water solutions of soluble metallic hydroxides are commonly referred to as *aqueous bases.* They are the most useful basic solutions. Aqueous ammonia solutions are traditionally called *ammonium hydroxide,* NH_4OH. This molecular species probably does not exist in water solutions except possibly through the formation of hydrogen bonds between some NH_3 and H_2O molecules. These solutions are more appropriately called ammonia-water solutions, or simply NH_3(aq) (aqueous ammonia). The most common basic solutions used in the laboratory are those of NaOH, KOH, $Ca(OH)_2$, and NH_3(aq). Chemists refer to them as being *alkaline* in behavior.

In the Brønsted definition, an acid is a *proton donor.* Accordingly, *a* ***base*** *is a proton acceptor.* In this sense, the hydroxide ion, OH^-, is the most common base. However, many other species also combine with protons. The Brønsted use of the term *base* includes all species that accept protons in solution.

You know that hydrogen chloride ionizes in water solution as a result of the hydrating action of the solvent dipoles.

$$\underset{\text{acid}}{HCl} + \underset{\text{base}}{H_2O} \rightarrow \underset{\text{acid}}{\overset{\text{weaker}}{H_3O^+}} + \underset{\text{base}}{\overset{\text{weaker}}{Cl^-}}$$

Here the water molecule is the base. It accepts a proton from the HCl molecule, the acid. This reaction forms the H_3O^+ ion, which is a weaker acid than HCl, and the Cl^- ion, which is a weaker base than H_2O. The relative strengths of acids and bases are discussed further in section 15.12.

Earlier, the neutralization reaction between aqueous solutions of HCl and NaOH was described. Here the H_3O^+ ion acts as the acid since it, and not the HCl molecule, is the proton donor. The OH^- ion is, of course, the proton acceptor or base.

$$\underset{\text{acid}}{H_3O^+} + \underset{\text{base}}{OH^-} \rightarrow \underset{\text{acid}}{H_2O} + \underset{\text{base}}{H_2O}$$

When HCl gas is dissolved in liquid ammonia, the ammonia acts as the base.

$$\underset{\text{acid}}{HCl} + \underset{\text{base}}{NH_3} \rightarrow \underset{\text{acid}}{NH_4^+} + \underset{\text{base}}{Cl^-}$$

When NH_3 gas is dissolved in water, the water donates protons and therefore acts as an acid. Ammonia accepts protons and therefore is the base. A low concentration of NH_4^+ ions and OH^- ions is produced in the reversible reaction.

$$\underset{\text{base}}{NH_3} + \underset{\text{acid}}{H_2O} \rightleftarrows \underset{\text{acid}}{NH_4^+} + \underset{\text{base}}{OH^-}$$

Suppose open vessels of a saturated aqueous NH_3 solution (concentrated ammonium hydroxide) and a saturated aqueous HCl solution (concentrated hydrochloric acid) are placed near each other. A dense white cloud of NH_4Cl forms above the vessels. This result shows how HCl (the acid) and NH_3 (the base) can react without the intervention of water. See Figure 15-10.

The Brønsted proton-transfer system of acids and bases is more general in scope than the aqueous system of acids and

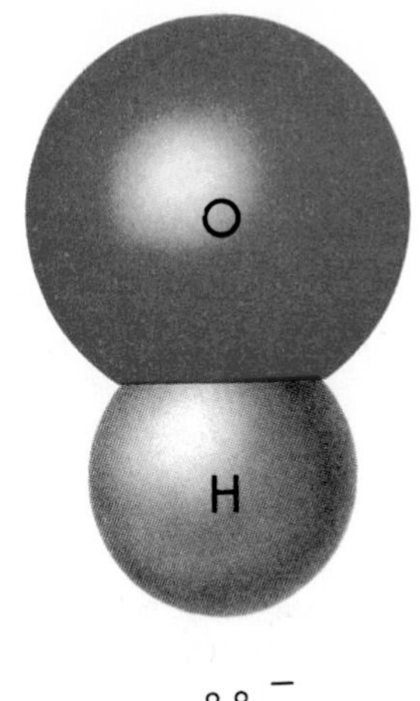

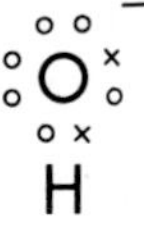

Figure 15-9. A space-filling model of the hydroxide ion and its electron-dot formula. A water molecule is formed when the hydroxide ion accepts a proton.

Figure 15-10. The acid HCl and the base NH_3 react to form the white cloud of NH_4Cl.

bases. It extends the acid-base concept to reactions in nonaqueous solutions. You will be concerned mainly with aqueous acids and aqueous bases, water solutions of H_3O^+ ions, and water solutions of OH^- ions.

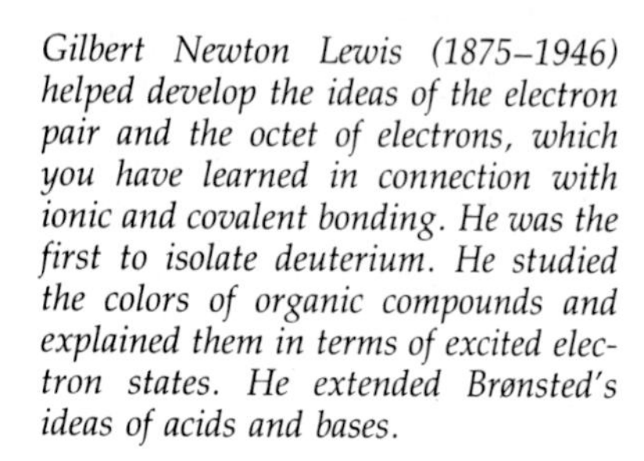
Gilbert Newton Lewis (1875–1946) helped develop the ideas of the electron pair and the octet of electrons, which you have learned in connection with ionic and covalent bonding. He was the first to isolate deuterium. He studied the colors of organic compounds and explained them in terms of excited electron states. He extended Brønsted's ideas of acids and bases.

An even more general concept of acids and bases was defined in 1938 by G. N. Lewis, an American chemist. According to the Lewis acid-base system, any species that acts as an *electron-pair acceptor* in a chemical reaction is an *acid*. An *electron-pair donor* is a *base*. The Brønsted system extends the proton transfer concept of acid-base reactions beyond the aqueous system to non-aqueous solutions. The Lewis acid-base system extends the concept of acid-base reactions to some reactions that do not involve proton transfers.

15.9 Properties of hydroxides 1. *Hydroxides of the active metals supply OH^- ions in solution.* Sodium and potassium hydroxides are very soluble in water. They are ionic compounds and are, therefore, completely ionized in water solution. Their solutions are *strongly alkaline* because of the high concentration of strongly basic OH^- ions. In speaking of such solutions, chemists often attach the property of the ions to the solution itself. Thus, they may speak of "strongly basic solutions."

$$Na^+ OH^- \rightarrow Na^+(aq) + OH^-(aq)$$
$$K^+ OH^- \rightarrow K^+(aq) + OH^-(aq)$$

Calcium and strontium hydroxides are also ionic compounds. Thus their water solutions are completely ionized. However, they are only slightly soluble in water and, therefore, only *moderately basic.*

$$Ca^{++}(OH^-)_2 \rightarrow Ca^{++}(aq) + 2OH^-(aq)$$
$$Sr^{++}(OH^-)_2 \rightarrow Sr^{++}(aq) + 2OH^-(aq)$$

The strength of the base depends on the *concentration* of OH^- ions *in solution*. It does not depend on the number of hydroxide ions per mole of the compound.

Ammonia-water solutions are *weakly basic* because they have a low concentration of OH^- ions. Ammonia, NH_3, is not a strong base and thus does not acquire very many protons from water molecules when in solution. Relatively few NH_4^+ ions and OH^- ions are present.

$$NH_3(aq) + H_2O \rightleftharpoons NH_4^+(aq) + OH^-(aq)$$

2. *Soluble hydroxides have a bitter taste.* Possibly you have tasted limewater and know that it is bitter. Soapsuds also taste bitter because of the presence of hydroxide ions. Strongly basic solutions are very caustic (they chemically burn the skin), and the metallic ions in them are sometimes poisonous.

3. *Solutions of hydroxides feel slippery.* The very soluble hydroxides, such as sodium hydroxide, attack the skin and can produce severe caustic burns. Dilute solutions have a soapy, slippery feel when rubbed between the thumb and fingers.

4. *Soluble hydroxides affect indicators.* The basic OH^- ions in solutions of the soluble hydroxides cause *litmus* to turn from *red* to *blue.* This is just the opposite of the color change caused by H_3O^+ ions of acid solutions. In a basic solution, *phenolphthalein* turns *red,* and *methyl orange* changes to *yellow.* Many insoluble hydroxides, on the other hand, produce too few hydroxide ions to affect indicators.

5. *Hydroxides neutralize acids.* The neutralization of HNO_3 by KOH can be represented empirically by the equation

$$KOH + HNO_3 \rightarrow KNO_3 + H_2O$$

Of course, ionic KOH dissociates in water solution and exists as hydrated K^+ ions and OH^- ions.

$$K^+OH^- \rightarrow K^+ + OH^-$$

In water solution, the covalent HNO_3 is ionized and exists as hydrated protons and nitrate ions.

$$HNO_3 + H_2O \rightarrow H_3O^+ + NO_3^-$$

The complete ionic equation for this neutralization reaction is

$$H_3O^+ + NO_3^- + K^+ + OH^- \rightarrow K^+ + NO_3^- + 2H_2O$$

With the removal of the spectator ions, the net ionic equation becomes

$$H_3O^+ + OH^- \rightarrow 2H_2O$$

This represents the only chemical reaction during the neutralization process. The hydrated K^+ and NO_3^- ions join as ionic crystals of the salt KNO_3 only when the water is evaporated.

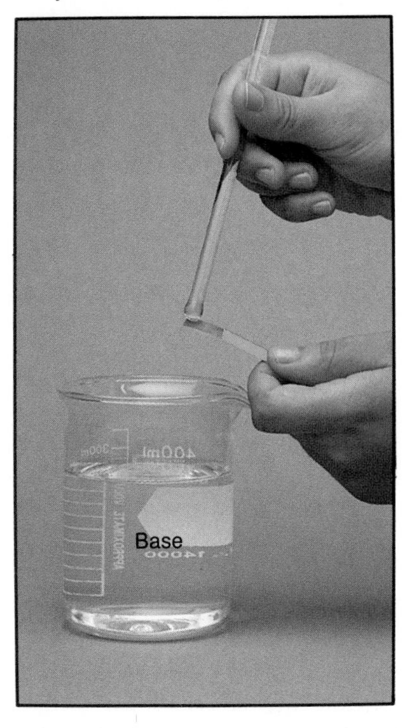

Figure 15-11. Red litmus indicator changes to blue in an aqueous base.

6. *Hydroxides react with the oxides of nonmetals.* Such reactions form salts and sometimes water. For example, sodium hydroxide reacts with carbon dioxide in different ways. The products may be either carbonate *or* hydrogen carbonate ions, depending on the relative quantities of reactants. Two moles of NaOH per mole of CO_2 forms sodium carbonate, Na_2CO_3, and H_2O.

$$\mathbf{CO_2 + 2NaOH \rightarrow Na_2CO_3 + H_2O}$$

Common names for sodium hydrogen carbonate, $NaHCO_3$, are "sodium bicarbonate" and "baking soda."

One mole of NaOH per mole of CO_2 forms only sodium hydrogen carbonate, $NaHCO_3$.

$$\mathbf{CO_2 + NaOH \rightarrow NaHCO_3}$$

The net reactions are

$$\mathbf{CO_2 + 2OH^- \rightarrow CO_3^{--} + H_2O}$$

and

$$\mathbf{CO_2 + OH^- \rightarrow HCO_3^-}$$

Carbon dioxide is the acid anhydride of carbonic acid. Therefore, these reactions are essentially neutralization reactions between carbonic acid and sodium hydroxide.

7. *Certain hydroxides are amphoteric.* The metallic hydroxides, except those of the active metals, are practically insoluble in water. They are weakly basic in the presence of strong acids. They are weakly acidic in the presence of strong bases. *Substances that have acidic or basic properties under appropriate conditions are said to be* ***amphoteric.*** Amphoteric hydroxides, while insoluble in water, are soluble in acids and bases. Aluminum is a metal whose ions form an amphoteric hydroxide. The amphoteric nature of such metallic hydroxides is related to reactions involving hydrated ions.

Aluminum hydroxide separates as a white jelly-like precipitate when hydroxide ions (as NaOH, for example) are added to a solution of a soluble aluminum salt.

$$\mathbf{Al(H_2O)_6^{+++}(aq) + 3OH^-(aq) \rightarrow Al(H_2O)_3(OH)_3(s) + 3H_2O(l)}$$

The OH^- ions are strongly basic. They accept protons from three water molecules of each hydrated aluminum ion to form the insoluble hydrated aluminum hydroxide, $Al(H_2O)_3(OH)_3$.

When an excess of the base is added, the high concentration of hydroxide ions causes the aluminum hydroxide precipitate to dissolve. The negative aluminate ions are formed.

$$\mathbf{Al(H_2O)_3(OH)_3(s) + OH^-(aq) \rightarrow Al(H_2O)_2(OH)_4^-(aq) + H_2O(l)}$$

The OH^- ion removes a proton from one of the three water molecules in the hydrated aluminum hydroxide structure. Observe the soluble aluminate ion formed has two molecules of water of hydration. In this reaction, the hydrated aluminum hydroxide *acts as an acid in the presence of the strong base.*

Table 15-4
SOLUBILITY OF METALLIC HYDROXIDES

Hydroxide	*Solubility* (g/100 g H_2O at 20 °C)
soluble (> 1 g/100 g H_2O)	
KOH	112
NaOH	109
LiOH	12.4
$Ba(OH)_2$	3.89
slightly soluble (>0.1 g/100 g H_2O)	
$Ca(OH)_2$	0.173
insoluble (<0.1 g/100 g H_2O)	
$Pb(OH)_2$	0.016
$Mg(OH)_2$	0.0009
$Sn(OH)_2$	0.0002
$Zn(OH)_2$	negligible
$Cu(OH)_2$	negligible
$Al(OH)_3$	negligible
$Cr(OH)_3$	negligible
$Fe(OH)_3$	negligible

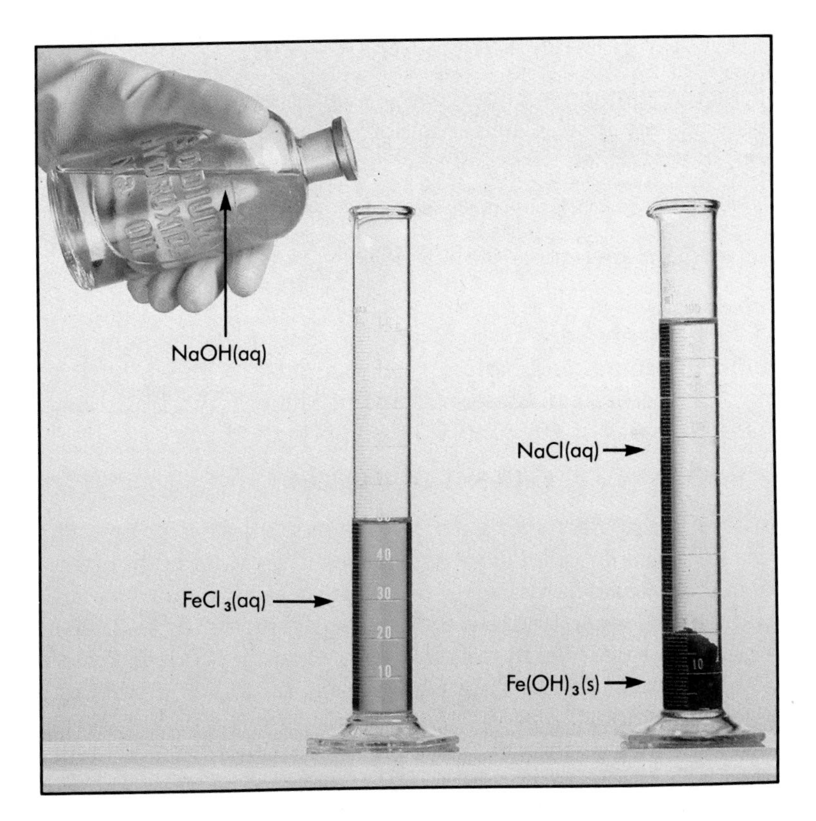

Figure 15-12. The hydroxides of heavy metals are practically insoluble as shown in this reaction.

The amphoteric aluminum hydroxide also dissolves in an excess of hydronium ions. Hydrated aluminum ions and water are the products.

$$Al(H_2O)_3(OH)_3(s) + 3H_3O^+ \rightarrow Al(H_2O)_6{}^{+++}(aq) + 3H_2O(l)$$

This reaction is the reverse of the reaction above in which the hydrated aluminum hydroxide is precipitated. The three hydroxide ions accept protons from the three hydronium ions to form the three water molecules. In this reaction, the hydrated aluminum hydroxide *acts as a base in the presence of the strong acid.* It illustrates proton-transfer reactions, or protolysis, in which protons are transferred from an acid to a base.

The hydroxides of zinc, lead(II), chromium(III), and antimony(III) are other common amphoteric hydroxides. Iron(III) hydroxide is not amphoteric. It is not soluble in an excess of OH^- ions. See Figure 15-12.

In the Brønsted system of acids and bases, water is an amphoteric substance. When a water molecule accepts a proton from hydrogen chloride, it acts as a base. On the other hand, when a water molecule donates a proton to ammonia, it acts as an acid. Indeed, in the slight ionization of water, one water molecule donates a proton to another water molecule. Thus some of the water molecules behave as an acid while others behave as a base.

$$H_2O + H_2O \rightleftarrows H_3O^+ + OH^-$$

Liquid ammonia undergoes a similar ionization, but to a lesser extent than water. Ammonium ions and amide ions, NH_2^-, are formed.

$$NH_3 + NH_3 \rightleftarrows NH_4^+ + NH_2^-$$

Ammonia is therefore amphoteric.

15.10 Basic anhydrides. Oxides of sodium, potassium, calcium, strontium, and barium react vigorously with water. You may have seen a plasterer *slaking* quicklime, CaO, by adding water to it. *Slaked lime,* $Ca(OH)_2$, was being formed.

$$CaO + H_2O \rightarrow Ca(OH)_2$$

Basic anhydrides are oxides of active metals. See Section 12.18(3).

Oxides that react with water to produce solutions containing the basic OH^- *ions are called* ***basic anhydrides.*** The oxides of the active metals are basic anhydrides. (In contrast, acid anhydrides are oxides of nonmetals. They are covalent compounds and have molecular structures. In reactions with water they form oxyacids with one or more O—H groups capable of donating protons. Refer to Section 15.7.)

The oxides of the active metals are ionic in structure. Like other ionic substances, they are solids at room temperature. They contain O^{--} ions. When placed in water, O^{--} ions react with the water to form basic OH^- ions.

$$O^{--} + H_2O \rightarrow 2OH^-$$

If the metallic hydroxide is soluble in water, the solution is basic because of the presence of OH^- ions.

15.11 Hydroxides and periodic trends In general, the hydroxides of the active metals are strongly basic. The O—H groups are present as ions, and the compounds are usually quite soluble. Other hydroxide compounds may be weakly basic, amphoteric, or acidic. The higher the oxidation state of the atom combined with the O—H group, the more covalent is the bond between this atom and the O—H group. Further, the more covalent the bond, the more difficult it is to remove the OH^- ion. For example, chromium(II) hydroxide is basic, chromium(III) hydroxide is amphoteric, and chromium(VI) forms an oxyacid.

With amphoteric hydroxides, it appears that the O—H bond is as easily broken as the bond between the metal and the O—H group. A strong base acquires a proton by breaking the O—H bond. An acid, on the other hand, donates a proton to the O—H group. In doing so, it breaks the bond between the O—H group and the metal atom.

Hydroxides of atoms having high electronegativity and high oxidation states are acidic. The O—H bond is more easily broken than the bond between the O—H group and the central atom. For example, O_3ClOH is strongly acidic. The chlorine atom is in the +7 oxidation state. It is also the central atom to which three oxygen atoms and one O—H group are bonded. In water solution, the molecule donates a proton from its O—H group to form the negative O_3ClO^- ion.

$$O_3ClOH + H_2O \rightarrow H_3O^+ + O_3ClO^-$$

If the formula is written in the conventional form of acids, O_3ClOH becomes $HClO_4$. You can recognize this formula as *perchloric* acid, an oxyacid. The equation now looks more familiar.

$$HClO_4 + H_2O \rightarrow H_3O^+ + ClO_4^-$$

Figure 15-8 shows the electron-dot configuration from which the formula O_3ClOH is easily recognized.

It may not be apparent from the formula of a substance whether it has acidic or basic properties. You have observed that oxides that react with water form oxygen-hydrogen groups. In general, the oxygen-hydrogen groups formed by ionic oxides (metal oxides) are OH^- ions. Their compounds are hydroxides, and their solutions are basic.

Oxygen-hydrogen groups formed by molecular oxides (nonmetal oxides) are not ionic. Instead, they are bonded covalently to another atom in the product molecule. Such groups are identified as hydroxyl groups in Section 15.7. Hydroxyl groups donate protons in aqueous solution. The compounds have acid properties and are oxyacids.

Figure 15-13 shows electron-dot formulas for several molecular substances that contain hydroxyl groups. Their formulas can be written $PO(OH)_3$, $SO_2(OH)_2$, CH_3COOH, and C_2H_5OH. However, none has the basic properties characteristic of OH^- ions in water solution. The first three compounds are oxyacids. This acidic character is recognized by writing their formulas as H_3PO_4, H_2SO_4, and $HC_2H_3O_2$. The fourth compound, ethanol, lacks either acidic or basic properties in aqueous solution. Observe that its formula can be written as HC_2H_5O.

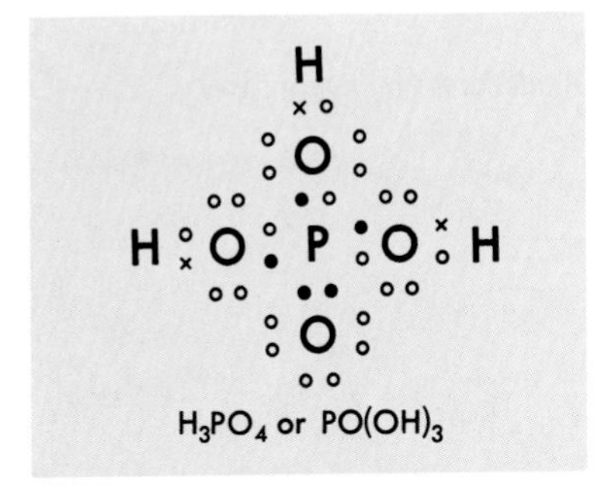

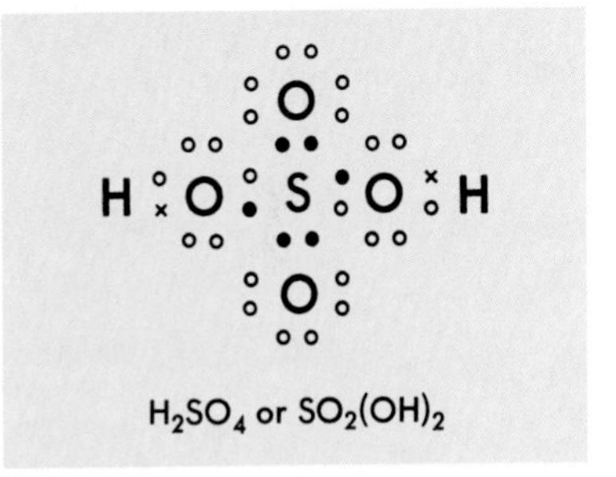

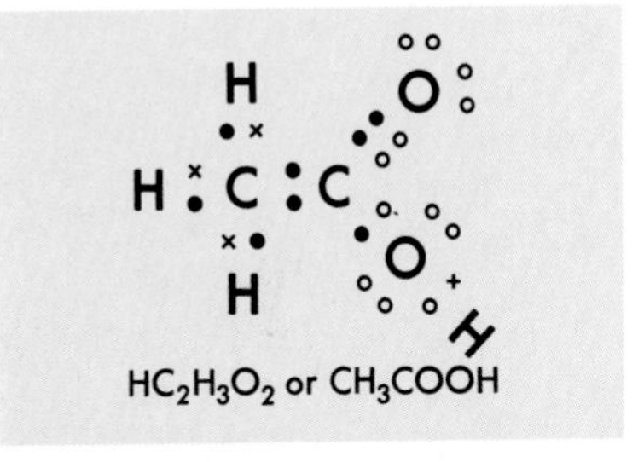

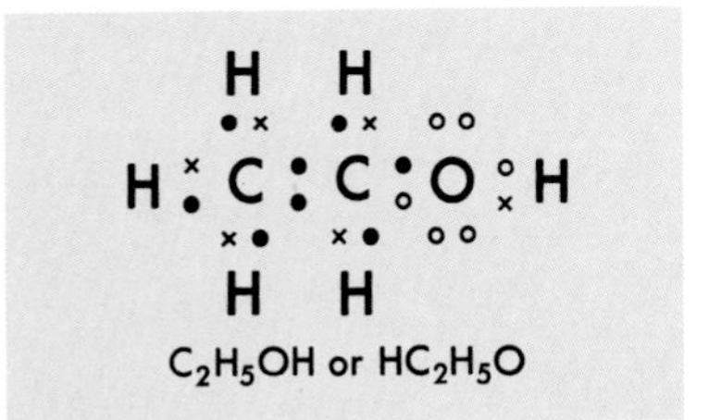

Figure 15-13. Molecular formulas alone will not identify acidic or basic properties.

15.12 Relative strengths of acids and bases

The Brønsted definition of acids and bases provides a basis for the study of protolysis, or proton-transfer reactions. Any molecule or ion capable of donating a proton is considered to be an acid. Any molecule or ion that can accept the proton is a base.

Suppose that an acid (in the Brønsted sense) gives up a proton. The remainder of the acid particle itself is then capable of accepting a proton. Therefore, this remaining particle may be considered to be a base; it is called a *conjugate base. A* ***conjugate base*** *is the species that remains after an acid has given up a proton.*

An aqueous solution of sulfuric acid contains H_3O^+ ions and HSO_4^- ions. With further dilution, the HSO_4^- ions may give up protons to H_2O molecules.

$$\underset{\text{acid}}{HSO_4^-} + \underset{\text{base}}{H_2O} \rightarrow \underset{\text{acid}}{H_3O^+} + \underset{\text{base}}{SO_4^{--}}$$

Acids have conjugate bases.

The SO_4^{--} ion is the remainder of the HSO_4^- ion after its proton has been removed. It is the conjugate base of the acid HSO_4^-.

The SO_4^{--} ion, as a base, can accept a proton from the H_3O^+ ion. When this occurs, the acid HSO_4^- ion is formed.

$$\underset{\text{acid}}{H_3O^+} + \underset{\text{base}}{SO_4^{--}} \rightarrow \underset{\text{acid}}{HSO_4^-} + \underset{\text{base}}{H_2O}$$

Bases have conjugate acids.

The HSO_4^- ion can be called the *conjugate acid* of the base SO_4^{--}. *A* ***conjugate acid*** *is the species formed when a base takes on a proton.* Thus in the example given, the HSO_4^- ion and the SO_4^{--} ion are a *conjugate acid-base pair.*

The H_2O molecule also acts as a base in this reaction. It receives the proton given up by the HSO_4^- ion to form the H_3O^+ ion. Thus the H_3O^+ ion is the conjugate acid of the base H_2O.

Similarly, the acidic H_3O^+ ion gives up a proton to the basic SO_4^{--} ion to form the H_2O molecule. Thus the H_2O molecule is the conjugate base of the H_3O^+ ion. The H_2O molecule and the H_3O^+ ion make up the second conjugate acid-base pair in the reaction.

In the foregoing reaction, each reactant and product is labeled either as an acid or a base. There are two conjugate acid-base pairs.

Conjugate acid-base pair	HSO_4^-	SO_4^{--}
	acid	**base**
Conjugate acid-base pair	H_2O	H_3O^+
	base	**acid**

Each acid has one more proton than its conjugate base.

Similar reasoning can be applied to the equation for the initial ionization of H_2SO_4. Here the HSO_4^- ion is the *conjugate base* of H_2SO_4.

$$\underset{\text{acid}}{H_2SO_4} + \underset{\text{base}}{H_2O} \rightarrow \underset{\text{acid}}{H_3O^+} + \underset{\text{base}}{HSO_4^-}$$

The HSO_4^- ion is *both* the conjugate acid of the base SO_4^{--} and the conjugate base of the acid H_2SO_4. Thus the HSO_4^- ion is amphoteric.

You know that HCl is highly ionized even in concentrated aqueous solutions. The hydrogen chloride molecule gives up protons readily and is therefore a strong acid. It follows that the Cl^- ion, the conjugate base of this acid, has little tendency to retain the proton. It is, consequently, a weak base.

This observation suggests an important statement that follows naturally from the Brønsted definition of acids and bases: *the stronger an acid, the weaker its conjugate base; the stronger a base, the weaker its conjugate acid.*

The stronger the acid, the weaker its conjugate base. The stronger the base, the weaker its conjugate acid.

An aqueous solution of the strong acid $HClO_4$ is highly ionized. The reaction to the right is practically complete even in concentrated solutions.

$$\underset{\text{acid}}{\overset{\text{stronger}}{HClO_4}} + \underset{\text{base}}{\overset{\text{stronger}}{H_2O}} \rightleftarrows \underset{\text{acid}}{\overset{\text{weaker}}{H_3O^+}} + \underset{\text{base}}{\overset{\text{weaker}}{ClO_4^-}}$$

The ClO_4^- ion is the conjugate base of $HClO_4$. It is too weak a base to compete successfully with the base H_2O in acquiring a proton. The H_3O^+ ion is the conjugate acid of H_2O. It is too weak an acid to compete successfully with the acid $HClO_4$ in donating a proton. Thus there is little tendency for the reaction to proceed to the left to re-form the $HClO_4$ and H_2O molecules.

Now examine the situation in an aqueous solution of acetic acid.

$$\underset{\text{acid}}{\overset{\text{weaker}}{HC_2H_3O_2}} + \underset{\text{base}}{\overset{\text{weaker}}{H_2O}} \rightleftarrows \underset{\text{acid}}{\overset{\text{stronger}}{H_3O^+}} + \underset{\text{base}}{\overset{\text{stronger}}{C_2H_3O_2^-}}$$

The H_3O^+ ion concentration is quite low even in dilute solutions. This fact indicates that $HC_2H_3O_2$ is indeed a weak acid. It does not compete very successfully with H_3O^+ ions in donating protons to a base. The H_2O molecules do not compete very successfully with $C_2H_3O_2^-$ ions in accepting protons. The H_3O^+ ion is the stronger acid and the $C_2H_3O_2^-$ ion is the stronger base. Thus the reaction tends to proceed to the left.

Observe that in each example the stronger acid had the weaker conjugate base and the stronger base had the weaker conjugate acid. Also note that, in each reversible situation, the reaction tended to proceed toward the weaker acid and base.

These observations suggest a second important statement that follows naturally from the Brønsted definition: *Protolysis reactions favor the production of the weaker acid and the weaker base.*

Protolysis occurs when a proton donor and a proton acceptor are brought together in a solution. The extent of this protolysis

Table 15-5
RELATIVE STRENGTHS OF ACIDS AND BASES

Acid	Formula	Conjugate base	Formula
perchloric	$HClO_4$	perchlorate ion	ClO_4^-
hydrogen chloride	HCl	chloride ion	Cl^-
nitric	HNO_3	nitrate ion	NO_3^-
sulfuric	H_2SO_4	hydrogen sulfate ion	HSO_4^-
hydronium ion	H_3O^+	water	H_2O
hydrogen sulfate ion	HSO_4^-	sulfate ion	SO_4^{--}
phosphoric	H_3PO_4	dihydrogen phosphate ion	$H_2PO_4^-$
hydrogen fluoride	HF	fluoride ion	F^-
acetic	$HC_2H_3O_2$	acetate ion	$C_2H_3O_2^-$
carbonic	H_2CO_3	hydrogen carbonate ion	HCO_3^-
hydrogen sulfide	H_2S	hydrosulfide ion	HS^-
ammonium ion	NH_4^+	ammonia	NH_3
hydrogen carbonate ion	HCO_3^-	carbonate ion	CO_3^{--}
water	H_2O	hydroxide ion	OH^-
ammonia	NH_3	amide ion	NH_2^-
hydrogen	H_2	hydride ion	H^-

Decreasing Acid Strength (downward arrow) — *Decreasing Base Strength* (upward arrow)

depends on the relative strengths of the acids and bases involved. For a proton-transfer reaction to approach completion, the reactants must be much stronger as an acid and a base than the products.

Table 15-5 shows the relative strengths of several Brønsted acids and their conjugate bases. Observe that the strongest acid listed, $HClO_4$, has the weakest conjugate base listed, ClO_4^-. The weakest acid, H_2, has the strongest conjugate base, the hydride ion, H^-. A violent protolysis could result from bringing together the strongest acid and the strongest base in certain proportions. Such a reaction would be highly exothermic and dangerous.

SALTS

15.13 Nature of salts Common table salt, NaCl, is only one of a large class of compounds that chemists refer to as *salts*. The solution of an aqueous acid contains H_3O^+ ions and negatively charged nonmetal ions (anions). These particles result from ionization of the acid in water. The solution of an aqueous base contains positively charged metal ions (cations) and OH^- ions. These particles result from dissociation of the ionic metallic hydroxide in water. A neutralization reaction between two such solutions removes almost all the H_3O^+ and OH^- ions. They unite to form water, which is only very slightly ionized. The cations of the hydroxide and the anions of the acid are spectator ions. They have no part in the neutralization reaction.

As an example, consider the neutralization reaction between aqueous solutions of HCl and KOH. Aqueous ions are present in the acid solution because of ionization.

Table 15-6
SOLUBILITY OF SALTS

1. Common sodium, potassium, and ammonium compounds are soluble in water.
2. Common nitrates, acetates, and chlorates are soluble.
3. Common chlorides are soluble except silver, mercury(I), and lead. [Lead(II) chloride is soluble in hot water.]
4. Common sulfates are soluble except calcium, barium, strontium, and lead.
5. Common carbonates, phosphates, and silicates are insoluble except sodium, potassium, and ammonium.
6. Common sulfides are insoluble except calcium, barium, strontium, magnesium, sodium, potassium, and ammonium.

$$HCl(g) + H_2O \rightarrow H_3O^+(aq) + Cl^-(aq)$$

Aqueous ions are present in the hydroxide solution because of dissociation.

$$K^+OH^-(s) \rightarrow K^+(aq) + OH^-(aq)$$

Neutralization occurs when the proper quantities of the two solutions are mixed.

$$H_3O^+(aq) + Cl^-(aq) + K^+(aq) + OH^-(aq) \rightarrow K^+(aq) + Cl^-(aq) + 2H_2O$$

Removing all spectator ions, the net equation for the neutralization reaction becomes

$$H_3O^+(aq) + OH^-(aq) \rightarrow 2H_2O$$

As solvent water is evaporated, K^+ cations and Cl^- anions no longer remain separated from each other by water dipoles. They form a characteristic ionic crystalline structure and separate from solution as the *salt* KCl. *A compound composed of the positive ions of an aqueous base and the negative ions of an aqueous acid is called a* ***salt.*** All salts, by this definition, are ionic substances. They vary in solubility in water, but their aqueous solutions are solutions of hydrated ions. See Table 15-6 on the solubility of salts.

ROUGH RULES OF SOLUBILITY

soluble:	*>1 g/100 g H_2O*
slightly soluble:	*>0.1 g/100 g H_2O*
insoluble:	*<0.1 g/100 g H_2O*

15.14 Salt-producing reactions There are several ways of forming salts, but not all of these ways apply to the formation of every salt.

1. *Direct union of the elements.* Some metals react directly with certain nonmetals to form a salt. For example, burning sodium in an atmosphere of chlorine gas produces sodium chloride.

$$2Na + Cl_2 \rightarrow 2NaCl$$

2. *Reaction of a metal with an acid.* Many metals replace hydrogen in an aqueous acid to form the corresponding salt. Zinc reacts with hydrochloric acid to form zinc chloride and hydrogen gas.

$$Zn + 2HCl \rightarrow ZnCl_2 + H_2(g)$$

3. *Reaction of a metallic oxide with an aqueous acid.* The oxides of some metals react with an acid to form a salt. Magnesium oxide treated with hydrochloric acid forms magnesium chloride and water.

$$MgO + 2HCl \rightarrow MgCl_2 + H_2O$$

If calcium oxide is substituted for magnesium oxide in this reaction, the salt formed is calcium chloride.

$$CaO + 2HCl \rightarrow CaCl_2 + H_2O$$

4. *Reaction of a nonmetallic oxide with a base.* The oxides of some nonmetals react with a soluble hydroxide to form a salt. Carbon dioxide gas passed into limewater (saturated calcium hydroxide solution) forms insoluble calcium carbonate and water.

$$CO_2 + Ca(OH)_2 \rightarrow CaCO_3(s) + H_2O$$

Additional carbon dioxide converts the calcium carbonate to soluble calcium hydrogen carbonate.

$$CO_2 + H_2O + CaCO_3 \rightarrow Ca(HCO_3)_2$$

The reactions of sulfur dioxide in limewater are similar. Insoluble calcium sulfite, $CaSO_3$, is first formed.

$$SO_2 + Ca(OH)_2 \rightarrow CaSO_3(s) + H_2O$$

When excess sulfur dioxide gas is bubbled through limewater, soluble calcium hydrogen sulfite is formed.

$$SO_2 + H_2O + CaSO_3 \rightarrow Ca(HSO_3)_2$$

5. *Acid-base neutralization.* When an acid neutralizes a soluble hydroxide, a salt may be recovered from the water solvent. This salt corresponds to the metallic ion of the base and nonmetallic ion of the acid. Many different salts can be prepared by neutralization. An example is the reaction between hydrochloric acid and sodium hydroxide mixed in chemically equivalent quantities. When the solvent water is evaporated, sodium chloride remains.

$$NaOH + HCl \rightarrow NaCl + H_2O$$

6. *Ionic reaction.* Two salts may be prepared in ionic reactions if one of them is practically insoluble. The equation for the reaction between solutions of sodium sulfate and barium chloride is

$$BaCl_2 + Na_2SO_4 \rightarrow 2NaCl + BaSO_4(s)$$

Since both reactants are dissociated in water solution, the ionic equation is more useful.

$$Ba^{++}(aq) + 2Cl^-(aq) + 2Na^+(aq) + SO_4^{--}(aq) \rightarrow 2Na^+(aq) + 2Cl^-(aq) + BaSO_4(s)$$

In this reaction, barium sulfate is only very slightly soluble and precipitates readily from the solution. Since precipitates always

Table 15-7
SALT NOMENCLATURE

Formula	Stock name
$CuCl$	copper(I) chloride
$CuCl_2$	copper(II) chloride
FeO	iron(II) oxide
Fe_2O_3	iron(III) oxide
Fe_3O_4	iron(II, III) oxide
$MnCl_2$	manganese(II) chloride
$MnCl_4$	manganese(IV) chloride
PtO_2	platinum(IV) oxide
$Cr_2(SO_3)_3$	chromium(III) sulfite
$CoCO_3$	cobalt(II) carbonate
$Co_2(SO_4)_3$	cobalt(III) sulfate
Cu_2SO_4	copper(I) sulfate
$CuSO_4$	copper(II) sulfate
$Fe_3(PO_4)_2$	iron(II) phosphate
$Hg(NO_3)_2$	mercury(II) nitrate
$KCaPO_4$	potassium calcium phosphate
$NaHCO_3$	sodium hydrogen carbonate

separate from saturated solutions, the solvent water contains a very low saturation concentration of Ba^{++} ions and SO_4^{--} ions in addition to the spectator ions Na^+ and Cl^-. If sodium chloride is recovered by evaporating the solvent water, it will necessarily contain some barium sulfate.

7. *Reaction of an acid with a carbonate.* A salt can be obtained from this reaction because the other products are water and carbon dioxide gas. If hydrochloric acid is added to a solution of sodium carbonate, the following reaction occurs:

$$2HCl + Na_2CO_3 \rightarrow 2NaCl + H_2O + CO_2(g)$$

Carbon dioxide bubbles out of the solution as a gas. Sodium chloride can be recovered by evaporation.

8. *Reaction of a metallic oxide with a nonmetallic oxide.* An oxygen-containing salt can be formed by the reaction between a basic oxide and an acidic oxide. Water is not involved in this process. Instead, the dry oxides are mixed and heated. Metallic carbonates and phosphates are typical of the salts produced.

$$MgO + CO_2 \rightarrow MgCO_3$$
$$CaO + CO_2 \rightarrow CaCO_3$$
$$6CaO + P_4O_{10} \rightarrow 2Ca_3(PO_4)_2$$

15.15 Naming salts Salts are generally named by combining the names of the ions of which they are composed. For example, the name of $Ba(NO_3)_2$ is *barium nitrate.* By agreement the positive ion, in this case the Ba^{++} ion, is named first. The name of the negative ion, in this case the NO_3^- ion, follows.

Table 15-8
ACID-SALT NOMENCLATURE

Formula	*Name of acid*	*Name of salt anion*
HF	hydrofluoric	fluoride
HBr	hydrobromic	bromide
HI	hydriodic	iodide
HCl	hydrochloric	chloride
$HClO$	hypochlorous	hypochlorite
$HClO_2$	chlorous	chlorite
$HClO_3$	chloric	chlorate
$HClO_4$	perchloric	perchlorate
HIO_3	iodic	iodate
H_2MnO_4	manganic	manganate
$HMnO_4$	permanganic	permanganate
H_2S	hydrosulfuric	sulfide
H_2SO_3	sulfurous	sulfite
H_2SO_4	sulfuric	sulfate
HNO_2	nitrous	nitrite
HNO_3	nitric	nitrate
H_2CO_3	carbonic	carbonate
H_3PO_3	phosphorous	phosphite
H_3PO_4	phosphoric	phosphate

Over the years, many difficulties have arisen in the naming of salts. For example, many outdated names of salts have carried over into the present naming system. These old names do not provide for a simple translation from name to formula or from formula to name. In 1940, the International Union of Pure and Applied Chemistry recommended a more logical system for naming inorganic compounds. It is called the *Stock system,* named after Alfred Stock (1876–1946), an inorganic chemist.

Alfred Stock was born in Poland and educated in Germany. He received the Ph.D. degree in chemistry at the University of Berlin in 1899, graduating magna cum laude. He is best known for his work on the hydrides of boron and silicon. For many years, Stock was a victim of chronic mercury poisoning. When he finally recognized its cause, he studied the disease and its dangers and effectively warned others about it.

The Stock system is used in this book for naming salts that contain metals *with variable oxidation states.* Several examples of Stock names of salts are given in Table 15-7. Observe that in *double salts* the more electropositive cation is named first. *A* ***double salt*** *is one in which two different kinds of metallic ions are present.*

The names of anions (negative ions) take the same root and prefix as the acid in which they occur. However, the acid ending *-ic* is changed to *-ate,* and the ending *-ous* is changed to *-ite.* Salts derived from binary acids take the ending *-ide.* Table 15-8 shows the names of the anions of salts produced by the reactions of various acids.

Salt anions that are polyatomic and include metallic atoms with variable oxidation states may have rather complex names in the Stock system. For example, polyatomic MnO_4^- anion is the *tetraoxomanganate(VII) ion* in this system. However, its potassium salt, $KMnO_4$, is well known as *potassium permanganate;* it will not likely become *potassium tetraoxomanganate(VII).*

SUMMARY

Acids and bases are described historically in terms of their water solutions. In this aqueous acid-base system, a substance is an acid if it produces hydronium ions in water solution. Similarly, a substance is a base if it contributes hydroxide ions to its water solution. The properties of acids and bases, their reactions with other substances, and the neutralization reaction between them are the properties of traditional acids and bases in aqueous solutions.

In the aqueous acid-base system, an acid contains hydrogen which, through ionization in water solution, is transferred as a proton (hydrogen ion) to a water molecule to form the hydronium ion. This aqueous system has been broadened by the Brønsted definition of acids and bases. By this definition, all proton donors are acids, and all proton acceptors are bases. The Brønsted proton-transfer system extends the acid-base concept of reactions to nonaqueous solutions. Hydrogen chloride gas is a Brønsted acid when dissolved in ammonia. An HCl molecule donates a proton to an NH_3 molecule. Ammonia is a Brønsted base. It accepts a proton from the HCl molecule.

Other acid-base systems have been defined. The Lewis acid-base system defines an acid as an electron-pair acceptor. A proton-transfer may not necessarily be involved in a Lewis acid-base reaction.

The hydroxides of certain metals in the middle region of the periodic table are amphoteric. They are insoluble in water but dissolve in acidic and basic solutions. Amphoteric hydroxides behave as acids in the presence of strong bases. They behave as bases in the presence of strong acids.

In a proton-transfer reaction, the acid gives up a proton to the base. The species that remains is a base, the conjugate base of the acid donor. The base that accepts the proton in the reaction becomes an acid. It is the conjugate acid of the base. Together they constitute an acid-base pair. The stronger the acid, the weaker is its conjugate base. Such proton-transfer reactions tend to produce weaker acids and also weaker bases.

Some nonmetallic oxides combine with water to form oxyacids. These oxides are called acid anhydrides. Oxides of active metals react with water to form hydroxides. These oxides are called basic anhydrides.

Salts are compounds composed of the positive ion of an aqueous base and the negative ion of an aqueous acid. They are ionic compounds. Salts vary in their solubility in water; their aqueous solutions are solutions of hydrated ions. Generally, the names of salts follow the Stock system for naming inorganic compounds in this textbook.

VOCABULARY

acid
acid anhydride
alkaline
amphoteric
aqueous
base
basic anhydride
binary acid
conjugate acid
conjugate base
diprotic
double salt
electron-pair acceptor
electron-pair donor
hydrochloric acid
hydronium ion
hydroxide ion
hydroxyl group
inorganic acid
litmus
methyl orange
mineral acid
monoprotic
neutralization
nitric acid
organic acid
oxyacid
phenolphthalein
phosphoric acid
pickling
protolysis
proton acceptor
proton donor
salt
sulfuric acid
triprotic

QUESTIONS

GROUP A

1. Distinguish between organic and inorganic acids on the basis of origin. *A1, A4a*
2. (*a*) What specific group of atoms identifies an organic acid? (*b*) Write the structural configuration of this group. *A1, A3a*
3. (*a*) Identify the four most important industrial acids in the order of their importance. (*b*) What are their major uses? *A3b*
4. The acidic properties of aqueous acid solutions are characterized by the presence of what aqueous ion? *A2b, A3c*
5. Upon what basis can an aqueous acid be described as a proton donor? *A2c*
6. Referring to Question 5, what generic (group) name characterizes the "proton acceptor?" *A2c*
7. Compile a list of eight properties that aqueous acids have in common. *A3c*
8. (*a*) Describe the scheme for naming binary acids. (*b*) Give the name of the aqueous acid formed by dissolving hydrogen sulfide in water. *A6*
9. Give both the *name of the pure substance* and the *name of the aqueous acid* for each of the following oxyacids: (*a*) $HBrO$; (*b*) $HBrO_2$; (*c*) $HBrO_3$. *A6*
10. Define: (*a*) acid anhydride; (*b*) basic anhydride; (*c*) Give an example of each. *A4c*
11. A dilute aqueous solution of hydrogen chloride and an aqueous solution of calcium hydroxide are mixed in chemically equivalent quantities. (*a*) Write the empirical equation for the neutralization reaction. (*b*) Write the ionic equation. (*c*) Write the net ionic equation. *A7b*
12. How can you interpret the Brønsted definition of a base as a proton acceptor? *A2c*
13. Compile a list of seven properties that aqueous solutions of the hydroxides of active metals have in common. *A3c*
14. Aluminum hydroxide is said to be *amphoteric*. Explain. *A9c*
15. Characterize the composition and structure of salts. A5a
16. Describe the scheme for naming salts. *A6*

QUESTIONS

GROUP B

17. (*a*) Describe a method you would use to prepare a small quantity of barium sulfate in the laboratory, and give your reasons for the method chosen. (*b*) Write the ionic equation for the reaction. (*c*) Write the net ionic equation. *A8*
18. Would silver chloride be a suitable source of Cl^- ions for an ionic reaction with another salt? Explain. *A8*
19. An aqueous solution of hydrogen chloride has distinctly acidic properties. However, the pure liquid hydrogen chloride does not have these acidic properties and is not an acid in the traditional sense. Explain. *A2b, A5b, A7a*
20. Dilute sulfuric acid is a nonoxidizing acid. Write the equation for the reaction between zinc and dilute sulfuric acid: (*a*) as an ionic equation; (*b*) as a net ionic equation. *A7b, A9b*
21. Hot concentrated sulfuric acid is an oxidizing acid. When zinc reacts with this acid, zinc sulfate, hydrogen sulfide gas, and water are formed. Write the equation: (*a*) as an ionic equation; (*b*) as a net ionic equation. *A7b, A9b*
22. Dilute nitric acid (an oxidizing acid) reacts with copper to form copper (II) nitrate, nitrogen monoxide gas, and water. Write the equation: (*a*) as an ionic equation; (*b*) as a net ionic equation. *A7b, A9b*
23. Concentrated nitric acid (an oxidizing acid) reacts with copper to form copper(II) ni-

trate, nitrogen dioxide, and water. Write the equation: *(a)* as an ionic equation; *(b)* as a net ionic equation. *A7b, A9b*

24. Hydrogen chloride is described as an acid when it is dissolved in ammonia. *(a)* Justify this description of hydrogen chloride. *(b)* How should the ammonia be described? *(c)* Write the equation for this reaction. *(d)* Write this equation using electron-dot formulas. *A2c*

25. Ammonia is described as a base when it is dissolved in water. *(a)* Justify this description of ammonia. *(b)* How should the water be described? *(c)* Write the equation for this reaction. *(d)* Write this equation using electron-dot formulas. *A2c, A5b*

26. *(a)* When the H_2O molecule acts as an acid, what is its conjugate base? *(b)* When the H_2O molecule acts as a base, what is its conjugate acid? *(c)* When the NH_3 molecule acts as an acid, what is its conjugate base? *(d)* When the NH_3 molecule acts as a base, what is its conjugate acid? *A2c, A5b*

27. Hydrogen fluoride gas is slightly ionized in water solution. Write the ionization equation and identify two conjugate acid-base pairs. *A7a*

28. Write two equations that show collectively the amphoteric character of the HSO_4^- ion. *A7a, A7d*

29. *(a)* Define *amphoteric hydroxide.* *(b)* Complete and balance the following equations that demonstrate the amphoteric character of water-insoluble hydrated aluminum hydroxide. *A2d, A9c*

$Al(H_2O)_3(OH)_3(s) + OH^-(aq) \rightarrow$

$Al(H_2O)_3(OH)_3(s) + H_3O^+ \rightarrow$

30. Water-insoluble hydrated zinc hydroxide, $Zn(OH)_2(H_2O)_2$, dissolves in an excess of aqueous OH^- ions to form soluble zincate ions, $Zn(OH)_4^{--}$, and water. Write the net ionic equation. *A8c*

31. The hydrated zinc hydroxide of Question 30 also dissolves in an excess of aqueous H_3O^+ ions to form hydrated zinc ions, $Zn(H_2O)_4^{++}$, and water. Write the net ionic equation. *A8c*

32. Write the equation for the slight ionization of water and identify the two conjugate acid-base pairs to show that water is amphoteric. *A5b*

33. Classify the following compounds as *soluble, slightly soluble,* or *insoluble* on the basis of their solubilities in water: *(a)* $Al(OH)_3$; *(b)* $(NH_4)_2S$; *(c)* $CaCO_3$; *(d)* $Cu_3(PO_4)_2$; *(e)* $FeCl_2$; *(f)* $PbCrO_4$; *(g)* MgI_2; *(h)* $AgCl$; *(i)* $SrSO_4$; *(j)* $Zn(NO_3)_2$.

34. Name the following compounds: *(a)* $Mg(MnO_4)_2$; *(b)* $Na_2Cr_2O_7$; *(c)* $CuSO_4$; *(d)* HIO_3; *(e)* $NaClO$; *(f)* $K_4Fe(CN)_6$; *(g)* NH_4NO_3; *(h)* $Ba(NO_2)_2$. *A6*

PROBLEMS

Group A

1. Nitric acid can be prepared in the laboratory by the reaction of sodium nitrate with sulfuric acid. Sodium hydrogen sulfate is also formed. *(a)* How many grams of sulfuric acid are required to produce $10\bar{0}$ g of nitric acid? *(b)* How many grams of sodium hydrogen sulfate are formed?

2. How many liters of dry carbon dioxide can be collected at 21 °C and 745 mm pressure from a reaction between 32.5 g of calcium carbonate and an excess of hydrochloric acid?

3. What quantity of calcium silicate, $CaSiO_3$, can be prepared by heating a mixture of 60.0 g of calcium oxide and 75.0 g of silicon dioxide?

4. Suppose 50.0 L of dry carbon dioxide gas, measured at 25.0 °C and 755 mm pressure, are available to convert hot calcium oxide to calcium carbonate. *(a)* What quantity of calcium oxide is required? *(b)* is produced?

Gerhard Herzberg (Canadian) was awarded the 1971 Nobel Prize in chemistry for determining the structures of molecules and fragments called free radicals. The method he used to study many diatomic and polyatomic molecules was spectroscopy, which is based on the fact that atoms and molecules absorb and emit electromagnetic radiation at discrete wavelengths. He also discovered the spectra of certain free radicals, which are unstable intermediates in many reactions. He even used spectroscopy to identify molecules in the atmospheres of other planets, in comets, and in interstellar space.

Acid-Base Titration and pH

GOALS

In this chapter you will gain an understanding of: • the difference between solution concentration expressed as molality and molarity • chemical equivalents of acids, bases, elements, and salts • solving problems involving equivalents • normality and normal solutions • pH in both operational and mathematical terms • solving problems involving the calculation of pH and H_3O^+ concentration • acid-base titration and indicators • solving titration problems involving molar and normal solutions • expressing solution concentration in terms of molarity and normality

SOLUTION CONCENTRATIONS

16.1 Molar solutions In your studies of the influence of solutes on the freezing points and boiling points of solvents (Chapter 13), you learned that the ratio of solute-to-solvent molecules is the significant consideration. The freezing-point depression and boiling-point elevation *of a solvent* are related to the *number* of solute particles dissolved in a given *quantity* of solvent. The concentration of such a solution is expressed in terms of *molality*. Solutions of known molality are prepared when it is important to know the ratio of solute-to-solvent molecules.

Recall that *molality, m,* is an expression of solution concentration in terms of *moles of solute* and *kilograms of solvent*. For example, *a 1-molal* (1-*m*) solution contains *1 mole of solute per kilogram of solvent*. For any given solvent, two solutions of equal molality have the same ratio of solute-to-solvent molecules. Chemists are able to determine molecular weights of many soluble substances by observing the effects of their dilute solutions of known molalities on the freezing points of their solvents.

In this chapter, we are interested in the quantity of solute participating as a reactant in a chemical reaction occurring in a solution environment. Because the volume of a solution is easily measured, it is convenient to know the quantity of a solute in terms of the volume of its solution. In this approach, solution concentration is expressed in terms of a *known quantity of a solute in a given volume of solution*. When a solution is prepared in this way, any required quantity of solute can be selected by measuring out the volume of solution containing that quantity of solute.

Figure 16-1. A volumetric flask for measuring precise volumes of liquids. When filled to the mark on the neck of the flask with a liquid at the temperature etched on the flask, it contains 500 ± 0.20 mL of the liquid being measured.

If the quantity of solute is stated in *moles* and the volume of solution in *liters*, the concentration of a solution is expressed in *molarity*. The symbol for molarity is *M*. *The* ***molarity*** *of a solution is an expression of the number of moles of solute per liter of solution. A one-molar* (1-*M*) solution contains *1 mole of solute per liter of solution*. Solutions of the same molarity have the same mole concentration of solutes.

A mole of sodium chloride, NaCl, has a mass of 58.5 g, its gram-formula weight. This quantity of NaCl dissolved in enough water to make exactly 1 liter of solution gives a 1-*M* solution. Half this quantity of NaCl in 1 liter of solution forms a 0.5-*M* solution. Twice this quantity per liter of solution yields a 2-*M* solution.

A *volumetric flask* like the one in Figure 16-1 is commonly used in preparing solutions of known molarity. A measured quantity of solute is dissolved in a portion of solvent in the flask. Then more solvent is added to fill the flask to the mark on the neck. Thus the quantity of solute and the volume of solution are known, and the molarity of the solution is easily calculated.

As an example, suppose you wish to prepare a 1-*M* solution of potassium chromate. The formula weight of K_2CrO_4 is 194. Thus a mole of K_2CrO_4 has a mass of 194 g. To prepare 1 L of the 1-*M* solution, 194 g of K_2CrO_4 *must be dissolved in enough water to make 1 liter of solution*. To prepare $10\bar{0}$ mL of the 1-*M* solution, 19.4 g of K_2CrO_4 must be dissolved in enough water to make 100 mL of solution. Similarly, a 0.5-*M* solution requires 0.5 mole (97.0 g) of K_2CrO_4 per liter of solution. A 0.05-*M* solution requires 0.05 mole (9.70 g) of K_2CrO_4 per liter of solution.

Molarity: *moles solute per liter of solution.*

Molality: *moles solute per kilogram of solvent.*

Observe that solution *molarity* is based on the *volume of solution*. (Solution *molality*, on the other hand, is based on the *mass of solvent*.) Equal volumes of solutions of the same molarity have equal mole quantities of solutes. Molarity is preferred when measuring volumes of solutions. For very dilute solutions, the distinction between molality and molarity is not significant.

16.2 Chemical equivalents of acids and bases Solution concentrations can be expressed in a way that allows *chemically equivalent quantities* of different solutes to be measured very simply. These quantities of solutes are called *equivalents* (equiv). ***Equivalents*** *are the quantities of substances that have the same combining capacity in chemical reactions.*

Consider the following equations. They show that 36.5 g (1 mole) of HCl and 49 g (½ mole) of H_2SO_4 are chemically equivalent in neutralization reactions with basic KOH.

$$\begin{array}{ccccc} HCl & + \; KOH & \rightarrow & KCl & + \; H_2O \\ 1 \text{ mole} & 1 \text{ mole} & \rightarrow & 1 \text{ mole} & 1 \text{ mole} \\ 36.5 \text{ g} & 56 \text{ g} & & & \end{array}$$

$$H_2SO_4 + 2KOH \rightarrow K_2SO_4 + 2H_2O$$

or,

$$\tfrac{1}{2}H_2SO_4 + KOH \rightarrow \tfrac{1}{2}K_2SO_4 + H_2O$$

$\tfrac{1}{2}$ mole — 1 mole — $\tfrac{1}{2}$ mole — 1 mole

49 g — 56 g

The equations also reveal that both 36.5 g of HCl and 49 g of H_2SO_4 are chemically equivalent to 56 g (1 mole) of KOH in these neutralization reactions.

Suppose KOH is replaced with $Ca(OH)_2$ in one of the prior reactions. The new equation shows that 37 g (½ mole) of $Ca(OH)_2$ is equivalent to 56 g (1 mole) of KOH.

$$2HCl + Ca(OH)_2 \rightarrow CaCl_2 + 2H_2O$$

or,

$$HCl + \tfrac{1}{2}Ca(OH)_2 \rightarrow \tfrac{1}{2}CaCl_2 + H_2O$$

1 mole — $\tfrac{1}{2}$ mole — $\tfrac{1}{2}$ mole — 1 mole

36.5 g — 37 g

In proton-transfer reactions, *one equivalent of an acid is the quantity, in grams, that supplies one mole of protons.* HCl and H_2SO_4 are the acids in the neutralization reactions just considered. The quantity representing *one equivalent* of each acid in these reactions can be determined as follows:

An acid equivalent donates 1 mole of protons.

$$1 \text{ equiv HCl} = \frac{1 \text{ mole HCl}}{1 \text{ mole } H_3O^+} \times \frac{36.5 \text{ g}}{\text{mole}} = \frac{36.5 \text{ g HCl}}{\text{mole } H_3O^+}$$

$$1 \text{ equiv } H_2SO_4 = \frac{1 \text{ mole } H_2SO_4}{2 \text{ moles } H_3O^+} \times \frac{98 \text{ g}}{\text{mole}} = \frac{49 \text{ g } H_2SO_4}{\text{mole } H_3O^+}$$

One equivalent of a base is the quantity, in grams, that accepts one mole of protons or supplies one mole of OH^- ions. For KOH and $Ca(OH)_2$ in the above reactions,

A base equivalent accepts 1 mole of protons or supplies 1 mole of OH^- ions.

$$1 \text{ equiv KOH} = \frac{1 \text{ mole KOH}}{1 \text{ mole } OH^-} \times \frac{56 \text{ g}}{\text{mole}} = \frac{56 \text{ g KOH}}{\text{mole } OH^-}$$

$$1 \text{ equiv } Ca(OH)_2 = \frac{1 \text{ mole } Ca(OH)_2}{2 \text{ moles } OH^-} \times \frac{74 \text{ g}}{\text{mole}} = \frac{37 \text{ g } Ca(OH)_2}{\text{mole } OH^-}$$

HCl, HNO_3, and $HC_2H_3O_2$ are monoprotic acids. One mole of each can supply 1 mole of H_3O^+ ions. Therefore, *1 equivalent of a monoprotic acid is the same as 1 mole of the acid.* One mole of a diprotic acid such as H_2SO_4 can supply 2 moles of H_3O^+ ions. Thus, *for complete neutralization, 1 equivalent of a diprotic acid is the same as ½ mole of the acid.* H_3PO_4 is triprotic and can furnish 3 moles of H_3O^+ ions per mole of acid. *When completely neutralized, 1 equivalent of a triprotic acid is the same as ⅓ mole of the acid.*

A diprotic acid has 2 atoms of ionizable hydrogen per molecule.

A triprotic acid has 3 atoms of ionizable hydrogen per molecule.

A similar relationship exists between chemical equivalents and moles of bases. One mole of KOH supplies 1 equivalent of OH^- ions. One mole of $Ca(OH)_2$ supplies 2 equivalents of OH^-

ions. Therefore, 1 equivalent of KOH is the same as 1 mole of KOH, and 1 equivalent of $Ca(OH)_2$ is the same as ½ mole of $Ca(OH)_2$.

In many chemical reactions, a diprotic or triprotic acid is not completely neutralized. In such a case, the number of moles of protons supplied per mole of acid is determined by the reaction it undergoes. For example, suppose a solution containing 1 mole of H_2SO_4 is added to a solution containing 1 mole of NaOH. The salt sodium hydrogen sulfate is then recovered by evaporating the water solvent. Observe that the neutralization of H_2SO_4 is *not* complete.

$$\begin{array}{cccc} \mathbf{H_2SO_4} + & \mathbf{NaOH} \rightarrow & \mathbf{NaHSO_4} + & \mathbf{H_2O} \\ \mathbf{1\ mole} & \mathbf{1\ mole} & \mathbf{1\ mole} & \mathbf{1\ mole} \\ \mathbf{98\ g} & \mathbf{4\bar{0}\ g} & & \end{array}$$

One mole of H_2SO_4 supplies 1 mole of protons to the base to form an "acid salt" containing HSO_4^- ions. Therefore 1 equivalent of H_2SO_4 is the same as 1 mole of the acid, 98 g, *in this reaction*.

$$\mathbf{1\ equiv\ H_2SO_4 = \frac{1\ mole\ H_2SO_4}{1\ mole\ H_3O^+} \times \frac{98\ g}{mole} = \frac{98\ g\ H_2SO_4}{mole\ H_3O^+}}$$

Now, suppose a solution containing 1 mole of H_3PO_4 is added to one containing 1 mole of NaOH. The reaction is

$$\begin{array}{cccc} \mathbf{H_3PO_4} + & \mathbf{NaOH} \rightarrow & \mathbf{NaH_2PO_4} + & \mathbf{H_2O} \\ \mathbf{1\ mole} & \mathbf{1\ mole} & \mathbf{1\ mole} & \mathbf{1\ mole} \\ \mathbf{98.0\ g} & \mathbf{40.0\ g} & & \end{array}$$

The salt sodium dihydrogen phosphate can be recovered by evaporation. One mole of H_3PO_4 supplies 1 mole of protons to the base to form a salt containing $H_2PO_4^-$ ions. Thus one equivalent of H_3PO_4 is the same as 1 mole of the acid, 98.0 g, *in this reaction*.

$$\mathbf{1\ equiv\ H_3PO_4 = \frac{1\ mole\ H_3PO_4}{1\ mole\ H_3O^+} \times \frac{98.0\ g}{mole} = \frac{98.0\ g\ H_3PO_4}{mole\ H_3O^+}}$$

An equivalent of H_3PO_4 depends on the reaction.

Suppose the basic solution in the reaction above contained 2 moles of NaOH. The salt recovered by evaporation would be Na_2HPO_4. One equivalent of H_3PO_4 in this reaction is the same as ½ mole of the acid, 49.0 g. If, in another reaction, the neutralization of the triprotic acid is complete, 1 equivalent of H_3PO_4 is the same as ⅓ mole of the acid, 32.7 g.

16.3 Chemical equivalents of elements Now consider a reactant in an electron-transfer reaction. A chemical equivalent of such a reactant is *the quantity, in grams, that supplies or acquires 1 mole of electrons in a chemical reaction.* In the following reaction, a mole of sodium atoms (23 g) loses 1 mole of electrons to form 1 mole of Na^+ ions.

$$\begin{array}{llll} \mathbf{Na} & \rightarrow \mathbf{Na^+} & + & \mathbf{e^-} \\ \mathbf{1\ mole} & \mathbf{1\ mole} & & \mathbf{1\ mole} \\ \mathbf{23\ g} & \mathbf{23\ g} & & \end{array}$$

Thus 1 equivalent of sodium is the same as 1 mole of sodium atoms, 23 g.

A mole of calcium atoms ($4\bar{0}$ g) supplies 2 moles of electrons when Ca^{++} ions are formed. A mole of aluminum atoms (27 g) supplies 3 moles of electrons when Al^{+++} ions are formed. Thus 1 equivalent of calcium is the mass of ½ mole of calcium atoms, $2\bar{0}$ g. One equivalent of aluminum is the mass of ⅓ mole of aluminum atoms, 9.0 g.

$$\begin{array}{lll} \frac{1}{2}\,\mathbf{Ca} & \rightarrow \frac{1}{2}\,\mathbf{Ca^{++}} & + \ \mathbf{e^-} \\ \frac{1}{2}\,\mathbf{mole} & \frac{1}{2}\,\mathbf{mole} & \mathbf{1\ mole} \\ \mathbf{2\bar{0}\ g} & \mathbf{2\bar{0}\ g} & \end{array}$$

and

$$\begin{array}{lll} \frac{1}{3}\,\mathbf{Al} & \rightarrow \frac{1}{3}\,\mathbf{Al^{+++}} & + \ \mathbf{e^-} \\ \frac{1}{3}\,\mathbf{mole} & \frac{1}{3}\,\mathbf{mole} & \mathbf{1\ mole} \\ \mathbf{9.0\ g} & \mathbf{9.0\ g} & \end{array}$$

These relationships can be summarized as follows:

$$\mathbf{1\ equiv\ Na} = \frac{\mathbf{1\ mole\ Na}}{\mathbf{1\ mole\ e^-}} \times \frac{\mathbf{23\ g}}{\mathbf{mole}} = \frac{\mathbf{23\ g\ Na}}{\mathbf{mole\ e^-}}$$

$$\mathbf{1\ equiv\ Ca} = \frac{\mathbf{1\ mole\ Ca}}{\mathbf{2\ moles\ e^-}} \times \frac{\mathbf{4\bar{0}\ g}}{\mathbf{mole}} = \frac{\mathbf{2\bar{0}\ g\ Ca}}{\mathbf{mole\ e^-}}$$

$$\mathbf{1\ equiv\ Al} = \frac{\mathbf{1\ mole\ Al}}{\mathbf{3\ moles\ e^-}} \times \frac{\mathbf{27\ g}}{\mathbf{mole}} = \frac{\mathbf{9.0\ g\ Al}}{\mathbf{mole\ e^-}}$$

Observe that when Na^+ ions are formed, the numerical *change* in oxidation state for sodium atoms is 1. The change for calcium atoms is 2. For aluminum atoms it is 3. Such numbers can ordinarily be used to determine one equivalent of an element for a given reaction. The mass of 1 mole of atoms of the element (1 g-at wt) is divided by the *change in oxidation state* these atoms undergo in a chemical reaction.

Oxidizing and reducing agents with several common oxidation states are given special attention in Chapter 22.

$$\mathbf{1\ equiv\ Na} = \frac{\mathbf{23\ g}}{\mathbf{1}} = \mathbf{23\ g}$$

$$\mathbf{1\ equiv\ Ca} = \frac{\mathbf{4\bar{0}\ g}}{\mathbf{2}} = \mathbf{2\bar{0}\ g}$$

$$\mathbf{1\ equiv\ Al} = \frac{\mathbf{27\ g}}{\mathbf{3}} = \mathbf{9.0\ g}$$

16.4 Chemical equivalents of salts A similar method can ordinarily be used to find the mass of 1 equivalent of a salt. The mass of 1 mole of the salt is divided by the *total* positive (or negative) ionic charge indicated by its formula. This total positive charge is determined by multiplying the number of cations, which is shown in the formula of the salt, by the charge on each cation.

$$\textbf{1 equiv (salt)} = \frac{\textbf{mass of 1 mole of salt}}{\textbf{total positive charge}}$$

The formula for sodium sulfate is Na_2SO_4, and the mass of 1 mole is 142 g. Each of the two Na^+ ions has a +1 charge. The total positive charge is 2. (Observe that the total negative charge is also 2.)

$$\textbf{1 equiv } \mathbf{Na_2SO_4} = \frac{\textbf{142 g}}{\textbf{2}} = \textbf{71.0 g}$$

One mole of calcium phosphate, $Ca_3(PO_4)_2$, has a mass of $31\bar{0}$ g. Each of the three Ca^{++} ions has a +2 charge. The total positive charge is 6. (Observe that the total negative charge is also 6.)

$$\textbf{1 equiv } \mathbf{Ca_3(PO_4)_2} = \frac{\mathbf{31\bar{0}}\textbf{ g}}{\textbf{6}} = \textbf{51.7 g}$$

16.5 Normal solutions Solution concentration based on the volume of solution can now be expressed in a second way by stating the quantity of solute in equivalents. This method is called *normality, N. The* ***normality*** *of a solution expresses the number of equivalents of solute per liter of solution.* A one-normal (1-*N*) solution contains 1 equivalent of solute *per liter of solution.* For a solution to be 0.25 *N*, it must contain 0.25 equivalent of solute per liter of solution. *Equal volumes of solutions of the same normality are chemically equivalent.* The expressions for solution concentration are summarized in Table 16-1.

Normality: *equivalents solute per liter solution.*

A mole of the monoprotic hydrogen chloride has a mass of 36.5 g. As a reactant, this quantity of HCl can furnish 1 mole of protons. Thus 1 mole of HCl in 1 liter of aqueous solution provides 1 *equivalent* (equiv) of protons as H_3O^+ ions. The concentration of this solution is 1 *N*.

Suppose a solution of HCl that can supply 0.100 mole of

Table 16-1
METHODS OF EXPRESSING CONCENTRATION OF SOLUTIONS

Name	*Symbol*	*Solute unit*	*Solvent unit*	*Dimensions*
molality	*m*	mole	kilogram solvent	$\frac{\text{mole solute}}{\text{kg solvent}}$
molarity	*M*	mole	liter solution	$\frac{\text{mole solute}}{\text{liter solution}}$
normality	*N*	equivalent	liter solution	$\frac{\text{equiv solute}}{\text{liter solution}}$

H_3O^+ ions per liter (a 0.100-N HCl solution) is required. This solution must consist of 3.65 g of HCl dissolved in water and diluted to a 1.00-liter volume. However, *the solute is 3.65 g of anhydrous hydrogen chloride in one liter of solution*, not 3.65 g of the concentrated hydrochloric acid ordinarily found in the laboratory. The volume of concentrated hydrochloric acid containing 3.65 g of hydrogen chloride must be diluted to 1-liter volume. How can the required volume of concentrated HCl solution be correctly determined?

First, the mass percentage of HCl in the concentrated solution and the density of the concentrated solution must be known. This information is printed on the label of the concentrated hydrochloric acid container, as shown in Figure 16-2. Representative values for common laboratory acids are listed in Table 16-2 on the next page.

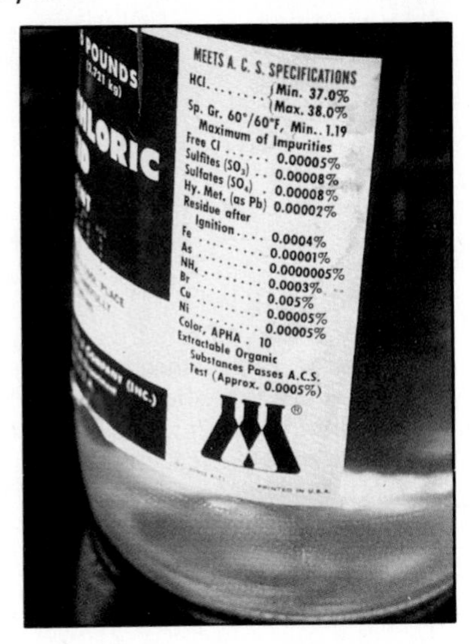

Figure 16-2. The manufacturer's label on a reagent bottle provides important analysis information for the chemist.

$$\textbf{solute HCl} = \textbf{37.5\%}$$

$$\textbf{solution density} = \textbf{1.19 g/mL}$$

Thus $10\bar{0}$ g of concentrated solution contains 37.5 g of HCl. Knowing the density of the concentrated solution to be 1.19 g/mL, the mass of HCl per milliliter of concentrated solution can be determined.

$$\frac{37.5 \text{ g HCl}}{10\bar{0} \text{ g conc soln}} \times \frac{1.19 \text{ g}}{\text{mL}} = \frac{0.446 \text{ g HCl}}{\text{mL conc soln}}$$

For 1.00 L of 0.100-N solution, 3.65 g of anhydrous HCl is required, since

$$\frac{0.100 \text{ equiv HCl}}{\text{L}} \times \frac{36.5 \text{ g}}{\text{equiv}} \times 1.00 \text{ L} = 3.65 \text{ g HCl}$$

The mass of HCl per milliliter of concentrated solution and the mass of HCl required are now known. From these data, the volume of concentrated hydrochloric acid solution can be finally calculated.

$$3.65 \text{ g HCl} \times \frac{\text{mL conc soln}}{0.446 \text{ g HCl}} = 8.18 \text{ mL conc soln}$$

Thus 8.18 mL of concentrated HCl solution diluted to 1.00 L with distilled water gives a 0.100-N solution of HCl.

You have already learned that 1 mole of H_2SO_4 contains 2 equivalents of that substance. A 1-M solution contains 98 g of H_2SO_4 per liter of solution. However, a 1-N solution contains 49 g (98 g ÷ 2) of H_2SO_4 per liter of solution. A 5-N solution contains 245 g (49 g × 5) of H_2SO_4 per liter. Similarly, 0.01-N H_2SO_4 contains 0.49 g (49 g ÷ 100) of H_2SO_4 per liter of solution. Concentrated sulfuric acid is usually 95%–98% H_2SO_4 and the acid has a density of about 1.84 g/mL. Dilutions to desired normalities are calculated as shown previously for HCl. See Table 16-2.

Table 16-2
CONCENTRATIONS OF COMMON ACIDS

	(Average values for freshly opened bottles)			
	Acetic	*Hydrochloric*	*Nitric*	*Sulfuric*
formula	$HC_2H_3O_2$	HCl	HNO_3	H_2SO_4
molecular weight	60.03	36.46	63.02	98.08
density of concentrated reagent, g/cm^3	1.06	1.19	1.42	1.84
percentage assay concentrated reagent	99.5	37.5	69.5	96.0
grams active ingredient/mL reagent	1.055	0.446	0.985	1.76
normality of concentrated reagent	17.6	12.2	15.6	35.9
mL concentrated reagent/liter *N* solution	56.9	81.8	64.0	27.9
molarity of concentrated reagent	17.6	12.2	15.6	17.95
mL concentrated reagent/liter *M* solution	56.9	81.8	64.0	55.8

Compounds containing water of crystallization present special problems in preparing solutions. For example, crystalline copper(II) sulfate has the empirical formula

$$\mathbf{CuSO_4 \cdot 5H_2O}$$

The formula weight is 249.5. One mole of this hydrate, 249.5 g, contains 1 mole of $CuSO_4$, 159.5 g. This fact must be recognized when moles or equivalents of crystalline hydrates are being measured.

A 1-*M* $CuSO_4$ solution contains 159.5 g of $CuSO_4$ per liter of solution. This 1-*M* solution is also a 2-*N* solution because 1 mole of $CuSO_4$ contains 2 equivalents. A 1-*N* solution requires 79.75 g of $CuSO_4$ per liter. Of course, this solution is 0.5 *M*.

If a mole of a solute is also 1 equivalent, the molarity and normality of the solution *are the same.* A 1-*M* HCl solution is also a 1-*N* solution. If a mole of solute is two equivalents, a 1-*M* solution is 2 *N*. A 0.01-*M* H_2SO_4 solution is therefore 0.02 *N*. Similarly, a 0.01-*M* H_3PO_4 solution is 0.03 *N* if it is completely neutralized. Solutions of equal normality are chemically equivalent, volume for volume.

16.6 Ion concentration in water Water is very weakly ionized by self-ionization, as illustrated in Figure 16-3. This process is sometimes referred to as *autoprotolysis*. The very poor conductivity of pure water results from the slight ionization of water itself. This fact can be demonstrated by testing water that has been highly purified by several different techniques.

Electric conductivity measurements of pure water show that at 25 °C it is very slightly ionized to H_3O^+ and OH^- ions. Concentrations of these ions in pure water are only

$$\frac{\text{1 mole } H_3O^+}{10^7 \text{ L } H_2O} \quad \text{and} \quad \frac{\text{1 mole } OH^-}{10^7 \text{ L } H_2O}$$

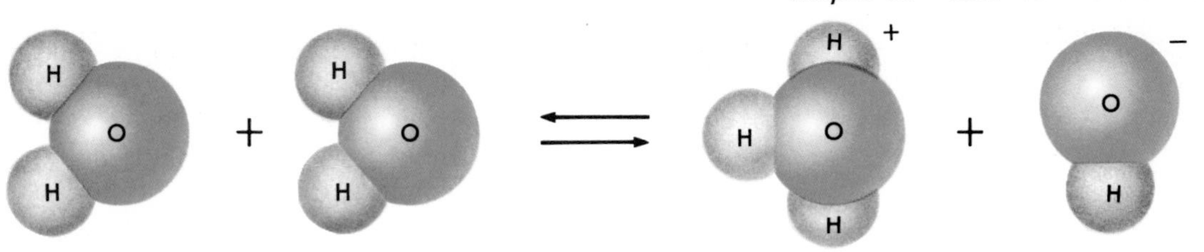

Figure 16-3. Water is weakly ionized by a process of self-ionization.

One liter of water has a mass of 997 g at 25 °C (1 L = 1000 g at 4 °C). The mass of 1 mole of water is 18.0 g. Using these quantities, 1 liter contains 55.4 moles of water at 25 °C.

$$\frac{\mathbf{997\ g}}{\mathbf{L}} \times \frac{\mathbf{1\ mole}}{\mathbf{18.0\ g}} = \mathbf{55.4\ moles/L}$$

The extent of the ionization can be stated as a percentage if the concentration of H_3O^+ ions (and also OH^- ions) is expressed in moles of ions per mole of water.

Pure water is very slightly ionized.

$$\frac{\mathbf{1\ mole\ H_3O^+}}{\mathbf{10^7\ L\ H_2O}} \times \frac{\mathbf{1\ L}}{\mathbf{55.4\ moles}} = \frac{\mathbf{2 \times 10^{-9}\ mole\ H_3O^+}}{\mathbf{mole\ H_2O}}$$

This result shows that water is about 0.0000002% ionized at 25 °C.

It is more useful to express ion concentration as *moles per liter* than as moles per 10,000,000 liters. This change is accomplished by dividing both terms in the expression "moles per 10^7 liters" by 10^7.

$$\frac{\mathbf{1\ mole\ H_3O^+ \div 10^7}}{\mathbf{10^7\ L\ H_2O \div 10^7}} = \frac{\mathbf{10^{-7}\ mole\ H_3O^+}}{\mathbf{L\ H_2O}}$$

Thus the concentration of H_3O^+ ions (and OH^- ions) in water at 25 °C is 10^{-7} mole per liter of H_2O.

Chemists use a standard notation to represent concentration in terms of *moles per liter*. The symbol or formula of the particular ion or molecule is enclosed in brackets, [ ]. For example, $[H_3O^+]$ means *hydronium ion concentration in moles per liter*. For the ionic concentrations in water at 25 °C,

$$\mathbf{[H_3O^+] = 10^{-7}\ mole/L}$$

and

$$\mathbf{[OH^-] = 10^{-7}\ mole/L}$$

or

$$\mathbf{[H_3O^+] = [OH^-] = 10^{-7}\ mole/L}$$

Pure water is neutral because $[H_3O^+] = [OH^-]$.

Because the H_3O^+ ion concentration and the OH^- ion concentration are equal, water is neutral. It is neither acidic nor basic. This neutrality prevails in any solution in which $[H_3O^+] = [OH^-]$.

Reminder: 10^{-5} *is a larger number than* 10^{-7}.

If the H_3O^+ ion concentration in a solution exceeds 10^{-7} mole/liter, the solution is acidic. For example, a solution containing 10^{-5} mole H_3O^+ ion per liter is acidic. If the OH^- ion concentration exceeds 10^{-7} mole per liter, the solution is basic, or alkaline. Thus a solution containing 10^{-4} mole OH^- ion per liter is basic.

In water and in dilute solutions $[H_3O^+] \times [OH^-] = 10^{-14}$ *is a constant.*

It is also true that the *product* of the $[H_3O^+]$ and $[OH^-]$ remains constant in water and dilute aqueous solutions as long as the temperature does not change. Recall that by Le Chatelier's principle, an increase in concentration of either of these ionic species in an aqueous mixture at equilibrium causes a decrease in concentration of the other species. In water and dilute aqueous solutions at 25 °C,

$$[H_3O^+] \times [OH^-] = \text{a constant}$$

$$[H_3O^+]\,[OH^-] = (1 \times 10^{-7} \text{ mole/L})^2$$

$$[H_3O^+]\,[OH^-] = 1 \times 10^{-14} \text{ mole}^2/\text{L}^2$$

The ionization of water increases slightly with temperature increases. The ion product $[H_3O^+]\,[OH^-]$ of 1×10^{-14} mole²/L² is commonly used as a *constant* within the ordinary range of room temperatures.

16.7 The pH of a solution The range of solution concentrations encountered by chemists is great. It varies from about 10 *M* to perhaps 10^{-15} *M*. However, concentrations of less than 1 *M* are most commonly used.

As stated above, the product of $[H_3O^+]$ and $[OH^-]$ is a constant at a given temperature. Therefore, if the concentration of either ionic species is known, the concentration of the other species can be determined. For example, the OH^- ion concentration of a 0.01-*M* NaOH solution is 0.01 or 10^{-2} mole/liter. The H_3O^+ ion concentration of this solution is calculated as follows:

Reminder:
$a^m \times a^n = a^{m+n}$
$10^6 \times 10^3 = 10^{6+3} = 10^9$
$10^5 \times 10^{-2} = 10^{5+(-2)} = 10^3$
$a^m \div a^n = a^{m-n}$
$10^6 \div 10^3 = 10^{6-3} = 10^3$
$10^5 \div 10^{-2} = 10^{5-(-2)} = 10^7$

$$[H_3O^+]\,[OH^-] = 1 \times 10^{-14} \text{ mole}^2/\text{L}^2$$

$$[H_3O^+] = \frac{1 \times 10^{-14} \text{ mole}^2/\text{L}^2}{[OH^-]}$$

$$[H_3O^+] = \frac{1 \times 10^{-14} \text{ mole}^2/\text{L}^2}{1 \times 10^{-2} \text{ mole/L}}$$

$$[H_3O^+] = 1 \times 10^{-12} \text{ mole/L}$$

Sample Problem 16-1 follows.

SAMPLE PROBLEM 16-1

A 0.0001-*M* solution of HNO_3 has been prepared for a laboratory experiment. *(a)* Calculate the $[H_3O^+]$ of this solution. *(b)* Calculate the $[OH^-]$.

SOLUTION

(a) HNO_3 is a monoprotic acid giving 1 mole of H_3O^+ ions per mole of HNO_3 when completely ionized in aqueous solution.

For 0.0001-*M* aqueous HNO_3

$$[H_3O^+] = 0.0001 \text{ mole/L} = 10^{-4} \text{ mole/L}$$

(b) $[H_3O^+]\,[OH^-] = 10^{-14} \text{ mole}^2/L^2$

$$[OH^-] = \frac{10^{-14} \text{ mole}^2/L^2}{[H_3O^+]} = \frac{10^{-14} \text{ mole}^2/L^2}{10^{-4} \text{ mole/L}}$$

$$[OH^-] = 10^{-10} \text{ mole/L}$$

The acidity or alkalinity of a solution can be expressed in terms of its hydronium ion concentration. An $[H_3O^+]$ *higher* than 10^{-7} mole/liter (a *smaller* negative exponent) indicates an acid solution. An $[H_3O^+]$ *lower* than 10^{-7} mole/liter (a *larger* negative exponent) indicates an alkaline solution.

Expressing acidity or alkalinity in this way can become cumbersome, especially in dilute solutions, whether using decimal or scientific notations. Because it is more convenient, chemists use a quantity called pH to indicate the hydronium ion concentration of a solution.

pH is called the hydronium ion index.

Numerically, the pH of a solution is the common logarithm of the number of liters of solution that contains one mole of H_3O^+ ions. The number of liters of solution is equal to the *reciprocal* of the H_3O^+ ion concentration. This concentration is given in moles of H_3O^+ ions per liter of solution. The reciprocal expression is

$$\frac{1}{[H_3O^+]}$$

Thus, the ***pH of a solution*** *is defined as the common logarithm of the reciprocal of the hydronium ion concentration. The pH is expressed by the equation*

$$pH = \log \frac{1}{[H_3O^+]}$$

The common logarithm of a number is the power to which 10 must be raised to give the number. Thus 0.0000001 is 10^{-7} and its reciprocal is 10,000,000, or 10^7. The logarithm of 10^7 is 7.

pH of neutral solutions = 7.

Pure water is slightly ionized, and at 25 °C contains 0.0000001 or 10^{-7} mole of H_3O^+ per liter. The pH of water is therefore

$$pH = \log \frac{1}{0.0000001}$$

$$pH = \log \frac{1}{10^{-7}}$$

$$pH = \log 10^7$$

$$pH = 7$$

Table 16-3
APPROXIMATE pH OF SOME COMMON MATERIALS (at 25 °C)

Material	*pH*
1.0-*N* HCl	0.1
1.0-*N* H_2SO_4	0.3
0.1-*N* HCl	1.1
0.1-*N* H_2SO_4	1.2
gastric juice	2.0
0.01-*N* H_2SO_4	2.1
lemons	2.3
vinegar	2.8
0.1-*N* $HC_2H_3O_2$	2.9
soft drinks	3.0
apples	3.1
grapefruit	3.1
oranges	3.5
cherries	3.6
tomatoes	4.2
bananas	4.6
bread	5.5
potatoes	5.8
rainwater	6.2
milk	6.5
pure water	7.0
eggs	7.8
0.1-*N* $NaHCO_3$	8.4
seawater	8.5
milk of magnesia	10.5
0.1-*N* NH_3	11.1
0.1-*N* Na_2CO_3	11.6
0.1-*N* $NaOH$	13.0
1.0-*N* $NaOH$	14.0
1.0-*N* KOH	14.0

Suppose the H_3O^+ ion concentration in a solution is *higher* than that in pure water. Then the number of liters required to provide 1 mole of H_3O^+ ions is *smaller*. Consequently, the pH is a *smaller* number than 7. Such a solution is *acidic*. On the other hand, suppose the H_3O^+ ion concentration is *lower than* that in pure water. The pH is then a *larger* number than 7. Such a solution is *basic*.

The range of pH values usually falls between 0 and 14. The pH system is particularly useful in describing the acidity or alkalinity of solutions that are not far from neutral. This includes many food substances and fluids encountered in physiology. The pH of some common substances is given in Table 16-3.

There are two basic types of pH problems that will concern you. They are

1. The calculation of pH when the $[H_3O^+]$ of a solution is known.

2. The calculation of $[H_3O^+]$ when the pH of a solution is known.

These two calculation methods are examined in Sections 16.8 and 16.9.

16.8 Calculation of pH In the simplest pH problems, the $[H_3O^+]$ of the solution is an integral power of 10, such as 1 *M* or 0.01 *M*. These problems can be solved *by inspection*. The pH equation based on the definition stated in Section 16.7 is

$$pH = \log \frac{1}{[H_3O^+]}$$

Since

$$\log \frac{1}{[H_3O^+]} = -\log [H_3O^+]$$

the first equation can be written in a more useful form to solve for pH.

$$pH = -\log [H_3O^+]$$

In an aqueous solution in which $[H_3O^+] = 10^{-6}$ mole/liter, the pH = 6.

pH of acidic solutions < 7.

pH of basic solutions > 7.

$pH = -\log[H_3O^+]$.

Chemical solutions with pH values below 0 and above 14 can be prepared. For example, the pH of 6-*M* H_2SO_4 is between 0 and −1. The pH of 3-*M* KOH is near 14.5. However, only common pH values in the 0–14 range will be considered. Observe that the *pH of a solution is the exponent of the hydronium ion concentration with the sign changed.* Sample Problems 16-2 and 16-3 further illustrate this fact.

SAMPLE PROBLEM 16-2

Determine the pH of a 0.001-*M* HCl solution.

SOLUTION

$$\mathrm{pH} = -\log [\mathrm{H_3O^+}]$$

$$[\mathrm{H_3O^+}] = 0.001 \text{ mole/L} = 10^{-3} \text{ mole/L}$$

$$\mathrm{pH} = -\log 10^{-3} = -(-3)$$

$$\mathrm{pH} = 3$$

Notice that if the magnitude of the $[H_3O^+] = 10^{-3}$, the pH = 3.

SAMPLE PROBLEM 16-3

What is the pH of a 0.001-*M* NaOH solution?

SOLUTION

$$\mathrm{pH} = -\log [\mathrm{H_3O^+}]$$

$$[\mathrm{H_3O^+}]\,[\mathrm{OH^-}] = 10^{-14} \text{ mole}^2/\text{L}^2$$

$$[\mathrm{H_3O^+}] = \frac{10^{-14} \text{ mole}^2/\text{L}^2}{[\mathrm{OH^-}]}$$

$$[\mathrm{OH^-}] = 0.001 \text{ mole/L} = 10^{-3} \text{ mole/L}$$

$$[\mathrm{H_3O^+}] = \frac{10^{-14} \text{ mole}^2/\text{L}^2}{10^{-3} \text{ mole/L}} = 10^{-11} \text{ mole/L}$$

$$\mathrm{pH} = -\log 10^{-11} = -(-11)$$

$$\mathrm{pH} = 11$$

Observe that if the magnitude of the $[H_3O^+] = 10^{-11}$, the pH = 11.

The preceding problems have hydronium ion concentrations that are integral powers of ten. They are easily solved by inspection. However, many problems involve hydrogen ion concentrations that are not integral powers of ten. Solving such problems requires some basic knowledge of logarithms and exponents. The common logarithms of numbers are listed in Appendix Table 17. The following brief review of logarithms and the use of the table of logarithms may be helpful at this point.

The common logarithm of a number is the exponent or the power to which 10 must be raised in order to obtain the given number. A logarithm is composed of two parts: the *characteristic,* or integral part; and the *mantissa,* or decimal part. The characteristic of the logarithm of any whole or mixed number is one less than the number of digits to the left of its decimal point. The characteristic of the logarithm of a decimal fraction is always negative and is numerically one greater than the number of zeros immediately to the right of the decimal point. Mantissas are read from tables such as Appendix Table 17. Mantissas are always positive. In determining the mantissa, the decimal point in the original number is ignored since its position is indicated by the characteristic.

Logarithms are exponents and follow the laws of exponents. Specifically, the logarithm of a product equals the sum of the logarithms of the factors.

To find the number whose logarithm is given, determine the digits in the number from the table of mantissas. The characteristic indicates the position of the decimal point.

As the logarithm of a number is the power to which 10 must be raised to give the number, the logarithm of 10^5 is 5 and the logarithm of 10^{-5} is -5. But suppose you require the logarithm of 2.5×10^{-3}. The characteristic of the logarithm of 2.5 is 0, one less than the number of digits to the left of the decimal point. From Appendix Table 17, the mantissa of the logarithm of 2.5 is .40 (to two significant figures). The complete logarithm of 2.5 is 0.40. The logarithm of 10^{-3} is -3. The logarithm of a product is the sum of the logarithms of the factors. Therefore,

$$\log 2.5 \times 10^{-3} = \log 2.5 + \log 10^{-3} = 0.40 + (-3) = -2.60$$

Suppose you must express the number whose logarithm is -9 in scientific notation. The number whose logarithm is -9 is 10^{-9}. Or, stated in another way, the antilog of $(-9) = 10^{-9}$.

Now express the number whose logarithm is -11.30 in scientific notation. The number whose logarithm is -11.30 is $10^{-11.30}$. To convert this expression to scientific notation, the exponent must first be changed to the sum of a positive decimal fraction and a negative whole number:

$$-11.30 = 0.70 + (-12)$$

As the sum of exponents indicates a product:

$$10^{0.70 + (-12)} = 10^{0.70} \times 10^{-12}$$

From Appendix Table 17, 0.70 is the logarithm of 5.0 (to 2 significant figures).

$$10^{0.70} = 5.0$$

$$10^{0.70} \times 10^{-12} = 5.0 \times 10^{-12}$$

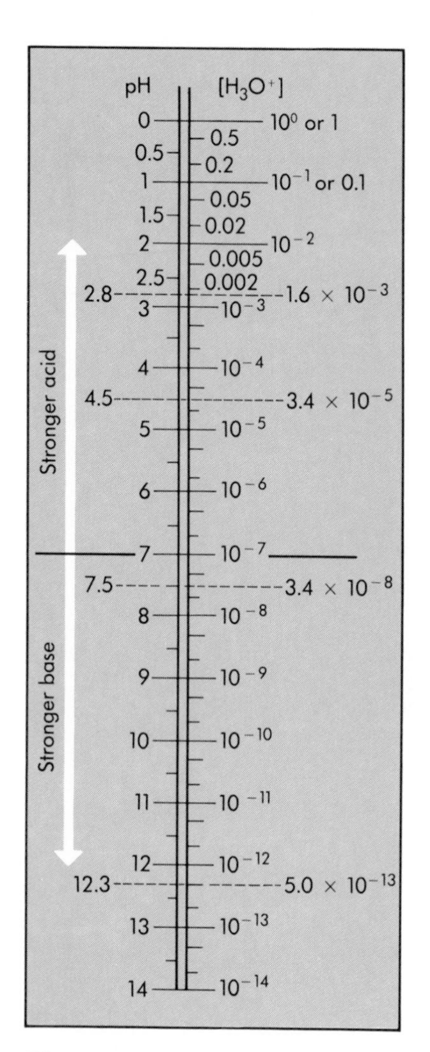

Figure 16-4. A comparison of the numerical pH scale and corresponding hydronium ion concentrations, $[H_3O^+]$.

The relationship between the pH and $[H_3O^+]$ is shown on the scale of Figure 16-4. This scale can be used to estimate the pH from a known $[H_3O^+]$ value or the $[H_3O^+]$ from a known pH value. For example, the $[H_3O^+]$ of a solution is 3.4×10^{-5} mole/L. Observe that 3.4×10^{-5} lies between 10^{-4} and 10^{-5} on the $[H_3O^+]$ scale of Figure 16-4. Thus the pH of the solution must be between 4 and 5. The value can be estimated to a second significant figure as 4.5 on the pH scale of Figure 16-4. Calculations are required to obtain a more precise pH value. However, this reliable estimate of pH helps prevent errors that commonly occur in such calculations. See Sample Problem 16-4.

SAMPLE PROBLEM 16-4

What is the pH of a solution if $[H_3O^+]$ is 3.4×10^{-5} mole/liter?

SOLUTION

$$\mathbf{pH = -\log [H_3O^+]}$$
$$\mathbf{pH = -\log (3.4 \times 10^{-5})}$$

The logarithm of a product is equal to the sum of the logarithms of each of the factors. Thus,

$$\mathbf{pH = -(\log 3.4 + \log 10^{-5})}$$

The log of $10^{-5} = -5$, and from the table of logarithms (Appendix Table 17), the log of 3.4 is found to be 0.53.

$$\mathbf{pH = -(0.53 - 5)}$$

Therefore,

$$\mathbf{pH = 4.47}$$

Compare this result with that obtained from Figure 16-4.

16.9 Calculation of $[H_3O^+]$ The pH of a solution has been calculated knowing its hydronium ion concentration. Now suppose the pH of a solution is known. How can its hydronium ion concentration be determined?

The equation for the pH in terms of the $[H_3O^+]$ is

$$\mathbf{pH = -\log [H_3O^+]}$$

Remember that the base of common logarithms is 10. This equation can be restated in terms of $[H_3O^+]$ as follows:

$$\mathbf{\log [H_3O^+] = -\ pH}$$
$$\mathbf{[H_3O^+] = antilog\ (-pH)}$$
$$\mathbf{[H_3O^+] = 10^{-pH}}$$

$[H_3O^+]$ = antilog (−pH).

For an aqueous solution having a pH of 2, the $[H_3O^+]$ is equal to 10^{-2} mole/L. When the pH is 0, the $[H_3O^+]$ is 1 mole/L. Recall that $10^0 = 1$. Sample Problem 16-5 has a pH value that is a positive integer.

SAMPLE PROBLEM 16-5

Determine the hydronium ion concentration of an aqueous solution that has a pH of 4.

SOLUTION

$$pH = -\log [H_3O^+]$$
$$\log [H_3O^+] = -pH$$
$$[H_3O^+] = \text{antilog } (-pH) = \text{antilog } (-4)$$
$$\text{antilog } (-4) = 10^{-4}$$
$$[H_3O^+] = 10^{-4} \text{ mole/L}$$

Observe that if the pH = 4, the magnitude of the $[H_3O^+] = 10^{-4}$. The pH value in Sample Problem 16-6 is not an integral number.

SAMPLE PROBLEM 16-6

The pH of a solution is found to be 7.52. *(a)* What is the hydronium ion concentration? *(b)* What is the hydroxide ion concentration? *(c)* Is the solution acidic or basic?

SOLUTION

(a) The $[H_3O^+]$ is the number whose logarithm is -7.52. Therefore, the antilog of -7.52 will give the hydronium ion concentration.

$$pH = -\log [H_3O^+]$$

Solving for $[H_3O^+]$

$$\log [H_3O^+] = -pH$$
$$[H_3O^+] = \text{antilog } (-pH)$$
$$[H_3O^+] = \text{antilog } (-7.52)$$

But

$$\text{antilog } (-7.52) = \text{antilog } (0.48 - 8)$$

Thus

$$[H_3O^+] = \text{antilog } (0.48 - 8)$$
$$[H_3O^+] = \text{antilog } (0.48) \times \text{antilog } (-8)$$

The antilog of (0.48) is found from the table of logarithms to be 3.0. The antilog of (−8) is 10^{-8}. Therefore,

$$[H_3O^+] = 3.0 \times 10^{-8} \text{ mole/L}$$

Compare this result with that obtained from Figure 16-4.

(b) $$[H_3O^+][OH^-] = 1.0 \times 10^{-14} \text{ mole}^2/\text{L}^2$$

$$[OH^-] = \frac{1.0 \times 10^{-14} \text{ mole}^2/\text{L}^2}{3.0 \times 10^{-8} \text{ mole/L}}$$

$$[OH^-] = 0.33 \times 10^{-6} \text{ mole/L} = 3.3 \times 10^{-7} \text{ mole/L}$$

(c) Solution is slightly basic as pH > 7 and $[OH^-] > [H_3O^+]$.

Practice Problems

1. What is the pH of an aqueous solution whose $[H_3O^+]$ is 2.7×10^{-4} mole per liter? *ans.* 3.57
2. The $[H_3O^+]$ of an aqueous solution of a weak acid is 3.54×10^{-5} mole per liter. Find its pH. *ans.* 4.451
3. The pH of an aqueous solution is measured as 1.5. *(a)* Calculate the $[H_3O^+]$. *(b)* What is the $[OH^-]$ of the solution? *ans. (a)* 3.2×10^{-2} mole/L; *(b)* 3.1×10^{-13} mole/L

Table 16-4 shows the relationship between the hydronium ion and hydroxide ion concentrations, the product of these concentrations, and the pH for several solutions of typical molarities. Since KOH is a soluble ionic compound, its aqueous solutions are completely ionized. The molarity of each KOH solution indicates directly the $[OH^-]$. The product $[H_3O^+][OH^-]$ is constant, 10^{-14} mole^2 per liter^2 at 25 °C. Therefore, the $[H_3O^+]$ can be calculated. If the $[H_3O^+]$ is known, the pH can be determined as $-\log [H_3O^+]$.

Any aqueous solution of HCl that has a concentration below 1-*M* can be considered to be completely ionized. Thus the molarity of a 0.001-*M* HCl solution indicates directly the $[H_3O^+]$.

The weakly ionized $HC_2H_3O_2$ solution presents a different problem. Information about the concentrations of $HC_2H_3O_2$

Table 16-4
RELATIONSHIP OF $[H_3O^+]$ TO $[OH^-]$ AND pH (at 25 °C)

Solution	*$[H_3O^+]$*	*$[OH^-]$*	*$[H_3O^+][OH^-]$*	*pH*
0.02-*M* KOH	5.0×10^{-13}	2.0×10^{-2}	1.0×10^{-14}	12.3
0.01-*M* KOH	1.0×10^{-12}	1.0×10^{-2}	1.0×10^{-14}	12.0
pure H_2O	1.0×10^{-7}	1.0×10^{-7}	1.0×10^{-14}	7.0
0.001-*M* HCl	1.0×10^{-3}	1.0×10^{-11}	1.0×10^{-14}	3.0
0.1-*M* $HC_2H_3O_2$	1.3×10^{-3}	7.7×10^{-12}	1.0×10^{-14}	2.9

molecules, H_3O^+ ions, and $C_2H_3O_2^-$ ions in the equilibrium mixture in the aqueous solution may be lacking. However, the hydronium ion concentration can be determined by measuring the pH of the solution experimentally. The $[H_3O^+]$ is then determined as the antilog (−pH).

$$[H_3O^+] = \text{antilog } (-pH)$$

16.10 The neutralization reaction In a neutralization reaction, the basic OH^- ion acquires a proton from the H_3O^+ ion to form a molecule of water.

$$H_3O^+ + OH^- \rightarrow 2H_2O$$

One mole of H_3O^+ ions (19 g) and 1 mole of OH^- ions (17 g) are chemically equivalent. Neutralization occurs when H_3O^+ ions and OH^- ions are supplied in equal numbers. A liter of water at room temperature has an $[H_3O^+]$ and $[OH^-]$ of 10^{-7} *M* each. Furthermore, the product $[H_3O^+]$ $[OH^-]$ of 10^{-14} mole²/liter² is a constant for water and all dilute aqueous solutions at 25 °C.

In aqueous solutions: if $[H_3O^+]$ increases, $[OH^-]$ decreases; if $[OH^-]$ increases, $[H_3O^+]$ decreases.

If 0.1 mole of gaseous HCl is dissolved in the liter of water, the H_3O^+ ion concentration rises to 0.1 or 10^{-1} *M*. Since the product $[H_3O^+]$ $[OH^-]$ remains at 10^{-14}, the $[OH^-]$ obviously must decrease from 10^{-7} to 10^{-13} *M*. The OH^- ions are removed from solution by combining with H_3O^+ ions to form H_2O molecules. Almost 10^{-7} mole of H_3O^+ ions is also removed in this way. However, this quantity is only a small portion (0.0001%) of the 0.1 mole of H_3O^+ ions present in the liter of solution.

Now suppose 0.1 mole (4 g) of solid NaOH is added to the liter of 0.1-*M* HCl solution. Imagine, also, that the hydroxide and hydronium ions are somehow temporarily prevented from reacting with each other. The NaOH dissolves and supplies 0.1 mole of OH^- ions to the solution. Both $[H_3O^+]$ and $[OH^-]$ are now high, and their product is much greater than the constant value 10^{-14} for the dilute aqueous solution.

Suppose the chemical reaction is now allowed to begin. The ion-removal reaction will be as before except that this time there are as many OH^- ions as H_3O^+ ions to be removed. H_3O^+ and OH^- ions combine until the product $[H_3O^+]$ $[OH^-]$ returns to the constant value 10^{-14}, and

$$[H_3O^+] = [OH^-] = 10^{-7}\ M$$

Figure 16-5. The neutralization reaction. Hydronium ions and hydroxide ions form very slightly ionized water molecules.

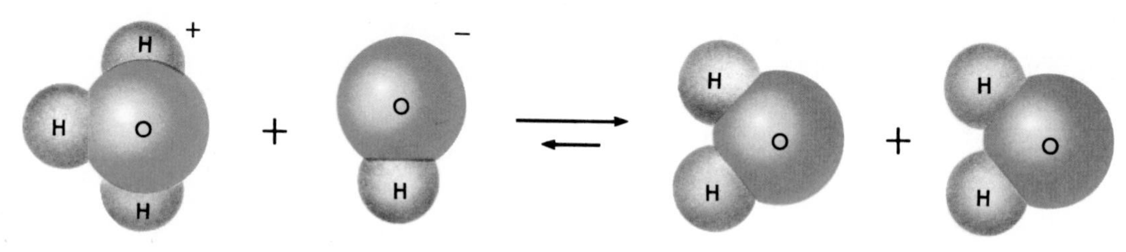

The resulting solution is neither acidic nor basic, but is neutral. The process is one in which chemically equivalent quantities of H_3O^+ ions and OH^- ions combine. A neutralization reaction has taken place.

Figure 16-6. An acid-base titration using an indicator for detecting the end point of the reaction and a magnetic stirrer for mixing the reactants during the titration.

16.11 Acid-base titration The preceding examples should help you understand the nature of the chemical reaction that occurs between acids and bases as a solution of one is progressively added to a solution of the other. This progressive addition of an acid to a base (or a base to an acid) in order to compare their concentrations is called *titration*. ***Titration*** *is the controlled addition of the measured amount of a solution of known concentration required to react completely with a measured amount of a solution of unknown concentration.*

Titration provides a sensitive means of determining the relative volumes of acidic and basic solutions that are chemically equivalent. If the concentration of one solution is known, the concentration of the other solution can be calculated. Titration is an important laboratory procedure and is often used in analytical chemistry. See Figure 16-6.

Suppose successive additions of an aqueous base are made to a measured volume of an aqueous acid. Eventually the acid is neutralized. With continued addition of base, the solution becomes distinctly basic. The pH has now changed from a low to a high numerical value. The change in pH occurs slowly at first, then rapidly through the neutral point, and slowly again as the solution becomes basic. Typical pH curves for strong acid–strong base and weak acid–strong base titrations are shown in Figure 16-7.

The very rapid change in pH occurs in the region where equivalent quantities of H_3O^+ and OH^- ions are present. Any method that shows this abrupt change in pH can be used to detect the *end point*, or *equivalence point*, of the titration.

End point: the point in a titration at which the reaction is just complete.

Many dyes have colors that are sensitive to pH changes. Some change color within the pH range in which an end point

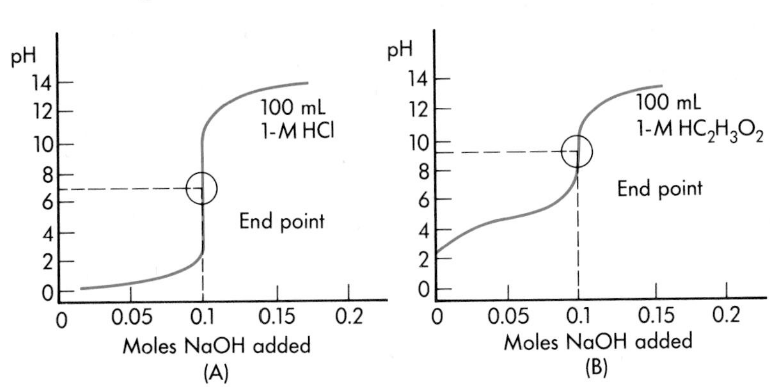

Figure 16-7. Acid-base titration curves: (A) strong acid–strong base; (B) weak acid–strong base.

occurs. Such dyes may serve as *indicators* in the titration process. Several indicators are listed in Table 16-5.

In order to actually know the concentration of a "known" solution, it is compared to a *standard solution* whose concentration is well established. The known solution is first prepared, and its volume is adjusted to the desired concentration. This concentration information is then refined by titrating the solution against a carefully measured quantity of a highly purified compound known as a *primary standard*. The actual concentration of the known solution becomes that established by this standardizing procedure.

16.12 Titration with molar solutions Burets like those shown in Figure 16-8 are commonly used in titration to measure solution volumes with good precision. Suppose an aqueous solution of NaOH of unknown concentration is added in successive small amounts to 10.0 mL of 0.01-*M* aqueous HCl containing a few drops of a suitable indicator until the end point is reached. The base buret readout shows that 20.0 mL of the NaOH solution was used. How can these titration data indicate the *molarity* of the basic solution?

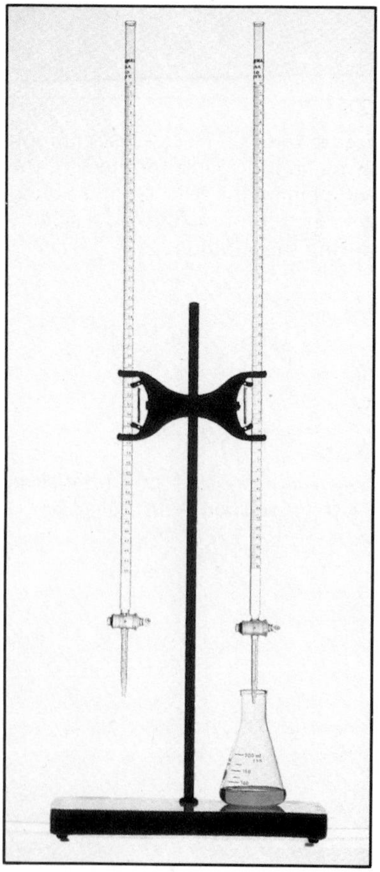

Figure 16-8. A common laboratory titration stand.

The empirical equation for the neutralization reaction is

$$\mathbf{HCl + NaOH \rightarrow NaCl + H_2O}$$

The volume and molarity of the HCl solution are known. From these data, the quantity, in moles, of HCl used when titrating can be determined.

$$\frac{\mathbf{10.0\ mL}}{\mathbf{1000\ mL/L}} \times \frac{\mathbf{0.01\ mole\ HCl}}{\mathbf{L}} = \mathbf{0.0001\ mole\ HCl\ used}$$

The balanced equation shows that *1 mole* of NaOH is used for *1 mole* of HCl. In other words, NaOH and HCl show chemical equivalence, mole for mole, in the reaction. Therefore, the quantity of NaOH used in the titration is also 0.0001 mole. This quantity was furnished by 20.0 mL of NaOH solution. The molarity of the NaOH solution is obtained as follows:

$$\frac{\mathbf{0.0001\ mole\ NaOH}}{\mathbf{20.0\ mL}} \times \frac{\mathbf{1000\ mL}}{\mathbf{L}} = \mathbf{0.005\ mole\ NaOH/L}$$

or

$$\mathbf{0.005\text{-}}\boldsymbol{M}\ \mathbf{NaOH}$$

Suppose the titration is repeated with the same "unknown" NaOH solution, but this time against 10.0 mL of a 0.01-*M* solution of diprotic H_2SO_4 as the "known" acid solution. When titrated to the end point, the base buret readout shows that 40.0

mL of the NaOH solution was used. The empirical equation for this reaction is

$$H_2SO_4 + 2NaOH \rightarrow Na_2SO_4 + 2H_2O$$

$$\frac{10.0 \text{ mL}}{1000 \text{ mL/L}} \times \frac{0.01 \text{ mole } H_2SO_4}{L} = 0.0001 \text{ mole } H_2SO_4 \text{ used}$$

The equation shows that *2 moles* of NaOH are required for *1 mole* of H_2SO_4. Therefore, 0.0002 mole NaOH is used in the titration as the chemical equivalent of 0.0001 mole H_2SO_4. The molarity of the NaOH solution is obtained as follows:

$$\frac{0.0002 \text{ mole NaOH}}{40.0 \text{ mL}} \times \frac{1000 \text{ mL}}{L} = 0.005 \text{ mole NaOH/L}$$

or

$$0.005\text{-}M \text{ NaOH}$$

To summarize, the molarity of an aqueous base (or acid) of unknown concentration can be determined by titrating against an aqueous acid (or base) of known concentration. The following steps are involved:

1. *Determine the moles solute of known solution used during the titration.*
2. *Determine the ratio—moles unknown solute/moles known solute—from a balanced equation.*
3. *Determine the moles solute of unknown solution used during the titration.*
4. *Determine the molarity of the unknown solution.*

Sample Problem 16-7 illustrates the titration process.

SAMPLE PROBLEM 16-7

In a titration, 27.4 mL of a standard solution of $Ba(OH)_2$ is added to a 20.0-mL sample of an HCl solution. The concentration of the standard solution is 0.0154 M. What is the molarity of the acid solution?

SOLUTION

The equation for this reaction is

$$2HCl + Ba(OH)_2 \rightarrow BaCl_2 + 2H_2O$$

The quantity, in moles, of $Ba(OH)_2$ used in the reaction can be found from the molarity of the standard solution and the volume used.

$$\frac{27.4 \text{ mL}}{1000 \text{ mL/L}} \times \frac{0.0154 \text{ mole } Ba(OH)_2}{L} = 0.000422 \text{ mole } Ba(OH)_2 \text{ used}$$

The equation shows that *2 moles* of HCl are used for *1 mole* of $Ba(OH)_2$. Therefore, 0.000844 mole of HCl is used since this is the chemical equivalent of 0.000422 mole of $Ba(OH)_2$. If 20.0 mL of the unknown solution contains 0.000844 mole of HCl, the concentration is

$$\frac{\textbf{0.000844 mole HCl}}{\textbf{20.0 mL}} \times \frac{\textbf{1000 mL}}{\textbf{L}} = \textbf{0.0422 mole/L}$$

or

$$\textbf{0.0422-}M\textbf{ HCl}$$

Practice Problems

1. A 15.5-mL sample of 0.215-*M* KOH solution required 21.2 mL of aqueous acetic acid in a titration experiment. Calculate the molarity of the acetic acid solution. *ans.* 0.157 *M*
2. By titration, 17.6 mL of aqueous H_2SO_4 just neutralized 27.4 mL of 0.0165-*M* LiOH solution. What was the molarity of the aqueous acid? *ans.* 0.0128 *M*

16.13 Titration with normal solutions Chemists sometimes prefer to express solution concentrations in terms of *normality*. The advantage in doing so is that concentrations are expressed directly in terms of equivalents of solute. Solutions of the same normality are chemically equivalent, milliliter for milliliter. Recall that the normality of a given solution is a whole number times its molarity, the factor depending on the substance and the reaction in which it is involved.

A very simple relationship exists between volumes and normalities of solutions used in titration. For example, a titration required 50.0 mL of a 0.100-*N* solution of NaOH to reach an end point with 10.0 mL of vinegar, a water solution of acetic acid ($HC_2H_3O_2$). The chemical equivalent of $HC_2H_3O_2$ used is the product of the volume of acid solution used, V_a, and the acid's normality, N_a.

$$V_a \times N_a = \textbf{equiv}_a$$

Similarly, the chemical equivalent of NaOH used is the product of the volume of base solution used, V_b, and the base's normality, N_b.

$$V_b \times N_b = \textbf{equiv}_b$$

At the end point in the titration,

$$\textbf{equiv}_a = \textbf{equiv}_b$$

Therefore,

$$V_aN_a = V_bN_b$$

In this example, the normality of the vinegar, N_a, is to be determined. Solving the above equation for N_a

$$N_a = \frac{V_b N_b}{V_a} = \frac{50.0 \text{ mL} \times 0.100\ N}{10.0 \text{ mL}}$$

$$N_a = 0.500\ N$$

In this simple numerical example, it is apparent that 5 times as much base solution was used in the titration as vinegar. Clearly, the vinegar is 5 times more concentrated than the base.

The acidity of vinegar is due to the presence of acetic acid. A 1.0-*N* acetic acid solution contains 1.0 equivalent per liter of solution. In this case, 1.0 equivalent equals 1.0 mole or $6\bar{0}$ g of $HC_2H_3O_2$ per liter of solution. The 0.50-*N* solution must contain $3\bar{0}$ g of $HC_2H_3O_2$ per liter. A liter of vinegar has a mass of about 1000 g. Thus, the sample of vinegar used will contain 3.0% acetic acid.

NaOH is a strong base, and $HC_2H_3O_2$ is a weak acid. The pH curve for this titration, Figure 16-7(B), differs from the curve for a strong acid-strong base titration. The end point occurs at a higher pH because the sodium acetate solution formed in the titration is slightly basic.

Aqueous solutions of some salts may be acidic or basic depending on the composition of the salt. If the anions are sufficiently basic, protons are removed from some water molecules and the $[OH^-]$ increases. If the cations are slightly acidic, protons are donated to some of the water molecules, and the $[H_3O^+]$ increases.

16.14 Indicators in titration Chemists have a wide choice of indicators for use in titration. They are able to choose one that changes color over the correct pH range for a particular reaction. Why is it not always suitable to use an indicator that changes color at a pH of 7?

Solutions of soluble hydroxides and acids mixed in chemically equivalent quantities are neutral only if both solutes are ionized to the same degree. The purpose of the indicator is to show, by a change in color, that the end point has been reached, that is, to show when equivalent quantities of the two solutes are together. Table 16-5 gives the color changes of several common indicators used in acid-base titrations. The pH range over which an indicator color change occurs is referred to as its *transition interval*. Observe the variations in the transition intervals for the different indicators. The choice of indicator is based on the suitability of its transition interval for a given acid-base reaction.

There are four possible types of acid-base combinations. In titration procedures these combinations may have end points occurring in different pH ranges as follows:

1. *Strong acid—strong base:* pH is about 7. Litmus is a suitable indicator, but the color change is not sharply defined. Bromothymol blue performs more satisfactorily. Its color change is shown in Figure 16-9.

2. *Strong acid–weak base:* pH is lower than 7. Methyl orange is a suitable indicator.

Table 16-5
INDICATOR COLORS

Indicator	*Color*			*Transition interval (pH)*
	Acid	*Transition*	*Base*	
methyl violet	yellow	aqua	blue	0.0– 1.6
methyl yellow	red	orange	yellow	2.9– 4.0
bromophenol blue	yellow	green	blue	3.0– 4.6
methyl orange	red	orange	yellow	3.2– 4.4
methyl red	red	buff	yellow	4.8– 6.0
litmus	red	pink	blue	5.5– 8.0
bromothymol blue	yellow	green	blue	6.0– 7.6
phenol red	yellow	orange	red	6.6– 8.0
phenolphthalein	colorless	pink	red	8.2–10.6
thymolphthalein	colorless	pale blue	blue	9.4–10.6
alizarin yellow	yellow	orange	red	10.0–12.0

3. *Weak acid–strong base:* pH is higher than 7. Phenolphthalein is a suitable indicator.

4. *Weak acid–weak base:* pH may be either higher or lower than 7, depending on which reactant is stronger. No indicator is satisfactory.

16.15 pH measurements Indicators used to detect end points in neutralization reactions are organic compounds. They have the characteristics of weak acids. When added to a solution in suitable form and concentration, an indicator gives the solution a characteristic color. If the pH of the solution is changed enough, as in titration, the indicator changes color over a definite pH range, the *transition interval.*

Because indicators are color sensitive to the pH of a solution, they are used to give information about the $[H_3O^+]$ of the solution. The color of un-ionized indicator molecules is different from that of indicator ions. In solutions of high hydronium ion

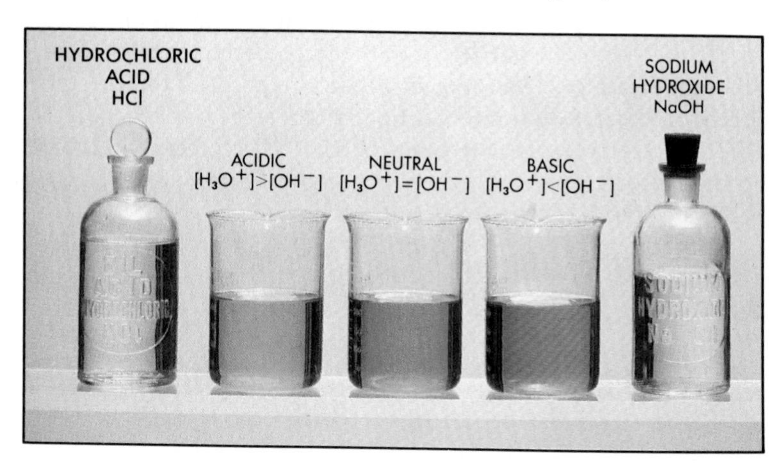

Figure 16-9. The acid, transition, and base colors of bromothymol blue.

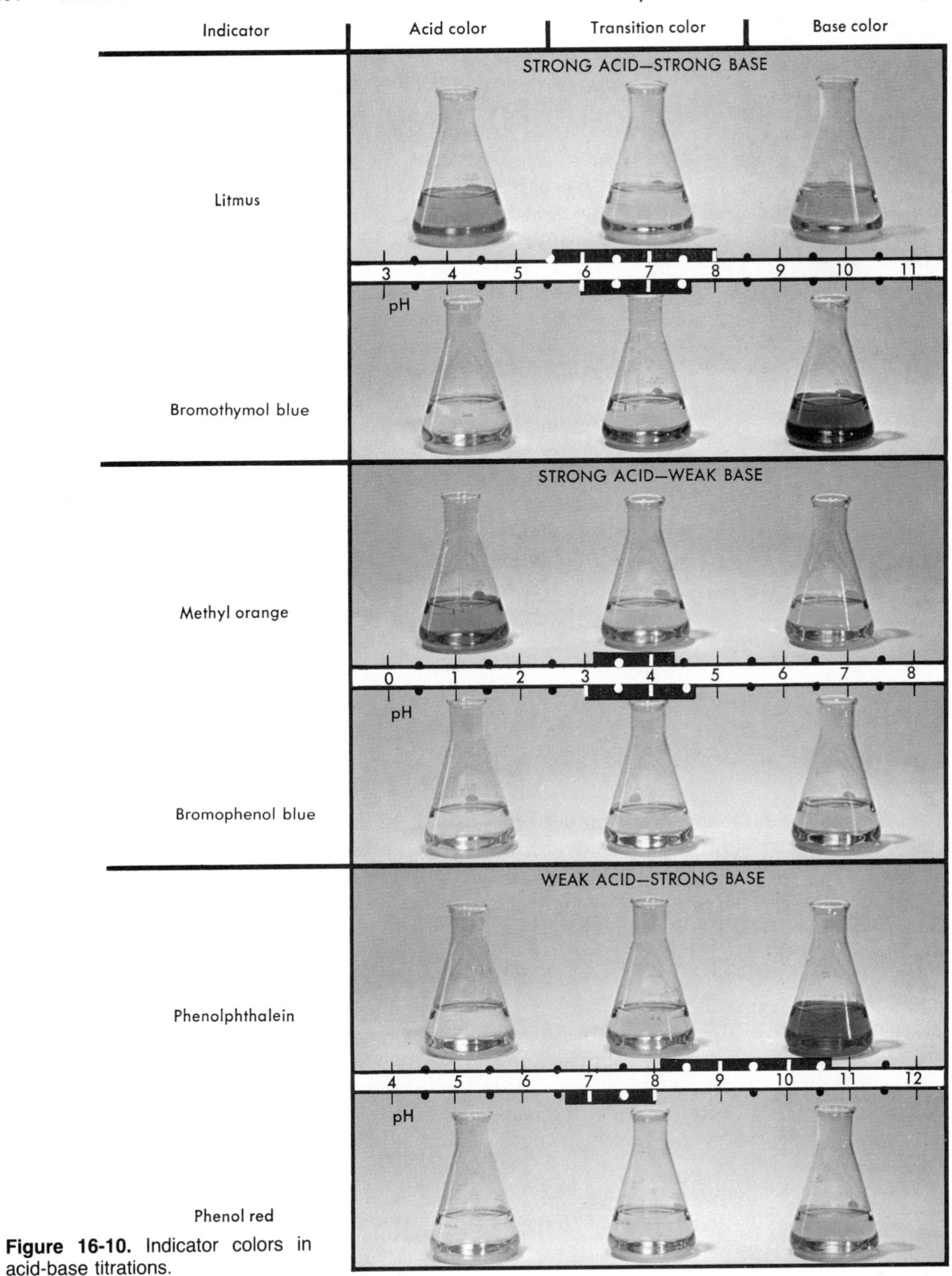

Figure 16-10. Indicator colors in acid-base titrations.

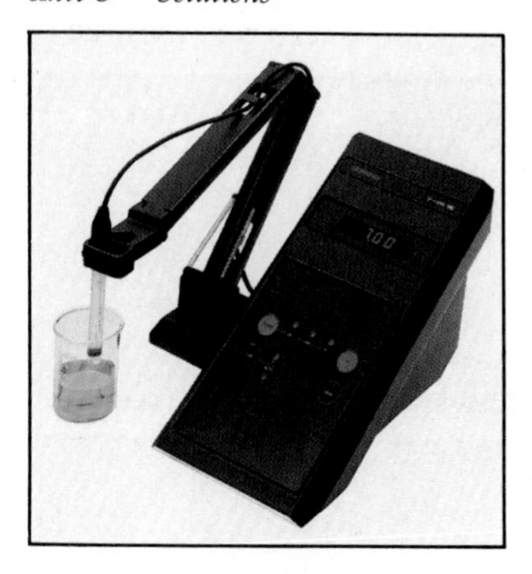

Figure 16-11. A pH meter.

concentration (low pH), the color of the molecular species prevails. Solutions of low hydronium ion concentrations (high pH) have the color of the ionic species. The acid, base, and transition colors of several indicators are shown in Figure 16-10.

An indicator added to different solutions may show the same *transition color*. If so, the solutions are considered to have the same pH. This is the basis for the common *colorimetric* determination of pH. A measured volume of a suitable indicator is added to each solution whose pH is to be determined. The color is then compared with that of the same indicator in solutions of known pH. By careful color comparison, the pH of a solution can be estimated with a precision of about 0.1 pH interval.

End-point determinations in titrations and the pH of solutions can be found in ways other than by the use of indicators. Instrumental methods are generally used by chemists to make rapid titrations and pH determinations. A pH meter, shown in Figure 16-11, provides a convenient way of measuring the pH of a solution. The pH meter measures the voltage difference between a special electrode and a reference electrode placed in the solution. The voltage changes as the H_3O^+ ion concentration of the solution changes. In an acid-base titration, a large change in voltage occurs when equivalent quantities of the two reactants are present in the solution. This voltage change is related to the sharp color change of an indicator near the end point of the titration process.

SUMMARY

Solution concentrations are expressed in terms of molality, molarity, or normality. Molar solutions are prepared using moles of solute per liter of solution. Normal solutions are prepared using chemical equivalents of solute per liter of solution.

Chemical equivalents of substances have the same combining capacities in reactions. One equivalent of an acid supplies one mole of protons. One equivalent of a base supplies one mole of OH^- ions or accepts one mole of protons. For diprotic and triprotic acids, the reactions they undergo determine what their chemical equivalencies will be.

Water is weakly ionized by autoprotolysis. At 25 °C, the hydronium ion concentration and hydroxide ion concentration are each 10^{-7} mole/liter. The product of these ion concentrations remains constant at 10^{-14} mole^2/liter^2 for water and aqueous solutions as long as the temperature remains at 25 °C. The ionization of water increases slightly with a rise in temperature.

The pH scale is used to indicate the hydronium ion concentration in aqueous solutions. In numerical notation, the pH of a solution is the common logarithm of the number of liters that contain one mole of H_3O^+ ions. Since this number of liters is the reciprocal of the H_3O^+ ion concentration of the solution, the pH is also the common logarithm of the reciprocal of the hydronium ion concentration. If the hydronium ion concentration of a solution is known, the pH can be calculated. Conversely, the hydronium ion concentration can be calculated if the pH of a solution is known. The pH of pure water and neutral aqueous solutions is 7. Acidic solutions have pH values lower than 7. Basic solutions have pH values higher than 7.

The concentrations of acids and bases are compared by a technique called titration. A solution of a base (or acid) of unknown concentration is added to a solution of an acid (or base) of known concentration until equal chemical equivalents of acid and base are present in the solution. An acid-base indicator is used to show when the end point in the titration is reached. Because the concentration of one solution and the volumes of both solutions used to reach the end point are known, the concentration of the other solution can be calculated.

Indicators change color over characteristic short pH ranges called their transition intervals. The end point is reached at a pH of about 7 for titrations of strong acids and strong bases. Indicators with transition intervals around a pH of 7 are used. Indicators with transition intervals below a pH of 7 are used in titration of strong acids and weak bases. Weak acid–strong base titrations require indicators with transition intervals above a pH of 7.

VOCABULARY

antilog
autoprotolysis
characteristic
colorimetric determination of pH
common logarithm
end point
equivalent
indicator
mantissa
molality
molarity
normality
pH
primary standard
standard solution
titration
transition color
transition interval
volumetric flask

QUESTIONS

Group A

1. Define: *(a)* mole; *(b)* equivalent. *A1*
2. Define the following terms and give the identifying symbol for each: *(a)* molality; *(b)* molarity; *(c)* normality. *A1*
3. On what basis is the mass of 1 equivalent determined: *(a)* for an acid; *(b)* for a base? *A1*
4. Determine the mass of 1 equivalent of each of the following: *(a)* Zn; *(b)* Li; *(c)* H_3O^+; *(d)* OH^-; *(e)* HCO_3^-; *(f)* KCl; *(g)* $HC_2H_3O_2$; *(h)* H_3PO_4; *(i)* $ZnSO_4$; *(j)* $Ca(NO_3)_2$. *A1*
5. Determine the number of equivalents per mole of each of the following: *(a)* $HClO_3$; *(b)* $Ba(NO_3)_2$; *(c)* HBr; *(d)* $AuCl_3$; *(e)* $KC_2H_3O_2$; *(f)* $CuSO_4$; *(g)* $Mg(OH)_2$; *(h)* $(NH_4)_2S$; *(i)* $FePO_4$; *(j)* K_2CrO_4. *A2b, A3c*
6. *(a)* Write equations representing two partial ionizations of aqueous sulfuric acid that occur with increasing dilution. *(b)* Write the equation for the complete ionization of aqueous sulfuric acid in very dilute solution. *A2a*
7. What is the meaning of the notation $[C_2H_3O_2^-]$? *A3c*
8. *(a)* The HSO_4^- ion is the conjugate base of what acid? *(b)* Write the ionization equation. *(c)* The HSO_4^- ion is the conjugate acid of what base? *(d)* Write the ionization equation for the reaction.
9. *(a)* Write the equation for the ionization of acetic acid in aqueous solution. *(b)* Identify the two conjugate acid-base pairs in this equation. *A3n*
10. *(a)* Define the pH of a solution. *(b)* What is the usual range of the pH scale? *A1*
11. The hydronium ion concentration in a certain aqueous solution is determined to be 1×10^{-4} mole per liter. *(a)* What is the pH of the solution? *(b)* What is the hydroxide ion concentration of this solution? *A2h*

12. The hydroxide ion concentration in a certain aqueous solution is 1×10^{-6} mole per liter. *(a)* What is the hydronium ion concentration of the solution? *(b)* What is the pH of this solution? *A2h*

13. Neither bromphenol blue nor methyl orange is a suitable indicator for the titration of 0.02-*N* acetic acid with sodium hydroxide. *(a)* Explain. *(b)* Identify two indicators that are suitable for this titration. *A2l*

QUESTIONS

Group B

14. Hydrogen carbonate, H_2CO_3, has 2 equivalents per mole and hydrogen chloride, HCl, has 1 equivalent per mole. Nevertheless, hydrochloric acid is described as a strong acid and carbonic acid as a weak acid. Explain. *A4c*
15. Test your saliva with litmus paper. *(a)* Is it acidic or basic? *(b)* Would you expect a toothpaste to be acidic or basic? Test those in your home. *A4d*
16. Devise a simple litmus-paper test for the soil in your lawn or garden. *(a)* Is the soil acidic or basic? *(b)* What can be added to reduce the acidity of soil? (See Sections 25.10 and 25.11.) *(c)* What can be added to reduce the alkalinity of soil? (See Section 25.12.) *A4d*
17. When a solution of potassium hydroxide, KOH, is neutralized with hydrochloric acid, HCl, the K^+ ions and Cl^- ions are called *spectator* ions. *(a)* Explain. *(b)* How can potassium chloride, KCl, be recovered? *A6*
18. *(a)* What indicator should be used to identify the end point of the neutralization reaction of Question 17? *(b)* Explain the reason for your indicator selection. *A5*
19. How many moles of sodium hydroxide are required for the complete neutralization of *(a)* 2 moles of hydrochloric acid; *(b)* 2 moles of sulfuric acid; *(c)* 2 moles of phosphoric acid? *(d)* Write the empirical equation for each reaction.
20. What mass of calcium hydroxide is required to prepare *(a)* 1.0 liter of 0.0010-*N* solution; *(b)* 1.0 liter of 0.0010-*M* solution? *A3c*
21. Ten million liters (10^7 L) of water at 25 °C contains what number of *(a)* moles of H_3O^+ ions; *(b)* equivalents of H_3O^+ ions; *(c)* grams of H_3O^+ ions; *(d)* moles of OH^- ions; *(e)* equivalents of OH^- ions; *(f)* grams of OH^- ions? *A2d*
22. What is the normality of the following: *(a)* a 0.0030-*M* solution of copper(II) nitrate; *(b)* a 0.12-*M* solution of sodium hydroxide; *(c)* a 1.5-*M* solution of sulfuric acid? *A2c*
23. What is the molarity of the following: *(a)* a 0.001-*N* solution of hydrobromic acid; *(b)* a 0.048-*N* solution of aluminum sulfate; *(c)* a 0.024-*N* solution of barium hydroxide? *A2c*

PROBLEMS

Group A

1. *(a)* What quantity of sodium hydroxide is required to neutralize 21.9 g of hydrogen chloride in aqueous solution? *(b)* How many moles of each reactant are involved in the reaction? *B9*
2. What quantity of potassium nitrate is required to prepare a 0.400-*m* solution in $75\bar{0}$ g of water?
3. How many grams of sugar, $C_{12}H_{22}O_{11}$, are contained in $20\bar{0}$ mL of a 0.250-*M* aqueous solution? *B3*
4. Determine the molarity of a solution prepared by dissolving 36.5 g of H_2SO_4 in enough water to make 2.00 L of solution. *B4*
5. What quantity of potassium iodide is required to make $50\bar{0}$ mL of 0.125-*M* solution? *B3*

6. Determine the molarity of a $CuBr_2$ solution that contains 145 g of solute in 2.50 L of solution. *B4*
7. How many grams of solute are required to make $15\bar{0}$ mL of a 0.250-*M* solution of $Al_2(SO_4)_3 \cdot 18H_2O$? *B4*
8. Calculate the mass of one equivalent of each of the following substances: *(a)* K; *(b)* Ca; *(c)* NaCl; *(d)* $CuSO_4$; *(e)* $KMnO_4$; *(f)* $Al_2(SO_4)_3$; *(g)* $ZnCrO_4$; *(h)* $Ni(NO_3)_2 \cdot 6H_2O$; *(i)* $Na_2CO_3 \cdot 10H_2O$; *(j)* $FeCl_3 \cdot 6H_2O$. *B2*
9. What is the normality of a solution that contains 5.4 g of Na_2SO_4 per liter of solution? *B6*
10. Determine the normality of a solution containing 473 g of $Al(NO_3)_3$ in 12.0 L of solution. *B6*
11. What quantity of $CuSO_4 \cdot 5H_2O$ is required to prepare $25\bar{0}$ mL of 0.200-*N* solution? *B5*
12. What quantity of $FeCl_3 \cdot 6H_2O$ is required to prepare $60\bar{0}$ mL of 0.160-*N* solution? *B5*
13. How many milliliters of 0.150-*N* solution of a metallic hydroxide are required to neutralize 30.0 mL of 0.500-*N* solution of an acid? *B10*
14. A chemistry student finds that it takes 34 mL of 0.50-*N* acid solution to neutralize $1\bar{0}$ mL of a sample of household ammonia. What is the normality of the ammonia-water solution? *B11*
15. *(a)* What is the pH of a 0.01-*M* solution of HCl, assuming complete ionization? *(b)* What is the $[OH^-]$ of a 0.01-*M* solution of NaOH? *(c)* What is the pH of this NaOH solution? *B7, B8*

PROBLEMS

Group B

16. How many solute molecules are contained in each milliliter of a 0.1-*M* solution of a nonelectrolyte?
17. An excess of zinc reacts with $50\bar{0}$ mL of an aqueous solution of HCl, and 2.61 L of H_2 gas are collected over water at 25.0 °C and 745.0 mm. Determine the molarity of the acid. *B4*
18. A certain batch of concentrated hydrochloric acid has a density of 1.19 g/mL and contains 37.2% HCl by mass. How many milliliters of concentrated hydrochloric acid are required to prepare *(a)* 1.00 L of 1.00-*M* HCl solution; *(b)* 2.50 L of 3.00-*M* HCl solution; *(c)* 2.00 L of 0.100-*N* HCl solution; *(d)* $50\bar{0}$ mL of 0.200-*N* HCl solution? *B4*
19. A stockroom supply of concentrated sulfuric acid is 95.0% H_2SO_4 by mass and has a density of 1.84 g/mL. How many milliliters of concentrated sulfuric acid are required to prepare *(a)* 1.00 L of 1.00-*M* H_2SO_4 solution; *(b)* $20\bar{0}$ mL of 0.200-*M* H_2SO_4 solution; *(c)* 4.00 L of 1.00−*N* H_2SO_4 solution; *(d)* 2.50 L of 0.200−*N* H_2SO_4 solution? *B3, B5*
20. In a laboratory titration, 15.0 mL of 0.275-*M* H_2SO_4 solution neutralizes 24.3 mL of NaOH solution. What is the molarity of the NaOH solution? *B9, B11*
21. A 10.0-mL sample of vinegar is diluted to $10\bar{0}$ mL with distilled water and titrated against 0.100-*M* sodium hydroxide solution using a suitable indicator. When the titration was completed, 30.0 mL of the diluted vinegar and 24.7 mL of the base had been withdrawn from the burets. Calculate the percentage of acetic acid, $HC_2H_3O_2$, in the vinegar sample. *B9, B11*
22. Determine the pH of a 0.02-*M* LiOH solution. *B7*
23. What is the pH of a 0.054-*M* solution of HCl? *B7*
24. A solution is determined experimentally to have a pH of 2.9. *(a)* Calculate the $[H_3O^+]$. *(b)* Calculate the $[OH^-]$. *B8*
25. Suppose you have mixed 20.0 mL of 0.150-*M* NaOH and 40.0 mL of 0.100-*M* HCl solutions in the laboratory. Calculate the pH of the resulting solution. *B7*

In 1923 the Nobel Prize in chemistry was given to Fritz Pregl (Austrian) for his method of microanalyzing organic substances. He needed to devise new analytical techniques when he began research on physiological substances available only in very small amounts. He refined his method, made it automatic, and used it to analyze enzymes, sera, and bile acids. Advantages of the method were that it required significantly smaller amounts of the substances, achieved higher accuracy, and involved less time than other techniques. Pregl's work led to the development of ultramicroanalysis.

Carbon and Its Oxides

17

GOALS

In this chapter you will gain an understanding of: • the chemical importance of carbon; organic chemistry • the structure and properties of carbon atoms • allotropy; diamond, graphite, and carbynes • the production, properties, and uses of various forms of carbon • the occurrence, preparation, molecular structure, properties, uses, and action on the human body of carbon dioxide and carbon monoxide

CARBON

17.1 Abundance and importance of carbon Charcoal and soot are forms of carbon that have been known from earliest times. But the elemental nature of carbon, its occurrence in charcoal, and its existence as diamond and graphite were not discovered until the late eighteenth century.

Carbon ranks about seventeenth in abundance by weight among the elements in the earth's crust. In importance, carbon ranks far higher. Carbon is present in body tissue and in the foods you eat. It is found in coal, petroleum, natural gas, and limestone and in all living things. In addition to its natural occurrence, chemists have synthesized hundreds of thousands of carbon compounds in the laboratory.

The study of carbon compounds is so important that it forms a separate branch of chemistry called *organic chemistry*. Originally, organic chemistry was defined as the study of materials derived from living organisms, and inorganic chemistry was the study of materials derived from mineral sources. Chemists have known for over 150 years that this is not a clear distinction. Many substances identical to those produced in living things can also be made from mineral materials. As a result, ***organic chemistry*** *today includes the study of carbon compounds whether or not they are produced by living organisms.*

In most substances containing carbon, the carbon is present in the *combined* form. It is usually united with hydrogen or with hydrogen and oxygen. This chapter first describes carbon in its *free* or *uncombined* forms and then considers carbon dioxide and carbon monoxide.

Figure 17-1. Coal is an important source of carbon and of carbon compounds.

17.2 Characteristics of carbon atoms Carbon is the element with atomic number 6 and electron configuration $1s^22s^22p^2$. On the periodic table, it is in the second period, midway between the active metal lithium and the active nonmetal fluorine. The two 1*s* electrons are tightly bound to the nucleus. The two 2*s* electrons and the two 2*p* electrons are the valence electrons. Carbon atoms show a very strong tendency to share electrons and form covalent bonds. This electron sharing usually has the effect of producing a stable highest-energy-level octet about a carbon atom. Having four valence electrons makes it possible for a carbon atom to form four covalent bonds. These bonds are directed in space toward the four vertices of a regular tetrahedron. This arrangement of carbon valence bonds is explained by sp^3 hybridization. The nucleus of the atom is at the center of the tetrahedron. See Figure 17-2.

Hybridization is explained in Section 6.17.

The property of forming covalent bonds is so strong in carbon atoms that they join readily with other elements. They also link together with other carbon atoms in chains, rings, plates, and networks. Depending on the number of atoms and the way they are bonded, each molecule thus formed is a unique carbon compound. The variety of ways in which carbon atoms can be linked explains why there are so many carbon compounds.

17.3 Allotropic forms of carbon In Section 9.11, the allotropic forms of the element oxygen—oxygen molecules, O_2, and ozone molecules, O_3—were described. Allotropy is the existence of an element in two or more forms in the same physical phase. The reason for oxygen's allotropy is the existence of two kinds of molecules, each with different numbers of atoms.

Allotropy occurs for either of two reasons: (1) An element has two or more kinds of molecules, each with different numbers of atoms, which exist in the same phase; or (2) an element has two or more different arrangements of atoms or molecules in a crystal.

Carbon occurs in two well-known solid allotropic forms: *diamond* is a hard crystalline form; *graphite* is a soft, grayish-black crystalline form. The reason for carbon's allotropy is the existence of different arrangements of atoms and different bonding between atoms in a crystal. A third allotrope, a *carbyne,* is one of a set of recently discovered crystalline varieties of carbon in which the carbon atoms are linked by both single and triple covalent bonds. Carbynes range in hardness from soft to almost as hard as diamond. They have been found in meteorites and have been produced synthetically from carbon monoxide.

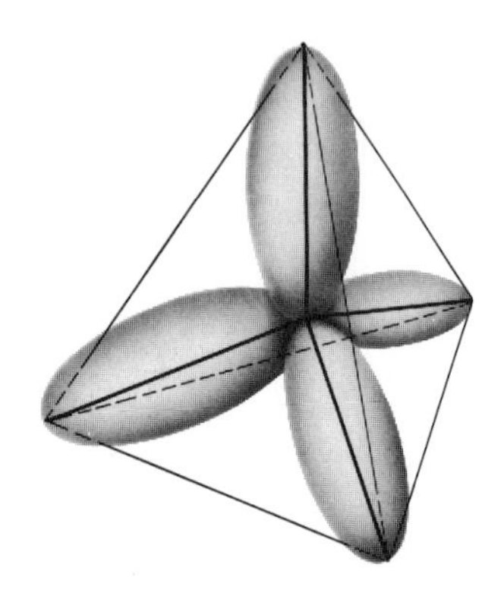

Figure 17-2. In sp^3 hybridization, the four covalent bonds of a carbon atom are directed in space toward the four vertices of a regular tetrahedron. The nucleus of the atom is at the center of the tetrahedron.

When substances that contain combined carbon are decomposed by heat, they leave black residues. These residues are sometimes collectively called *amorphous carbon* because they seem to have no definite crystalline shape. Examples of amorphous carbon are *coke, charcoal, boneblack,* and *carbon black.* Studies of these substances' structures, made by X-ray diffraction, reveal that the various forms of so-called amorphous carbon contain regions in which the carbon atoms have an orderly arrangement. In carbon black, for example, the carbon atoms are arranged somewhat as they are in a layer of graphite.

17.4 Diamond The most famous diamond mines in the world are located in South Africa. A significant new diamond mine is being developed in northern Australia. Diamonds in both these regions usually occur in the shafts of extinct volcanoes. It is believed that they were formed slowly under extreme heat and pressure. Diamonds, as they are mined, do not have the shape or sparkle of gem stones. The art of cutting and polishing gives them their brilliant appearance.

Figure 17-3. The crystal structure of diamond. In the diamond structure, each carbon atom is bonded to four tetrahedrally oriented carbon atoms.

Synthetic diamonds are chemically identical to natural diamonds but are produced in the laboratory. They are prepared by subjecting graphite and a metal that acts as solvent and catalyst to extremely high pressure (55,000 atm) and high temperature (2000 °C) for nearly a day.

A solvent is a dissolving medium: a catalyst is a substance that increases the rate of a chemical reaction without itself being permanently changed.

Diamond is the hardest material. It is the densest form of carbon, about 3.5 times as dense as water. Both the hardness (resistance to wear) and density are explained by its structure. Figure 17-3 shows that carbon atoms in diamond are covalently bonded in a strong, compact fashion. The distances between the carbon nuclei are 1.54 Å. Note that each carbon atom is tetrahedrally oriented to its four nearest neighbors. This type of structure gives the crystal its great strength in all three dimensions.

The rigidity of its structure gives diamond its hardness. The compactness, resulting from the small distances between nuclei, gives diamond its high density. The covalent network structure of diamond accounts for its extremely high melting point, about 3700 °C. Since all the valence electrons are used in forming covalent bonds, none can migrate. This explains why diamond is a nonconductor of electricity. Because of its extreme hardness, diamond is used for cutting, drilling, and grinding. A diamond is used as a long-lasting phonograph needle.

Diamond is the best conductor of heat. A perfect single diamond crystal conducts heat more than five times better than silver or copper. Silver and copper are the best metallic conductors. In diamond, heat is conducted by the transfer of energy of vibration from one carbon atom to the next. In a perfect single diamond crystal, this process is very efficient. The carbon atoms have a small mass. The forces binding the atoms together are strong and can easily transfer vibratory motion among the atoms.

Diamond is insoluble in ordinary solvents. In 1772, the French chemist Lavoisier burned a clear diamond in pure oxygen and obtained carbon dioxide as a product. This experiment proved to him that diamond contains carbon. The English chemist Smithson Tennant repeated the experiment in 1797. He weighed the diamond and the carbon dioxide it produced. The mass of carbon dioxide showed that diamond is pure carbon.

Smithson Tennant (1761–1815) was the discoverer of the transition elements osmium, atomic number 76, and iridium, atomic number 77. He discovered them in impure platinum.

17.5 Graphite Natural graphite deposits are found throughout the world. The major producers are the Republic of Korea, Austria, North Korea, and the Soviet Union.

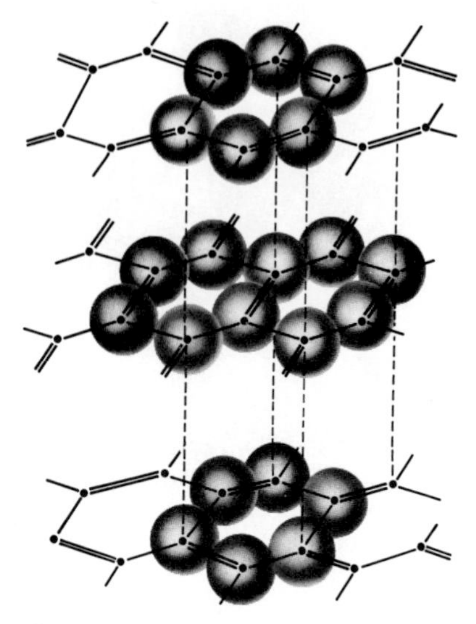

Figure 17-4. The crystal structure of graphite. Within each layer, each carbon atom is bonded to three other carbon atoms 120° apart. The distance between the layers has been exaggerated to show the structure of each layer more clearly.

Figure 17-5. A breakthrough in aerodynamics. Engineers weave lightness and strength into sail planes and airplanes by using carbon fibers or carbon-epoxy composites.

More than 70% of the graphite used in the United States is synthetic graphite. Most of this synthetic graphite is produced from petroleum coke. The process involves heating petroleum coke to about 2800 °C in special furnaces.

Graphite is nearly as remarkable for its softness as diamond is for its hardness. It is easily crumbled and has a greasy feel. Graphite crystals are hexagonal (six-sided) in cross section, with a density of about 2.26 g/cm^3. Graphite conducts electricity.

The structure of graphite readily explains these properties. The carbon atoms in graphite are arranged in layers of thin hexagonal plates, as shown in Figure 17-4. The distance between the centers of adjacent carbon atoms within a layer is 1.42 Å. This distance is less than the distance between adjacent carbon atoms in diamond. However, the distance between the centers of atoms in adjacent layers is 3.35 Å.

Figure 17-4 shows the bonding within a layer of graphite. Each carbon atom in a layer is bonded to only three other carbon atoms in that layer. This bonding consists of single and double covalent bonds between carbon atoms. When represented in this fashion, three *different* equivalent patterns appear. In each of these, some carbon-carbon bonds are single and others are double. There is, however, no experimental evidence that the bonds in a layer of graphite are of these two distinct types. On the contrary, the evidence indicates that the bonds are all the same. The layers of graphite have a resonance structure in which the carbon-carbon bonds are intermediate in character between single and double bonds. Each layer in graphite is a strongly bonded covalent network structure. As with diamond, this structure gives graphite a high melting point, about 3600 °C. The strong bonds between the carbon atoms within a layer make graphite difficult to pull apart in the direction of the layer. As a result, carbon fibers, in which the carbon is in the form of graphite, are very strong.

The layers of carbon atoms in graphite are too far apart for the formation of covalent bonds between them. They are held together by weak dispersion interaction forces. These forces result from electron motion within the layers. The weak attraction between layers accounts for the softness of graphite and its greasy feel as one layer slides over another.

On the average, the carbon atoms in graphite are farther apart than they are in diamond, so graphite has a lower density. The mobile electrons in each carbon-atom layer make graphite a fairly good conductor of electricity, even though it's a nonmetal. Like diamond, graphite does not dissolve in any ordinary solvent. Similarly, it forms carbon dioxide when burned in oxygen.

17.6 Uses of graphite The largest single use of natural graphite is for coating the molds used in metal casting. It is also used to increase the carbon content of steel and to make clay-graphite crucibles in which steel and other metals are melted. All these applications take advantage of the high melting point of graphite. Graphite is a good lubricant. It is sometimes mixed with petroleum jelly or motor oil to form graphite lubricants. It can be used for lubricating machine parts that operate at temperatures too high for the usual petroleum lubricants. Graphite leaves a gray streak or mark when it is drawn across a sheet of paper. In making "lead" pencils, graphite is powdered, mixed with clay, and then formed into sticks. The hardness of a pencil depends on the relative amount of clay that is used.

The most important use of synthetic graphite is in electrodes for electric-arc steelmaking furnaces. Synthetic graphite electrodes are also used in the electrolysis of salt water for making chlorine and sodium hydroxide. Graphite does not react with acids, bases, and organic and inorganic solvents. These properties make it useful in equipment for a variety of processes in the food, chemical, and petroleum industries. Graphite is also used in nuclear reactors, as described in Chapter 31.

If certain synthetic fibers are combined with plastic resins and heated under pressure, they become carbon fibers. As mentioned in Section 17.5, the form of carbon in these fibers is graphite. Carbon fibers are less dense than steel but are stronger and stiffer. They are used in aircraft floor decking, wings, and wing flaps and in weather and communication satellites. In sporting goods, carbon fibers are used to make golf club shafts, tennis rackets, fishing rods, and bicycle frames. Carbon fibers have been implanted in severely torn ligaments and tendons to help the natural process of reconstruction.

Figure 17-6. The luxuriant growth of vegetation during the carboniferous age is the source of our coal deposits today. Bottom photo shows fossils of that age.

17.7 Coal Coal is a solid, rocklike material that burns readily. It contains at least 50% carbon and varying amounts of moisture, volatile (easily vaporized) materials, and noncombustible mineral matter (ash).

Geologists believe that coal was formed about 3×10^8 years ago during the carboniferous age. At that time plants grew more luxuriantly than they do today. Possibly there was more carbon dioxide in the atmosphere, and it stimulated the growth of vegetation. Tree ferns, giant club mosses, and other vegetation growing in swamps supplied the material for the coal deposits.

Peat bogs were probably formed first. Extensive peat bogs are found in Pennsylvania, Michigan, Wisconsin, and other states. The bogs contain mosses, sedges, and other forms of vegetation that have undergone partial decomposition in swampy land in almost complete absence of air. Peat burns with a smoky flame, and its heat content is rather low. Peat contains a high percentage of moisture. In the United States, peat is used mainly as a soil conditioner, not as fuel. Comparative data for peat and representative samples of coal are given in Table 17-1.

In some areas, upheavals of parts of the earth's crust buried

Table 17-1
ANALYSIS OF PEAT AND REPRESENTATIVE COALS

	Peat	*Lignite*	*Subbituminous*	*Bituminous*	*Anthracite*
carbon, %		37.4	51.0	74.5	86.4
hydrogen, %		7.4	6.2	5.2	2.7
oxygen, %		48.4	34.5	9.2	3.6
nitrogen, %		0.7	1.0	1.5	0.9
sulfur, %	0.1	0.6	0.3	0.9	0.6
moisture, %	66.9	43.0	25.9	2.8	2.2
ash, %	4.8	5.5	6.9	8.8	5.9
energy value, kcal/kg	1680	3488	4784	7441	7683

thick masses of vegetable matter. Once buried, this material was subjected to increased temperatures and pressures over many millions of years. At a fairly early stage in these changes, lignite was formed. Lignite is sometimes called brown coal because of its brownish-black color. It is common in some of the western states, as the map in Figure 17-7 shows. Lignite burns with a smoky flame, and although its heat content is higher than that of peat, it is still significantly below that of the other coals. The moisture content of lignite is high.

Subbituminous coal, which is found extensively in Wyoming and other western states, may be considered the next stage of coal formation. The percentage of carbon is higher than in lignite, and the moisture content is lower. Subbituminous coal also has a higher heat content than lignite. It can be used for industrial heating, as in electric generating stations.

Bituminous coal appears to have been subjected to greater heat and pressure. The percentage of carbon is high; the mois-

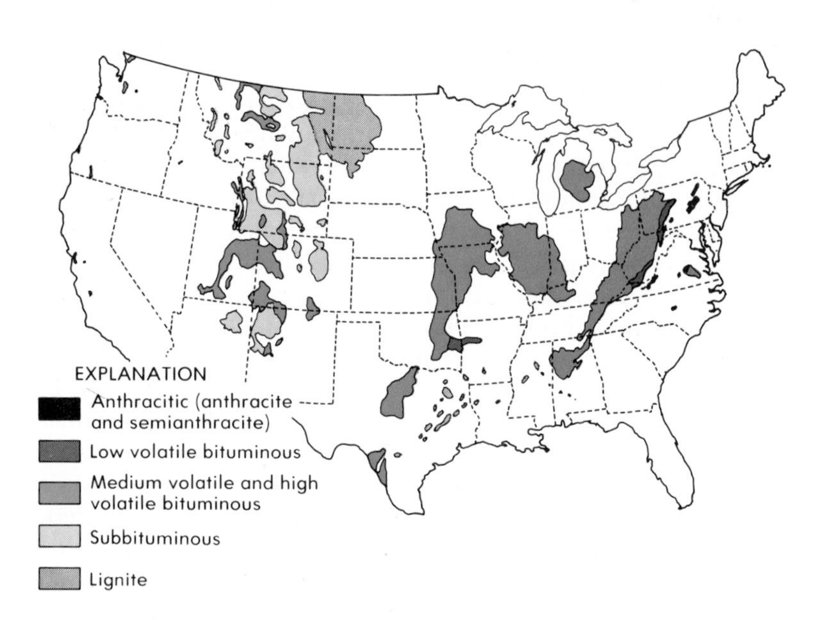

Figure 17-7. The various forms of coal are widely distributed in the United States.

ture content is low. Bituminous coal has a high heat content. It is widely distributed throughout the Appalachian region, the Midwest, and the Rocky Mountain states. It is used for home and industrial heating and for making coke and coal by-products.

Anthracite is believed to have been subjected to greater temperatures and pressures than other coals. Anthracite contains the least amount of volatile matter of all coals. It has the highest percentage of carbon. Most anthracite is found in Pennsylvania.

Coal is our most plentiful fossil fuel. The amount of coal in the United States that can be recovered using known techniques is presently estimated to be about 2×10^{11} metric tons. The total amount of coal mined each year is about 7×10^{8} metric tons. Three-fourths of this coal is used to generate electricity.

17.8 Destructive distillation Suppose a complex carbon-containing material, such as wood, is *heated in a closed container without access to air or oxygen. The complex material decomposes into simpler substances.* This process is known as ***destructive distillation.*** Coke, charcoal, and boneblack are prepared by destructive distillation of coal, wood, and bones, respectively.

17.9 Coke When bituminous coal is heated in a hard-glass test tube, a flammable gas escapes. This gas burns readily if mixed with air and ignited. Also, a tarlike liquid condenses on the upper walls of the tube. If the heating is continued until all the volatile material is driven off, coke is left as a residue.

Commercially, coke is prepared by destructive distillation of bituminous coal in by-product coke ovens. The volatile products are separated into *coal gas, ammonia,* and *coal tar.* Coal gas can be used as a fuel. Ammonia is used in making fertilizers. Coal tar can be separated by distillation into materials used to make drugs, dyes, and explosives. The black pitch that remains after distillation of coal tar is used to surface roads.

About 60,000,000 tons of coke are produced each year in the United States. Coke is a gray, porous solid that is harder and denser than charcoal. It burns with little flame, has a high heat content, and is a valuable fuel.

Figure 17-8. Coke is produced by destructive distillation of bituminous coal in by-product coke ovens.

Coke is an excellent reducing agent. It is widely used in obtaining the metals from the ores of iron, tin, copper, and zinc. These ores are either oxides or are converted into oxides. Coke readily reduces the metals in these oxides. It also has great structural strength and is free from volatile impurities.

17.10 Charcoal Destructive distillation of wood yields several gases that can be burned. It also produces methanol (wood alcohol), acetic acid, and other volatile products. The residue that remains is charcoal. Charcoal is prepared commercially by heating wood in *retorts.* These are closed containers in which substances are distilled or decomposed by heat. The burnable gases that result provide supplementary fuel. The other volatile products may be condensed and sold as by-products.

Adsorption is defined in Section 9.18 and further explained in Section 13.2.

Charcoal is a porous, black, brittle solid. It is odorless and tasteless. It is denser than water, but it often adsorbs enough gas to make it float on water. This ability to *adsorb* a large quantity of gas is the most remarkable physical property of charcoal. One cubic centimeter of freshly prepared willow charcoal adsorbs about 90 cubic centimeters of ammonia gas. Charcoal has been used to adsorb toxic waste products from the blood of persons suffering from liver failure, from drug overdose, and from accidental swallowing of poisonous carbon-hydrogen compounds by very young children. It is also used to remove ozone from the cabin atmosphere of jet airplanes.

At ordinary temperatures, charcoal is inactive and insoluble in all ordinary solvents. It is a good reducing agent because it unites with oxygen at a high temperature. Charcoal is also a good fuel, but it is more expensive than other common fuels.

17.11 Boneblack Animal charcoal, or *boneblack,* is produced by the destructive distillation of bones. The by-products of the process include bone oil and pyridine. These products may be added to ethanol (ethyl alcohol) to make it unfit for drinking. Boneblack usually contains calcium phosphate as an impurity. This can be removed by treating the boneblack with an acid.

17.12 Activated carbon Activated carbon is a form of carbon that is prepared in a way that gives it a very large internal surface area. This large surface area makes activated carbon useful for the adsorption of liquid or gaseous substances. Activated carbon can be made from a variety of carbon-containing materials. Such a material is first destructively distilled. The carbon produced is treated with steam or carbon dioxide at about 100 °C. These two processes produce a very porous form of carbon. This porosity creates a very large internal surface area. The surface area of a portion of activated carbon may be as high as 2000 m^2/g.

Activated carbon used for adsorption of gases must have a small pore structure. Coconut and other nut shells are the best sources of this type of activated carbon. Gas-adsorbent activated carbon is used in gas masks. It is also used for the recovery of volatile solvent vapors and the removal of impurities from gases in industrial processes. This form of activated carbon can remove odors from the air circulated by large air conditioning systems.

Activated carbon used for adsorption from the liquid phase comes from both animal and vegetable sources. Boneblack was first used for this purpose, but now such activated carbon is also made from coal, peat, and wood. Inorganic impurities are removed from activated carbon that is to be used in processing food or chemical products. Acids such as dilute hydrochloric or sulfuric react with these impurities to form soluble products that are washed away with water. Liquid-adsorbent activated carbon is used in the refining of cane sugar, beet sugar, and corn syrup. It is also used in municipal and industrial water treatment to adsorb harmful materials as well as objectionable odor and taste.

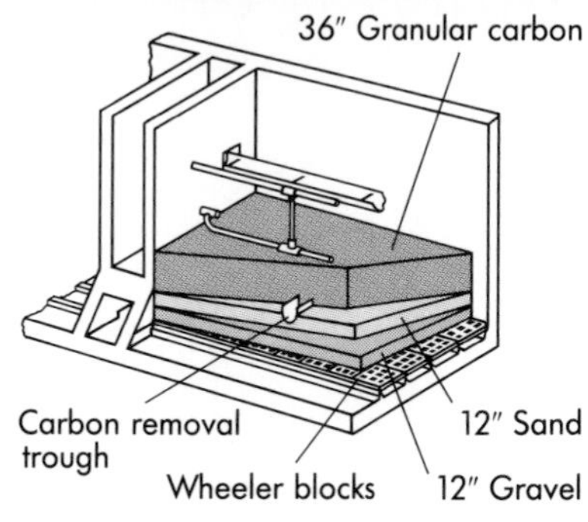

Figure 17-9. Activated carbon is used to remove objectionable color and odor and other impurities from water to be used in public water supplies. Bottom diagram shows a system used for filtering water through activated carbon.

17.13 Other forms of amorphous carbon Finely divided particles of carbon are set free when liquid or gaseous fuels composed of carbon and hydrogen are burned in an insufficient supply of air. These particles are commonly called *soot.* Soot is an example of the form of amorphous carbon called *carbon black.* Commercially, the production of carbon black involves making soot under carefully controlled conditions.

The most important method of making carbon black is called the *furnace process.* The furnace is made of materials that have high melting points, such as fire brick. In making carbon black, three materials are introduced into the furnace:

1. A spray of liquid fuel or the gaseous fuel vapor.
2. An additional fuel, such as petroleum-refinery gas, which is the fuel-gas produced when petroleum is refined.
3. Air, as a source of oxygen.

The supply of oxygen is so low that the fuels are only partially burned in the furnace. But enough fuel burns to provide the energy needed for decomposing the rest of the fuel. The carbon black is collected from the combustion products.

Over 90% of all carbon black produced is used in natural and synthetic rubber. It adds bulk to the rubber and acts as a reinforcing agent. Most of this rubber is used in tires. Carbon black helps to preserve the rubber and makes the tire wear longer. The second largest use of carbon black is in printer's ink. Other uses are in paints, phonograph records, and in coloring plastics.

An oil residue remains after the refining of crude petroleum. When destructively distilled, this residue produces a form of amorphous carbon called *petroleum coke.* Rods of petroleum coke are converted to synthetic graphite for use as electrodes. Such electrodes are used as the positive electrodes in dry cells. They are also used in the production of aluminum by electrolysis.

Carbon Dioxide

17.14 Occurrence Carbon dioxide comprises only 0.034% of the earth's atmosphere by volume. But it is a very important component of the air. The water of rivers, lakes, and oceans contains about 60 times as much dissolved carbon dioxide as the atmosphere. Both the decay of organic matter and the burning of fossil fuels produce carbon dioxide. So do the respiration processes of living things.

The amount of carbon dioxide in the earth's atmosphere is slowly increasing. It is believed this increase is produced by the burning of carbon-containing fuels and the destruction of forests throughout the world. An increase in the amount of carbon dioxide in the atmosphere enables more of the sun's heat to be held by the earth. If the amount of carbon dioxide in the atmosphere continues to increase, the increased heat absorption could eventually raise the average temperature around the earth. Such a temperature increase could change local conditions for growing crops, cause the polar ice caps to melt, and the ocean levels to rise.

Carbon dioxide is somewhat denser than air. It sometimes gathers in relatively large amounts in low-lying areas such as bogs, swamps, and marshes. It may also collect in mines, caves, and caverns. Some natural gases contain significant amounts of carbon dioxide.

The atmosphere of the planet Venus is 96% carbon dioxide by volume, with an atmospheric pressure more than 90 times that on the earth. On Mars, carbon dioxide is also the most abundant component of the atmosphere. It is 95% by volume of the Martian atmosphere. But the atmospheric pressure on Mars is only about $\frac{1}{135}$ of that on the earth.

17.15 Preparation of carbon dioxide 1. *By burning material that contains the element carbon.* Carbon dioxide is one of the products of the complete combustion in oxygen or air of any material that contains carbon. If air is used, the carbon dioxide prepared in this way is mixed with other gases. But if these gases do not interfere with the intended use of the carbon dioxide, this method is by far the cheapest and easiest.

$$\text{C (combined)} + O_2 \rightarrow CO_2$$

2. *By reaction of steam and natural gas.* Natural gas is usually a mixture of several gaseous compounds of carbon and hydrogen. The principal component of natural gas is methane, CH_4. Methane undergoes a series of reactions with steam in the presence of metallic oxide catalysts at temperatures between 500 °C and 1000 °C. The end products of these reactions are carbon dioxide and hydrogen. The equation for the overall reaction is

$$CH_4 + 2H_2O \rightarrow 4H_2 + CO_2$$

The primary purpose of this reaction is to prepare hydrogen for making synthetic ammonia. The carbon dioxide is a by-product. The carbon dioxide is separated from the hydrogen by dissolving the carbon dioxide in cold water under high pressure.

3. *By fermentation of molasses.* The enzymes of *zymase* are produced by yeast. They catalyze the fermentation of the sugar, $C_6H_{12}O_6$, in molasses. This fermentation produces ethanol (ethyl alcohol) and carbon dioxide. Although the process is complex, the overall reaction is

$$C_6H_{12}O_6(aq) \rightarrow 2C_2H_5OH(aq) + 2CO_2(g)$$

This equation represents a method by which some industrial alcohol is produced. The process is also an important source of carbon dioxide.

4. *By heating a carbonate.* When calcium carbonate (as limestone, marble, or shells) is heated strongly, calcium oxide and carbon dioxide are the products.

$$CaCO_3 \rightarrow CaO + CO_2$$

Calcium oxide, known as *quicklime,* is used for making plaster and mortar. The carbon dioxide is a by-product.

5. *By the action of an acid on a carbonate.* This is the usual laboratory method of preparing carbon dioxide. The gas-generating bottle in Figure 17-10 contains a few pieces of marble, $CaCO_3$. If dilute hydrochloric acid is poured through the funnel tube, carbon dioxide is given off rapidly. Calcium chloride may be recovered from the solution in the bottle.

This reaction proceeds in two stages. *First,* the marble and hydrochloric acid undergo an exchange reaction:

$$CaCO_3(s) + 2HCl(aq) \rightarrow CaCl_2(aq) + H_2CO_3(aq)$$

Second, carbonic acid is unstable and decomposes:

$$H_2CO_3(aq) \rightarrow H_2O(l) + CO_2(g)$$

The equation that summarizes these two reactions is

$$CaCO_3(s) + 2HCl(aq) \rightarrow CaCl_2(aq) + H_2O(l) + CO_2(g)$$

Even though carbon dioxide is soluble in water, it may be collected by water displacement if it is generated rapidly. It may also be collected by displacement of air. In this case, the receiver must be kept *mouth upward* because carbon dioxide is denser than air.

This reaction is an example of the general reaction of an acid and a carbonate. Almost any acid may be used, even a weak one

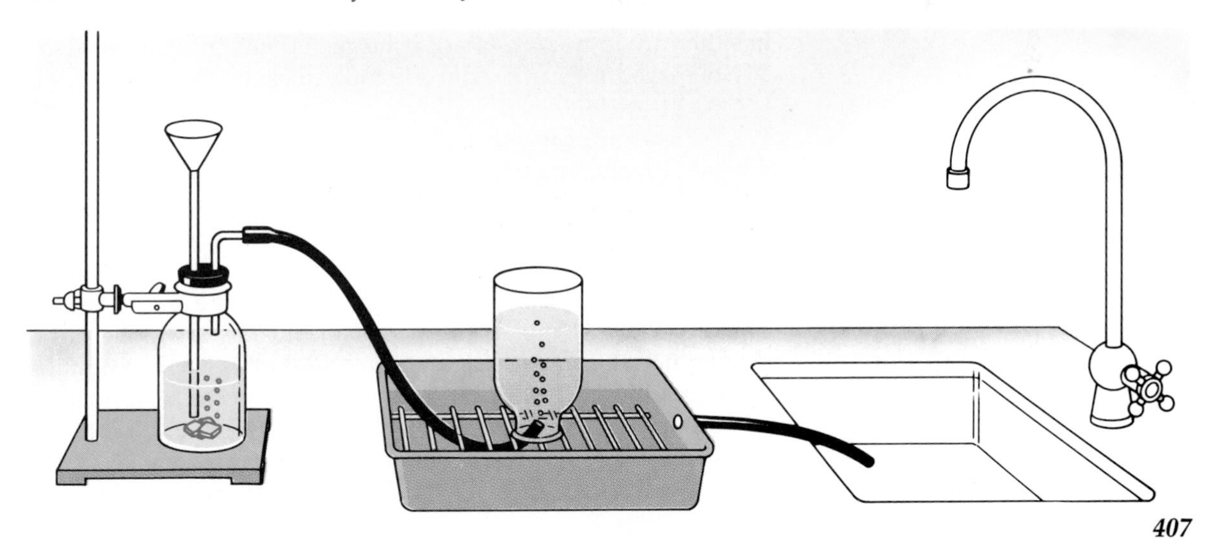

Figure 17-10. Carbon dioxide is prepared in the laboratory by the reaction of dilute hydrochloric acid on marble chips, calcium carbonate. Carbon dioxide is collected by water displacement.

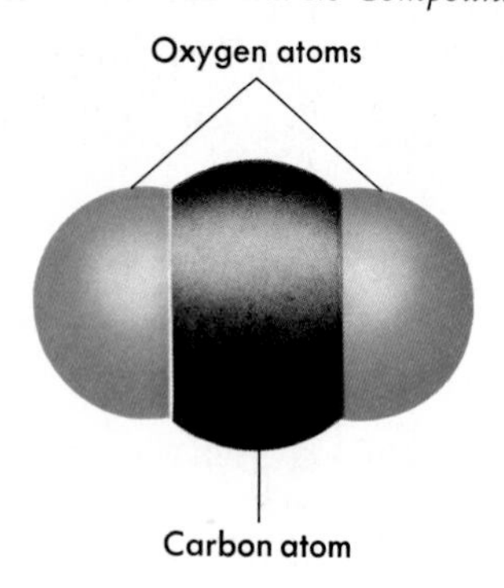

Figure 17-11. A carbon dioxide molecule is linear and consists of one carbon atom and two oxygen atoms.

such as the acetic acid in vinegar. Almost any carbonate may also be used, provided its cation does not have an interfering reaction with the anion of the acid. The ionic equation for the reaction is

$$CO_3^{--} + 2H_3O^+ \rightarrow 3H_2O + CO_2(g)$$

6. *By respiration and decay.* This process is a natural method of preparing carbon dioxide. The foods you eat contain compounds of carbon. Oxygen from the air you inhale is used in oxidizing this food. This oxidation supplies energy to maintain body temperature, move muscles, synthesize new compounds in the body, and transmit nerve impulses. Carbon dioxide, which you exhale into the air, is one of the products of this oxidation. All living things give off carbon dioxide during respiration.

When plants and animals die, decay begins and carbon dioxide is produced. This gas eventually finds its way into the surrounding air, or dissolves in surface or underground streams.

17.16 Structure of carbon dioxide molecules Carbon dioxide molecules are linear, with the two oxygen atoms bonded on opposite sides of the carbon atom. See Figure 17-11. The carbon-oxygen bonds in the molecule are somewhat polar. This polarity is explained by the electronegativity difference between carbon and oxygen. However, the arrangement of these bonds, exactly opposite one another, causes the molecule to be nonpolar.

Using the electron-dot symbols for carbon and oxygen, carbon dioxide molecules might be assigned the electron-dot formula

:Ö::C::Ö:

However, this formula is not strictly accurate. The carbon-oxygen bond distance predicted by it is larger than that actually found in carbon dioxide molecules. Carbon dioxide molecules are believed to be *resonance hybrids of four electron-dot structures.* Each of these structures, which are shown at the left, contributes about equally to the actual structure. Such a resonance hybrid has the carbon-oxygen bond distance and energy that are actually observed for carbon dioxide molecules.

{ :O̤::C::Ö: $^-$:Ö̤:C:::O:$^+$
  :Ö::C::O̤: $^+$:O:::C:Ö̤:$^-$ }

17.17 Physical properties of carbon dioxide Carbon dioxide is a gas at room temperature. This fact supports the theory that it has a simple, nonpolar molecular structure. Carbon dioxide is colorless with a faintly irritating odor and a slightly sour taste. The molecular weight of carbon dioxide is 44. Thus, its density is about 1.5 times that of air at the same temperature and pressure. The large, heavy molecules of carbon dioxide gas move more slowly than the smaller, lighter molecules of gaseous oxygen or hydrogen. Because of its high density and slow rate of diffusion, carbon dioxide can be poured from one vessel to another.

Carbon dioxide has a critical temperature of 31.0 °C and a critical pressure of 72.8 atm (Section 12.7). The liquid carbon dioxide in a fire extinguisher is under a pressure of 60 atm.

At high pressures, the molecules of a gas are much closer together than they are at a pressure of one atmosphere. At room temperature and a pressure of about 70 atmospheres, molecules of carbon dioxide attract each other strongly enough to condense to a liquid. If this liquid is permitted to evaporate rapidly under atmospheric pressure, part of it changes into a gas. This process absorbs heat from the remaining liquid, which is thus cooled until it solidifies in the form called *dry ice.*

Solid carbon dioxide has a high vapor pressure. Many molecules of solid carbon dioxide possess enough energy to escape from the surface of the solid into the air. The vapor pressure of solid carbon dioxide equals atmospheric pressure at −78.5 °C. As a result, solid carbon dioxide under atmospheric pressure sublimes at this temperature. Liquid carbon dioxide does not exist at atmospheric pressure. It can exist only at low temperatures with pressures higher than 5 atmospheres.

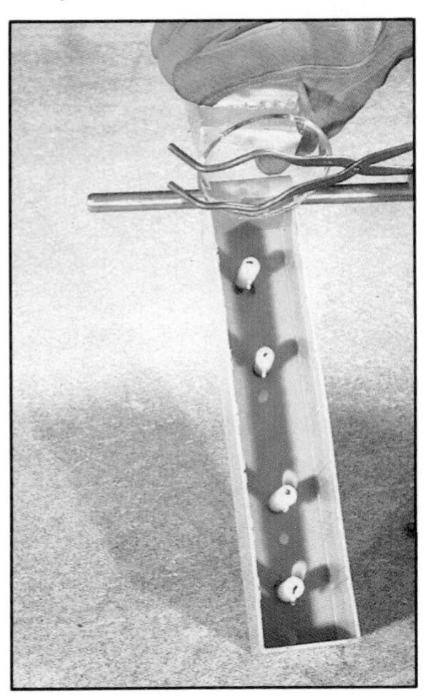

Figure 17-12. Carbon dioxide can be poured down this stair of candles because of its high density and slow rate of diffusion. The candles will be extinguished because carbon dioxide gas does not support combustion.

17.18 Chemical properties of carbon dioxide Carbon dioxide is a stable gas. It neither burns nor supports combustion. However, burning magnesium is hot enough to decompose carbon dioxide. A piece of burning magnesium ribbon continues to burn in a bottle of the gas. The magnesium unites vigorously with the oxygen set free by the decomposition. Carbon is produced, as shown by a coating of soot on the inside of the bottle.

$$2Mg(s) + CO_2(g) \rightarrow 2MgO(s) + C(s)$$

Carbon dioxide dissolves readily in cold water. A few of the dissolved molecules also unite with the water to form carbonic acid. Thus, carbon dioxide is the acid anhydride of carbonic acid.

An acid anhydride is the compound formed by removal of water from an acid.

$$H_2O + CO_2 \rightleftarrows H_2CO_3$$

Almost all the H_2CO_3 molecules ionize. Carbonic acid exists in water solution principally as ions.

$$H_2O + H_2CO_3 \rightleftarrows H_3O^+ + HCO_3^-$$

$$H_2O + HCO_3^- \rightleftarrows H_3O^+ + CO_3^{--}$$

Carbonic acid is a weak acid because of the slight reaction between CO_2 and H_2O, even though the few H_2CO_3 molecules formed ionize extensively. Carbonic acid is easily decomposed by heat since CO_2 is less soluble at higher temperatures. The reduction of the concentration of CO_2 in the water causes all the equilibria to shift to the left. This shift further decreases the H_2CO_3 and H_3O^+ concentrations.

You may find it helpful at this point to review Le Chatelier's principle, Section 12.5, and protolysis reactions, Section 15.12.

When carbon dioxide is passed into a water solution of a hydroxide, it reacts to form a carbonate.

$$CO_2 + 2OH^- \rightarrow CO_3^{--} + H_2O$$

If the positive ion of the hydroxide forms an insoluble carbonate, it is precipitated when carbon dioxide passes through the hydroxide solution.

A *test for carbon dioxide* is to bubble the gas through a saturated solution of $Ca(OH)_2$, called limewater. A white calcium carbonate precipitate indicates carbon dioxide's presence.

$$Ca^{++} + 2OH^{-} + CO_2 \rightarrow Ca^{++}CO_3^{--}(s) + H_2O$$

If excess carbon dioxide gas is bubbled through the solution, the precipitate disappears. The excess CO_2 gas reacts with the precipitate and water to form soluble calcium hydrogen carbonate.

$$Ca^{++}CO_3^{--} + H_2O + CO_2 \rightarrow Ca^{++} + 2HCO_3^{-}$$

As stated earlier (Section 17.14), normal air contains 0.034% carbon dioxide by volume. The air in a crowded, poorly ventilated room may contain as much as 1% carbon dioxide by volume. A concentration of from about 0.1% to 1% brings on a feeling of drowsiness and a headache. Concentrations of 8% to 10% or more cause death from lack of oxygen.

17.19 Uses of carbon dioxide 1. *It is necessary for photosynthesis.* ***Photosynthesis*** *means "putting together by means of light."* It is a complex process by which green plants manufacture carbohydrates with the aid of sunlight. *Chlorophyll,* the green coloring matter of plants, acts as a catalyst. Carbon dioxide from the air and water from the soil are the raw materials. Glucose, a simple sugar, $C_6H_{12}O_6$, is one of the products. The following simplified equation gives only the *reactants* and the *final products.*

$$6CO_2 + 12H_2O \rightarrow C_6H_{12}O_6 + 6O_2 + 6H_2O$$

The sugar may then be converted into a great variety of other plant products. The oxygen is given off to the atmosphere. Photosynthesis and the various other natural and artificial methods of producing atmospheric carbon dioxide comprise the *oxygen-carbon dioxide cycle.*

2. *Carbon dioxide is used as a refrigerant.* The most important commercial use of carbon dioxide is as a refrigerant for producing and storing frozen foods. Carbon dioxide vapor can be liquefied and carbon dioxide liquid can be made to evaporate at the required low temperatures (below −20 °C) by simply changing the pressure. These properties, coupled with the chemical stability of carbon dioxide, make it a good refrigerant for large installations in food and industrial plants. Nearly half of the carbon dioxide used commercially is used for refrigeration.

A small amount of solid carbon dioxide, dry ice, is used as a refrigerant. It is convenient because it has a very low temperature and sublimes at atmospheric pressure. The temperature of dry ice is so low that it *must never be handled with bare hands* because serious frostbite can result.

3. *Carbon dioxide improves petroleum recovery.* Carbon dioxide is injected into older underground oil fields where normal production has decreased. The carbon dioxide forces oil from the

oil-bearing deposits and makes possible the recovery of about 10–15% more petroleum from the field.

4. *Carbonated beverages contain carbon dioxide in solution.* Soft drinks are carbonated by forcing the gas into the beverages under pressure. When the bottles are opened, the excess pressure is released. Bubbles of carbon dioxide then escape rapidly from the liquid. The carbonation of beverages accounts for one-fourth of the commercial use of carbon dioxide.

5. *Leavening agents produce carbon dioxide.* Yeast is a common leavening agent. It is mixed with the flour and other ingredients used in making dough for bread. The living yeast plants produce enzymes that ferment the starches and sugars in the dough. This fermentation reaction produces ethanol and carbon dioxide. The carbon dioxide forms bubbles in the soft dough, causing it to "rise," or decrease in density. The ethanol is vaporized and driven off during the baking process.

Baking powder differs from baking soda. Baking soda is the compound sodium hydrogen carbonate. Baking powder is not a compound, but a dry *mixture* of compounds. It contains baking soda, which can yield the carbon dioxide. It also contains some powder that forms an acid when water is added. The acid compound varies with the kind of baking powder used. Cornstarch is used in baking powders to keep them dry until they are used.

6. *Carbon dioxide is used in fire extinguishers.* When the *soda-acid type* of fire extinguisher is inverted, sulfuric acid reacts with sodium hydrogen carbonate solution.

Sodium hydrogen carbonate is commonly called baking soda. This is the origin of the word soda *in the expression "soda-acid fire extinguisher."*

$$2NaHCO_3(aq) + H_2SO_4(aq) \rightarrow Na_2SO_4(aq) + 2H_2O(l) + 2CO_2(g)$$

The pressure of the gas forces a stream of liquid a considerable distance. The carbon dioxide dissolved in the liquid helps put out the fire, but water is the main extinguishing agent.

Liquid carbon dioxide fire extinguishers are very efficient and widely used. When the valve is opened, the nozzle directs a stream of carbon dioxide "snow" against the flame. Such an extinguisher is effective against oil fires and electric fires.

Figure 17-13. A liquid carbon dioxide fire extinguisher is effective in putting out oil fires and electric fires. It cools the burning material and shuts off the supply of oxygen needed for combustion.

Carbon Monoxide

17.20 The occurrence of carbon monoxide Carbon monoxide is found in samples of the atmosphere all over the world. The amounts vary from 0.04 ppm over the South Pacific Ocean to 360 ppm at street level in a crowded city on a calm day. (Ppm means parts per million. 1 ppm = 0.0001%.) A safe amount of carbon monoxide is somewhat under 50 ppm.

In congested areas, the carbon monoxide in the air comes mainly from incompletely burned fuels. The sources range from the chimney gases of improperly fired coal-burning furnaces to the exhaust gases from automobile engines. In the United States, as much as 90% of the people-caused carbon monoxide in the air

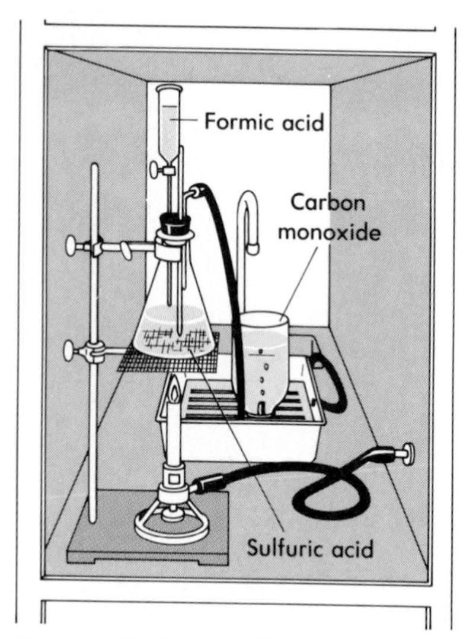

Figure 17-14. Carbon monoxide can be prepared in the laboratory by decomposing formic acid with hot, concentrated sulfuric acid.

may come from gasoline engines. Decaying plants and live algae add significant amounts of carbon monoxide to the air. It has been estimated that the oceans may release up to 85% as much carbon monoxide to the air as incompletely burned fuels do.

Specialists estimate that the carbon monoxide produced on a given day remains in the air for one to three months. Research is being carried out to determine what happens to carbon monoxide in the air, since no worldwide increase in its concentration has yet been detected. One explanation under study is that certain soil fungi convert carbon monoxide in the air to carbon dioxide. Another possible explanation is that carbon monoxide reacts with OH groups in the lower stratosphere to form CO_2 and H_2.

Traces of carbon monoxide have been found in the atmospheres of Venus, Mars, Jupiter, and Saturn's moon, Titan. Carbon monoxide is widespread in the interstellar gas clouds of our galaxy, especially in regions where stars are being formed. Carbon monoxide has also been detected in other galaxies.

17.21 Preparation of carbon monoxide 1. *By reducing carbon dioxide.* If carbon dioxide comes into contact with white-hot carbon or coke, it is reduced to carbon monoxide.

$$CO_2(g) + C(s) \rightarrow 2CO(g)$$

2. *By action of steam on hot coke.* Passing steam over white-hot coke produces a mixture called *water gas,* consisting mainly of carbon monoxide and hydrogen. This industrial method produces both carbon monoxide and hydrogen for use as fuel gases or as a synthesis gas for making organic compounds.

$$C(s) + H_2O(g) \rightarrow CO(g) + H_2(g)$$

The two gases can be separated by cooling and compression. This process liquefies the carbon monoxide but not the hydrogen. This reaction is fundamental to the production of fuel gases from coal. It may be carried out above ground in reacting vessels or below ground by burning damp seams of coal or lignite in air or oxygen forced down from the surface.

3. *By decomposing formic acid.* This decomposition reaction is the usual laboratory method for preparing carbon monoxide. Formic acid, HCOOH, is introduced one drop at a time into hot, concentrated sulfuric acid. Carbon monoxide is produced as each drop strikes the hot acid. See Figure 17-14. Concentrated sulfuric acid is an excellent dehydrating agent. It removes a molecule of water from each molecule of formic acid, leaving CO.

$$HCOOH(l) \xrightarrow{H_2SO_4} H_2O(l) + CO(g)$$

CAUTION: *When using this method, be sure the connections are tight so that the carbon monoxide does not escape.* It is preferable to prepare carbon monoxide in a hood.

Figure 17-15. A carbon monoxide molecule is a slightly polar molecule consisting of one carbon atom and one oxygen atom.

17.22 Structure of the carbon monoxide molecule Carbon monoxide molecules consist of one carbon atom and one oxygen atom covalently bonded. See Figure 17-15. The distance between the nuclei is 1.13 Å. The molecule is slightly polar, with *the carbon atom somewhat negative.* It is uncertain what accounts for these properties, but the carbon monoxide molecule is believed to be a resonance hybrid of the four structures shown at the right. Unlike carbon dioxide, however, these four structures do not contribute equally. The hybrid is estimated to be 10 percent $^{+}{:}\mathrm{C}{:}\ddot{\underset{\cdot\cdot}{\mathrm{O}}}{:}^{-}$, 20 percent each ${:}\mathrm{C}{::}\ddot{\mathrm{O}}{:}$ and ${:}\mathrm{C}{::}\underset{\cdot\cdot}{\mathrm{O}}{:}$, and 50 percent $^{-}{:}\mathrm{C}{:::}\mathrm{O}{:}^{+}$. The electronegativity difference discussed in Chapter 6 indicates that the oxygen in carbon monoxide would be negative. However, you must remember that those data apply only to *single* bonds between elements. The structure of carbon monoxide is complicated. Carbon monoxide's stability at ordinary temperatures is explained by the effect of the high percentage of triple-bonded structure. It also gives the carbon atom the slight negative charge in the polar molecule.

$$\left\{\begin{matrix} ^{+}{:}\mathrm{C}{:}\ddot{\underset{\cdot\cdot}{\mathrm{O}}}{:}^{-} & {:}\mathrm{C}{::}\ddot{\mathrm{O}}{:} \\ {:}\mathrm{C}{::}\underset{\cdot\cdot}{\mathrm{O}}{:} & ^{-}{:}\mathrm{C}{:::}\mathrm{O}{:}^{+} \end{matrix}\right\}$$

17.23 Physical properties of carbon monoxide Carbon monoxide is a colorless, odorless, tasteless gas. It is slightly less dense than air and only slightly soluble in water. Carbon monoxide has a low critical temperature, −140.23 °C, and a high critical pressure, 34.53 atm (atmospheres). These data indicate that the attractive forces between carbon monoxide molecules are low. Consequently, carbon monoxide is not easily liquefied. Neither is it readily adsorbed by charcoal. However, charcoal can be treated with certain metallic oxides that oxidize carbon monoxide to carbon dioxide. Gas masks containing treated charcoal protect wearers against carbon monoxide concentrations up to 2%. In atmospheres containing more than 2% carbon monoxide, a self-contained breathing apparatus is needed. This equipment, often used by firefighters, furnishes air from a cylinder of compressed air to the wearer through a mask.

17.24 Chemical properties and uses of carbon monoxide
1. *As a reducing agent.* Carbon monoxide is used in the production of iron, copper, and other metals from their oxides.

$$\mathbf{Fe_2O_3(s) + 3CO(g) \rightarrow 2Fe(s) + 3CO_2(g)}$$

2. *As a fuel.* Carbon monoxide burns with a blue flame. Many fuel gases contain carbon monoxide mixed with other gases that can be burned. Coal gas and water gas always contain some carbon monoxide.

3. *For synthesizing organic compounds.* Methanol, CH_3OH, is made from carbon monoxide and hydrogen under pressure. A mixture of oxides of copper, zinc, and chromium is used as a catalyst.

$$CO + 2H_2 \rightarrow CH_3OH$$

Carbon monoxide is also used in the synthesis of many other organic compounds.

Figure 17-16. Firefighters wearing self-contained breathing apparatus. This equipment furnishes air from the cylinder worn on the back through a hose to the face mask. With this equipment a firefighter can work safely in an atmosphere containing carbon monoxide.

17.25 The action of carbon monoxide on the human body

Carbon monoxide is poisonous because it unites very readily with *hemoglobin* molecules. Hemoglobin molecules in red blood cells serve as an oxygen carrier in the body's circulatory system. The attraction of hemoglobin for carbon monoxide is about 300 times greater than for oxygen. If carbon monoxide unites with hemoglobin, the hemoglobin is not available for carrying oxygen. When this happens, cells that need a lot of oxygen function less efficiently. Such cells are found in the heart, the skeletal muscles, and the central nervous system. Low levels of carbon monoxide thus impair a person's vision and reflexes. When persons breathe a large enough amount of carbon monoxide, they collapse because of oxygen starvation.

As stated earlier, Section 17.20, the safe level of carbon monoxide in the air is less than 50 ppm (parts per million). A person breathing air with as little as one part of carbon monoxide per thousand parts of air (1000 ppm) experiences nausea and headache in less than one hour. One part of carbon monoxide in one hundred parts of air (10,000 ppm) may prove fatal in 10 min.

To repeat: Carbon monoxide is colorless, odorless, and tasteless, and it induces drowsiness before actual collapse and death. Therefore, *extreme care must be taken to see that it does not contaminate the air in closed areas.* If a coal-burning furnace is not properly operated, carbon monoxide may escape and mix with the air in living or sleeping rooms. An unvented gas heater (one without a chimney to the outside) is also a potential source of carbon monoxide in a home. Carbon monoxide is a component of some fuel gases. Leaking gas lines are dangerous because of the poisonous nature of the gas as well as the fire hazard. Carbon monoxide is present in the exhaust of internal combustion engines. Thus, the engine of an automobile should never be left running in a closed garage. Similarly, an automobile should not be kept running to provide heat in a car parked with the windows closed. The smoke from burning tobacco contains up to 200 ppm of carbon monoxide. Thus, the blood of smokers contains several times the normal amount of hemoglobin-carbon monoxide.

SUMMARY

In combined form, carbon occurs in all living things. Foods, fuels, and carbonates contain combined carbon. In addition, hundreds of thousands of carbon compounds have been synthesized. Organic chemistry is the study of carbon compounds.

A carbon atom, with four valence electrons, forms covalent bonds with atoms of other elements or with other carbon atoms. Carbon atoms can link together in chains, rings, plates, and networks. The variety of ways in which carbon atoms can be linked accounts for the tremendous number of carbon compounds.

Uncombined carbon commonly occurs in the crystalline allotropic forms, graphite and diamond. Amorphous carbon includes coke, charcoal, boneblack, carbon black, and petroleum coke. The forms of amorphous carbon seem to have no definite shape, but contain regions in which the carbon atoms are arranged in an orderly way.

At ordinary temperatures all forms of carbon are inactive, but at higher temperatures they react with oxygen to form carbon dioxide.

Carbon dioxide is present in the air, although only in small amounts, as a result of decay, combustion, and the respiration of living things. The amount of carbon dioxide in the air is slowly increasing and may ultimately affect climatic conditions. Carbon dioxide can be prepared by burning carbon or its compounds, by reacting steam and natural gas, by fermenting molasses, by heating a carbonate, by the action of an acid on a carbonate, and by respiration and decay.

Carbon monoxide can be prepared by reducing carbon dioxide, by the action of steam on hot coke, and by decomposing formic acid. Carbon monoxide is a colorless, odorless, tasteless gas that is exceedingly poisonous. It burns with a blue flame.

VOCABULARY

amorphous carbon
anthracite
bituminous coal
boneblack
carbon black
carbyne
charcoal
chlorophyll
coal
coke
destructive distillation
diamond
dry ice
enzyme
fermentation
graphite
hemoglobin
leavening agent
lignite
organic chemistry
oxygen–carbon dioxide cycle
peat
petroleum coke
photosynthesis
respiration
subbituminous coal
sublime

QUESTIONS

Group A

1. *(a)* What is *organic chemistry? (b)* Why is organic chemistry a separate branch of chemistry? *A1, A4a*
2. How are the four covalent bonds of a carbon atom oriented in sp^3 hybridization? *A2b*
3. There are several times as many carbon compounds as noncarbon compounds. What property of carbon atoms makes this possible? *A2b*
4. *(a)* Define: *allotropy. (b)* Name three allotropic forms of carbon. *(c)* Why is amorphous carbon not considered a separate allotropic form of carbon? *A1, A3*
5. What property of diamond determines most of its industrial uses? *A3e*
6. Diamond is an excellent conductor of heat but a nonconductor of electricity. Why?
7. Graphite is soft, yet carbon fibers in which the carbon is in the form of graphite are very strong. Why? *A3e*
8. *(a)* Draw structural formulas for the three types of graphite layers. *(b)* How is the actual structure of graphite an example of resonance? *A2b*
9. List several properties of graphite that make it useful as a lubricant. *A3e*
10. What is coal? *A1*
11. *(a)* Explain how geologists believe coal was formed. *(b)* In what sequence were the various types of coal probably formed? *(c)* What conditions probably determined this sequence? *A4c*
12. *(a)* Define: *destructive distillation. (b)* Comment on the descriptive accuracy of this term. *A1, A3h*

13. After the destructive distillation of bituminous coal or wood, a form of carbon, coke or charcoal, remains. What property of carbon is responsible? *A3e, A3f*

14. *(a)* Define: *adsorption.* *(b)* Distinguish between adsorption and absorption. *A1*

15. *(a)* What is an important impurity in boneblack? *(b)* How can it be removed? *(c)* Why? Use information from Appendix Table 12 to support your answer. *A3c, A3h*

16. Why is activated carbon a useful adsorbent? *A1, A3c, A3g*

17. *(a)* How many steps are needed to prepare activated carbon? *(b)* Describe them. *A3c, A3i*

18. *(a)* In the manufacture of what product is most carbon black used? *(b)* What two purposes does the carbon black serve? *A3g*

19. For what is petroleum coke used? *A3g*

20. Explain why carbon dioxide is an important component of the atmosphere even though it occurs to only 0.034% by volume. *A3a*

21. *(a)* What change is occurring in the amount of carbon dioxide in the earth's atmosphere? *(b)* What are the probable causes? *A3a*

22. What will probably be the effect on the earth's climate if the present change in the amount of carbon dioxide in the atmosphere continues? *A3a*

23. *(a)* List four commercial methods for preparing carbon dioxide. *(b)* Write a balanced chemical equation for each method. *A3i, A5a*

24. Write a balanced chemical equation for the usual laboratory method of preparing carbon dioxide. *A3i, A5a*

25. What is the function of the enzymes of zymase during the fermentation of $C_6H_{12}O_6$ to C_2H_5OH and CO_2? *A1, A3i*

26. What are the disadvantages of collecting carbon dioxide *(a)* by water displacement; *(b)* by air displacement? *A3i*

27. List five physical properties of carbon dioxide. *A3e*

28. List five chemical properties of carbon dioxide. *A3f*

29. *(a)* Write a series of three equations which show the reactions that occur when carbon dioxide dissolves in cold water. *(b)* What is the name of the product formed? *(c)* Is this product a strong or weak acid? *(d)* Why? *A5b*

30. Describe the test for carbon dioxide. *A3f*

31. The oxygen–carbon dioxide cycle consists of two opposing parts. What are they? *A3g*

32. For what purpose is carbon dioxide used in petroleum recovery? *A3g*

33. *(a)* What is a leavening agent? *(b)* What compound is the source of carbon dioxide in most leavening agents? *(c)* How is the carbon dioxide released? *A3g*

34. *(a)* Write a balanced formula equation for the reaction that occurs when a soda-acid fire extinguisher is discharged. *(b)* Write the ionic and net ionic equations for this reaction. *A5c*

35. Describe two helpful actions that enable a liquid carbon dioxide fire extinguisher to put out a fire. *A3e, A3f*

36. What are the sources of the carbon monoxide in the atmosphere? *A3a*

37. In the atmosphere of what planets besides Earth is *(a)* carbon dioxide found; *(b)* carbon monoxide found? *A3a*

38. Why is sulfuric acid needed for the preparation of carbon monoxide from formic acid? *A3i*

39. *(a)* On the basis of their molecular weights, list oxygen, hydrogen, carbon dioxide, and carbon monoxide in order of increasing density. *(b)* What is the relationship between molecular weights and densities of gases that makes such a listing possible?

40. List three uses of carbon monoxide. *A3g*

PROBLEMS

Group A

1. The jewelers' mass unit for diamond is the carat. By definition, 1 carat equals exactly 200 mg. What is the volume of a 1.00-carat diamond? Obtain density data from Appendix Table 16.

2. For 100.0 g of carbon dioxide, calculate *(a)* the number of moles; *(b)* the number of molecules; *(c)* the volume at STP.
3. A complex oxide of carbon has the molecular formula $C_{12}O_9$. Determine its percentage composition.
4. *(a)* How many moles of iron(III) oxide can be reduced by the carbon in 5.00 moles of carbon monoxide, according to the equation: $Fe_2O_3 + 3CO \rightarrow 2Fe + 3CO_2$? *(b)* How many moles of iron are produced? *(c)* How many moles of carbon dioxide are produced?
5. How many grams of carbon monoxide are needed to react with 15.0 g of zinc oxide to produce elemental zinc? $ZnO + CO \rightarrow Zn + CO_2$
6. In Problem 5, *(a)* How many grams of zinc are produced? *(b)* What is the volume in liters at STP of the carbon dioxide gas produced?
7. How many grams of H_2SO_4 are required for the reaction with $90\bar{0}$ g of sodium hydrogen carbonate in a soda-acid fire extinguisher? $2NaHCO_3 + H_2SO_4 \rightarrow Na_2SO_4 + 2H_2O + 2CO_2$
8. Calculate the number of liters of carbon dioxide at STP given off during the discharge of the fire extinguisher of Problem 7.

Group B

9. On a mythical planet in a far-off galaxy, the element carbon consists of equal numbers of atoms of C-12, atomic mass exactly 12 *u*; C-13, atomic mass, 13.00335 *u*; and C-14, atomic mass, 14.00324 *u*. What is the atomic weight of carbon on this planet?
10. The stable oxides of carbon have the molecular formulas CO, CO_2, C_3O_2, and $C_{12}O_9$. What is the oxidation number of carbon in each oxide?
11. How many grams of carbon monoxide can be obtained by the dehydration of $25\bar{0}$ g of formic acid by sulfuric acid?
12. How many liters of dry carbon monoxide are produced in Problem 11 if the temperature is 27 °C and the pressure is $75\bar{0}$ mm?
13. How many liters of carbon dioxide result from the combustion of carbon monoxide in Problem 12 if the product is restored to 27°C and $75\bar{0}$ mm pressure?
14. A gaseous compound contains 52.9% carbon and 47.1% oxygen. One volume of this gas reacts with two volumes of oxygen and yields three volumes of carbon dioxide. Knowing that oxygen molecules are diatomic, determine the molecular formula of this compound.
15. How many grams of calcium carbonate are needed to prepare 2.50 L of dry carbon dioxide at 22 °C and 735 mm pressure by the reaction between calcium carbonate and hydrochloric acid?
16. How many milliliters of concentrated hydrochloric acid must be diluted with water to provide the HCl needed for the reaction of Problem 15? Concentrated hydrochloric acid is 38.0% HCl by mass and has a density of 1.20 g/mL.
17. If the acid of Problem 16 is diluted with water to make $10\bar{0}$ mL of solution, what is the molarity of the solution?
18. The pH of a saturated solution of carbon dioxide in water is 3.8. What is the $[H_3O^+]$?

Robert Burns Woodward (American) was awarded the Nobel Prize in chemistry in 1965 for preparing numerous biochemicals in the laboratory. Among them were cholesterol and chlorophyll, which had once been thought to be produced only by living things. His first major synthesis was the preparation of quinine from simple organic compounds, each of which could be synthesized from its elements rather than obtained from living matter. Woodward also synthesized cortisone, strychnine, reserpine, tetracycline, vitamin B_{12}, and many other complicated molecules.

Hydrocarbons

18

Goals

In this chapter you will gain an understanding of: • the abundance of carbon compounds • structural formulas for simple organic compounds; isomers • differences between organic and inorganic compounds • hydrocarbons; classification, structures, production, properties • natural gas and petroleum; petroleum substitutes • natural and synthetic rubber

An Introduction To Organic Compounds

18.1 Abundance of carbon compounds The number of possible carbon compounds is almost unlimited. Over 3,000,000 are known and about 100,000 new ones are isolated or synthesized each year. Why are there so many carbon compounds and how do they differ from noncarbon compounds? These questions will be answered in this section and in Section 18.4. Then you will learn about natural gas and petroleum, sources of many simple organic compounds. Most of these simple organic compounds are ***hydrocarbons,*** *compounds composed only of carbon and hydrogen.* You will next find out what specific hydrocarbons occur in natural gas and petroleum and how these compounds can be changed into other fundamental hydrocarbon compounds. Since the known supply of petroleum in the earth is rapidly being exhausted, some possible substitutes for petroleum as a source of organic compounds must be considered. In Chapter 19, you will read about a variety of organic compounds derived from hydrocarbons that are important in everyday life.

There are two reasons for so many carbon compounds:

1. *Carbon atoms link together by means of covalent bonds.* In Section 17.2, you learned how carbon atoms readily form covalent bonds with other carbon atoms. This makes possible the existence of molecules in which thousands of carbon atoms are bonded one to another. The molecules of some organic compounds are principally long carbon-atom chains with carbon-atom groups attached. Other carbon-compound molecules have carbon atoms linked together in rings. Still others may consist of several such rings joined together. Not only are carbon atoms linked by single covalent bonds, but they are sometimes linked by double or triple covalent bonds.

2. *The same atoms may be arranged in several different ways*. One of the substances in petroleum is a hydrocarbon called *octane*. Its molecular formula is C_8H_{18}. A molecule of octane consists of 8 carbon atoms and 18 hydrogen atoms. A carbon atom may form four single covalent bonds while a hydrogen atom forms only one single covalent bond. The straight-chain electron-dot structure for an octane molecule and its model are shown in the top-left photo of Figure 18-1. But there are other ways in which these same atoms can be arranged. For instance, three branched-chain formulas and their models are also shown in Figure 18-1.
All these formulas represent arrangements of 8 carbon atoms and 18 hydrogen atoms. Each carbon atom shares four electrons, and each hydrogen atom shares one electron. In addition to these four structures for octane, there are 14 others, making a total of 18 possible structures for octane. Each of these 18 structures has the same molecular formula. However, the different arrangements of the atoms in the molecules give each molecule

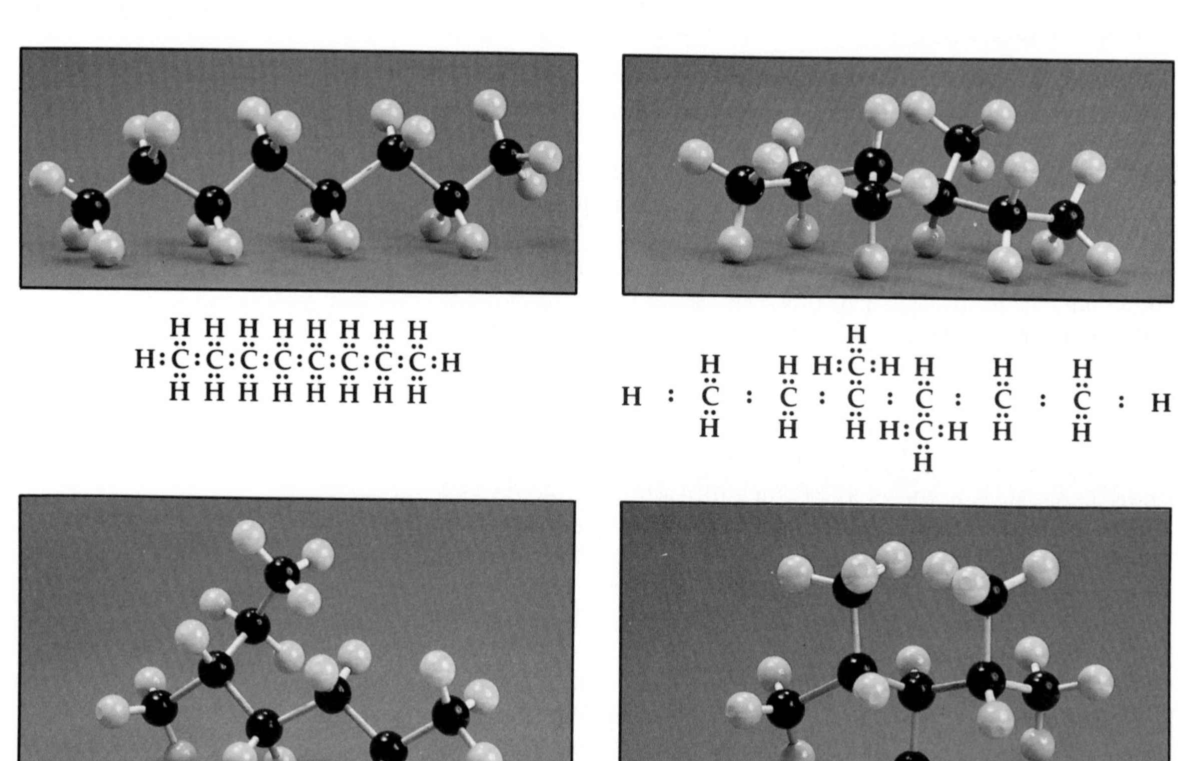

Figure 18–1. Models of the molecules of the four isomers of octane whose electron-dot structures are given below each isomer. Ball-and-stick models are used here instead of space-filling models so that you see the structures more clearly.

slightly different properties. Thus, each of these molecular arrangements represents a separate chemical compound. *Compounds having the same molecular formula but different structures are called* ***isomers.***

Figure 18-2. The first oil well in the United States was drilled in 1859 near Titusville, Pennsylvania, by Edwin Drake.

18.2 Structural formulas for organic compounds The formula H_2SO_4 for sulfuric acid gives enough information for most purposes in inorganic chemistry. But a molecular formula such as C_8H_{18} is not at all satisfactory in organic chemistry. It has already been noted that there are 18 different isomers of this compound. To indicate clearly a particular isomer, the organic chemist uses a ***structural formula.*** *Such a formula not only indicates what kinds of atoms and how many of each are involved but also shows how they are arrranged in the molecule.* Electron-dot formulas have been used to illustrate the isomers of octane. However, for routine equation work such formulas are tedious to draw. Organic chemists often substitute a dash (—) for the pair of shared electrons forming a covalent bond. Using the dash, the straight-chain structural formula for octane can be represented

```
    H   H   H   H   H   H   H   H
    |   |   |   |   |   |   |   |
H — C — C — C — C — C — C — C — C — H
    |   |   |   |   |   |   |   |
    H   H   H   H   H   H   H   H
```

For practice, on a separate sheet of paper, write the structural formulas for the three branched-chain isomers of octane given in Section 18.1.

When structural formulas are written, there must be no dangling bonds. Each dash must represent an electron pair that forms the covalent bond linking two atoms.

18.3 Determination of an organic structural formula There are two different organic compounds that consist of carbon, 52.2%, hydrogen, 13.0%, and oxygen, 34.8%. They have the same molecular weight, 46, and thus are isomers. One compound is a colorless liquid that boils at 78 °C. The other is a colorless gas that condenses to a liquid at −25 °C under one atmosphere pressure. Each has its own distinctive odor. How can their structural formulas be determined?

From the percentage composition, you can calculate the empirical formula, which is C_2H_6O. Since this empirical formula has a formula weight of 46, it must also be the molecular formula of each compound. From what you have already learned about bonding, there are only two ways in which two carbon atoms, six hydrogen atoms, and a single oxygen atom can combine, as shown in the right margin.

```
    H   H
    |   |
H — C — C — O — H
    |   |
    H   H
```

Structure A

```
    H       H
    |       |
H — C — O — C — H
    |       |
    H       H
```

Structure B

Now the problem is to match these structures to the two compounds. If each compound is tested for reaction with metallic sodium, only the liquid reacts. In the reaction, hydrogen is given off. The amount of hydrogen given off is equal to one-sixth of the hydrogen that the compound contains. This evidence indicates that in the molecules of the liquid, one of the six hydrogen atoms is bonded differently from the others. Structure *A* is indicated.

Next you discover that the liquid reacts with phosphorus trichloride and gives a product with the molecular formula C_2H_5Cl. In this reaction, chlorine has replaced both a hydrogen atom and an oxygen atom. Only one structural formula for C_2H_5Cl can be written, as shown in the left margin. You may assume that the chlorine atom occupies the same position as the oxygen and hydrogen atoms that it replaced. Structure *A* is again indicated. You might continue further, because much more evidence can be found to indicate that the liquid does indeed have Structure *A*. This liquid substance is ethanol, or ethyl alcohol. The gaseous substance has the other structural formula and is dimethyl ether.

$$\begin{array}{ccccc} & H & & H & \\ & | & & | & \\ H- & C & - & C & -Cl \\ & | & & | & \\ & H & & H & \end{array}$$

Note that structural formulas such as

$$\begin{array}{ccccc} & H & & H & \\ & | & & | & \\ H- & C & - & C & -H \\ & | & & | & \\ & H & & Cl & \end{array} \text{ and } \begin{array}{ccccc} & Cl & & H & \\ & | & & | & \\ H- & C & - & C & -H \\ & | & & | & \\ & H & & H & \end{array}$$

are the same as the one given above.

Methods similar to those just described can be used to determine the structural formulas of other simple organic compounds. Complicated molecules are generally broken down into simpler molecules. From the structures of these simpler molecules, the structure of the complex molecule can be determined. Sometimes simple molecules of known structure are combined to produce a complex molecule. A comparison of chemical and physical properties of compounds of unknown structure with those of known structure is sometimes helpful.

18.4 Differences between organic and inorganic compounds

The basic laws of chemistry are the same for organic and inorganic compounds. However, the behavior of organic compounds is somewhat different from that of inorganic compounds. Some of the most important differences are:

1. *Most organic compounds do not dissolve in water.* The majority of inorganic compounds do dissolve more or less readily in water. Organic compounds generally dissolve in such organic liquids as alcohol, chloroform, ether, or carbon tetrachloride.

The approximate rule that "like dissolves like" is explained in Section 13.3.

2. *Organic compounds are decomposed by heat more easily than most inorganic compounds.* The decomposition (charring) of sugar when it is heated moderately is familiar. Such charring on heating is often a test for organic substances. By contrast, many inorganic compounds, such as common salt (sodium chloride), can be vaporized at a red heat without decomposition.

3. *Organic reactions generally proceed at much slower rates.* Such reactions often require hours or even days for completion. However, organic reactions in living cells may take place with great speed. Most inorganic reactions occur almost as soon as solutions of the reactants are brought together.

4. *Organic reactions are greatly affected by reaction conditions.* Many inorganic reactions follow well-known patterns. This makes it possible for you to learn about many inorganic reactions by studying the general types of reactions explained in Chapter 8. There are some general types of organic reactions, too. But

changing the temperature or pressure or the nature of a catalyst can alter the identity of the products formed. The same organic reactants can form different products, depending on reaction conditions.

5. *Organic compounds exist as molecules consisting of atoms joined by covalent bonds.* Many inorganic compounds have ionic bonds. **CAUTION:** *Many organic compounds are* ***flammable*** *and* ***poisonous.*** *Quite a few cause or are suspected of causing cancer. Some organic reactions are rapid and highly exothermic. A student should not perform any experiments with organic compounds without detailed laboratory directions, and then only with adequate ventilation and other safety precautions and under the supervision of an experienced instructor.*

Figure 18–3. The search for new sources of petroleum continues beneath off-shore waters.

Hydrocarbon Series

18.5 Natural gas and petroleum Natural gas is a mixture of hydrocarbon gases and vapors found in porous formations in the earth's crust. Natural gas is mostly methane, CH_4. Frequently natural gas occurs with petroleum, or crude oil. Petroleum is a complex mixture of hydrocarbons. This hydrocarbon mixture varies greatly in composition from place to place. The hydrocarbon molecules in petroleum contain from one to more than 50 carbon atoms.

Natural gas and petroleum were probably formed by the decay of plants and animals living millions of years ago. Because of changes in the earth's surface, these plant and animal residues were trapped in rock formations where they slowly decomposed in the absence of atmospheric oxygen.

Natural gas and petroleum are the most common sources of energy. But, more important, they are a source of hydrocarbon chemical raw materials that is rapidly being exhausted. Discoveries of new sources of natural gas and petroleum in the United States are not keeping pace with the amounts being consumed. Estimates of the world petroleum situation indicate that demand will probably exceed available supplies in ten to twenty years.

18.6 The processing of natural gas Up to about 97% of natural gas is methane, CH_4. Mixed with the methane are other hydrocarbons whose molecules contain between two and seven carbon atoms. These different hydrocarbons have different boiling points and can be separated on this basis. This method of separation is called *fractional distillation*. The separation of nitrogen from oxgyen in liquid air is an example of this method.

Fractional distillation is a method of separating the components of a mixture on the basis of differences in their boiling points.

The separation of nitrogen from oxygen in liquid air is described in Section 9.4(5).

The hydrocarbons having 3 or 4 carbon atoms per molecule are sometimes separated and used as "bottled gas" for fuel. The hydrocarbons having 5 to 7 carbon atoms per molecule are liquids at ordinary temperature. Their vapors can be condensed and used as a solvent or in gasoline.

18.7 The refining of petroleum Petroleum is refined by separating crude oil into portions with properties suitable for certain uses. The method used is fractional distillation. No attempt is made to separate the petroleum into individual hydrocarbons. Instead, portions that distill between certain temperature ranges are collected in separate receivers. Table 18-1 summarizes the characteristics of the portions obtained from the fractional distillation of petroleum.

Petroleum refining is carried out in a *pipe still* and a *fractionating tower.* See Figure 18-4. The crude oil is heated to about 370 °C in the pipe still. At this temperature, nearly all the components of the crude oil are vaporized. The hot vapors are then discharged into the fractionating tower at a point near its base. Here, the portions with the highest condensation temperatures condense and are drawn off to collecting vessels. Portions with lower condensation temperatures continue to rise in the tower. As they rise, they are gradually cooled. In this way, the various portions reach their condensation temperatures at different levels. As they condense, the liquids collect in shallow troughs that line the inside of the tower. Pipes lead off the overflow of condensed liquids from the troughs. The gasoline fraction, together with the more volatile portions of the petroleum, passes as a gas from the top of the tower. The gasoline fraction is then liquefied in separate condensers. The uncondensed gases may be piped to

Table 18-1
SUMMARY OF FRACTIONAL DISTILLATION OF PETROLEUM

Portion	*No. of C atoms per molecule*	*Boiling point range (°C)*	*Uses*
gas	C_1 to C_5	−164-30	fuel; making carbon black, hydrogen, gasoline by alkylation
petroleum ether	C_5 to C_7	20-100	solvent; dry cleaning
gasoline	C_5 to C_{12}	30-200	motor fuel
kerosene	C_{12} to C_{16}	175-275	fuel
fuel oil diesel oil	C_{15} to C_{18}	250-400	furnace fuel; diesel engine fuel; cracking
lubricating oils greases petroleum jelly	C_{16}	350	lubrication
paraffin wax	C_{20}	melts 52-57	candles; waterproofing; home canning
pitch tar		residue	road constuction
petroleum coke		residue	electrodes

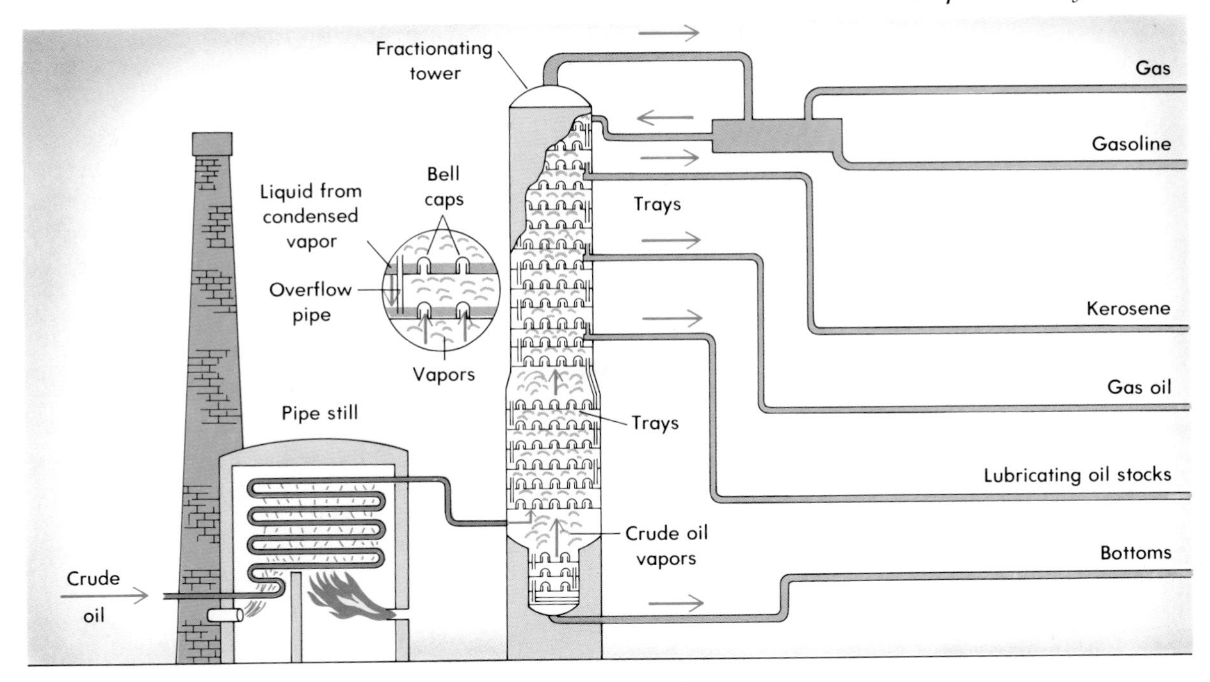

Figure 18–4. A cross section of a pipe still and fractionating tower used in refining petroleum.

the refinery. The liquid and gaseous fractions are subjected to other processes depending on what products are required.

18.8 Classification of hydrocarbons Compounds composed of only the two elements hydrogen and carbon are called *hydrocarbons.* The hydrocarbons are studied first because they are the basic structures from which other organic compounds are derived. Hydrocarbons are grouped into several different series of compounds. These groupings are based mainly on the type of bonding between carbon atoms.

1. The *alkanes* (al-*kaynes*) are straight-chain or branched-chain hydrocarbons. Their carbon atoms are connected by *single* covalent bonds only:

```
   H  H
   |  |
H—C—C—H
   |  |
   H  H
```

Alkanes are the most abundant hydrocarbons in natural gas and petroleum.

2. The *alkenes* (al-*keens*) are straight- or branched-chain hydrocarbons in which two carbon atoms in each molecule are connected by a *double* covalent bond:

```
H       H
 \     /
  C=C
 /     \
H       H
```

Figure 18–5. A petroleum refinery. The tower-like structure is a fractionating tower.

3. The *alkynes* (al-*kynes*) are straight- or branched-chain hydrocarbons in which two carbon atoms in each molecule are connected by a *triple* covalent bond:

$$H—C \equiv C—H$$

4. The *alkadienes* (al-kah-*dy*-eens) are straight- or branched-chain hydrocarbons that have *two double* covalent bonds between carbon atoms in each molecule:

```
H       H   H      H
 \      |   |     /
  C ═ C ─ C ═ C
 /                \
H                  H
```

5. The *aromatic hydrocarbons* have resonance structures. These structures are sometimes represented by alternate single and double covalent bonds in six-membered carbon *rings:*

```
⎧        H                      H        ⎫
⎪        |                      |        ⎪
⎪ H      C       H     H        C      H ⎪
⎪  \   /   \\   /       \    //   \   /  ⎪
⎪    C       C            C         C    ⎪
⎨    ‖       |            |         ‖    ⎬
⎪    C       C            C         C    ⎪
⎪  /   \   //   \       /    \\   /   \  ⎪
⎪ H      C       H     H        C      H ⎪
⎪        |                      |        ⎪
⎩        H                      H        ⎭
```

Petroleum from most sources contains some aromatic hydrocarbons. The amount varies from a few percent to as high as 40%.

18.9 The alkane series This series of organic compounds is sometimes called the *paraffin series* because paraffin wax is a mixture of hydrocarbons of this series. The word *paraffin* means *little attraction.* Compared to the other hydrocarbon series, the alkanes have low chemical reactivity. This stability results from their single covalent bonds. Because they have only single covalent bonds in each molecule, the alkanes are known as *saturated hydrocarbons.* Saturated bonding occurs when each carbon atom in the molecule forms four single covalent bonds with other atoms.

Table 18-2 lists a few members of the alkane series. The names of the first four members of this series follow no system. However, beginning with pentane, the first part of each name is a Greek or Latin numerical prefix. This prefix indicates the number of carbon atoms. The name of each member ends in *-ane,* the same as the name of the series. The letter prefix "*n*" for "normal" indicates the straight-chain isomer.

If you examine the formulas for successive alkanes, you will

Table 18-2
SOME MEMBERS OF THE ALKANE SERIES

Name	*Formula*	*Melting point* (°C)	*Boiling point* (°C)
methane	CH_4	−182	−164
ethane	C_2H_6	−183	−89
propane	C_3H_8	−190	−42
n-butane	C_4H_{10}	−138	0
2-methylpropane	C_4H_{10}	−159	−12
n-pentane	C_5H_{12}	−130	36
2-methylbutane	C_5H_{12}	−160	28
2,2-dimethylpropane	C_5H_{12}	−17	10
n-hexane	C_6H_{14}	−95	69
n-heptane	C_7H_{16}	−91	98
n-octane	C_8H_{18}	−57	126
n-nonane	C_9H_{20}	−51	151
n-decane	$C_{10}H_{22}$	−30	174
***			
n-eicosane	$C_{20}H_{42}$	37	343
***			
n-hexacontane	$C_{60}H_{122}$	99	

see a clear pattern. Each member of the series differs from the preceding one by the group $\mathbf{CH_2}$,

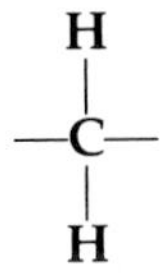

Compounds that differ in this fashion belong to a *homologous series*. *A* **homologous series** *is one in which adjacent members differ by a constant unit*. It is not necessary to remember the formulas for all members of a homologous series. A general formula, such as C_nH_{2n+2} for the alkanes, can be derived. Suppose a member of this series has 30 carbon atoms in its molecules. To find the number of hydrogen atoms, multiply 30 by 2, then add 2. The formula is $C_{30}H_{62}$.

18.10 Structures of the lower alkanes Each of the first three alkanes can have only *one* molecular structure. The formulas for these structures are given below. See the models in Figure 18-6.

```
    H            H   H            H   H   H
    |            |   |            |   |   |
  H—C—H      H—C—C—H      H—C—C—C—H
    |            |   |            |   |   |
    H            H   H            H   H   H
```

methane **ethane** **propane**

Butane, the alkane with four carbon atoms and ten hydrogen atoms, has *two* isomers. The straight-chain molecule is

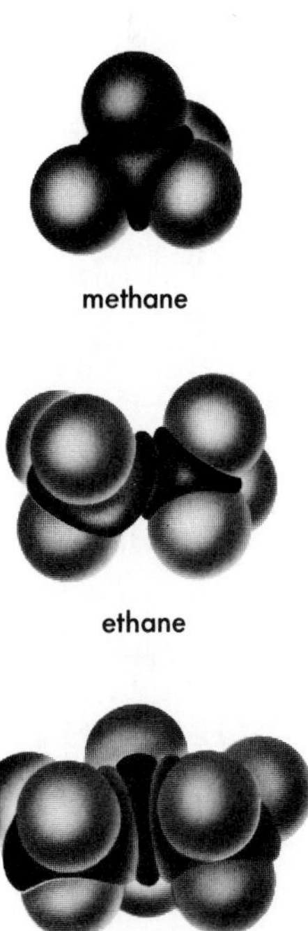

Figure 18–6. Models of molecules of the first three members of the alkane series of hydrocarbons.

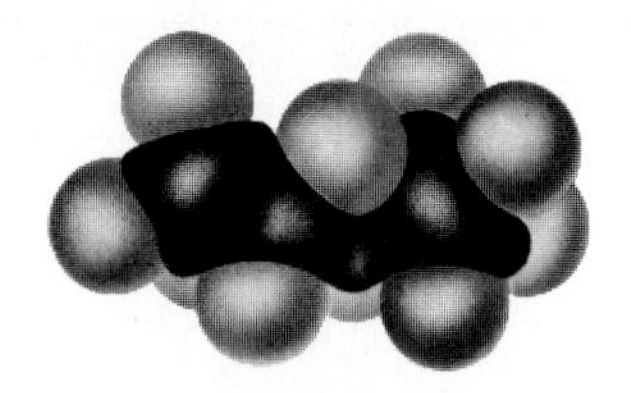

n-butane

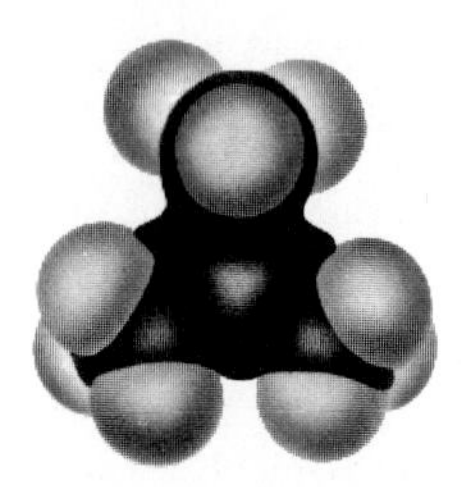

2-methylpropane

Figure 18–7. Models of molecules of the two isomers of butane.

named *n*-butane (*n* for *normal*). The branched-chain molecule is named 2-methylpropane. Their melting and boiling points are given in Table 18-2. Examine the models in Figure 18-7.

```
                          H      H      H
    H  H  H  H            |1     |2     |3
    |  |  |  |         H—C——————C——————C—H
 H—C—C—C—C—H             |      |      |
    |  |  |  |            |   H—C—H    |
    H  H  H  H            H      |      H
                                 H
```

***n*-butane** **2-methylpropane**

The name 2-methylpropane is derived from the structure of the molecule. The longest continuous carbon chain in the molecule is three carbon atoms long, as in propane. One hydrogen atom attached to the second carbon atom in propane is replaced by the CH_3— group. This is a *substitution group.* It is called the *methyl group.*

```
 |
H—C—H
 |
 H
```

is methane with one of the hydrogen atoms removed. Carbon atoms in the main chain of the molecule are numbered. This numbering begins at the end of the molecule giving the carbon atoms with substitution groups the smallest numbers. Thus in the name **2-methylpropane,** the 2 refers to the number of the carbon atom on which there is a substitution. *Methyl* is the substitution group. *Propane* is the parent hydrocarbon.

There are *three* possible pentanes (C_5H_{12}) whose names are: *n*-pentane, 2-methylbutane, and 2,2-dimethylpropane. Their melting and boiling points are given in Table 18-2. Their structural formulas are shown in the left margin. Why is the name **2-methylbutane** used rather than 3-methylbutane?

```
    H  H  H  H  H
    |  |  |  |  |
 H—C—C—C—C—C—H
    |  |  |  |  |
    H  H  H  H  H
```

***n*-pentane**

```
          H
     H    |     H     H
     |1   C2    |3    |4
  H—C————|————C—————C—H
     |  H—C—H  |      |
     H    |     H     H
          H
```

2-methylbutane

```
            H
            |
         H—C—H
      H     |     H
      |1    |2    |3
   H—C—————C—————C—H
      |     |     |
      H     |     H
         H—C—H
            |
            H
```

2,2-dimethylpropane

In the name **2,2-dimethylpropane,** the *2,2* refers to the position of both substitutions. The prefix *dimethyl* shows that the *two* substitutions are both *methyl* groups. *Propane* is the parent hydrocarbon.

Just as the CH_3— group derived from methane is the *methyl* group, C_2H_5— derived from ethane is the *ethyl* group. C_3H_7— derived from propane is the *propyl* group. C_4H_9— derived from butane is the *butyl* group. C_5H_{11}— is usually called the *amyl* group rather than the pentyl group. Other groups are given names following the general rule of dropping the *-ane* suffix and adding *-yl.* Any such group derived from an *alkane* is an *alkyl* group. The symbol **R**— is frequently used to represent an *alkyl* group in a formula.

18.11 Preparation of the alkanes Alkanes are generally found in petroleum and natural gas. It is fairly easy to separate the lower members of the alkane series individually from petroleum and natural gas by fractional distillation. However, the alkanes with higher boiling points are usually separated into mixtures with similar boiling points.

Methane is a colorless, nearly odorless gas that forms about 90% of natural gas. Pure methane can be separated from the other components of natural gas. Chemists sometimes prepare small amounts of methane in the laboratory by heating soda lime (which contains sodium hydroxide) with sodium acetate. See Figure 18-8.

$$NaC_2H_3O_2(s) + NaOH(s) \rightarrow CH_4(g) + Na_2CO_3(s)$$

Figure 18–8. A small quantity of methane can be prepared in the laboratory by heating a mixture of sodium acetate and soda lime. Methane is then collected by water displacement.

Ethane is a colorless gas that occurs in natural gas and is a product of petroleum refining. It has a higher melting point and boiling point than methane. These properties are related to ethane's higher molecular weight.

Methane and ethane are minor components of the atmosphere of the planet Jupiter. Some gaseous methane is present around Saturn and Saturn's moon Titan, and it is abundant about Uranus and Neptune. Solid methane has been detected on the surface of Pluto. Methane has also been found in interstellar space.

18.12 Reactions of the alkanes *1. Combustion.* Because the alkanes make up a large proportion of our gaseous and liquid fuels, their most important reaction is combustion. Methane burns with a bluish flame.

$$CH_4 + 2O_2 \rightarrow CO_2 + 2H_2O$$

Ethane and other alkanes also burn in air to form carbon dioxide and water vapor.

$$2C_2H_6 + 7O_2 \rightarrow 4CO_2 + 6H_2O$$

2. Substitution. The alkanes react with halogens such as chlorine or bromine. In reactions such as these, one or more atoms of a halogen are substituted for one or more atoms of hydrogen. Therefore, the products are called *substitution products.*

$$\begin{array}{c}H\\|\\H-C-H\\|\\H\end{array} + Br_2 \rightarrow \begin{array}{c}H\\|\\H-C-Br\\|\\H\end{array} + HBr$$

Figure 18–9. Methane and bromine undergo a substitution reaction, thereby producing substitution products.

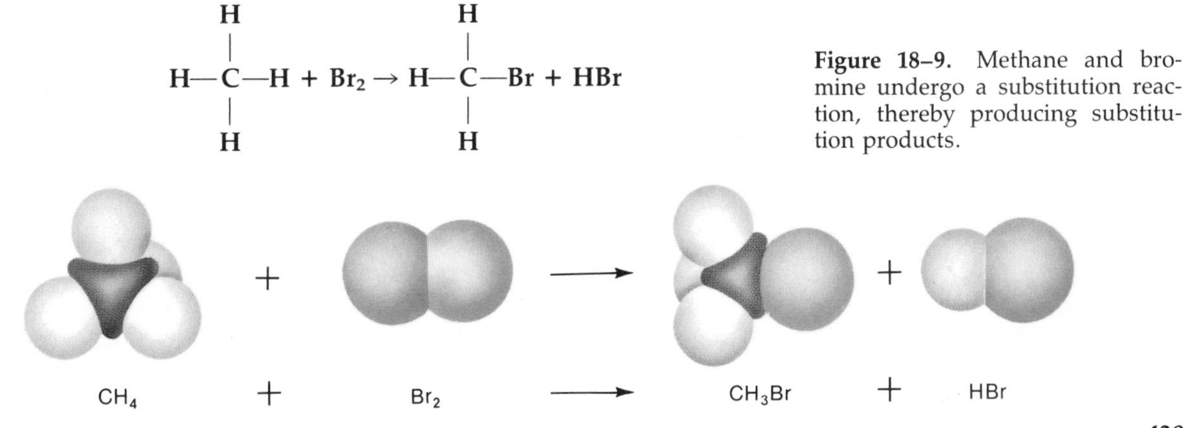

By supplying additional molecules of the halogen, a halogen atom can be substituted for each of the hydrogen atoms. Carbon tetrachloride, dichlorodifluoromethane, and Teflon are examples of halogen-substituted alkanes.

3. *Preparation of hydrogen.* Propane reacts with steam in the presence of a nickel catalyst at a temperature of about 850 °C.

Compare this reaction with the ones in Sections 9.17(4) and 17.15(2).

$$C_3H_8 + 6H_2O \rightarrow 3CO_2 + 10H_2$$

To separate the carbon dioxide from the hydrogen, the carbon dioxide under pressure is dissolved in water.

18.13 Alkene series The alkenes, sometimes called the *olefin series,* are distinguished by a double covalent bond between two carbon atoms. Thus, the simplest alkene must have two carbon atoms. Its structural formula is

```
H       H
 \     /
  C=C
 /     \
H       H
```

Its name is ethene. The name of an alkene comes from the name of the alkane with the same number of carbon atoms. Simply substitute the suffix *-ene* for the suffix *-ane.* Since eth*ane* is the alk*ane* with two carbon atoms, the alk*ene* with two carbon atoms is named eth*ene.* (This substance is also commonly called *ethylene.*) The general formula for the alkenes is C_nH_{2n}.

Ethene is the hydrocarbon commercially produced in greatest quantity in the United States. It is also the organic chemical industry's most important starting material.

In 1983, ethene ranked fifth in production behind sulfuric acid, nitrogen, lime, and oxygen. The amount produced in the United States was nearly thirteen million metric tons.

18.14 Preparations of alkenes 1. *Cracking alkanes.* The commercial method of producing alkenes is by *cracking* petroleum. ***Cracking*** *is a process by which complex organic molecules are broken up into simpler molecules. This process involves the action of heat or the action of heat and a catalyst.*

Cracking that uses heat alone is known as *thermal cracking.* During cracking, alkanes decompose in several ways that produce a variety of alkenes. A simple example is the thermal cracking of propane, which proceeds in either of two ways in nearly equal proportions.

Important reaction conditions are sometimes written near the yields sign.

$$C_3H_8 \xrightarrow{460\ ^\circ C} C_3H_6 + H_2$$

$$C_3H_8 \xrightarrow{460\ ^\circ C} C_2H_4 + CH_4$$

Alkenes (especially ethene), smaller alkanes, and hydrogen are typical products of the thermal cracking of alkanes. They can be separated, purified, and used as starting materials for making other organic compounds.

The cracking process that involves heat and a catalyst is *catalytic cracking*. The high-boiling fractions from petroleum distillation are catalytically cracked to produce smaller, lower-boiling hydrocarbons useful in gasoline. The catalysts used are mostly oxides of silicon and aluminum. The cracking reactions produce smaller alkanes and alkenes with highly branched structures. Aromatic hydrocarbons are also formed.

2. *Dehydration of alcohols*. Ethene can be prepared in the laboratory by dehydrating (removing water from) ethyl alcohol. Hot concentrated sulfuric acid is used as the dehydrating agent.

$$C_2H_5OH \xrightarrow[170\ ^\circ C]{H_2SO_4} C_2H_4 + H_2O$$

18.15 Reactions of alkenes 1. *Addition*. An organic compound that has one or more double or triple covalent bonds in each molecule is said to be *unsaturated*. It is chemically possible to add other atoms directly to such molecules to form molecules of a new compound. For example, hydrogen atoms can be added to an alkene in the presence of a finely divided nickel catalyst. This reaction produces the corresponding alkane. Breaking the carbon-carbon double bond provides two new bond positions for the hydrogen atoms.

$$\begin{array}{c}H \qquad\quad H\\ \diagdown\ \ \diagup\\ C{=}C\\ \diagup\ \ \diagdown\\ H \qquad\quad H\end{array} + H_2 \xrightarrow{Ni} \begin{array}{c}\ \ H\ \ \ H\\ \ \ |\ \ \ \ |\\ H{-}C{-}C{-}H\\ \ \ |\ \ \ \ |\\ \ \ H\ \ \ H\end{array}$$

Halogen atoms can be added readily to alkene molecules. For example, two bromine atoms added directly to ethene form 1,2-dibromoethane. See Figure 18-10.

$$\begin{array}{c}H \qquad\quad H\\ \diagdown\ \ \diagup\\ C{=}C\\ \diagup\ \ \diagdown\\ H \qquad\quad H\end{array} + Br_2 \rightarrow \begin{array}{c}\ \ Br\ \ Br\\ \ \ |\ \ \ \ |\\ H{-}C{-}C{-}H\\ \ \ |\ \ \ \ |\\ \ \ H\ \ \ H\end{array}$$

Figure 18-10. Ethene and bromine undergo an addition reaction.

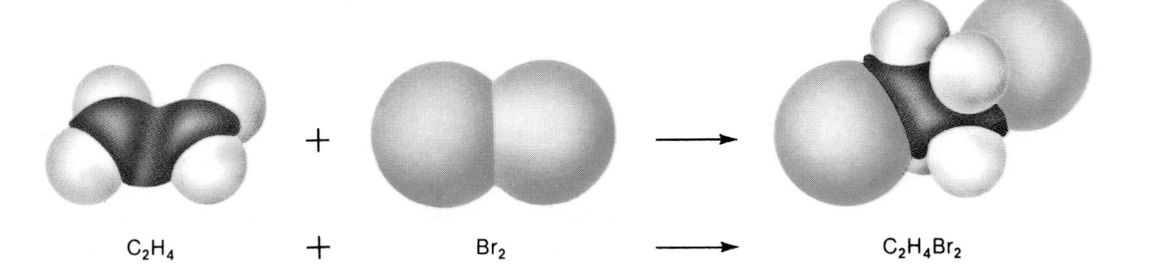

When the double bond between the carbon atoms breaks, there is one bond position available for each bromine atom.

The name of the product is 1,2-dibromoethane. The basic part of the name, *ethane,* is that of the related alkane with two carbon atoms. *Dibromo-* refers to the two bromine atoms that have been substituted for hydrogen atoms in ethane. *1,2-* means that one bromine atom is bonded to the first carbon atom and the other is bonded to the second carbon atom. An isomer, 1,1-dibromoethane, has the formula

$$\begin{array}{ccccccc} & & \mathbf{Br} & & \mathbf{H} & & \\ & & | & & | & & \\ \mathbf{Br} & — & \mathbf{C} & — & \mathbf{C} & — & \mathbf{H} \\ & & | & & | & & \\ & & \mathbf{H} & & \mathbf{H} & & \end{array}$$

A molecule of a hydrogen halide, such as hydrogen bromide, can be added to an alkene molecule. What is the name of the product?

$$\begin{array}{ccc} \mathbf{H} & & \mathbf{H} \\ & \diagdown \quad \diagup & \\ & \mathbf{C{=}C} & \\ & \diagup \quad \diagdown & \\ \mathbf{H} & & \mathbf{H} \end{array} + \mathbf{HBr} \rightarrow \begin{array}{ccccccc} & & \mathbf{H} & & \mathbf{Br} & & \\ & & | & & | & & \\ \mathbf{H} & — & \mathbf{C} & — & \mathbf{C} & — & \mathbf{H} \\ & & | & & | & & \\ & & \mathbf{H} & & \mathbf{H} & & \end{array}$$

2. *Polymerization.* Molecules of ethene join together, or *polymerize,* at about 15 °C and 20 atmospheres pressure in the presence of a catalyst. The resulting large molecules have molecular weights of about 30,000. This polymerized material is called *polyethylene.* It is made up of many single units called *monomers.*

$$\begin{array}{ccccc} & \mathbf{H} & & \mathbf{H} & \\ & | & & | & \\ — & \mathbf{C} & — & \mathbf{C} & — \\ & | & & | & \\ & \mathbf{H} & & \mathbf{H} & \end{array}$$

Many monomers join together to make the ***polymer*** (many units). Polyethylene is used in film for packaging, in electric insulation, in a variety of containers, and in tubing.

3. *Alkylation.* In alkylation, gaseous alkanes and alkenes are combined. For example, 2-methylpropane and 2-methylpropene combine in the presence of a sulfuric acid or anhydrous hydrogen fluoride catalyst. The result is a highly branched octane, 2,2,4-trimethylpentane, an important component of gasoline.

4. *Combustion.* The alkenes burn in oxygen. For example,

$$\mathbf{C_2H_4 + 3O_2 \rightarrow 2CO_2 + 2H_2O}$$

18.16 Alkyne series The alkynes are distinguished by a triple covalent bond between two carbon atoms. This series is sometimes called the *acetylene series* because the simplest alkyne has the common name *acetylene.* It has two carbon atoms, with the structural formula

$$H—C{\equiv}C—H$$

The names of the alk*ynes* are derived from the names of the alk*anes* that have the same number of carbon atoms. The suffix *-yne* is substituted for *-ane*. Hence the chemical name for acetylene, the simplest alkyne, is *ethyne*. The general formula for the alkynes is C_nH_{2n-2}.

Ethyne, C_2H_2, is a minor component of the atmospheres of Jupiter and Saturn. Ethyne molecules have also been detected in interstellar space.

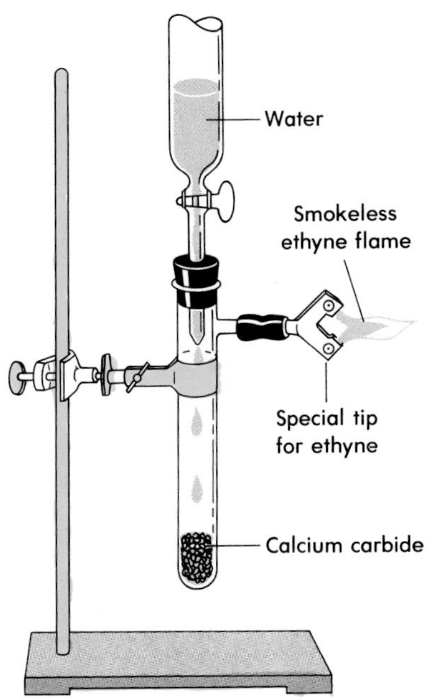

Figure 18–11. A convenient method of preparing a small quantity of ethyne (acetylene) in the laboratory by the action of water on calcium carbide.

18.17 Preparations of ethyne 1. *From calcium carbide.* Ethyne, a colorless gas, can be prepared by the action of water on calcium carbide, CaC_2. Calcium carbide is made from limestone, $CaCO_3$, in a series of operations. First, the limestone is heated in a kiln (oven). CaO is produced.

$$CaCO_3(s) \rightarrow CaO(s) + CO_2(g)$$

The calcium oxide is then heated with coke at 2000 °C in an electric resistance furnace.

$$CaO(s) + 3C(s) \rightarrow CaC_2(s) + CO(g)$$

Calcium carbide is an ionic compound with the electron-dot structure

$$Ca^{++}$$
$$^{-}{:}C{:::}C{:}^{-}$$

When it reacts with water, hydrogen replaces the calcium in the calcium carbide structure. The bonding of hydrogen and carbon in ethyne is covalent.

$$CaC_2(s) + 2H_2O(l) \rightarrow C_2H_2(g) + Ca(OH)_2(aq)$$

The preparation of ethyne from calcium carbide is both a commercial and laboratory method. See Figure 18-11.

2. *By partial oxidation of methane.* This is an alternate commercial preparation of ethyne. Methane can be partially oxidized under controlled conditions to yield ethyne, carbon monoxide, and hydrogen.

$$6CH_4 + O_2 \xrightarrow{1500\ °C} 2C_2H_2 + 2CO + 10H_2$$

The source of the methane is petroleum refining. The carbon monoxide and some of the hydrogen can be used to produce methanol. The balance of the hydrogen can be used as a fuel to maintain the temperature required for the reaction.

The preparation of methanol from carbon monoxide and hydrogen is described in Section 17.24(3).

18.18 Reactions of ethyne 1. *Combustion.* Ethyne burns in air with a very smoky flame. Carbon, carbon dioxide, and water vapor are the products of combustion. With special burners, the combustion to carbon dioxide and water vapor is complete.

$$2C_2H_2 + 5O_2 \rightarrow 4CO_2 + 2H_2O$$

The oxyacetylene welding torch burns ethyne in the presence of oxygen.

2. *Halogen addition.* Ethyne is more unsaturated than ethene because of the triple bond. It is chemically possible to add two molecules of bromine to an ethyne molecule to form 1,1,2,2-tetrabromoethane.

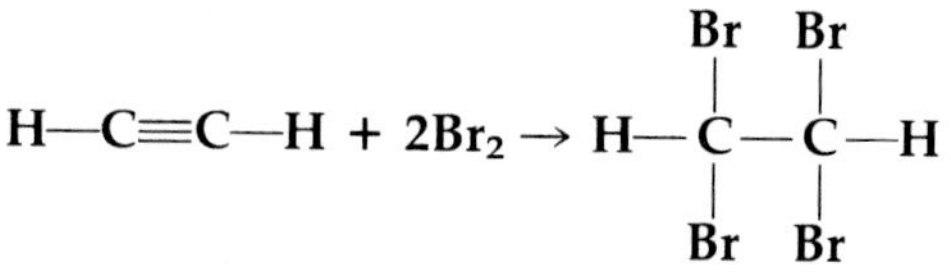

$$H{-}C{\equiv}C{-}H + 2Br_2 \rightarrow H{-}\underset{\substack{|\\Br}}{\overset{\substack{Br\\|}}{C}}{-}\underset{\substack{|\\Br}}{\overset{\substack{Br\\|}}{C}}{-}H$$

You have now learned three related words: monomer, dimer, and polymer. Monomer means one unit of a compound that can undergo polymerization. Dimer means a polymer formed from two units of a monomer. Polymer means a compound consisting of many repeating structural units.

3. *Dimerization.* Two molecules of ethyne may combine and form the *dimer* (two units), vinylacetylene. This dimerization is brought about by passing ethyne through a water solution of copper(I) chloride and ammonium chloride. The solution acts as a catalyst.

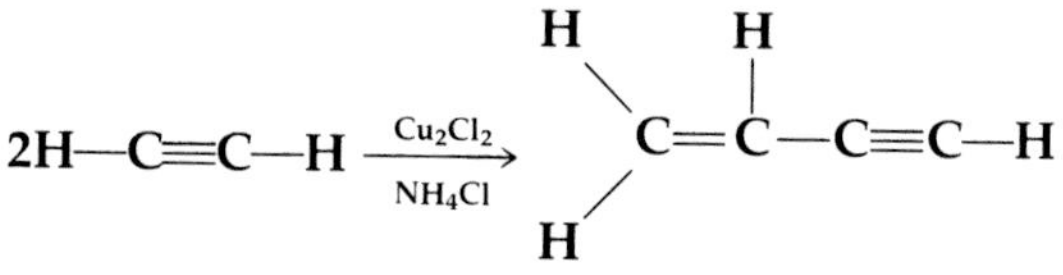

$$2H{-}C{\equiv}C{-}H \xrightarrow[NH_4Cl]{Cu_2Cl_2} H_2C{=}\overset{\substack{H\\|}}{C}{-}C{\equiv}C{-}H$$

vinylacetylene

The $CH_2{=}CH{-}$ group is the *vinyl* group. Vinylacetylene is the basic raw material for producing Neoprene, a synthetic rubber.

18.19 Butadiene: an important alkadiene Alkadienes have two double covalent bonds in each molecule. The *-ene* suffix indicates a double bond. The *-diene* suffix indicates two double bonds. The names of the alkadienes are derived in much the same way as those of the other hydrocarbon series. Butadiene must, therefore, have four carbon atoms and contain two double bonds in each molecule. See the structure in the left margin.

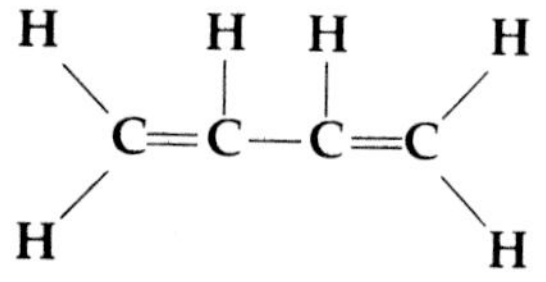

$$H_2C{=}\overset{\substack{H\\|}}{C}{-}\overset{\substack{H\\|}}{C}{=}CH_2$$

Actually, this structure is 1,3-butadiene, since the double bonds follow the first and third carbon atoms. However, 1,2-butadiene, its isomer, is so uncommon that 1,3-butadiene is usually called simply butadiene.

Butadiene is prepared by cracking petroleum fractions containing butane. It is used in the manufacture of *SBR* rubber, the most common type of synthetic rubber.

18.20 The aromatic hydrocarbons The aromatic hydrocarbons are generally obtained from coal tar and petroleum. Benzene, the best known aromatic hydrocarbon, has the molecular formula C_6H_6. It may be represented by the following resonance formula, in which the two structures contribute equally.

The bonds in benzene are neither single bonds nor double bonds. Instead, each bond is a *resonance hybrid bond.* All the carbon-carbon bonds in the molecule are the same. As a result, benzene and other aromatic hydrocarbons are not as unsaturated as the alkenes.

Because of the resonance structure of benzene, the benzene ring is sometimes abbreviated as

The $\mathbf{C_6H_5}$— group derived from benzene is the *phenyl* group.

Benzene is produced commercially from petroleum. It is a flammable liquid that is used as a solvent. Benzene is used in manufacturing many other chemicals, including dyes, drugs, and explosives. Benzene has a strong, yet fairly pleasant, aromatic odor. It is less dense than water and only very slightly soluble in water. *Benzene is poisonous.* The vapors are harmful to breathe and are very flammable. *It is now recommended that benzene not be used in school laboratories.*

Exposure to benzene can result in the destruction of the bone marrow's ability to make blood cells. It may also cause leukemia.

18.21 Reactions of benzene 1. *Halogenation.* Benzene reacts with bromine in the presence of iron to form the substitution product bromobenzene, or phenyl bromide.

$$+ \ Br_2 \xrightarrow{Fe} \quad + \ HBr$$

Further treatment causes the successive substitution of other bromine atoms for hydrogen atoms. With complete substitution, hexabromobenzene is produced.

benzene, C_6H_6

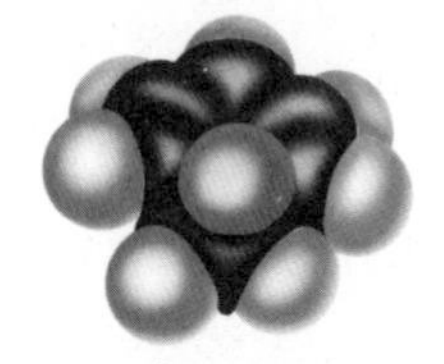

toluene, C_7H_8

naphthalene, $C_{10}H_8$

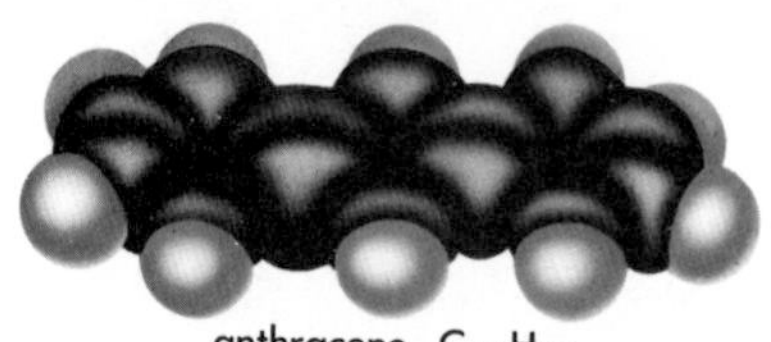

anthracene, $C_{14}H_{10}$

Figure 18–12. Molecular models of common aromatic hydrocarbons.

2. *Nitration.* Nitrobenzene is produced by treating benzene with concentrated nitric and sulfuric acids.

$$C_6H_6 + HNO_3 \xrightarrow{H_2SO_4} C_6H_5NO_2 + H_2O$$

3. *Sulfonation.* Benzenesulfonic acid is produced at room temperature by treating benzene with fuming sulfuric acid. (Fuming sulfuric acid contains an excess of sulfur trioxide.)

$$C_6H_6 + (HO)_2SO_2 \xrightarrow{SO_3} C_6H_5SO_2OH + H_2O$$

4. *Friedel-Crafts reaction.* An alkyl group can be introduced into the benzene ring by using an alkyl halide in the presence of anhydrous aluminum chloride.

$$C_6H_6 + RCl \xrightarrow{AlCl_3} C_6H_5R + HCl$$

18.22 Other aromatic hydrocarbons Toluene, or methyl benzene, is obtained from petroleum.

$$C_6H_5CH_3$$

Toluene is used to make benzene.

The xylenes or dimethylbenzenes, $C_6H_4(CH_3)_2$, are a mixture of three liquid isomers. The xylenes are used as starting materials for the production of polyester fibers, films, and bottles.

Ethylbenzene is produced by the Friedel-Crafts reaction of benzene and ethene in the presence of hydrogen chloride.

What is the original raw material from which both the benzene and ethene needed for this reaction are derived?

$$C_6H_6 + C_2H_4 \xrightarrow[HCl]{AlCl_3} C_6H_5C_2H_5$$

Ethylbenzene is treated with a catalyst of mixed metallic oxides to eliminate hydrogen and produce styrene. Styrene is used along with butadiene in making *SBR* synthetic rubber.

$$C_6H_5C_2H_5 \xrightarrow[650\ °C]{catalyst} C_6H_5CH{=}CH_2 + H_2$$

Styrene may be polymerized to polystyrene, a tough, transparent plastic. Polystyrene molecules in this form have molecular weights of about 500,000. Styrofoam, a porous form of polystyrene, is used as a packaging and insulating material and for making throwaway beverage tumblers.

Naphthalene, $C_{10}H_8$, is a coal-tar or petroleum product that crystallizes in white, shining scales. Naphthalene molecules have a structure made up of two benzene rings joined by a common side.

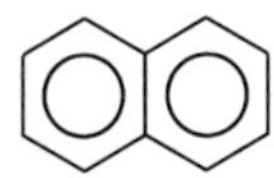

Naphthalene is used as a raw material for the manufacture of some resins and dyes.

Anthracene, $C_{14}H_{10}$, has a structure made up of three benzene rings joined together.

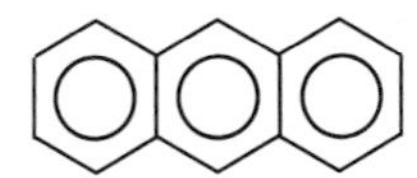

Like naphthalene, anthracene forms a whole series of hydrocarbons. They differ from the compounds related to benzene in that there is more than one ring. Anthracene is obtained commercially from coal tar and used to produce synthetic dyes.

Figure 18–13. This plant produces over 400,000 metric tons of styrene annually.

18.23 Octane number of gasoline *Knocking* occurs in an automobile engine when the mixture of gasoline vapor and air in the cylinders explodes spontaneously rather than burning at a uniform rate. Knocking causes a loss of power and may harm the engine. It can be prevented by using a gasoline that resists this tendency to explode spontaneously as the temperature and pressure increase within the cylinder.

Air mixtures of gasoline that consists mostly of straight-chain hydrocarbons tend to knock badly in automobiles. Air mixtures of hydrocarbons with branched-chains and rings resist exploding spontaneously as the temperature and pressure in the cylinder increase. Thus, branched-chain and ring hydrocarbons have less tendency to knock than straight-chain hydrocarbons. To improve the antiknock qualities, refiners produce a gasoline mixture that contains a large proportion of branched-chain and ring hydrocarbon molecules.

Certain compounds, when added to gasoline in small amounts, improve the antiknock properties. The best known of these is lead tetraethyl, $Pb(C_2H_5)_4$. The use of lead tetraethyl in gasoline is currently being reduced because it contributes to air pollution.

The octane rating of a gasoline is a number that indicates the tendency of a gasoline to knock in a high-compression engine. The higher the number, the less the tendency of the gasoline to

Figure 18–14. A technician determining the octane rating of a sample of gasoline.

Figure 18–15. Oil shale, found in some western states, such as Colorado, is another source of a petroleum-like liquid.

knock. To determine the octane rating, the gasoline is burned in a standard test engine. Its performance in the test engine is compared with a fuel of known octane rating. The fuels used as standards are *n*-heptane and 2,2,4-trimethylpentane, which is also named iso-octane. Iso-octane has excellent antiknock properties. It has an arbitrary octane rating of 100. Normal heptane knocks badly and has a rating of zero. A gasoline with the characteristics of a mixture of 90% iso-octane and 10% *n*-heptane has an octane rating of 90.

Straight-chain alkanes have low octane ratings. Saturated ring-type hydrocarbons have intermediate octane ratings. Highly branched alkanes and aromatic (benzene ring-shaped) hydrocarbons have high octane ratings.

18.24 Petroleum substitutes It is estimated that about 73% of the energy used in the United States comes from natural gas and petroleum. Within a few decades the known supplies of these important chemical raw materials may be exhausted. Continued wasteful use of natural gas and petroleum as fuels threatens the future supply of these raw materials for other important products. Alternate fuels or sources of energy should be developed. Energy will need to be used more efficiently as the sources of energy becomes scarcer.

Coal is by far the most widespread and plentiful fuel remaining in the United States. It is used to supply nearly 20% of our energy. At the present rate of consumption, the amount of recoverable coal in the United States is expected to last 300 years. This period of time may be shortened if the rate of consumption increases. Both the mining and burning of coal present environmental problems that need to be resolved.

The only commercial-sized plants currently using this process are in South Africa.

There are two well-known methods by which coal can be converted to fuel gases and petroleumlike liquids. The Fischer-Tropsch process starts with the mixture of CO and H_2 formed by reacting steam and hot coke. These gases then react at suitable pressures and temperatures in the presence of a metallic catalyst to form a variety of straight-chain hydrocarbons. The amount of unsaturated products varies with the catalyst used. The Bergius process involves reacting finely powdered coal suspended in oil with hydrogen at very high pressure and high temperature, usually in the presence of a metallic catalyst. The complex many-ring compounds in coal are broken down. Liquid hydrocarbons are formed. Neither process is economical in the United States.

These processes provided aviation and other fuels for Germany during World War II.

Oil shale, found in some western states, is another source of a petroleumlike liquid. In one process, the oil shale is mined and brought to the surface. There it is heated in large ovens to drive out the oil, which is then refined. Another method involves the underground heating and collection of the oil. More research is needed to develop economical production methods. Here, too, environmental problems must be resolved.

Nuclear fission (see Chapter 31) already accounts for the production of about 13% of the electricity in the United States. This, however, is only about 4% of our total energy use. Nuclear fission is a promising source of energy for the next several decades if we resolve environmental and safety problems associated with the location and operation of nuclear power plants.

The remaining energy used in the United States, about 3% of the total, is furnished mostly by water power.

Researchers are seeking ways to obtain large quantities of energy from underground steam, the sun, ocean water temperature differences, nuclear fusion, and biomass.

Biomass is the collective term for living materials, or matter derived from living things, that could be used to provide energy. It includes wood and agricultural products and their wastes, algae, animal wastes, sewage, etc.

Rubber

18.25 Nature of rubber *Rubber* is an elastic hydrocarbon material obtained from rubber trees. Each tree daily yields about one ounce of a milky fluid called *latex.* Latex contains about 35% rubber in colloidal suspension. When formic acid is added to latex, the rubber separates out as a curd-like mass. After being washed and dried, it is shipped to market as large sheets of crude rubber.

The simplest formula for rubber is $(C_5H_8)_x$. The structural formula for a single C_5H_8 unit is thought to be the one shown in the margin. The C_5H_8 unit is a monomer of rubber. The "x" is believed to be a large number. Thus, rubber is a polymer made up of a large number of C_5H_8 units. The units are joined in a zigzag chain. This arrangement accounts for rubber's elasticity.

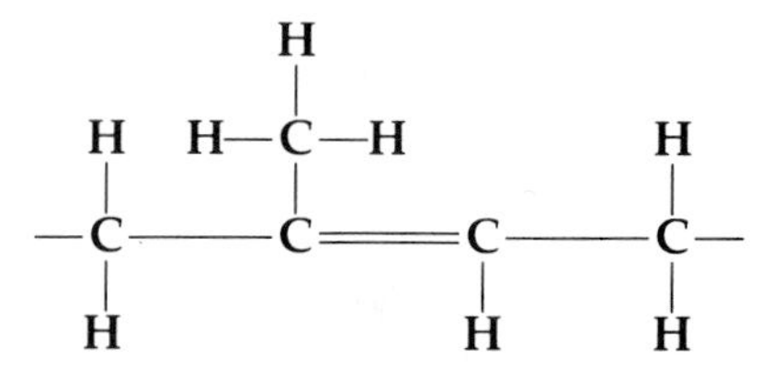

18.26 Compounding of rubber For commercial processing, raw rubber is mixed with a number of other materials in large batches. The ingredients in these batches vary according to the products to be made. Sulfur, however, is always one of the ingredients. Automobile tires contain considerable amounts of carbon black. This substance adds to the bulk and increases the wearing qualities of the tires.

After mixing, the product is shaped either in a mold or from thin sheets. The whole mass is then vulcanized. ***Vulcanization*** *involves heating the rubber mixture to a definite temperature for a definite time.* It gives the article a permanent shape, makes the rubber more elastic, and causes it to lose its sticky qualities. The changes that occur during vulcanization are many and complex. It is believed, however, that the sulfur atoms form bonds between adjacent rubber molecules. Organic catalysts are added to speed the process. Other organic chemicals are added to reduce the effect of oxygen from the air. This treatment slows the aging process that makes the rubber become hard and brittle.

Figure 18–16. Radial automobile tires are made from a mixture, half of which is natural rubber and half synthetic rubber.

18.27 Neoprene, a synthetic rubber About the year 1910, scientists first produced hydrocarbon synthetic rubber by polymerizing isoprene, C_5H_8. This synthetic product cost too much and was inferior to natural rubber from plantations in the East Indies. In 1931, a successful synthetic rubber called *neoprene* appeared on the market.

The dimerization of ethyne to form vinylacetylene is described in Section 18.18(3). From what materials can ethyne be prepared?

Hydrogen chloride adds to vinylacetylene to yield chloroprene:

```
H  H                               H  H     H
|  |                               |  |     |
C=C—C≡C—H + HCl → C=C—C=C
|                                  |     |  |
H                                  H     Cl H
```

vinylacetylene **chloroprene**

```
  H  Cl  H  H
  |  |   |  |
—C—C=C—C—
  |         |
  H         H
```

The catalytic polymerization of chloroprene yields the neoprene unit, as shown in the left margin. Oils and greases cause natural rubber to swell and rot. They have little effect on neoprene. Hence, neoprene is used in gasoline delivery hoses and in other objects that must be flexible while resisting the action of hydrocarbons.

18.28 *SBR,* a synthetic rubber for tires

SBR, Styrene Butadiene Rubber, is a good all-purpose synthetic rubber. It can replace natural rubber for most purposes. However, radial automobile tires are made from a mixture of about half natural rubber and half synthetic rubber. *SBR* is used for automobile tire treads because it resists wear better than other synthetic rubbers. It is made by churning *butadiene* (Section 18.19) and *styrene* (Section 18.22) together in soapy water. The churning is carried out at 5 °C, using a catalyst. This causes the chemicals to polymerize and form *SBR.* The addition of an acid causes the rubber to separate in curdlike masses, which are washed and dried. A possible structural unit is shown in the structural formula below:

What is the original raw material from which SBR is derived?

```
  H  H  H  H  H  H  H  H  H  H  H  H
  |  |  |  |  |  |  |  |  |  |  |  |
—C—C—C—C=C—C—C—C—C—C=C—C—
  |  |  |        |  |  |  |        |
  H  |  H        H  H  |  H        H
     |                 C—H
   (C6H5)              ||
                     H—C—H
```

SBR **structural unit**

SUMMARY

The number of possible carbon compounds is almost unlimited. There are many carbon compounds because: (1) carbon atoms link together by means of covalent bonds; (2) the same atoms may be arranged in several different ways. Isomers are compounds having the same molecular formula but different structures. Structural formulas indicate what kinds of atoms, how many of each, and how they are arranged in the molecule.

Most organic compounds do not dissolve in water. Organic compounds are decomposed by heat more easily than most inorganic compounds. Organic reactions generally proceed at much slower rates than inorganic reactions. Organic reactions are greatly affected by reaction conditions. Organic compounds exist as molecules having atoms joined by covalent bonds.

Hydrocarbons are compounds composed of only carbon and hydrogen. Their sources are natural gas and petroleum. Some pure hydrocarbons, but more commonly mixtures of hydrocarbons, are separated from natural gas and petroleum by fractional distillation. Fractional distillation is a method of separating the components of a mixture on the basis of differences in their boiling points.

Hydrocarbons are grouped into different series based mainly on the type of bonding between carbon atoms: (1) Alkanes—straight-chain or branched-chain hydrocarbons with carbon atoms connected by only single covalent bonds; (2) Alkenes—straight- or branched-chain hydrocarbons in which two carbon atoms in each molecule are connected by a double covalent bond; (3) Alkynes—straight- or branched-chain hydrocarbons in which two carbon atoms in each molecule are connected by a triple covalent bond; (4) Alkadienes—straight- or branched-chain hydrocarbons having two double covalent bonds between carbon atoms in each molecule; (5) Aromatic hydrocarbons—resonance structures represented by alternate single and double covalent bonds in six-membered carbon rings.

Alkanes are found in natural gas and petroleum. Alkenes are produced from petroleum by cracking. Cracking is a process by which complex organic molecules are broken up into simpler molecules by the action of heat and sometimes a catalyst. Ethyne, the simplest alkyne, is prepared from calcium carbide or by the partial oxidation of methane from petroleum refining. Butadiene, an alkadiene, is prepared by cracking butane separated from petroleum. Simple aromatic hydrocarbons are obtained from petroleum; more complex aromatic hydrocarbons come from coal tar.

Important types of hydrocarbon reactions: Alkanes—combustion, substitution, cracking. Alkenes—addition, polymerization, alkylation, combustion. Alkynes—combustion, addition, dimerization. Benzene—halogenation, nitration, sulfonation, addition of alkyl groups.

Substitutes for natural gas and petroleum can be made from coal. The processes for this conversion are the Fischer-Tropsch process and the Bergius process. Another source of a petroleum-like liquid is oil shale.

Natural rubber is an elastic hydrocarbon material obtained from rubber trees. It is a polymer of a large number of C_5H_8 units joined in a zig-zag chain. Rubber is compounded by mixing it with sulfur, sometimes carbon black, and other ingredients. The mixture is vulcanized by heating it to a definite temperature for a definite time. This gives the rubber article a permanent shape, makes it more elastic, and causes it to lose its sticky qualities. Neoprene and *SBR* are two kinds of synthetic rubber.

VOCABULARY

addition
alkadienes
alkanes
alkenes
alkyl group
alkylation
alkynes
aromatic hydrocarbons
cracking
dimer
dimerization
Friedel-Crafts reaction
halogenation
homologous series
hydrocarbon
isomer
knocking
monomer
nitration
octane rating
olefin series
paraffin series
petroleum
polymer
polymerization
saturated
structural formula
substitution group
sulfonation
unsaturated
vinyl group
vulcanization

QUESTIONS

GROUP A

1. Why are there so many carbon compounds? *A2a*
2. *(a)* What is a hydrocarbon? Give the formula of *(b)* the simplest saturated hydrocarbon; *(c)* the unsaturated hydrocarbon having the fewest atoms per molecule. *A1, A6a*
3. *(a)* In an electron-dot formula, what does **:** represent? *(b)* In a structural formula, what does — represent? *A7a*
4. *(a)* What are isomers? *(b)* Illustrate using structural formulas for the simplest alkene having isomers. *A1, A6b, A7b*
5. What information is provided by a properly written structural formula? *A1*
6. List five important differences between organic and inorganic compounds. *A2c*
7. Describe the composition of *(a)* natural gas; *(b)* petroleum. *A1*
8. Define *(a) distillation; (b) fractional distillation. (c)* How effectively are the components of petroleum separated by fractional distillation? *A1, A2d*
9. What use is made of the petroleum fraction *(a)* having 5 to 12 carbon atoms per molecule; *(b)* having a melting point range of 52 °C to 57 °C? *A3c*
10. What is the general formula for the members of the *(a)* alkane series; *(b)* alkene series; *(c)* alkyne series? *A6a*
11. A hydrocarbon molecule contains seven carbon atoms. Give its empirical formula if it is *(a)* an alkane; *(b)* an alkene; *(c)* an alkyne. *A6a*
12. *(a)* What is a homologous series? *(b)* Name the first four members of the alkane series. *(c)* How are the other members of the alkane series named? *A3d, A5a, A6a*
13. *(a)* What does the formula RH represent? *(b)* What does the symbol R— represent? *A2e*
14. Write the balanced formula equation for the laboratory preparation of methane. *A4a*
15. Write a balanced formula equation for the complete combustion of *(a)* methane; *(b)* ethene; *(c)* ethyne; *(d)* butadiene. *A4b*
16. Write equations for the step-by-step substitution of each of the hydrogen atoms in methane by chlorine. *A4b, A5c*
17. *(a)* Which hydrocarbon is commercially produced in greatest quantity in the United States? *(b)* What method is used for this commercial production.? *A2f, A3i*
18. Define: *cracking, thermal cracking, catalytic cracking. A1, A2f*
19. Butene reacts with hydrogen in the presence of a nickel catalyst. Write a structural-formula equation for the reaction. *A4b*
20. What is *(a)* a monomer; *(b)* a dimer; *(c)* a polymer; *(d)* polymerization? *A1*
21. Give two uses for ethyne. *A3h*
22. *(a)* What do the terms *saturated* and *unsaturated* mean when applied to hydrocarbons? *(b)* What other meaning do these terms have in chemistry? *A1*
23. Calcium carbide is sold in airtight metal cans. Why? *A2f, A2g, A3f*
24. By drawing the structural formulas, determine the number of isomers of pentadiene. *A7b*
25. How are naphthalene and anthracene molecules related structurally to benzene molecules? *A2h*
26. *(a)* What causes knocking in an automobile engine? *(b)* How is gasoline formulated to prevent knocking? *A3k, A3l*
27. What does it mean if a gasoline has an octane rating of 89? *A2i, A3m*
28. *(a)* What two general uses are made of natural gas and petroleum in the United States? *(b)* Which use is more important and should be protected by conservation measures? *A2j*
29. Describe two methods of obtaining oil from oil shale. *A2k*
30. How is rubber separated out from latex? *A2l*
31. Why is rubber elastic? *A2l*
32. How is rubber compounded? *A2m*
33. *(a)* What kinds of additives are used in making rubber goods? *(b)* Why? *A2m*
34. Define: *vulcanization. A1*

35. What advantage does neoprene have over natural rubber? *A2l*

Group B

36. Why are organic compounds with covalent bonds usually less stable to heat than inorganic compounds with ionic bonds? *A2c*
37. How does the action shown in Figure 12-6 compare with that on a single trough in Figure 18-4? *A2d*
38. Draw structural formulas for the five isomers of hexane. *A7b*
39. When decane, $C_{10}H_{22}$, burns completely, carbon dioxide and water vapor are formed. Write the balanced formula equation for this reaction. *A4b*
40. The compound 1,3-dichloropropane is a colorless liquid that boils at 120.4 °C under standard pressure. Draw the structural formula for this compound. *A5c, A7a*
41. Draw structural formulas for any three isomers of trichloropentane, and name the isomers whose formulas you draw. *A5c*
42. The element that appears in the greatest number of compounds is hydrogen. The element forming the second greatest number of compounds is carbon. Why are there more hydrogen compounds than carbon compounds? *A2a*
43. The organic compounds 2-methylpropane and 2-methylpropene react to form 2,2,4-trimethylpentane. Write the structural formula equation for the reaction. *A4b*
44. Ethyne molecules dimerize to form vinylacetylene. Write the structural formula equation. *A4b*
45. *(a)* Is it geometrically possible for the four hydrogen atoms attached to the end carbon atoms in the 1,3-butadiene molecule to lie in the same plane? *(b)* If carbon-hydrogen bonds on adjacent singly bonded carbon atoms tend to repel each other, would it be likely for all six hydrogen atoms to lie in the same plane? *(c)* If they do, what is their relation to the plane of the carbon atoms? *A7a*

PROBLEMS

Group A

1. What is the mass in kilograms of 12.0 gallons of gasoline? Assume the gasoline is isooctane, which has a density of 0.692 g/mL.
2. What is the mass in grams of 5.00 moles of pentane?
3. Calculate the percentage of sulfur in benzenesulfonic acid.
4. What volume of carbon dioxide is produced by the complete combustion of 15.0 L of propane? The volumes of carbon dioxide and of propane are measured under the same experimental conditions.
5. How many grams of calcium carbide are required for the production of 2.5 L of ethyne at STP?

Group B

6. As the number of carbon atoms in an alkane molecule increases, does the percentage of hydrogen increase, decrease, or remain the same?
7. A compound consists of 60.0% carbon, 26.7% oxygen, and 13.3% hydrogen. Its molecular weight is $6\bar{0}$. What are the possible structures for molecules of this chemical compound?
8. Calculate the energy change of the substitution reaction between one mole of methane and one mole of bromine molecules. Use the bond energies given in Table 6-4.
9. A volume of ethene (115 mL) is collected by water displacement at 22 °C and 738 mm pressure. What is the volume of the dry ethene at STP?
10. A compound consists of 93.75% carbon and 6.25% hydrogen. The substance dissolves in benzene, and 6.14 g of it lowers the freezing point of $10\bar{0}$ g of benzene 2.45 C°. *(a)* What is the empirical formula of the compound? *(b)* What is its molecular weight? *(c)* What is its molecular formula? See Table 13-5 for necessary data.

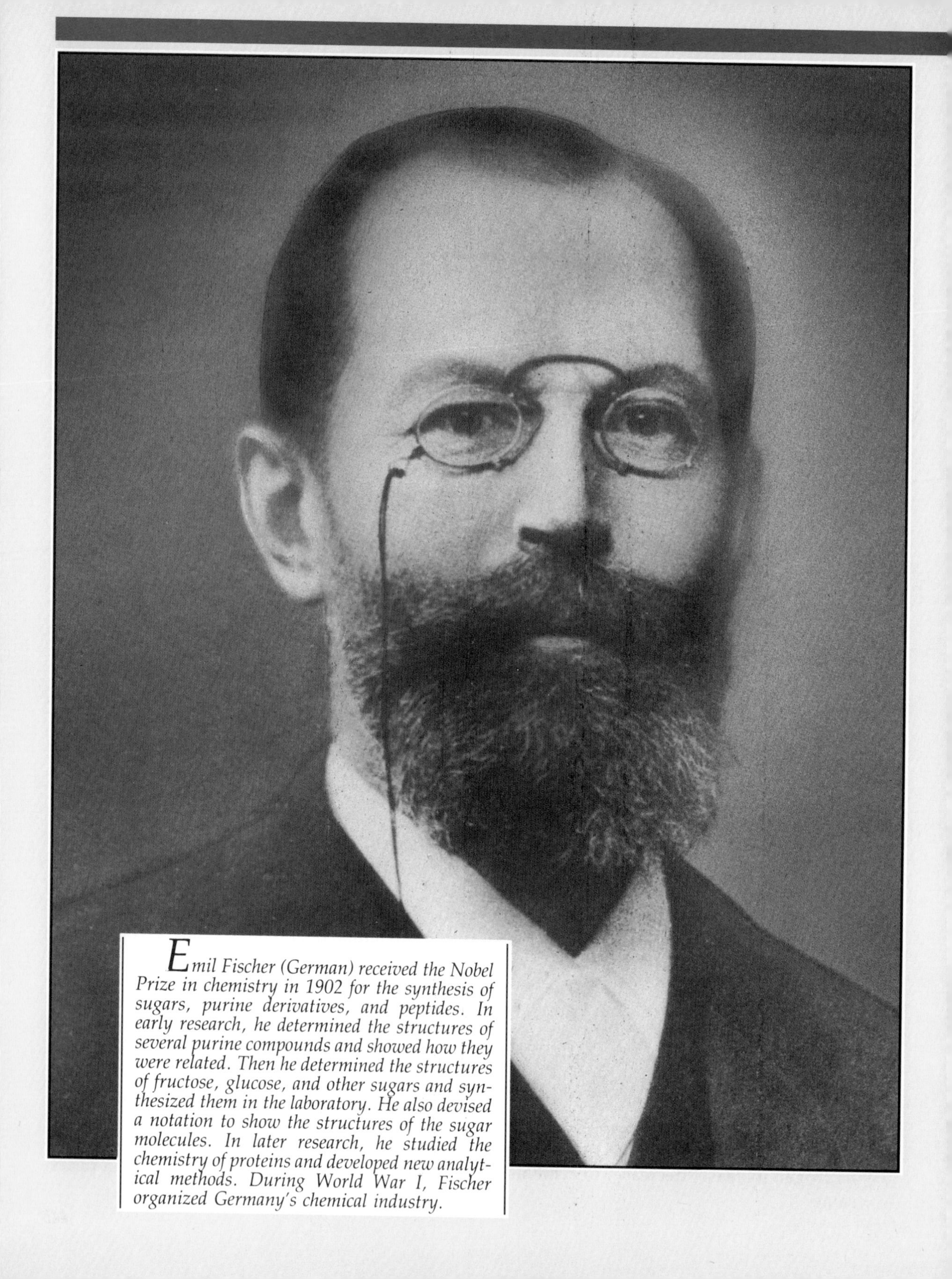

Emil Fischer (German) received the Nobel Prize in chemistry in 1902 for the synthesis of sugars, purine derivatives, and peptides. In early research, he determined the structures of several purine compounds and showed how they were related. Then he determined the structures of fructose, glucose, and other sugars and synthesized them in the laboratory. He also devised a notation to show the structures of the sugar molecules. In later research, he studied the chemistry of proteins and developed new analytical methods. During World War I, Fischer organized Germany's chemical industry.

Hydrocarbon Substitution Products

19

GOALS

In this chapter you will gain an understanding of: • the general structural formulas for the alkyl halides, alcohols, ethers, aldehydes, ketones, carboxylic acids, and esters • the preparations, properties, and uses of these organic compounds • esterification and saponification reactions

HALOGEN SUBSTITUTION PRODUCTS

19.1 Preparation of alkyl halides *An **alkyl halide** is an alkane in which a halogen atom*—fluorine, chlorine, bromine, or iodine—*is substituted for a hydrogen atom.* Since **R—** is often used to represent an alkyl group and **—X** any halogen, an alkyl halide can be represented as **RX.**

Many halogenated hydrocarbons are toxic and affect the liver, the kidneys, or the heart muscle. Some are believed to cause cancer.

1. *Direct halogenation.* In Section 18.12, you learned that the halogens react with alkanes to form substitution products. For example, under suitable conditions halogen atoms can be substituted for each of the four hydrogen atoms in methane. This reaction occurs in four stages.

$$CH_4 + X_2 \rightarrow CH_3X + HX$$

$$CH_3X + X_2 \rightarrow CH_2X_2 + HX$$

$$CH_2X_2 + X_2 \rightarrow CHX_3 + HX$$

$$CHX_3 + X_2 \rightarrow CX_4 + HX$$

2. *From alkenes and alkynes.* As discussed in Sections 18.15 and 18.18, alkenes and alkynes react with halogens or hydrogen halides to form alkyl halides.

3. *From alcohols.* Alcohols are alkanes in which the hydroxyl group, **—OH,** has been substituted for hydrogen. Hence, an alcohol has the type formula **ROH.** The reaction of an alcohol with a hydrogen halide, HCl, HBr, or HI, yields the corresponding alkyl halide.

$$ROH + HX \rightarrow RX + H_2O$$

19.2 Reactions of the alkyl halides Alkyl halides react with many molecules and ions. One result of these changes is the substitution of another atom or group of atoms for the halogen atom.

For example, alkyl halides react with water solutions of strong hydroxides to yield alcohols and the halide ion. The hydroxyl group from the hydroxide is substituted for the halogen atom in the alkyl halide.

$$RX + OH^- \rightarrow ROH + X^-$$

The Friedel-Crafts reaction [Section 18.21(4)] is an alkyl halide substitution reaction. Here the phenyl group, C_6H_5—, is substituted for the halogen atom in the alkyl halide.

$$RCl + C_6H_6 \xrightarrow{AlCl_3} C_6H_5R + HCl$$

19.3 Specific alkyl halides Tetrachloromethane, CCl_4, is commonly called carbon tetrachloride. It is a colorless, volatile, nonflammable liquid. It is an excellent nonpolar solvent. Carbon tetrachloride is sometimes used for dry-cleaning fabrics, removing grease from metals, and extracting oils from seeds. Its vapors are poisonous. Therefore, *there must be good ventilation when carbon tetrachloride is used. Its use in high school laboratories should be avoided.* Its most important use is in the preparation of Freon-type compounds. Carbon tetrachloride is prepared commercially by the direct chlorination of methane.

$$CH_4 + 4Cl_2 \rightarrow CCl_4 + 4HCl$$

Dichlorodifluoromethane, CCl_2F_2, is the most important member of a family of compounds named Freon. It is used as a refrigerant in mechanical refrigerators and air conditioners and as a bubble-making agent in the manufacture of plastic foams. This particular Freon is an odorless, nontoxic, nonflammable, easily liquefied gas. It is prepared from carbon tetrachloride and hydrofluoric acid with antimony compounds as catalysts.

$$CCl_4 + 2HF \xrightarrow{catalyst} CCl_2F_2 + 2HCl$$

Freon-type compounds, released into the air, do rise into the stratosphere. There, it is believed, ultraviolet radiation from the sun decomposes these compounds, releasing free halogen atoms. The free halogen atoms then react with the ozone molecules. Scientists believe the ozone molecules become oxygen molecules by a series of reactions. What happens to the halogen atoms after these reactions is unclear. One theory is that the halogen atoms start the decomposition of more ozone. In this way, a small number of halogen atoms can decompose a large number of ozone molecules. Another theory is that the halogen atoms become combined in a compound with nitrogen and oxy-

The presence of ozone in the stratosphere and its role in protecting the earth's surface from an excess of ultraviolet radiation from the sun are described in Section 9.9.

gen and do not react with more ozone. A decrease in the concentration of ozone in the stratosphere would allow more ultraviolet radiation to reach the earth. Because this form of radiation is potentially harmful, the production of Freons has been limited, and the use of Freons as propellants in aerosol cans has been discontinued in the United States. In 1984, atmospheric research indicated that the concentration of ozone in the upper atmosphere is less likely to be changed by human activities than had been believed earlier.

Tetrafluoroethene, C_2F_4, can be polymerized. The product has a structure in which the following unit occurs again and again:

$$\begin{array}{cc} \mathrm{F} & \mathrm{F} \\ | & | \\ -\mathrm{C}- & \mathrm{C}- \\ | & | \\ \mathrm{F} & \mathrm{F} \end{array}$$

This material is called Teflon. Teflon is a very inactive, flexible substance that is stable up to about 325 °C. It is made into fibers for weaving chemical-resistant fabrics or into rods from which small parts may be formed. Teflon has a very low coefficient of friction. It is useful where heat-resistant, nonlubricated moving parts are needed. It is also used to coat metals of cooking utensils and give them a "nonsticking" surface. As an adhesive, Teflon is used to bond artificial diamond grit to aluminum for grinding wheels used in cutting ceramics and metals.

1,2-dichloroethane, $ClCH_2CH_2Cl$, commonly called ethylene dichloride, is made by chlorinating ethene, as described in Section 18.15. This liquid is used as a solvent and for making adhesives. Most of it, however, is converted to vinyl chloride.

Figure 19-1. Teflon is used in modern architecture, and to make tubing, wire insulation, and a variety of molded containers, as well as a nonstick coating for cooking utensils.

The vinyl group is $CH_2{=}CH{-}$. See Section 18.18.

$$ClCH_2{-}CH_2Cl \xrightarrow{\text{heat}} CH_2{=}CHCl + HCl$$

Vinyl chloride is then polymerized to form polyvinyl chloride, which has the structural unit

$$-CH_2-CHCl-$$

Polyvinyl chloride has a molecular weight of about 1,500,000. It is used for plastic pipe, plastic siding for buildings, film, wire insulation, and phonograph records.

Alcohols

19.4 General preparations of alcohols Alcohols are alkanes in which one or more *hydroxyl* groups, —OH, has been substituted for a like number of hydrogen atoms. (The covalent bonded hydroxyl group must not be confused with the ionic bonded hydroxide ion.)

1. *Hydration of alkenes.* Alkenes react with water in the presence of hydronium ions that are furnished by H_2SO_4 or H_3PO_4 as a catalyst; alcohols are formed. The reaction of ethene used commercially is

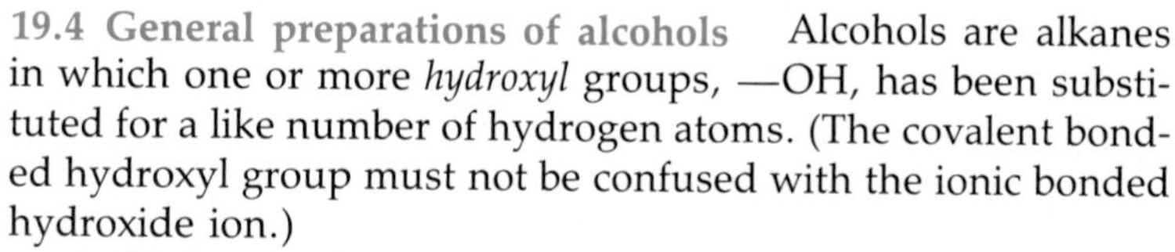

The effect is the addition of water across the ethene double bond. The product, C_2H_5OH, is called ethanol or ethyl alcohol.

2. *From alkyl halides.* You learned in Section 19.2 that alkyl halides react with water solutions of strong hydroxides to yield alcohols.

19.5 Preparation, properties, and uses of specific alcohols 1. *Methanol.* Methanol is prepared, by means of a catalyst, from carbon monoxide and hydrogen under pressure. See Section 17.24. Methanol is a colorless liquid with a rather pleasant odor. It has a low density and boils at 64.7 °C. *It is very poisonous, even when used externally.* If taken internally in small quantities, it causes blindness by destroying the cells of the optic nerve. Larger quantities cause death. Methanol is a good fuel, burning with a hot, smokeless flame. It is used as a solvent and is added

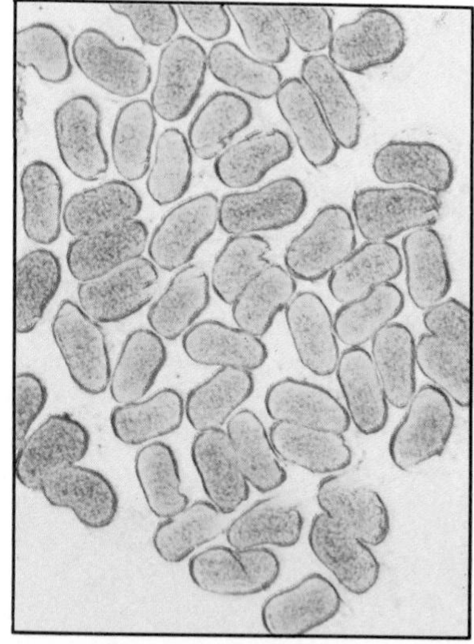

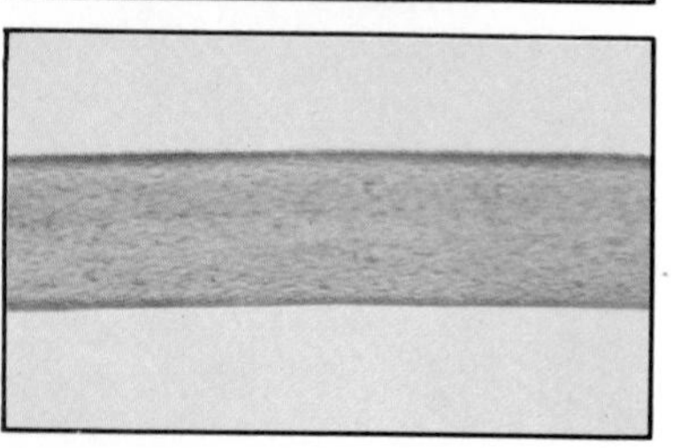

Figure 19-2. Teflon is a fluorocarbon. The fibers are formed of long-chain carbon molecules in which available bonds are saturated with fluorine.

Figure 19-3. Large quantities of methanol are produced in this plant (left photo). Methanol is prepared, by means of a catalyst, from carbon monoxide and hydrogen under pressure. Large quantities of ethanol are produced by hydrating ethene. Another method is fermentation. Both methanol and ethanol can be added to unleaded gasoline to form gasohol, a fuel that is used chiefly in automobile and truck engines (right photo).

to ethanol to make it unfit for drinking. It serves as a starting material for preparing other organic compounds such as formaldehyde and acetic acid. Methanol is also used in automobile gas tank de-icers. The methanol in these products prevents water that has condensed in the tank from freezing there as well as in the gas line and carburetor. Review Section 13.12.

2. *Ethanol.* Large quantities of ethanol are produced by hydrating ethene. Another method is fermentation. If yeast is added to a dilute solution of sugar or molasses at room temperature, chemical action soon occurs. The yeast plants secrete the enzymes sucrase and zymase. These enzymes act as catalysts in changing the sugar into alcohol and carbon dioxide.

$$C_{12}H_{22}O_{11} + H_2O \rightarrow 4CO_2 + 4C_2H_5OH$$

Both processes are used in producing industrial alcohol. However, hydration of ethene is a less expensive method. Another ethanol-preparation process now being developed involves catalytically reacting methanol with carbon monoxide and hydrogen. By this method, ethanol can be prepared starting with coal rather than petroleum.

Ethanol is a colorless liquid. It has a characteristic odor and a sharp, biting taste. It boils at 78 °C, freezes at −115 °C, and burns with a nearly colorless flame. Ethanol is a good solvent for many organic compounds that are insoluble in water. Accordingly, it is used for making a variety of solutions for medicinal use. Ethanol is also used for making ether and acetaldehyde.

Ethanol is the alcohol in alcoholic beverages. It affects the central nervous system and can damage the liver. *Denatured alcohol* is a mixture composed principally of ethanol. But other poisonous and nauseating materials have been added to it to make it unfit for beverage purposes. It can still be used for industrial purposes, however.

In ethylene glycol and other alcohols with more than one hydroxyl group, the hydroxyl groups are usually bonded to different carbon atoms. Two hydroxyl groups attached to the same carbon atom are mostly unstable. They may produce water and either a formyl group (Section 19.8) or a carbonyl group (Section 19.10).

Molecules of methanol and ethanol have been detected in interstellar space.

3. *Ethylene glycol.* The compound ethylene glycol, $C_2H_4(OH)_2$, has the structural formula

$$\begin{array}{ccccccc} & & H & & H & & \\ & & | & & | & & \\ H & — & C & — & C & — & H \\ & & | & & | & & \\ & & O & & O & & \\ & & | & & | & & \\ & & H & & H & & \end{array}$$

Ethylene glycol is an alcohol containing two hydroxyl groups. It may be prepared by first catalytically oxidizing ethene to ethene oxide.

$$\begin{array}{ccccccc} & & H & & H & & \\ & & | & & | & & \\ H & — & C & = & C & — & H \end{array} + \tfrac{1}{2}O_2 \xrightarrow{Ag} \begin{array}{ccccccc} & & H & & H & & \\ & & | & & | & & \\ H & — & C & — & C & — & H \\ & & \diagdown & O & \diagup & & \end{array}$$

Then ethene oxide is reacted with water in the presence of an acid to yield ethylene glycol.

$$\begin{array}{ccccccc} & & H & & H & & \\ & & | & & | & & \\ H & — & C & — & C & — & H \\ & & \diagdown & O & \diagup & & \end{array} + H_2O \xrightarrow{acid} \begin{array}{ccccccc} & & H & & H & & \\ & & | & & | & & \\ H & — & C & — & C & — & H \\ & & | & & | & & \\ & & O & & O & & \\ & & | & & | & & \\ & & H & & H & & \end{array}$$

Ethylene glycol is used extensively as a "permanent" antifreeze in automobile radiators. Its boiling point (198 °C) is so much higher than that of water that it does not readily evaporate or boil away. Ethylene glycol is also used in the production of one type of polyester fiber and film. Ethylene glycol is poisonous.

4. *Glycerol.* Glycerol, or glycerin, $C_3H_5(OH)_3$, has the structural formula

$$\begin{array}{ccccccc} & & H & & & & \\ & & | & & & & \\ H & — & C & — & O & — & H \\ & & | & & & & \\ H & — & C & — & O & — & H \\ & & | & & & & \\ H & — & C & — & O & — & H \\ & & | & & & & \\ & & H & & & & \end{array}$$

Hygroscopic materials are described in Section 12.19.

It is a colorless, odorless, slow-flowing liquid with a sweet taste. It has a low vapor pressure and is hygroscopic. It is used in making synthetic resins for paints and in cigarettes to keep the tobacco moist. It is also used in the manufacture of cellophane, nitroglycerin, and some toilet soaps. Glycerol is an ingredient of cosmetics and drugs and is used in many foods and beverages. It is a by-product of soap manufacture. Large amounts of glycerol are synthesized from propene, a product of petroleum cracking.

Figure 19-4. Glycerol is often an ingredient of cosmetics. Large quantities of glycerol are synthesized from propene.

19.6 Reactions of alcohols 1. *With sodium.* Sodium reacts vigorously with ethanol, releasing hydrogen. A second product of the reaction is sodium ethoxide, C_2H_5ONa. This compound is recovered as a white solid after the excess ethanol is evaporated. This reaction is similar to the reaction of sodium with water.

$$2C_2H_5OH + 2Na \rightarrow 2C_2H_5ONa + H_2$$

2. *With HX.* Alcohols react with concentrated water solutions of hydrogen halides, particularly hydrobromic, HBr, and hydriodic, HI, acids. These reactions produce alkyl halides. Sulfuric acid is used as a dehydrating agent.

$$ROH + HBr \rightarrow RBr + H_2O$$

3. *Dehydration.* Depending on reaction conditions, ethanol dehydrated by hot, concentrated sulfuric acid yields either diethyl ether, $C_2H_5OC_2H_5$, or ethene, C_2H_4.

$$2C_2H_5OH \rightarrow C_2H_5OC_2H_5 + H_2O$$

$$C_2H_5OH \rightarrow C_2H_4 + H_2O$$

4. *Oxidation.* Alcohols that have the hydroxyl group attached to the end carbon can be oxidized by hot copper(II) oxide. The product of such an oxidation is an aldehyde, RCHO.

$$RCH_2OH + CuO \rightarrow RCHO + H_2O + Cu$$

Low-molecular-weight alcohols are flammable and burn readily in air.

$$2CH_3OH + 3O_2 \rightarrow 2CO_2 + 4H_2O$$

5. *Sulfation of long-chain alcohols.* 1-dodecanol, $C_{12}H_{25}OH$, is commonly called lauryl alcohol. It is obtained from coconut oil by hydrogenation and partial decomposition. Lauryl alcohol may be *sulfated* by treatment with sulfuric acid and then neutralized with sodium hydroxide. This process yields sodium lauryl sulfate, $C_{12}H_{25}OSO_2ONa$, a very effective cleaning agent.

$$C_{12}H_{25}OH + H_2SO_4 \rightarrow C_{12}H_{25}OSO_2OH + H_2O$$

$$C_{12}H_{25}OSO_2OH + NaOH \rightarrow C_{12}H_{25}OSO_2ONa + H_2O$$

Do not confuse the similar-sounding terms sulfonation *[Section 18.21(3)] and* sulfation. *Sulfonation is the substitution of an —SO_2OH group for a hydrogen atom bonded to a carbon atom. The carbon-hydrogen bond is replaced by a carbon-sulfur bond. On the other hand, sulfation is the substitution of an —SO_2OH group for a hydrogen atom bonded to an oxygen atom. The oxygen-hydrogen bond is replaced by an oxygen-sulfur bond.*

ETHERS

19.7 Ethers: organic oxides Ethers have the general formula **ROR′**. **R** and **R′** represent the same or different alkyl groups. Ethers may be thought of as structurally similar to water, but both hydrogen atoms are replaced by alkyl groups. Ethers can be prepared by dehydrating alcohols as described in Section 19.6. Diethyl ether is commonly known as *ether*. It is a volatile, very flammable, colorless liquid of characteristic odor. It can be made by heating ethanol and sulfuric acid to 140 °C.

$$2C_2H_5OH \xrightarrow{H_2SO_4} C_2H_5OC_2H_5 + H_2O$$

Ether slows down the operation of the central nervous system; this accounts for its former use as an anesthetic. It is also used as a solvent for fats and oils. Ether molecules have been found in outer space.

One of the methyl butyl ethers, $(CH_3)_3COCH_3$, is used in gasoline to increase the octane rating. It has the advantage of being lead-free. This ether is made by catalytically reacting methanol with 2-methylpropene, a gas produced during petroleum cracking.

ALDEHYDES

19.8 Preparations of aldehydes An aldehyde is a compound that has a hydrocarbon group and one or more *formyl* groups, $-C(=O)H$. The general formula for an *aldehyde* is **RCHO.**

The general method of preparing aldehydes was mentioned in Section 19.6. A common process of this type involves passing methanol vapor and a regulated amount of air over heated copper. Formaldehyde, HCHO, is produced.

$$2Cu + O_2 \rightarrow 2CuO$$

$$CH_3OH + CuO \rightarrow HCHO + H_2O + Cu$$

The commercial preparation of formaldehyde involves a silver or an iron-molybdenum oxide catalyst. At room temperature, formaldehyde is a gas with a strangling odor. Dissolved in water, it makes an excellent disinfectant. It is also used to preserve biological or medical specimens. By far the largest uses of formaldehyde are in making certain types of adhesives for plywood and particle board and resins for plastics.

Acetaldehyde, CH_3CHO, can be made by the oxidation of ethanol. Commercially, it is produced directly from ethene by reacting ethene with oxygen, water, and hydrochloric acid in the presence of a $PdCl_2$–$CuCl_2$ catalyst. Acetaldehyde is a stable liquid used in preparing acetic acid and other organic compounds. Both formaldehyde and acetaldehyde molecules are found in outer space.

1. *Oxidation.* The mild oxidation of an aldehyde produces the organic acid having the same number of carbon atoms. The oxidation of acetaldehyde to acetic acid is typical.

$$CH_3CHO + O \text{ (from oxidizing agent)} \rightarrow CH_3COOH$$

2. *Fehling's test.* Fehling's solution A is copper(II) sulfate solution. Fehling's solution B is sodium hydroxide and sodium tartrate solution. If these are mixed with an aldehyde and heated, the aldehyde is oxidized. The copper(II) ion is reduced to copper(I) and precipitated as brick-red copper(I) oxide.

$$RCHO + 2CuSO_4 + 5NaOH \rightarrow RCOONa + Cu_2O + 2Na_2SO_4 + 3H_2O$$

3. *Hydrogen addition.* Adding hydrogen to aldehydes in the presence of finely divided nickel or platinum produces alcohols.

$$RCHO + H_2 \xrightarrow{Ni} RCH_2OH$$

This reaction is the reverse of the oxidation of alcohols to aldehydes, as mentioned in Section 19.6.

KETONES

Organic compounds containing the *carbonyl* group, $\rangle C{=}O$, and having the general formula **RCOR′** are *ketones*.

Ketones can be prepared from alcohols that do *not* have the hydroxyl group attached to an end-carbon atom. For example, acetone, CH_3COCH_3, is prepared by the mild oxidation of 2-propanol, $CH_3CHOHCH_3$. Potassium dichromate, $K_2Cr_2O_7$, in water solution is the oxidizing agent.

$$\underset{\text{2-propanol}}{H{-}\overset{H}{\underset{H}{C}}{-}\overset{O{-}H}{\underset{H}{C}}{-}\overset{H}{\underset{H}{C}}{-}H} + O \xrightarrow{\text{(from oxidizing agent)}} \underset{\text{acetone}}{H{-}\overset{H}{\underset{H}{C}}{-}\overset{O}{\overset{\|}{C}}{-}\overset{H}{\underset{H}{C}}{-}H} + H_2O$$

Acetone is a colorless, volatile liquid. It is widely used as a solvent in the manufacture of acetate rayon. Storage tanks for ethyne gas are loosely filled with asbestos saturated with acetone. The ethyne dissolves in the acetone and, by so doing, occupies less volume. This procedure increases the amount of ethyne that can safely be compressed into the tank. Acetone and other

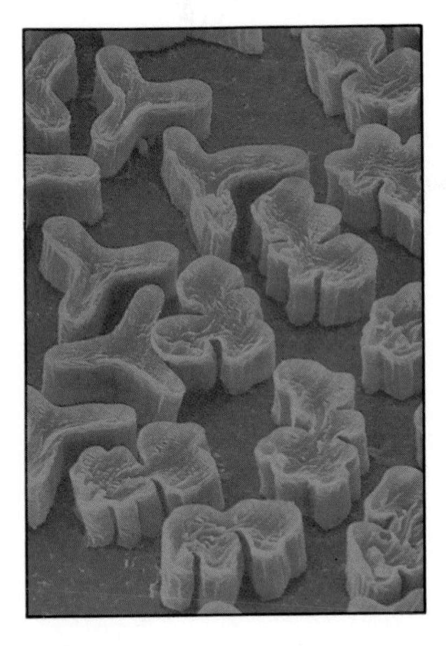

Figure 19-5. Cross-sections of cellulose acetate fibers. Acetone is used as a solvent in the production of cellulose acetate (acetate rayon) fibers.

ketones are used for cleaning metals, removing stains, and preparing synthetic organic chemicals. Acetone is a digestive product of untreated diabetics.

Carboxylic Acids and Esters

19.11 Preparations of carboxylic acids Many organic acids and their salts occur naturally. They are found in sour milk, unripe fruits, rhubarb, sorrel, and other plants. All organic acids contain the *carboxyl* group,

$$-C\overset{\displaystyle O}{\underset{\displaystyle O-H}{\gtrless}}$$

The general formula for a *carboxylic acid* is **RCOOH.**

1. *Oxidation of alcohols or aldehydes.* The oxidation of alcohols to aldehydes and of aldehydes to carboxylic acids was described in Sections 19.6 and 19.9. Acetic acid can be produced by the catalytic oxidation of acetaldehyde. A new method of preparing very pure acetic acid is by catalytically reacting methanol with carbon monoxide at 180 °C and 30–40 atm pressure.

$$CH_3OH + CO \xrightarrow{catalyst} CH_3COOH$$

Concentrated acetic acid is a colorless liquid that is a good solvent for some organic chemicals. It is used for making cellulose acetate, a basic material of many fibers and films.

Cider vinegar is made from apple cider that has fermented to hard cider. The ethanol in hard cider is slowly oxidized by the oxygen of the air. This oxidation produces acetic acid. The reaction is catalyzed by enzymes from certain bacteria.

$$C_2H_5OH + O_2 \rightarrow CH_3COOH + H_2O$$

Vinegar contains from 4% to 6% acetic acid.

2. *Formic acid.* Formic acid, $HCOOH$, is prepared from sodium hydroxide solution and carbon monoxide under pressure. This reaction yields sodium formate, $HCOONa$.

$$NaOH + CO \rightarrow HCOONa$$

If sodium formate is carefully heated with sulfuric acid, formic acid distills off.

$$HCOONa + H_2SO_4 \rightarrow HCOOH + NaHSO_4$$

Formic acid is found in nature in stinging nettle plants and in certain ants. It is one of the molecules found in outer space. Formic acid is used as an acidifying agent in the textile industry.

Resonance is described in Section 6.22.

Formic acid molecules, as shown in the top margin at right,

contain two carbon-oxygen bonds. One of them, the double bond, is 1.23 Å long. The other, the single bond, is 1.36 Å long. This evidence supports the idea that these bonds are different. In sodium formate, an ionic compound, the formate ion also has two carbon-oxygen bonds. But these bonds have the same length, 1.27 Å. What explanation is there for this difference? Because the carbon-oxygen bonds in the formate ion are the same length, they should be similar. The formate ion must be a resonance hybrid of two structures, as shown in the margin at right, and have carbon-oxygen bonds midway in character between single and double covalent bonds. The anions of other carboxylic acids have similar resonance hybrid structures.

$$H{-}C({=}O){-}O{-}H$$

$$\left\{ H{-}C({=}O){-}O^- \qquad H{-}C(-O^-){=}O \right\}$$

19.12 Reactions of carboxylic acids *1. Ionization.* This reaction involves the one hydrogen atom bonded to an oxygen atom in the carboxyl group. This hydrogen atom ionizes in water solution, giving carboxylic acids their acid properties. The hydrogen atoms bonded to carbon atoms in these acids *never* ionize in water solution.

$$HCOOH + H_2O \rightleftarrows H_3O^+ + HCOO^-$$

$$CH_3COOH + H_2O \rightleftarrows H_3O^+ + CH_3COO^-$$

These equilibria yield low H_3O^+ ion concentrations. Therefore, carboxylic acids are generally weak acids.

2. *Neutralization.* An organic acid may be neutralized by a hydroxide. A salt is formed in a reaction similar to that of inorganic acids.

$$CH_3COOH + NaOH \rightarrow CH_3COONa + H_2O$$

3. *Esterification. An* **ester** *is produced when an acid reacts with an alcohol.* Such reactions are called *esterification reactions.* For example, ethyl acetate is the ester formed when ethanol and acetic acid react. The general formula for an ester is **RCOOR′.**

$$CH_3COOH \quad + \quad C_2H_5OH \xrightarrow{H_2SO_4} CH_3COOC_2H_5 \quad + H_2O$$

$$H{-}\underset{H}{\overset{H}{C}}{-}C({=}O){-}O{-}H \; + \; H{-}O{-}\underset{H}{\overset{H}{C}}{-}\underset{H}{\overset{H}{C}}{-}H \xrightarrow{H_2SO_4} H{-}\underset{H}{\overset{H}{C}}{-}C({=}O){-}O{-}\underset{H}{\overset{H}{C}}{-}\underset{H}{\overset{H}{C}}{-}H \; + H_2O$$

All such reactions between acids and alcohols are reversible. The achievement of equilibrium is slow, and a small amount of sulfuric acid is used as a catalyst. Experiments with alcohols containing oxygen-18 show that the oxygen of the water product comes from the acid.

Figure 19-6. The flavor of pineapples is due mostly to the presence of ethyl butyrate in the juice.

19.13 Esters It is also possible to prepare esters by the reaction of alcohols with inorganic acids. Glyceryl trinitrate, known as nitroglycerin, is an example.

$$C_3H_5(OH)_3 + 3HNO_3 \xrightarrow{H_2SO_4} C_3H_5(NO_3)_3 + 3H_2O$$

Esters give fruits their characteristic flavors and odors. Amyl acetate, $CH_3COOC_5H_{11}$, has an odor somewhat resembling bananas. Commonly called "banana oil," this ester is used as the carrier for some aluminum paints. Ethyl butyrate, $C_3H_7COOC_2H_5$, has an odor and flavor that resemble pineapples. Ripe pineapples contain some of this ester and smaller amounts of other esters.

Esters can be decomposed by hydrolysis into the alcohol and acid from which they were derived. This hydrolysis may occur in the presence of dilute acid or metallic hydroxide solutions.

19.14 Saponification Fats and oils from plants or animals are esters of glycerol and long-carbon-chain acids. The carbon chains of the acids usually contain from 12 to 20 carbon atoms. The structure of a fat or oil can be represented as

$$\begin{array}{c} RCOOCH_2 \\ | \\ R'COOCH \\ | \\ R''COOCH_2 \end{array}$$

R, R′, and **R″** are saturated or unsaturated long-chain-hydrocarbon groups.

The only difference between a fat and an oil is the physical phase of each at room temperature. Oils are liquids at room temperature; fats are solids. Long-carbon-chain acids with double bonds produce esters with lower melting points. Hence, oils usually contain more unsaturated hydrocarbon chains than fats do.

Saponification *is the hydrolysis of a fat using a solution of a strong hydroxide.* Alkaline hydrolysis produces the sodium salt of the long-chain carboxylic acid instead of the acid itself.

$$\begin{array}{c} RCOOCH_2 \\ | \\ R'COOCH \\ | \\ R''COOCH_2 \end{array} + 3NaOH \rightarrow RCOONa + R'COONa + R''COONa + C_3H_5(OH)_3$$

Soaps are generally made by hydrolyzing fats and oils with water heated to about 250 °C in a closed container. The water is under a pressure of about 50 atmospheres. At this pressure, water does not boil despite the temperature, which is well above its normal boiling point. The long-chain carboxylic acids thus produced are neutralized with sodium hydroxide. This neutralization yields a mixture of sodium salts that makes up soap.

If the acid chains are unsaturated, a soft soap results. Soaps with hydrocarbon chains of 10 to 12 carbon atoms are soluble in water and produce a large-bubble lather. Soaps containing hydrocarbon chains of 16 to 18 carbon atoms are less soluble in water and give a longer-lasting, small-bubble lather. Soap that is a mixture of potassium salts, rather than sodium salts, is generally more soluble in water.

SUMMARY

Hydrocarbon substitution products are those in which an atom such as chlorine or a group such as hydroxyl is substituted for one or more of the hydrogen atoms in a hydrocarbon.

An alkyl halide is an alkane in which a halogen atom is substituted for a hydrogen atom. The general formula for an alkyl halide is RX. Alkyl halides are prepared by the reaction of halogens with alkanes, alkenes, and alkynes and by the reaction of hydrogen halides with alkenes, alkynes, and alcohols. Many reactions of alkyl halides are substitution reactions. Examples of alkyl halides are carbon tetrachloride, the Freons, Teflon, and vinyl chloride.

Alcohols are alkanes in which one or more hydroxyl groups have been substituted for a like number of hydrogen atoms. The general formula is ROH. They are prepared by the hydration of alkenes and from alkyl halides by substitution. Important alcohols are methanol, ethanol, ethylene glycol, and glycerol. Alcohols react with sodium and with hydrogen halides. They also undergo dehydration, oxidation, and sulfation reactions.

Ethers are organic oxides with the general formula ROR′. They can be prepared by dehydrating alcohols. Diethyl ether is an example.

Aldehydes have a hydrocarbon group and one or more formyl groups. The general formula is RCHO. They are prepared by the controlled oxidation of alcohols having a hydroxyl group on an end carbon. Formaldehyde and acetaldehyde are examples. Aldehydes undergo oxidation to acids and reduction with hydrogen to alcohols. They give positive Fehling's tests.

Ketones are organic compounds containing the carbonyl group. They have the general formula RCOR′. Ketones are prepared by oxidizing alcohols that do not have the hydroxyl group attached to the end-carbon atom. Acetone is an important ketone.

The general formula for carboxylic acids is RCOOH. They can be prepared by the oxidation of alcohols and aldehydes. Formic acid and acetic acid are familiar examples. Organic acids undergo ionization, neutralization, and esterification reactions. An ester is produced when an acid reacts with an alcohol. The general formula is RCOOR′. Esters give fruits their characteristic flavors and odors. Saponification is the hydrolysis of a fat using a solution of a strong hydroxide, such as sodium hydroxide. This hydrolysis reaction yields a mixture of sodium salts that makes up soap.

VOCABULARY

alcohol
aldehyde
alkyl halide
carbonyl group
carboxyl group
carboxylic acid
denatured alcohol
ester
esterification
ether
fat
formyl group
hydroxyl group
ketone
oil
saponification

QUESTIONS

Group A

1. Write the general formula for a(n) *(a)* alkyl halide; *(b)* alcohol; *(c)* ether; *(d)* aldehyde; *(e)* ketone; *(f)* carboxylic acid; *(g)* ester; *(h)* fat. *A5*
2. Define *(a) sulfation; (b) ionization; (c) neutralization; (d) esterification; (e) saponification.* *A1*
3. *(a)* What is an important physical property of carbon tetrachloride? *(b)* What uses of carbon tetrachloride depend on this property? *A3a*
4. *(a)* What precautions must be taken when using carbon tetrachloride? *(b)* Why? *A3c*
5. Draw a structural formula showing the portion of a polymer formed from three molecules of vinyl chloride. *A6*
6. During 1983–1984, when this book was being written, there was much controversy over the danger of exposure to 1,2-dibromoethane (ethylene dibromide, EDB) used to protect fruit and grain from insects. What action has been taken on this problem since then?
7. Alcohols and metallic hydroxides both contain **—OH** groups. Why are their chemical properties so different? *A3a*
8. How does methanol affect the body? *A3c*
9. What is *denatured alcohol?* *A1*
10. Why is glycerol an ingredient in many moisturizing skin lotions? *A3a*
11. Write the balanced formula equation for the reaction of sodium with *(a)* water; *(b)* methanol. *A4c*
12. How many molecules of oxygen are required for the complete combustion of one molecule of 2-propanol, $CH_3CHOHCH_3$? This compound is commonly called isopropyl alcohol. *A4c*
13. Give the two most important uses for formaldehyde. *A3a*
14. *(a)* Describe the Fehling's test. *(b)* What organic group gives a positive Fehling's test? *A3a*
15. Give five uses for acetone. *A3a*
16. How is vinegar made from apple cider? *A4b*
17. *(a)* Oxalic acid is a dicarboxylic acid with the structural formula $\mathbf{H{-}O{-}\overset{\overset{\Large O}{\|}}{C}{-}\overset{\overset{\Large O}{\|}}{C}{-}O{-}H}$. Write equations for the step-by-step complete ionization of oxalic acid. *(b)* How many moles of sodium hydroxide are required for the complete neutralization of three moles of oxalic acid? *A4c*
18. What is the source of the hydrogen and oxygen atoms of the water eliminated during an esterification reaction? *A4c*
19. List the types of reactions given in this chapter in which sulfuric acid is used as a dehydrating agent. *A4*
20. *(a)* How are fats and oils alike? *(b)* How do they differ? *(c)* Why do they differ? *A1*

Group B

21. Draw structural formulas for *(a)* trichloromethane (chloroform); *(b)* 1,3-dihydroxypropane; *(c)* diethyl ether; *(d)* acetaldehyde; *(e)* methylethylketone; *(f)* acetic acid; *(g)* ethyl formate. *A5, A6*
22. Write an equation for the preparation of ethyl bromide starting with *(a)* ethane; *(b)* ethene; *(c)* ethanol. *A4a*
23. Describe the types of alkyl halide substitution reactions given in this chapter. *A4c*
24. Starting with methane, chlorine, and hydrogen fluoride, write equations for the preparation of dichlorodifluoromethane. *A4b*
25. *(a)* What is believed to be the effect of Freon-type compounds in the stratosphere? *(b)* Consult current sources to learn the extent to which the concentration of ozone in the upper atmosphere has changed recently and the reason scientists give for the change. *A2a*
26. Using structural formula equations, show how ethanol may be prepared from *(a)* ethene; *(b)* ethyl chloride; *(c)* sugar ($C_{12}H_{22}O_{11}$); *(d)* methanol. *A4b*

27. On the basis of molecular weight and boiling point, compare the advantages of methanol, ethanol, and ethylene glycol as antifreezes. *A3a*

28. Why is a 50%–50% mixture of ethylene glycol and water better than water alone as a radiator fluid for extremely hot summer driving conditions? *A3a*

29. Write an equation for the preparation of propyl iodide from propyl alcohol. *A4a*

30. Write a formula equation for the preparation of propyl ether from propyl alcohol. *A4a*

31. *(a)* Write an equation for the mild oxidation of 2-butanol, $CH_3CHOHCH_2CH_3$. *(b)* Name the product formed. *A4c*

32. Write a formula equation showing the preparation of acetic acid from *(a)* methanol; *(b)* ethanol. *A4b*

33. *(a)* How do the two carbon-oxygen bonds in formic acid compare? *(b)* How do the two carbon–oxygen bonds in sodium formate compare? *(c)* Account for the difference. *A1, A3a*

34. The ester *n*-butyl acetate is to be prepared from *n*-butyl alcohol, C_4H_9OH, and acetic acid. Write the equation for the preparation. *A4a*

35. Write a balanced formula equation for the saponification with NaOH of a fat having the formula $(C_{17}H_{35}COO)_3C_3H_5$. *A4b*

PROBLEMS

Group A

1. Calculate the molecular weight of trichlorofluoromethane.

2. Sodium hydroxide solution (45.0 mL) exactly neutralizes 35.0 mL of 0.150-*N* formic acid solution. What is the normality of the sodium hydroxide solution?

3. Calculate the number of grams of glycerol that must be dissolved in 0.250 kg of water in order to prepare a 0.400-*m* solution.

4. Calculate the freezing point and boiling point of the water of Problem 3.

5. How many grams of diprotic oxalic acid, $(COOH)_2$, are required to prepare 1.75 L of 0.200-*N* acid?

Group B

6. What is the percentage composition of $C_{12}H_{25}OSO_2ONa$, sodium lauryl sulfate?

7. A compound is found to contain 54.5% carbon, 9.1% hydrogen, and 36.4% oxygen. *(a)* Determine the empirical formula. *(b)* The compound is soluble in ethanol. If 11.0 g of the compound is dissolved in $25\bar{0}$ g of ethanol, the ethanol boils at 78.87 °C. What is the molecular weight of the compound? *(c)* What is the molecular formula? Consult Table 13-5 for additional data.

8. The hydronium ion concentration in 0.05-*M* acetic acid is 9.4×10^{-4} mole/L. What is the pH of this solution?

9. What volume of ethanol must be diluted with water to prepare $50\bar{0}$ mL of 0.750-*M* solution? The density of ethanol is given as 0.789 g/mL.

10. The density of ethylene glycol is 1.432 g/mL. *(a)* Calculate the theoretical freezing point of the water in a 50% by volume solution of ethylene glycol. *(b)* The actual freezing point of such a solution is about −37 °C. Account for any difference from your calculation.

In 1977, the Nobel Prize in chemistry was given to Ilya Prigogine (Belgian) for his contributions to nonequilibrium thermodynamics. He was born in Moscow, studied at the University of Brussels in Belgium, and then taught in Belgium and the United States. Prigogine's research in physical chemistry has covered several complex areas, such as the molecular theory of solutions, and the thermodynamics of irreversible phenomena, as well as nonequilibrium statistical mechanics.

Reaction Energy and Reaction Kinetics

20

GOALS

In this chapter you will gain an understanding of: • the term *reaction mechanism* • the relationship between the heat of reaction and the terms *endothermic* and *exothermic,* and *stable* and *unstable* • solving problems to determine the heat of formation, heat of combustion, and heat of reaction • writing thermochemical equations for reactions • the factors that contribute to the driving force of chemical reactions • the terms *free energy, entropy,* and *enthalpy* • reaction mechanisms and the role of the collision theory • the difference between activation energy and activated complex • the five major factors that influence reaction rates • the reaction rate law

20.1 Introduction Chemical equations are often written to represent reactions between different molecular species. The equations show the initial reactants and the final products. They do not show how, during the reaction process, reactant molecules become product molecules. For example, gaseous iodine and hydrogen may react chemically to form gaseous hydrogen iodide. The equation for this reaction is

$$H_2(g) + I_2(g) \rightarrow 2HI(g)$$

This equation makes it easy to visualize molecules of hydrogen and iodine colliding in just the right way and then separating as hydrogen iodide molecules. During the extremely brief encounter, H—H and I—I bonds must be broken, and H—I bonds must be formed.

Chemists are interested in learning how reacting molecules change as a reaction proceeds. New techniques have been developed for studying molecular changes, even those occurring in fast reaction systems. These methods explore the rates of chemical reactions and the pathways along which they occur. The branch of chemistry that is concerned with *reaction rates* and *reaction pathways* is called *chemical kinetics.* The pathway from reactants to products may be a sequence of simple steps called the *reaction mechanism.*

When reaction systems are investigated, the role of energy in the systems is of primary interest. Every substance has a characteristic internal energy because of its structure and physical state. Definite quantities of energy are released or absorbed when new substances are formed from reacting substances. Energy changes occur even when products and reactants are both at the same temperature. This energy is related to the breaking and forming of bonds and to the strengths of these bonds.

The first part of this chapter deals with *thermochemistry*, the changes in heat energy that accompany chemical reactions. Later in the chapter, the modern theories of reaction pathways and reaction rates are introduced.

ENERGY OF REACTION

20.2 Heat of reaction Chemical reactions are either exothermic or endothermic processes. During a reaction, a certain amount of chemical binding energy is changed into thermal (internal) energy, or vice versa. Usually the energy change can be measured as heat released or absorbed during the reaction. That is, the energy change can be measured as the *change in heat content* of the substances reacting.

Heat is thermal energy in the process of being added to or removed from a substance during a reaction.

Chemical reactions are generally carried out in open vessels. Volumes may change in such vessels, but pressures remain constant. The heat content of a substance under constant pressure is often called *enthalpy* of the substance. The symbol for enthalpy (and for heat content at constant pressure) is H. If a process is exothermic, the total heat content of the products is *lower* than that of the reactants. The products of an endothermic reaction must have a *higher* heat content than the reactants.

One mole of a substance has a characteristic heat content, just as it has a characteristic mass. The heat content is a measure of the internal energy stored in the substance during its formation. This stored heat content *cannot* be measured *directly*. However, the *change* in heat content that occurs during chemical reaction *can* be measured. This quantity is the heat released during an exothermic change or the heat absorbed during an endothermic change. It is called *heat of reaction. The* ***heat of reaction*** *is the quantity of heat released or absorbed during a chemical reaction.*

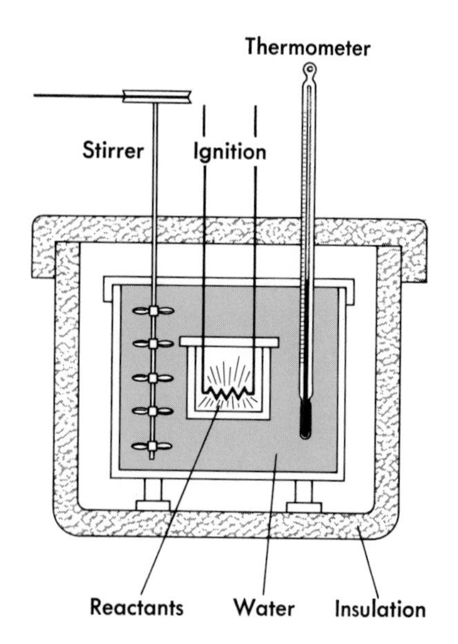

Figure 20-1. Schematic diagram of an ignition calorimeter used for measuring nutritional calories.

The heat of reaction is measured when the final state of a system is brought to the same temperature as that of the initial state. Unless otherwise stated, the reaction is assumed to be at 25 °C under standard atmospheric pressure. In addition, each substance is assumed to be in its usual (standard) state at these conditions. For this reason, the phases of reactants and products should be shown along with their formulas in thermochemical equations.

If a mixture of hydrogen and oxygen is ignited, water is formed and heat energy is released. Because this reaction occurs as an explosion, it is carried out in a special *calorimeter*, as shown in Figure 20-1. Known quantities of reactants are sealed in the reaction chamber, which is immersed in a known quantity of water in an insulated vessel. The heat given off (or absorbed) during the reaction is determined from the temperature change of the known mass of water.

1 kcal of heat is required to change the temperature of 1 kg of water 1 C°.

The quantity of heat given up is proportional to the quantity of water formed in the reaction. No heat is supplied externally, except to ignite the mixture. For this reason, the heat content of the product water must be less than the heat content of the reactants before ignition. The equation for this reaction is ordinarily written as

$$2H_2 + O_2 \rightarrow 2H_2O$$

and the following fact becomes evident: when 2 moles of hydrogen gas at room temperature is burned, 1 mole of oxygen gas is used, and 2 moles of water vapor is formed.

After the product water is brought back to room temperature (the temperature of the initial state), the reaction heat given up by the system is found to be 136.64 kcal. The thermochemical equation can now be written as

$$2H_2(g) + O_2(g) \rightarrow 2H_2O(l) + 136.64 \text{ kcal}$$

The letter symbols (g), (l), *and* (s) *identify gas, liquid, and solid phases, respectively, in an equation.*

Heats of reaction are usually expressed in terms of *kilocalories per mole* of substance. From the previous equation, the following equality can be stated.

$$\begin{matrix}\text{heat content} & & \text{heat content} & & \text{heat content} & & \\ \text{of 1 mole of} & + & \text{of } \frac{1}{2} \text{ mole of} & = & \text{of 1 mole of} & + & 68.32 \text{ kcal} \\ \text{hydrogen gas} & & \text{oxygen gas} & & \text{liquid water} & & \end{matrix}$$

The thermochemical equation can now indicate the heat of reaction in kcal/mole of product:

$$H_2(g) + \tfrac{1}{2}O_2(g) \rightarrow H_2O(l) + 68.32 \text{ kcal}$$

This equation shows that one mole of the product (liquid) water has a heat content 68.32 kcal *lower* than that of the two gaseous reactants before the reaction occurred. To decompose one mole of water to produce one mole of hydrogen and one-half mole of oxygen, this quantity of energy must be supplied from an external source. The decomposition reaction, which is the reverse of the composition reaction described earlier, is written

$$H_2O(l) + 68.32 \text{ kcal} \rightarrow H_2(g) + \tfrac{1}{2}O_2(g)$$

Here the products, 1 mole of hydrogen and ½ mole of oxygen, have a total heat content 68.32 kcal *higher* than that of the 1 mole of water before the reaction. The two reactions can be represented by the reversible equation

$$H_2(g) + \tfrac{1}{2}O_2(g) \rightleftarrows H_2O(l) + 68.32 \text{ kcal}$$

If the *heat content* of a substance is symbolized by the letter H, then the *change in heat content* can be represented by ΔH. The

Greek letter Δ (delta) signifies "change in." Thus the change in heat content, ΔH, during a reaction is the difference between the heat content of products and the heat content of reactants.

$$\Delta H = \text{heat content of products} - \text{heat content of reactants}$$

Exothermic processes: ΔH is negative.

Endothermic processes: ΔH is positive.

In this notation scheme, the ΔH for an exothermic reaction has a *negative* sign. Thus, in the preceding reaction

$$\Delta H = -68.32 \text{ kcal/mole}$$

The thermochemical equation

$$H_2(g) + \tfrac{1}{2}O_2(g) \rightarrow H_2O(l) + 68.32 \text{ kcal}$$

has the same meaning as

$$H_2(g) + \tfrac{1}{2}O_2(g) \rightarrow H_2O(l) \qquad \Delta H = -68.32 \text{ kcal}$$

The ΔH for an endothermic reaction is signified by using a *positive* value. This sign convention is an arbitrary one, but it is logical since the heat of reaction is said to be *negative* when the heat content of the system is *decreasing* (exothermic reaction). It is said to be *positive* when the heat content of the system is *increasing* (endothermic reaction). See Figure 20-2.

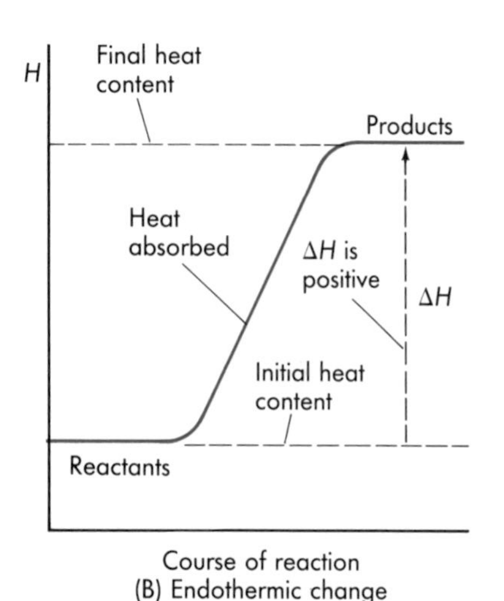

Figure 20-2. Change in heat content during a chemical reaction.

20.3 Heat of formation Chemical reactions in which elements combine to form compounds are generally exothermic. In these composition reactions the product compounds have lower heat contents than their separate elements, the reactants. The products are also more stable than the uncombined reactants. The formation of water from hydrogen and oxygen illustrates this change in stability.

Elemental hydrogen and oxygen exist as nonpolar diatomic molecules. Water molecules are covalent structures. They are polar because they are bent molecules with an electronegativity difference between the hydrogen and oxygen atoms.

Energy was given up when the covalent bonds of the diatomic hydrogen and oxygen molecules were originally formed. Therefore, energy is required to break these bonds if the hydrogen and oxygen atoms are to combine. On the other hand, more energy is released when the two polar bonds of the water molecule are formed. Heat is given off during this composition reaction. The heat of reaction is quite high ($\Delta H = -68.32$ kcal/mole of water formed), suggesting that water molecules are more stable than hydrogen and oxygen molecules.

The heat released or absorbed in a composition reaction is a useful indicator of product stability. It is referred to as the *heat of formation* of the compound. *The heat of reaction released or absorbed when 1 mole of a compound is formed from its elements is called the* ***molar heat of formation*** *of the compound.* By convention, each element in its standard state is said to have a heat content of zero. Then the ΔH during the formation of one mole of a com-

Table 20-1
HEAT OF FORMATION

ΔH_f = heat of formation of the substance from its elements. All values of ΔH_f are expressed as kcal/mole at 25 °C. Negative values of ΔH_f indicate exothermic reactions. (s) = solid, (l) = liquid, (g) = gas.

Substance	*Formula*	ΔH_f
ammonia (g)	NH_3	−11.02
barium nitrate (s)	$Ba(NO_3)_2$	−237.11
benzene (l)	C_6H_6	+11.71
calcium chloride (s)	$CaCl_2$	−190.2
carbon (diamond) (s)	C	+0.45
carbon (graphite) (s)	C	0.00
carbon dioxide (g)	CO_2	−94.05
copper(II) sulfate (s)	$CuSO_4$	−184.36
ethyne (acetylene) (g)	C_2H_2	+54.19
hydrogen chloride (g)	HCl	−22.06
water (l)	H_2O	−68.32
nitrogen dioxide (g)	NO_2	+7.93
ozone (g)	O_3	+34.1
sodium chloride (s)	NaCl	−98.23
sulfur dioxide (g)	SO_2	−70.94
zinc sulfate (s)	$ZnSO_4$	−234.9

pound from its elements (all in their standard states) is its standard heat of formation. Table 20-1 gives the heats of formation of some common compounds. A more complete list is given in Appendix Table 14.

The sign convention and ΔH notation adopted for heat of reaction apply to heat of formation as well. Heat of formation is merely one category of reaction heats. To distinguish a particular reaction heat as a heat of formation, the more specific notation ΔH_f is used. Heats of formation have negative values for exothermic composition reactions. They have positive values for endothermic composition reactions.

Observe that most of the heats of formation given in Appendix Table 14 are negative. Only a few compounds, such as hydrogen iodide and carbon disulfide, have positive heats of formation.

20.4 Stability and heat of formation A large amount of energy is released when a compound with a high negative heat of formation is formed. The same amount of energy is required to decompose such a compound into its separate elements. This energy must be supplied to the reaction from an external source. *Such compounds are very stable.* The reactions forming them proceed spontaneously, once they start, and are usually vigorous. Carbon dioxide has a high heat of formation. The ΔH_f of carbon dioxide is −94.05 kcal per mole of gas produced.

Compounds with positive and low negative values of heats of formation are generally unstable. Hydrogen sulfide, H_2S, has a heat of

The stability of a compound is related to its ΔH.

formation of −4.82 kcal per mole. It is not very stable and decomposes when heated. Hydrogen iodide, HI, has a low positive heat of formation, +6.33 kcal/mole. It is a colorless gas that decomposes somewhat when stored at room temperature. As it decomposes, violet iodine vapor becomes visible throughout the container of the gas.

A compound with a high positive heat of formation is very unstable and may react or decompose explosively. For example, ethyne reacts explosively with oxygen. Nitrogen tri-iodide and mercury fulminate decompose explosively. The formation reactions of such compounds store a great deal of energy within them. Mercury fulminate, $HgC_2N_2O_2$, has a heat of formation of +64 kcal/mole. Its instability makes it useful as a detonator for explosives.

Fuels are energy-rich substances.

20.5 The heat of combustion Fuels, whether for the furnace, automobile, or rocket, are energy-rich substances. The products of their combustion are energy-poor substances. In these combustion reactions the energy yield may be very high. The products of the chemical action may be of little importance compared to the quantity of heat energy given off.

The combustion of 1 mole of pure carbon (graphite) yields 94.05 kcal of heat energy.

$$C(s) + O_2(g) \rightarrow CO_2(g) \qquad \Delta H = -94.05 \text{ kcal}$$

The heat of reaction released by the complete combustion of 1 mole of a substance is called the ***heat of combustion*** *of the substance.* Observe that the heat of combustion is defined in terms of *1 mole of reactant.* The heat of formation, on the other hand, is defined in terms of *1 mole of product.* The general heat of reaction notation, ΔH, applies to heats of combustion as well. However, the ΔH_c notation may be used to refer specifically to heat of combustion. See Table 20-2. A more extensive list is in Appendix Table 15.

In some cases, a substance cannot be formed directly from its elements in a rapid composition reaction. The heat of formation of such a compound can be found by using the heats of reaction of a series of related reactions. Heats of combustion are sometimes useful in these calculations when used according to the following equation:

$$\text{heat of formation of compound X} = \text{sum of heats of formation of products of combustion of compound X} - \text{heat of combustion of compound X}$$

CO_2 and H_2O are the products of complete combustion of many organic compounds. Their heats of formation are known. The heats of formation of the organic compounds can be calculated indirectly according to the preceding expression. For example, the heat of combustion, ΔH_c, of methane is determined to be −212.79 kcal/mole.

ΔH_f can be determined indirectly.

$$CH_4 + 2O_2 \rightarrow CO_2 + 2H_2O + 212.79 \text{ kcal}$$

Table 20-2
HEAT OF COMBUSTION

ΔH_c = heat of combustion of the given substance. All values of ΔH_c are expressed as kcal/mole of substance oxidized to H_2O (l) and/or CO_2 (g) at constant pressure and 25 °C. (s) = solid, (l) = liquid, (g) = gas.

Substance	*Formula*	ΔH_c
hydrogen (g)	H_2	−68.32
carbon (graphite) (s)	C	−94.05
carbon monoxide (g)	CO	−67.64
methane (g)	CH_4	−212.79
ethane (g)	C_2H_6	−372.81
propane (g)	C_3H_8	−530.57
butane (g)	C_4H_{10}	−687.98
pentane (g)	C_5H_{12}	−845.16
hexane (l)	C_6H_{14}	−995.01
heptane (l)	C_7H_{16}	−1149.9
octane (l)	C_8H_{18}	−1302.7
ethene (ethylene) (g)	C_2H_4	−337.23
propene (propylene) (g)	C_3H_6	−490.2
ethyne (acetylene (g)	C_2H_2	−310.61
benzene (l)	C_6H_6	−780.96
toluene (l)	C_7H_8	−934.2

or

$$CH_4 + 2O_2 \rightarrow CO_2 + 2H_2O \qquad \Delta H_c = -212.79 \text{ kcal}$$

The combustion of 1 mole of CH_4 forms 1 mole of CO_2 and 2 moles of H_2O. From Table 20-1, ΔH_f for CO_2 is −94.05 kcal/mole and for H_2O is −68.32 kcal/mole. The heat of formation for methane can be calculated from these data as follows:

$$\Delta H_f(CH_4) = \Delta H_f(CO_2) + 2\Delta H_f(H_2O) - \Delta H_c(CH_4)$$

$$\Delta H_f(CH_4) = -94.05 \frac{\text{kcal}}{\text{mole}} + 2\left(-68.32 \frac{\text{kcal}}{\text{mole}}\right) - \left(-212.79 \frac{\text{kcal}}{\text{mole}}\right)$$

$$\Delta H_f(CH_4) = -17.90 \text{ kcal/mole}$$

When carbon is burned in a limited supply of oxygen, carbon monoxide is produced. In this reaction carbon is probably first oxidized to CO_2. Some of the CO_2 may in turn be reduced to CO by hot carbon. The result is an uncertain mixture of the two gases.

CO is formed indirectly.

$$C(s) + O_2(g) \rightarrow CO_2(g)$$

$$C(s) + CO_2(g) \rightarrow 2CO(g)$$

Because of this uncertainty, the heat of formation of CO cannot be determined by measuring directly the heat given off during the reaction. However, both carbon and carbon monoxide can be burned completely to carbon dioxide. The heat of formation of CO_2 and the heat of combustion of CO are then known (Tables 20-1 and 20-2). From these reaction heats, the heat of formation of CO is determined using the equality stated previously.

For the combustion of carbon,

$$C(s) + O_2(g) \rightarrow CO_2(g) + 94.05 \text{ kcal}$$

Thus

$$\Delta H_f \text{ of } (CO_2) = -94.05 \text{ kcal/mole}$$

For the combustion of carbon monoxide,

$$2CO(g) + O_2(g) \rightarrow 2CO_2(g) + 135.28 \text{ kcal}$$

Recall that heat of combustion is defined in terms of 1 mole of a reactant.

Rewriting this equation for 1 mole of the reactant CO,

$$CO(g) + \tfrac{1}{2}O_2(g) \rightarrow CO_2(g) + 67.64 \text{ kcal}$$

Thus

$$\Delta H_c \text{ of } CO = -67.64 \text{ kcal/mole}$$

But

$$\Delta H_f(CO) = \Delta H_f(CO_2) - \Delta H_c(CO)$$

Then, by substitution,

$$\Delta H_f(CO) = -94.05 \text{ kcal/mole} - (-67.64 \text{ kcal/mole})$$

$$\Delta H_f(CO) = -26.41 \text{ kcal/mole}$$

The heat of formation of CO can be added to the heat of combustion of CO to verify the heat of formation of CO_2 determined previously.

$$C(s) + \tfrac{1}{2}O_2(g) \rightarrow CO(g) \qquad \Delta H_f = -26.41 \text{ kcal}$$
$$CO(g) + \tfrac{1}{2}O_2(g) \rightarrow CO_2(g) \qquad \Delta H_c = -67.64 \text{ kcal}$$
$$\overline{C(s) + O_2(g) \rightarrow CO_2(g) \qquad \Delta H_f = -94.05 \text{ kcal}}$$

A thermochemical equation for the formation of CO can now be derived directly from its elements. To do so, the equations for the oxidation of carbon and the reduction of carbon dioxide are combined.

$$C(s) + O_2(g) \rightarrow CO_2(g) \qquad \text{(oxidation of C)}$$
$$C(s) + CO_2(g) \rightarrow 2CO(g) \qquad \text{(reduction of } CO_2\text{)}$$
$$\overline{2C(s) + O_2(g) \rightarrow 2CO(g) \qquad \text{(net reaction)}}$$

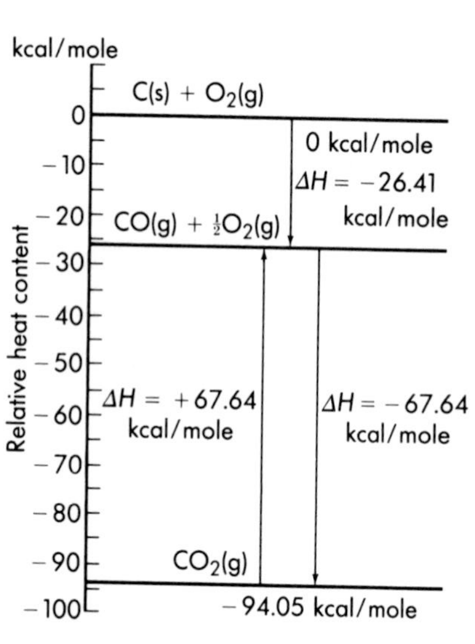

Figure 20-3. Heat of formation diagram for carbon dioxide and carbon monoxide.

Energy is conserved during a chemical reaction; the heat absorbed in decomposing a compound must be equal to the heat released in its formation under the same conditions. Thus, if there is a reason to write the equation for a reaction in reverse form, the sign of ΔH for the reaction must be reversed.

Suppose these principles are applied to the thermochemical equation for the combustion of carbon monoxide:

$$CO(g) + \tfrac{1}{2}O_2(g) \rightarrow CO_2(g) \qquad \Delta H_c = -67.64 \text{ kcal}$$

Writing the reverse of this reaction:

$$CO_2(g) \rightarrow CO(g) + \tfrac{1}{2}O_2(g) \qquad \Delta H_c = +67.64 \text{ kcal}$$

ΔH for a given overall reaction is independent of the reaction pathway.

Observe that the sign of ΔH has been reversed to express the change in heat content for a reverse reaction. This principle is actually part of a more general one: *The heat of a given overall reaction is the same regardless of the intermediate steps involved.* The heat of formation of CO calculated from the following equations illustrates this additivity principle.

$$\begin{array}{ll} C(s) + O_2(g) \rightarrow CO_2(g) & \Delta H_f = -94.05 \text{ kcal} \\ CO_2(g) \rightarrow CO(g) + \frac{1}{2}O_2(g) & \Delta H_c = +67.64 \text{ kcal} \\ \hline C(s) + \frac{1}{2}O_2(g) \rightarrow CO(g) & \Delta H_f = -26.41 \text{ kcal} \end{array}$$

Referring to the endothermic reaction between carbon and steam first mentioned in Section 2.14, this "water-gas" reaction occurs spontaneously at the temperature of white-hot carbon. It produces a mixture of CO and H_2 that can be used as a gaseous fuel. What is the thermochemistry of this fuel?

The heat of formation of water is normally expressed in terms of the change in heat content between liquid water at 25 °C and its separate elements at the same temperature. Water vapor must give up 10.52 kcal/mole in order to condense to its liquid phase at 25 °C. This condensation is a physical change.

$$H_2O(g) \rightarrow H_2O(l) + 10.52 \text{ kcal}$$

Thus the steam in the reaction has a higher heat content (and a lower heat of formation) by 10.52 kcal than liquid water.

The following thermochemical equation shows the heat of formation of the product water as a gas from its composition reaction.

$$H_2(g) + \frac{1}{2}O_2(g) \rightarrow H_2O(g) + 57.80 \text{ kcal}$$

The complete relationship is shown by the following series of equations.

$$\begin{array}{ll} H_2(g) + \frac{1}{2}O_2(g) \rightarrow H_2O(g) & \Delta H = -57.80 \text{ kcal} \\ H_2O(g) \rightarrow H_2O(l) & \Delta H = -10.52 \text{ kcal} \\ \hline H_2(g) + \frac{1}{2}O_2(g) \rightarrow H_2O(l) & \Delta H = -68.32 \text{ kcal} \end{array}$$

The water-gas reaction is endothermic and spontaneous.

Experiments have established the quantity of heat absorbed in the water-gas reaction as 31.39 kcal per mole of carbon used. The thermochemical equation is

$$H_2O(g) + C(s) + 31.39 \text{ kcal} \rightarrow CO(g) + H_2(g)$$

The heat of reaction is absorbed; the heat content of the products CO and H_2 is greater than the heat content of the reactants C and H_2O.

When the product gases are burned as fuel, two combustion reactions occur. Carbon dioxide is the product of one, and water vapor is the product of the other. Both are familiar reactions shown earlier in this section.

$$CO(g) + \frac{1}{2}O_2(g) \rightarrow CO_2(g) + 67.64 \text{ kcal}$$

$$H_2(g) + \frac{1}{2}O_2(g) \rightarrow H_2O(g) + 57.80 \text{ kcal}$$

These reactions are exothermic, and the heats of combustion, ΔH_c, have negative values.

There are now three thermochemical equations representing the formation of the water gas and its combustion as a fuel. Suppose these equations are arranged in a series. The net additive result is shown below.

$$\begin{array}{ll} H_2O(g) + C(s) \rightarrow CO(g) + H_2(g) & \Delta H = +31.39 \text{ kcal} \\ CO(g) + \frac{1}{2}O_2(g) \rightarrow CO_2(g) & \Delta H = -67.64 \text{ kcal} \\ H_2(g) + \frac{1}{2}O_2(g) \rightarrow H_2O(g) & \Delta H = -57.80 \text{ kcal} \\ \hline C(s) + O_2(g) \rightarrow CO_2(g) & \Delta H = -94.05 \text{ kcal} \end{array}$$

The combined heat of combustion of the gases, CO and H_2, is −125.44 kcal. Observe that this value is higher than that of carbon (−94.05 kcal). However, it is higher only by the amount of heat energy put into the first reaction (+31.39 kcal).

20.6 Bond energy and reaction heat The change in heat content of a reaction system is related to (1) the change in the number of bonds breaking and forming, and (2) the strengths of these bonds as the reactants form products. The reaction for the formation of water gas can be used to test these relationships.

The two oxygen-to-hydrogen bonds of each steam molecule must be broken, as must the carbon-to-carbon bonds of the graphite. Carbon-to-oxygen and hydrogen-to-hydrogen bonds must be formed. Recall that energy is absorbed when bonds are broken and energy is released when bonds are formed.

A possible reaction mechanism for the water-gas reaction in which there is an intermediate stage of free atoms is illustrated in Figure 20-4. Observe that a heat input of 390 kcal is required to break the bonds of 1 mole of graphite and 1 mole of steam. The formation of bonds in the final state releases 358 kcal of heat energy. The net effect is that 32 kcal of heat must be supplied to the system from an external source. This quantity agrees closely with the experimental value of 31.39 kcal of heat input per mole of carbon used.

20.7 Factors that drive reactions Whether a reaction occurs and how it occurs are questions that have always concerned chemists. The answers to these questions involve a study of reaction mechanisms and reaction kinetics conducted within the framework of the laws of thermodynamics. This is the realm of *physical chemistry* and *physics*. Here, the concepts that chemists label collectively as the "driving force" of chemical reactions will be examined. Some of these concepts were discussed briefly in Section 2.14.

Thermodynamics is the experimental study of the relationship between heat energy and other forms of energy.

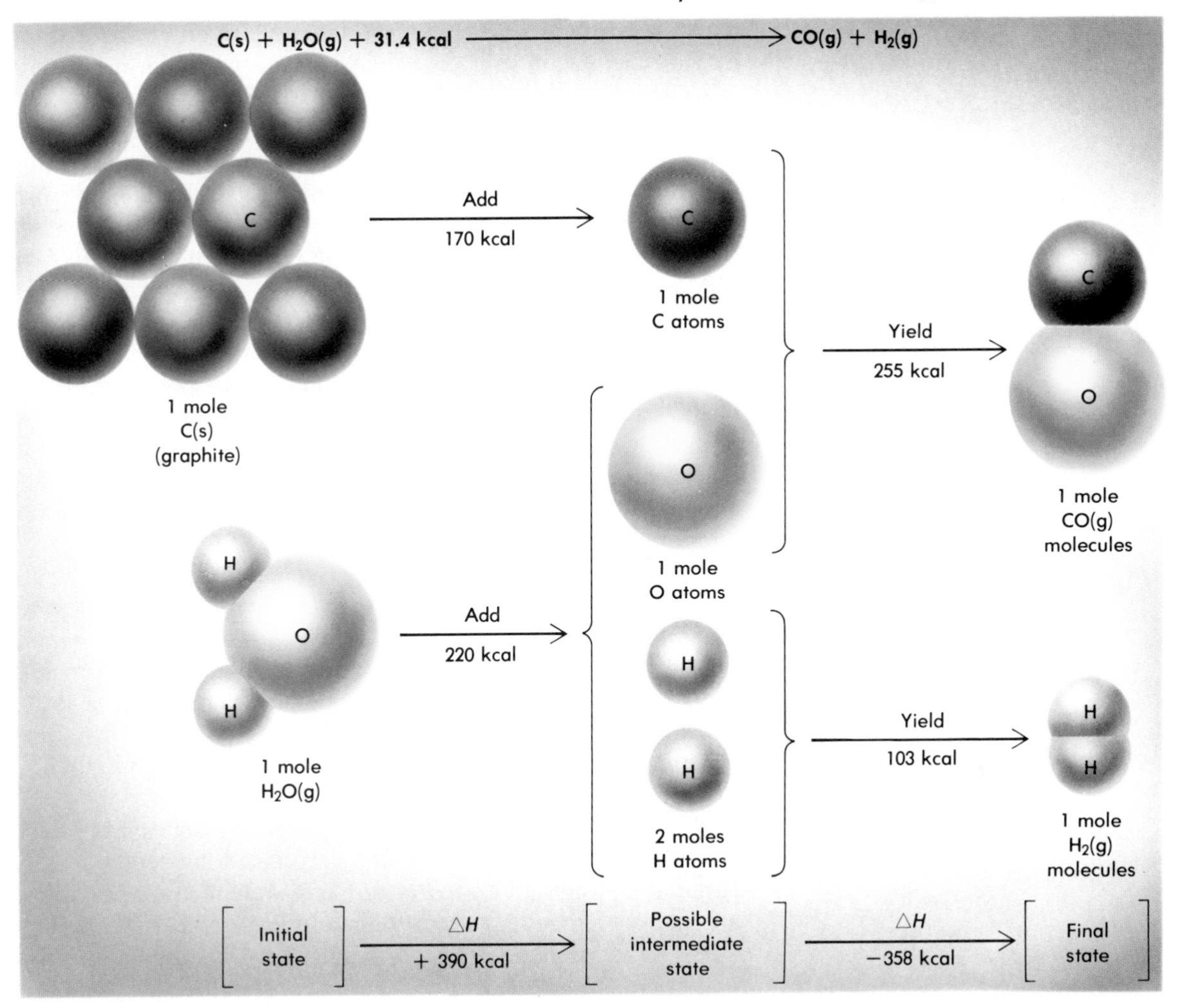

Figure 20-4. A possible mechanism for the water-gas reaction.

By observation, it is recognized that most reactions occurring spontaneously in nature are exothermic; they give off energy and lead to lower energy states and more stable structural configurations. The quantity H, the energy content of a system, is a function of the state of the system. The drive toward a favorable change in energy content, ΔH, is responsible for this *tendency for processes to occur that lead to the lowest possible energy state.*

In its principal connotation, "state" refers to the "condition" of a substance or system.

If the driving force depended on this energy-change tendency alone, no endothermic reaction could take place spontaneously. One could predict that only exothermic reactions are spontaneous. A great deal of evidence shows that most reactions do release energy. In fact, the greater the quantity of energy given up, the more vigorous such reactions tend to be.

The disturbing fact is that endothermic reactions *do* take place spontaneously. They occur simply as a result of mixing reactants. The production of water gas involves such a reaction. Steam is passed into white-hot coke (impure carbon), and the

Reaction systems that absorb energy are endothermic.

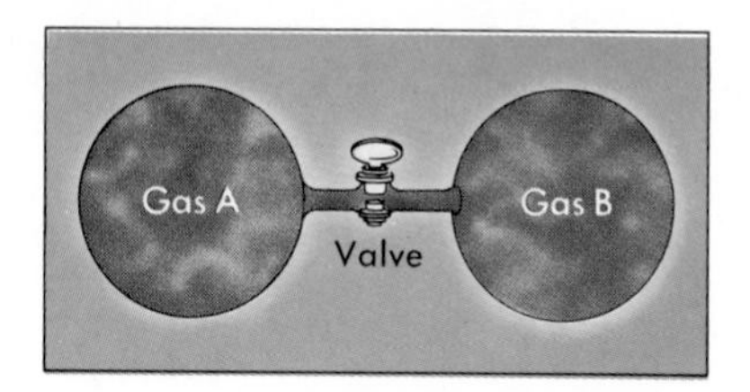

Figure 20-5. The mixing of gases may occur without an energy change.

reaction proceeds spontaneously. It is not driven by any activity outside the reacting system. The process is endothermic: heat is absorbed, and the reaction system cools. Therefore, more heat must be supplied by periodically blowing air into the coke to cause some combustion. From Section 20.5 it is recognized that the product gases, CO and H_2, collectively have a higher heat content than the reactants, steam and carbon. The energy change is positive. This unfavorable energy change cannot be the factor that drives the reaction.

$$H_2O(g) + C(s) \rightarrow CO(g) + H_2(g) \qquad \Delta H = +31.39 \text{ kcal}$$

To see how an endothermic reaction can occur spontaneously, consider a simple physical process that proceeds by and of itself with no energy change. Figure 20-5 shows two identical flasks connected by a valve. One flask is filled with ideal gas **A** and the other with ideal gas **B.** The entire system is at room temperature.

When the valve is opened, the two gases mix until they are distributed evenly throughout the two containers. The gases will remain in this state indefinitely, with no tendency to separate. The temperature remains constant throughout the process. Thus, the total heat content cannot have changed to a lower level. Clearly, the self-mixing process must be caused by a driving force other than the energy-change tendency. What, then, is the driving force for this process?

Entropy describes the state of disorder.

The concept of *a tendency for processes to occur that lead to the highest possible state of disorder* was introduced in Section 2.14. In general, a system that can go from one state to another without an energy change will adopt the more disordered state, as illustrated in Figure 20-5. There is a quantity *S*, known as *entropy*, which is a function of the state of a system. In the water-gas reaction, the final mixed system of gases represents a more disordered state than the initial pure-gas system. In other words, a favorable entropy change has occurred. The entropy of the mixed-gas system is higher than the entropy of the pure-gas system. The factor that drives the reaction is the *tendency for the disorder (entropy) of the system to increase.*

Natural processes tend toward (1) minimum energy and toward (2) maximum disorder.

Thus the property of a system that drives the reaction depends on two factors, the tendency toward *lowest energy* and the tendency toward *highest entropy.* A new function of the state of a reaction system has been defined to relate the energy and entropy factors at a given temperature. It is called the *free energy*, *G*, of the system. It is the function that simultaneously assesses both the energy-change and entropy-change tendencies. In this sense, a reaction system proceeds spontaneously in the way that *lowers its free energy.*

A process is spontaneous in the direction of lower free energy.

The net driving force is called the *free-energy change*, ΔG, of the system. When the energy change and the entropy change oppose each other, the direction of the net driving force is determined by the factor having the larger influence. The free-energy change, ΔG, indicates the quantity of energy that is *free* to perform useful work.

At a constant pressure and temperature, *the* ***free-energy change,*** ΔG, is a property of the reaction system *defined as the difference between the change in heat content,* ΔH, *and the product of the temperature and the entropy change,* $T\Delta S$.

$$\Delta G = \Delta H - T\Delta S$$

Here ΔG is the change in free energy of the system, ΔH is the change in the heat content, T is the temperature in °K, and ΔS is the change in entropy. (ΔG, ΔH, and the product $T\Delta S$ all have the same dimensions, usually kilocalories per mole. What are the dimensions of ΔS?)

A chemical reaction proceeds spontaneously if it is accompanied by a decrease in free energy. That is, it proceeds spontaneously if the free energy of the products is lower than that of the reactants. In such a case, the *free energy change,* ΔG, in the system is *negative.*

In exothermic reactions ΔH has a negative value. In endothermic reactions its value is positive. From the expression for ΔG, observe that the more negative ΔH is, the more negative ΔG is likely to be. Reaction systems that change from a high-energy state to a low one tend to proceed spontaneously. Also observe that the more positive ΔS is, the more negative ΔG is likely to be. Thus systems that change from well-ordered states to highly disordered states also tend to proceed spontaneously. Both of these tendencies in a given reaction system are assessed simultaneously in terms of the free-energy change, ΔG.

In some processes ΔH is negative and ΔS is positive. Here, the process should proceed spontaneously because ΔG is negative regardless of the relative magnitudes of ΔH and ΔS. Processes in which ΔH and ΔS have the same signs are more common. Note that the sign of ΔG can be either positive or negative depending on the temperature T. The temperature of the system is the dominant factor that determines the relative importance of the tendency toward lower energy and the tendency toward higher entropy.

ΔG is negative for spontaneous reactions or processes.

Consider again the water-gas reaction,

$$H_2O(g) + C(s) \rightarrow CO(g) + H_2(g) \qquad \Delta H = +31.39 \text{ kcal}$$

Here, a negative free-energy change would require that $T\Delta S$ be positive and greater than 31.39 kcal. This can be so if T is large, or if ΔS is large and positive, or both. Consider first the magnitude and sign of ΔS.

Recall that the solid phase is well-ordered and the gaseous phase is random. One of the reactants, carbon, is a solid, and the other, steam, is a gas. However, both products are gases. The change from the orderly solid phase to the random gaseous phase involves an *increase* in entropy. In general, a change from a solid to a gas will proceed with an increase in entropy. If all reactants and products are gases, an increase in the number of product particles increases entropy.

At low temperatures, whether ΔS is positive or negative, the product $T\Delta S$ may be small compared to ΔH. In such cases, the reaction proceeds as the energy change predicts.

Careful measurements show that ΔS for the water-gas reaction is +0.0320 kcal/mole K at 25 °C (298 K). Thus

$$T\Delta S = 298\ \text{K} \times 0.0320\ \frac{\text{kcal}}{\text{mole K}} = 9.54\ \frac{\text{kcal}}{\text{mole}}$$

$$\Delta G = \Delta H - T\Delta S = +31.39\ \frac{\text{kcal}}{\text{mole}} - 9.54\ \frac{\text{kcal}}{\text{mole}}$$

and

$$\Delta G = +21.85\ \frac{\text{kcal}}{\text{mole}}$$

Since ΔG has a positive value, the reaction is not spontaneous at a temperature of 25 °C.

Increases in temperature tend to favor increases in entropy. When ΔS is positive, a high temperature gives $T\Delta S$ a large positive value. Certainly, at a high enough temperature $T\Delta S$ will be larger than ΔH, and ΔG will have a negative value.

Higher temperature—higher entropy.

The water-gas reaction occurs at the temperature of white-hot carbon, approximately 900 °C. The values of ΔH and ΔS are different at different temperatures. However, if they are assumed to remain about the same at the reaction temperature of 900 °C, an approximate value of ΔG can be calculated for this temperature. Retaining the value of ΔH as +31.39 kcal/mole, $T\Delta S$ is determined at 1173 K (900 °C) to be

$$T\Delta S = 1173\ \text{K} \times 0.0320\ \frac{\text{kcal}}{\text{mole K}} = 37.5\ \frac{\text{kcal}}{\text{mole}}$$

$$\Delta G = \Delta H - T\Delta S = +31.39\ \frac{\text{kcal}}{\text{mole}} - 37.5\ \frac{\text{kcal}}{\text{mole}}$$

$$\Delta G = -6.1\ \text{kcal/mole}$$

The free-energy change for the water-gas reaction is negative at 900 °C, and the reaction is spontaneous at this temperature.

Reaction Mechanisms

20.8 Reaction pathways Chemical reactions involve breaking existing chemical bonds and forming new ones. The relationships and arrangements of atoms in the products of a reaction are different from those in the reactants. Colorless hydrogen gas consists of pairs of hydrogen atoms bonded together as diatomic molecules, H_2. Violet-colored iodine vapor is also diatomic, consisting of pairs of iodine atoms bonded together as I_2 molecules. A chemical reaction between these two gases at elevated temperatures produces hydrogen iodide, HI, a colorless gas. Hydrogen iodide molecules, in turn, tend to decompose and re-form hydrogen and iodine molecules. Their reaction equations are

$$H_2(g) + I_2(g) \rightarrow 2HI(g)$$

and

$$2HI(g) \rightarrow H_2(g) + I_2(g)$$

These equations indicate only what molecular species disappear as a result of the reactions and what species are produced. They do not show the pathway by which either reaction proceeds. That is, they do not show the step-by-step sequence of reactions by which the overall chemical change occurs. Such a sequence, when known, is called the *reaction pathway* or *reaction mechanism*.

A chemical system can usually be examined before the reaction occurs or after the reaction is over. However, such an examination reveals nothing about the pathway by which the action proceeded. See Figure 20-6. For most chemical reactions, only the reactants that disappear and the final products that appear are known. In other words, only the net chemical change is directly observable.

Sometimes chemists are able to devise experiments that reveal a sequence of steps in a reaction pathway. They attempt to learn how the speed of a reaction is affected by various factors. Such factors may include temperature, concentrations of reactants and products, and the effects of catalysts. Radioactive tracer techniques are sometimes helpful.

A chemical reaction might occur in a single step or in a sequence of steps. Each reaction step is usually a relatively simple process. Complicated chemical reactions take place in a sequence of simple steps. Even a reaction that appears from its balanced equation to be a simple process may actually occur in a sequence of steps.

The reaction between hydrogen gas and bromine vapor to produce hydrogen bromide gas is an example of a *homogeneous reaction*. The reaction system involves a single phase—the gas phase. This reaction system is also an example of a *homogeneous chemical system*. In such a system, all reactants and products in all intermediate steps are in the same phase. The overall chemical equation for this reaction is

$$H_2(g) + Br_2(g) \rightleftarrows 2HBr(g)$$

It might appear that one hydrogen molecule reacts with one bromine molecule to form the product in a simple, one-step process. However, chemists have found through kinetic studies that the reaction follows a more complex pathway. The initial forward reaction (and the terminating reverse reaction) is

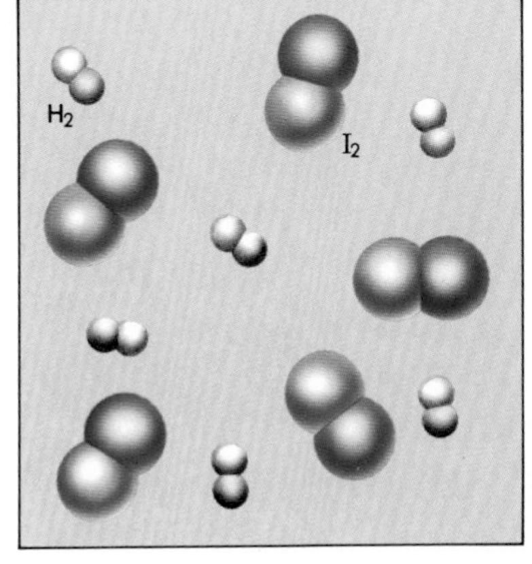

Initial state

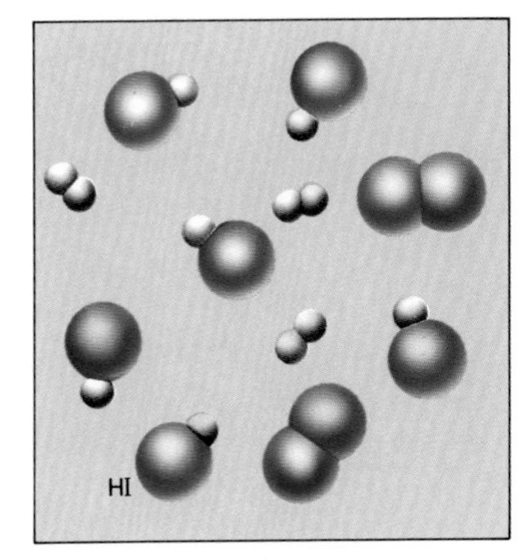

Final state

Figure 20-6. The initial and final states of a mixture of hydrogen gas and iodine vapor which react to form hydrogen iodide gas.

$$Br_2 \rightleftarrows 2Br \quad (1)$$

This initial step provides bromine atoms that, in turn, initiate a sequence of two steps which can repeat themselves.

$$Br + H_2 \rightleftarrows HBr + H \quad (2)$$

$$H + Br_2 \rightleftarrows HBr + Br \quad (3)$$

Observe that the sum of steps 2 and 3 gives the net reaction as shown in the overall equation above.

The equation

$$H_2(g) + I_2(g) \rightleftarrows 2HI(g)$$

represents another homogeneous chemical system. Figure 20-6 illustrates the initial reactants and final products identified by this equation. For many years chemists thought this reaction was a simple, one-step process. They assumed it involved two molecules, $H_2 + I_2$, in the forward direction and two molecules, HI + HI, in the reverse reaction.

Recent experiments have shown that the reaction $H_2 + I_2$ does not take place, and chemists have proposed alternative mechanisms for the reaction. The first has a two-step pathway:

$$I_2 \rightleftarrows 2I \quad (1)$$

$$2I + H_2 \rightleftarrows 2HI \quad (2)$$

The second possible mechanism has a three-step pathway:

$$I_2 \rightleftarrows 2I \quad (1)$$

$$I + H_2 \rightleftarrows H_2I \quad (2)$$

$$H_2I + I \rightleftarrows 2HI \quad (3)$$

It appears that, no matter how simple the balanced equation, the reaction pathway may be complicated and difficult to determine. Some chemists believe that simple, one-step reaction mechanisms are unlikely. They suggest that the only simple reactions are those that have not been thoroughly studied.

20.9 Collision theory In order for reactions to occur between substances, their particles (molecules, atoms, or ions) must collide. Further, these collisions must result in interactions. Chemists use this *collision theory* to interpret many facts they observe about chemical reactions.

From the kinetic theory you know that the molecules of gases are continuously in random motion. The molecules have kinetic energies ranging from very low to very high values. The energies of the greatest portion are near the average for all the molecules in the system. When the temperature of a gas is raised, the average kinetic energy of the molecules is increased. Therefore, their speed is increased as is shown in Figure 20-7.

Consider what happens on a molecular scale in one step of a homogeneous reaction system. A good example is the decomposition of hydrogen iodide. The reaction for the first step in the decomposition pathway is

$$HI + HI \rightarrow H_2 + 2I$$

According to the collision theory, the two gas molecules must collide in order to react. Further, they must collide while favorably oriented and with enough energy to disrupt the bonds of the molecules. If they do so, a reshuffling of bonds leads to the formation of the new particle species of the products.

A collision may be too gentle. Here, the distance between the colliding molecules is never small enough to disrupt old bonds or form new ones. The two molecules simply rebound from each other unchanged. This effect is illustrated in Figure 20-8(A).

Similarly, a collision in which the reactant molecules are poorly oriented has little effect. The distance between certain atoms is never small enough for new bonds to form. The colliding molecules rebound without changing. A poorly oriented collision is shown in Figure 20-8(B).

However, a collision may be suitably oriented and violent enough to cause an interpenetration of the electron clouds of the colliding molecules. Then the distance between the reactant particles *does* become small enough for new bonds to form. See Figure 20-8(C). The minimum energy required to produce this effective collision is called the *activation energy* for the reaction.

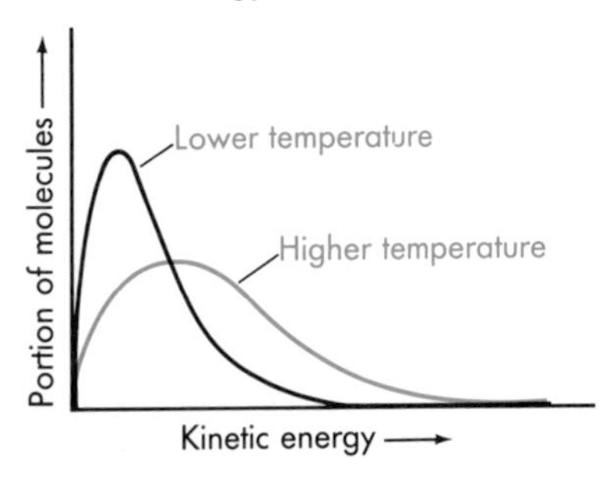

Figure 20-7. Energy distribution among gas molecules at two different temperatures.

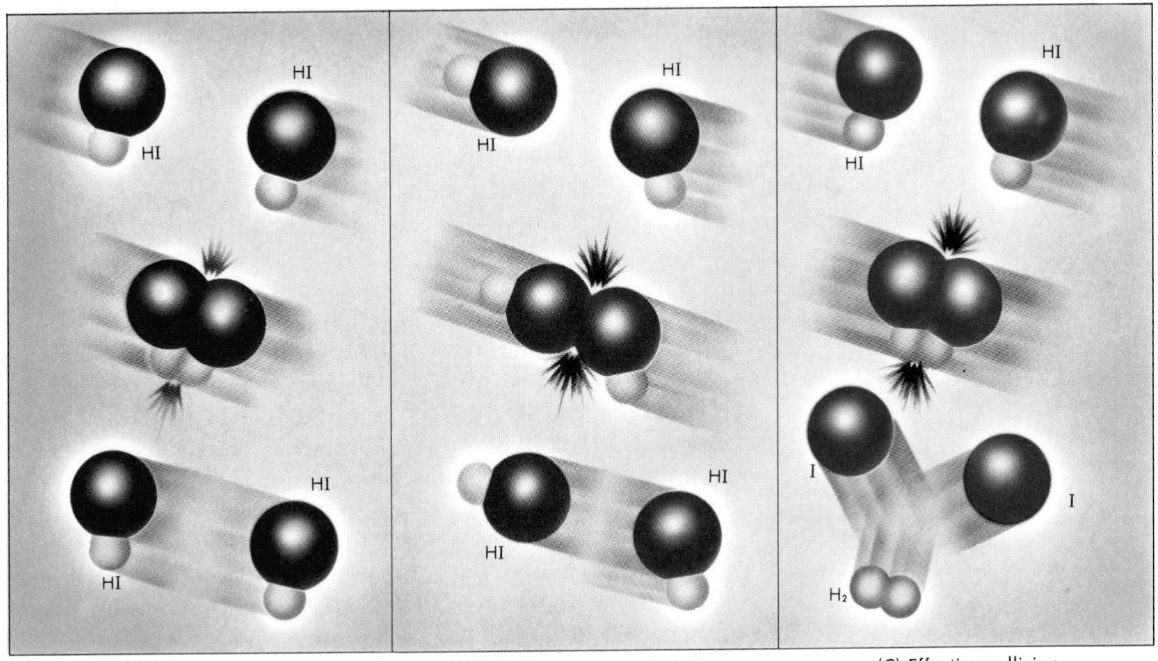

Figure 20-8. Possible collision patterns for HI molecules.

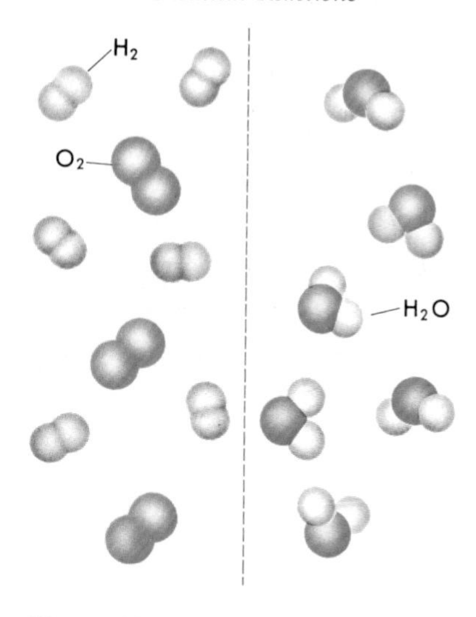

Figure 20-9. A mixture of hydrogen and oxygen molecules will form very stable water molecules when properly activated.

Thus, collision theory provides two reasons why a collision between reactant molecules may fail to produce a new chemical species: (1) *the collision is not energetic enough to supply the required activation energy,* and (2) *the colliding molecules are not oriented in a way that enables them to react with each other.* What then is the reaction pathway of an effective collision that produces a new chemical species?

20.10 Activation energy Consider the reaction for the formation of water from the diatomic gases oxygen and hydrogen. The heat of formation is quite high; $\Delta H = -68.3$ kcal/mole at 25 °C. The free energy change is also large; $\Delta G = -56.7$ kcal/mole. Why then do hydrogen and oxygen not combine spontaneously to form water when they are mixed at room temperature?

Hydrogen and oxygen gases exist as diatomic molecules. By some reaction mechanism, the bonds of these molecular species must be broken. Then new bonds between oxygen and hydrogen atoms must be formed. Bond breaking is an endothermic process, and bond forming is exothermic. Even though the net process is exothermic, it appears that an initial energy "kick" is needed to start the reaction.

It might help to think of the reactants as lying in an energy "trough." They must be lifted from this trough before they can react to form water, even though the energy content of the product is lower than that of the reactants. Once the reaction is started, the energy released is enough to sustain the reaction by activating other molecules. Thus the reaction rate keeps increasing. It is finally limited only by the time required for reactant particles to acquire the energy and make contact.

The energy needed to lift the reactants from the energy trough is called *activation energy.* It is the energy required to loosen bonds in molecules so they can become reactive. Energy from a flame, a spark discharge, or the energy associated with high temperatures or radiations may start reactants along the pathway of reaction. The activation-energy concept is illustrated in Figure 20-10.

The reverse reaction is the decomposition of water molecules. The product water lies in an energy trough deeper than the one from which the reactants were lifted. The water molecules must be lifted from this deeper energy trough before they can decompose to form oxygen and hydrogen. The activation energy needed to start this endothermic reaction is greater than that required for the original exothermic change. The difference equals the amount of energy of reaction, ΔH, released in the original reaction. See Figure 20-11.

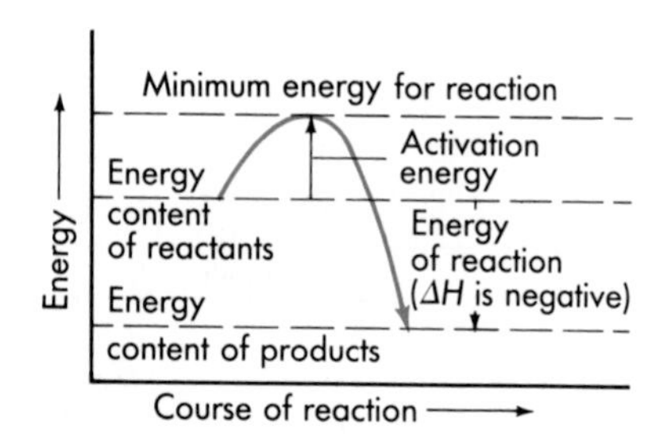

Figure 20-10. Pathway of an exothermic reaction.

This situation can be compared to two mountain valleys separated by a high mountain pass. One valley is lower than the other. Still, to get to it from the upper valley, one must climb over the high pass. The return trip to the upper valley can be made only by climbing over the same high pass again. For this trip, however, one must climb up a greater height since the starting point is lower.

In Figure 20-12, the difference in height of the two valley floors is compared to the energy of reaction, ΔH. Energies E_a and E_a' represent the activation energies of the forward and reverse reactions, respectively. These quantities are compared to the heights of the pass above the high and low valleys.

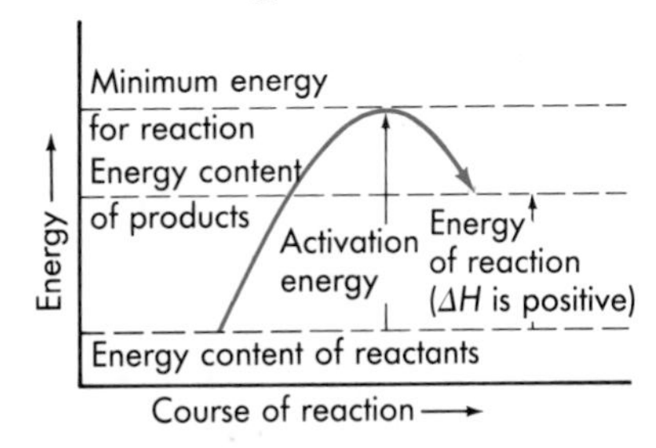

Figure 20-11. Pathway of an endothermic reaction.

20.11 The activated complex The possession of high motion energy (kinetic energy) does not make molecules unstable. However, when molecules collide, some of this energy is converted into internal (potential) energy within the colliding molecules. If enough energy is converted, the molecules may be activated.

When particles collide with energy at least equal to the activation energy for the species involved, their interpenetration disrupts existing bonds. New bonds can then form. In the brief interval of bond disruption and bond formation, the *collision complex* is said to be in a *transition state.* Some sort of partial bonding exists in this state. *A transitional structure results from an effective collision. This structure persists while old bonds are breaking and new bonds are forming. It is called the* ***activated complex.***

Effective collisions produce transitional complexes.

An activated complex is formed when an effective collision raises the internal energies of the reactants to their *minimum-energy-for-reaction level.* See Figure 20-12. Both forward and reverse reactions go through the same activated complex. Suppose a bond is in the process of being broken in the activated complex for the forward reaction. The same bond is in the process of being formed in the activated complex for the reverse reaction. Observe that the activated complex occurs at the maximum energy position along the reaction pathway. In this sense the activated complex defines the activation energy for the system. ***Activation energy*** *is the energy required to transform the reactants into the activated complex.*

In its brief existence, the activated complex has partial bonding of both reactant and product. In this state, it may respond to either of two possibilities: (1) It may re-form the original bonds and separate into the reactant particles or, (2) It may form new bonds and separate into product particles. Usually the formation of products is just as likely as the formation of reactants. *Do not confuse the activated complex with the intermediate products of different steps of a reaction mechanism.* The activated complex is a molecular complex in which bonds are in the process of being broken and formed.

A possible configuration of the activated complex in the hydrogen iodide reaction is shown in Figure 20-13. The broken lines represent some sort of partial bonding in the particle. This transitional structure may produce two HI molecules or an H_2 molecule and two I atoms.

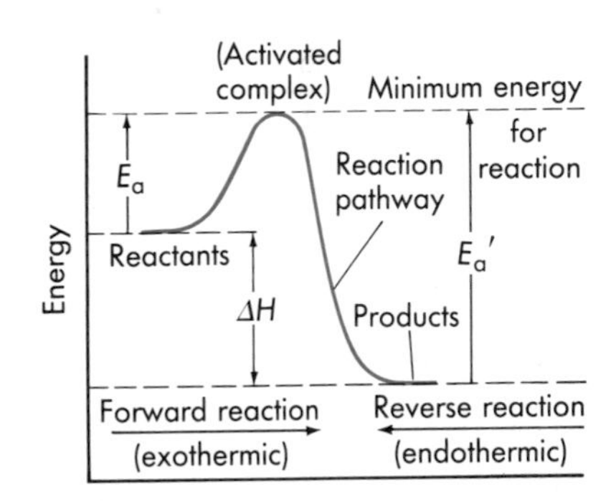

Figure 20-12. Activation energies for the forward reaction E_a and the reverse reaction E_a', and the change in internal energy ΔH in a reversible reaction.

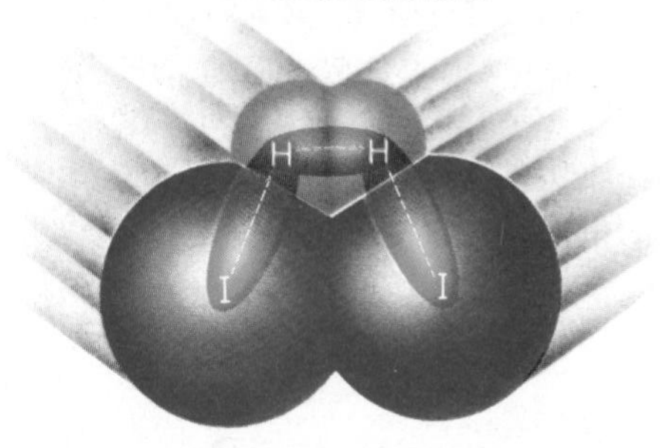

Figure 20-13. Possible activated complex configuration which could form either 2HI or H_2 + 2I.

Figure 20-14 shows the energy profile for the HI decomposition, which occurs at elevated temperatures of 400–500 °C. Of the 43.8-kcal activation energy, 40.8 kcal are available for excitation of the H_2 and I_2 molecules. Of this amount, 35.5 kcal are used in producing iodine atoms.

20.12 Rate-influencing factors The rates of chemical reactions vary widely. Some reactions are over in an instant, and others take months or years to complete. *The rate of reaction is measured by the amount of reactants converted to products in a unit of time.* Two conditions are necessary for reactions (other than simple decompositions) to occur at all. First, particles must come in contact. Second, this contact must result in interaction. Thus the *rate* of a reaction depends on the *collision frequency* of the reactants and on the *collision efficiency*. (An efficient collision is one with enough energy for activation and in which the reactant molecules are favorably oriented.)

Reaction rates depend on (1) collision frequency and (2) collision efficiency.

Changing conditions may affect either the frequency of collisions or the collision efficiency. Any such change influences the reaction rate. Five important factors influence the rate of chemical reaction.

1. *Nature of the reactants.* Hydrogen combines vigorously with chlorine under certain conditions. Under the same conditions it may react only feebly with nitrogen. Sodium and oxygen combine much more rapidly than iron and oxygen under similar conditions. Platinum and oxygen do not combine directly. Atoms, ions, and molecules are the particles of substances that react. Bonds are broken and other bonds are formed in reactions. The rate of reaction depends on the particular bonds involved.

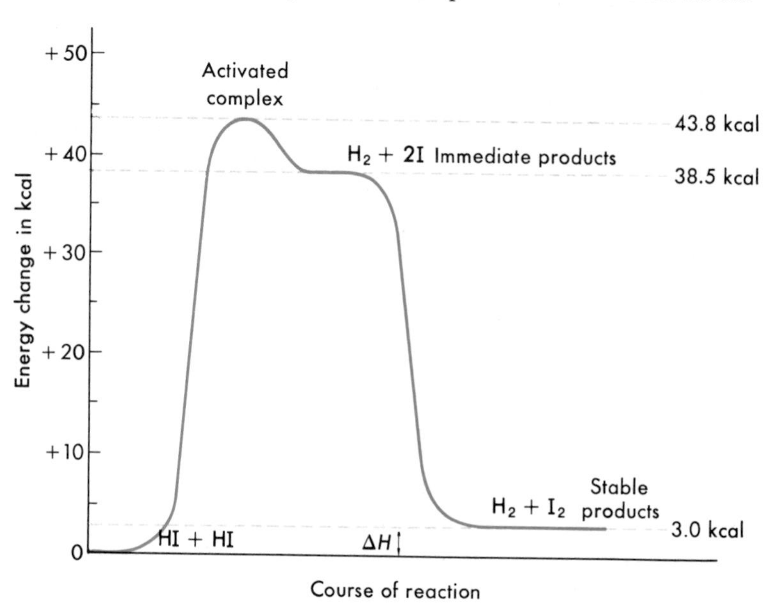

Figure 20-14. An energy profile—experimental results of kinetic studies of hydrogen iodide decomposition at elevated temperatures.

2. *Amount of surface.* The solution rate for a crystalline solid in water is increased if the crystals are first broken down into small pieces. A cube of solute measuring 1 cm on each edge presents only 6 cm^2 of contact area to the solvent. This same cube when ground to a fine powder might provide a contact area of the order of 10^4 times the original area. Consequently, the solution rate of the powdered solid is greatly increased.

A lump of coal burns slowly when kindled in air. The rate of burning can be increased by breaking the lump into smaller pieces, exposing new surfaces. If the piece of coal is powdered and ignited while suspended in air, it burns explosively. Nickel in large pieces shows no noticeable oxidation in air, but finely powdered nickel reacts vigorously and spectacularly.

These reactions, as mentioned above, are examples of *heterogeneous reactions. Heterogeneous reactions involve reactants in two different phases.* Such reactions can occur only when the two phases are in contact. Thus the amount of surface of a solid (or liquid) reactant is an important *rate* consideration. *Gases* and *dissolved* particles do not have surfaces in the sense just described. *In heterogeneous reactions the reaction rate is proportional to the area of contact of the reacting substances.*

Some chemical reactions between gases actually take place at the walls of the container. Others occur at the surface of a solid or liquid catalyst. If the products are also gases, such a reaction system presents a problem of language. It is a *heterogeneous* reaction because it takes place between two phases. On the other hand, it is a *homogeneous* chemical system because all the reactants and all the products are in one phase.

3. *Effect of concentration.* Suppose a small lump of charcoal is heated in air until combustion begins. If it is then lowered into a bottle of pure oxygen, the reaction proceeds at a much faster rate. A substance that oxidizes in air reacts more vigorously in pure oxygen, as shown in Figure 20-15. The partial pressure of oxygen in air is approximately one-fifth of the total pressure. Pure oxygen at the same pressure as the air has five times the *concentration* of oxygen molecules.

This charcoal oxidation is a heterogeneous reaction system in which one reactant is a gas. Not only does the reaction rate depend on the amount of exposed charcoal surface, *it depends on the concentration of the gas as well.*

Homogeneous reactions may involve reactants in liquid or gaseous solutions. The concentration of gases changes with pressure according to Boyle's law. In liquid solutions, the concentration of reactants changes if either the quantity of solute or the quantity of solvent is changed. Solids and liquids are practically noncompressible. Thus it is not possible to change the concentration of pure solids and pure liquids to any measurable extent.

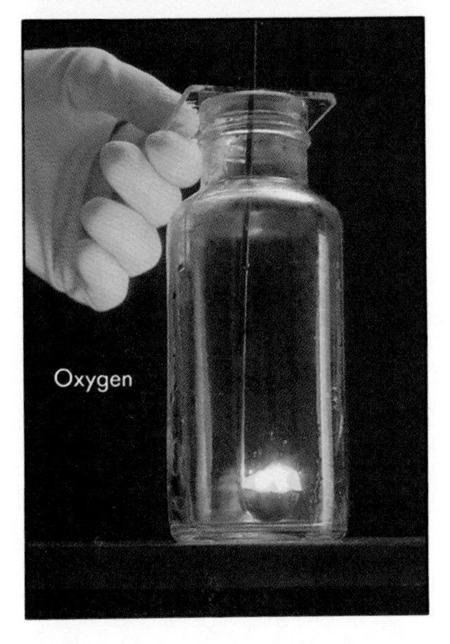

Figure 20-15. Carbon burns faster in oxygen than in air because of the higher concentration of oxygen molecules.

Reaction rate: effect of concentration of reactants is determined experimentally.

In homogeneous reaction systems, reaction rates depend on the concentration of the reactants. From collision theory, a rate increase might be expected if the concentration of one or more of the reactants is increased. Lowering the concentration should have the opposite effect. However, the specific effect of concentration changes *must be determined experimentally.*

Increasing the concentration of substance **A** in reaction with substance **B** could increase the reaction rate, decrease it, or have no effect on it. The effect depends on the particular reaction. *One cannot tell from the balanced equation for the net reaction how the reaction rate is affected by a change in concentration of reactants.* Chemists account for these differences in behavior in terms of the reaction mechanisms.

Reaction rate: slowest step is the rate-determining step.

Complex chemical reactions may take place in a *series* of simple steps. Instead of a single activated complex, there may be several activated complexes in sequence along the reaction pathway. Of these steps, the one that proceeds at the slowest rate determines the overall reaction rate. When this slowest-rate step can be identified, it is called the *rate-determining step* for the chemical reaction.

4. *Effect of temperature.* The average kinetic energy of the particles of a substance is proportional to the temperature of the substance. Collision theory explains why a rise in temperature increases the rate of chemical reaction. According to this theory, a decrease in temperature lowers the reaction rate for both exothermic and endothermic reactions.

At room temperature, the rates of many reactions roughly double or triple with a 10 C° rise in temperature. However, this rule must be used with caution. The actual rate increase with a given rise in temperature *must be determined experimentally.*

Large increases in reaction rate are caused partly by the increase in collision frequency of reactant particles. However, for a chemical reaction to occur, the particles must also collide with enough energy to cause them to react. At higher temperatures more particles possess enough energy to form the activated complex when collisions occur. In other words, more particles have the necessary activation energy. Thus a rise in temperature produces an increase in collision energy and collision frequency.

5. *Action of catalysts.* Some chemical reactions proceed quite slowly. Frequently their reaction rates can be increased dramatically by the presence of a *catalyst. A* ***catalyst*** *is a substance, or substances, that increases the rate of a chemical reaction without itself being permanently consumed.* Catalytic activity is called *catalysis.* Catalysts do not appear among the final products of reactions they accelerate. They may participate in one step along a reaction pathway and be regenerated in a later step. In large-scale and cost-sensitive reaction systems, the catalysts are recovered and reused.

Enzymes are biochemical catalysts. Each enzyme accelerates a specific digestive process, thus enabling the chemical reaction to be completed in a few hours. The same uncatalyzed process might require several weeks for completion.

A catalyst that is in the same phase as all the reactants and products in a reaction system is called a *homogeneous catalyst*. When its phase is different from that of the reactants, it is called a *heterogeneous catalyst*. Metals are often used as heterogeneous catalysts. The catalysis of the reaction is promoted by adsorption of reactants on the metal surfaces, which has the effect of increasing the concentration (and therefore the contacts) of the reactants.

Such catalytic actions are hindered by the presence of substances called *inhibitors*. These inhibitors, once referred to as negative catalysts, interfere with the surface chemistry of heterogeneous catalysts and thus *inhibit* and eventually destroy the catalytic action. An example is the "poisoning" of metallic oxide catalysts in the catalytic converters of automobile exhaust systems by leaded gasoline.

Catalysts may provide alternate reaction pathways.

Chemists believe that a catalyst provides an alternate pathway or reaction mechanism in which the potential energy barrier between reactants and products is lowered. The catalyst may be effective in forming an alternate activated complex that requires a lower activation energy, as suggested in the energy profiles of Figure 20-16.

20.13 Reaction rate law The influence of reactant concentrations on reaction rate in a given chemical system has been mentioned. Measuring this influence is a challenging part of kinetics. Chemists determine the relationship between the rate of a reaction and the concentration of one reactant by first keeping the

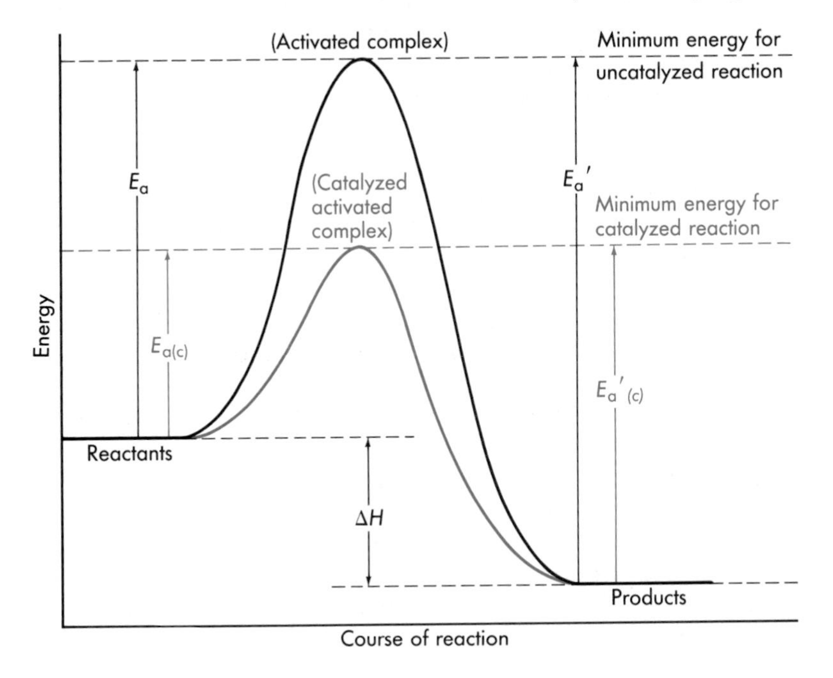

Figure 20-16. Possible difference in potential-energy change along alternate reaction pathways, one catalyzed and the other uncatalyzed.

concentrations of other reactants and the temperature of the system constant. Then the reaction rate is measured for various concentrations of the reactant in question. A series of such experiments reveals how the concentration of each reactant affects the reaction rate.

The following homogeneous reaction is used as an illustration. The reaction is carried out in a vessel of constant volume and at an elevated *constant* temperature.

$$2H_2(g) + 2NO(g) \rightarrow N_2(g) + 2H_2O(g)$$

Here, four moles of reactant gases produce three moles of product gases. Thus the pressure of the system diminishes as the reaction proceeds. The rate of the reaction is determined by measuring the change of pressure in the vessel with time.

Suppose a series of experiments is conducted using the same initial concentration of nitrogen monoxide but different initial concentrations of hydrogen. The initial reaction rate is found to vary directly with the hydrogen concentration. That is, doubling the concentration of H_2 doubles the rate, and tripling the concentration of H_2 triples the rate. Therefore,

$$R \propto [H_2]$$

Here R is the reaction rate and $[H_2]$ is the molecular concentration of hydrogen in moles per liter. The $\propto$ is a proportionality symbol.

Now suppose the same initial concentration of hydrogen is used but the initial concentrations of nitrogen monoxide are varied. The initial reaction rate is found to increase four times when the NO concentration is doubled and nine times when the concentration of NO is tripled. Thus the reaction rate varies directly with the *square* of the nitrogen monoxide concentration.

$$R \propto [NO]^2$$

Since R is proportional to $[H_2]$ and to $[NO]^2$, it is proportional to their product.

$$R \propto [H_2][NO]^2$$

By introducing an appropriate proportionality constant k, the expression becomes an equality.

$$R = k[H_2][NO]^2$$

This equation relating the reaction rate and concentrations of reactants is called the *rate law* for the reaction. It is constant for a specific reaction at a given temperature. A rise in temperature causes an increase in the rate of nearly all reactions. Thus the value of k usually increases as the temperature increases.

Collision theory indicates that the number of collisions between particles increases as the concentration of these particles is raised. The reaction rate *for any step* in a reaction pathway

Figure 20-17. Dr. Reatha Clark King is a chemist and educator who has been a pioneer in thermochemistry research and in university administration. Since 1977 she has served as president of Metropolitan State University in St. Paul, Minnesota. Dr. King is keenly interested in educational opportunities for women and minorities, and has spoken frequently on this subject in the U.S. and other countries.

is directly proportional to the frequency of collisions between the particles involved. It is also directly proportional to the collision efficiency.

Consider a single step in a reaction pathway. Suppose one molecule of gas **A** collides with one molecule of gas **B** to form two molecules of substance **C.** The equation for the step is

$$\mathbf{A + B \rightarrow 2C}$$

One particle of each reactant is involved in each collision. Thus doubling the concentration of either reactant will double the collision frequency. It will also double the reaction rate *for this step.* Therefore, the rate is directly proportional to the concentration of **A** and of **B.** The rate law for the step becomes

$$R = k[\mathrm{A}][\mathrm{B}]$$

Now suppose the reaction is reversible. In this reverse step, two molecules of **C** must collide to form one molecule of **A** and one of **B.**

$$\mathbf{2C \rightarrow A + B}$$

Thus the reaction rate *for this reverse step* is directly proportional to [C] × [C]. The rate law for the step becomes

$$R = k[\mathrm{C}]^2$$

A simple relationship *may* exist between the chemical equation *for a step in a reaction pathway* and *the rate law for that step.* Observe that the *power* to which the molar concentration of each reactant or product is raised in the rate laws above corresponds to the coefficient for the reactant or product in the balanced equation.

This relationship *does not always hold.* It holds if the reaction follows a simple one-step pathway. A reaction may proceed in a series of steps, however, and in such reactions *the rate law is simply that for the slowest (rate-determining) step.* Thus the rate law for a reaction must be determined experimentally. *It cannot be written from the balanced equation for the net reaction.*

Again consider the hydrogen–nitrogen monoxide reaction. The balanced equation for the net reaction is

$$2H_2(g) + 2NO(g) \rightarrow N_2(g) + 2H_2O(g)$$

Suppose this reaction occurs in a single step involving a collision between two H_2 molecules and two NO molecules. Doubling the concentration of either reactant should quadruple the collision frequency and the rate. The rate law should then be proportional to $[H_2]^2$ as well as $[NO]^2$. It would take the form

$$R = k[H_2]^2[NO]^2$$

This equation is *not* in agreement with experimental results. Thus the assumed reaction mechanism cannot be correct.

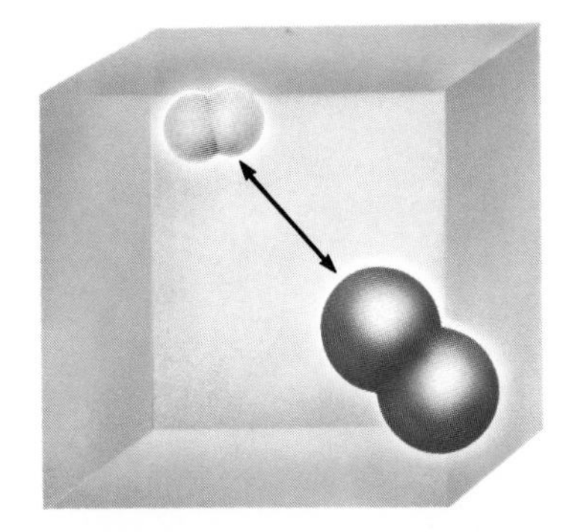

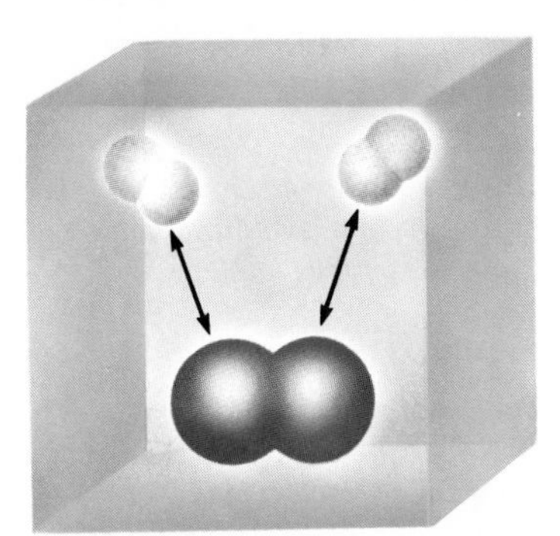

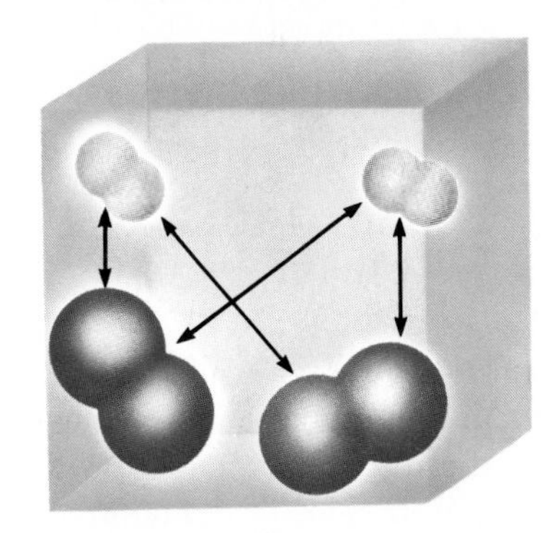

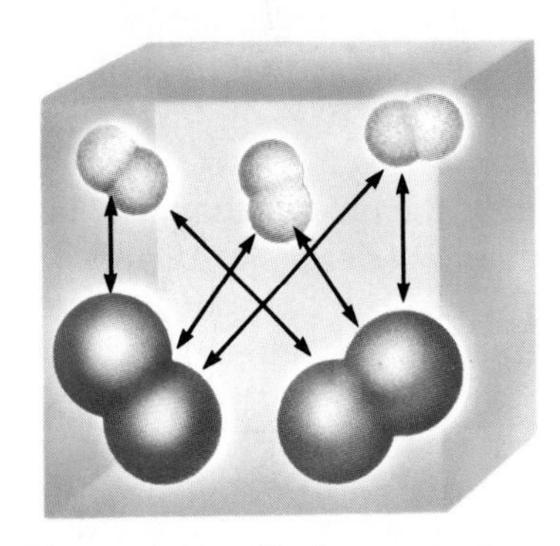

Figure 20-18. Under constant conditions, the collision frequency increases with the concentration of each reactant.

The experimental rate law for the net reaction is actually

$$R = k[H_2][NO]^2$$

Therefore, the reaction pathway must consist of more than one step. The slowest step is rate-determining. The rate law found experimentally for the net reaction is the rate expression of this step.

The general form for a rate law is

$$R = k[A]^n[B]^m\text{-----}$$

Here R is the reaction rate, k is the rate constant, and **[A]** and **[B]-----** represent the molar concentrations of reactants. The n and m are the respective powers to which the concentrations are raised, *based on experimental data.* Again it should be emphasized that one *cannot* assume that the coefficients in the balanced equation for a net reaction are the exponents in the rate law for the reaction.

In the rate law for a reaction, the exponents of reactant concentrations are determined experimentally.

The dependence of reaction rate on concentration of reactants was first recognized as a general principle by two Norwegian chemists, Guldberg and Waage, in 1867. It was stated as the ***law of mass action:*** *the rate of a chemical reaction is directly proportional to the product of the concentrations of reacting substances, each raised to the appropriate power.* This principle is interpreted in terms of the rate law for a chemical system in modern reaction kinetics.

Cato Maximilian Guldberg (1836–1902) was professor of applied mathematics and technology and Peter Waage (1833–1900) was professor of chemistry at the University of Oslo. They collaborated on the mathematical expression of the law of mass action.

SUMMARY

Heat energy is released or absorbed during chemical reactions as the heat of reaction. An exothermic reaction has products with a total heat content lower than that of the reactants. An endothermic reaction has products with a heat content higher than that of the reactants. Inclusion of the heat of reaction information in the chemical equation for a reaction is customary when the energy change is important. Such an equation is called a thermochemical equation.

The heat content of a substance is represented by the letter H and is usually expressed in kilocalories per mole. The change in heat content, the difference between the heat content of products and reactants, is represented as ΔH. For endothermic reactions, ΔH is a positive quantity; for exothermic reactions, it is negative. Changes in heat content in reaction systems are related to changes in the number of bonds breaking and forming and to the strengths of these bonds as reactants form products.

The heat of reaction released or absorbed during the formation of one mole of a compound from its elements is called its heat of formation. The stability of a compound is related to its heat of formation. The heat of reaction released during the complete combustion of one mole of a substance is called its heat of combustion. The heat of formation of a compound not formed directly from its elements may be determined from its heat of combustion and the heats of formation of its combustion products.

The driving force of reaction systems consists of tendencies for processes to occur that lead to lower energy and to higher entropy (state of disorder). The assessment of these two tendencies in a reaction system is expressed in terms of a change in the free energy of the system, ΔG. If the reaction process results in lower free energy, it tends to proceed spontaneously. At low temperatures, the change in heat content toward a lower energy state is the driving force that dominates reaction systems. At high temperatures, the change in entropy toward a higher entropy state dominates reaction systems. Processes that do not proceed spontaneously at low temperatures may become spontaneous at some higher temperature.

A chemical reaction may occur in a sequence of simple steps called the reaction pathway or reaction mechanism. Chemists use the collision theory to help them interpret their observations of reaction processes. According to the collision theory, for reactions to occur, particles must collide and these collisions must result in interactions between the particles. The minimum energy required to produce effective collisions is the activation energy for the reaction. If a collision is energetic enough and the colliding molecules are suitably oriented, an activated complex is formed for a brief interval. It is a molecular complex in which bonds are being broken and formed.

Factors that influence the rate of a chemical reaction are (1) the nature of the reactants, (2) the amount of contact area, (3) the concentration of the reactants, (4) temperature, and (5) catalysis. The rate of a reaction is determined by the amount of reactants converted to products per unit of time. An equation that relates the reaction rate and the concentration of reactants is called the rate law for the reaction. The rate law for a reaction is the rate law for the slowest step in the reaction pathway. This slowest step is the rate-determining step for the reaction. The rate law for a reaction must be determined experimentally. It cannot be written from the balanced equation for the overall reaction.

VOCABULARY

activated complex
activation energy
bond energy
calorimeter
catalysis
chemical kinetics
collision theory
enthalpy
entropy
free energy
free energy change
heat content
heat of combustion
heat of formation
heat of reaction
heterogeneous catalyst
heterogeneous reaction
homogeneous
homogeneous catalyst
homogeneous chemical system
homogeneous reaction
inhibitor
intermediate product
law of mass action
molar heat of formation
rate-determining step
rate law
reaction mechanism
reaction pathway
reaction rate
stable
thermochemical equation
thermochemistry
unstable

QUESTIONS

Group A

1. What evidence can be cited to illustrate the fact that a substance has a characteristic internal energy? *A3i*
2. How does the heat content of the products of a reaction system compare with the heat content of the reactants when the reaction is *(a)* endothermic; *(b)* exothermic? *A3b*
3. Name two factors that collectively determine the driving force of chemical reactions. *A2f*
4. Changes of phase in the direction of the solid phase favor what kind of entropy change? *A2f*
5. What is the effect on the entropy of a system when temperature is raised? *A2f*
6. *(a)* Define the molar heat of formation of a

compound. (b) What specific notation represents "heat of formation"? A1, A3a

7. What is the basis for assigning a negative value to the change in heat content, ΔH, in an exothermic system? A1, A2d
8. To what does the term *activated complex* refer? A1
9. Define *activation energy* in terms of the activated complex of a reaction. A1
10. In a reversible reaction, how does the activation energy required for the exothermic change compare with the activation energy for the endothermic change? A2c, A2d
11. Give two reasons why a collision between reactant molecules may not be effective in producing new chemical species. A1, A2i

QUESTIONS

GROUP B

12. Using the energy profile of Figure 20-14, what is the activation energy for the reaction that produces hydrogen iodide? A1h
13. Considering the structure and phase of substances in a reacting system, to what is the energy change in the reaction related? A2i
14. A compound is found to have a heat of formation ΔH_f of +87.3 kcal/mole. What does this imply about its stability? Explain. A2b
15. Given: two flasks, each containing a different gas at room temperature. The flasks are connected so that the two gases mix. What kind of evidence would indicate that the gases experienced no change in energy content during the mixing? A3d
16. Explain the mixing tendency of Question 15. A2f
17. Explain the circumstances under which an endothermic reaction is spontaneous. A2g
18. Explain the circumstances under which an exothermic reaction does not proceed spontaneously. A2g
19. (a) How can you justify calling the reaction pathway that is shown in Figure 20-12 the minimum energy pathway for reaction? (b) What significance is associated with the maximum energy region of this minimum energy pathway? A2l
20. The balanced equation for a homogeneous reaction between two gases, $4A + B \rightarrow 2C + 2D$, shows that 4 molecules of A react with 1 molecule of B to form 2 molecules of C and 2 molecules of D. The simultaneous collision of 4 molecules of one reactant with 1 molecule of the other reactant is extremely improbable. Recognizing this, what would you assume about the nature of the reaction mechanism for this reaction system? A3h, A4, A3
21. The decomposition of nitrogen dioxide
$$2NO_2 \rightarrow 2NO + O_2$$
occurs in a two-step sequence at elevated temperatures. The first step is
$$NO_2 \rightarrow NO + O$$
Predict a possible second step that, when combined with the first step, gives the complete reaction. A4, A3
22. For each of the following reactions, predict whether the reaction, once started, is likely to proceed rapidly or slowly. State your reason for each prediction. A2m
    (a) $H_2(g) + Cl_2(g) \rightarrow 2HCl(g)$
    (b) $Ag^+(aq) + Cl^-(aq) \rightarrow AgCl(s)$
    (c) $Fe(chunk) + S(l) \rightarrow FeS(s)$
23. What property would you measure in order to determine the reaction rate for the following reaction? Justify your choice. A2o, A3k
$$2NO_2(g) \rightarrow N_2O_4(g)$$
24. Ozone decomposes according to the following equation:
$$2O_3 \rightarrow 3O_2$$

The reaction proceeds in two steps. Propose a possible two-step mechanism. *A1, A3h, A4c*

25. In a thought experiment, suppose 2 moles of hydrogen gas and 1 mole of iodine vapor are passed simultaneously into a 1-liter flask. The rate law for the forward reaction is $R = k[I_2][H_2]$. What is the effect on the rate of the forward reaction if *(a)* the temperature is increased; *(b)* 1 mole of iodine vapor is added; *(c)* 1 mole of hydrogen is removed; *(d)* the volume of the flask is reduced (assume this is possible); *(e)* a catalyst is introduced into the flask? *A1h*

PROBLEMS

GROUP A

1. Write the thermochemical equation for the complete combustion of 1 mole of propane gas. Calculate the heat of formation of propane from the heat of combustion and product heats of formation data. *A6b, B1*
2. Write the thermochemical equation for the complete combustion of 1 mole of ethane gas. Calculate the heat of formation of ethane from the heat of combustion and product heats of formation data. *A6b, B1*
3. Write the thermochemical equation for the complete combustion of 1 mole of ethyne (acetylene) and calculate its heat of formation. *A6b, B1*
4. Write the thermochemical equation for the complete combustion of 1 mole of benzene and calculate its heat of formation. *A6b*
5. Using heats of formation data, calculate the heat of combustion of 1 mole of hydrogen gas. *A6b, B2*
6. Calculate the heat of combustion of 1 mole of sulfur. *A6b, B2*
7. The heat of formation of ethanol (C_2H_5OH) is −66.20 kcal/mole at 25 °C. Calculate the heat of combustion of 1 mole of ethanol.

PROBLEMS

GROUP B

8. The concentration of reactant A changes from 0.0375 *M* to 0.0250 *M* in the reaction time interval 0.0 min–18.3 min. What is the reaction rate during this time interval *(a)* per minute; *(b)* per second? *B3*
9. Calculate the heat of formation of H_2SO_4 (1) from the ΔH for the combustion of sulfur to SO_2(g), the oxidation of SO_2 to SO_3(g), and the solution of SO_3 in H_2O(1) (ht of soln: −31.65 kcal) to give H_2SO_4(1) at 25 °C. *A6b, B1*
10. A chemical reaction is expressed by the balanced chemical equation

$$A + B \rightarrow C$$

Three reaction rate experiments yield the following data:

Experiment number	Initial [A]	Initial [B]	Initial rate of formation of C
1	0.20 *M*	0.20 *M*	2.0×10^{-4} *M*/min
2	0.20 *M*	0.40 *M*	8.0×10^{-4} *M*/min
3	0.40 *M*	0.40 *M*	1.6×10^{-3} *M*/min

(a) Determine the rate law for the reaction. *(b)* Calculate the value of the specific rate constant. *(c)* If the initial concentrations of both A and B are 0.30 *M*, at what initial rate is C formed? *B3, B4*

Wilhelm Ostwald (German) was awarded the Nobel Prize in chemistry in 1909 for his studies of catalysis, chemical equilibrium, and rates of reactions. He was a pioneer in the development of physical chemistry as a separate branch of chemistry, but he refused to believe in the existence of atoms. He began his research by studying reactions in solution. His studies in catalysis indicated that catalysts accelerate reactions that otherwise proceed slowly. He also discovered a method for converting ammonia to nitric acid by using a platinum catalyst; this method later became important in industry.

Chemical Equilibrium

21

GOALS

In this chapter you will gain an understanding of: • the nature of reversible reactions • physical and chemical equilibria • the equilibrium constant and its calculation • factors that disturb equilibrium, and predict their influence using Le Chatelier's principle • the three conditions that cause an ionic reaction to run to completion • the common ion effect • ionization constants—weak acids and water • hydrolysis of salts: cation and anion hydrolysis • solubility: solubility-product constant, calculating solubilities, predicting precipitates

21.1 Reversible reactions The products formed in a chemical reaction may, in turn, react to re-form the original reactants. Such chemical reactions are said to be reversible. All chemical reactions are considered to be reversible under suitable conditions. Theoretically at least, every reaction pathway can be traversed in both directions. Some reverse reactions occur or are driven less easily than others. For example, a temperature near 3000 °C is required for water vapor to decompose into measurable quantities of hydrogen and oxygen. In some cases the conditions for the reverse reactions are not known. An example is the reaction in which potassium chlorate decomposes into oxygen and potassium chloride. Chemists do not know how to reverse this reaction in a single process.

Mercury(II) oxide decomposes when heated strongly.

$$2HgO(s) \rightarrow 2Hg(l) + O_2(g)$$

Mercury and oxygen combine to form mercury(II) oxide when heated gently.

$$2Hg(l) + O_2(g) \rightarrow 2HgO(s)$$

Suppose mercury(II) oxide is heated in a closed container from which neither the mercury nor the oxygen can escape. Once decomposition has begun, the mercury and oxygen released can recombine to form mercury(II) oxide again. Thus both reactions proceed at the same time. Under these conditions, the rate of the composition reaction will eventually equal that of the decomposition reaction. Mercury and oxygen combine to form mercury(II) oxide just as fast as mercury(II) oxide decomposes into mercury and oxygen. The amounts of mercury(II) oxide,

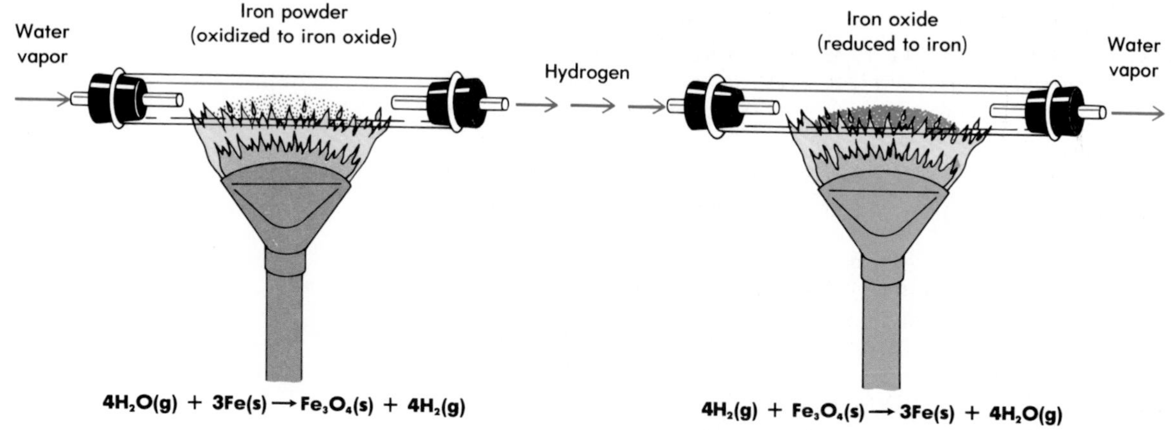

Figure 21-1. A reversible reaction. Water vapor (steam) passed over the hot iron is reduced to hydrogen, and the iron is oxidized to iron(II, III) oxide. Hydrogen passed over the hot iron oxide is oxidized to water vapor, and the iron oxide is reduced to iron.

mercury, and oxygen can be expected to remain constant as long as these conditions persist. A state of *equilibrium* has been reached between the two chemical reactions. *Both reactions continue, but there is no net change in the composition of the system.* The equilibrium may be written as

$$2HgO(s) \rightleftarrows 2Hg(l) + O_2(g)$$

Chemical equilibrium *is a state of balance in which the rates of opposing reactions are exactly equal.*

Equilibrium: opposing processes proceeding at the same rate.

A reaction system in equilibrium shows equal tendencies to proceed in the forward and reverse directions. In Section 20.7, two factors that constitute the driving force for reactions are distinguished: tendency toward higher entropy and tendency toward lower energy. *At equilibrium, the driving force of the energy-change factor is balanced by the driving force of the entropy-change factor.* Under different circumstances, conditions may favor one reaction direction over the other, and a net reaction may result.

21.2 Equilibrium, a dynamic state Equilibrium systems consist of opposing processes occurring at the same time and at the same rate. The evaporation of a liquid in a closed vessel and the condensation of its saturated vapor proceed at equal rates. The resulting equilibrium vapor pressure is a characteristic of the liquid at the prevailing temperature.

If an excess of sugar is placed in water, sugar molecules go into solution. Some of these molecules in turn separate from solution and rejoin the crystals. At saturation, molecules of sugar are crystallizing at the same rate that crystal molecules are dissolving. See Section 12.3.

The preceding are examples of *physical equilibria.* The opposing physical processes occur at exactly the same rate. *Equilibrium is a dynamic state in which two opposing processes proceed simultaneously at the same rate.*

Solution equilibrium is an example of physical equilibrium.

Electrovalent compounds, such as sodium chloride, are composed of ions. Their aqueous solutions consist of hydrated solute ions dispersed throughout the water solvent. When an excess of sodium chloride is placed in water, a saturated solution eventually results. A *solution equilibrium* is established in which the rate of association of ions re-forming the crystals equals the rate of dissociation of ions from the crystals. This equilibrium is shown in the ionic equation

$$Na^+Cl^-(s) \rightleftharpoons Na^+(aq) + Cl^-(aq)$$

Equilibrium is a dynamic state.

The dynamic character of this equilibrium system is easily demonstrated. Suppose an irregularly shaped crystal of sodium chloride is placed in a saturated solution of the salt. The *shape* of the crystal gradually changes, becoming more regular as time passes. However, the *mass* of the crystal does not change.

Polar compounds, such as acetic acid, are very soluble in water. Molecules of acetic acid in water solution ionize, forming H_3O^+ and $C_2H_3O_2^-$ ions. Pairs of these ions tend to rejoin, forming acetic acid molecules in the solution. This tendency is very strong. Even in fairly dilute solutions, equilibrium is quickly established between un-ionized molecules in solution and their hydrated ions. This system is an example of *ionic equilibrium*. The ionic equilibrium of acetic acid in water solution is represented by the equation

$$HC_2H_3O_2(aq) + H_2O(l) \rightleftharpoons H_3O^+(aq) + C_2H_3O_2^-(aq)$$

Many chemical reactions are reversible under ordinary conditions of temperature and concentration. They may reach a state of equilibrium unless at least one of the substances involved escapes or is removed. In some cases, however, the forward reaction is nearly completed before the reverse reaction rate becomes high enough to establish equilibrium. *Here the products of the forward reaction (→) are favored.* This kind of reaction is referred to as the *reaction to the right* because the convention for writing chemical reactions is that *left-to-right* is forward and *right-to-left* is reverse.

An equilibrium established late in the reaction process favors the products of the forward reaction.

$$HBr + H_2O \rightleftharpoons H_3O^+ + Br^-$$

In other cases, the forward reaction is barely under way when the rate of the reverse reaction becomes equal to that of the forward reaction, and equilibrium is established. *In these cases, the products of the reverse reaction (←), the original reactants, are favored.* This kind of chemical reaction is referred to as the *reaction to the left*.

An equilibrium established early in the reaction process favors the products of the reverse reaction.

$$H_2CO_3 + H_2O \rightleftharpoons H_3O^+ + HCO_3^-$$

In still other cases, both forward and reverse reactions occur to nearly the same extent before chemical equilibrium is thus established. *Neither reaction is favored; considerable concentrations of both reactants and products are present at equilibrium.*

$$H_3PO_4 + H_2O \rightleftarrows H_3O^+ + H_2PO_4^-$$

Chemical reactions are ordinarily used to convert available reactants into more desirable products. Chemists try to produce as much of these products as possible from the reactants used. Chemical equilibrium may seriously limit the possibilities of a seemingly useful reaction. In dealing with equilibrium systems, it is important to recognize the conditions that influence reaction rates. Factors that determine the rate of chemical action are discussed in Section 20.12. They are (1) *the nature of the reactants,* (2) *the temperature,* (3) *the presence of a catalyst,* (4) *the surface area, and* (5) *the concentration of reactants.*

Heterogeneous reactions involve two or more phases of matter.

In homogeneous reactions, reactants and products are in the same phase.

In heterogeneous reactions, the chemical reaction takes place at the surfaces where the reactants in different phases meet. Thus the surface area presented by solid and liquid reactants is important in rate considerations. Homogeneous reactions occur between gases and between substances dissolved in liquid solvents. Here the concentration of each reactant is an important rate factor.

21.3 The equilibrium constant Many chemical reactions seem likely to yield useful products. After they are started, however, they *appear* to slow down and finally stop without having run to completion. Such reactions are reversible and happen to *reach a state of equilibrium* before the reactants are completely changed into products. Both forward and reverse processes continue at the same rate. The concentrations of products and reactants remain constant.

The time required for reaction systems to reach equilibrium varies widely. It may be a fraction of a second or a great many years, depending on the system and the conditions. Ionic reactions in solution usually reach equilibrium very quickly.

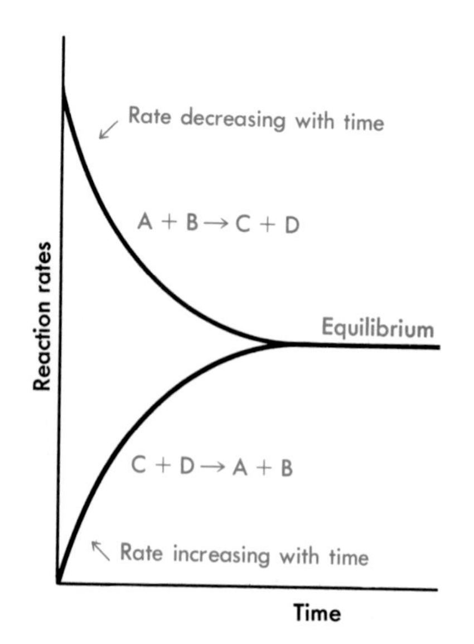

Figure 21-2. Reaction rates for the hypothetical equilibrium reaction system **A + B ⇄ C + D**. The rate of the forward reaction is represented by curve **A + B**. Curve **C + D** represents the rate of the reverse reaction. At equilibrium, the two rates are equal.

Suppose two substances, **A** and **B,** react to form products **C** and **D,** as in Figure 21-2. In turn, **C** and **D** react to produce **A** and **B.** Under certain conditions, equilibrium occurs in this reversible reaction. This hypothetical equilibrium reaction equation is

$$A + B \rightleftarrows C + D$$

Initially, the concentrations of **C** and **D** are zero, and those of **A** and **B** are maximum. With time, the rate of the forward reaction *decreases* as **A** and **B** are used up. Meanwhile, the rate of the reverse reaction increases as **C** and **D** are formed. As these two reaction rates become equal, equilibrium is established. The individual concentrations of **A, B, C,** and **D** undergo no further change if conditions remain the same.

At equilibrium, the ratio of the product **[C]** × **[D]** *to the product*

[A] × [B] *has a definite value at a given temperature.* It is known as the ***equilibrium constant*** of the reaction and is designated by the letter *K*. Thus,

$$\frac{[C] \times [D]}{[A] \times [B]} = K$$

Observe that the concentrations of substances on the right side of the chemical equation are given in the numerator. These substances are the *products* of the forward reaction. The concentrations of substances on the left side of the chemical equation are in the denominator. These substances are the *reactants* of the forward reaction. Concentrations of reactants and products are given in *moles per liter*. The constant *K* is independent of the initial concentrations. It is, however, dependent on the fixed temperature of the system.

Equilibrium constant: *K has a unique value for each equilibrium system at a specific temperature.*

The value of *K* for a given equilibrium reaction is important to chemists. It shows them the extent to which the reactants are converted into the products of the reaction. If *K* is equal to 1, the products of the concentrations in the numerator and denominator have the same value. If the value of *K* is very small, the forward reaction occurs only very slightly before equilibrium is established. A large value of *K* indicates an equilibrium in which the original reactants are largely converted to products. The numerical value of *K* for a particular equilibrium system is obtained experimentally. The chemist must analyze the equilibrium mixture and determine the concentrations of all substances.

If $K < 1$, *reactants of the forward reaction are favored.*

If $K > 1$, *products of the forward reaction are favored.*

Suppose the balanced equation for an equilibrium system has the general form

$$nA + mB + \text{-----} \rightleftarrows xC + yD + \text{-----}$$

The constant relationship at equilibrium (for this generalized reaction system) becomes

$$K = \frac{[C]^x[D]^y\text{-----}}{[A]^n[B]^m\text{-----}}$$

The ***equilibrium constant*** *K is the ratio of the product of the concentrations of substances formed at equilibrium to the product of the concentrations of reacting substances, each concentration being raised to the power that is the coefficient of that substance in the chemical equation.* This equation for *K* is sometimes referred to as the *chemical equilibrium law* or as the *mass action expression.*

To illustrate, suppose a reaction system at equilibrium is shown by the equation

$$3A + B \rightleftarrows 2C + 3D$$

The equilibrium constant *K* is given by the expression

$$K = \frac{[C]^2[D]^3}{[A]^3[B]}$$

The reaction between H_2 and I_2 in a closed flask at an elevated temperature is easy to follow; the rate at which the violet color of the iodine vapor diminishes is simply observed. Suppose this reaction runs to completion with respect to iodine. If so, the color must disappear entirely since the product, hydrogen iodide, is a colorless gas.

The color does not disappear entirely because the reaction is reversible. Hydrogen iodide decomposes to re-form hydrogen and iodine. The rate of this reverse reaction increases as the concentration of hydrogen iodide builds up. Meanwhile, the concentrations of hydrogen and iodine decrease as they are used up. The rate of the forward reaction decreases accordingly.

As the rates of the opposing reactions become equal, an equilibrium is established. A constant intensity of the violet color indicates that equilibrium exists among hydrogen, iodine, and hydrogen iodide. The net chemical equation for the reaction system at equilibrium is

$$H_2(g) + I_2(g) \rightleftarrows 2HI(g)$$

Thus,

$$K = \frac{[HI]^2}{[H_2][I_2]}$$

Concentrations of substances in an equilibrium system are determined experimentally.

Chemists have carefully measured the concentrations of H_2, I_2, and HI in equilibrium mixtures at various temperatures. In some experiments, the flasks were filled with hydrogen iodide at known pressure. The flasks were held at fixed temperatures until equilibrium was established. In other experiments, hydrogen and iodine were the substances introduced.

Experimental data together with the calculated values for K are listed in Table 21-1. Experiments 1 and 2 began with hydrogen iodide. Experiments 3 and 4 began with hydrogen and iodine. Note the close agreement obtained for numerical values of the equilibrium constant.

The equilibrium constant K for this equilibrium reaction system at 425 °C has the average value of 54.34. This value for K should hold for any system of H_2, I_2, and HI at equilibrium *at this*

Table 21-1
TYPICAL EQUILIBRIUM CONCENTRATIONS OF H_2, I_2, AND HI IN MOLE/LITER AT 425 °C

Exp.	*$[H_2]$*	*$[I_2]$*	*[HI]*	$K = \frac{[HI]^2}{[H_2][I_2]}$
(1)	0.4953×10^{-3}	0.4953×10^{-3}	3.655×10^{-3}	54.46
(2)	1.141×10^{-3}	1.141×10^{-3}	8.410×10^{-3}	54.33
(3)	3.560×10^{-3}	1.250×10^{-3}	15.59×10^{-3}	54.62
(4)	2.252×10^{-3}	2.336×10^{-3}	16.85×10^{-3}	53.97
			Average	54.34

Equilibrium constants are computed from experimental data.

temperature. If the calculation for K yields a different result, there must be a reason. Either the H_2, I_2, and HI system has not reached equilibrium or the temperature of the system is not 425 °C. Sample Problem 21-1 is a typical example.

SAMPLE PROBLEM 21-1

An equilibrium mixture of H_2, I_2, and HI gases at 425 °C is determined to consist of 4.5647×10^{-3} mole/liter of H_2, 0.7378×10^{-3} mole/liter of I_2, and also 13.544×10^{-3} mole/liter of HI. What is the equilibrium constant for the system at this temperature?

SOLUTION

The balanced equation for the equilibrium system is

$$\mathbf{H_2(g) + I_2(g) \rightleftarrows 2HI(g)}$$

$$K = \frac{[HI]^2}{[H_2][I_2]}$$

$$K = \frac{[13.544 \times 10^{-3}]^2}{[4.5647 \times 10^{-3}][0.7378 \times 10^{-3}]} = 54.47$$

This value for K is in close agreement with the average of the four experimental values given in Table 21-1.

Practice Problems

1. An equilibrium mixture at 425 °C is determined to consist of 1.83×10^{-3} mole/L of H_2, 3.13×10^{-3} mole/L of I_2, and also 17.7×10^{-3} mole/L of HI. Calculate the equilibrium constant K. *ans.* 54.7
2. An equilibrium mixture at 425 °C consists of 4.79×10^{-4} mole/L of H_2, 4.79×10^{-4} mole/L of I_2, and 3.53×10^{-3} mole/L of HI. Determine the equilibrium constant. *ans.* 54.3

The balanced chemical equation for the equilibrium system yields the expression for the equilibrium constant. The data in Table 21-1 show that this expression is an experimental fact. The equilibrium concentrations of reactants and product are determined experimentally. The values of K are calculated from these concentrations. No information concerning the kinetics of the reacting systems is required.

However, you know an equilibrium system involves forward and reverse reactions proceeding at equal rates. Thus the rate laws for the forward and reverse reactions should yield the same equilibrium constant as does the balanced chemical equation.

Suppose an equilibrium system is expressed by the chemical equation

$$A_2 + B_2 \rightleftarrows 2C$$

Assume that both forward and reverse reactions are simple, one-step processes. The rate law for the forward reaction becomes

$$R_f = k_f[A_2][B_2]$$

***Rate law:** review Section 20.13.*

The rate law for the reverse reaction becomes

$$R_r = k_r[C]^2$$

At equilibrium,

$$R_f = R_r$$

Therefore,

$$k_f[A_2][B_2] = k_r[C]^2$$

Rearranging terms,

$$\frac{k_f}{k_r} = \frac{[C]^2}{[A_2][B_2]} = K$$

The expression is the same as the mass action equation derived from the balanced chemical equation (Section 20.13).

Most reaction mechanisms are complex and proceed by way of a sequence of simple steps. The exponents in the *rate equation* for such a reaction cannot be taken from the balanced equation. They must be determined experimentally.

21.4 Factors that disturb equilibrium In systems that have attained chemical equilibrium, opposing reactions are proceeding at equal rates. Any change that alters the rate of either reaction *disturbs the original equilibrium.* The system then seeks a new equilibrium state. By shifting an equilibrium in the desired direction, chemists can often increase production of important industrial chemicals.

Le Chatelier's principle is a "must know."

Le Chatelier's principle (Section 12.5) provides a means of predicting the influence of disturbing factors on equilibrium systems. It may be helpful at this point to restate Le Chatelier's principle: *If a system at equilibrium is subjected to a stress, the equilibrium is shifted in the direction that relieves the stress.* This principle holds for all kinds of dynamic equilibria, physical and ionic as well as chemical. In applying Le Chatelier's principle to chemical equilibrium, three important stresses will be considered.

1. *Change in concentration.* From collision theory, an increase in the concentration of a reactant causes an increase in collision frequency. A reaction resulting from these collisions should then proceed at a faster rate. Consider the hypothetical reaction

$$A + B \rightleftarrows C + D$$

An increase in the concentration of **A** will shift the equilibrium to the *right*. Both **A** and **B** will be used up faster, and more of **C** and **D** will be formed. The equilibrium will be reestablished with a lower concentration of **B.** *The equilibrium has shifted in such direction as to reduce the stress caused by the increase in concentration.*

A change in concentration of a reactant will place a stress on an equilibrium system.

Similarly, an increase in the concentration of **B** drives the reaction to the *right*. An increase in either **C** or **D** shifts the equilibrium to the *left*. A *decrease* in the concentration of either **C** or **D** has the same effect as an *increase* in the concentration of **A** or **B.** That is, it will shift the equilibrium to the *right*.

Changes in concentration have no effect on the value of the equilibrium constant. All concentrations give the same numerical ratio for the equilibrium constant when equilibrium is reestablished. Thus Le Chatelier's principle leads to the same predictions for changes in concentration as does the equilibrium constant.

2. *Change in pressure.* A change in pressure can affect only equilibrium systems in which *gases* are involved. According to Le Chatelier's principle, *if the pressure on an equilibrium system is increased, the reaction is driven in the direction that relieves the pressure.*

The Haber process for catalytic synthesis of ammonia from its elements illustrates the influence of pressure on an equilibrium system.

$$N_2(g) + 3H_2(g) \rightleftarrows 2NH_3(g)$$

The equation indicates that 4 molecules of the reactant gases form 2 molecules of ammonia gas. Suppose the equilibrium mixture is subjected to an increase in pressure. This pressure can be relieved by the reaction that produces fewer gas molecules and, therefore, a smaller volume. Thus the stress is *lessened* by the formation of ammonia. Equilibrium is shifted toward the right, favoring the production of ammonia. High pressure is desirable in this industrial process. Figure 21-3 illustrates the effect of pressure on this equilibrium system.

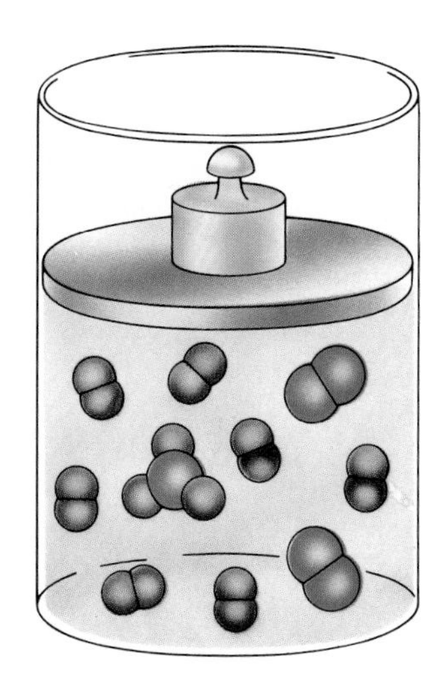

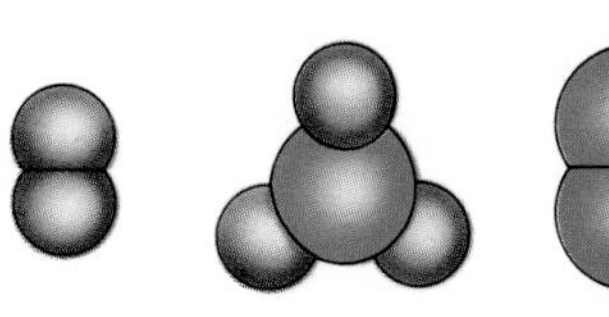

Figure 21-3. Increased pressure results in a higher yield of ammonia because the equilibrium shifts in the direction that produces fewer molecules.

An increase in the partial pressure of H_2 or N_2 in the reaction vessel shifts the equilibrium to the right as this increase in pressure effects an increase in concentration of the resultant gas [Section 21.4 (*1*)]. Similarly, a decrease in the partial pressure of NH_3 gas in the reaction vessel shifts the equilibrium to the right. A change in the partial pressure (and thus the concentration) of at least one of the three reactant gases is necessary for the equilibrium position to be affected.

The introduction of an inert gas such as helium into the reaction vessel for the synthesis of ammonia would increase the total pressure in the vessel. However, the partial pressures (and the concentrations) of the reactant gases present would not be altered by this kind of pressure change, and it cannot affect the equilibrium position of the reaction system.

In the Haber process, the ammonia produced is continuously removed by condensation to a liquid. This condensation

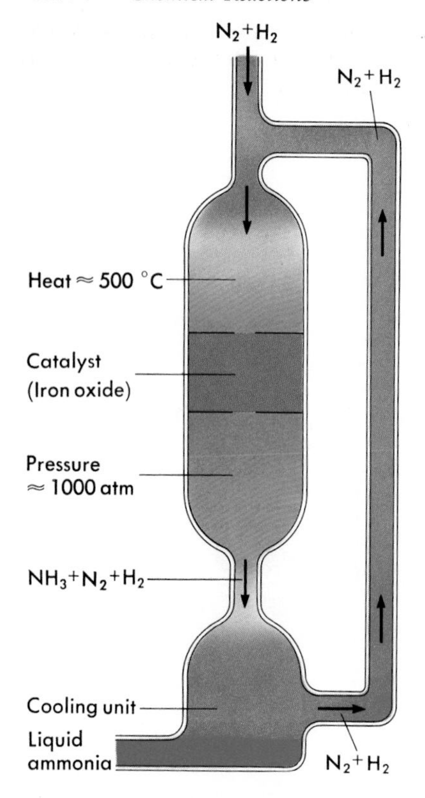

Figure 21-4. The Haber process for the production of ammonia. Application of Le Chatelier's principle to the equilibrium system, **$N_2(g)$ + $3H_2(g) \rightleftarrows 2NH_3(g)$ + 22 kcal,** suggests that high pressure and low temperature favor the yield of ammonia. Why, then, is a moderately high temperature used in this industrial process?

removes most of the product from the (gas) phase in which the reaction occurs. The change in concentration also tends to shift the equilibrium to the right.

Many chemical processes involve heterogeneous reactions in which the reactants and products are in different phases. The *concentrations* in equilibrium systems of pure substances in solid and liquid phases are not changed by adding or removing quantities of the substances. The equilibrium constant expresses a relationship between *relative* concentrations of reactants and products. Therefore, a pure substance in a condensed phase can be removed from the expression for the equilibrium constant by substituting the number "1" for its concentration, which remains at unity value in the equilibrium system.

Consider the equilibrium system given by the equation

$$CaCO_3(s) \rightleftarrows CaO(s) + CO_2(g)$$

Carbon dioxide is the only substance in the system subject to changes in concentration. Since it is a gas, the forward (decomposition) reaction is favored by a *low* pressure. The expression for the equilibrium constant is

$$K = \frac{[CaO][CO_2]}{[CaCO_3]} = \frac{[1][CO_2]}{[1]} = [CO_2]$$

In the reaction

$$CO(g) + H_2O(g) \rightleftarrows CO_2(g) + H_2(g)$$

there are equal numbers of molecules of gaseous reactants and gaseous products. Pressure change cannot produce a shift in equilibrium. Thus, *pressure change has no effect* on the position of equilibrium in this reaction.

Obviously, an increase in pressure on confined gases has the same effect as an increase in the concentrations of these gases. Recall that changes in concentration have no effect on the value of the equilibrium constant. Thus *changes in pressure do not affect the value of the equilibrium constant.*

3. *Change in temperature.* Chemical reactions are either exothermic or endothermic. Reversible reactions are exothermic in one direction and endothermic in the other. The effect of changing the temperature of an equilibrium mixture depends on which of the opposing reactions is endothermic.

According to Le Chatelier's principle, *addition* of heat shifts the equilibrium so that heat is absorbed. This favors the *endothermic* reaction. The *removal* of heat favors the *exothermic* reaction. A rise in temperature increases the rate of any reaction. In an equilibrium system, the rates of the opposing reactions are raised *unequally.* Thus *the value of the equilibrium constant for a given system is affected by the temperature.*

The synthesis of ammonia is *exothermic.*

$$N_2(g) + 3H_2(g) \rightleftarrows 2NH_3(g) + 22 \text{ kcal}$$

A high temperature is not desirable as it favors the decomposition of ammonia, the *endothermic* reaction. However, at ordinary temperatures, the forward reaction is too slow to be commercially useful. The temperature used represents a compromise between kinetic and equilibrium requirements. It is high enough that equilibrium is established rapidly but low enough that the equilibrium concentration of ammonia is significant. The reactions of the system are also accelerated by a suitable catalyst. Moderate temperature (about 500 °C) and very high pressure (700–1000 atmospheres) produce a satisfactory yield of ammonia.

In an equilibrium system, one reaction is exothermic, and the other reaction is endothermic.

The numerical values of equilibrium constants range from very large to very small numbers. You have learned that they are independent of changes in concentrations but not of changes in temperature. The addition of a catalyst accelerates both forward and reverse reactions equally. Therefore, it does not affect the value of ***K***. The time required for a system to reach equilibrium can be reduced dramatically when the system is catalyzed.

***K** changes if temperature changes.*

Several equilibrium systems are listed in Table 21-2. The table also gives numerical values of the equilibrium constants at various temperatures. A very small ***K*** value means that the equilibrium mixture consists mainly of the substances on the left of the equation. If the value of ***K*** is large, the equilibrium mixture consists mainly of the substances on the right of the equation.

21.5 Reactions that run to completion Many reactions are easily reversible under suitable conditions. A state of equilibrium may be established unless one or more of the products escapes or is removed. An equilibrium reaction may be driven in the preferred direction by applying Le Chatelier's principle.

Table 21-2
EQUILIBRIUM CONSTANTS

Equilibrium system	*Value of K*	*Temp. (°C)*
$N_2(g) + 3H_2(g) \rightleftharpoons 2NH_3(g)$	2.66×10^{-2}	350°
$N_2(g) + 3H_2(g) \rightleftharpoons 2NH_3(g)$	6.59×10^{-3}	450°
$N_2(g) + 3H_2(g) \rightleftharpoons 2NH_3(g)$	2.37×10^{-3}	727°
$2H_2(g) + S_2(g) \rightleftharpoons 2H_2S(g)$	9.39×10^{-5}	477°
$H_2(g) + CO_2(g) \rightleftharpoons H_2O(g) + CO(g)$	4.40	1727°
$2H_2O(g) \rightleftharpoons 2H_2(g) + O_2(g)$	5.31×10^{-10}	1727°
$2CO(g) + O_2(g) \rightleftharpoons 2CO_2(g)$	2.24×10^{22}	727°
$H_2(g) + I_2(g) \rightleftharpoons 2HI(g)$	66.9	350°
$H_2(g) + I_2(g) \rightleftharpoons 2HI(g)$	54.4	425°
$H_2(g) + I_2(g) \rightleftharpoons 2HI(g)$	45.9	490°
$C(s) + CO_2(g) \rightleftharpoons 2CO(g)$	14.1	1123°
$Cu(s) + 2Ag^+(aq) \rightleftharpoons Cu^{++}(aq) + 2Ag(s)$	2×10^{15}	25°
$I_2(g) \rightleftharpoons 2I(g)$	3.76×10^{-5}	727°
$2O_3(g) \rightleftharpoons 3O_2(g)$	2.54×10^{12}	1727°
$N_2(g) \rightleftharpoons 2N(g)$	1.31×10^{-31}	1000°

Figure 21-5. Death Valley Devil's Golf Course salt formations.

Some reactions appear to go to completion in the forward direction. No one has found a method of recombining potassium chloride and oxygen directly once potassium chlorate decomposes. Sugar is decomposed into carbon and water by the application of heat. Yet no single-step method of recombining these products is known.

Many compounds are formed by the interaction of ions in solutions. If solutions of two electrolytes are mixed, two pairings of ions are possible. These pairings may or may not occur. If dilute solutions of sodium chloride and potassium bromide are mixed, no reaction occurs. The resulting solution merely contains a mixture of Na^+, K^+, Cl^-, and Br^- ions. Association of ions occurs only if enough water is evaporated to cause crystals to separate from solution. The yield is then a mixture of NaCl, KCl, NaBr, and KBr.

With some combinations of ions, reactions do occur. *Such reactions appear to run to completion in the sense that the ions are almost completely removed from solution.* Chemists can predict that certain ion reactions will run practically to completion. The extent to which the reacting ions are removed from solution depends on (1) *the solubility of the compound formed and* (2) *the degree of ionization, if the compound is soluble.* Thus a product that *escapes as a gas or is precipitated as a solid, or is only slightly ionized* effectively removes the bulk of the reacting ions from solution. Consider some specific examples of such reactions.

1. *Formation of a gas.* Unstable substances formed as products of ionic reactions decompose spontaneously. An example is carbonic acid, which yields a gas as a decomposition product.

$$H_2CO_3 \rightarrow H_2O + CO_2(g)$$

Carbonic acid: *aqueous solutions called "carbonic acid" are mainly solutions of CO_2 in water. The relatively few H_2CO_3 molecules formed are ionized, making the solution weakly acidic.*

Carbonic acid is produced in the reaction between sodium hydrogen carbonate and hydrochloric acid, as shown by the equation

$$NaHCO_3 + HCl \rightarrow NaCl + H_2CO_3$$

The aqueous solution of HCl contains H_3O^+ and Cl^- ions. A careful examination of the reaction mechanism suggests that the HCO_3^- ion, a Brønsted base, acquires a proton from the H_3O^+ ion, a Brønsted acid; Na^+ and Cl^- ions are merely spectator ions. The following equations may be more appropriate for this chemical reaction.

$$H_3O^+ + HCO_3^- \rightarrow H_2O + \cancel{H_2CO_3}$$

$$\cancel{H_2CO_3} \rightarrow H_2O + CO_2(g)$$

or, simply,

$$H_3O^+ + HCO_3^- \rightarrow 2H_2O + CO_2(g)$$

The reaction runs practically to completion because one of the products escapes as a gas. Of course, the sodium ions and chloride ions produce sodium chloride crystals when the water solvent is evaporated.

The reaction between iron(II) sulfide and hydrochloric acid goes practically to completion because a gaseous product escapes. The reaction equation is

$$\mathbf{FeS + 2HCl \rightarrow FeCl_2 + H_2S(g)}$$

The net ionic equation is

$$\mathbf{FeS(s) + 2H_3O^+(aq) \rightarrow Fe^{++}(aq) + H_2S(g) + 2H_2O(l)}$$

The hydrogen sulfide formed is only moderately soluble and is largely given off as a gas. The iron(II) chloride is formed on evaporation of the water.

2. *Formation of a precipitate.* When solutions of sodium chloride and silver nitrate are mixed, a white precipitate of silver chloride immediately forms.

$$\mathbf{Na^+ + Cl^- + Ag^+ + NO_3^- \rightarrow Na^+ + NO_3^- + Ag^+Cl^-(s)}$$

If chemically equivalent amounts of the two solutes are used, Na^+ ions and NO_3^- ions remain in solution in appreciable quantities. All but a meager portion of the Ag^+ ions and Cl^- ions combine and separate from the solution as a precipitate of AgCl. Silver chloride is very sparingly soluble in water. Like any other substance of either low or high solubility, it separates by precipitation from a *saturated solution* of its particles. *The reaction effectively runs to completion because an "insoluble" product is formed.*

The only reaction is between the silver ions and chloride ions. Omitting the spectator ions, Na^+ and NO_3^-, the net ionic equation is simply

$$\mathbf{Ag^+(aq) + Cl^-(aq) \rightarrow Ag^+Cl^-(s)}$$

Crystalline sodium nitrate is recovered by evaporation of the water solvent.

3. *Formation of a slightly ionized product.* Neutralization reactions between H_3O^+ ions from aqueous acids and OH^- ions from aqueous bases result in the formation of water molecules. A reaction between HCl and NaOH illustrates this process. Aqueous HCl supplies H_3O^+ ions and Cl^- ions to the solution, and aqueous NaOH supplies Na^+ ions and OH^- ions.

$$\mathbf{H_3O^+ + Cl^- + Na^+ + OH^- \rightarrow Na^+ + Cl^- + 2H_2O}$$

Neglecting the spectator ions, the net ionic equation is simply

$$\mathbf{H_3O^+(aq) + OH^-(aq) \rightarrow 2H_2O(l)}$$

Water is only slightly ionized and exists almost entirely as covalent molecules. Thus hydronium ions and hydroxide ions are almost entirely removed from the solution. *The reaction effectively runs to completion because the product is only slightly ionized.* Sodium chloride separates on evaporation of the water.

In proton-transfer systems of acids and bases, the H_3O^+ ion is an acid, the OH^- ion is a base.

21.6 Common ion effect Suppose hydrogen chloride gas is bubbled into a saturated solution of sodium chloride. As the hydrogen chloride dissolves, sodium chloride separates as a precipitate. The mass action principle applies to this example, since chloride ions are *common* to both solutes. The concentration of chloride ions is increased; that of sodium ions is not. As sodium chloride crystals form, the concentration of sodium ions in the solution is lowered. Thus, increasing the concentration of chloride ions has the effect of decreasing the concentration of sodium ions. This phenomenon is known as the *common ion effect*.

Common ion effect: *add a common ionic species to a saturated solution of a salt, and (salt) solute precipitates.*

Eventually, the rate of dissociation of sodium chloride crystals equals the rate of association of sodium and chloride ions. An equilibrium then exists.

$$Na^+Cl^-(s) \rightleftarrows Na^+(aq) + Cl^-(aq)$$

Further additions of hydrogen chloride disturb this equilibrium and drive the reaction to the *left*. *By forcing the reaction to the left*, more sodium chloride separates. This further reduces the concentration of the sodium ions in solution.

The common ion effect is also observed when *one* ion species of a weak electrolyte is added in excess to a solution. Acetic acid is such an electrolyte. A 0.1-*M* $HC_2H_3O_2$ solution is about 1.4% ionized. The ionic equilibrium is shown by the equation

Common ion effect: *add a salt of a weak acid to a solution of the weak acid, and the* $[H_3O^+]$ *decreases.*

$$HC_2H_3O_2 + H_2O \rightleftarrows H_3O^+ + C_2H_3O_2^-$$

Sodium acetate, an ionic salt, is completely dissociated in water solution. Small additions of sodium acetate to a solution containing acetic acid greatly increase the acetate ion concentration. The equilibrium shifts in the direction that uses acetate ions. More molecules of acetic acid are formed and the concentration of hydronium ions is reduced. In general, *the addition of a salt with an ion common to the solution of a weak electrolyte reduces the ionization of the electrolyte*. A 0.1-*M* $HC_2H_3O_2$ solution has a pH of 2.9. A solution containing 0.1-*M* concentrations of both acetic acid and sodium acetate has a pH of 4.6.

Common ion effect: *add a salt of a weak base to a solution of the weak base, and the* $[OH^-]$ *decreases.*

21.7 Ionization constant of a weak acid About 1.4% of the solute molecules in a 0.1-*M* acetic acid solution are ionized at room temperature. The remaining 98.6% of the $HC_2H_3O_2$ molecules are un-ionized. Thus the water solution contains three species of particles in equilibrium: $HC_2H_3O_2$ molecules, H_3O^+ ions, and $C_2H_3O_2^-$ ions.

At equilibrium, the *rate* of the forward reaction

$$HC_2H_3O_2 + H_2O \rightarrow H_3O^+ + C_2H_3O_2^-$$

is equal to the *rate* of the reverse reaction

$$HC_2H_3O_2 + H_2O \leftarrow H_3O^+ + C_2H_3O_2^-$$

The equilibrium equation is

$$HC_2H_3O_2 + H_2O \rightleftarrows H_3O^+ + C_2H_3O_2^-$$

The equilibrium constant for this system expresses the *equilibrium ratio of ions to molecules.*

From the equilibrium equation for the ionization of acetic acid,

$$K = \frac{[H_3O^+][C_2H_3O_2^-]}{[HC_2H_3O_2][H_2O]}$$

Water molecules are greatly in excess of acetic acid molecules at the 0.1-M concentration. Without introducing a measurable error, the mole concentration of H_2O molecules can be assumed to remain constant in such a solution. Thus the product $K[H_2O]$ is constant.

Weak acids are slightly ionized in aqueous solution.

$$K[H_2O] = \frac{[H_3O^+][C_2H_3O_2^-]}{[HC_2H_3O_2]}$$

By setting $K[H_2O] = K_a$,

$$K_a = \frac{[H_3O^+][C_2H_3O_2^-]}{[HC_2H_3O_2]}$$

In this expression, K_a is called the *ionization constant* of the weak acid. The concentration of water molecules in pure water and in dilute solutions is about 55 moles per liter. (At 25 °C it is 55.4 moles/liter.) Therefore, the ionization constant of a weak acid, K_a, is about 55 times larger than the equilibrium constant K.

Ionization constant: $K_a = K[H_2O] = 55.4K$

The equilibrium equation for the typical weak acid **HB** is

$$HB(aq) + H_2O(l) \rightleftarrows H_3O^+(aq) + B^-(aq)$$

From this equation, the expression for K_a can be written in the general form

$$K_a = \frac{[H_3O^+][B^-]}{[HB]}$$

How can the numerical value of the ionization constant K_a for acetic acid at a specific temperature be determined? First, the equilibrium concentrations of H_3O^+ ions, $C_2H_3O_2^-$ ions, and $HC_2H_3O_2$ molecules must be known. The ionization of a molecule of $HC_2H_3O_2$ in water yields one H_3O^+ ion and one $C_2H_3O_2^-$ ion. Therefore, these concentrations can be found experimentally by measuring the pH of the solution.

Suppose that precise measurements in an experiment show the pH of a 0.1000-M solution of acetic acid to be 2.876 at 25 °C. The numerical value of K_a for $HC_2H_3O_2$ at 25 °C can be determined as follows:

$[H_3O^+] = 10^{-pH}$

$$[H_3O^+] = [C_2H_3O_2^-] = 10^{-2.876}\ \frac{\text{mole}}{\text{liter}}$$

$$\text{antilog}\ (-2.876) = \text{antilog}\ (0.124 - 3) = 1.33 \times 10^{-3}$$

$$[H_3O^+] = [C_2H_3O_2^-] = 1.33 \times 10^{-3}$$

$$[HC_2H_3O_2] = 0.1000 - 0.00133 = 0.0987$$

$$K_a = \frac{[H_3O^+][C_2H_3O_2^-]}{[HC_2H_3O_2]}$$

$$K_a = \frac{(1.33 \times 10^{-3})^2}{9.87 \times 10^{-2}} = 1.79 \times 10^{-5}$$

Ionization data and constants for some dilute acetic acid solutions at 25 °C are given in Table 21-3.

An increase in temperature causes the equilibrium to shift according to Le Chatelier's principle. Thus K_a has a new value for each temperature. An increase in the concentration of $C_2H_3O_2^-$ ions, through the addition of $NaC_2H_3O_2$, also disturbs the equilibrium. This disturbance causes a decrease in $[H_3O^+]$ and an increase in $[HC_2H_3O_2]$. Eventually, the equilibrium is reestablished with the same value of K_a. However, there is a higher concentration of un-ionized acetic acid molecules and a lower concentration of H_3O^+ ions. Changes in the hydronium ion concentration and changes in pH go together. In this example, the reduction in $[H_3O^+]$ means an increase in the pH of the solution.

K_a changes if temperature changes.

A solution containing both a weak acid and a salt of the acid can react with either an acid or a base. The pH of the solution remains nearly constant even when small additions of acids or bases are present. Suppose an acid is added to a solution of acetic acid and sodium acetate. Acetate ions react with the added hydronium ions to form un-ionized acetic acid molecules.

$$C_2H_3O_2^-(aq) + H^+(aq) \rightarrow HC_2H_3O_2(aq)$$

The hydronium-ion concentration and the pH of the solution remain practically unchanged.

Suppose a small amount of a base is added to the solution. The OH^- ions of the base remove hydronium ions as un-ionized water molecules. However, acetic acid molecules ionize and restore the equilibrium concentration of hydronium ions.

$$HC_2H_3O_2(aq) \rightarrow H^+(aq) + C_2H_3O_2^-(aq)$$

Table 21-3
IONIZATION CONSTANT OF ACETIC ACID

Molarity	*% ionized*	*$[H_3O^+]$*	*$[HC_2H_3O_2]$*	*K_a*
0.1000	1.33	0.00133	0.09867	1.79×10^{-5}
0.0500	1.89	0.000945	0.04906	1.82×10^{-5}
0.0100	4.17	0.000417	0.009583	1.81×10^{-5}
0.0050	5.86	0.000293	0.004707	1.81×10^{-5}
0.0010	12.60	0.000126	0.000874	1.72×10^{-5}

The pH of the solution again remains practically unchanged.

A solution of a weak base containing a salt of the base behaves in a similar manner. The hydroxide-ion concentration (and the pH) of the solution remain essentially constant with small additions of acids or bases. Suppose a base is added to an aqueous solution of ammonia that also contains ammonium chloride. Ammonium ions remove the added hydroxide ions as un-ionized water molecules.

$$NH_4^+(aq) + OH^-(aq) \rightarrow NH_3(aq) + H_2O$$

If a small amount of an acid is added to the solution, hydroxide ions from the solution remove the added hydronium ions as un-ionized water molecules. Ammonia molecules in the solution ionize and restore the equilibrium concentration of hydronium ions and the pH of the solution.

$$NH_3(aq) + H_2O \rightarrow NH_4^+(aq) + OH^-(aq)$$

The common ion salt in each of these solutions acts as a "buffer" against significant changes in pH in the solution. The solutions are referred to as *buffered* solutions. They are buffered against changes in pH.

Buffers help stabilize the pH of chemical solutions.

Buffer action has many important applications in chemistry and physiology. Human blood is naturally buffered to maintain a pH of about 7.3. Certain physiological functions require slight variations in pH. However, large changes would lead to serious disturbances of normal body functions or even death.

21.8 Ionization constant of water Pure water is a very poor conductor of electricity because it is very slightly ionized. According to the proton-transfer system of acids and bases, some water molecules donate protons, acting as an acid. Other water molecules, which accept these protons, act as a base.

$$H_2O + H_2O \rightleftarrows H_3O^+ + OH^-$$

The degree of ionization is slight. Equilibrium is quickly established with a very low concentration of H_3O^+ and OH^- ions.

Conductivity experiments with pure water at 25 °C show that the concentrations of H_3O^+ and OH^- ions are both 10^{-7} mole per liter. The expression for the equilibrium constant is

$$K = \frac{[H_3O^+][OH^-]}{[H_2O]^2}$$

A liter of water at 25 °C contains

$$\frac{997 \text{ g } H_2O}{18.0 \text{ g/mole}} = 55.4 \text{ moles } H_2O$$

This concentration of water molecules remains practically the same in all dilute solutions.

Thus both $[H_2O]^2$ and K in the above equilibrium expression

In water and aqueous solutions at 25 °C, $[H_3O^+][OH^-] = 10^{-14}$.

are constants. Their product is the constant K_w, *the ion-product constant for water.* It is equal to the product of the molar concentrations of the H_3O^+ and OH^- ions.

$$K_w = [H_3O^+][OH^-]$$

At 25 °C,

$$K_w = 10^{-7} \times 10^{-7} = 10^{-14}$$

The product, K_w, *of the molar concentrations of* H_3O^+ *and* OH^- *ions has this constant value not only in pure water but in all water solutions at 25 °C.* An acid solution with a pH of 4 has a $[H_3O^+]$ of 10^{-4} mole per liter and a $[OH^-]$ of 10^{-10} mole per liter. An alkaline solution with a pH of 8 has a $[H_3O^+]$ of 10^{-8} mole per liter and a $[OH^-]$ of 10^{-6} mole per liter.

$10^{-7} \times 10^{-7} = 10^{-14}$
$10^{-4} \times 10^{-10} = 10^{-14}$
$10^{-8} \times 10^{-6} = 10^{-14}$

21.9 Hydrolysis of salts When a salt is dissolved in water, the solution might be expected to be neutral. The aqueous solutions of some salts, such as NaCl and KNO_3, are neutral; their solutions have a pH of 7.

Some other salts form aqueous solutions that are either acidic or basic. As salts, they are completely dissociated in aqueous solution; but unlike salts whose aqueous solutions are neutral, their ions may react chemically with the water solvent. The ions of a salt that do react with the solvent are said to *hydrolyze* in water solution. ***Hydrolysis*** *is a reaction between water and ions of a dissolved salt.*

Some ions that do not hydrolyze in aqueous solution: Ba^{++}, Ca^{++}, K^+, Na^+, HSO_4^-, NO_3^-.

Four general categories of salts relating to hydrolysis reactions can be described.

1. *Salts of strong acids and strong bases.* Recall that a salt is one product of an aqueous acid-base neutralization reaction. The salt consists of the positive ions *(cations)* of the base and the negative ions *(anions)* of the acid. For example,

$$HCl(aq) + NaOH(aq) \rightarrow Na^+Cl^-(aq) + H_2O(l)$$

HCl(aq) is a *strong* acid, and NaOH(aq) is a *strong* base. Equivalent quantities of these two reactants yield a neutral aqueous solution of Na^+ and Cl^- ions; the pH of the solution is 7. Neither the Na^+ *cation* (the positive ion of the strong base) nor the Cl^- *anion* (the negative ion of the strong acid) undergoes hydrolysis in water solutions.

Similarly, KNO_3 is the salt of the strong acid HNO_3 and the strong base KOH. Measurements show that the pH of an aqueous KNO_3 solution is always very close to 7. *Neither the cations of a strong base nor the anions of a strong acid hydrolyze appreciably in aqueous solutions.*

2. *Salts of weak acids and strong bases.* Their aqueous solutions are *alkaline. Anions* of the dissolved salt are hydrolyzed in the water solvent and the pH of the solution is *raised,* indicating that the *hydroxide-ion concentration has increased.* Observe that the *anions* of the salt which undergo hydrolysis are the negative ions

from the *weak acid*. The *cations* of the salt, the positive ions from the *strong base*, do not hydrolyze appreciably.

3. *Salts of strong acids and weak bases*. Their aqueous solutions are *acidic*. *Cations* of the dissolved salt are hydrolyzed in the water solvent, and the pH of the solution is *lowered*, indicating that the *hydronium-ion concentration has increased*. Observe that in this case the *cations* of the salt which undergo hydrolysis are the positive ions from the *weak base*. The *anions* of the salt, the negative ions from the *strong acid*, do not hydrolyze appreciably.

4. *Salts of weak acids and weak bases*. Both ions of the dissolved salt are hydrolyzed extensively in the water solvent, and the solution can be either neutral, acidic, or basic depending on the salt dissolved.

Figure 21-6. A polarized light photomicrograph of sodium carbonate crystals ($9\frac{1}{2}$x). Sodium carbonate, $Na_2CO_3 \cdot 10H_2O$, is commonly called washing soda.

If both ions of a salt in aqueous solution are hydrolyzed equally, the solution remains neutral. Ammonium acetate is such a salt. In cases in which both the acid and the base are very weak indeed, the salt may undergo essentially complete decomposition to hydrolysis products. When aluminum sulfide is placed in water, both a precipitate and a gas are formed as hydrolysis products. The reaction is

$$\mathbf{Al_2S_3 + 6H_2O \rightarrow 2Al(OH)_3(s) + 3H_2S(g)}$$

Both products are very sparingly soluble in water and are effectively removed from solution.

Hydrolysis often has important effects on the properties of solutions. Sodium carbonate, washing soda, is widely used as a cleaning agent because of the alkaline properties of its water solution. Sodium hydrogen carbonate, baking soda, forms a mildly alkaline solution in water and has many practical uses. Through the study of hydrolysis, you can understand why the end point of a neutralization reaction (Section 16.14) can occur at a pH other than 7.

21.10 Anion hydrolysis

Review Section 15.12.

The *cations* of the salt of a weak acid and a strong base do not hydrolyze in water. Only *anion* hydrolysis occurs in an aqueous solution of such a salt. In the Brønsted sense, the anion of the salt is the *conjugate base* (proton acceptor) of the weak acid. These ions are basic enough to remove protons from some water molecules (proton donors) to form OH^- ions. An equilibrium is established in which the net effect of the anion hydrolysis is an increase in the hydroxide-ion concentration, $[OH^-]$, of the solution. Using the arbitrary symbol B^- for the anion, this hydrolysis reaction is represented by the general anion hydrolysis equation that follows:

Salts of weak acids and strong bases form basic solutions.

$$\mathbf{B^-(aq) + H_2O(l) \rightleftarrows HB(aq) + OH^-(aq)}$$

In the reaction to the right, the anion B^- acquires a proton from the water molecule to form the weak acid HB and the hydroxide ion OH^-.

Some anions that hydrolyze in aqueous solutions: $C_2H_3O_2^-$, CO_3^{--}, CN^-, PO_4^{---}, S^{--}.

The equilibrium constant for a hydrolysis reaction is called the *hydrolysis constant* K_h. From this general anion hydrolysis equation, the expression for K_h is

$$K_h = \frac{[HB][OH^-]}{[B^-]}$$

This hydrolysis constant K_h is also expressed by the ratio of the ion-product constant K_w for water to the ionization constant K_a for the weak acid.

$$K_h = \frac{K_w}{K_a}$$

The validity of this expression can be demonstrated in the following way. From Sections 21.7 and 21.8,

$$K_w = [H_3O^+][OH^-] \quad \text{and} \quad K_a = \frac{[H_3O^+][B^-]}{[HB]}$$

Thus,

$$K_h = \frac{K_w}{K_a} = \frac{[H_3O^+][OH^-]}{\frac{[H_3O^+][B^-]}{[HB]}} = \frac{[HB][OH^-]}{[B^-]}$$

Suppose sodium carbonate is dissolved in water and the solution is tested with litmus papers. The solution turns red litmus blue. The solution contains a higher concentration of OH^- ions than pure water and is alkaline.

Sodium ions, Na^+, do not undergo hydrolysis in aqueous solution, but carbonate ions, CO_3^{--}, react as a Brønsted base. A CO_3^{--} anion acquires a proton from a water molecule to form the slightly ionized hydrogen carbonate ion, HCO_3^-, and the OH^- ion.

$$CO_3^{--} + H_2O \rightleftarrows HCO_3^- + OH^-$$

The OH^- ion concentration *increases* until equilibrium is established. Consequently, the H_3O^+ ion concentration *decreases* since the product $[H_3O^+][OH^-]$ remains equal to the ionization constant K_w of water at the temperature of the solution. Thus the pH is *higher* than 7, and the solution is *alkaline*. In general, *salts formed from weak acids and strong bases hydrolyze in water to form alkaline solutions.*

Anion hydrolysis: *pH > 7, solution alkaline.*

21.11 Cation hydrolysis The *anions* of the salt of a strong acid and a weak base do not hydrolyze in water. Only *cation* hydrolysis occurs in an aqueous solution of such a salt.

An aqueous solution of ammonium chloride, NH_4Cl, turns blue litmus paper red. This color change indicates that hydrolysis has occurred and the solution contains an excess of H_3O^+ ions. Chloride ions, the *anions* of the salt, show no noticeable tendency to hydrolyze in aqueous solution. Ammonium ions, the *cations* of the salt, donate protons to water molecules according to the equation

$$NH_4^+ + H_2O \rightleftarrows H_3O^+ + NH_3$$

Equilibrium is established with an *increased* H_3O^+ ion concentration. The pH is *lower* than 7, and the solution is *acidic.*

The metallic cations of some salts are hydrated in aqueous solution. These hydrated cations may donate protons to water molecules. Because of this cation hydrolysis, the solution becomes acidic. For example, aluminum chloride forms the hydrated cations

Salts of strong acids and weak bases form acidic solutions.

$$Al(H_2O)_6^{+++}$$

Copper(II) sulfate in aqueous solution forms the light-blue hydrated cations

Some cations that hydrolyze in aqueous solutions: Al^{+++}, NH_4^+, Cr^{+++}, Cu^{++}, Fe^{+++}, Sn^{++++}.

$$Cu(H_2O)_4^{++}$$

These ions react with water to form hydronium ions as shown in the equations that follow:

$$Al(H_2O)_6^{+++} + H_2O \rightleftarrows Al(H_2O)_5OH^{++} + H_3O^+$$

$$Cu(H_2O)_4^{++} + H_2O \rightleftarrows Cu(H_2O)_3OH^+ + H_3O^+$$

Cations such as $Cu(H_2O)_3OH^+$ ions may experience a secondary hydrolysis to a slight extent

$$Cu(H_2O)_3OH^+ + H_2O \rightleftarrows Cu(H_2O)_2(OH)_2 + H_3O^+$$

Expressions for the hydrolysis constant K_h for cation acids are similar to those of anion bases. In general, *salts formed from strong acids and weak bases hydrolyze in water to form acidic solutions.*

Hydrated copper(II) hydroxide is not very soluble. The scummy appearance of reagent bottles in which copper(II) salt solutions are stored for a long time is caused by the slight secondary hydrolysis of $Cu(H_2O)_3OH^+$ cations. This hydrolysis is aided by the formation of a slightly soluble product.

21.12 Solubility product A *saturated* solution contains the maximum amount of solute possible at a given temperature in equilibrium with an undissolved excess of the substance. A saturated solution is *not necessarily* a concentrated solution. The concentration may be high or low, depending on the solubility of the solute.

Solubility guide:

soluble	> 1.0 *g/100 g* H_2O
slightly soluble	> 0.1 *g/100 g* H_2O
insoluble	< 0.1 *g/100 g* H_2O

A rough rule is often used to express solubilities qualitatively. By this rule, a substance is said to be *soluble* if the solubility is greater than 1 g per 100 g of water. It is said to be *insoluble* if the solubility is less than 0.1 g per 100 g of water. Solubilities that fall between these limits are described as *slightly soluble.*

Substances usually referred to as insoluble are, in fact, very *sparingly soluble.* An extremely small quantity of such a solute saturates the solution. Equilibrium is established with the undissolved excess remaining in contact with the solution. Equilibria between sparingly soluble solids and their saturated solutions are especially important in analytical chemistry.

You have observed that silver chloride precipitates when Ag^+ and Cl^- ions are placed in the same solution. Silver chloride is so sparingly soluble in water that it is described as insoluble.

"Insoluble" salts are actually very sparingly soluble in water.

The solution reaches saturation at a very low concentration of its ions. All Ag^+ and Cl^- ions in excess of this concentration eventually separate as an AgCl precipitate.

The equilibrium principles developed in this chapter apply to all saturated solutions of sparingly soluble salts. Consider the equilibrium system in a saturated solution of silver chloride containing an excess of the solid salt. The equilibrium equation is

$$\mathbf{AgCl(s) \rightleftarrows Ag^+(aq) + Cl^-(aq)}$$

The equilibrium constant is expressed as

$$K = \frac{[Ag^+][Cl^-]}{[AgCl]}$$

Here, [AgCl] refers to concentration of undissolved AgCl.

Because the concentration of a pure substance in the solid or liquid phase remains constant (Section 21.4), adding more solid AgCl to this equilibrium system does not change the *concentration* of the undissolved AgCl present. Thus [AgCl] in the equation is a constant. By combining the two constants, this equation becomes

$$K[AgCl] = [Ag^+][Cl^-]$$

The product $\boldsymbol{K}$**[AgCl]** is also a constant. It is called the *solubility-product constant* $\boldsymbol{K_{sp}}$.

$$K_{sp} = K[AgCl]$$

Therefore,

$$K_{sp} = [Ag^+][Cl^-]$$

Thus the solubility-product constant $\boldsymbol{K_{sp}}$ of AgCl *is the product of the molar concentrations of its ions in a saturated solution.*

Calcium fluoride is another sparingly soluble salt. The equilibrium system in a saturated CaF_2 solution is given by the following equation

$$\mathbf{CaF_2(s) \rightleftarrows Ca^{++}(aq) + 2F^-(aq)}$$

The solubility-product constant is given by

$$K_{sp} = [Ca^{++}][F^-]^2$$

Notice how this constant differs from the solubility-product constant for AgCl. For CaF_2, $\boldsymbol{K_{sp}}$ is the product of the molar concentration of Ca^{++} ions and the molar concentration of F^- ions *squared*.

Similar equations apply to any sparingly soluble salt having the general formula M_aX_b. The equilibrium system in a saturated solution is shown by

$$\mathbf{M_aX_b \rightleftarrows aM^{+b} + bX^{-a}}$$

The solubility-product constant is expressed by

$$K_{sp} = [M^{+b}]^a[X^{-a}]^b$$

The ***solubility-product constant*** *of a substance is the product of the molar concentrations of its ions in a saturated solution, each raised to the appropriate power.*

From solubility data listed in Appendix Table 13, 1.94×10^{-4} g of AgCl saturates $10\bar{0}$ g of water at 20 °C. One mole of AgCl has a mass of 143.4 g. Therefore, the saturation concentration (solubility) of AgCl can be expressed in moles per liter.

$$\frac{1.94 \times 10^{-4}\ \text{g}}{10.0^2\ \text{g}} \times \frac{10^3\ \text{g}}{\text{L}} \times \frac{\text{mole}}{1.434 \times 10^2\ \text{g}} = 1.35 \times 10^{-5}\ \text{mole/L}$$

The equilibrium equation is

$$AgCl(aq) \rightleftarrows Ag^+(aq) + Cl^-(aq)$$

Silver chloride dissociates in solution, contributing equal numbers of aqueous Ag^+ and Cl^- ions. Therefore, the ion concentrations in the saturated solution are

$$[Ag^+] = 1.35 \times 10^{-5}$$

$$[Cl^-] = 1.35 \times 10^{-5}$$

and

$$K_{sp} = [Ag^+][Cl^-]$$

$$K_{sp} = (1.35 \times 10^{-5})(1.35 \times 10^{-5})$$

$$K_{sp} = (1.35 \times 10^{-5})^2$$

$$K_{sp} = 1.82 \times 10^{-10}$$

This result is the solubility-product constant of AgCl at 20 °C.

From Appendix Table 13, the solubility of CaF_2 at 25 °C is 1.7×10^{-3} g/$10\bar{0}$ g of water. Expressed in moles per liter as before, this concentration becomes 2.2×10^{-4} mole/L.

The equilibrium equation for a saturated solution of CaF_2 is

$$CaF_2 \rightleftarrows Ca^{++} + 2F^-$$

CaF_2 dissociates in solution to yield *twice as many* F^- ions as Ca^{++} ions. The ion concentrations in the saturated solution are

$$[Ca^{++}] = 2.2 \times 10^{-4}$$

$$[F^-] = 2(2.2 \times 10^{-4})$$

and

$$K_{sp} = [Ca^{++}][F^-]^2$$

$$K_{sp} = (2.2 \times 10^{-4})(4.4 \times 10^{-4})^2$$

$$K_{sp} = (2.2 \times 10^{-4})(1.9 \times 10^{-7})$$

$$K_{sp} = 4.2 \times 10^{-11}$$

Thus the solubility-product constant of CaF_2 is 4.2×10^{-11} at 25 °C.

If the ion product $[Ca^{++}][F^-]^2$ is *less* than the value for K_{sp} at

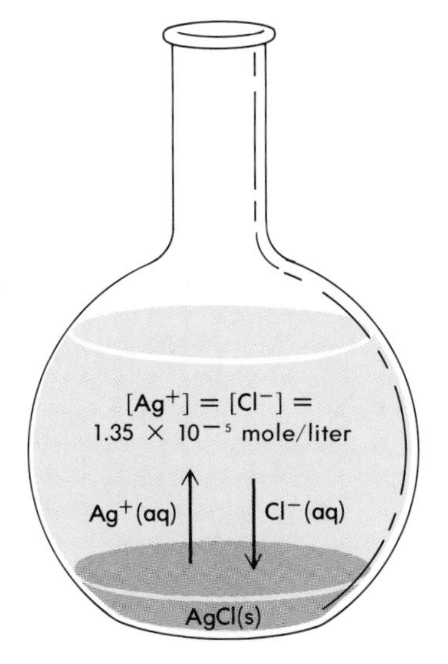

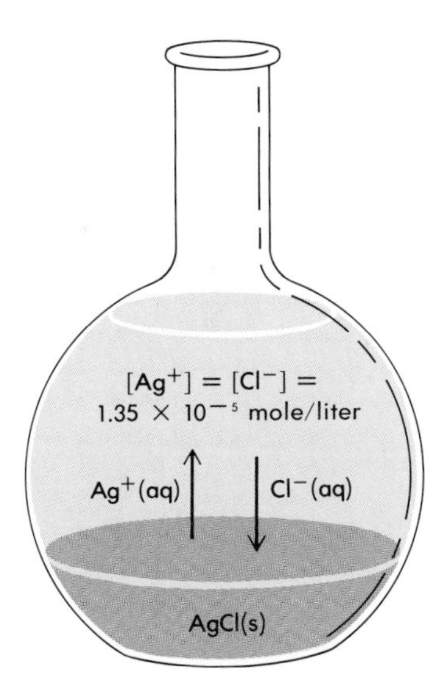

Figure 21-7. The heterogeneous equilibrium in a saturated solution is not disturbed by the addition of more of the solid phase.

Salts precipitate from their saturated solutions.

a particular temperature, the solution is *unsaturated*. If the ion product is *greater* than the value for K_{sp}, CaF_2 precipitates. This precipitation reduces the concentrations of Ca^{++} and F^- ions until equilibrium is established. The solubility equilibrium is then

$$CaF_2(s) \rightleftarrows Ca^{++}(aq) + 2F^-(aq)$$

It is difficult to measure very small concentrations of a solute with precision. For this reason, solubility data from different sources may result in slightly different values of K_{sp} for a substance. Thus calculations of K_{sp} ordinarily should be limited to two significant figures. Representative values of K_{sp} at 25 °C for some sparingly soluble compounds are listed in Table 21-4.

Table 21-4
SOLUBILITY-PRODUCT CONSTANTS K_{sp} at 25 °C

Salt	*Ion product*	K_{sp}
$AgC_2H_3O_2$	$[Ag^+][C_2H_3O_2^-]$	2.5×10^{-3}
$AgBr$	$[Ag^+][Br^-]$	5.0×10^{-13}
Ag_2CO_3	$[Ag^+]^2[CO_3^{--}]$	8.1×10^{-12}
$AgCl$	$[Ag^+][Cl^-]$	1.8×10^{-10}
AgI	$[Ag^+][I^-]$	8.3×10^{-17}
Ag_2S	$[Ag^+]^2[S^{--}]$	1.6×10^{-49}
$Al(OH)_3$	$[Al^{+++}][OH^-]^3$	1.3×10^{-33}
$BaCO_3$	$[Ba^{++}][CO_3^{--}]$	1.2×10^{-8}
$BaSO_4$	$[Ba^{++}][SO_4^{--}]$	1.1×10^{-10}
Bi_2S_3	$[Bi^{+++}]^2[S^{--}]^3$	1.0×10^{-97}
CdS	$[Cd^{++}][S^{--}]$	8.0×10^{-27}
$CaCO_3$	$[Ca^{++}][CO_3^{--}]$	1.4×10^{-8}
CaF_2	$[Ca^{++}][F^-]^2$	4.2×10^{-11}
$Ca(OH)_2$	$[Ca^{++}][OH^-]^2$	5.5×10^{-6}
$CaSO_4$	$[Ca^{++}][SO_4^{--}]$	9.1×10^{-6}
CoS	$[Co^{++}][S^{--}]$	3.0×10^{-26}
Co_2S_3	$[Co^{+++}]^2[S^{--}]^3$	2.6×10^{-124}
$CuCl$	$[Cu^+][Cl^-]$	1.2×10^{-6}
Cu_2S	$[Cu^+]^2[S^{--}]$	2.5×10^{-48}
CuS	$[Cu^{++}][S^{--}]$	6.3×10^{-36}
FeS	$[Fe^{++}][S^{--}]$	6.3×10^{-18}
Fe_2S_3	$[Fe^{+++}]^2[S^{--}]^3$	1.4×10^{-85}
$Fe(OH)_3$	$[Fe^{+++}][OH^-]^3$	1.5×10^{-36}
HgS	$[Hg^{++}][S^{--}]$	1.6×10^{-52}
$MgCO_3$	$[Mg^{++}][CO_3^{--}]$	3.5×10^{-8}
$Mg(OH)_2$	$[Mg^{++}][OH^-]^2$	1.5×10^{-11}
MnS	$[Mn^{++}][S^{--}]$	2.5×10^{-13}
NiS	$[Ni^{++}][S^{--}]$	3.2×10^{-19}
$PbCl_2$	$[Pb^{++}][Cl^-]^2$	1.9×10^{-4}
$PbCrO_4$	$[Pb^{++}][CrO_4^{--}]$	2.8×10^{-13}
$PbSO_4$	$[Pb^{++}][SO_4^{--}]$	1.8×10^{-8}
PbS	$[Pb^{++}][S^{--}]$	8.0×10^{-28}
SnS	$[Sn^{++}][S^{--}]$	1.2×10^{-25}
$SrSO_4$	$[Sr^{++}][SO_4^{--}]$	3.2×10^{-7}
ZnS	$[Zn^{++}][S^{--}]$	1.6×10^{-24}

Practice Problems

Determine the solubility-product constant, K_{sp}, from solubility data for each of the following substances.

1. Copper(I) chloride. Solubility: 1.08×10^{-2} g/100 g H_2O at 25 °C. *ans.* 1.19×10^{-6}
2. Lead(II) chloride. Solubility: 1.0 g/10$\bar{0}$ g H_2O at 25 °C. *ans.* 1.86×10^{-4}

21.13 Calculating solubilities Solubility-product constants are computed from very careful measurements of solubilities and other solution properties. Once known, the solubility product is very helpful in determining the solubility of a sparingly soluble salt.

Suppose you wish to know how much barium carbonate, $BaCO_3$, can be dissolved in one liter of water at 25 °C. From Table 21-4, K_{sp} for $BaCO_3$ has the numerical value 1.2×10^{-8}. The solubility equation is written as follows:

$$BaCO_3(s) \rightleftarrows Ba^{++}(aq) + CO_3^{--}(aq)$$

Knowing the value for K_{sp},

$$K_{sp} = [Ba^{++}][CO_3^{--}] = 1.2 \times 10^{-8}$$

Therefore, $BaCO_3$ dissolves until the product of the molar concentrations of Ba^{++} and CO_3^{--} equals 1.2×10^{-8}.

The solubility equilibrium equation shows that Ba^{++} ions and CO_3^{--} ions enter the solution in equal numbers as the salt dissolves. Thus,

$$[Ba^{++}] = [CO_3^{--}] = [BaCO_3] \text{ dissolved}$$

$$K_{sp} = 1.2 \times 10^{-8} = [BaCO_3]^2$$

$$[BaCO_3] = \sqrt{1.2 \times 10^{-8}}$$

$$[BaCO_3] = 1.1 \times 10^{-4} \text{ mole/L}$$

The solubility of $BaCO_3$ is 1.1×10^{-4} mole/liter. Thus the solution concentration is 1.1×10^{-4} *M* for Ba^{++} ions and for CO_3^{--} ions, 1.1×10^{-4} *M*.

$K_{sp} = [Ba^{++}][CO_3^{--}]$

Practice Problems

1. Calculate the concentration of $AgC_2H_3O_2$ in mole/L from the solubility-product constant K_{sp} as listed in Table 21-4. *ans.* 5.0×10^{-2} mole/L
2. Verify that the value for K_{sp} you computed as the answer to Practice Problem Number 1, Section 21.12, yields the concentration of CuCl given in the problem. *ans.* 1.0×10^{-2} g CuCl/10$\bar{0}$ g H_2O

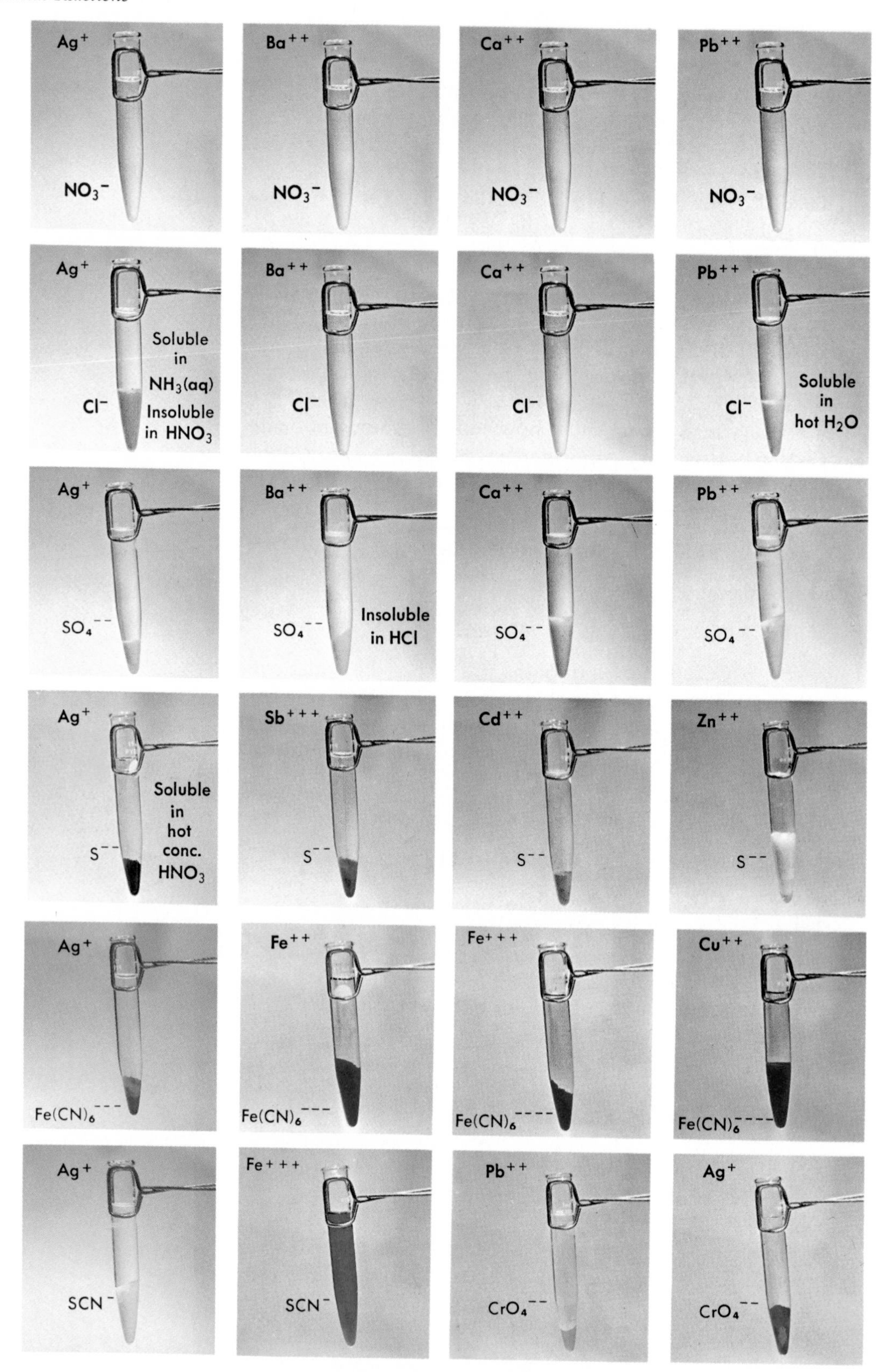

Ag+
NO3−
Ba++
NO3−
Ca++
NO3−
Pb++
NO3−
Ag+
Cl−
Soluble in NH3(aq) Insoluble in HNO3
Ba++
Cl−
Ca++
Cl−
Pb++
Cl−
Soluble in hot H2O
Ag+
SO4−−
Ba++
SO4−−
Insoluble in HCl
Ca++
SO4−−
Pb++
SO4−−
Ag+
S−−
Soluble in hot conc. HNO3
Sb+++
S−−
Cd++
S−−
Zn++
S−−
Ag+
Fe(CN)6−−−
Fe++
Fe(CN)6−−−
Fe+++
Fe(CN)6−−−−
Cu++
Fe(CN)6−−−
Ag+
SCN−
Fe+++
SCN−
Pb++
CrO4−−
Ag+
CrO4−−

21.14 Precipitation calculations In the example used in Section 21.13, the $BaCO_3$ served as the source of both Ba^{++} and CO_3^{--} ions. Thus the concentrations of the two ions were equal. However, the equilibrium condition does not require the two ion concentrations to be equal. It requires only that the *ion product* $[Ba^{++}][CO_3^{--}]$ *not* exceed the value of K_{sp} for the system.

Suppose unequal quantities of $BaCl_2$ and $CaCO_3$ are added to water. A high concentration of Ba^{++} ions and a low concentration of CO_3^{--} ions might result. If the ion product $[Ba^{++}][CO_3^{--}]$ exceeds the K_{sp} of $BaCO_3$, a precipitate of $BaCO_3$ forms. Precipitation would continue until the ion concentrations decreased to equilibrium values. The photos on the left show the behavior of some negative ions in the presence of certain metallic ions.

The solubility product can be used to predict whether a precipitate forms when two solutions are mixed. Calculations involved in such a prediction are given in Sample Problem 21-2.

SAMPLE PROBLEM 21-2

Will a precipitate form if 20.0 mL of 0.010-*M* $BaCl_2$ solution is mixed with 20.0 mL of 0.0050-*M* Na_2SO_4 solution?

SOLUTION

The two possible new pairings of ions are NaCl and $BaSO_4$. Of these, $BaSO_4$ is a sparingly soluble salt. It will then precipitate from the resulting solution if the ion product $[Ba^{++}][SO_4^{--}]$ exceeds the value of the solubility-product constant K_{sp} for $BaSO_4$. From the table of solubility products (Table 21-4), the K_{sp} is found to be 1.1×10^{-10}

The solubility equilibrium equation is

$$BaSO_4(s) \rightleftarrows Ba^{++}(aq) + SO_4^{--}(aq)$$

and the equilibrium condition is

$$K_{sp} = [Ba^{++}][SO_4^{--}] = 1.1 \times 10^{-10}$$

If the ion product $[Ba^{++}][SO_4^{--}]$ exceeds 1.1×10^{-10}, precipitation of $BaSO_4$ is predicted. Mole quantities of Ba^{++} and SO_4^{--} ions:

$$0.020 \text{ liter} \times \frac{0.010 \text{ mole } Ba^{++}}{\text{liter}} = 0.00020 \text{ mole } Ba^{++}$$

$$0.020 \text{ liter} \times \frac{0.0050 \text{ mole } SO_4^{--}}{\text{liter}} = 0.00010 \text{ mole } SO_4^{--}$$

Total volume of solution containing Ba^{++} and SO_4^{--} ions:

$$0.020 \text{ liter} + 0.020 \text{ liter} = 0.040 \text{ liter}$$

Ba^{++} and SO_4^{--} ion concentrations:

$$\frac{0.00020 \text{ mole } Ba^{++}}{0.040 \text{ liter}} = 5.0 \times 10^{-3} \text{ mole } Ba^{++}/\text{liter}$$

$$\frac{0.00010 \text{ mole } SO_4^{--}}{0.040 \text{ liter}} = 2.5 \times 10^{-3} \text{ mole } SO_4^{--}/\text{liter}$$

Trial value of ion product:

$$[Ba^{++}][SO_4^{--}] = (5.0 \times 10^{-3})(2.5 \times 10^{-3}) = 1.2 \times 10^{-5}$$

The ion product is much greater than $K_{sp}(K_{sp} = 1.1 \times 10^{-10})$, so precipitation occurs.

Practice Problems

1. Does a precipitate form when 10$\bar{0}$ mL of 0.0025-*M* $AgNO_3$ and 15$\bar{0}$ mL of 0.0020-*M* NaBr solutions are mixed?
   ans. AgBr ion product $= 1.2 \times 10^{-6}$; $K_{sp}(AgBr) = 5.0 \times 10^{-13}$; AgBr precipitates.
2. Does a precipitate form when 2$\bar{0}$ mL of 0.038-*M* $Pb(NO_3)_2$ and 3$\bar{0}$ mL of 0.018-*M* KCl solutions are mixed?
   ans. $PbCl_2$ ion product $= 1.8 \times 10^{-6}$;
   $K_{sp}(PbCl_2) = 1.9 \times 10^{-4}$; $PbCl_2$ does not precipitate.

The solubility-product principle can be very useful when applied to solutions of sparingly soluble substances. It *cannot* be applied to solutions of moderately soluble or very soluble substances. Many solubility-product constants are known only roughly because of difficulties involved in solubility measurements. Sometimes, as with the hydrolysis of an ion in solution, it is necessary to consider two equilibria simultaneously. The solubility product is sensitive to changes in solution temperature to the extent that the solubility of the dissolved substance is affected by such changes.

SUMMARY

Many chemical reactions are reversible. A reaction system in which the forward and reverse reactions occur simultaneously at the same rate is said to be in equilibrium. Both reactions continue, but there is no net change in the composition of the system.

Equilibrium is a dynamic state involving physical as well as chemical processes. At equilibrium, the energy-change and the entropy-change factors are equal. The relative concentrations of reactants and products in a reaction system at equilibrium are expressed in terms of an equilibrium constant. At equilibrium, the ratio of the product of the mole concentrations of substances formed to the product of the mole concentrations of reactants has a definite numerical value at a given temperature. This numerical ratio is the equilibrium constant K. For values of K greater than 1, the products of the forward reaction are favored. For values of K less than 1, the reactants of the forward reaction are favored. The equilibrium constant for an equilibrium system varies only with the system's temperature.

Any change that alters the rate of either the forward or reverse reaction disturbs the equilibrium of the system. The equilibrium is subjected to a stress. According to Le Chatelier's principle, the equilibrium is shifted in the direction that relieves the stress. Only when a change in temperature causes the stress is the numerical value of K changed when equilibrium is reestablished. A forward reaction shows little tendency to reach an equilibrium with the reverse reaction when the conditions that encourage each are very different.

The common ion effect is recognized when a solute containing ions like those of a reactant in an equilibrium system is added to the system. Le Chatelier's principle explains the response of the system to the stress caused by the increase in concentration of common ions.

The equilibrium expressions of weak acids can be modified to give an expression for the ionization constant of the acid. This ionization constant is derived from the fact that the equation for the equilibrium constant includes $[H_2O]$, which is also a constant. The ionization constant is then set equal to the product of these two constants, $K_a = K[H_2O]$. When the equilibrium expression for the slight ionization of water is considered, the ionization constant for water, K_w, becomes $K_w = K[H_2O]^2 = [H_3O^+][OH^-]$. At 25 °C, $K_w = 1 \times 10^{-14}$ for water and all aqueous solutions.

When salts are dissolved in water, the resulting solutions are not always neutral. Salts formed from strong bases and weak acids have aqueous solutions that are basic. The anions of the salt act as a base. They acquire protons from water molecules and increase the $[OH^-]$ of the solution. The process is called anion hydrolysis.

Salts formed from strong acids and weak bases have aqueous solutions that are acidic. The cations of the salt act as an acid. They donate protons to water molecules and increase the $[H_3O^+]$ of the solution. This process is called cation hydrolysis.

Salts formed from strong acids and strong bases do not hydrolyze in water, and their solutions are neutral. Salts formed from weak acids and weak bases may hydrolyze completely in water solution.

Salts that are usually described as being insoluble in water are, in fact, very sparingly soluble. Their ions form saturated aqueous solutions at extremely low concentrations. Equilibrium is quickly established between undissolved solute and dissolved solute ions. Because the concentration of undissolved salt is constant, the equilibrium expression yields a useful constant, K_{sp}, called the solubility-product constant. The numerical value of K_{sp} is determined by the product of the mole concentrations of solute ions in the saturated solution. Solubility-product constants are useful analytical values.

VOCABULARY

anion
anion hydrolysis
buffered solution
cation
cation hydrolysis
chemical equilibrium
chemical equilibrium law
common ion effect
dynamic
equilibrium constant
hydrolysis
hydrolysis constant
insoluble
ionic equilibrium
ionization constant
ionization constant of water
mass action expression
physical equilibrium
precipitate
reversible reaction
slightly soluble
solubility-product constant
soluble
sparingly soluble

QUESTIONS

GROUP A

1. Evaluate the statement: As a reversible reaction system approaches equilibrium, the rates of the opposing reactions decrease, and at equilibrium the reactions stop. *A1*
2. What is the meaning of the term *dynamic* as applied to an equilibrium system? *A1*
3. Give two examples of physical equilibrium. *A1, A3c*
4. Using polar compounds recognized as weak electrolytes, write ionic equations for three examples of ionic equilibrium. *A5b*
5. State the law of mass action. *A1, A4a*
6. *(a)* State Le Chatelier's principle. *(b)* To what kinds of equilibria does it apply? *A1*
7. A combustion reaction proceeding in air under standard pressure is transferred to an atmosphere of pure oxygen under the same pressure. *(a)* What effect would you observe? *(b)* How can you account for this effect? *A3g*
8. *(a)* What three factors can disturb, or shift, an equilibrium? *(b)* Which of these factors affects the value of the equilibrium constant? *A2b, A2e*
9. Identify the three conditions under which ionic reactions involving ionic substances can run to completion. Write an equation for each. *A2i, A3j*
10. What is the solubility characteristic of substances involved in solubility equilibrium systems? *A3o*
11. Define the solubility-product constant. *A1, A4e*

GROUP B

12. Natural crystals of NaCl have a regular cubic shape. Suppose an irregularly shaped crystal fragment of NaCl is suspended in a saturated aqueous solution of this salt. With time, the shape of the crystal is observed to gradually change, becoming more nearly cubic but with no change in mass. *(a)* What inference can you draw from these observations regarding the character of the equilibrium system? *(b)* Write the ionic equation for the equilibrium process. *A1, A4a*
13. The reaction between steam and iron is reversible. Steam passed over hot iron produces magnetic iron oxide (Fe_3O_4) and hydrogen. Hydrogen passed over hot magnetic iron oxide reduces it to iron and forms steam. Suggest a method by which this reversible reaction could be brought to a state of equilibrium. **CAUTION:** *Just propose a possible method; do not attempt to construct and test one.* *A2a*
14. According to the following equilibrium reaction $\mathbf{CO + 2H_2 \rightleftharpoons CH_3OH + 24\ kcal}$, methanol is produced synthetically as a gas by the reaction between carbon monoxide and hydrogen in the presence of a catalyst. Write the expression for the equilibrium constant of this reaction. *A4a*
15. Suggest a way to regulate the temperature of the equilibrium mixture of CO, H_2, and CH_3OH of Question 14 in order to increase the yield of methanol. Explain. *A2g*
16. How would you propose to regulate the pressure on the equilibrium mixture of Question 14 to increase the yield of methanol? Explain. *A2d*
17. In the reaction $\mathbf{A + B \rightleftharpoons C}$, the concentrations of **A**, **B**, and **C** in the equilibrium mixture were found to be 2.0, 3.0, and 1.0 moles per liter, respectively. What is the equilibrium constant of this reaction? *A4a*
18. Write the net ionic equations for the following reactions in aqueous solution. If no visible reaction occurs, write NO REACTION. Omit all *spectator ions.* Label precipitates (s) and gases (g). Refer to solubility data in Appendix Table 13 as required. Use a separate sheet of paper. *Do not write in this book.* *A5b*

    (a) $\mathbf{BaCO_3 + HNO_3 \rightarrow}$
    (b) $\mathbf{Pb(NO_3)_2 + NaCl \rightarrow}$
    (c) $\mathbf{CuSO_4 + HCl \rightarrow}$
    (d) $\mathbf{Ca_3(PO_4)_2 + NaNO_3 \rightarrow}$

(e) $Ba(NO_3)_2 + H_2SO_4 \rightarrow$
(f) $FeS + NaCl \rightarrow$
(g) $AgC_2H_3O_2 + HCl \rightarrow$
(h) $Na_3PO_4 + CuSO_4 \rightarrow$
(i) $BaCl_2 + Na_2SO_4 \rightarrow$
(j) $CuO + H_2SO_4 \rightarrow$

19. The pH of a solution containing both acetic acid and sodium acetate is higher than that of a solution containing the same concentration of acetic acid alone. Explain. *A2j*
20. Referring to Table 21-2, write the expression for the equilibrium constant for each equilibrium system listed. *A4a*
21. Refer to Table 21-1. Observe that $[H_2] = [I_2]$ in the first two experiments, but are not equal in the last two experiments. Explain. *A5b*
22. What is the effect of changes in pressure on the $H_2 + I_2 \rightleftarrows 2HI$ equilibrium? Explain. *A2d*
23. (a) From the development of K_a in Section 21.7, show how you would express an ionization constant K_b for the weak base NH_3. (b) In this case, $K_b = 1.8 \times 10^{-5}$. What is the meaning of this numerical value? *A4b*
24. Given a hydrolysis reaction $B^- + H_2O \rightleftarrows HB + OH^-$, demonstrate that $K_h = K_w/K_a$. *A4d*
25. The ionization constant K_a for acetic acid is 1.8×10^{-5} at 25 °C. Explain the meaning of this value. *A4b*
26. Using a separate sheet of paper, construct a table similar to the model shown below. Complete all columns for each pH value given in the left column, using a pH of 7 as your guide. Do not write in the table in this book. *A4c*

pH	*$[H_3O^+]$ (mole/liter)*	*$[OH^-]$ (mole/liter)*	*$[H_3O^+][OH^-]$*	*Property*
0				
1				
3				
5				
7	$10^{-7} = 0.0000001$	$10^{-7} = 0.0000001$	10^{-14}	Neutral
9				
11				
13				
14				

PROBLEMS

Group A

1. The H_3O^+ ion concentration of a solution is 0.00045 mole per liter. This quantity can be expressed as $[H_3O^+] = 4.5 \times 10^{-4}$ mole per liter. What is the pH of the solution? *B2*
2. What is the pH of a 0.0024-*M* solution of HCl? (At this concentration HCl is completely ionized.) *B2*
3. Find the pH of 0.025-*M* of KOH. *B2*
4. Given a 250-mL volumetric flask, distilled water, and NaOH, (a) state how you would prepare 250 mL of 0.30-*M* NaOH solution. (b) What is the normality of the solution?
5. (a) What quantity of copper(II) sulfate pentahydrate is required to prepare $75\overline{0}$ mL of 1.50-*M* solution? (b) What is the normality of the solution?

Luis Federico Leloir (Argentine) received the 1970 Nobel Prize in chemistry for discovering chemical compounds that affect the storage of chemical energy in living organisms. In his early research, he investigated the oxidation of fatty acids. Later, by studying how complex sugars are broken down into simpler carbohydrates, he discovered sugar nucleotides and their important role in the biosynthesis of carbohydrates. He also proved that certain liver enzymes are essential for synthesizing glycogen in the body.

Oxidation-Reduction Reactions

22

Goals

In this chapter you will gain an understanding of: • the processes of oxidation and reduction • oxidation-reduction reactions • balancing oxidation-reduction equations: oxidation-number method and ion-electron method • oxidizing and reducing agents • electrochemical reactions: electrochemical and electrolytic cells • electrolysis • electrode potentials

22.1 Oxidation and reduction processes Reactions in which the atoms or ions of an element attain a more positive (or less negative) oxidation state are recognized in Section 6.5 as *oxidation processes.* An element species that undergoes a change in oxidation state in a positive sense is said to be *oxidized.* Reactions in which the atoms or ions of an element attain a more negative (or less positive) oxidation state are recognized as *reduction processes.* An element species that undergoes a change in oxidation state in a negative sense is said to be *reduced.* The rules for assigning oxidation numbers are given in Section 6.25 and are summarized in Table 22-1.

1. *Oxidation.* The combustion of metallic sodium in an atmosphere of chlorine gas illustrates the oxidation process. The product is sodium chloride, and the sodium-chlorine bond is ionic. The empirical equation for this reaction is

$$2Na(s) + Cl_2(g) \rightarrow 2NaCl(s)$$

In this reaction each sodium atom loses an electron and becomes a sodium ion.

$$Na \rightarrow Na^+ + e^-$$

The oxidation state of sodium has changed from the 0 state of the atom (Rule 1) to the more positive +1 state of the ion (Rule 2). The sodium atom is said to be *oxidized* to the sodium ion. This change in oxidation state is indicated by the oxidation numbers placed above the symbol of the atom and the ion. For each sodium atom,

$$\overset{0}{Na} \rightarrow \overset{+1}{Na^+} + e^-$$

A second example of an oxidation process occurs when hydrogen burns in chlorine to form molecular hydrogen chloride

Table 22-1
RULES FOR ASSIGNING OXIDATION NUMBERS

1. The oxidation number of an atom of a free element is zero.
2. The oxidation number of a monatomic ion is equal to its charge.
3. The algebraic sum of the oxidation numbers of the atoms in the formula of a compound is zero.
4. The oxidation number of combined hydrogen is +1, except in metallic hydrides where it is −1.
5. The oxidation number of combined oxygen is −2. A common exception is in peroxides where it is −1. (In compounds with fluorine, the oxidation number of oxygen is +2.)
6. In combinations of nonmetals, the oxidation number of the less electronegative element is positive and of the more electronegative element is negative.
7. The algebraic sum of the oxidation numbers of the atoms in the formula of a polyatomic ion is equal to its charge.

gas. The hydrogen-chlorine bond is covalent. The formula equation for this reaction is

$$H_2(g) + Cl_2(g) \rightarrow 2HCl(g)$$

By Rule 1, the oxidation state of *each* hydrogen atom in the hydrogen molecule is 0. By Rule 4, the oxidation state of the combined hydrogen in the HCl molecule is +1. Because the hydrogen atom has changed from the 0 to the more positive +1 oxidation state, this change is an *oxidation process.*

$$\overset{0}{H_2} + Cl_2 \rightarrow 2\overset{+1}{H}Cl$$

2. *Reduction.* How shall the behavior of chlorine be regarded in these reactions with sodium and hydrogen? With sodium, each chlorine atom acquires an electron and becomes a chloride ion. The oxidation state of chlorine changes from the 0 state of the chlorine atom (Rule 1) to the more negative −1 state of the chloride ion (Rule 2). The chlorine atom is *reduced* to the chloride ion. For each chlorine atom,

$$\overset{0}{Cl} + e^- \rightarrow \overset{-1}{Cl^-}$$

In the reaction of chlorine with hydrogen, the pair of electrons shared by the hydrogen and chlorine atoms in the hydrogen chloride molecule is not shared equally. Rather, the pair of electrons is more strongly attracted to the chlorine atom because of its higher electronegativity. By Rules 4 and 3, chlorine is assigned an oxidation number of −1. Thus the chlorine atom is changed from the 0 to the −1 oxidation state, a *reduction process.*

$$H_2 + \overset{0}{Cl_2} \rightarrow 2H\overset{-1}{Cl}$$

Oxidation numbers are assigned to the atoms of covalent molecular species in this arbitrary way to indicate their oxidation states. For the hydrogen chloride molecule, the oxidation number of the hydrogen atom is +1, and that of the chlorine atom is −1.

Oxidation-reduction reactions are called "redox" reactions.

$$\overset{0}{H_2} + \overset{0}{Cl_2} \rightarrow 2\overset{+1\,-1}{HCl}$$

22.2 Oxidation and reduction occur simultaneously If *oxidation* occurs during a chemical reaction, then *reduction* must occur simultaneously. Furthermore, the *amount* of oxidation must match the *amount* of reduction. *Any chemical process in which elements undergo a change in oxidation number is an* ***oxidation-reduction reaction.*** The name is often shortened to *"redox" reaction.* The processes of oxidation and reduction are *equivalent* in every oxidation-reduction reaction.

Observe that as metallic sodium burns in chlorine gas, *two* sodium atoms are oxidized to Na^+ ions as *one* diatomic chlorine molecule is reduced to *two* Cl^- ions. Elemental sodium and elemental chlorine are chemically equivalent atom for atom and ion for ion.

$$2\overset{0}{Na} \rightarrow 2\overset{+1}{Na^+} + 2e^- \quad \text{(oxidation)}$$

$$\overset{0}{Cl_2} + 2e^- \rightarrow 2\overset{-1}{Cl^-} \quad \text{(reduction)}$$

$$2\overset{0}{Na} + \overset{0}{Cl_2} \rightarrow 2\overset{+1}{Na^+}\overset{-1}{Cl^-} \quad \text{(combined)}$$

Suppose a scheme is devised to recover metallic aluminum from an aluminum salt. In this scheme, sodium atoms are oxidized to Na^+ ions, and Al^{+++} ions are reduced to aluminum atoms. For these oxidation and reduction processes to be equivalent, three Na^+ ions are formed for each Al^{+++} ion reduced.

$$3\overset{0}{Na} \rightarrow 3\overset{+1}{Na^+} + 3e^- \quad \text{(oxidation)}$$

$$\overset{+3}{Al^{+++}} + 3e^- \rightarrow \overset{0}{Al} \quad \text{(reduction)}$$

$$3\overset{0}{Na} + \overset{+3}{Al^{+++}} \rightarrow 3\overset{+1}{Na^+} + \overset{0}{Al} \quad \text{(combined)}$$

Observe that in the combined equation for each example, both electric charge and atoms are conserved, and that the oxidation and reduction processes are indeed equivalent.

Many of the reactions studied in elementary chemistry involve oxidation-reduction processes. Of the examples considered, the first and second are composition reactions. The third example is a replacement reaction. Not all composition reactions are oxidation-reduction reactions, however. Sulfur dioxide gas,

Figure 22-1. The combustion of antimony in chlorine is an oxidation-reduction reaction.

Figure 22-2. As $KMnO_4$ solution is added to an acidic solution of $FeSO_4$, Fe^{++} ions are oxidized to Fe^{+++} ions, and the red MnO_4^- ions are changed to colorless Mn^{++} ions. When all Fe^{++} ions are oxidized, MnO_4^- ions are no longer changed to colorless Mn^{++} ions. Thus the first faint appearance of the MnO_4^- color indicates the end point of the titration.

SO_2, dissolves in water to form an acidic solution containing a low concentration of sulfurous acid, H_2SO_3.

$$\overset{+4\,-2}{SO_2} + \overset{+1\,-2}{H_2O} \rightarrow \overset{+1\,+4\,-2}{H_2SO_3}$$

Observe that the oxidation states of all elemental species remain unchanged in this composition reaction. It is not an oxidation-reduction reaction.

When a solution of sodium chloride is added to a solution of silver nitrate, an ion-exchange reaction occurs, and a white precipitate of silver chloride separates.

$$\overset{+1}{Na^+} + \overset{-1}{Cl^-} + \overset{+1}{Ag^+} + \overset{+5-2}{NO_3^-} \rightarrow \overset{+1}{Na^+} + \overset{+5\,-2}{NO_3^-} + \overset{+1}{Ag^+}\ \overset{-1}{Cl^-}(s)$$

Or more simply by omitting spectator ions,

$$\overset{+1}{Ag^+}(aq) + \overset{-1}{Cl^-}(aq) \rightarrow \overset{+1}{Ag^+}\ \overset{-1}{Cl^-}(s)$$

The oxidation state of each species of monatomic ion remains unchanged. This reaction is not an oxidation-reduction reaction.

Oxidation-reduction reactions sometimes involve polyatomic ions. An example is the *permanganate ion,* MnO_4^-, in aqueous potassium permanganate. Under certain conditions this MnO_4^- ion is changed to the *manganese(II) ion,* Mn^{++}. See Figure 22-2. Under other conditions, the MnO_4^- ion is changed to the *manganate ion,* MnO_4^{--}.

When writing the equation for either of these reactions, all changes in elemental oxidation states must be known. The ionic charge alone is of little help if the ion consists of two or more different elements. Assigning the proper oxidation number to each atom present in a polyatomic ion simplifies the task of balancing oxidation-reduction equations. This assignment is accomplished by following the appropriate rules of Table 22-1.

The oxidation state of an atom of an element is 0. This is true whether the atom exists as a free particle or as one of several in a molecule of the element. The oxidation state of each hydrogen atom in the molecule H_2 is 0. It is also 0 for monatomic hydrogen. The oxidation state of sulfur is 0 in S, S_2, S_4, S_6, and S_8—all of which exist. All atoms in their elemental forms have oxidation numbers of 0. They are written as C, H_2, N_2, Pt, and S_8.

In the following electronic equations, the substances on the left are *oxidized.*

$$\overset{0}{Na} \rightarrow \overset{+1}{Na^+} + e^-$$

$$\overset{0}{Fe} \rightarrow \overset{+2}{Fe^{++}} + 2e^-$$

$$\overset{+2}{Fe^{++}} \rightarrow \overset{+3}{Fe^{+++}} + e^-$$

$$2\overset{-1}{Cl^-} \rightarrow \overset{0}{Cl_2} + 2e^-$$

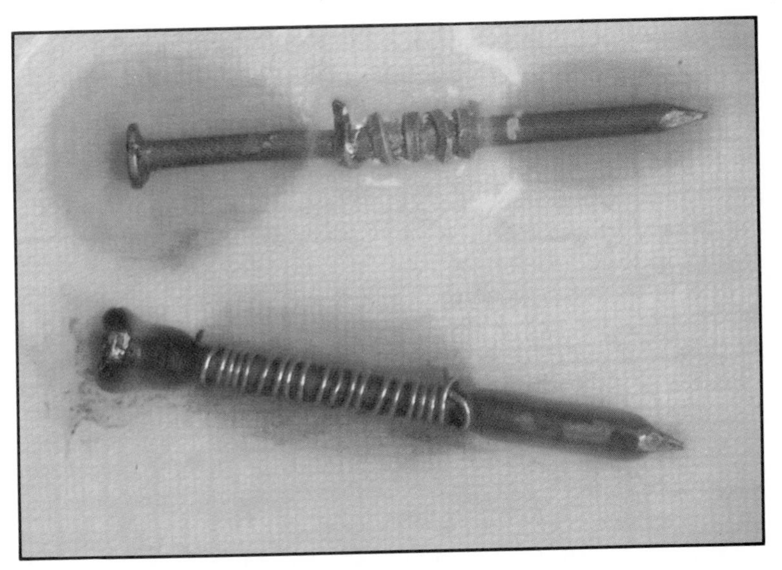

Figure 22-3. (Right Photo) An oxidation-reduction reaction occurs when a strip of zinc is immersed in an aqueous solution of copper(II) sulfate. What substance is oxidized? What is reduced? (Left Photo) Oxidation-Reduction. Zinc wrapped around iron nail—iron reduced; Copper wrapped around iron nail—iron oxidized.

The number above each symbol is the oxidation number of that particle. (Ionic charges are shown as right superscripts where appropriate.) The difference between oxidation numbers indicates the change in oxidation state. The chloride ion is assigned the oxidation number −1, a *negative* oxidation state. Oxidation results in the change of an oxidation number *in a positive sense.* The change from −1 to 0 is a change in a positive sense, as is the change from 0 to +1, from +1 to +2, or from −3 to −2.

In the electronic equations that follow, the substances on the left are *reduced.*

$$\overset{+1}{Na^+} + e^- \rightarrow \overset{0}{Na}$$

$$\overset{0}{Cl_2} + 2e^- \rightarrow 2\overset{-1}{Cl^-}$$

$$\overset{+3}{Fe^{+++}} + e^- \rightarrow \overset{+2}{Fe^{++}}$$

$$\overset{+2}{Cu^{++}} + 2e^- \rightarrow \overset{0}{Cu}$$

$$\overset{0}{Br_2} + 2e^- \rightarrow 2\overset{-1}{Br^-}$$

In the third equation the iron(III) ion, oxidation number +3, is reduced to the iron(II) state with oxidation number +2. Reduction results in the change of an oxidation number *in a negative sense.* The change from +3 to +2 is a change in a negative sense, as is the change from 0 to −1, from +1 to 0, from +1 to −1, or from −2 to −4.

The oxidation state of each monatomic ion in a binary salt is the same as the ionic charge. In binary covalent compounds, shared electrons are arbitrarily assigned to the more electronegative element.

Oxidation numbers are usually placed above the symbols to avoid having them mistaken for ionic charges.

In the two compounds H_2SO_4 and H_2SO_3, oxygen is assigned the oxidation number −2 and hydrogen +1. In the H_2SO_4 molecule, the total contribution of the 4 atoms of oxygen is 4(−2) = −8. The total contribution of the 2 atoms of hydrogen is 2(+1) = +2. Of course the H_2SO_4 molecule is electrically neutral; therefore the oxidation number of the single sulfur atom must be +6. The appropriate oxidation number can now be placed *above* each symbol in the formula.

$$\overset{+1}{H_2}\ \overset{+6}{S}\ \overset{-2}{O_4}$$

$$2(+1) + 1(+6) + 4(-2) = 0$$

In the sulfurous acid molecule, H_2SO_3, the total contribution of oxygen is 3(−2) = −6. For hydrogen it is 2(+1) = +2. Thus the oxidation number of the single atom of sulfur must be +4.

$$\overset{+1}{H_2}\ \overset{+4}{S}\ \overset{-2}{O_3}$$

$$2(+1) + 1(+4) + 3(-2) = 0$$

Suppose these rules are applied to a salt containing a polyatomic ion. Potassium permanganate is composed of potassium ions, K^+, and polyatomic permanganate ions, MnO_4^-. The empirical formula is $KMnO_4$. Oxygen has the oxidation number −2, as before. The total contribution of the 4 oxygen atoms is 4(−2) = −8. The K^+ ion is assigned the oxidation number that equals its ionic charge, +1. The oxidation number of the manganese atom in the polyatomic ion must then be +7. The formula showing these oxidation states of the elements can be written

$$\overset{+1}{K^+}[\overset{+7}{Mn}(\overset{-2}{O})_4]^-$$

or more simply,

$$\overset{+1}{K}\overset{+7}{Mn}\overset{-2}{O_4}$$

Manganese has several important oxidation states. This oxidation number, +7, represents its highest oxidation state.

22.3 Balancing oxidation-reduction equations A principal use of oxidation numbers is in balancing equations for oxidation-reduction reactions. These reactions are recognized by the fact that changes in oxidation states occur among reactants that undergo oxidation or reduction during a chemical reaction. In the equation-balancing process, some accommodation is made for the *conservation of electrons* as well as for the *conservation of atoms*. The use of a pair of electronic equations to achieve electron balance in an oxidation-reduction equation is demonstrated in Section 22.2.

Several methods are used to balance "redox" equations. Two of these methods are illustrated here. The first is the *oxidation-number* or *electron-transfer* method. This basic scheme is simply an extension of the procedure you are now using. The second scheme, the *ion-electron* or *half-reaction* method, is commonly used with ionic equations.

1. Oxidation-number method. In all but the simplest reactions, the electron shift is balanced for the particles oxidized and the particles reduced by balancing the pair of electronic equations that represents this shift. Then the coefficients of all reactants and products are adjusted in the usual way (by inspection) to achieve a balanced equation.

The oxidation-number method of balancing oxidation-reduction equations includes the steps listed below. Observe that Steps 2, 3, and 4 represent additions to the steps you are accustomed to using in equation balancing.

Step 1: Write the skeleton equation for the reaction. To do this, the reactants and products must be known and each represented by the correct formula.

Step 2: Assign oxidation numbers to all elements and determine what is oxidized and what is reduced.

Some redox equations are simple enough to be balanced by inspection.

Step 3: Write the electronic equation for the oxidation process and the electronic equation for the reduction process.

Step 4: Adjust the coefficients in both electronic equations so that the number of electrons lost equals the number gained.

Step 5: Place these coefficients in the skeleton equation.

Step 6: Supply the proper coefficients for the rest of the equation to satisfy the conservation of atoms.

These steps are illustrated by applying them first to a very simple oxidation-reduction reaction that is easily balanced by inspection. Gaseous hydrogen sulfide burns in air to form sulfur dioxide and water. These facts are used to write the skeleton equation.

Step 1:

$$H_2S + O_2 \rightarrow SO_2 + H_2O$$

Oxidation numbers are now assigned. Changes in oxidation numbers indicate that sulfur is oxidized from the −2 state to the +4 state; oxygen is reduced from the 0 state to the −2 state. The oxidation number of hydrogen remains the same; it plays no part in the primary action of oxidation-reduction.

Step 2:

$$\overset{+1}{H_2}\overset{-2}{S} + \overset{0}{O_2} \rightarrow \overset{+4}{S}\overset{-2}{O_2} + \overset{+1}{H_2}\overset{-2}{O}$$

The change in oxidation state of sulfur requires the loss of 6 electrons: (−2) − (+4) = −6. The change in oxidation state of oxygen requires the gain of 2 electrons: (0) − (−2) = +2. The electronic equations for these two reactions are

Step 3:

$$\overset{-2}{S} \rightarrow \overset{+4}{S} + 6e^- \quad \textbf{(oxidation)}$$

$$\overset{0}{O} + 2e^- \rightarrow \overset{-2}{O} \quad \textbf{(reduction)}$$

Free oxygen is diatomic. Thus 4 electrons must be gained during the reduction of a molecule of free oxygen.

$$\overset{0}{O_2} + 4e^- \rightarrow 2\overset{-2}{O}$$

The coefficients of the two electronic equations are now adjusted so that the number of electrons lost in the oxidation of sulfur is equal to the number gained in the reduction of oxygen. The smallest number of electrons common to both equations is 12. To show the gain and loss of 12 electrons in the two equations, the oxidation equation is multiplied by 2, and the reduction equation is multiplied by 3.

Step 4:

$$2\overset{-2}{S} \rightarrow 2\overset{+4}{S} + 12e^-$$

$$3\overset{0}{O_2} + 12e^- \rightarrow 6\overset{-2}{O}$$

Hence the coefficients of H_2S and SO_2 are both 2, and the coefficient of O_2 is 3. Observe that the $6\overset{-2}{O}$ is divided between the two products SO_2 and H_2O. The coefficient 6 is accounted for by the coefficient 2 in front of each formula. These coefficients are transferred to the skeleton equation.

Step 5:

$$2H_2S + 3O_2 \rightarrow 2SO_2 + 2H_2O$$

When a redox equation is complex enough that balancing by inspection would be tedious, the redox sequence should be used.

The coefficients of the equation are adjusted in the usual way to satisfy the law of conservation of atoms. In this case, no further adjustments are needed; the equation is balanced.

Step 6:

$$2H_2S + 3O_2 \rightarrow 2SO_2 + 2H_2O$$

As a second example, an oxidation-reduction equation that is slightly more difficult to balance will be used. This reaction occurs between manganese dioxide and hydrochloric acid.

Water, manganese(II) chloride, and chlorine gas are formed. The skeleton equation is

$$\overset{+4\ -2}{MnO_2} + \overset{+1\ -1}{HCl} \rightarrow \overset{+1\ -2}{H_2O} + \overset{+2\ -1}{MnCl_2} + \overset{0}{Cl_2}$$

Oxidation numbers are assigned to the elements in the reaction. They show that $\overset{+4}{Mn}$ is reduced to $\overset{+2}{Mn^{++}}$, and some of the $\overset{-1}{Cl^-}$ is oxidized to $\overset{0}{Cl}$. Hydrogen and oxygen do not take part in the oxidation-reduction reaction. The electronic equations are

$$2\overset{-1}{Cl^-} \rightarrow \overset{0}{Cl_2} + 2e^- \quad \textbf{(oxidation)}$$

$$\overset{+4}{Mn} + 2e^- \rightarrow \overset{+2}{Mn^{++}} \quad \textbf{(reduction)}$$

The number of electrons lost and gained is the same. The coefficients are transferred to the skeleton equation, which becomes

$$MnO_2 + 2HCl \rightarrow H_2O + MnCl_2 + Cl_2$$

The complete equation can now be balanced by inspection. Two additional molecules of HCl are needed to supply the two Cl^- ions of the $MnCl_2$. This balancing requires 2 molecules of water. These water molecules also account for the 2 oxygen atoms of the MnO_2. The final equation reads

$$MnO_2 + 4HCl \rightarrow 2H_2O + MnCl_2 + Cl_2$$

The equations for both of these examples can be balanced without using electronic equations to balance electron shifts. The step process is now applied to a more complicated oxidation-reduction reaction in Sample Problem 22-1.

SAMPLE PROBLEM 22-1

The oxidation-reduction reaction between hydrochloric acid and potassium permanganate yields water, potassium chloride, manganese(II) chloride, and chlorine gas. Write the balanced equation.

SOLUTION

First, write the skeleton equation. Be careful to show the correct formula for each reactant and each product. Appropriate oxidation numbers are placed above the symbols of the elements.

$$\overset{+1\ -1}{HCl} + \overset{+1\ +7\ -2}{KMnO_4} \rightarrow \overset{+1\ -2}{H_2O} + \overset{+1\ -1}{KCl} + \overset{+2\ -1}{MnCl_2} + \overset{0}{Cl_2}$$

Some chloride ions are oxidized to chlorine atoms. The manganese of the permanganate ions is reduced to manganese(II) ions. Electronic equations for these two reactions are

$$2\overset{-1}{Cl^-} \rightarrow \overset{0}{Cl_2} + 2e^-$$

$$\overset{+7}{Mn} + 5e^- \rightarrow \overset{+2}{Mn^{++}}$$

The electron shift must involve an equal number of electrons in these two equations. This number is 10. The first equation is multiplied by 5 and the second by 2.

$$10\overset{-1}{Cl^-} \rightarrow 5\overset{0}{Cl_2} + 10e^-$$

$$2\overset{+7}{Mn} + 10e^- \rightarrow 2\overset{+2}{Mn^{++}}$$

These coefficients are transferred to the skeleton equation, which becomes

$$10HCl + 2KMnO_4 \rightarrow H_2O + KCl + 2MnCl_2 + 5Cl_2$$

By inspection, $2KMnO_4$ produces 2KCl and $8H_2O$. Now 2KCl and $2MnCl_2$ call for 6 additional molecules of HCl. The balanced equation becomes

$$16HCl + 2KMnO_4 \rightarrow 8H_2O + 2KCl + 2MnCl_2 + 5Cl_2$$

Practice Problem

A pure solution of nitric acid is colorless but becomes yellow on standing because of slight decomposition that forms brown nitrogen dioxide, oxygen, and water. Using the oxidation-number method, balance the redox equation for this decomposition reaction.

ans.

$$H\overset{+5}{N}\overset{-2}{O_3}(aq) \rightarrow \overset{+4}{N}O_2(g) + \overset{0}{O_2}(g) + H_2O(l)$$

$$4\overset{+5}{N} + 4e^- \rightarrow 4\overset{+4}{N} \qquad 2\overset{-2}{O} \rightarrow \overset{0}{O_2} + 4e^-$$

$$4HNO_3(aq) \rightarrow 4NO_2(g) + O_2(g) + 2H_2O(l)$$

2. ***The ion-electron method.*** In this alternate scheme the participating species are determined as before by assigning oxidation numbers. Separate oxidation and reduction equations are then balanced for both atoms and charge (adjusted for equal numbers of electrons gained and lost) and added together to give a balanced net ionic equation.

To compare the two balancing schemes, suppose the ion-electron method is used to balance the same equation previously used to illustrate the oxidation-number method.

Step 1: Write the skeleton equation.

$$HCl(aq) + KMnO_4(aq) \rightarrow H_2O(l) + KCl(aq) + MnCl_2(aq) + Cl_2(g)$$

Step 2: Write the ionic equation.

$$H^+(aq) + Cl^- + K^+ + MnO_4^- \rightarrow H_2O + K^+ + Cl^- + Mn^{++} + 2Cl^- + Cl_2$$

Step 3: Assign oxidation numbers and retain only those species that include an element that changes oxidation state.

$H^+(aq)$ is equivalent to H_3O^+ for representing the hydronium ion.

$$\overset{-1}{Cl^-} + \overset{+7}{Mn}O_4^- \rightarrow \overset{+2}{Mn^{++}} + \overset{0}{Cl_2}$$

Step 4: Write the equation for the reduction.

$$\overset{+7}{Mn}O_4^- \rightarrow \overset{+2}{Mn^{++}}$$

Observe from the initial equation that oxygen forms water. Account for the 4 oxygen atoms by adding $4H_2O$ on the right.

$$MnO_4^- \rightarrow Mn^{++} + 4H_2O$$

This addition now requires that $8H^+(aq)$ be added on the left.

$$\underset{(-1)}{MnO_4^-} + \underset{(+8)}{8H^+(aq)} \rightarrow \underset{(+2)}{Mn^{++}} + \underset{(0)}{4H_2O}$$

The equation is now balanced for atoms but not for charge; $5e^-$ must be added on the left side to balance the charge.

$$\underset{(-1)}{MnO_4^-} + \underset{(+8)}{8H^+(aq)} + \underset{(-5)}{5e^-} \rightarrow \underset{(+2)}{Mn^{++}} + \underset{(0)}{4H_2O}$$

The reduction equation is now balanced for both atoms and charge.

Step 5: Write the equation for the oxidation and balance for atoms.

$$\underset{(-2)}{2\overset{-1}{Cl^-}} \rightarrow \underset{(0)}{\overset{0}{Cl_2}}$$

To balance charge, $2e^-$ must be added on the right.

$$\underset{(-2)}{2Cl^-} \rightarrow \underset{(0)}{Cl_2} + \underset{(-2)}{2e^-}$$

Step 6: Add the balanced reduction equation to the balanced oxidation equation. The reduction equation must be multiplied by 2 and the oxidation equation by 5 so that the number of electrons gained in reduction is equal to that lost in oxidation.

$$2[MnO_4^- + 8H^+(aq) + 5e^- \rightarrow Mn^{++} + 4H_2O]$$
$$5[2Cl^- \rightarrow Cl_2 + 2e^-]$$
$$\overline{2MnO_4^- + 16H^+(aq) + 10Cl^- + 10e^- \rightarrow 2Mn^{++} + 8H_2O + 5Cl_2 + 10e^-}$$

The $10e^-$ cancel, and the ionic equation is balanced.

$$2MnO_4^- + 16H^+(aq) + 10Cl^- \rightarrow 2Mn^{++} + 8H_2O + 5Cl_2$$

Step 7: To write the formula equation, one K^+ ion must be included for each MnO_4^- ion, and six more Cl^- ions must be added so there is one for each $H^+(aq)$ ion. The formula equation may then be stated as follows:

$$2KMnO_4 + 16HCl \rightarrow 2MnCl_2 + 2KCl + 8H_2O + 5Cl_2$$

Practice Problems

Balance the equations for the following oxidation-reduction reactions using *(a)* the oxidation-number method and *(b)* the ion-electron method.

1. Copper reacts with hot, concentrated sulfuric acid to form copper(II) sulfate, sulfur dioxide, and water.

ans. (a)

$$\overset{+6}{S} + 2e^- \rightarrow \overset{+4}{S}$$

$$\overset{0}{Cu} \rightarrow \overset{+2}{Cu} + 2e^-$$

$$Cu + 2H_2SO_4 \rightarrow CuSO_4 + SO_2 + 2H_2O$$

(b)

$$4H^+(aq) + \overset{+6}{S}O_4^{--} + 2e^- \rightarrow \overset{+4}{S}O_2 + 2H_2O$$

$$\overset{0}{Cu} \rightarrow \overset{+2}{Cu} + 2e^-$$

$$Cu + 2H_2SO_4 \rightarrow CuSO_4 + SO_2 + 2H_2O$$

2. The products of a reaction between nitric acid and potassium iodide are potassium nitrate, iodine, nitrogen monoxide, and water.

ans. (a)

$$2\overset{+5}{N} + 6e^- \rightarrow 2\overset{+2}{N}$$

$$6\overset{-1}{I} \rightarrow 3\overset{0}{I_2} + 6e^-$$

$$8HNO_3 + 6KI \rightarrow 6KNO_3 + 3I_2 + 2NO + 4H_2O$$

(b)

$$8H^+(aq) + 2\overset{+5}{N}O_3^- + 6e^- \rightarrow 2\overset{+2}{N}O + 4H_2O$$

$$6\overset{-1}{I^-} \rightarrow 3\overset{0}{I_2} + 6e^-$$

$$8H^+(aq) + 2NO_3^- + 6I^- \rightarrow 3I_2 + 2NO + 4H_2O$$

or

$$8HNO_3 + 6KI \rightarrow 6KNO_3 + 3I_2 + 2NO + 4H_2O$$

A strong oxidizing agent easily acquires electrons and becomes a weak reducing agent reluctant to give up electrons.

22.4 Oxidizing and reducing agents An ***oxidizing agent*** *attains a more negative oxidation state* during an oxidation-reduction reaction. A ***reducing agent*** *attains a more positive oxidation state.*

Table 22-2
OXIDATION-REDUCTION TERMINOLOGY

Term	*Change in oxidation number*	*Change in electron population*
oxidation	in a positive sense	loss of electrons
reduction	in a negative sense	gain of electrons
oxidizing agent	in a negative sense	acquires electrons
reducing agent	in a positive sense	supplies electrons
substance oxidized	in a positive sense	loses electrons
substance reduced	in a negative sense	gains electrons

These terms are defined in Section 6.5. It follows that the substance oxidized is also the reducing agent, and the substance reduced is the oxidizing agent. An oxidized substance becomes a potential oxidizing agent. Similarly, a reduced substance is a potential reducing agent. Study the oxidation-reduction terms presented in Table 22-2.

The relatively large atoms of the Sodium Family of metals make up Group I of the periodic table. These atoms have weak attraction for their valence electrons and form positive ions readily. They are *very active reducing agents.* According to electrochemical measurements, the lithium atom is the most active reducing agent of all the common elements. The lithium ion, on the other hand, is the weakest oxidizing agent of the common ions. The electronegativity scale suggests that Group I metals starting with lithium should become progressively more active reducing agents. Except for lithium, this is true. A possible basis for the unusual activity of lithium is discussed in Chapter 24.

Atoms of the Halogen Family, Group VII of the periodic table, have a strong attraction for electrons. They form negative ions readily and are *very active oxidizing agents.* The fluorine atom is the most highly electronegative atom. It is also the most active oxidizing agent among the elements. Because of its strong attraction for electrons, the fluoride ion is the weakest reducing agent.

In Table 22-3 many familiar substances are arranged according to their activity as oxidizing and reducing agents. The left column shows the relative abilities of some metals to displace other metals from their compounds. Such displacements are oxidation-reduction processes. Zinc, for example, appears above copper. Thus zinc is the more active reducing agent and displaces copper ions from solutions of copper compounds.

$$\mathbf{Zn(s) + Cu^{++}(aq) \rightarrow Zn^{++}(aq) + Cu(s)}$$

$$\overset{0}{\mathbf{Zn}} \rightarrow \overset{+2}{\mathbf{Zn}}^{++} + 2e^- \qquad \textbf{(oxidation)}$$

$$\overset{+2}{\mathbf{Cu}}^{++} + 2e^- \rightarrow \overset{0}{\mathbf{Cu}} \qquad \textbf{(reduction)}$$

Table 22-3
RELATIVE STRENGTH OF OXIDIZING AND REDUCING AGENTS

	Reducing agents	*Oxidizing agents*	
Strong ↓	Li	Li^+	Weak ↑
	K	K^+	
	Ca	Ca^{++}	
	Na	Na^+	
	Mg	Mg^{++}	
	Al	Al^{+++}	
	Zn	Zn^{++}	
	Cr	Cr^{+++}	
	Fe	Fe^{++}	
	Ni	Ni^{++}	
	Sn	Sn^{++}	
	Pb	Pb^{++}	
	H_2	H_3O^+	
	H_2S	S	
	Cu	Cu^{++}	
	I^-	I_2	
	MnO_4^{--}	MnO_4^-	
	Fe^{++}	Fe^{+++}	
	Hg	Hg_2^{++}	
	Ag	Ag^+	
	NO_2^-	NO_3^-	
	Br^-	Br_2	
	Mn^{++}	MnO_2	
	SO_2	H_2SO_4 (conc)	
	Cr^{+++}	$Cr_2O_7^{--}$	
	Cl^-	Cl_2	
	Mn^{++}	MnO_4^-	
Weak	F^-	F_2	Strong

The copper(II) ion, on the other hand, is a more active oxidizing agent than the zinc ion.

Nonmetals and some important ions are included in the series. Any reducing agent is oxidized by the oxidizing agents below it. Observe that F_2 displaces Cl^-, Br^-, and I^- ions from their solutions. Cl_2 displaces Br^- and I^- ions; Br_2 displaces I^- ions. Accordingly, in the equation

$$\mathbf{Cl_2 + 2Br^-(aq) \rightarrow 2Cl^-(aq) + Br_2}$$

$$\mathbf{2\overset{-1}{Br^-} \rightarrow \overset{0}{Br_2} + 2e^-} \qquad \textbf{(oxidation)}$$

$$\mathbf{\overset{0}{Cl_2} + 2e^- \rightarrow 2\overset{-1}{Cl^-}} \qquad \textbf{(reduction)}$$

Permanganate ions, MnO_4^-, and dichromate ions, $Cr_2O_7^{--}$, are important oxidizing agents. They are used mainly in the form of their potassium salts. In neutral or mildly basic solutions, $\overset{+7}{Mn}$ in permanganate ions is reduced to $\overset{+4}{Mn}$ in MnO_2. If a solution is strongly basic, manganate ions, MnO_4^{--}, containing $\overset{+6}{Mn}$ are formed. In acid solutions, the $\overset{+7}{Mn}$ in permanganate ions is reduced to manganese(II) ion, Mn^{++}, and $\overset{+6}{Cr}$ in dichromate ions is reduced to chromium(III) ion, Cr^{+++}.

Peroxide ions, O_2^{--}, have a relatively unstable covalent bond between the two oxygen atoms. The electron-dot formula is then written as

$$\left[:\ddot{\underset{\times\bullet}{O}}:\ddot{\underset{\times\bullet}{O}}:\right]^{--}$$

This structure represents an intermediate state of oxidation between free oxygen and oxides. The oxidation number of oxygen in the peroxide structure is -1.

Hydrogen peroxide, H_2O_2, decomposes by the oxidation and reduction of its oxygen. The products are water and molecular oxygen.

$$\mathbf{H_2\overset{-1}{O_2} + H_2\overset{-1}{O_2} \rightarrow 2H_2\overset{-2}{O} + \overset{0}{O_2}(g)}$$

Autooxidation: H_2O_2 is both an oxidizing and a reducing agent; it undergoes both oxidation and reduction.

In this decomposition, half of the oxygen in the peroxide is *reduced* to the oxide, forming water, and half is *oxidized* to gaseous oxygen. The process is called *autooxidation*. Impurities in a water solution of hydrogen peroxide may catalyze this process.

Oxygen in H_2O_2 is *oxidized* from the -1 to the 0 oxidation state when oxygen is formed. It is *reduced* from the -1 to the -2 state when water is formed. Thus H_2O_2 can be either an oxidizing agent or a reducing agent.

22.5 Chemical equivalents of oxidizing and reducing agents

In Section 16.2, the chemical equivalent (equiv) of a reactant is described as the *mass of the reactant that loses or acquires the Avogadro number of electrons* in a chemical reaction. If a reactant is oxidized, the mass that loses the Avogadro number of electrons is one equivalent (1 equiv) of that reactant. If a reactant is reduced, the mass that acquires the Avogadro number of electrons is one equivalent of the reactant. Thus 1 equiv of any reducing agent will always react with 1 equiv of any oxidizing agent.

Quantities of reactants are often expressed in terms of chemical equivalents. To do so for oxidizing and reducing agents, *the particular oxidation-reduction reaction must be known.* For example, one atom of iron loses 2 electrons when oxidized to iron(II), Fe^{++}.

$$\overset{0}{\mathbf{Fe}} \rightarrow \overset{+2}{\mathbf{Fe}}^{++} + \mathbf{2e^-}$$

One mole of iron, 55.8 g, gives up 2 times the Avogadro number of electrons when oxidized to the +2 oxidation state. Thus the mass of iron that releases the Avogadro number of electrons in such a reaction is *one-half mole.* It follows that one equivalent is 55.8 g Fe ÷ 2 = 27.9 g Fe.

One atom of iron loses 3 electrons when oxidized to iron(III), Fe^{+++}.

$$\overset{0}{\mathbf{Fe}} \rightarrow \overset{+3}{\mathbf{Fe}}^{+++} + \mathbf{3e^-}$$

One mole of iron oxidized to the +3 oxidation state gives up 3 times the Avogadro number of electrons. Thus, one equivalent of iron oxidized to the +3 state is 55.8 g Fe ÷ 3 = 18.6 g Fe.

Fe^{+++} ions in an iron(III) chloride solution are reduced to the +2 oxidation state by the addition of a tin(II) chloride solution. The Sn^{++} ions are the reducing agent. Here, one mole of Fe^{+++} ions acquires one Avogadro number of electrons and is reduced to Fe^{++} ions. One equivalent of iron in this reaction is 55.8 g ÷ 1 = 55.8 g. One mole of the reducing agent, Sn^{++} ions, loses 2 times the Avogadro number of electrons. One equivalent of tin in this reaction is 118.7 g ÷ 2 = 59.35 g. These relationships are shown clearly in the electronic equations for this oxidation-reduction reaction.

$$\mathbf{2}\overset{+3}{\mathbf{Fe}}^{+++} + \mathbf{2e^-} \rightarrow \mathbf{2}\overset{+2}{\mathbf{Fe}}^{++}$$

$$\overset{+2}{\mathbf{Sn}}^{++} \rightarrow \overset{+4}{\mathbf{Sn}}^{++++} + \mathbf{2e^-}$$

$$\textbf{1 equiv Fe}^{+++} = \frac{\mathbf{2Fe^{+++}}}{\mathbf{e^-}\textbf{ gained}} = \frac{\mathbf{2 \times 55.8\ g}}{\mathbf{2}} = \mathbf{55.8\ g}$$

$$\textbf{1 equiv Sn}^{++} = \frac{\mathbf{Sn^{++}}}{\mathbf{e^-}\textbf{ lost}} = \frac{\mathbf{118.7\ g}}{\mathbf{2}} = \mathbf{59.35\ g}$$

22.6 Electrochemical reactions 1. *Electrochemical cells.* Oxidation-reduction reactions involve a transfer of electrons from the substance oxidized to the substance reduced. If such reactions occur *spontaneously,* they can be used as sources of electric energy. If the reactants are in contact, the energy released during the electron transfer is in the form of heat. If the reactants are

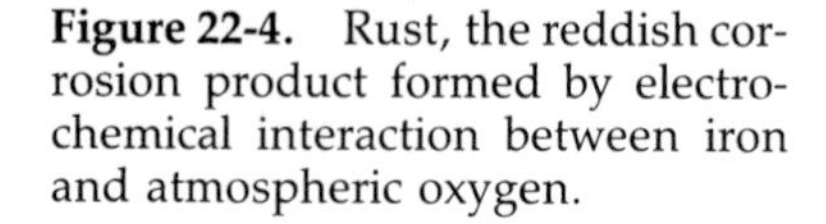

Figure 22-4. Rust, the reddish corrosion product formed by electrochemical interaction between iron and atmospheric oxygen.

separated in an electrolytic solution and are connected externally by a wire conductor, the transfer of electrons takes place through the wire. Such an arrangement is known as an *electrochemical cell.* This transfer of electrons through a metallic conductor constitutes an electric current. Under these conditions, only part of the energy released during the electron transfer appears as heat; the remainder is available as electric energy.

The dry cell is a common source of electric energy in the laboratory. Small dry cells are familiar as flashlight batteries. A zinc cup serves as the negative electrode. A carbon rod serves as the positive electrode. The carbon electrode is surrounded by a mixture of manganese dioxide and powdered carbon. The electrolyte is a moist paste of ammonium chloride containing some zinc chloride. These components of the dry cell are illustrated in the schematic diagram of Figure 22-5.

When the external circuit is closed, *zinc atoms are oxidized* at the *negative* (zinc) *electrode.*

$$\overset{0}{Zn} \rightarrow \overset{+2}{Zn}^{++} + 2e^-$$

Electrons flow through the external circuit to the *positive* (carbon) *electrode.* If manganese dioxide were not present, hydrogen gas would be formed at the carbon electrode by the reduction reaction shown in the following equation.

$$2N\overset{+1}{H}_4^+ + 2e^- \rightarrow 2NH_3 + \overset{0}{H}_2(g)$$

However, manganese oxidizes the hydrogen gas to form water. Thus *manganese* rather than hydrogen is reduced at the positive (carbon) electrode.

$$2\overset{+4}{Mn}O_2 + 2NH_4^+ + 2e^- \rightarrow \overset{+3}{Mn}_2O_3 + 2NH_3 + H_2O$$

The ammonia reacts with Zn^{++} ions to form the complex $Zn(NH_3)_4^{++}$ ions.

In electrochemistry, *the electrode at which oxidation occurs is commonly called the* ***anode,*** and *the electrode at which reduction occurs is called the* ***cathode.*** In this context, because *oxidation* occurs at the *negative electrode* of an electrochemical cell (spontaneous reactions), *the negative electrode is an anode.* Because *reduction* occurs at the *positive electrode, the positive electrode is a cathode.* The zinc electrode (oxidation) of Figure 22-5 is the *anode,* and the carbon electrode (reduction) is the *cathode* of the zinc-carbon dry cell.

Observe that the electrochemical definitions of *anode* and *cathode* lead to different results for electrolytic cells (driven reactions) in the following sections.

2. *Electrolytic cells.* Some oxidation-reduction reactions *do not occur spontaneously.* However, such reactions can be driven by means of electric energy. *The process whereby an electric current is used to drive oxidation-reduction reactions is called* ***electrolysis.*** Basically, the *electrolytic cell* consists of a pair of electrodes and an electrolytic solution in a suitable container. A battery or other direct-current source is connected across the cell. One electrode, which is connected to the negative terminal of the battery, acquires a negative charge. The other electrode, connected to the positive terminal of the battery, acquires a positive charge.

Ion migration in the cell is responsible for the transfer of electric charge between electrodes. Positively charged ions *migrate* toward the negative electrode. Negatively charged ions migrate toward the positive electrode.

In an electrolytic cell, *reduction* occurs at the *negative electrode.* Electrons are removed from the electrode in this process. Oxidation occurs at the positive electrode, which acquires electrons in the process. The chemical reactions at the electrodes complete the electric circuit between the battery and the cell. The closed-loop path for electric current allows energy to be transferred from the battery to the electrolytic cell. This energy drives the electrode reactions in the cell.

Because *oxidation* occurs at the positive electrode of an electrolytic cell (driven reactions), *the positive electrode is an anode.* Because *reduction* occurs at the *negative electrode, the negative electrode is a cathode.* In Figure 22-6, observe that oxygen gas is recovered (oxidation) at the anode and hydrogen gas is recovered (reduction) at the cathode of the electrolysis-in-water cell.

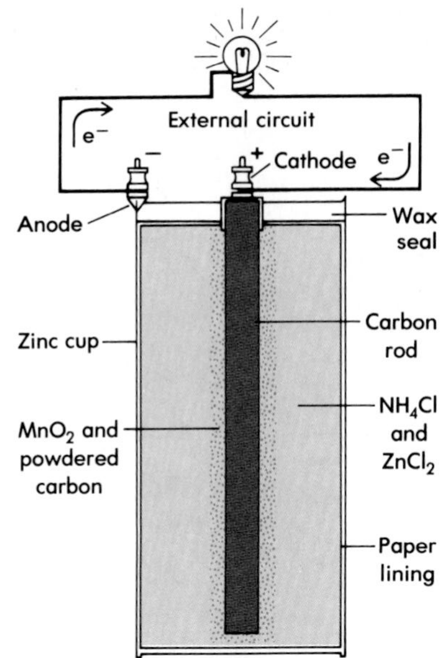

Figure 22-5. In the common dry cell, zinc is oxidized to Zn^{++} at the negative electrode and manganese(IV) is reduced to manganese(III) at the positive electrode.

22.7 Electrolysis of water In the decomposition of water by electrolysis, energy is transferred from the energy source to the decomposition products. The reaction is endothermic, and the amount of energy required is 68.32 kcal/mole of water decomposed. The overall reaction is

$$2H_2O(l) + 136.64 \text{ kcal} \rightarrow 2H_2(g) + O_2(g)$$

Review molar heat relationships for water, Section 20.2.

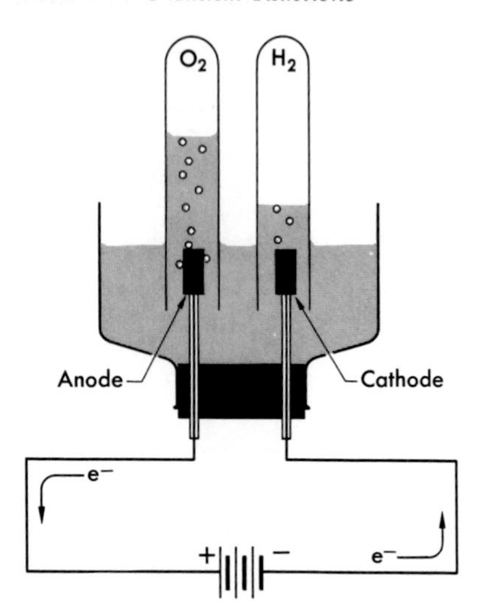

Figure 22-6. The electrolysis of water. In electrolytic cells, reduction occurs at the negative electrode, and oxidation occurs at the positive electrode.

Hydrogen gas is given up at the cathode, and oxygen gas is given up at the anode.

A suitable electrolysis cell is shown in Figure 22-5. It consists of two inert electrodes made of platinum immersed in water. A very small amount of an electrolyte, such as H_2SO_4, is added to provide adequate conductivity. The electrodes are connected to a battery that supplies the electric energy to drive the decomposition reaction forward.

The electric current in the external circuit consists of a flow of electrons. The electrode of the cell connected to the negative electrode of the battery acquires an excess of electrons and becomes the *cathode* of the electrolytic cell. The other electrode of the cell, which is connected to the positive electrode of the battery, loses electrons to the battery and becomes the *anode* of the electrolytic cell. The battery can be thought of as an electron pump supplying electrons to the cathode of the electrolytic cell and recovering electrons from the anode of the cell at equal rates.

Reduction occurs at the *cathode;* hydrogen gas is the product. Because the amount of sulfuric acid added to improve conductivity is small, its contribution of hydronium ions to the water is negligible. It is believed that, at very low concentrations, water molecules are reduced by acquiring electrons directly from the cathode.

cathode reaction: (reduction)

$$2\overset{+1}{H_2}O + 2e^- \rightarrow 2OH^- + \overset{0}{H_2}(g)$$

The OH^- ion concentration in the solution around the cathode rises.

Oxidation occurs at the *anode;* oxygen gas is the product. There are SO_4^{--} ions, OH^- ions, and water molecules in the region about the anode. The OH^- ion concentration is quite low, and they are not likely to appear in the anode reaction. The SO_4^{--} ions, also at low concentration, are more difficult to oxidize than water molecules. For these reasons, chemists believe that water molecules are oxidized by giving up electrons directly to the anode.

anode reaction: (oxidation)

$$6H_2\overset{-2}{O} \rightarrow 4H_3O^+ + \overset{0}{O_2}(g) + 4e^-$$

The H_3O^+ ion concentration in the solution around the anode rises.

The overall cell reaction is the sum of the net cathode and anode reactions. The equation for the cathode reaction is doubled because oxidation and reduction processes are equivalent.

cathode: $4H_2O + 4e^- \rightarrow 4OH^- + 2H_2(g)$
anode: $6H_2O \rightarrow 4H_3O^+ + O_2(g) + 4e^-$
cell: $10H_2O \rightarrow 2H_2(g) + O_2(g) + 4H_3O^+ + 4OH^-$

The solution around the cathode becomes basic because of the production of OH^- ions. The solution around the anode becomes acidic because of the production of H_3O^+ ions. Because of the ordinary mixing tendency in the solution, these ions can be expected to eventually diffuse together and form water. If this neutralization is complete, the net electrolysis reaction is

$$10H_2O \rightarrow 2H_2(g) + O_2(g) + 4H_3O^+ + 4OH^-$$
$$4H_3O^+ + 4OH^- \rightarrow 8H_2O$$
net: $2H_2O \rightarrow 2H_2(g) + O_2(g)$

Reduction at the cathode lowers the oxidation state of hydrogen from +1 to 0; the oxidation state of oxygen at the cathode remains −2. At the anode, oxidation raises the oxidation state of oxygen from −2 to 0; the oxidation state of hydrogen at the anode remains +1.

22.8 Electrolysis of aqueous salt solutions The electrode products from the electrolysis of aqueous salt solutions are determined by the relative ease with which the different particles present can be oxidized or reduced. In the case of aqueous NaCl, for example, Na^+ ions are more difficult to reduce at the cathode than H_2O molecules or H_3O^+ ions. Since the solution of NaCl is neutral, the H_3O^+ ion concentration remains very low (10^{-7} mole/liter). Therefore, H_2O molecules are the particles reduced at the cathode.

cathode reaction: (reduction)

$$2\overset{+1}{H_2}O + 2e^- \rightarrow 2OH^- + \overset{0}{H_2}(g)$$

Thus Na^+ ions remain in solution, and hydrogen gas is released.

In general, metals that are easily oxidized form ions that are difficult to reduce. Thus hydrogen gas, not sodium metal, is the cathode product in the preceding reaction. On the other hand, metals such as copper, silver, and gold are difficult to oxidize, and they form ions that are easily reduced. Aqueous solutions of their salts give up the metal at the cathode.

The choice for anode reaction in the aqueous NaCl electrolysis lies between Cl^- ions and H_2O molecules. The Cl^- ions are more easily oxidized, so Cl_2 gas is produced at the anode.

Electrolysis of concentrated NaCl solutions yields Cl_2 at the anode.

anode reaction: (oxidation)

$$2\overset{-1}{Cl^-} \rightarrow \overset{0}{Cl_2}(g) + 2e^-$$

Adding the cathode and anode equations gives the net reaction for the cell.

$$\text{net: } 2H_2O + 2Cl^- \rightarrow 2OH^- + H_2(g) + Cl_2(g)$$

Dilute NaCl solutions may yield both Cl_2 and O_2 at the anode.

Very dilute NaCl solutions yield O_2 at the anode.

The electrolytic solution gradually changes from aqueous NaCl to aqueous NaOH as the electrolysis continues, as long as the Cl_2 gas is continuously removed from the cell.

Aqueous Br^- ions and I^- ions are oxidized electrolytically in the same way as Cl^- ions. Their solutions give the free halogen at the anode. On the other hand, aqueous solutions of negative ions that do not participate in the oxidation reaction (NO_3^- ions, for example) give O_2 gas at the anode.

22.9 Electroplating Note that inactive metals form ions that are more easily reduced than hydrogen. This fact makes possible an electrolytic process called *electroplating*.

An electroplating cell contains a solution of a salt of the plating metal. It has an object to be plated (the cathode) and a piece of the plating metal (the anode). A silverplating cell, illustrated in Figure 22-7, contains a solution of a soluble silver salt and a silver anode. The cathode is the object to be plated. The silver anode is connected to the positive electrode of a battery or other source of direct current. The object to be plated is connected to the negative electrode. *Silver ions are reduced at the cathode* of the cell when electrons flow through the circuit.

$$\overset{+1}{Ag^+} + e^- \rightarrow \overset{0}{Ag}$$

Silver atoms are oxidized at the anode.

$$\overset{0}{Ag} \rightarrow \overset{+1}{Ag^+} + e^-$$

Silver ions are removed from the solution at the cathode and deposited as metallic silver. Meanwhile, metallic silver is removed from the anode as ions. This action maintains the Ag^+ ion concentration of the solution. Thus, in effect, silver is transferred from the anode to the cathode of the cell.

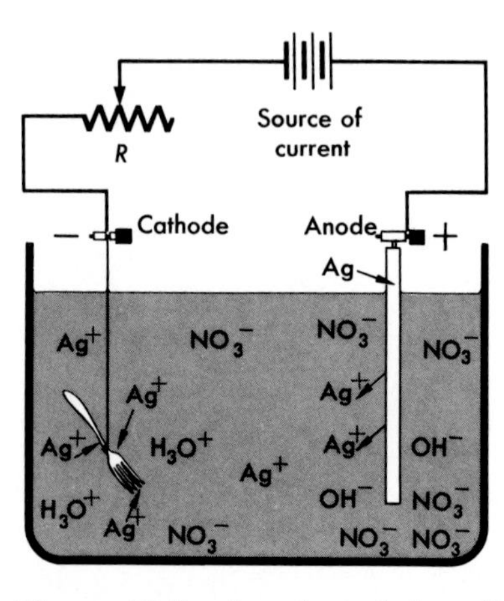

Figure 22-7. An electrolytic cell used for silver plating.

In these discussions of electrochemical and electrolytic cells, the name (*anode* or *cathode*) is determined by the redox half-reaction occurring at that electrode. In this scheme, commonly used by chemists, the electrode at which *oxidation* occurs is the *anode*, and the electrode at which *reduction* occurs is the *cathode*. The names of the electrodes are reversed with respect to the electrode charge in electrochemical cells and electrolytic cells, as illustrated in Figure 22-8.

22.10 Rechargeable cells A rechargeable cell, in effect, combines the oxidation-reduction chemistry of both the electrochemical cell and the electrolytic cell. When the rechargeable cell is

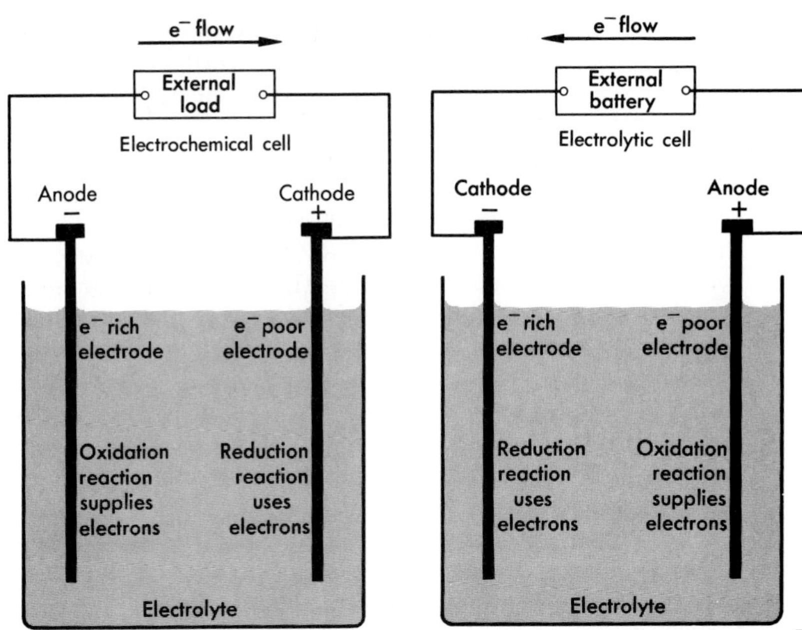

Figure 22-8. A comparison of electrochemical and electrolytic cells.

used as a source of electric energy (discharge cycle), chemical energy is converted to electric energy by the oxidation-reduction reactions of an electrochemical cell. When the cell is being recharged (charge cycle), electric energy is converted to chemical energy by the oxidation-reduction reactions of an electrolytic cell.

The standard twelve-volt automobile battery consists of six lead(IV) oxide-lead-sulfuric acid rechargeable cells connected in series. The battery is charged by the conversion of electric energy from an external source to chemical energy stored in the battery by an oxidation-reduction reaction in which each cell acts as an *electrolytic* cell. While the battery is used as a source of electric energy, chemical energy stored during the charging cycle is converted to electric energy by the reverse oxidation-reduction reaction in which each cell acts as an *electrochemical* cell. These charge and discharge modes are regulated automatically to maintain the state of charge in the battery while the automobile engine is in operation.

22.11 Electrode potentials Oxidation-reduction systems are the result of two distinct reactions: *oxidation,* in which electrons are supplied to the system, and *reduction,* in which electrons are acquired from the system. In electrochemical cells, these reactions take place at the separate electrodes. The oxidation-reduction reaction is the sum of these two separate reactions. As said in Section 22.6, in electrochemical cells, oxidation occurs at the negative electrode and reduction at the positive electrode.

As the cell reaction begins, a difference in *electric potential* develops between the electrodes. This potential difference can be measured by a voltmeter connected across the two electrodes. It is a measure of the energy required to move a certain electric charge between the electrodes. Potential difference is measured in *volts*.

Consider the electrochemical cell shown in Figure 22-9. A strip of zinc is placed in a solution of $ZnSO_4$, and a strip of copper is placed in a solution of $CuSO_4$. The two solutions are separated by a porous partition that permits ions to pass but otherwise prevents mixing of the solutions. Such an arrangement is called a *voltaic cell*. It is capable of generating a small electron current in an external circuit connected between the electrodes.

In the two electrode reactions, the zinc electrode acquires a negative charge relative to the copper. The copper electrode becomes positively charged relative to the zinc. This reaction shows that zinc atoms have a stronger tendency to enter the solution as ions than do copper atoms. Zinc is said to be more active, or more easily oxidized, than copper.

In the cell shown in Figure 22-9, then, the reaction at the surface of the zinc electrode is an oxidation.

$$\overset{0}{Zn}(s) \rightarrow \overset{+2}{Zn}^{++}(aq) + 2e^-$$

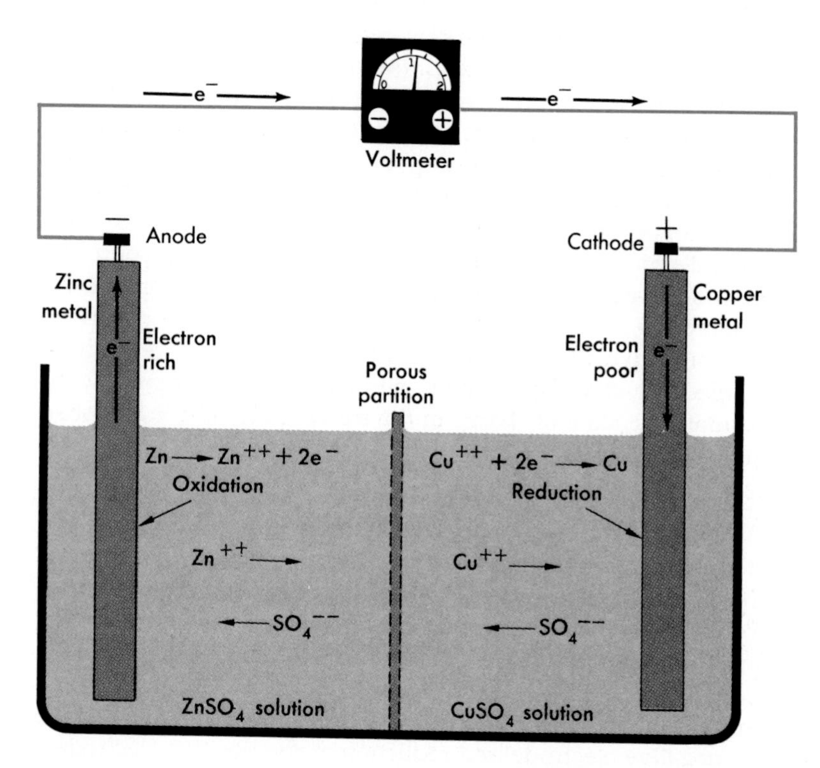

Figure 22-9. A Zn-Cu voltaic cell. An electrochemical cell: oxidation occurs at the negative electrode, and reduction occurs at the positive electrode.

The reaction at the surface of the copper electrode is a reduction.

$$\overset{+2}{Cu^{++}}(aq) + 2e^- \rightarrow \overset{0}{Cu}(s)$$

As Zn^{++} ions form, electrons accumulate on the zinc electrode, giving it a negative charge. As Cu atoms form, electrons are removed from the copper electrode, giving it a positive charge. Electrons also flow through the external circuit from the zinc electrode to the copper electrode. Here they replace the electrons removed as Cu^{++} ions undergo reduction to Cu atoms. Thus, in effect, electrons are transferred from Zn atoms through the external circuit to Cu^{++} ions. The overall reaction can be written as

$$Zn + Cu^{++} \rightarrow Zn^{++} + Cu$$

A voltmeter connected across the Cu-Zn voltaic cell measures the potential difference. This difference is about 1.1 volts when the solution concentrations of Zn^{++} and Cu^{++} ions are each 1 *m*.

A voltaic cell consists of two metal electrodes, each in contact with a solution of its ions. Each of these portions is called a *half-cell*. The reaction taking place at each electrode is called a *half-reaction*.

The potential difference between an electrode and its solution in a half-reaction is known as its **electrode potential.** The sum of the electrode potentials for the two half-reactions roughly equals the potential difference measured across the complete voltaic cell.

The potential difference across a voltaic cell is easily measured. However, there is no way to measure an individual electrode potential. The electrode potential of a half-reaction can be determined by using a *standard half-cell* along with it as a reference electrode. An arbitrary potential is assigned to the standard reference electrode. Relative to this potential, a specific potential can be determined for the other electrode of the complete cell. Electrode potentials are expressed as reduction (or oxidation) potentials. They provide a reliable indication of the tendency of a substance to undergo reduction (or oxidation).

Chemists use a *hydrogen electrode* immersed in a molar solution of $H^+(aq)$ ions as a standard reference electrode. This practice provides a convenient way to examine the relative tendencies of metals to react with aqueous hydrogen ions [$H^+(aq)$ or H_3O^+]. It is responsible for the activity series of metals listed in Table 8-2. A hydrogen electrode is shown in Figure 22-10. It consists of a platinum electrode dipping into an acid solution of 1-M concentration and surrounded by hydrogen gas at 1 atmosphere pressure. This *standard hydrogen electrode is assigned a potential of zero volt.* The half-cell reaction is

$$\overset{0}{H_2}(g) \rightleftarrows 2\overset{+1}{H^+}(aq) + 2e^-$$

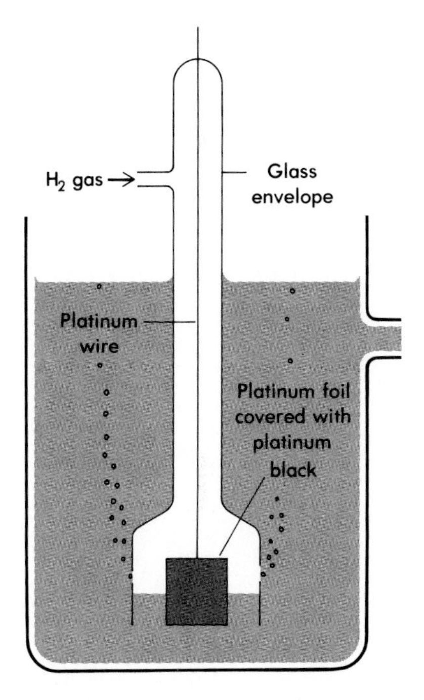

Figure 22-10. Hydrogen electrode, the standard reference electrode for measuring electrode potentials. A layer of hydrogen adsorbed on the surface of the platinum black is the actual electrode surface in contact with the solution.

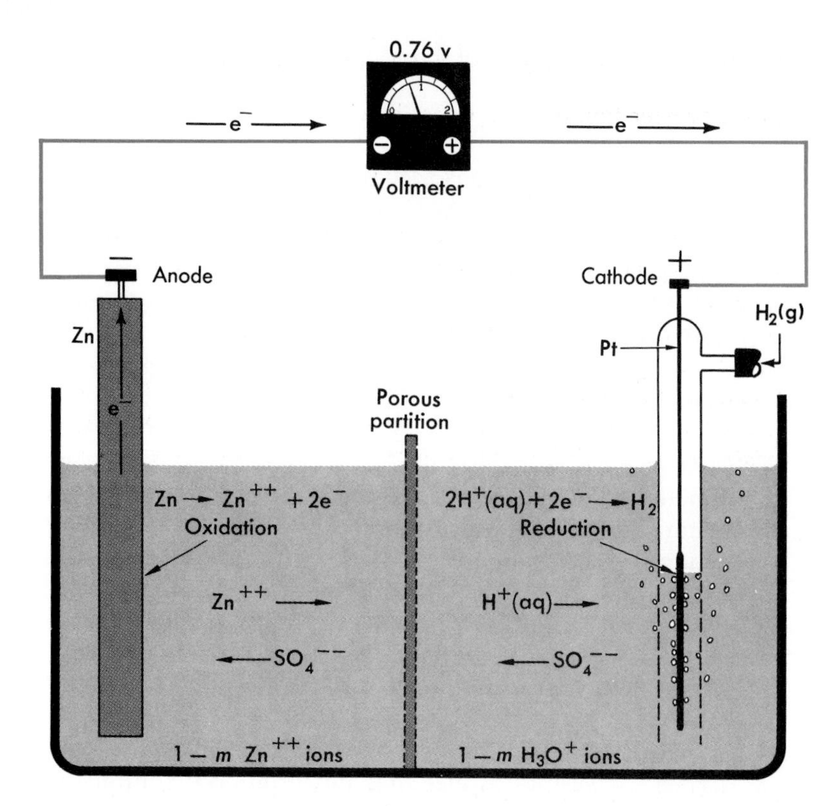

Figure 22-11. The electrode potential of the zinc half-cell is measured by coupling it with a standard hydrogen electrode.

The hydrogen reference electrode is arbitrarily assigned a standard potential of zero volt.

To repeat, the potential of the hydrogen electrode is arbitrarily set at zero volt. Therefore, the potential difference across the complete cell is attributed entirely to the electrode of the other half-cell.

Suppose a complete cell consists of a zinc half-cell and a standard hydrogen half-cell, as in Figure 22-11. The potential difference across the cell measures the electrode potential of the zinc electrode relative to the hydrogen electrode (the zero reference electrode). It is found to be −0.76 volt. *The standard electrode potential, E^0, of an electrode is given a negative value if this electrode has a negative charge relative to the standard hydrogen electrode.* Electrons flow through the external circuit from the zinc to the hydrogen electrode. There, $H^+(aq)$ ions are reduced to H_2 gas.

This reaction indicates that the tendency for Zn^{++} ions to be reduced to Zn atoms is 0.76 volt less than the tendency for $H^+(aq)$ ions to be reduced to H_2. The half-reaction (as a reduction) is

$$\overset{+2}{\mathbf{Zn}}^{++} + \mathbf{2e^-} \rightarrow \overset{0}{\mathbf{Zn}} \qquad E^0 = -0.76 \text{ v}$$

This reaction has less tendency to occur than

$$\mathbf{2}\overset{+1}{\mathbf{H}}{}^{+}\mathbf{(aq)} + \mathbf{2e^-} \rightarrow \overset{0}{\mathbf{H}}_2\mathbf{(g)}$$

by 0.76 volt. This statement also means that the half-reaction (as an oxidation)

$$\overset{0}{\text{Zn}} \rightarrow \overset{+2}{\text{Zn}^{++}} + 2e^- \qquad E^0 = +0.76 \text{ v}$$

has a greater tendency to occur by 0.76 volt than

$$\overset{0}{\text{H}_2}(g) \rightarrow 2\overset{+1}{\text{H}^+}(aq) + 2e^-$$

Observe that the sign of the electrode potential is reversed when the Zn half-cell reaction is written as an oxidation.

A copper half-cell coupled with the standard hydrogen electrode gives a potential difference measurement of +0.34 volt. This measurement indicates that $Cu^{++}(aq)$ ions are more readily reduced than $H^+(aq)$ ions. *The standard electrode potential, E^0, of an electrode is given a positive value if this electrode has a positive charge relative to the standard hydrogen electrode.* The half-reaction for copper (as a reduction) is

$$\overset{+2}{\text{Cu}^{++}} + 2e^- \rightarrow \overset{0}{\text{Cu}} \qquad E^0 = +0.34 \text{ v}$$

This reaction has a greater tendency to occur than

$$2\overset{+1}{\text{H}^+}(aq) + 2e^- \rightarrow \overset{0}{\text{H}_2}(g)$$

by 0.34 volt.

Two observations can be made from these measurements: (1) Zinc has a greater tendency to yield electrons than hydrogen by 0.76 volt. (2) Hydrogen has a greater tendency to yield electrons than copper by 0.34 volt. Taken together, these potentials indicate that zinc has a greater tendency toward oxidation than copper by 1.10 volts (0.76 v + 0.34 v).

How do these electrode potentials apply to the Zn-Cu voltaic cell of Figure 22-9? The potential difference across the complete cell is obtained by adding the electrode potentials of the two half-reactions; the Zn half-reaction is written as an oxidation and the Cu half-reaction as a reduction.

$$\begin{array}{ll} \overset{0}{\text{Zn}} \rightarrow \overset{+2}{\text{Zn}^{++}} + 2e^- & E^0 = +0.76 \text{ v} \\ \overset{+2}{\text{Cu}^{++}} + 2e^- \rightarrow \overset{0}{\text{Cu}} & E^0 = +0.34 \text{ v} \\ \hline \text{Zn} + \text{Cu}^{++} \rightarrow \text{Zn}^{++} + \text{Cu} & E^0 = +1.10 \text{ v} \end{array}$$

The positive sign of the potential difference shows that the reaction proceeds spontaneously to the right.

Half-reactions for some common electrodes and their standard electrode potentials are listed in Table 22-4. These reactions are arranged according to their standard electrode potentials, E^0, relative to a standard hydrogen reference electrode. All electrode reactions are written as *reduction reactions* to the right. Electrode potentials are given as *reduction potentials*. Half-reactions with *positive* reduction potentials occur spontaneously to the right as

Table 22-4 STANDARD ELECTRODE POTENTIALS (as reduction potentials)	
Half-reaction	*Electrode potential* (E^0)
$Li^+ + e^- \rightleftharpoons Li$	−3.04 v
$K^+ + e^- \rightleftharpoons K$	−2.92 v
$Ba^{++} + 2e^- \rightleftharpoons Ba$	−2.90 v
$Ca^{++} + 2e^- \rightleftharpoons Ca$	−2.76 v
$Na^+ + e^- \rightleftharpoons Na$	−2.71 v
$Mg^{++} + 2e^- \rightleftharpoons Mg$	−2.38 v
$Al^{+++} + 3e^- \rightleftharpoons Al$	−1.71 v
$Zn^{++} + 2e^- \rightleftharpoons Zn$	−0.76 v
$Cr^{+++} + 3e^- \rightleftharpoons Cr$	−0.74 v
$S + 2e^- \rightleftharpoons S^{--}$	−0.51 v
$Fe^{++} + 2e^- \rightleftharpoons Fe$	−0.41 v
$Cd^{++} + 2e^- \rightleftharpoons Cd$	−0.40 v
$Co^{++} + 2e^- \rightleftharpoons Co$	−0.28 v
$Ni^{++} + 2e^- \rightleftharpoons Ni$	−0.23 v
$Sn^{++} + 2e^- \rightleftharpoons Sn$	−0.14 v
$Pb^{++} + 2e^- \rightleftharpoons Pb$	−0.13 v
$Fe^{+++} + 3e^- \rightleftharpoons Fe$	−0.04 v
$2H^+(aq) + 2e^- \rightleftharpoons H_2$	0.00 v
$S + 2H^+(aq) + 2e^- \rightleftharpoons H_2S(aq)$	+0.14 v
$Cu^{++} + e^- \rightleftharpoons Cu^+$	+0.16 v
$Cu^{++} + 2e^- \rightleftharpoons Cu$	+0.34 v
$I_2 + 2e^- \rightleftharpoons 2I^-$	+0.54 v
$MnO_4^- + e^- \rightleftharpoons MnO_4^{--}$	+0.56 v
$Fe^{+++} + e^- \rightleftharpoons Fe^{++}$	+0.77 v
$Hg_2^{++} + 2e^- \rightleftharpoons 2Hg$	+0.80 v
$Ag^+ + e^- \rightleftharpoons Ag$	+0.80 v
$Hg^{++} + 2e^- \rightleftharpoons Hg$	+0.85 v
$Br_2 + 2e^- \rightleftharpoons 2Br^-$	+1.06 v
$MnO_2 + 4H^+(aq) + 2e^- \rightleftharpoons Mn^{++} + 2H_2O$	+1.21 v
$Cr_2O_7^{--} + 14H^+(aq) + 6e^- \rightleftharpoons 2Cr^{+++} + 7H_2O$	+1.33 v
$Cl_2 + 2e^- \rightleftharpoons 2Cl^-$	+1.36 v
$Au^{+++} + 3e^- \rightleftharpoons Au$	+1.42 v
$MnO_4^- + 8H^+(aq) + 5e^- \rightleftharpoons Mn^{++} + 4H_2O$	+1.49 v
$F_2 + 2e^- \rightleftharpoons 2F^-$	+2.87 v

reduction reactions. Half-reactions with *negative* reduction potentials occur spontaneously to the left as *oxidation reactions.* When a half-reaction is written as an oxidation reaction, the sign of the electrode potential is changed. The potential then becomes an *oxidation potential.*

Standard electrode potentials describe the relative reaction tendencies of reactants under standardized conditions. As concentrations or temperatures are changed, reaction tendencies change as predicted by Le Chatelier's principle.

The magnitude of the electrode potential measures the tendency of the *reduction half-reaction* to occur as the equation is written in Table 22-4. The half-reaction at the top of the column has the *least* tendency toward reduction (adding electrons). Stated in another way, it has the *greatest* tendency to occur as an oxidation (yielding electrons). The half-reaction at the bottom of the column has the *greatest* tendency to occur as a reduction. Thus it has the *least* tendency to occur as an oxidation.

The *lower* a half-reaction is in the column, the *greater* is the tendency for its *reduction reaction* to occur. The *higher* a half-reaction is in the column, the *greater* is the tendency for the *oxidation reaction* to occur. For example, potassium has a large negative electrode potential and a strong tendency to form K^+ ions. Thus, potassium is a strong reducing agent. Fluorine has a large positive electrode potential and a strong tendency to form F^- ions. Fluorine, then, is a strong oxidizing agent. Compare the listings in Table 22-3 with those in Table 22-4.

SUMMARY

The atoms or ions of an element that attain a more positive oxidation state in a chemical reaction are said to be oxidized. Those particles that attain a more negative oxidation state are reduced. The oxidation state of an element is indicated by an oxidation number. Oxidation results in a change of oxidation number in a positive sense. Reduction results in a change of oxidation number in a negative sense. Both oxidation and reduction occur simultaneously and in equivalent amounts in redox reactions.

The equations for simple oxidation-reduction reactions can be balanced by inspection. Equations for more complicated oxidation-reduction reactions are more easily balanced if the electrons gained and lost by elements that change oxidation numbers are first made equivalent. These equations can be successfully balanced by following a sequence of steps.

In redox reactions, the substance reduced acts as an oxidizing agent because it acquires electrons from the substance oxidized. The substance oxidized becomes a reducing agent because it supplies the electrons to the substance reduced. A strong oxidizing agent easily acquires electrons and becomes a weak reducing agent reluctant to give up electrons. A strong reducing agent easily gives up electrons and becomes a weak oxidizing agent.

One equivalent of a reactant oxidized or reduced is the quantity that loses or gains one mole (the Avogadro number) of electrons. For a chemical equivalent of a substance involved in a redox reaction to be determined, the particular oxidation-reduction process must be known.

Some oxidation-reduction reactions occur spontaneously and may be used as sources of electric energy when arranged in electrochemical cells. These redox processes are called electrochemical reactions. Other oxidation-reduction reactions that are not spontaneous can be driven by an external source of electric energy. These redox processes are called electrolysis reactions.

The oxidation or reduction reaction between one electrode and its electrolyte in an electrochemical cell is called a half-reaction. The potential difference between the electrode and its solution is called the electrode potential. The sum of electrode potentials of the two half-reactions of a cell roughly is its potential difference.

When measured under standard conditions using a standard hydrogen half-cell as a reference, the measured value is the standard electrode potential of that electrode. These standard electrode potentials indicate the relative strengths of electrode substances as oxidizing and reducing agents.

VOCABULARY

anode
autooxidation
cathode
charging cycle
chemical equivalent
conservation of atoms
conservation of electrons
discharge cycle
electrochemical cell
electrochemical process
electrode potential
electrolysis
electrolytic cell
electrolytic process
electron-transfer method
electroplating
half-cell
half-reaction method
ion-electron method
ion migration
negative oxidation state
oxidation
oxidation-reduction reaction
oxidizing agent
positive oxidation state
potential difference
redox reaction
reducing agent
reduction
standard electrode potential
storage battery
volt
voltaic cell

QUESTIONS

Group A

1. Distinguish between the processes of oxidation and reduction. *A1*
2. An oxidation process and a reduction process occur simultaneously. Explain. *A2b*
3. Elementary particles that acquire electrons during a chemical reaction undergo what change in oxidation state? *A1*
4. Identify the oxidation-reduction reactions among the following: *A2a*
   (*a*) $2Na + Cl_2 \rightarrow 2NaCl$
   (*b*) $C + O_2 \rightarrow CO_2$
   (*c*) $2H_2O \rightleftarrows 2H_2 + O_2$
   (*d*) $NaCl + AgNO_3 \rightarrow AgCl + NaNO_3$
   (*e*) $NH_3 + HCl \rightarrow NH_4^+ + Cl^-$
   (*f*) $2KClO_3 \rightarrow 2KCl + 3O_2$
   (*g*) $H_2 + Cl_2 \rightarrow 2HCl$
   (*h*) $2H_2 + O_2 \rightarrow 2H_2O$
   (*i*) $H_2SO_4 + 2KOH \rightarrow K_2SO_4 + 2H_2O$
   (*j*) $Zn + CuSO_4 \rightarrow ZnSO_4 + Cu$
5. For each oxidation-reduction reaction in Question 4, identify: (*a*) the substance oxidized; (*b*) the substance reduced; (*c*) the oxidizing agent; (*d*) the reducing agent. *A2b*
6. What is the oxidation number of manganese in (*a*) the permanganate ion, MnO_4^-; (*b*) the manganate ion, MnO_4^{--}? *A4, A5a*
7. Determine the oxidation number of each element in the following compounds: (*a*) $Ca(ClO_3)_2$; (*b*) Na_2HPO_4; (*c*) K_2CO_3; (*d*) H_3PO_3; (*e*) $Fe(OH)_3$. *A4, A5a*
8. Assign oxidation numbers to all elements in the following compounds: (*a*) $PbSO_4$; (*b*) H_2O_2; (*c*) $K_2Cr_2O_7$; (*d*) H_2SO_3; (*e*) $HClO_4$. *A4, A5a*
9. List the essential components of a copper-plating cell. *A2k*
10. Define: (*a*) electrode potential; (*b*) half-reaction; (*c*) half-cell. *A1*
11. The standard hydrogen electrode is assigned an electrode potential of 0.00 volt. Explain why this voltage is assigned. *A2n*
12. List in proper sequence the six steps involved in balancing oxidation-reduction equations by the oxidation-number method. *B1a*
13. Complete the first four steps listed in Question 12 for each of the following: *B1a*
    (*a*) zinc + hydrochloric acid → zinc chloride + hydrogen.
    (*b*) iron + copper(II) sulfate → iron(II) sulfate + copper.
    (*c*) potassium dichromate + sulfur + water → sulfur dioxide + potassium hydroxide + chromium(III) oxide.
    (*d*) bromine + water → hydrobromic acid + hypobromous acid.

GROUP B

14. Zinc reacts with nitric acid to form zinc nitrate, ammonium nitrate, and water. Balance the redox equation by the oxidation-number method. *B1a*

15. Zinc reacts with sodium chromate in a sodium hydroxide solution to form Na_2ZnO_2, $NaCrO_2$, and H_2O. Balance the redox equation using the oxidation-number method. *B1a*

16. Balance the following oxidation-reduction equation by the oxidation-number method. *B1a*

$$K_2Cr_2O_7 + HCl \rightarrow KCl + CrCl_3 + H_2O + Cl_2$$

17. Potassium carbonate and bromine react to form potassium bromide, potassium bromate, and carbon dioxide. Balance the equation using the oxidation-number method. *B1a*

18. Potassium permanganate, sodium sulfite, and sulfuric acid react to form potassium sulfate, manganese(II) sulfate, sodium sulfate, and water. Balance the equation for this oxidation-reduction reaction by the oxidation-number method. *B1a*

19. Concentrated nitric acid reacts with copper to form copper(II) nitrate, nitrogen dioxide, and water. Balance the equation using *(a)* the oxidation-number method; *(b)* the ion-electron method. *B1a, B1b*

20. Dilute nitric acid reacts with copper to form copper(II) nitrate, nitrogen monoxide, and water. Balance the equation using *(a)* the oxidation-number method; *(b)* the ion-electron method. *B1a, B1b*

21. Hot, concentrated sulfuric acid reacts with zinc to form zinc sulfate, hydrogen sulfide, and water. Balance the equation using *(a)* the oxidation-number method; *(b)* the ion-electron method. *B1a, B1b*

22. Nitric acid and hydrogen sulfide react to form nitrogen monoxide, sulfur, and water. Balance the equation by *(a)* the oxidation-number method; *(b)* the ion-electron method. *B1a, B1b*

23. Balance the equation for the following redox reaction by the ion-electron method. *B1b*

zinc + nitric acid →
zinc nitrate + nitrogen monoxide + water

24. Balance the equation for the following redox reaction by the ion-electron method. *B1b*

$$KMnO_4 + FeSO_4 + H_2SO_4 \rightarrow K_2SO_4 + MnSO_4 + Fe_2(SO_4)_3 + H_2O$$

25. Referring to Table 22-3, the active metals down to magnesium replace hydrogen from water. Magnesium and succeeding metals replace hydrogen from steam. Metals near the bottom of the list do not replace hydrogen from steam. How can the data given in the table help you explain the behavior of these metals? *A6b*

26. Using data from Table 22-4, write the equation for the half-reactions of a voltaic cell having Cu and Ag electrodes. *A7b*

27. *(a)* Determine the potential difference across the voltaic cell of Question 26. *(b)* Write the equation for the overall reaction of the cell in the direction that it proceeds spontaneously. *A7b, B3*

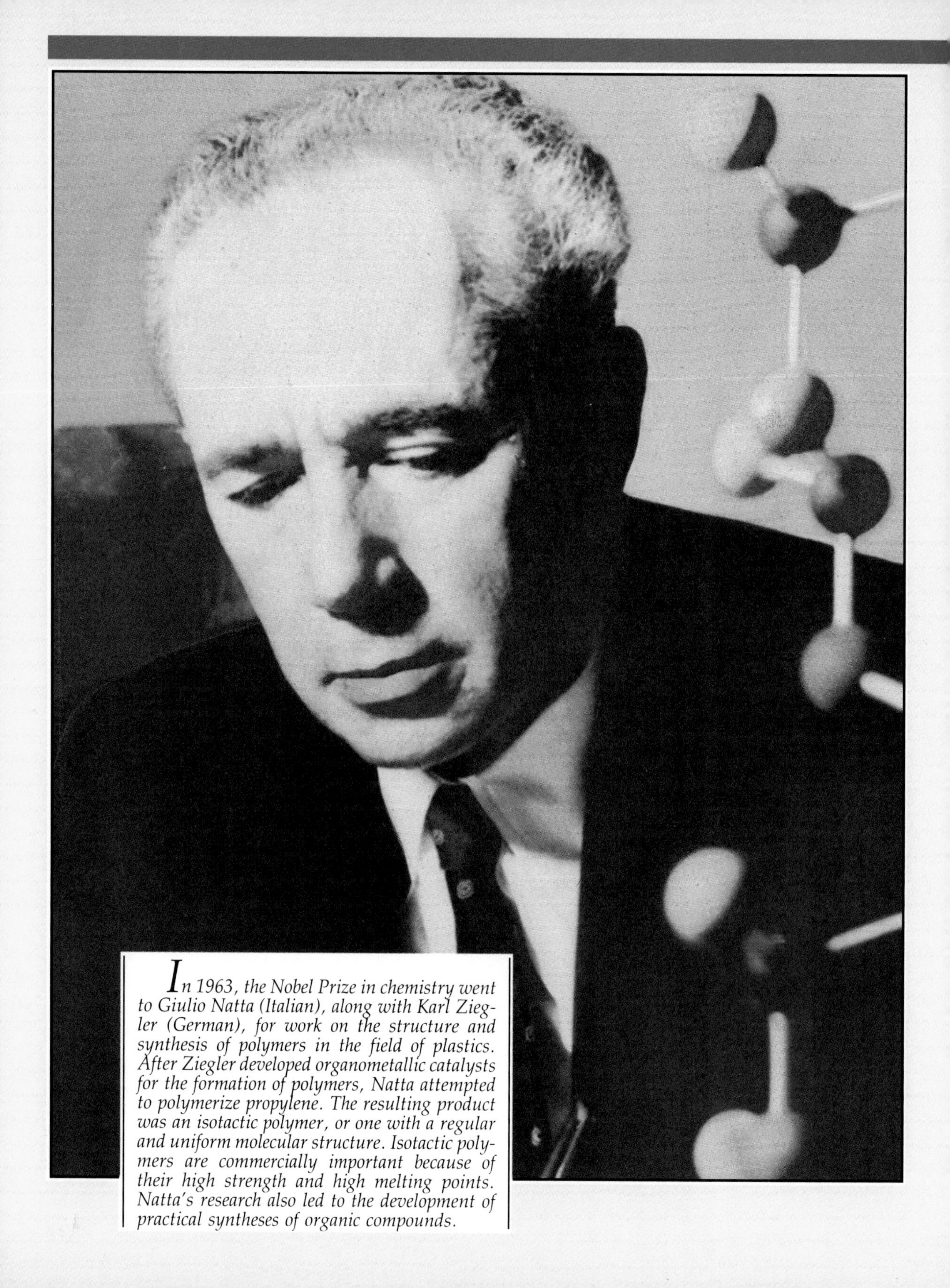

In 1963, the Nobel Prize in chemistry went to Giulio Natta (Italian), along with Karl Ziegler (German), for work on the structure and synthesis of polymers in the field of plastics. After Ziegler developed organometallic catalysts for the formation of polymers, Natta attempted to polymerize propylene. The resulting product was an isotactic polymer, or one with a regular and uniform molecular structure. Isotactic polymers are commercially important because of their high strength and high melting points. Natta's research also led to the development of practical syntheses of organic compounds.

Elements of Period Three

23

GOALS

In this chapter you will gain an understanding of: • the major physical and chemical properties of the elements of Period Three • the relationship between the variation in atomic structure and the different properties shown by the elements of Period Three • the properties of the oxides of Period Three elements • the properties of the binary hydrogen compounds of Period Three elements

23.1 General appearance In Chapter 5 you learned how the chemical elements are arranged by atomic number in a periodic table. In this table, the elements in a given column have similar chemical properties. This chemical similarity is explained by the similar arrangement of electrons in the highest energy level. Across a row of the periodic table, the properties of the elements vary. The first element is a highly active metal. Metals of decreasing activity come after it. They are followed by metalloids. Nonmetals of increasing activity come next. The last element in a row is a noble gas. This wide variation in chemical properties is due to the different highest-energy-level electron arrangements.

In this chapter the elements in Period Three will be discussed. Period Three is a short period containing eight elements. The first three elements are sodium, magnesium, and aluminum. All are silvery metals with a characteristic metallic luster. Sodium is soft enough to be cut easily. Magnesium and aluminum are somewhat harder, but can easily be scratched with a knife. Silicon, the fourth element, is still harder. It is gray with a metallic luster. Because silicon has properties between those of metals and nonmetals, it is classed as a *metalloid*. Phosphorus is the fifth element. It commonly exists as an active, pale yellow, waxy solid or as a more stable red powder. Phosphorus is a nonmetal. The next two elements, sulfur and chlorine, are also nonmetals. Sulfur is a brittle yellow solid, and chlorine is a greenish-yellow gas. Argon, the noble gas of this period, is colorless. Photographs of samples of the elements of Period Three are shown in Figure 5-4.

These elements illustrate the variation in properties that occurs within a period of the table. In later chapters, elements will be considered mainly within their family relationships.

23.2 Structure and physical properties Table 23-1 gives melting point, boiling point, density, and color data for each element of Period Three.

Figure 23-1. The beauty and brilliance of a fireworks display depend on the very rapid combustion of elements such as magnesium, phosphorus, and sulfur.

Metallic crystals are described in text Section 12.14(3).

Because sodium, magnesium, and aluminum are metals, they have certain physical properties in common. Each is a good conductor of heat and electricity, has a silvery luster, and is ductile and malleable. Sodium and aluminum form cubic crystals. Magnesium forms hexagonal crystals. The crystal lattice of each of these elements is made up of metallic ions in an electron "gas" of valence electrons. The good conductivity of these metals shows that their free electrons are highly mobile. Their silvery luster results from free electrons at the crystal surfaces. When light strikes these surfaces, the valence electrons absorb energy and begin to vibrate. These vibrations re-radiate the energy in all directions as light.

The binding force in metallic crystals is the attraction between positive ions and the negative electron "gas" that fills the lattice. This force is exerted equally in all directions. Its magnitude determines the degree of hardness, ductility, and mallea-

Table 23-1
PHYSICAL PROPERTIES OF ELEMENTS OF PERIOD THREE

Element	*Melting point* (°C)	*Boiling point* (°C)	*Density* (g/cm^3)	*Color*
sodium	97.8	882.9	0.971	silver
magnesium	649	1090	1.74	silver
aluminum	660.4	2467	2.70	silver
silicon	1410	2355	2.33	gray
phosphorus	44.1 (white)	280	1.82	pale yellow
sulfur	112.8 (rhombic)	444.7	2.07	yellow
chlorine	−101.0	−34.6	3.214 g/L	green-yellow
argon	−189.2	−185.7	1.784 g/L	colorless

bility of these metals. The magnitude of this binding force is further indicated by the melting and boiling points of the metals. Sodium is softer and has a lower melting point than magnesium and aluminum. These properties indicate that the forces holding sodium ions in fixed positions in the crystal lattice are weaker than those in magnesium and aluminum. Because of the rather wide temperature range over which the metals are liquids, the forces holding the mobile ions together in the liquid phases of these three metals must be quite strong.

Magnesium is denser than sodium, and aluminum is denser than magnesium. The denser elements have heavier, though smaller, atoms than the less dense elements. Besides being heavier, the smaller atoms tend to pack more closely together.

Silicon exists as a covalent network of atoms with a cubic crystal structure. The pattern of atomic arrangement in the silicon crystal is identical to diamond, shown in Figure 17-3. However, the distance between nuclei is greater in silicon, 2.35 Å, than in diamond, 1.54 Å. The electron configuration of carbon is $1s^2 2s^2 2p^2$ and hybridization produces four equivalent sp^3 orbitals. Carbon atoms form four equivalent covalent bonds in the diamond structure. The electron configuration of silicon is similar to carbon, $1s^2 2s^2 2p^6 3s^2 3p^2$. Silicon also undergoes hybridization in the third energy level, forming four equivalent sp^3 orbitals. Silicon, like diamond, has four equivalent covalent bonds. Unlike diamond, which is a nonconductor, silicon has a low electric conductivity. Some of the valence electrons in silicon crystals are free to move. Review Section 12.14(2).

Figure 23-2. The Papago Indian Solar Project in Gunsite, Arizona uses solar cells as a source of energy. Silicon is used to make these photovoltaic devices, which convert sunlight directly into electricity. The blue color of each cell is an antireflection coating that increases the amount of sunlight the cells absorb.

There is a great difference between the melting points of the two consecutive elements, aluminum and silicon. This difference is an indication that the two elements have different structures. Aluminum has a metallic structure, and silicon has a covalent network structure. Silicon is less dense than aluminum, even though silicon atoms are smaller and heavier than aluminum atoms. Hence, the structure of silicon must be less compact than that of aluminum.

Phosphorus has the electron configuration $1s^2 2s^2 2p^6 3s^2 3p^3$. With its three half-filled $3p$ orbitals, a phosphorus atom can form three covalent bonds. Elemental white phosphorus exists as separate P_4 molecules. Each atom of a P_4 molecule forms three covalent bonds. See Figure 23-3. White phosphorus has low melting and boiling points and is quite soft. These properties indicate that only weak dispersion interaction forces exist among the P_4 molecules. White phosphorus has no luster. It is a poor conductor of heat and a nonconductor of electricity. These typical properties of nonmetals are a sign that they lack free electrons.

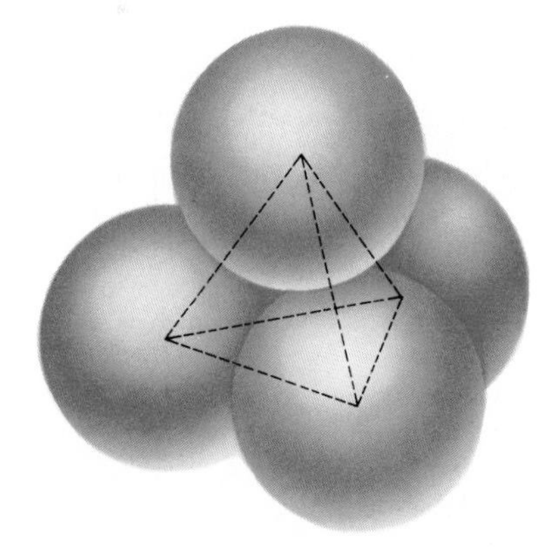

Figure 23-3. The structure of P_4 molecules of phosphorus.

The electron configuration of sulfur is $1s^2 2s^2 2p^6 3s^2 3p^4$. Sulfur atoms have two half-filled $3p$ orbitals and should form two covalent bonds. They do so in elemental sulfur. However, instead of forming a diatomic molecule like oxygen, sulfur atoms form a

Figure 23-4. The structure of S_8 molecules of sulfur.

molecule consisting of eight atoms joined together in a puckered ring. See Figure 23-4. The brittleness and relatively low melting point of sulfur indicate that there are only weak dispersion interaction forces between S_8 molecules. Sulfur has the common nonmetallic properties: no silvery luster, poor thermal conductivity, and nonconduction of electricity.

You have learned (Section 6.10) that chlorine exists as diatomic molecules, Cl_2. A single covalent bond joins the two atoms. Since chlorine is a gas at room temperature, the dispersion interaction forces between chlorine molecules must be very weak.

Argon exists as single atoms because it has no half-filled orbitals to use in bond formation. The very low boiling point of argon shows that the dispersion interaction forces between argon atoms are exceedingly weak.

23.3 Electron configurations, ionization energies, and oxidation states Table 23-2 lists the electron configuration, first ionization energy, oxidation states, and atomic radius for each of the Period Three elements.

When ionization energy was first discussed in Section 5.7, you learned that it generally increases across a period because of the increasing nuclear charge. But *decreases* in ionization energy occur in Period Three at aluminum and again at sulfur. The decrease between magnesium and aluminum indicates a distinct difference in their highest-energy-level electron arrangements. The highest-energy electron of aluminum is the first electron in the 3*p* sublevel. The highest-energy electrons of magnesium occupy the 3*s* sublevel. Less energy is required to remove the 3*p* electron because the 3*p* sublevel has higher energy than the 3*s* sublevel. A decrease in ionization energy also occurs between phosphorus and sulfur. It shows that the pairing of electrons in the 3*p* sublevel begins with sulfur. The half-filled 3*p* sublevel of phosphorus has three singly occupied orbitals. The 3*p* sublevel in sulfur has one filled and two singly occupied orbitals. The configuration of phosphorus is *more stable* than the configuration

Table 23-2
ATOMIC STRUCTURE AND RELATED PROPERTIES OF ELEMENTS OF PERIOD THREE

Element	*Electron configuration*	*First ionization energy* (kcal/mole atoms)	*Principal oxidation states*	*Atomic radius* (Å)
sodium	$1s^2 2s^2 2p^6 3s^1$	119	+1	1.54
magnesium	$1s^2 2s^2 2p^6 3s^2$	176	+2	1.36
aluminum	$1s^2 2s^2 2p^6 3s^2 3p^1$	138	+3	1.18
silicon	$1s^2 2s^2 2p^6 3s^2 3p^2$	188	+4	1.11
phosphorus	$1s^2 2s^2 2p^6 3s^2 3p^3$	242	+3, +5	1.06
sulfur	$1s^2 2s^2 2p^6 3s^2 3p^4$	239	−2, +4, +6	1.02
chlorine	$1s^2 2s^2 2p^6 3s^2 3p^5$	299	−1, +5, +7	0.99
argon	$1s^2 2s^2 2p^6 3s^2 3p^6$	363	0	0.98

of sulfur. In other words, the phosphorus configuration has lower energy. As a result, more energy is required to remove an electron from a phosphorus atom than from a sulfur atom.

The atomic radii decrease gradually across the period. This general decrease is easily explained. Successive electrons enter the same energy level, while the nuclear charge becomes successively greater and pulls the electrons closer to the nucleus. The relationship between atomic radius and ionic radius for sodium, magnesium, aluminum, sulfur, and chlorine was described in Section 6.8.

The oxidation states of sodium, +1, magnesium, +2, and aluminum, +3, were explained in Section 5.8. Sodium has one electron that can be removed at low energy. Magnesium has two and aluminum three such electrons. Silicon atoms undergo sp^3 hybridization and can form four covalent bonds. Its oxidation state is +4. Phosphorus, with the electron-dot symbol ·Ṗ̣:, can add three electrons to each atom to form P^{---} ions, as in the compounds of phosphorus known as phosphides. Here its oxidation number is −3. It is also possible for phosphorus atoms to share three or five electrons in forming covalent compounds. Such sharing gives oxidation numbers of +3 or +5.

Sulfur, with the electron-dot symbol ·Ṣ̈:, can add two electrons to each atom. This electron addition forms sulfide ions, S^{--}, which have oxidation number −2. Sulfur can also share two, four, or six electrons, giving oxidation states of +2, +4, or +6, respectively.

Chlorine, with the electron-dot symbol :C̣̈l:, commonly adds one electron to each atom, forming chloride ions. Here, chlorine has the oxidation number −1. However, it is also possible for chlorine to share one, three, five, or even all seven electrons in covalent bonding. In such cases the oxidation numbers are +1, +3, +5, or +7, respectively.

Since argon is not yet known to form compounds, it has an oxidation number of zero.

Figure 23-5. Photo shows a blue-green argon laser and a red helium-neon laser used in conjunction with a diffraction grating to create these multiple laser beams.

23.4 Properties of oxides of Period Three elements The elements of Period Three form a great variety of compounds with oxygen. Sodium, phosphorus, sulfur, and chlorine form two or more oxides. Only one oxide of each element will be described here. See Table 23-3.

Sodium oxide, Na_2O, and magnesium oxide, MgO, are white solids. The reaction of sodium with oxygen usually yields sodium peroxide, Na_2O_2. However, some sodium oxide can be prepared by heating sodium at about 180 °C in a limited amount of dry oxygen. Magnesium oxide is produced by burning magnesium in pure oxygen. Both sodium oxide and magnesium oxide are ionic compounds. Sodium oxide readily dissolves in water and forms a basic solution containing Na^+ and OH^- ions.

$$Na_2O + H_2O \rightarrow 2Na^+ + 2OH^-$$

You may find it helpful to review Section 15.7, acid anhydrides, and Section 15.10, basic anhydrides.

Table 23-3
OXIDES OF PERIOD THREE ELEMENTS

Oxide	*Melting point* (°C)	*Boiling point* (°C)	*Type of compound*	*Nature of reaction with water*
Na_2O	subl. 1275		ionic	forms OH^-
MgO	2852	3600	ionic	forms OH^-
Al_2O_3	2072	2980	ionic	none; hydroxide is amphoteric
SiO_2	1610	2230	covalent network	none
P_4O_{10}	subl. 300		molecular	forms H^+ (aq)
SO_2	−73	$-1\overline{0}$	molecular	forms H^+ (aq)
Cl_2O	−121	2	molecular	forms H^+ (aq)

Magnesium oxide is only slightly soluble in water, but it, too, forms a basic solution.

$$MgO + H_2O \rightarrow Mg^{++} + 2OH^-$$

Aluminum oxide, Al_2O_3, is a white, ionic compound. It can be produced by burning aluminum in pure oxygen. More commonly, it is prepared by heating aluminum hydroxide or the hydrated oxides of aluminum. Because aluminum oxide is virtually insoluble in water, the dehydration reaction is not reversible. Aluminum hydroxide is a white, insoluble, jellylike substance. It must be prepared indirectly from aluminum oxide. Aluminum hydroxide is amphoteric; it reacts as a base in the presence of hydronium ions and as an acid in the presence of hydroxide ions.

Compare these equations with the more detailed ones given in Section 15.9(7).

$$Al(OH)_3 + 3H_3O^+ \rightarrow Al(H_2O)_6^{+++}$$

$$Al(OH)_3 + OH^- \rightarrow Al(OH)_4^-$$

Silicon dioxide, SiO_2, exists widely in nature as quartz. White sand is mostly silicon dioxide. Silicon dioxide has a covalent network structure. It consists of silicon atoms tetrahedrally bonded to four oxygen atoms. Each oxygen atom, in turn, is bonded to another silicon atom tetrahedrally bonded to four oxygen atoms. This structure extends indefinitely. All the silicon-oxygen bond distances are equal. The silicon-oxygen bonds are covalent, but with some degree of ionic character.

Silicon dioxide is practically insoluble in water. It acts as an acid in hot, concentrated solutions containing hydroxide ions.

$$SiO_2 + 4OH^- \rightarrow SiO_4^{----} + 2H_2O$$

Silicon dioxide does not react with common acids, but does react with hydrofluoric acid.

$$SiO_2 + 4HF \rightarrow SiF_4(g) + 2H_2O$$

Figure 23-6. Sand dunes in Death Valley. White sand is mostly silicon dioxide.

In an abundant supply of oxygen, phosphorus burns and forms diphosphorus pentoxide, P_4O_{10}. (This formula is the cor-

Figure 23-7. Mining phosphate rock. From phosphate rock, elemental phosphorus and phosphoric acid are prepared. Phosphoric acid is used in manufacturing fertilizers.

rect molecular formula. The name is taken from the corresponding empirical formula, P_2O_5.) P_4O_{10} is a white, molecular solid. It reacts rapidly with water, forming a solution containing phosphoric acid, H_3PO_4.

In 1983 phosphoric acid ranked ninth in chemicals produced in the United States. Over 9 million metric tons were produced.

$$P_4O_{10} + 6H_2O \rightarrow 4H^+(aq) + 4H_2PO_4^-$$

Phosphoric acid is a moderately strong acid. The equilibrium constant for its first ionization is of the order of 10^{-2}.

The explanation of ionization constant of an acid is given in Section 21.7.

Sulfur burns in air or oxygen to form sulfur dioxide, SO_2. This compound is molecular. At room temperature it is a colorless gas with a choking odor. Sulfur dioxide readily dissolves in and reacts with water. This reaction forms a solution containing sulfurous acid, H_2SO_3.

$$SO_2 + H_2O \rightarrow H^+(aq) + HSO_3^-$$

This acid is also moderately strong. The equilibrium constant for its first ionization is of the order of 10^{-2}.

Dichlorine monoxide, Cl_2O, is prepared by the reaction of chlorine with freshly prepared mercury(II) oxide.

$$2Cl_2 + 2HgO \rightarrow HgO{\cdot}HgCl_2 + Cl_2O(g)$$

This compound is an unstable yellow-brown gas. It reacts with water to give a solution containing hypochlorous acid, HClO.

$$Cl_2O + H_2O \rightarrow 2HClO$$

$$2HClO \rightleftarrows 2H^+(aq) + 2ClO^-$$

Figure 23-8. Sulfur banks on Hawaii Island. The yellow crust on these deposits is elemental sulfur. The mass of sulfur gases put into the atmosphere by a volcanic eruption is estimated in the millions of metric tons.

Hypochlorous acid is a very weak acid that exists only in water solution. Its ionization constant is of the order of 10^{-8}.

23.5 Properties of the binary hydrogen compounds of Period Three elements Each of the elements of Period Three forms at least one binary compound with hydrogen. Silicon, phosphorus, and sulfur each form two or more binary hydrogen compounds. As with the oxides, only one binary hydrogen compound of each Period Three element will be described here. See Table 23-4.

Sodium hydride, NaH, is formed from sodium and hydrogen at moderately high temperatures. It is an ionic compound that crystallizes in a cubic system, as does sodium chloride. The negative ion in this compound is the hydride ion, H^-. Hydride ions are strong bases. They readily remove protons from water molecules to form hydrogen molecules and hydroxide ions. Sodium hydride reacts vigorously with water, yielding hydrogen gas and hydroxide ions.

$$NaH + H_2O \rightarrow H_2 + Na^+ + OH^-$$

Magnesium hydride, MgH_2, is a less stable compound than sodium hydride. It can be made directly by heating the elements under high pressure. Complex indirect methods are more satisfactory, however. It is a nonvolatile, colorless solid that reacts violently with water.

$$MgH_2 + 2H_2O \rightarrow Mg(OH)_2 + 2H_2$$

In a vacuum, magnesium hydride is stable to about 300 °C. Above this temperature, it decomposes into its elements. Magnesium hydride is an ionic compound that forms tetragonal crystals. The bonding has some covalent character, however.

Aluminum hydride, AlH_3, can be made by the action of lithium aluminum hydride and 100% sulfuric acid dissolved in a complex ether.

$$2LiAlH_4 + H_2SO_4 \xrightarrow[\text{solution)}]{\text{(ether}} 2AlH_3 + 2H_2 + Li_2SO_4$$

Aluminum hydride is a white solid having a covalent network structure, $(AlH_3)_x$. It reacts vigorously with water. It is stable in a vacuum up to about 100 °C. At higher temperatures, it decomposes to aluminum and hydrogen.

Table 23-4
BINARY HYDROGEN COMPOUNDS OF PERIOD THREE ELEMENTS

Compound	*Melting point* (°C)	*Boiling point* (°C)	*Type of compound*	*Nature of reaction with water*
NaH	d. 800		ionic	forms $OH^- + H_2$
MgH_2	d. 280		ionic	forms $OH^- + H_2$
$(AlH_3)_x$	d. 100		covalent network	forms $OH^- + H_2$
SiH_4	−185	−112	molecular	none
PH_3	−133	−88	molecular	none
H_2S	−86	−61	molecular	forms H^+ (aq)
HCl	−115	−85	molecular	forms H^+ (aq)

Monosilane, SiH_4, is a colorless, stable, but readily flammable gas. It consists of nonpolar covalent molecules. Monosilane may be prepared by reducing silicon dioxide with hydrogen at high pressure and moderate temperature.

$$SiO_2 + 4H_2 \xrightarrow[175\ ^\circ C]{400\ atm} SiH_4(g) + 2H_2O$$

Monosilane does not react with pure water, but reacts readily with basic solutions to yield hydrogen and silicon dioxide.

$$SiH_4 + 2H_2O \xrightarrow{(OH^-)} SiO_2 + 4H_2$$

Other known silicon hydrides are Si_2H_6, Si_3H_8, Si_4H_{10}, Si_5H_{12}, and Si_6H_{14}. Notice the similarity of these formulas to those of the alkanes. These compounds, however, are much more reactive than the alkanes. This greater reactivity indicates that silicon-silicon bonds and silicon-hydrogen bonds in the silicon hydrides are weaker than carbon-carbon and carbon-hydrogen bonds in the alkanes.

Recall that the general formula for the alkanes is C_nH_{2n+2}. See Section 18.9.

Phosphine, PH_3, is one of two or three known binary phosphorus-hydrogen compounds. It is a molecular substance in which the atoms are covalently bonded in an ammonia-type structure. At room temperature it is a colorless gas with an unpleasant odor like that of decayed fish. Phosphine is very poisonous and flammable. The simplest method of preparing phosphine is by treating calcium phosphide with dilute acid. Phosphine prepared by this method is spontaneously flammable in air because of impurities.

The pyramidal structure of the ammonia molecule was described in text Section 6.16.

$$Ca_3P_2 + 6H^+(aq) \rightarrow 3Ca^{++} + 2PH_3(g)$$

Phosphine is not very soluble in water; the dilute solution that is produced is neutral.

Hydrogen sulfide, H_2S, is a colorless, foul-smelling, very poisonous gas. Its odor resembles that of decayed eggs. It is a molecular compound, with polar covalent bonding within the molecule. It is usually prepared in the laboratory by the action of dilute hydrochloric acid on iron(II) sulfide.

$$FeS + 2H^+(aq) \rightarrow Fe^{++} + H_2S$$

Hydrogen sulfide is flammable. The products of combustion are water vapor and sulfur or sulfur dioxide, depending on combustion conditions.

Hydrogen sulfide dissolves in and reacts with water, forming a weakly acid solution of hydrosulfuric acid. The first ionization constant of hydrosulfuric acid is of the order of 10^{-7}.

$$H_2S \rightleftharpoons H^+(aq) + HS^-$$

The binary hydrogen compound of chlorine is hydrogen chloride, HCl. This compound is a colorless, sharp-odored, poisonous gas. Hydrogen chloride consists of highly polar covalent molecules. The gas can be prepared by direct combination of its elements or by heating sodium chloride with moderately concentrated sulfuric acid.

$$H_2 + Cl_2 \rightarrow 2HCl$$

$$NaCl + H_2SO_4 \rightarrow NaHSO_4 + HCl$$

Hydrogen chloride is very soluble in water. It is almost completely ionized by the water, yielding hydronium ions and chloride ions.

$$HCl \rightleftarrows H^+(aq) + Cl^-$$

This solution, called hydrochloric acid, is a strong acid.

SUMMARY

Some physical properties of the Period Three elements are described: hardness, color, melting point, boiling point, density, crystal structure, and conductivity. The chemical properties of the Period Three elements are briefly explained in terms of element type, atomic and molecular structure, and interatomic and intermolecular forces. Their first ionization energies, principal oxidation states, and atomic radii are compared.

The preparation and properties of the oxides and binary hydrogen compounds of the Period Three elements are described and compared. The reactions of these compounds with water are also described and compared.

QUESTIONS

Group A

1. Explain the variation in metallic-nonmetallic properties of the Period Three elements in terms of the variation in number of highest-energy-level electrons. *A2a*
2. *(a)* What physical properties do metals like sodium, magnesium, and aluminum have in common? *(b)* What physical properties do nonmetals like phosphorus, sulfur, and chlorine have in common? *A2b*
3. What is the nature of the binding force between *(a)* atoms of sodium; *(b)* atoms of silicon; *(c)* atoms of chlorine; *(d)* molecules of chlorine; *(e)* atoms of argon? *A1, A2b*
4. Aluminum and silicon have consecutive atomic numbers, yet the melting point of silicon is much higher than that of aluminum. Why? *A2b*
5. What is the difference between the interatomic forces of elements that are malleable and of those that are brittle? *A1, A2b*
6. *(a)* From Table 6-3, what is the radius in angstroms of a sodium atom and of a sodium ion? *(b)* Why is the sodium ion so much smaller? *A2e*
7. *(a)* What is the radius in angstroms of a chlorine atom and of a chloride ion? *(b)* Explain the difference in radius. *A2e*
8. How do the reactions of Na_2O, SiO_2, and Cl_2O with water compare? *A2g*
9. How do the reactions of NaH, SiH_4, and HCl with water containing OH^- ions compare? *A2i, A2m*
10. The bond energies of these single bonds are: C—C, 83 kcal/mole; C—H, 98 kcal/mole; Si—Si, 54 kcal/mole; Si—H, 77 kcal/mole. Why are silicon hydrides much more reactive than alkanes? *A2l*

Group B

11. Why do sodium, magnesium, and aluminum have a silvery luster? *A2b*
12. *(a)* What is the ratio of the density of aluminum to that of sodium? *(b)* What is the ratio

of the atomic weight of aluminum to that of sodium? *(c)* Why is the density ratio greater than the atomic weight ratio? *A2b*

13. Which Period Three element has *(a)* the lowest boiling point? *(b)* the highest boiling point? *(c)* How do the magnitudes of the binding forces between atoms of these elements as liquids compare? *A2b*
14. Diamond is a nonconductor of electricity; silicon has a low electric conductivity. Explain this difference even though diamond and silicon have the same crystal structure and type of bonding. *A2c*
15. Phosphorus atoms form three covalent bonds, sulfur atoms form two covalent bonds, and chlorine atoms form one covalent bond; yet phosphorus molecules are tetratomic, sulfur molecules are octatomic, and chlorine molecules are diatomic. Explain. *A2d*
16. Explain the observed change in first ionization energy *(a)* between magnesium and aluminum; *(b)* between phosphorus and sulfur. *A2e*
17. *(a)* Draw electron-dot symbols for the elements of Period Three. *(b)* Using these symbols, explain how the common oxidation states for each element are attained. *A2e*
18. Calculate the percentage of ionic character and predict the type of bonding in the oxides of the elements of Period Three. How do your predictions compare with the actual bonding in these compounds? *A2g*
19. Calculate the percentage of ionic character and predict the type of bonding in the binary hydrogen compounds of the elements of Period Three. How do your predictions compare with the actual bonding in these compounds? *A2i*
20. How do the structures of silicon dioxide and elemental silicon compare? *A2c, A2g*
21. *(a)* How do the structures of NH_3 and PH_3 molecules compare in shape, size, and type of bonding? *(b)* How do H_2S and H_2O molecules compare? *A2i*
22. Write the balanced formula equation for the reaction with water of *(a)* MgO; *(b)* SO_2. *A4b*
23. Write the balanced formula equation for the reaction with water of *(a)* MgH_2; *(b)* H_2S. *A4f*
24. Write the balanced formula equation for the reaction of aluminum hydride and water. *A4f*

PROBLEMS

Group A

1. What is the mass in grams of a block of aluminum 5.0 cm long, 3.0 cm wide, and 1.5 cm high?
2. How many magnesium atoms are in a length of magnesium ribbon of mass 0.725 g?
3. What is the percentage of chlorine in Cl_2O?
4. What volume of hydrogen at STP is formed by the reaction of $14\bar{0}$ mL of gaseous monosilane at STP with water containing hydroxide ions? *A4g*
5. How many grams of calcium phosphide must react with water containing hydronium ions to prepare 6.0 g of phosphine? *A4e*

Group B

6. What volume will 0.135 mole of argon occupy at 25.0 °C and 715 mm pressure?
7. Concentrated phosphoric acid is 85.5% H_3PO_4 by mass and has a density of 1.70 g/mL. How many milliliters of concentrated phosphoric acid are needed to prepare $10\bar{0}$ mL of 0.250-*M* H_3PO_4 solution?
8. A solution of 0.28 g of white phosphorus in 20.0 g of benzene lowers the freezing point of the benzene 0.58 C°. *(a)* Calculate the molecular weight of white phosphorus from these data. *(b)* What is the corresponding molecular formula? See Table 13-5 for additional data.
9. A saturated solution of $Mg(OH)_2$ is approximately 0.000031 *N*. What is its pH?

The 1914 Nobel Prize in chemistry was given to Theodore W. Richards (American) for determining accurate atomic weights of numerous elements. For his doctoral dissertation at Harvard, he redetermined the atomic weight of oxygen, which he found to be 15.869 ± 0.0017 compared to hydrogen. In later research, he corrected previously accepted atomic weights of many elements. He also corrected numerous heats of neutralization and combustion. Through the development of new methods of measurement, Richards was able to observe the isotopes of such elements as oxygen, hydrogen, and lead.

The Metals of Group I

24

GOALS

In this chapter you will gain an understanding of: • the structure and properties of the Group I metals • the occurrence, uses, and compounds of the Group I metals • the Downs cell • the Solvay process • spectroscopy

24.1 Structure and properties The group of elements at the left of the periodic table includes lithium, sodium, potassium, rubidium, cesium, and francium. These elements are all chemically active metals. They are known as the *alkali metals* and also as the Sodium Family of elements. Because of their great chemical reactivity, none exists in nature in the elemental state. They are found in natural compounds as monatomic ions with a +1 charge. Sodium and potassium are plentiful, lithium is fairly rare, and rubidium and cesium are rarer still. Francium does not exist as a stable element. Only trace quantities have been produced in certain nuclear reactions.

The Group I elements form hydroxides whose water solutions are extremely alkaline. Thus, the Group I elements are called "alkali metals."

The Group I elements possess metallic characteristics to a high degree. Each has a silvery luster, is a good conductor of electricity and heat, and is ductile and malleable. These metals are relatively soft and can be cut with a knife. The properties of the Group I metals are related to their characteristic metallic crystal lattice structures, which are described in Section 12.14(3).

The crystal lattice consists of metallic ions with a +1 charge. The lattice is built around a body-centered cubic unit cell in which each metallic ion is surrounded by eight nearest neighbors at the corners of the cube. A model of the body-centered cubic unit cell is shown in Figure 24-1. Valence electrons form an electron "gas" that permeates the lattice structure. These "free" electrons belong to the metallic crystal as a whole. The mobility of the free electrons in Group I metals accounts for their high thermal and electric conductivity and silvery luster as well.

The softness, ductility, and malleability of the alkali metals are explained by the binding force in the metallic crystal lattice. Their low melting points and densities, together with their softness, distinguish these elements from the more familiar common metals. See Table 24-1 for properties of the alkali metals.

Ductile: *capable of being drawn into a wire.*
Malleable: *capable of being rolled or pounded into a sheet.*

The atoms of each element in Group I have a single electron in their highest energy level. Lithium atoms have the electron

Table 24-1
PROPERTIES OF GROUP I ELEMENTS

Element	*Atomic number*	*Atomic weight*	*Electron configuration*	*Oxidation number*	*Melting point* (°C)	*Boiling point* (°C)	*Density* (g/cm³)	*Atomic radius* (Å)	*Ionic radius* (Å)
lithium	3	6.941	2,1	+1	180.5	1347	0.534	1.23	0.68
sodium	11	22.98977	2,8,1	+1	97.8	883	0.971	1.54	0.97
potassium	19	39.0983	2,8,8,1	+1	63.6	774	0.862	2.03	1.33
rubidium	37	85.4678	2,8,18,8,1	+1	38.9	688	1.532	2.16	1.47
cesium	55	132.9054	2,8,18,18,8,1	+1	28.4	678	1.873	2.35	1.67
francium	87	[223]	2,8,18,32,18,8,1	+1	27	677	—	—	—

configuration $1s^22s^1$. All other Group I elements have next-to-highest energy levels containing eight electrons. An ion formed by removing the single valence electron has the stable electron configuration of the preceding noble gas. For example, the sodium ion has the electron configuration $1s^22s^22p^6$. This matches the configuration of the neon atom. The electron configuration of the potassium ion, $1s^22s^22p^63s^23p^6$, matches the argon atom.

It is also possible to *add* an electron to a Group I atom to produce an anion having an electron pair in the *s* orbital of the highest energy level. An example is the Na^- anion which has the electron configuration $1s^22s^22p^63s^2$. Here the oxidation number of sodium is -1. Similar anions of potassium, rubidium, and cesium have also been produced.

24.2 Chemical activity The very active free metals of the Sodium Family are obtained by reducing the +1 ions in their natural compounds. The metals are vigorous reducing agents. They have a weak attraction for their valence electrons. Ionization energy decreases as the atom size increases going down the group. See Table 24-2. This decreasing energy with increasing size shows that it is easier to remove electrons physically from the highest energy levels of the heavier atoms. However, if you examine the net energy of reactions in which electrons are given up to other elements, you find that the lithium atom is the strongest reducing agent. Although it may be harder to remove an electron from the lithium atom, more energy is given back by the subsequent interaction of the lithium ion and its surroundings than by the larger alkali-metal ions with their surroundings.

Handling and storing the alkali metals is difficult because of their chemical activity. They are usually stored submerged in kerosene or some other liquid hydrocarbon because they react vigorously with water. See Figure 24-2. Reacting with H_2O, they release H_2 to form strongly basic hydroxide solutions.

$$2K(s) + 2H_2O(l) \rightarrow 2K^+(aq) + 2OH^-(aq) + H_2(g)$$

Figure 24-1. Body-centered cubic unit cell.

All of the ordinary compounds of the alkali metals are ionic, including their hydrides. Only lithium forms the oxide directly with oxygen. Sodium forms the peroxide instead, and the higher metals tend to form superoxides of the form $M^+O_2^-$. However, the ordinary oxides can be prepared indirectly. These oxides are basic anhydrides. In water they form hydroxides.

Nearly all the compounds of the alkali metals are quite soluble in water. The alkali-metal ions are colorless, have little tendency to hydrolyze in water solution, or to form polyatomic ions.

Compounds of the more important alkali metals are easily identified by *flame tests.* Their compounds impart characteristic colors to a Bunsen flame. Sodium compounds color the flame yellow. Lithium compounds give a flame a red (carmine) color. Potassium colors a flame violet; rubidium and cesium give reddish violet (magenta) flames.

Table 24-2
IONIZATION ENERGY AND THE SIZE OF ATOMS

Element	*Relative size of atoms*	*Ionization energy* (kcal/mole)
Li		124
Na		119
K		10$\bar{0}$
Rb		96.3
Cs		89.8
Fr	—	—

SODIUM

24.3 Occurrence of sodium Metallic sodium is never found free in nature. However, compounds containing the Na^+ ion and sodium complexes exist in soil, natural waters, and in plants and animals. Because of the solubility of sodium compounds, it is almost impossible to find a sodium-free material. Vast quantities of sodium chloride are present in sea water and rock salt deposits. There are important deposits of sodium nitrate in Chile and Peru. The carbonates, sulfates, and borates of sodium are found in dry lake beds.

24.4 Preparation of sodium Sir Humphry Davy (1778–1829) prepared metallic sodium in 1807 by the electrolysis of fused sodium hydroxide. Today sodium is prepared by the electrolysis of fused sodium chloride. An apparatus called the *Downs cell* (see Figure 24-3) is used in this process. Sodium chloride has a high melting point, 801 °C. Calcium chloride is mixed with it to lower the melting point to 580 °C. Liquid sodium is collected under oil. Chlorine gas is produced simultaneously. This gas is kept separate from the metallic sodium by an iron-gauze diaphragm.

The sodium is recovered by reducing Na^+ ions. This occurs at the negative electrode of the Downs cell. Each sodium ion acquires an electron from this electrode to form a neutral sodium atom.

$$Na^+ + e^- \rightarrow Na(l)$$

Chloride ions are oxidized at the positive electrode. Each chloride ion loses an electron to this electrode to form a neutral chlorine atom.

$$Cl^- \rightarrow Cl + e^-$$

But two chlorine atoms form the diatomic molecule of elemental chlorine gas. The net reaction at the positive electrode is

$$2Cl^- \rightarrow Cl_2(g) + 2e^-$$

Figure 24-2. Potassium, a Group I metal, like other members of this group reacts vigorously with water. The light for this photograph was produced by dropping a small amount of potassium into a beaker of water.

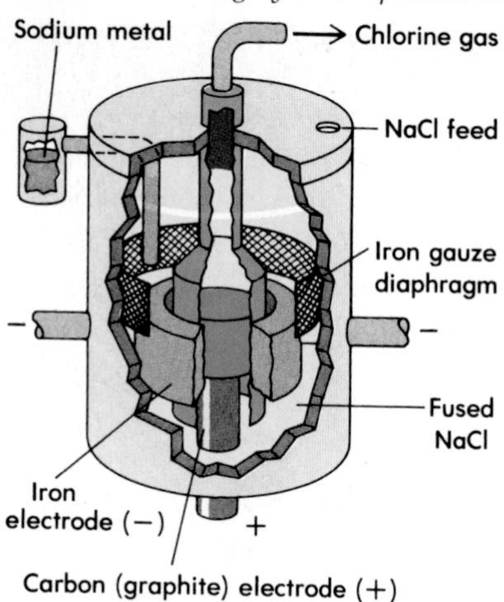

Figure 24-3. Elemental sodium is produced by the electrolysis of melted sodium chloride in a Downs cell. Chlorine is a by-product.

In the overall cell reaction, electron-transfer balance is maintained between the electrodes. Two Na^+ ions are reduced for each Cl_2 molecule formed. The cell reaction can be written

negative electrode:	$2Na^+ + 2e^- \rightarrow 2Na(l)$
positive electrode:	$2Cl^- \rightarrow Cl_2(g) + 2e^-$
cell:	$2Na^+ + 2Cl^- \rightarrow 2Na(l) + Cl_2(g)$

24.5 Properties and uses of sodium Sodium is a silvery-white, lustrous metal that tarnishes rapidly when exposed to air. It is very soft and has a lower density than water and a low melting point. A pellet of sodium dropped into water melts from the heat of the vigorous exothermic reaction that occurs. This reaction yields hydrogen gas and a strongly basic solution.

$$2Na(s) + 2H_2O(l) \rightarrow 2Na^+(aq) + 2OH^-(aq) + H_2(g)$$

When exposed to air, sodium unites with oxygen to form sodium peroxide, Na_2O_2. By supplying sodium in excess, some sodium oxide, Na_2O, can be produced along with the bulk product, Na_2O_2. This production of Na_2O is a result of the strong reducing character of sodium atoms. Sodium oxide can also be formed by heating NaOH with sodium.

$$2NaOH + 2Na \rightarrow 2Na_2O + H_2(g)$$

The superoxide of sodium, NaO_2, can be prepared indirectly. Superoxides have the O_2^- ions with one unpaired electron.

Sodium reacts with all aqueous acids. It burns in an atmosphere of chlorine gas, and directly forms sodium chloride.

A flame test of sodium compounds reveals a strong yellow color characteristic of vaporized sodium atoms. This yellow flame commonly identifies sodium. See Figure 24-4.

Most of the sodium produced in the United States is used in making tetraethyl lead, an antiknock additive for gasoline. Sodium is used as a heat-transfer agent and in making dyes and organic compounds. Another use is in sodium vapor lamps.

Figure 24-4. Sodium salts impart a strong yellow color to the Bunsen flame. This yellow cannot be seen through cobalt-blue glass. See Figure 24-8.

24.6 Sodium chloride Sodium chloride is found in sea water, in salt wells, and in deposits of rock salt. Rock salt is mined in many places in the world.

Pure sodium chloride is not deliquescent. Magnesium chloride is very deliquescent, however, and is usually present in sodium salt as an impurity. This explains why table salt becomes wet and sticky in damp weather.

Sodium chloride is essential in the diets of humans and animals and is present in certain body fluids. Perspiration contains considerable amounts of sodium chloride. People who perspire freely in hot weather may experience a significant loss of this essential mineral during periods of vigorous activity.

Sodium chloride is the cheapest compound of sodium and is used as a starting material in the production of most other sodium compounds. Some sodium compounds are most easily prepared directly from metallic sodium, however.

Figure 24-5. Native crystal formations of sodium chloride, known as halite, in Death Valley, California.

24.7 Sodium hydroxide Most commercial sodium hydroxide is produced by electrolysis of an aqueous sodium chloride solution. This electrolysis of aqueous NaCl is different from that of fused NaCl in the Downs cell. Chlorine gas is produced at the positive electrode, but hydrogen gas (instead of metallic sodium) is produced at the negative electrode. The solution, meanwhile, becomes aqueous NaOH. Evidence indicates that water molecules acquire electrons from the negative electrode and are reduced to H_2 gas and OH^- ions.

The cell reaction for the electrolysis of an aqueous NaCl solution is

electrode (−) $2H_2O + 2e^- \rightarrow H_2(g) + 2OH^-(aq)$

electrode (+) $2Cl^- \rightarrow Cl_2(g) + 2e^-$

cell: $2H_2O + 2Cl^- \rightarrow H_2(g) + Cl_2(g) + 2OH^-(aq)$

As the Cl^- ion concentration diminishes, the OH^- ion concentration increases. The Na^+ ion concentration remains unchanged during electrolysis. Thus the solution is converted from aqueous NaCl to aqueous NaOH.

Sodium hydroxide converts some types of animal and vegetable matter into soluble materials by chemical action. It is very *caustic* and has destructive effects on skin, hair, and wool.

Commercial NaOH is called lye or caustic soda.

Sodium hydroxide is a white crystalline solid. It is marketed in the form of flakes, pellets, and sticks. It is very deliquescent and dissolves in the water that it removes from the air. It reacts with carbon dioxide from the air, producing sodium carbonate. Its water solution is strongly basic.

The 1983 production of sodium hydroxide was over 9 million metric tons. It ranked seventh in amount among chemicals produced in the United States.

Sodium hydroxide reacts with fats, forming soap and glycerol. One of its important uses, therefore, is in making soap. It is also used in the production of rayon, cellulose film, and paper pulp and in petroleum refining.

24.8 The Solvay process Approximately three-fourths of the sodium carbonate produced in the United States comes from the mining of natural sodium carbonate deposits located in Wyoming or from the evaporation of brine from Searles Lake, California. The remaining sodium carbonate production in the United States and much of that produced in Europe is manufactured by the Solvay process. This process, a classic example of efficiency in chemical production, was developed in the 1860's by Ernest Solvay (1838–1922), a Belgian industrial chemist.

The raw materials for the Solvay process are common salt, limestone, and coal. The salt is pumped as brine from salt wells.

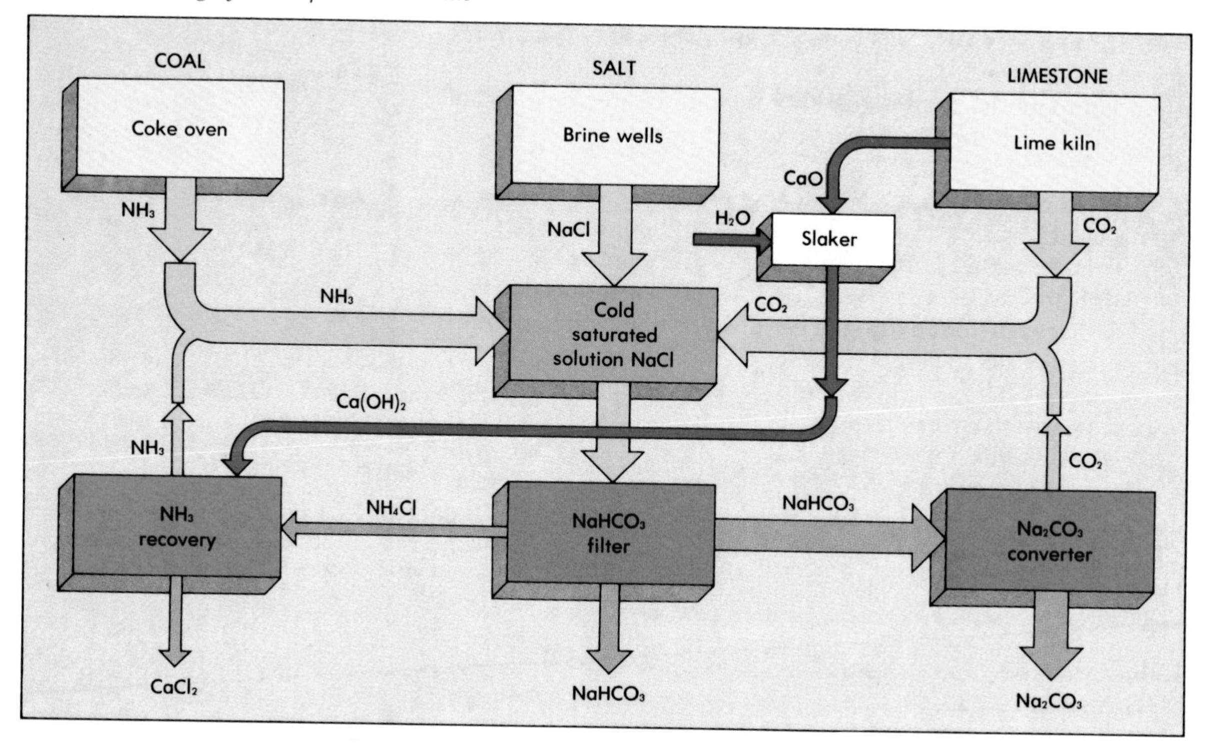

Figure 24-6. A flow diagram of the Solvay process.

The thermal decomposition of limestone yields the carbon dioxide and calcium oxide needed in the process.

$$CaCO_3(s) \rightarrow CaO(s) + CO_2(g) \qquad \text{(Equation 1)}$$

Coal is converted into coke, gas, coal tar, and ammonia by destructive distillation. The coke and gas are used as fuel in the plant. The ammonia is also used in the process, and the coal tar is sold as a useful by-product. See Figure 24-6.

To begin the process, a cold saturated solution of sodium chloride is further saturated with ammonia and carbon dioxide. The following reactions occur:

$$CO_2(g) + H_2O \rightarrow H_2CO_3$$

$$H_2CO_3 + NH_3(g) \rightarrow NH_4^+(aq) + HCO_3^-(aq)$$

$$\text{Net:} \quad CO_2(g) + NH_3(g) + H_2O \rightarrow NH_4^+(aq) + HCO_3^-(aq)$$

The HCO_3^- ions form in a solution that has a high concentration of Na^+ ions. Since sodium hydrogen carbonate is only slightly soluble in this cold solution, it precipitates.

$$Na^+(aq) + HCO_3^-(aq) \rightarrow NaHCO_3(s)$$

The solution that remains contains NH_4^+ ions and Cl^- ions. The precipitated sodium hydrogen carbonate is filtered and dried. It is either sold as *baking soda* or converted into sodium carbonate by thermal decomposition.

$NaHCO_3$: baking soda
Na_2CO_3: soda ash
$Na_2CO_3 \cdot 10H_2O$: washing soda

$$2NaHCO_3(s) \rightarrow Na_2CO_3(s) + H_2O(g) + CO_2(g)$$

The dried sodium carbonate is an important industrial chemical called *soda ash*.

The 1983 production of sodium carbonate was about 7.7 million metric tons. It ranked tenth in amount among chemicals produced in the United States.

The ammonia used in the process is more valuable than the sodium carbonate or sodium hydrogen carbonate. Hence, it must be recovered and used over again if the process is to be profitable. The calcium oxide from Equation 1 is slaked by adding water. Calcium hydroxide is formed in this reaction.

$$CaO(s) + H_2O(l) \rightarrow Ca(OH)_2(s)$$

The calcium hydroxide is added to the solution containing NH_4^+ ions and Cl^- ions to release the ammonia from the ammonium ion.

$$NH_4^+(aq) + OH^-(aq) \rightarrow NH_3(g) + H_2O$$

Table 24-3
REPRESENTATIVE SODIUM COMPOUNDS

Chemical name	*Common name*	*Formula*	*Color*	*Uses*
sodium tetraborate, decahydrate	borax	$Na_2B_4O_7 \cdot 10H_2O$	white	as a water softener; in making glass; as a flux
sodium carbonate, decahydrate	washing soda	$Na_2CO_3 \cdot 10H_2O$	white	as a water softener; in making glass
sodium hydrogen carbonate	baking soda	$NaHCO_3$	white	as a leavening agent in baking
sodium cyanide	prussiate of soda	$NaCN$	white	to destroy vermin; to extract gold from ores; in silver and gold plating; in case-hardening steel
sodium hydride	(none)	NaH	white	in cleaning scale and rust from steel forgings and castings
sodium nitrate	Chile saltpeter	$NaNO_3$	white, or colorless	as a fertilizer
sodium peroxide	(none)	Na_2O_2	yellowish white	as an oxidizing and bleaching agent; as a source of oxygen
sodium phosphate, decahydrate	TSP	$Na_3PO_4 \cdot 10H_2O$	white	as a cleaning agent; as a water softener
sodium sulfate, decahydrate	Glauber's salt	$Na_2SO_4 \cdot 10H_2O$	white, or colorless	in making glass; as a cathartic in medicine
sodium thiosulfate, pentahydrate	hypo	$Na_2S_2O_3 \cdot 5H_2O$	white, or colorless	as a fixer in photography; as an antichlor
sodium sulfide	(none)	Na_2S	colorless	in the preparation of sulfur dyes; for dyeing cotton; to remove hair from hides

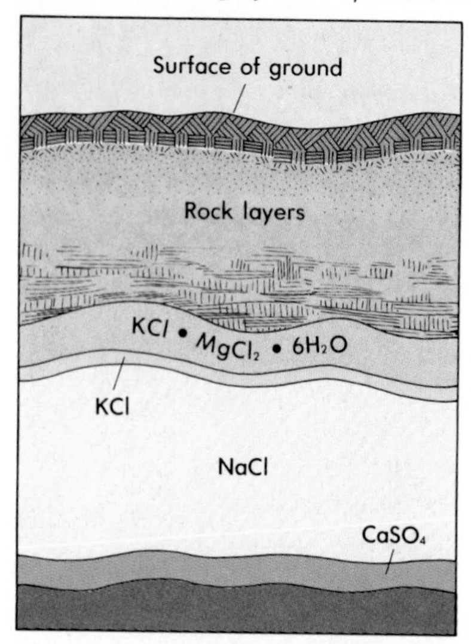

Figure 24-7. A cross section of a salt deposit, showing how the different minerals were deposited as the sea water evaporated.

The solution now contains mainly Ca^{++} ions and Cl^- ions. Neither of these ions is recycled into the process. As a by-product, calcium chloride has some use as an inexpensive dehydrating agent. The supply generally exceeds the demand, however.

Sodium hydrogen carbonate, as baking soda, is the main ingredient of baking powders. One important industrial use of sodium carbonate is in the production of glass.

The water solution of sodium carbonate is mildly basic because of hydrolysis of CO_3^{--} ions.

$$CO_3^{--} + H_2O \rightleftarrows HCO_3^- + OH^-$$

Potassium

24.9 Preparation and properties of potassium Potassium is abundant in nature and widely distributed, but only in combined form. Great deposits of combined potassium occur in the form of feldspar, a potassium aluminosilicate mineral. This mineral is part of all granitic rocks. It is insoluble and weathers very slowly. Large deposits of potassium chloride, crystallized with magnesium and calcium compounds, are found in Texas and New Mexico. See Figure 24-7. Some potassium compounds are taken from Searles Lake in California.

Potassium was first prepared by Sir Humphry Davy in 1807. His method involved the electrolysis of fused potassium hydroxide. The small annual production today is by the reaction

$$KCl + Na \rightarrow NaCl + K$$

The reaction proceeds to the right, and equilibrium is prevented by removing the potassium.

Potassium metal is soft and of low density. It has a silvery luster that quickly tarnishes bluish-gray when exposed to air. Potassium is more active than sodium. It floats on water and reacts with the water so rapidly that the hydrogen gas given off usually ignites.

Potassium imparts a fleeting violet color to a Bunsen flame. The color comes from vaporizing potassium atoms. The presence of sodium, however, masks the violet color of potassium in the flame. Potassium is detected in a mixture of sodium and potassium compounds by observing the colored flame through cobalt-blue glass. This glass filters out the yellow of the sodium flame, allowing the violet potassium flame to show clearly. A typical potassium flame is shown in Figure 24-8.

24.10 Compounds of potassium All common potassium compounds are soluble in water. Potassium hydroxide, which is prepared by the electrolysis of a solution of potassium chloride, has the typical properties of a strong alkali. Potassium nitrate is made by mixing hot, concentrated solutions of potassium chloride and sodium nitrate.

$$KCl + NaNO_3 \rightarrow KNO_3 + NaCl(s)$$

The solubility curves of Figure 13-11 can be helpful in understanding how potassium nitrate is recovered in this process. Examine the solubility curves of the four possible pairs of ions in the mixed solution. Sodium chloride is the least soluble. It has almost the same solubility over the liquid temperature range of water. Potassium nitrate is much less soluble at low temperature than at high temperature.

The solution is evaporated at high temperature, and sodium chloride first crystallizes. As evaporation continues, sodium chloride continues to separate, and the concentration of potassium and nitrate ions increases. The crystallized sodium chloride is removed, and the solution is allowed to cool. Very little sodium chloride separates as the solubility remains about the same. Potassium nitrate crystallizes very rapidly as the solution cools. It can be purified by recrystallization.

Sodium compounds are often used instead of potassium compounds because they are usually less expensive. Most glass is made with sodium carbonate, but potassium carbonate yields a more lustrous glass, which is preferred for optical uses. Potassium nitrate is not hygroscopic. For this reason, it is used instead of sodium nitrate in making black gunpowder.

One important use of potassium compounds has no substitute. Green plants must have these compounds to grow properly. Therefore, complete chemical fertilizers always contain an appropriate amount of potassium.

Figure 24-8. Potassium salts color the Bunsen flame violet. The potassium flame is masked by the presence of sodium salts but is purple when viewed through cobalt-blue glass. Thus, a cobalt-blue glass permits the flame test to detect elemental potassium in a sodium-potassium mixture.

Table 24-4
REPRESENTATIVE POTASSIUM COMPOUNDS

Chemical name	*Common name*	*Formula*	*Color*	*Uses*
potassium bromide	(none)	KBr	white	as a sedative; in photography
potassium carbonate	potash	K_2CO_3	white	in making glass; in making soap
potassium chlorate	(none)	$KClO_3$	white	as an oxidizing agent; in fireworks; in explosives
potassium chloride	(none)	KCl	white	as a source of potassium; as a fertilizer
potassium hydroxide	caustic potash	KOH	white	in making soft soap; as a battery electrolyte
potassium iodide	(none)	KI	white	in medicine; in iodized salt; in photography
potassium nitrate	saltpeter	KNO_3	white	in black gunpowder; in fireworks; in curing meats
potassium permanganate	(none)	$KMnO_4$	purple	as a germicide; as an oxidizing agent

SPECTROSCOPY

24.11 Use of a spectroscope One type of ***spectroscope*** consists of a glass prism and a *collimator tube* which focuses a narrow beam of light rays on the prism. It also has a small telescope for examining the light that passes through the prism. When white light passes through a triangular prism, a band of colors called a *continuous spectrum* appears. This effect is caused by the unequal bending of light of different wavelengths as it enters the prism and as it emerges from it. A continuous spectrum is shown in the top band of Figure 24-9.

A spectroscope is an optical instrument that separates the light entering it into the component wavelengths, which are then seen as the light spectrum.

Examination of a sodium flame by spectroscope reveals a characteristic bright-yellow line. Since this yellow line is always in the same relative place in the spectrum, it identifies sodium. Potassium produces both red and violet spectral lines. The characteristic color lines of several chemical elements are shown in Figure 24-9.

24.12 Origin of spectral lines A platinum wire held in a Bunsen flame becomes incandescent and emits white light. When the incandescent wire is viewed through a spectroscope, a continuous spectrum of colors is observed. The energy of the white light is distributed over a continuous range of light frequencies. This range includes the entire visible spectrum. White light is spread out, forming a spectrum as it passes through the prism of the spectroscope. Here the light rays of different frequencies are bent different amounts. Light energy of the shortest wavelengths (highest frequencies) is bent most, forming the deep violet color seen at one end of the visible spectrum. Light energy of the longest wavelengths (lowest frequencies) is bent least, forming the deep red color characteristic of the other end of the visible spectrum. Between these two extremes there is a gradual blending from one color to the next. It is possible to recognize six elementary colors: *red, orange, yellow, green, blue,* and *violet.* In general, incandescent solids and gases under high pressure give continuous spectra. Refer to the top band in Figure 24-9.

The relation between frequency and wavelength of electromagnetic radiation is explained in Section 4.2.

The bright-line spectra of elements were first described in Section 4.3.

Luminous (glowing) gases and vapors under low pressure give discontinuous spectra called *bright-line spectra.* These spectra consist of narrow lines of color that correspond to the light energy of certain wavelengths. The atoms of each element produce that element's characteristic line spectra.

Electrons in atoms are restricted to energies of only certain values. In unexcited atoms, electrons occupy the lowest energy levels available to them. The energy of an electron can change only as it moves from one energy level to another. A certain amount of energy is absorbed with each jump to a higher energy level; a certain amount is released with each jump to a lower level. The quantity of energy in each change is equal to the difference between the separate energy levels involved.

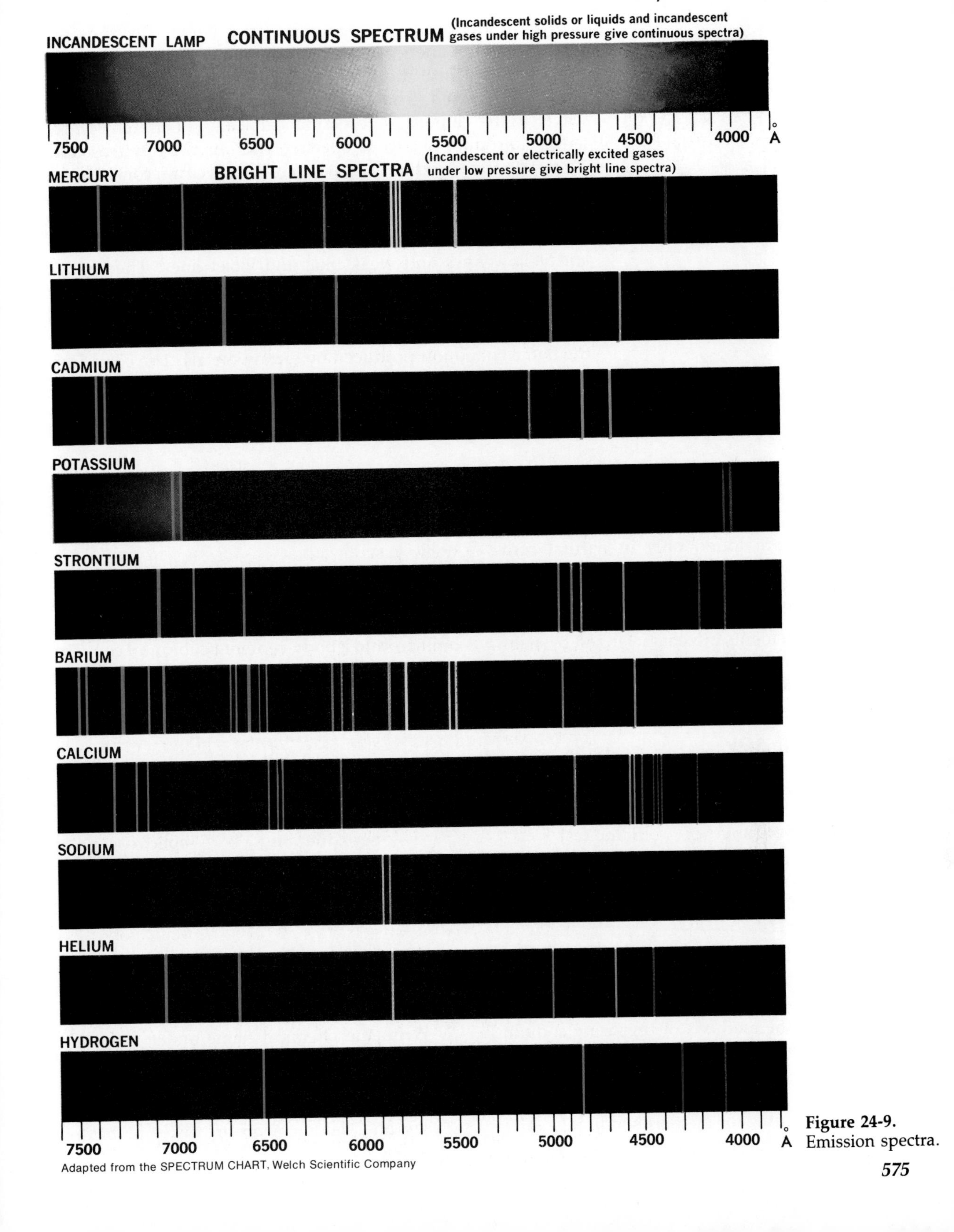

Figure 24-9. Emission spectra.

When substances are vaporized in a flame, electrons are raised to higher energy levels by heat energy. When these electrons fall back into the lower energy levels available to them, energy is released. The energy released by any substance has wavelengths characteristic of that substance. Different wavelengths produce different spectral lines in the spectroscope. Thus, vaporized sodium atoms produce a spectrum consisting of two narrow yellow lines very close together (seen in the ordinary spectroscope as a single yellow line). Potassium atoms produce two red lines and a violet line. Lithium atoms yield intense red and yellow lines and weak blue and violet lines. The spectra produced by excited atoms of different elements are as distinct as fingerprints.

Bright lines in the visible portion of the spectrum account for the flame coloration produced by certain metals. The color seen is the combination of emitted light of different wavelengths.

SUMMARY

The elements of Group I in the periodic table are among the most chemically active metals. Group I elements do not exist as free elements in nature; they are found only in natural compounds as monopositive ions. The elements of Group I are referred to collectively as the Sodium Family and as the alkali metals.

Group I elements are characterized by a single highest-energy-level electron added to the preceding noble-gas structure. These elements have relatively low ionization energies. When the outer *s* electron is removed, each alkali-metal ion has the stable electron configuration of a noble gas.

Group I elements are soft, silvery-white metals that are good conductors of heat and electricity. Their structure is a lattice of metallic ions built around a body-centered cubic unit cell. The valence electrons form an electron "gas" that permeates the lattice structure.

VOCABULARY

alkali metal
baking soda
bright-line spectrum
caustic
continuous spectrum
Downs cell
flame test
halite
incandescent
lye
peroxide ion
potash
soda ash
Solvay process
spectroscope
superoxide ion

QUESTIONS

Group A

1. How do the electron configurations of Group I atoms and ions differ? *A2c*
2. Describe the methods of preparing sodium and potassium. *A2g*
3. Write the ionic equation for the reaction of sodium and water. *A4a*
4. List three uses for metallic sodium. *A3j*
5. Distinguish *caustic* from *corrosive*. *A1*
6. Identify: *(a)* the raw materials used in the Solvay process; *(b)* the products and by-products of the process. *A2g*
7. Write the chemical name and formula for the following substances: *(a)* caustic soda; *(b)* washing soda; *(c)* baking soda. *A3k*
8. What leavening action occurs when molasses and baking soda are used in baking cookies? *A3k*

9. What are the principal sources of potassium compounds in the United States? *A2a*
10. *(a)* How are metallic sodium and potassium stored in the laboratory stockroom? *(b)* Why must they be stored in this manner? *A2f*
11. Write the ionic equation for the reaction of potassium and water. *A4a*
12. Describe the flame tests for sodium and potassium. *A3d*
13. Write three equations that represent the production of sodium carbonate in the Solvay process. *A2g*
14. Write three equations for the recovery of ammonia in the Solvay process. *A4b*
15. Describe: *(a)* the metallic crystal lattice of the alkali metals; *(b)* the unit cell upon which this lattice is built. *A2d*

GROUP B

16. The elements of Group I are described as soft, malleable metals with low melting and low boiling points. How can you account for these Group I properties? *A2b*
17. Why is NaCl necessary in the diet of many animals and people? *A3l*
18. Why is sodium chloride used as a starting material for preparing metallic sodium and other compounds of sodium? *A4b*
19. *(a)* Why are sodium compounds more frequently used than potassium compounds? *(b)* What are the purposes for which sodium compounds cannot be substituted for potassium compounds? *A3m*
20. What by-product of the Solvay process has such limited use and yet is produced in such quantity that disposal of it is actually a problem to the manufacturers? *A2g*
21. Why does table salt become sticky in damp weather, although pure sodium chloride is not deliquescent? *A2h*
22. Explain why potassium has a lower density than sodium, although it consists of heavier atoms. *A2e*
23. Write the equation for the net reaction when sodium hydroxide is exposed to air. *A3h*
24. In the Solvay process, the reaction between sodium chloride and ammonium hydrogen carbonate run to completion. Why? *A4b*
25. Why does a solution of sodium carbonate in water turn red litmus paper blue? *A1*
26. Suppose you had a tremendous quantity of acid that had to be neutralized and that NaOH, KOH, and LiOH were all available at the same price per kilogram. Which of these three would you use? Why?
27. A 0.1-*M* solution of $HC_2H_3O_2$ is found to have a pH of 2.9. A solution of 0.1-*M* $NaC_2H_3O_2$ is added to the acetic acid solution and the pH rises. Explain.

PROBLEMS

GROUP A

1. *(a)* How many grams of sulfuric acid in water solution can be neutralized by 15.0 g of sodium hydroxide; *(b)* 15.0 g of potassium hydroxide?
2. Given 1.50 kg of sodium nitrate and 1.50 kg of potassium chloride, how many kilograms of potassium nitrate can be produced by reacting these two substances?
3. If sodium carbonate decahydrate sells for 50.0 cents per kilogram, what is the anhydrous sodium carbonate worth per kilogram?
4. How many liters of carbon dioxide can be liberated from 50.0 g of each of the following: *(a)* Na_2CO_3; *(b)* $NaHCO_3$; *(c)* K_2CO_3; *(d)* $KHCO_3$?

GROUP B

5. How many kilograms of sodium chloride are required to produce 1.00 metric ton of anhydrous sodium carbonate?
6. How many cubic meters of carbon dioxide (at STP) are needed in Problem 5?
7. How many cubic meters of carbon dioxide gas at $15\bar{0}$ °C and 745 mm pressure will supply that needed in Problem 6?
8. A solution is prepared by dissolving 1.1 g NaOH in water and diluting it to $50\bar{0}$ mL. What is the pH of the solution?

Frederick Soddy (British) was awarded the 1921 Nobel Prize in chemistry for studying radioactive substances and the occurrence and nature of isotopes. Early in his research, he worked with Rutherford in developing a theory for the disintegration of radioactive substances. Later he discovered that products of radioactive transformation could not be separated from some ordinary elements (such as lead) by chemical means. Since these substances were indistinguishable except for their atomic weights, he proposed that they were the same elements but in different forms, which he called "isotopes."

The Metals of Group II

GOALS

In this chapter you will gain an understanding of: • the physical and chemical properties of the Group II metals • the major uses of Group II metals • the important compounds of Group II metals • soft water and hard water • methods of softening hard water

25.1 The Calcium Family The elements of Group II of the periodic table are members of the Calcium Family. They are the metals beryllium, magnesium, calcium, strontium, barium, and radium. These elements are also called the *alkaline-earth metals.* Like the alkali metals, they are never found as free elements in nature. The metals must be recovered from their natural compounds. Many of their compounds are insoluble or slightly soluble and are found in the earth's crust. Their carbonates, phosphates, silicates, and sulfates are the most important deposits.

*Early chemists used the term "earth" for nonmetallic substances, such as the Group II metal oxides, that were practically insoluble in water and unchanged by strong heating. The Group II metal oxides give an alkaline reaction with water. Thus, these metal oxides have historically been called **alkaline earths,** and the metals **alkaline-earth metals.***

Beryllium and magnesium are commercially important light metals. In their chemical behavior, they resemble the corresponding alkali metals, lithium and sodium. Radium is important because it is radioactive. Radioactivity and other properties of radium are discussed in Chapter 31. The remaining three elements of Group II—calcium, strontium, and barium—have similar properties. They are considered to be typical members of the Calcium Family.

Each alkaline-earth element has two valence electrons beyond the stable configuration of the preceding noble gas. All form doubly charged ions of the M^{++} type. Thus ions of the alkaline-earth metals have stable noble-gas structures. Their chemistry, like that of the alkali metals of Group I, is generally uncomplicated.

The attraction between the metal ions and the electron "gas" of the alkaline-earth metal crystals is stronger than in the alkali metals. Therefore, the Calcium Family metals are denser and harder than the corresponding Sodium Family metals, and they have higher melting and boiling points.

Group II metals are harder and stronger than Group I metals.

The atoms and ions of the alkaline-earth metals are smaller than those of the corresponding alkali metals because of their higher nuclear charge. For example, the magnesium ion, Mg^{++}, has the same electron configuration as the sodium ion, Na^{+}.

Figure 25-1. Beryllium being processed in Utah. This metal produces alloys that are extremely elastic.

This configuration is $1s^22s^22p^6$. However, Mg^{++} has a nuclear charge of +12, and Na^+ has a nuclear charge of +11. The higher nuclear charge of the Mg^{++} ion attracts electrons more strongly and results in a smaller ionic radius. Some properties of Group II metals are listed in Table 25-1.

In Section 24.2 you learned that ionization energies of Group I metals decrease down the group as atomic size increases. This same relationship is seen in Group II. The smaller atoms hold their outer electrons more securely than do the larger atoms.

Recall that ionization energy measures the tendency of an isolated atom to hold a valence electron. The *first* ionization energy relates to the removal of one electron from the neutral atom. The *second* ionization energy is that required to remove an electron from the ion with a +1 charge (the atom that has already had one valence electron removed). This second ionization energy is always higher than the first because the particle from which the electron is removed now has a positive charge.

If a third electron were to be removed from a Group II atom, it would come from the stable inner electron configuration (the noble-gas structure) of the +2 ion. The predicted energy requirement would be much higher. In fact, the energies required for a third level of ionization of alkaline-earth elements are very high. They exceed the energies usually available in chemical reactions. Thus +2 ions of these elements are the only ones observed. The first, second, and third ionization energies of the Group II elements are listed in Table 25-2.

Beryllium and its compounds are extremely toxic.

The alkaline-earth metals form hydrides, oxides or peroxides, and halides similar to those of the alkali metals. Almost all hydrides are ionic and contain H^- ions. An exception is BeH_2. Binary compounds of beryllium have fairly strong covalent bonds. They react with water to release hydrogen gas, and they form basic hydroxide solutions. Calcium hydride is often used as a laboratory source of hydrogen.

The oxides of the alkaline-earth metals have very high melting points. CaO and MgO are used as heat-resistant (refractory)

Table 25-1
PROPERTIES OF GROUP II ELEMENTS

Element	*Atomic number*	*Atomic weight*	*Electron configuration*	*Oxidation number*	*Melting point* (°C)	*Boiling point* (°C)	*Density* (g/cm³)	*Atomic radius* (Å)	*Ionic radius* (Å)
beryllium	4	9.01218	2,2	+2	1280	2970	1.85	0.89	0.35
magnesium	12	24.305	2,8,2	+2	649	1090	1.74	1.36	0.66
calcium	20	40.08	2,8,8,2	+2	839	1484	1.55	1.74	0.99
strontium	38	87.62	2,8,18,8,2	+2	769	1384	2.54	1.91	1.12
barium	56	137.33	2,8,18,18,8,2	+2	725	1640	3.5	1.98	1.34
radium	88	226.0254	2,8,18,32,18,8,2	+2	700	1140	5(?)	2.20	1.43

materials. Beryllium oxide is amphoteric; otherwise the oxides are basic. With the exception of beryllium, the oxides are essentially ionic, but with more covalent character than the alkali-metal oxides. Strontium and barium form peroxides with oxygen, probably because of the large size of their ions.

The hydroxides are formed by adding water to the oxides. Except for $Be(OH)_2$, which is amphoteric, the hydroxides dissociate in water solution to yield OH^- ions. Hydroxides above barium are only slightly soluble in water, the solubility increasing with the metallic ion's size. Solutions of these hydroxides have low concentrations of OH^- ions; they are weakly basic because of the slight solubility in water.

Table 25-2
IONIZATION ENERGIES OF GROUP II ATOMS

Element	*Ionization energy* *First* (kcal/mole)	*Second*	*Third*
Be	215	$\overline{420}$	3549
Mg	176	347	1848
Ca	141	274	1174
Sr	$\underline{131}$	254	1005(?)
Ba	$\overline{120}$	231	850(?)
Ra	122	234	—

MAGNESIUM

25.2 Occurrence of magnesium Magnesium compounds are widely distributed on land and in the sea. Magnesium sulfate is found in the earth's crust in many places. Important deposits occur in the state of Washington and in British Columbia, Canada. A double chloride of potassium and magnesium is mined from the salt deposits of Texas and New Mexico. See Figure 24-7. Sea water contains significant amounts of magnesium compounds. Talc and asbestos are silicates of magnesium. Their properties are related to their silicate structures.

Dolomite, $CaCO_3 \cdot MgCO_3$, is a double carbonate of magnesium and calcium that is found throughout the United States and Europe. It is an excellent building stone and is used for lining steel furnaces. Pulverized dolomite neutralizes soil acids and also supplies magnesium for plant growth.

Elemental magnesium, shown in Figure 25-2, was first prepared by Davy in 1808. This same series of experiments resulted in the isolation of sodium, calcium, and similar active metals.

Figure 25-2. (Top) Feathery crystals shown here are composed of magnesium, the eighth most abundant element. (Bottom) Magnesium sulfate crystals are photomicrographed (×8).

25.3 Extraction of magnesium Most commercial magnesium is produced by electrolytic reduction of molten magnesium chloride. The Mg^{++} ions may be obtained from sea water, brine, or minerals. Sea water is the most economical source.

About three kilograms of magnesium are recovered from each metric ton of sea water processed. The water is first treated with lime made from oyster shells, which are cheap and readily available. This treatment causes magnesium ions to precipitate as magnesium hydroxide. The reactions are shown by the following equations:

$$CaO(s) + H_2O(l) \rightarrow Ca(OH)_2(s)$$

$$Ca(OH)_2(s) \rightleftarrows Ca^{++}(aq) + 2OH^-(aq) \qquad K_{sp} = 5.5 \times 10^{-6}$$

$$Mg^{++}(aq) + 2OH^-(aq) \rightleftarrows Mg(OH)_2(s) \qquad K_{sp} = 1.5 \times 10^{-11}$$

The two values for K_{sp} show that $Mg(OH)_2$ is less soluble than $Ca(OH)_2$. The precipitation of $Mg(OH)_2$ lowers the concentration of OH^- ions in equilibrium with $Ca(OH)_2$, and more $Ca(OH)_2$ dissolves.

Magnesium hydroxide is separated from the water by filtration. The addition of hydrochloric acid converts the magnesium hydroxide to magnesium chloride.

$$\mathbf{Mg(OH)_2 + 2HCl \rightarrow MgCl_2 + 2H_2O}$$

Calcium chloride and sodium chloride are added to increase the conductivity and lower the melting point of the magnesium chloride. Magnesium metal is then recovered from the fused $MgCl_2$ by electrolysis.

$$\mathbf{Mg^{++} + 2Cl^- \rightarrow Mg(l) + Cl_2(g)}$$

Chlorine gas collected at the positive electrode is used to produce the hydrochloric acid required in the process. The Ca^{++} and Na^+ ions have higher reduction (electrode) potentials than Mg^{++} ions. Thus they are not reduced under the electrolysis conditions used. A flow diagram for this magnesium extraction process is shown in Figure 25-3.

25.4 Properties and uses of magnesium Magnesium is a silver-white metal with a density of 1.74 g/cm^3. When heated, it becomes ductile and malleable. Its tensile strength (resistance to being pulled apart) is not quite as great as that of aluminum.

Dry air does not affect magnesium. In moist air, however, a coating of basic magnesium carbonate forms on the surface of the metal. Because this coating is not porous, it protects the metal

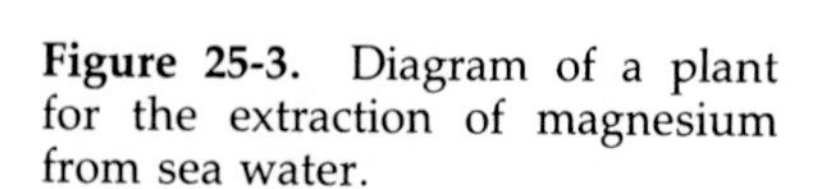
Figure 25-3. Diagram of a plant for the extraction of magnesium from sea water.

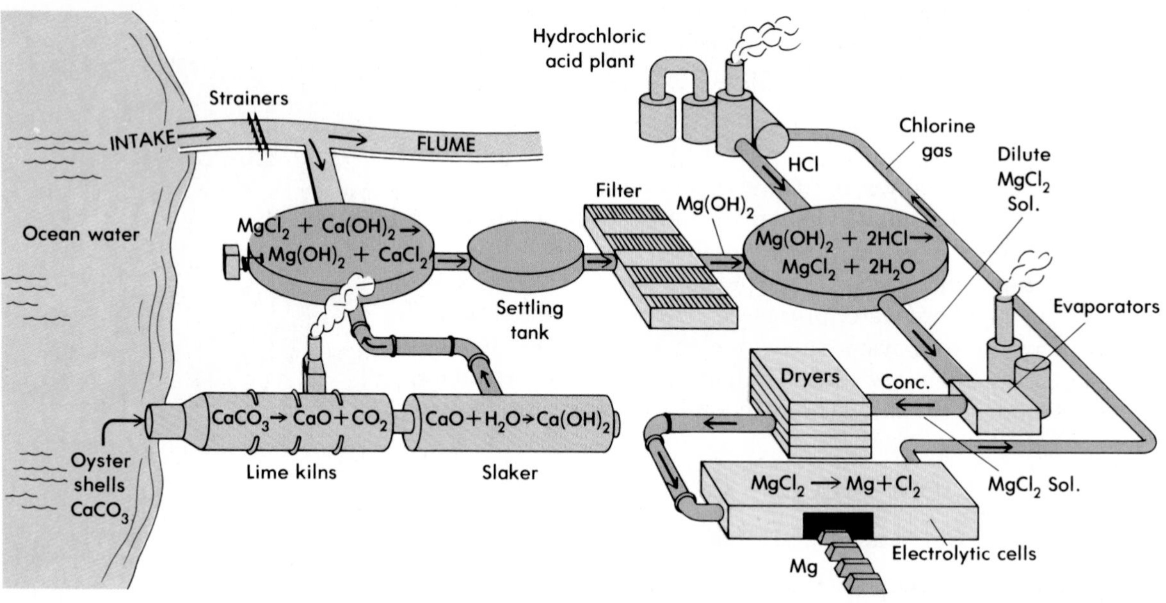

underneath from tarnishing. *A metal that forms a nonporous, non-scaling coat of tarnish is said to be a* ***self-protective metal.***

Aluminum and zinc are also self-protective metals; iron is not.

When heated in air to the kindling point, magnesium burns intensely and gives off a dazzling white light. See Figure 8-5. The combustion produces magnesium oxide, MgO, and magnesium nitride, Mg_3N_2. Once ignited, magnesium also burns in steam, in carbon dioxide, and in nitrogen. Magnesium is one of the few metals that combines directly with nitrogen. Boiling water reacts slowly with magnesium to form magnesium hydroxide and hydrogen. Common acids react with magnesium.

CAUTION: never look directly at burning magnesium for it can seriously damage the eyes.

The brilliant white light of burning magnesium makes it useful for flares and fireworks. Magnesium forms light, strong alloys with aluminum. Examples are *magnalium* and *Dow metal.* Other alloying metals are lithium, thorium, zinc, and manganese. Magnesium alloys are used for making tools and fixtures, beams of chemical balances, and for auto and airplane parts.

CALCIUM

25.5 Preparation of calcium Calcium ranks fifth in abundance by weight among the elements in the earth's crust, atmosphere, and surface waters. It is widely distributed in many rock and mineral forms. The best known of these mineral forms are the carbonate and the sulfate.

Since calcium occurs as combined Ca^{++} ions, the metal must be recovered by reduction. Electrolytic reduction involves fused calcium chloride.

$$Ca^{++} + 2Cl^- \rightarrow Ca + Cl_2(g)$$

Table 25-3
COMMON MAGNESIUM COMPOUNDS

Chemical name	*Common name*	*Formula*	*Appearance*	*Uses*
magnesium carbonate	(none)	$MgCO_3$	white, usually fluffy	for lining furnaces; in making the oxide
basic magnesium carbonate	magnesia alba	$Mg_4(OH)_2(CO_3)_3 \cdot 3H_2O$	soft, white powder	in tooth cleansers; for pipe coverings
magnesium chloride	(none)	$MgCl_2$	white, crystalline solid	with asbestos for stone flooring
magnesium hydroxide	milk of magnesia	$Mg(OH)_2$	white, milky suspension	as antacid; in laxatives
magnesium oxide	magnesia	MgO	white powder	as refractory; for lining furnaces
magnesium sulfate, heptahydrate	Epsom salts	$MgSO_4 \cdot 7H_2O$	white, crystalline solid	in laxatives, cathartics; in dye industry

Figure 25-4. Calcium salts impart a yellowish-red color to the Bunsen flame. Sodium contamination often changes the color to orange-yellow. The flame color is greenish when viewed through cobalt-blue glass. The calcium flame color is masked by the yellowish-green flame of barium salts. A calcium flame test must be used with caution.

Today chemical reduction is generally employed for recovering calcium. Calcium oxide is heated with aluminum in a vacuum retort.

$$3CaO + 2Al \rightarrow Al_2O_3 + 3Ca(g)$$

The reaction proceeds because the calcium metal is distilled off as a gas at the reaction temperature.

25.6 Properties of calcium Metallic calcium is silver-white in color, but a freshly cut piece tarnishes to bluish-gray within a few hours. Calcium is somewhat harder than lead. The density of calcium is 1.55 g/cm^3. If a piece of calcium is added to water, it reacts with the water to slowly give off hydrogen gas.

$$Ca + 2H_2O \rightarrow Ca(OH)_2 + H_2(g)$$

The calcium hydroxide produced is only slightly soluble and coats the surface of the calcium. This coating protects the metal from rapid interaction with the water.

Calcium is a good reducing agent. It burns in oxygen with a bright, yellowish-red flame. It also unites directly with chlorine. Calcium salts impart a yellowish-red color to a Bunsen flame, as illustrated in Figure 25-4. Certain limitations in the flame test for calcium are cited in the caption accompanying this figure.

Calcium is used in small amounts to reduce uranium tetrafluoride to uranium and thorium dioxide to thorium.

$$UF_4 + 2Ca \rightarrow 2CaF_2 + U$$
$$ThO_2 + 2Ca \rightarrow 2CaO + Th$$

Calcium is an effective deoxidizer for iron, steel, and copper. Some alloy steels contain small quantities of calcium. Lead-calcium alloys are used for bearings in machines. Calcium is also used to harden lead for cables and storage-battery plates.

25.7 Calcium carbonate Calcium carbonate, $CaCO_3$, is an abundant mineral found in many different forms.

1. *Limestone.* This is the most common form of calcium carbonate. It was formed in past geologic ages by great pressures on layers of seashells. Limestone is found in layers as a *sedimentary rock.* Pure calcium carbonate is white or colorless when crystalline. Most deposits, however, are gray because of impurities.

Limestone is used for making glass, iron, and steel and as a source of carbon dioxide. It is a building stone, and large quantities are used for building roads. Powdered or pulverized limestone is used to neutralize acid soils. Limestone is heated to produce calcium oxide, CaO, commonly called *lime.*

A mixture of limestone and clay is converted to *cement* by heating it strongly in a rotary kiln. Some limestone deposits, called *natural cement,* contain clay and calcium carbonate naturally mixed in about the right proportions for making cement.

Figure 25-5. Limestone is the most common form of calcium carbonate. It is found in layers as a sedimentary rock, as shown at White Elephant Terrace in the state of Wyoming.

2. *Calcite.* The clear, crystalline form of calcium carbonate is known as calcite. Transparent, colorless specimens are called *Iceland spar.*

3. *Marble.* This rock was originally limestone. Heat and pressure changed it into marble with a resulting increase in the size of the calcium carbonate crystals. Hence it is classed as a *metamorphic* (changed) *rock.*

4. *Shells.* The shells of such animals as snails, clams, and oysters consist largely of calcium carbonate. In some places, large masses of such shells have become cemented together to form rock called *coquina* (koh-*kee*-nuh). It is used as a building stone. Tiny marine animals called *polyps* deposit limestone as they build *coral* reefs. *Chalk,* as in the chalk cliffs of England, consists of the microscopic shells of small marine animals. Blackboard "chalk" is made of claylike material mixed with calcium carbonate; it should not be confused with natural chalk.

5. *Precipitated chalk.* This form of calcium carbonate is made by the reaction of sodium carbonate with calcium chloride.

$$Na_2CO_3 + CaCl_2 \rightarrow CaCO_3(s) + 2NaCl$$

It is soft and finely divided and thus forms a nongritty scouring powder suitable for toothpastes and tooth powders. Under the name of *whiting,* it is used in paints to fill the pores in the wood. When it is ground with linseed oil, it forms putty.

Figure 25-6. The shells of animals such as snails, clams, and oysters consist largely of calcium carbonate. Also, tiny marine animals called polyps deposit limestone as they build coral reefs. (Top photo) Gorgonia coral, (middle) fungia coral on soft coral, and (bottom) tree coral. All are found off the coast of Israel in the Red Sea.

25.8 Hardness in water Rainwater falling on the earth usually contains carbon dioxide in solution. As it soaks through the ground, it reaches deposits of limestone or dolomite. Some of the calcium carbonate and magnesium carbonate in these rocks reacts with the CO_2 solution. This reaction produces the soluble hydrogen carbonates of these metals.

$$CaCO_3(s) + CO_2(aq) + H_2O(l) \rightarrow Ca^{++}(aq) + 2HCO_3^-(aq)$$

$$MgCO_3(s) + CO_2(aq) + H_2O(l) \rightarrow Mg^{++}(aq) + 2HCO_3^-(aq)$$

The ground water now contains Ca^{++}, Mg^{++}, and HCO_3^- ions in solution. It is said to be *"hard"* water. This term indicates that it is "hard" to get a lather when soap is added to the water. Water that lathers readily with soap is called *"soft"* water. The terms are not precise, but they are in common use. Deposits of iron and other heavy metals in the ground may also produce hard water.

Water hardness is of two types. *Temporary hardness* results from the presence of HCO_3^- ions along with the metal ions. *Permanent hardness* results when other negative ions (usually SO_4^{--}) more stable than the HCO_3^- ion are present along with the metal ions.

The main ingredient or ordinary soap is water-soluble sodium stearate, $NaC_{18}H_{35}O_2$. When some soap is added to water containing Ca^{++} ions, the large $C_{18}H_{35}O_2^-$ ions react with the Ca^{++} ions. The reaction produces the insoluble stearate $Ca(C_{18}H_{35}O_2)_2$, which deposits as a gray scum.

$$Ca^{++} + 2C_{18}H_{35}O_2^- \rightarrow Ca(C_{18}H_{35}O_2)_2(s)$$

The metal ions in hard water continue to react with soap to form precipitates until all the ions are removed. No lasting lather is produced up to this point. Soft water does not contain these ions and lathers easily when soap is added. See Figure 25-7.

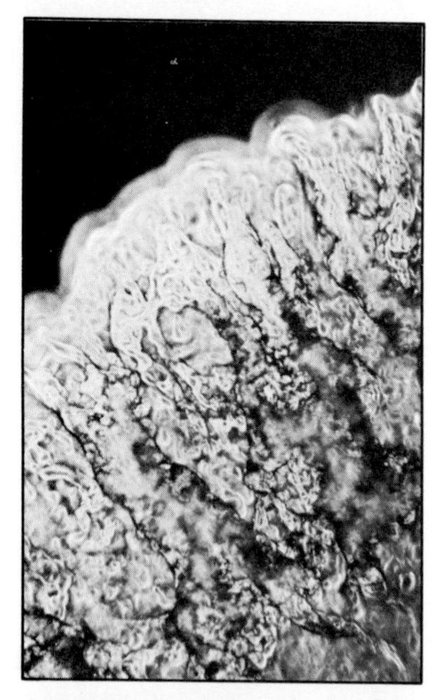

Figure 25-7. Photomicrographs of soap in the process of dissolving in hard water (top) and in soft water (bottom). In hard water the solution process is inhibited by a barrier of sticky soap curd; but in soft water, soap streamers extend into the water, forming a clear solution.

25.9 Softening hard water Hard water is a nuisance in laundering. The sticky precipitate wastes soap and collects on the fibers of the garments being laundered. In bathing, the hard water does not lather freely, and the precipitate forms a scum on the bathtub. In steam boilers, temporary hard water containing HCO_3^- ions presents a still more serious problem. When this water is boiled, calcium carbonate collects as a hard scale inside the boiler and the steam pipes. It may form a thick crust, acting as a heat insulator and preventing efficient transfer of heat. Thus, hard water should be softened before it is used.

There are several practical methods for softening water. Generally, the nature of the hardness and the quantity of soft water required determine the best method.

1. *Boiling* (for temporary hardness). Hard water containing HCO_3^- ions can be softened by boiling. The metal ions precipitate as carbonates according to the following equation:

$$Ca^{++} + 2HCO_3^- \rightarrow CaCO_3(s) + H_2O + CO_2(g)$$

This reaction is reversible except for the fact that boiling drives off the CO_2.

2. *Precipitation.* When sodium carbonate, Na_2CO_3, is added to hard water, Ca^{++} and Mg^{++} ions precipitate as insoluble carbonates. The Na^+ ions added to the water cause no difficulties with soap.

$$Ca^{++} + CO_3^{--} \rightarrow CaCO_3(s)$$

$$Mg^{++} + CO_3^{--} \rightarrow MgCO_3(s)$$

The addition of a basic solution such as $NH_3(aq)$ or limewater to temporary hard water supplies OH^- ions that neutralize the HCO_3^- ions and precipitate the "hard" metal ions as carbonates.

$$Ca^{++} + HCO_3^- + OH^- \rightarrow CaCO_3(s) + H_2O$$

$$Mg^{++} + HCO_3^- + OH^- \rightarrow MgCO_3(s) + H_2O$$

Other precipitating agents such as borax and sodium phosphate are sometimes used. Both produce basic solutions by hydrolysis; the phosphate salts of calcium and magnesium are insoluble.

3. *Ion exchange.* Certain natural minerals called *zeolites* have porous, three-dimensional networks of silicate-aluminate

groups. These networks are negatively charged. Mobile Na^+ ions are distributed throughout the porous network. If hard water is allowed to stand in contact with sodium zeolite, Na^+ ions are exchanged for the aqueous Ca^{++} and Mg^{++} ions. The process is an *ion exchange.* The zeolite is the *ion exchanger.*

$$Ca^{++}(aq) + 2Na\ zeolite \rightleftarrows Ca(zeolite)_2 + 2Na^+(aq)$$

The reaction can be reversed and the Na zeolite regenerated by adding a concentrated sodium chloride solution to the Ca(zeolite)$_2$. This solution has a high concentration of Na^+ ions. The Ca^{++} ions are replaced by Na^+ ions according to the mass-action principle.

The law of mass action is explained in Section 20.13.

A synthetic zeolite known as *permutit* reacts more rapidly than natural zeolites. Permutit is used today in many household water softeners. Sodium chloride, the cheapest source of Na^+ ions, is used to renew the exchanger in these softeners.

Chemists have developed *ion-exchange resins* far superior to the zeolites. See Figure 25-8. One type of resin is called an *acid-exchange resin* or a *cation exchanger.* This type has large negatively charged organic units whose neutralizing ions in water are H_3O^+ ions. A second type of resin is the *base-exchange resin* or *anion exchanger.* These resins have large positively charged organic units whose neutralizing ions in water are OH^- ions.

Used together, the two types of ion-exchange resins can remove all positive and negative ions from water. The cation exchanger removes metallic ions (cations) and replaces them with H_3O^+ ions. The water is then passed through the anion exchanger. Here, negative ions (anions) are removed and

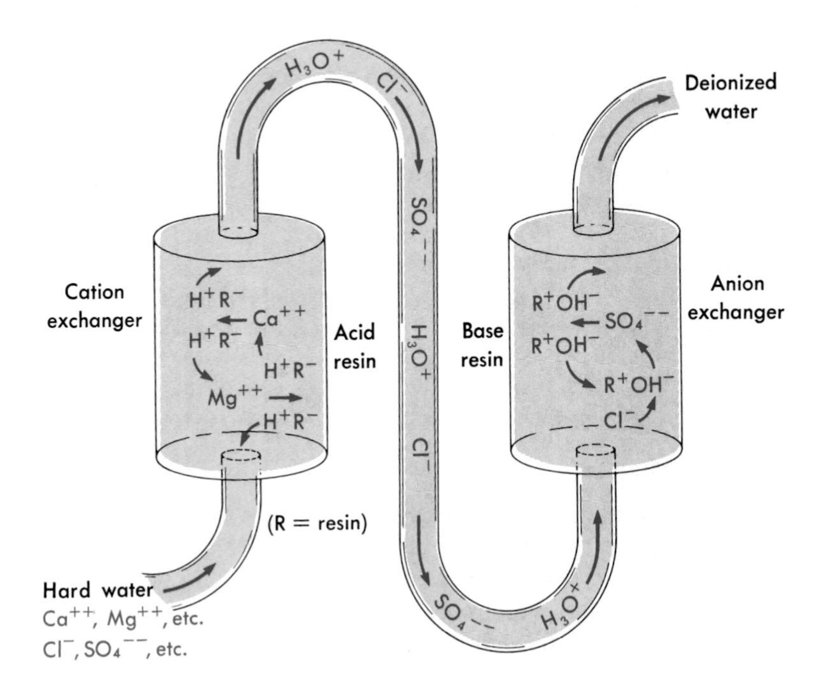

Figure 25-8. An ion-exchange system. All positive ions in the water are exchanged for hydronium ions in the acid resin. All negative ions are exchanged for hydroxide ions in the base resin. Water leaving the system is completely deionized.

replaced by OH^- ions. The H_3O^+ and OH^- ions form water by neutralization.

Natural water or a water solution of salts treated by combinations of ion-exchange resins is made *ion-free.* (Only the small equilibrium quantities of H_3O^+ and OH^- ions remain.) This treated water is called *deionized* or *demineralized* water. Deionized water is now used in many processes that once required distilled water. Deionized water is as free of ions as is carefully distilled water, but it may contain some dissolved carbon dioxide.

An acid-exchange resin can be renewed by running a strong acid through it. Similarly, a base-exchange resin can be renewed by using a basic solution.

Lime heated to incandescence gives off a brilliant white light—"limelight"—that once served as theater spotlight.

25.10 Calcium oxide Calcium oxide is a white, ionic solid commonly called *lime* or *quicklime.* It is *refractory* since it does not melt or vaporize below the temperature of the electric arc. It unites chemically with water to form calcium hydroxide.

$$CaO(s) + H_2O \rightarrow Ca(OH)_2(s) \qquad \Delta H = -16 \text{ kcal}$$

***$CaCO_3$:** limestone*
***CaO:** lime or quicklime*
***$Ca(OH)_2$:** slaked lime or hydrated lime*

During this process, called *slaking,* the mass swells, and a large amount of heat is released. The reaction is strongly exothermic. Because $Ca(OH)_2$ is prepared by this *slaking* process, its common name is *slaked lime.*

A lump of quicklime exposed to air gradually absorbs water. It swells, then cracks, and finally crumbles to a powder. It first forms calcium hydroxide and then slowly unites with carbon dioxide from the air to form calcium carbonate. Thus a mixture of calcium hydroxide and calcium carbonate is formed. Such a mixture is valuable for treating acidic soils. However, lime that has been air slaked cannot be used to make mortar and plaster.

*Historically, the thermal decomposition of calcium carbonate has been called **calcination.** The calcium carbonate is said to have been **calcined.***

Calcium oxide is produced by heating calcium carbonate to a high temperature in a kiln, as shown in Figure 25-9.

$$CaCO_3(s) \rightarrow CaO(s) + CO_2(g)$$

A high concentration of carbon dioxide would drive the reaction in the reverse direction according to Le Chatelier's principle. Such an equilibrium is avoided by removing the carbon dioxide from the kiln as it forms.

In 1983, the commercial production of lime in the United States was over 13 million metric tons. Lime ranked third, after sulfuric acid and nitrogen, in tonnage produced.

25.11 Calcium hydroxide Calcium hydroxide, or *slaked lime,* is a white solid that is sparingly soluble in water. Its water solution, called *limewater,* has basic properties. A suspension of calcium hydroxide in water is known as milk of lime. Mixed with flour paste or glue, it makes whitewash.

Calcium hydroxide is the cheapest hydroxide. It is used to remove hair from hides before they are tanned or made into leather. It is useful for reducing acidity in soils, for freeing ammonia from ammonium compounds, for softening temporary hard water, and for making mortar and plaster.

Lime mortar consists of slaked lime, sand, and water. Mortar of this type has been used for many centuries. The mortar first hardens as water evaporates. Crystals of $Ca(OH)_2$ form and cement the grains of sand together. Exposed to air, the mortar continues to get harder. The calcium hydroxide is slowly converted to the carbonate by action of atmospheric carbon dioxide.

$$Ca(OH)_2 + CO_2 \rightarrow CaCO_3 + H_2O$$

25.12 Calcium sulfate Calcium sulfate occurs as the mineral gypsum, a dihydrate $CaSO_4 \cdot 2H_2O$. Extensive deposits of gypsum occur in various regions of the United States. Large quantities are used in the construction industry and in agriculture. Gypsum is useful in treating alkaline soils.

When gypsum is heated gently, it partially dehydrates to form a white powder known as *plaster of paris*. The equation is

$$\underset{\text{gypsum}}{2CaSO_4 \cdot 2H_2O(s)} \rightarrow \underset{\text{plaster of paris}}{(CaSO_4)_2 \cdot H_2O(s)} + 3H_2O(g)$$

The formula for plaster of paris, $(CaSO_4)_2 \cdot H_2O$, is equivalent to $CaSO_4 \cdot \frac{1}{2}H_2O$, as it is sometimes written.

When plaster of paris is mixed with water, it hydrates back to the gypsum structure. The reaction is the reverse of the equation given above.

$$(CaSO_4)_2 \cdot H_2O + 3H_2O \rightarrow 2CaSO_4 \cdot 2H_2O$$

A thin paste (or slurry) of plaster of paris and water sets quickly and expands slightly. It forms a solid mass of interlacing crystals of gypsum. Because of the slight expansion, it gives remarkably faithful reproductions when cast in molds. The plaster of paris slurries are suited for making wallboard and rocklath, surgical casts, pottery molds, statuary, and many other uses.

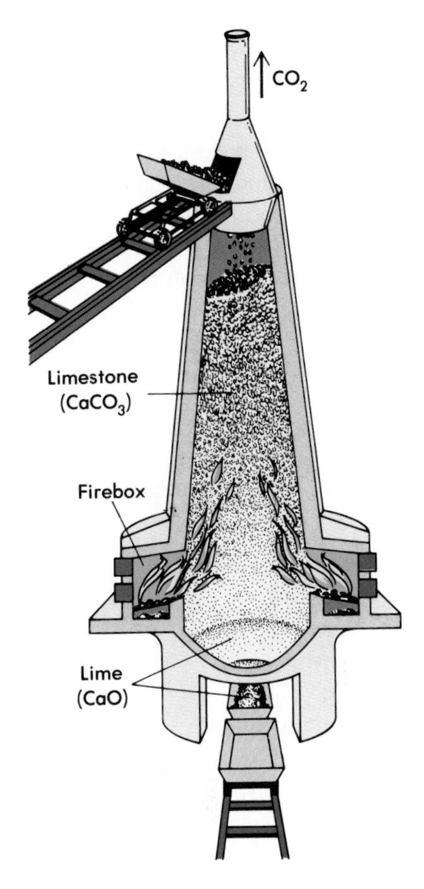

Figure 25-9. Calcination: an ancient process used by alchemists and still used today to convert limestone, $CaCO_3$, into lime, CaO.

SUMMARY

The elements of Group II are the metals beryllium, magnesium, calcium, strontium, barium, and radium. They are known as the alkaline-earth metals and as the Calcium Family of elements. They are not found as free elements in nature. The metals of Group II are recovered from their natural compounds.

The alkaline-earth metals are characterized by two valence electrons beyond the preceding noble-gas structure. Their ions have the electron configurations of the preceding noble gases. The metals are denser and harder than the corresponding members of the alkali metals of Group I and have higher melting points. Bonding shows some covalency. The chemistry of the elements of Group II, like that of the Group I elements, is uncomplicated.

Magnesium is recovered mainly by electrolysis of its chloride. Sea water is the most economical source of magnesium. Magnesium is a light, self-protective metal. It is used structurally when alloyed with other metals and, because it burns with a brilliant white light, for flares and fireworks. Soluble magnesium compounds contribute to the "hardness" of water.

Calcium is a low-density and relatively hard

metal. It is a good reducing agent. It reacts with water slowly and burns in oxygen with a yellow-red flame. Calcium is widely distributed in the earth's crust as insoluble compounds, mainly as carbonates and sulfates. It is recovered by electrolytic reduction from its chloride and by chemical reduction with aluminum.

Groundwater usually contains Ca^{++} ions contributed by soluble and slightly soluble calcium compounds. Water containing certain minerals in solution, mainly the hydrogen carbonates and sulfates of calcium and magnesium, is called hard water. The presence of these substances interferes with the use of water in laundering and in steam boilers and steam pipes. Water is softened by the removal of the hardening agents. Hard water is softened by boiling, precipitation, and ion-exchange methods.

The oxide, hydroxide, carbonate, and sulfate of calcium are important industrial substances. Calcium oxide is produced by heating calcium carbonate. Calcium hydroxide is formed by reacting calcium oxide with water. Calcium sulfate is mined mainly as the mineral gypsum.

VOCABULARY

alkaline-earth metals
calcination
calcite
cement
chalk
deionized water
dolomite
gypsum
hard water
Iceland spar
ion exchange
ion-exchange resin
lime
limestone
marble
permanent hardness
permutit
plaster of paris
precipitated chalk
quicklime
refractory
self-protective metal
slaked lime
slaking
soft water
temporary hardness
zeolite

QUESTIONS

Group A

1. *(a)* Write the equation for the production of quicklime from limestone. *(b)* How is an equilibrium avoided in this process? *A4a*
2. *(a)* What is dolomite? *(b)* For what purposes is it used? *A1*
3. Since magnesium is an active metal, why, unlike iron, do objects made from it not corrode rapidly? *A1*
4. Compare the preparation of elemental calcium with that of elemental sodium. *A2b*
5. What are the important uses of metallic calcium? *A3g*
6. In what forms is calcium carbonate found in nature? *A3i*
7. *(a)* What does the term *hard water* mean? *(b)* What does the term *soft water* mean? *(c)* What is the difference between temporary and permanent hardness of water? *A1*
8. What principle is employed to regenerate a zeolite water softener? *A1*
9. Distinguish *(a)* limestone; *(b)* quicklime; *(c)* slaked lime; *(d)* lime; *(e)* hydrated lime. *A3j*
10. For what gas is limewater used as a test solution? *A4d*
11. How could you demonstrate that a piece of coral is a carbonate?
12. Write an empirical equation to show the action of water containing dissolved carbon dioxide on limestone. *A2g*
13. How is cement manufactured from clay and limestone? *A2j*

Group B

14. Write three balanced equations to show the steps in the preparation of magnesium from sea water. *A2f, A4a*
15. How do beryllium compounds differ from those of the other Group II elements? *A3b*
16. Why are the members of the Calcium Family denser and harder than the corresponding members of the Sodium Family? *A2b*
17. *(a)* Write the equation for the slaking of lime. *(b)* Give a reason for the reaction's going to completion. *A4a*
18. Give two reasons why the reaction of calcium with water is not as vigorous as that of potassium with water. *A4b*
19. *(a)* What metallic ions cause water to be hard? *(b)* What negative ion causes temporary hardness? *(c)* What negative ion causes permanent hardness? *A2g*
20. What type of chemical reaction occurs between soap and hard water? *A2h*
21. Write an empirical equation to show the softening action of sodium carbonate on hard water containing *(a)* calcium sulfate; *(b)* magnesium hydrogen carbonate. *A2h*
22. *(a)* How is plaster of paris made? *(b)* Why does it harden? *A3j*
23. *(a)* What impurity may still remain in water that has been passed through both an acid-exchange resin and a base-exchange resin? *(b)* For what kinds of solutions would such water be unsuited? *A2h*
24. How can you account for the fact that temporary hard water, containing Ca^{++} ions, is softened by the addition of limewater, a solution containing Ca^{++} ions and OH^- ions? *A2h*

PROBLEMS

Group A

1. How many kilograms of calcium oxide can be produced from 1.000 metric ton of limestone that contains 12.5% of impurities?
2. How much mass will 86.0 kg of gypsum lose when it is converted into plaster of paris?
3. Calculate the percentage of beryllium in beryl, $Be_3Al_2Si_6O_{18}$.
4. What quantity of magnesium can be prepared from a metric ton of magnesium oxide, MgO?
5. A cubic kilometer of sea water contains in solution minerals that form about 4.0×10^6 kg of magnesium chloride. How much metallic magnesium could be obtained from this volume of sea water?
6. How many liters of hydrogen at STP can be prepared by reacting 8.10 g of magnesium with excess dilute hydrochloric acid?

Group B

7. If dolomite is 95.0% a double carbonate of calcium and magnesium, together with 5.0% of impurities such as iron and silica, what is the percentage of magnesium in the sample? (Compute to 3 significant figures.)
8. How many kilograms of carbon dioxide can be obtained from a metric ton of oyster shells that are 81.0% calcium carbonate?
9. How many liters will the carbon dioxide produced in Problem 8 occupy at STP?
10. If the carbon dioxide of Problem 9 is measured at $72\bar{0}$ mm pressure and 20 °C, what volume does it occupy?
11. The thermal decomposition of a charge of limestone produced $50\bar{0}$ m^3 of carbon dioxide when stored at 28 °C and $95\bar{0}$ mm pressure. How many moles of the gas were stored?

Alfred Werner (Swiss) received the Nobel Prize in chemistry in 1913 for his theory of the arrangement of atoms in molecules. He proposed that in organic compounds nitrogen had a tetrahedral structure similar to carbon's. He made geometric models to explain this isomerism. He predicted that metal complexes could exist as stereoisomers and that complexes of the same formula can exist in more than one form, exhibiting different colors and electrical and optical properties, depending on the arrangement of the atoms. He also suggested that transition metals could have multiple oxidation states.

The Transition Metals

GOALS

In this chapter you will gain an understanding of: • the general unique properties of the transition elements • the structural characteristics responsible for the unique nature of the transition elements • the members of the Iron Family • iron and steel production • the members of the Copper Family • the natural occurrence, general refining methods, and compounds of iron and copper

TRANSITION ELEMENTS

26.1 General properties The periodic table of the elements, Section 5.5, has short periods through the third row of elements. The fourth and succeeding rows are long periods. Beginning with Period 4, ten subgroups of elements intervene between Group II and Group III. These elements occupy the *transition region* of the periodic table. They are called the *transition elements* because, with increasing atomic number, there is an increase in the number of electrons in the next-to-highest energy level. This inner building of electronic structure can be interpreted as an interruption, between Group II and Group III, in the regular increase of highest-energy-level electrons from Group I to Group VIII across a period. See Figure 26-1 on the next page.

***Periodic table:** periods are horizontal rows; groups are vertical columns.*

The properties of the elements remain similar going down each main group. However, going horizontally across a short period from Group I to Group VII, the properties change steadily from metallic to nonmetallic. In the transition region, horizontal similarities may be as great as (in some cases even greater than) similarities going down a subgroup.

All transition elements are metals. The metallic character is related to the presence of no more than one or two electrons in the highest energy levels of their atoms. These metals are generally harder and more brittle than the Group I and Group II metals. Their melting points are higher. Mercury (at. no. 80) is a familiar exception. It is the only metal that is liquid at room temperature. The transition metals are good conductors of heat and electricity. Such conduction is characteristic of a metallic crystal lattice permeated by an electron gas.

In very hot weather, mercury may lose this distinction, as gallium (atomic number 31) and cesium (atomic number 55) have melting points of 29.8 °C and 28.4 °C, respectively.

The chemical properties of transition elements are varied. Electrons of the next-to-highest *d* sublevel as well as those of the

TRANSITION ELEMENTS

Period	II											III
3	24.305 **Mg** 12 (2, 8, 2)											26.98154 **Al** 13 (2, 8, 3)
4	40.08 **Ca** 20 (2, 8, 8, 2)	44.9559 **Sc** 21 (2, 8, 9, 2)	47.88 **Ti** 22 (2, 8, 10, 2)	50.9415 **V** 23 (2, 8, 11, 2)	51.996 **Cr** 24 (2, 8, 13, 1)	54.9380 **Mn** 25 (2, 8, 13, 2)	55.847 **Fe** 26 (2, 8, 14, 2)	58.9332 **Co** 27 (2, 8, 15, 2)	58.69 **Ni** 28 (2, 8, 16, 2)	63.546 **Cu** 29 (2, 8, 18, 1)	65.38 **Zn** 30 (2, 8, 18, 2)	69.72 **Ga** 31 (2, 8, 18, 3)
5	87.62 **Sr** 38 (2, 8, 18, 8, 2)	88.9059 **Y** 39 (2, 8, 18, 9, 2)	91.22 **Zr** 40 (2, 8, 18, 10, 2)	92.9064 **Nb** 41 (2, 8, 18, 12, 1)	95.94 **Mo** 42 (2, 8, 18, 13, 1)	[98] **Tc** 43 (2, 8, 18, 13, 2)	101.07 **Ru** 44 (2, 8, 18, 15, 1)	102.9055 **Rh** 45 (2, 8, 18, 16, 1)	106.42 **Pd** 46 (2, 8, 18, 18, 0)	107.8682 **Ag** 47 (2, 8, 18, 18, 1)	112.41 **Cd** 48 (2, 8, 18, 18, 2)	114.82 **In** 49 (2, 8, 18, 18, 3)
6	137.33 **Ba** 56 (2, 8, 18, 18, 8, 2)	Lanthanide Series; 174.967 **Lu** 71 (2, 8, 18, 32, 9, 2)	178.49 **Hf** 72 (2, 8, 18, 32, 10, 2)	180.9479 **Ta** 73 (2, 8, 18, 32, 11, 2)	183.85 **W** 74 (2, 8, 18, 32, 12, 2)	186.207 **Re** 75 (2, 8, 18, 32, 13, 2)	190.2 **Os** 76 (2, 8, 18, 32, 14, 2)	192.22 **Ir** 77 (2, 8, 18, 32, 15, 2)	195.08 **Pt** 78 (2, 8, 18, 32, 17, 1)	196.9665 **Au** 79 (2, 8, 18, 32, 18, 1)	200.59 **Hg** 80 (2, 8, 18, 32, 18, 2)	204.383 **Tl** 81 (2, 8, 18, 32, 18, 3)
7	226.0254 **Ra** 88 (2, 8, 18, 32, 18, 8, 2)	Actinide Series; [260] **Lr** 103 (2, 8, 18, 32, 32, 9, 2)	[261] **Unq** 104 (2, 8, 18, 32, 32, 10, 2)	[262] **Unp** 105 (2, 8, 18, 32, 32, 11, 2)	[263] **Unh** 106 (2, 8, 18, 32, 32, 12, 2)	[261?] **Uns** 107 (2, 8, 18, 32, 32, 13, 2)	[265] **Uno** 108 (2, 8, 18, 32, 32, 14, 2)	[266?] **Une** 109 (2, 8, 18, 32, 32, 15, 2)				

Figure 26-1. Transition metals are characterized by the buildup of *d*-sublevel electrons in the next-to-highest energy level.

highest energy level may become involved in the formation of some compounds. Most transition metals exhibit variable oxidation states in forming compounds, and most of their compounds have color. They have a pronounced tendency to form complex ions. Many of their compounds are attracted into a magnetic field (a property called paramagnetism).

26.2 Transition subgroups The electron configuration of each transition element is shown in Figure 26-1. Compare the electron configurations of the Group II metals with the corresponding metals of the first transition subgroup, the scandium (Sc) subgroup. What differences do you observe? Now compare the electron configurations across the first-row transition metals. Note the generally regular pattern of change in electron configuration with increasing atomic number. Notice also whether irregularities appear in the pattern.

The stepwise buildup of the electron configuration across the first-row transition metals results in the $3d$ sublevel being filled with 10 electrons. Following this $3d$ filling, the regular increase in $4p$ electrons occurs in Group III through Group VIII.

There are minor irregularities in the $3d$ electron buildup at Cr (at. no. 24) and at Cu (at. no. 29). The electron configuration for chromium is -----$3d^5 4s^1$ instead of -----$3d^4 4s^2$ as might have been predicted. That for copper is -----$3d^{10} 4s^1$ instead of the -----$3d^9 4s^2$ that might have been expected. These apparent irregularities are attributed to the extra stability associated with half-filled and completely filled sublevels.

In the fifth period following strontium (-----$4s^2 4p^6 5s^2$) there is a similar filling of the $4d$ sublevel before the buildup of the $5p$ sublevel occurs to complete the period. In the sixth period following barium (-----$4s^2 4p^6 4d^{10} 5s^2 5p^6 6s^2$) the filling of the $5d$ sub-

Figure 26-2. Potato-size nodules found on the Pacific Ocean floor contain nickel, copper, cobalt, and manganese. These nodules are a potentially new source of strategic transition metals.

level is interrupted by the buildup of the 4*f* sublevel. These two expansions complicate this period. There are seven 4*f* orbitals that can accommodate 14 electrons. In the *lanthanide series*, the number of 4*f* electrons increases with atomic number.

The lanthanide and actinide series of elements are described in Section 5.5.

In the seventh period, the 6*d* sublevel filling is similarly interrupted by the buildup of the 5*f* sublevel characteristic of the *actinide series* of elements. The placement of lawrencium (at. no. 103) in the scandium subgroup marks the resumption of the 6*d* buildup. Unnilquadium, unnilpentium, unnilhexium, unnilseptium, unniloctium, and unnilennium in the titanium, vanadium, chromium, manganese, iron, and cobalt subgroups, respectively, indicate a continuation of this 6*d* buildup of electrons.

The transition elements occupy the long periods in the periodic table. The choice of the first and last elements in the series depends somewhat on the definition used. Based on *d* and *s* electron configurations, the transition elements consist of the ten subgroups between Group II and Group III. The first-row transition elements are the most important, and they are the most abundant. Their electron configurations are in Table 26-1.

26.3 Oxidation states Variable oxidation states are common among the transition metals. Some form different compounds in which they exhibit several oxidation states. The energies of the outermost *d* and *s* electrons do not differ greatly. The energies to remove these electrons are relatively low.

The *d* sublevel has five available orbitals that can hold ten electrons when all are filled. The *s* sublevel has one orbital, and it can hold two electrons. Electrons occupy the *d* orbitals singly as long as unoccupied orbitals of similar energy exist. The *s* electrons and one or more *d* electrons can be used in bonding.

In the first-row transition metals, several 4*s* and 3*d* electrons may be transferred to or shared with other substances. Thus several oxidation states become possible. Remember that the

Table 26-1
ELECTRON CONFIGURATIONS OF FIRST-ROW TRANSITION ELEMENTS

Name	*Symbol*	*Atomic number*	*Number of electrons in sublevels*						
			1*s*	2*s*	2*p*	3*s*	3*p*	3*d*	4*s*
scandium	Sc	21	2	2	6	2	6	1	2
titanium	Ti	22	2	2	6	2	6	2	2
vanadium	V	23	2	2	6	2	6	3	2
chromium	Cr	24	2	2	6	2	6	5	1
manganese	Mn	25	2	2	6	2	6	5	2
iron	Fe	26	2	2	6	2	6	6	2
cobalt	Co	27	2	2	6	2	6	7	2
nickel	Ni	28	2	2	6	2	6	8	2
copper	Cu	29	2	2	6	2	6	10	1
zinc	Zn	30	2	2	6	2	6	10	2

maximum oxidation state is limited by the total number of 4*s* and 3*d* electrons present.

Table 26-2 gives the metals' common oxidation states and 3*d* and 4*s* electron configurations. It uses orbital notations introduced in Section 4.6. In these notations ___ represents an unoccupied orbital. The symbol ↑ represents an orbital occupied by one electron, and ↓↑ represents an orbital with an electron pair.

Observe that the maximum oxidation state increases to +7 for manganese and then decreases abruptly beyond manganese. The difficulty in forming the higher oxidation states increases toward the end of the row. This increase in difficulty is due to the general increase in ionization energy with atomic number. The higher oxidation states generally involve covalent bonding.

Manganese atoms lose two 4*s* electrons to become Mn^{++} ions. Higher oxidation states involve one or more of the 3*d* electrons. In the permanganate ion, MnO_4^-, manganese is in the +7 oxidation state, covalently bonded with the oxygen atoms.

When the electron configuration includes both 3*d* and 4*s* valence electrons, the 3*d* electrons are lower in energy than the 4*s* electrons. Thus the first electron removed in an ionizing reaction is the one most loosely held, a 4*s* electron. This fact is illustrated by the step-by-step ionization of titanium. The ground-state configurations of the valence electrons are:

$$\begin{array}{ll} \mathbf{Ti} & \mathbf{3d^2 4s^2} \\ \mathbf{Ti^+} & \mathbf{3d^2 4s^1} \\ \mathbf{Ti^{++}} & \mathbf{3d^2 4s^0} \\ \mathbf{Ti^{+++}} & \mathbf{3d^1} \\ \mathbf{Ti^{++++}} & \mathbf{3d^0} \end{array}$$

Table 26-2
OXIDATION STATES OF FIRST-ROW TRANSITION ELEMENTS

Element	*Electron configuration*			*Common oxidation states*
	$1s^22s^22p^63s^23p^6$	3*d*	4*s*	
scandium	Each transition	↑ __ __ __ __	↓↑	+3
titanium	element has all	↑ ↑ __ __ __	↓↑	+2 +3 +4
vanadium	these sublevels	↑ ↑ ↑ __ __	↓↑	+2 +3 +4 +5
chromium	filled in an	↑ ↑ ↑ ↑ ↑	↑	+2 +3 +6
manganese	argon	↑ ↑ ↑ ↑ ↑	↓↑	+2 +3 +4 +6 +7
iron	structure	↓↑ ↑ ↑ ↑ ↑	↓↑	+2 +3
cobalt		↓↑ ↓↑ ↑ ↑ ↑	↓↑	+2 +3
nickel		↓↑ ↓↑ ↓↑ ↑ ↑	↓↑	+2 +3
copper		↓↑ ↓↑ ↓↑ ↓↑ ↓↑	↑	+1 +2
zinc		↓↑ ↓↑ ↓↑ ↓↑ ↓↑	↓↑	+2

26.4 Color A striking property of many compounds of transition metals is their color. Not all compounds formed with transition metals are colored. On the other hand, most colored inorganic compounds involve elements of the transition region of the periodic table. See Figure 26-3. The color of the compounds and their solutions may vary depending on what other ions or polyatomic groups are associated with the transition element. See Table 26-3. In general, these colored compounds are thought to have some electrons that are easily excited by selectively absorbing part of the energy of white light.

The continuous spectrum of Figure 24-9 shows that white light is composed of red, orange, yellow, green, blue, and violet colors of light. If light energy of a discrete band of wavelengths corresponding to a color region is absorbed by a substance, the remaining light reflected or transmitted is no longer white. It is the *complement* of the color removed. The color of the substance that is seen is the complement of the color of light absorbed. *The energies of the wavelengths of white light that are absorbed giving these compounds color are the energies required to raise **d** electrons to higher energy states.*

Complementary colors are two colors that yield white light when combined. Examples of complementary colors are blue and yellow, red and bluish-green, and green and magenta.

Copper(II) sulfate pentahydrate crystals are blue when viewed in white light. Their aqueous solutions are also blue. They are blue because the energy of wavelengths corresponding to *yellow light* is absorbed. The light reflected (or transmitted) is *blue light*, the complement of yellow light. When these crystals are heated, the water of hydration is given up, and the anhydrous salt is neither blue nor crystalline. It is a white powder. If water is added to the anhydrous powder, it turns blue. Since many sulfate compounds are colorless, it must be the *hydrated* copper(II) ion, $Cu(H_2O)_4^{++}$, that is blue.

In $CuSO_4 \cdot 5H_2O$ crystals, four water molecules are close to the copper(II) ion. The fifth is more distant.

26.5 The formation of complex ions The NO_3^- ion, SO_4^{--} ion, and the NH_4^+ ion are examples of *polyatomic ions*. They are

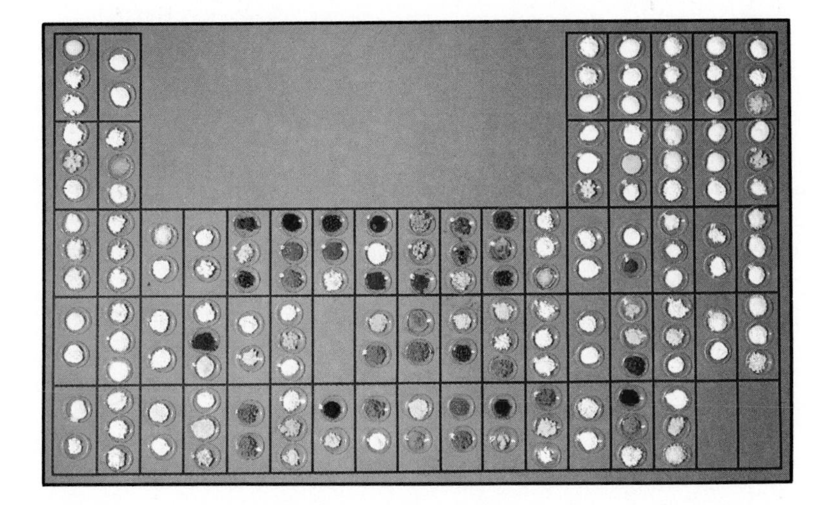

Figure 26-3. Most compounds of transition metals are colored. The color depends on the metal, its oxidation state, and the anion with which it is combined.

Table 26-3
COLORS OF COMPOUNDS OF TRANSITION METALS

Compound	*Formula*	*Color*
titanium(III) chloride	$TiCl_3$	violet
titanium(III) sulfate	$Ti_2(SO_4)_3$	green
titanium(IV) chloride	$TiCl_4$	yellow
vanadium(II) chloride	VCl_2	green
vanadium(II) sulfate, heptahydrate	$VSO_4 \cdot 7H_2O$	violet
vanadium(III) chloride	VCl_3	pink
vanadium(IV) chloride	VCl_4	red-brown
chromium(II) acetate	$Cr(C_2H_3O_2)_2$	red
chromium(II) sulfate, heptahydrate	$CrSO_4 \cdot 7H_2O$	blue
chromium(III) chloride	$CrCl_3$	violet
manganese(II) chloride	$MnCl_2$	pink
manganese(III) sulfate	$Mn_2(SO_4)_3$	green
manganese(IV) oxide	MnO_2	black
iron(II) chloride, dihydrate	$FeCl_2 \cdot 2H_2O$	green
iron(II) sulfate, heptahydrate	$FeSO_4 \cdot 7H_2O$	blue-green
iron(III) chloride, hexahydrate	$FeCl_3 \cdot 6H_2O$	brown-yellow
iron(III) sulfate	$Fe_2(SO_4)_3$	yellow
iron(III) thiocyanate	$Fe(SCN)_3$	red
cobalt(II) chloride, hexahydrate	$CoCl_2 \cdot 6H_2O$	red
cobalt(II) nitrate, hexahydrate	$Co(NO_3)_2 \cdot 6H_2O$	red
cobalt(II) sulfate, heptahydrate	$CoSO_4 \cdot 7H_2O$	pink
cobalt(III) sulfate	$Co_2(SO_4)_3 \cdot 18H_2O$	blue-green
nickel(II) hydroxide	$Ni(OH)_2$	green
nickel(II) nitrate, hexahydrate	$Ni(NO_3)_2 \cdot 6H_2O$	green
nickel(II) sulfate	$NiSO_4$	yellow
copper(I) carbonate	Cu_2CO_3	yellow
copper(I) oxide	Cu_2O	red
copper(II) oxide	CuO	black
copper(II) nitrate, hexahydrate	$Cu(NO_3)_2 \cdot 6H_2O$	blue
copper(II) sulfate, pentahydrate	$CuSO_4 \cdot 5H_2O$	blue

Polyatomic ions are described in Section 6.23.

charged particles made up of more than a single atom. These familiar ions are covalent structures. Their net ionic charge is the algebraic sum of the oxidation numbers of all the atoms present. Because of their small size and their stability, they behave just like single-atom ions, and go through many reactions unchanged.

The name *complex ion* is ordinarily restricted to an ionic species composed of a central metal ion combined with a specific number of polar molecules or ions. The charge on the complex ion is the sum of the charge on the central ion and that of all the attached units. Complex ions vary in stability, but none is as stable as the common polyatomic ions mentioned above.

The most common complex ions are formed by ions of the transition metals with chemical species such as chloride ions

(Cl^-), ammonia molecules (NH_3), water molecules (H_2O), and cyanide ions (CN^-). The transition cation is always the central ion on which the complex is formed. The number of units covalently bonded to, or *coordinated* with, the central ion is its *coordination number* for that complex ion. The Fe^{++} ion in the $Fe(CN)_6^{----}$ complex shown in Figure 26-4 has a coordination number of 6.

The hydrated Cu^{++} ion, $Cu(H_2O)_4^{++}$, discussed in Section 26.4, is a complex ion. The Cu^{++} ion is coordinated with 4 molecules of water of crystallization. It is the ionic species that crystallizes from an aqueous sulfate solution as the compound $CuSO_4 \cdot 5H_2O$.

When an excess of concentrated ammonia-water solution is added to an aqueous solution of Cu^{++} ions (a $CuSO_4$ solution, for example), the light-blue color of the solution changes to a deep-blue color. The soluble $Cu(NH_3)_4^{++}$ complex has been formed.

$$\underset{\text{light blue}}{Cu(H_2O)_4^{++}} + 4NH_3(aq) \rightarrow \underset{\text{deep blue}}{Cu(NH_3)_4^{++}} + 4H_2O$$

Zinc and cadmium form similar complex ions with ammonia. Table 26-4 gives some typical complex ions formed by transition metals.

If the ammonia-water solution is added to a solution of silver nitrate, the complex $Ag(NH_3)_2^+$ ion is formed.

$$Ag^+ + 2NH_3 \rightarrow Ag(NH_3)_2^+$$

$$Ag^+ + 2:\underset{\ddot{H}}{\overset{H}{\ddot{N}}}:H \rightarrow \left[H:\underset{H}{\overset{H}{\ddot{N}}}:Ag:\underset{H}{\overset{H}{\ddot{N}}}:H\right]^+$$

Brackets are used in this expression simply to enclose the electron-dot formula and indicate that the +1 charge applies to the ion as a whole. They do not *indicate a concentration expression.*

Because complex ions are unstable to some extent, the reaction is reversible. The more stable the complex ion, the more quickly

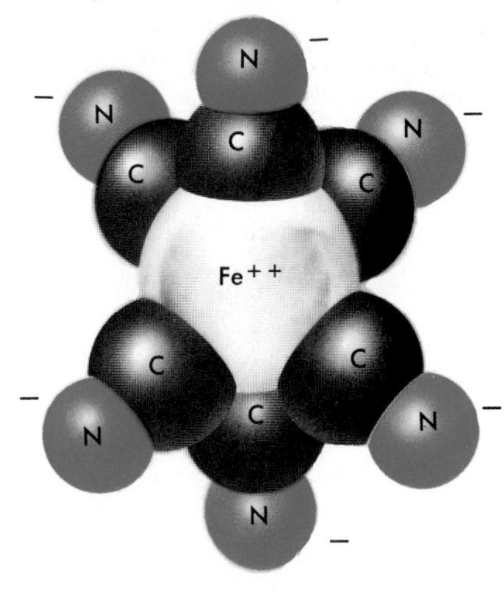

Figure 26-4. A possible space model of a complex ion, the hexacyanoferrate(II) ion, $Fe(CN)_6^{----}$. It is sometimes called the ferrocyanide ion.

Table 26-4
SOME COMMON COMPLEX IONS

Coordinating atom	*Coordinating group*	*Complex ions*	*Color*
N	H :N̈:H Ḧ	$Ag(NH_3)_2^+$ $Ni(NH_3)_4^{++}$ $Co(NH_3)_6^{+++}$	— blue blue
C	:C:::N:⁻	$Fe(CN)_6^{----}$ $Fe(CN)_6^{---}$	yellow red
S	:Ö: :S̈:S̈:Ö:⁻⁻ :Ö:	$Ag(S_2O_3)_2^{---}$	—
N	:S̈::C::N̈:⁻	$FeSCN^{++}$	red

equilibrium is established between the reactions involving the formation and dissociation of the complex ions. At equilibrium the equation is

$$Ag(NH_3)_2^+ \rightleftarrows Ag^+ + 2NH_3$$

Equilibrium constant: *see Section 21.3.*

The equilibrium constant has the usual form.

$$K = \frac{[Ag^+][NH_3]^2}{[Ag(NH_3)_2^+]} = 7 \times 10^{-8}$$

In cases of complex-ion equilibria, the equilibrium constant is called an *instability* constant. The *larger* the value of K, the more unstable is the complex ion.

When metallic ions in solution form complexes, they are, in effect, removed from the solution as simple ions. Thus the formation of complex ions increases the solubility of metallic ions. The solubility of a precipitate may also be increased if cations of the substance form complex ions. For example, the formation of chloride complexes may *increase* solubility where, by common ion effect, a *decrease* would be expected.

Common ion effect: *see Section 21.6.*

Consider the equilibrium of sparingly soluble AgCl.

$$AgCl(s) \rightleftarrows Ag^+(aq) + Cl^-(aq)$$

A small addition of Cl^- ions *does* shift the equilibrium to the left and lower the solubility as expected. However, if a large concentration of Cl^- ions is added, *the solid AgCl dissolves.* This apparent contradiction is explained by the formation of $AgCl_2^-$ complex ions.

The stability of complex ions is indicated by the small values of the equilibrium constants, K. See Table 26-5. The smaller the value of K, the more stable is the complex ion and the more effectively it holds the central metallic ion as part of the soluble complex.

Table 26-5
STABILITY OF COMPLEX IONS

Complex ion	*Equilibrium reaction*	*K*
$Ag(NH_3)_2^+$	$Ag(NH_3)_2^+ \rightleftharpoons Ag^+ + 2NH_3$	7×10^{-8}
$Co(NH_3)_6^{++}$	$Co(NH_3)_6^{++} \rightleftharpoons Co^{++} + 6NH_3$	1×10^{-5}
$Co(NH_3)_6^{+++}$	$Co(NH_3)_6^{+++} \rightleftharpoons Co^{+++} + 6NH_3$	2×10^{-34}
$Cu(NH_3)_4^{++}$	$Cu(NH_3)_4^{++} \rightleftharpoons Cu^{++} + 4NH_3$	3×10^{-13}
$Ag(CN)_2^-$	$Ag(CN)_2^- \rightleftharpoons Ag^+ + 2CN^-$	2×10^{-19}
$Au(CN)_2^-$	$Au(CN)_2^- \rightleftharpoons Au^+ + 2CN^-$	5×10^{-39}
$Cu(CN)_3^{--}$	$Cu(CN)_3^{--} \rightleftharpoons Cu^+ + 3CN^-$	1×10^{-35}
$Fe(CN)_6^{----}$	$Fe(CN)_6^{----} \rightleftharpoons Fe^{++} + 6CN^-$	1×10^{-35}
$Fe(CN)_6^{---}$	$Fe(CN)_6^{---} \rightleftharpoons Fe^{+++} + 6CN^-$	1×10^{-42}
$FeSCN^{++}$	$FeSCN^{++} \rightleftharpoons Fe^{+++} + SCN^-$	8×10^{-3}
$Ag(S_2O_3)_2^{---}$	$Ag(S_2O_3)_2^{---} \rightleftharpoons Ag^+ + 2S_2O_3^{--}$	6×10^{-14}

26.6 Paramagnetism *Substances that are weakly attracted into a magnetic field are said to be **paramagnetic.*** Paramagnetism among elements and their compounds indicates that there are unpaired electrons in the structure. Some transition metals have unpaired *d* electrons.

Manganese atoms, for example, have 5 unpaired 3*d* electrons as do Mn^{++} ions. Electron sharing in covalent complexes may reduce the number of unpaired electrons to the point where the substance is not paramagnetic. Most compounds of transition metals are paramagnetic in all but the highest oxidation states.

Figure 26-5. "Setting-out" the Cathedral of St. John the Divine. Zinc templates are used to trace the shape of each stone onto rough limestone blocks.

26.7 Transition metal similarities The number of electrons in the highest energy level of the transition elements remains fairly constant. It never exceeds two electrons as the number of electrons in the *d* sublevel increases across a period. Similarities often appear in a horizontal sequence of transition elements as well as vertically through a subgroup. In some cases, horizontal similarity may exceed vertical similarity. The first five subgroups are those headed by Sc, Ti, V, Cr, and Mn. Generally, similarities among these elements are recognized within each subgroup. The 6th, 7th, and 8th subgroups are headed by Fe, Co, and Ni. Here the similarities across each period are greater than those down each subgroup. Iron, cobalt, and nickel resemble each other more than do iron, ruthenium, and osmium, the members of the iron subgroup. The 9th and 10th subgroups are headed by Cu and Zn. Here again, the greatest similarity is within each subgroup.

The Iron Family

26.8 Members of the Iron Family The heavy metals, iron, cobalt, and nickel, are fourth-row transition elements. Each is the first member of the subgroup of transition metals that bears its name. See Figure 26-1. The similarities between iron, cobalt, and nickel are greater than those within each of the subgroups they head. Thus iron, cobalt, and nickel are commonly referred to as the Iron Family. The remaining six members of these three subgroups have properties similar to platinum and are considered to be members of the Platinum Family. They are called noble metals because they show little chemical activity. They are rare and expensive.

Family relationships in the 6th, 7th, and 8th subgroups of transition elements are horizontal rather than vertical, as in the other subgroups.

Iron is the most important member of the Iron Family. Alloys of iron, cobalt, and nickel are important structural metals. Some properties of the Iron Family are listed in Table 26-6.

The Iron Family is located in the middle of the transition elements. The next-to-highest energy level is incomplete. The 3*d* sublevels of iron, cobalt, and nickel contain only 6, 7, and 8 elec-

Table 26-6
THE IRON FAMILY

Element	*Atomic number*	*Atomic weight*	*Electron configuration*	*Oxidation numbers*	*Melting point* (°C)	*Boiling point* (°C)	*Density* (g/cm³)
iron	26	55.847	2,8,14,2	+2, +3	1535	2750	7.87
cobalt	27	58.9332	2,8,15,2	+2, +3	1495	2870	8.9
nickel	28	58.69	2,8,16,2	+2, +3	1453	2732	8.90

trons, respectively, instead of the full number, 10. Each member exhibits the +2 and +3 oxidation states. The Fe^{++} ion is easily oxidized to the Fe^{+++} ion by air and other oxidizing agents. The +3 oxidation state occurs only rarely in cobalt and nickel. The electron configurations of iron, cobalt, and nickel sublevels are shown in Table 26-7. In this table paired and unpaired electrons of the 3*d* and 4*s* sublevels are represented by paired and unpaired dots. Compare with Table 26-2.

The two 4*s* electrons are removed easily, as is usually the case with metals. This removal forms the Fe^{++}, Co^{++}, or Ni^{++} ion. In the case of iron, one 3*d* electron is also easily removed since the five remaining 3*d* electrons make a half-filled sublevel. When two 4*s* and one 3*d* electrons are removed, the Fe^{+++} ion is formed. It becomes increasingly more difficult to remove a 3*d* electron from cobalt and nickel. This increasing difficulty is partly explained by the higher nuclear charges. Another factor is that neither 3*d* sublevel would be left at the filled or half-filled stage. Neither Co^{+++} nor Ni^{+++} ions are common. However, cobalt atoms in the +3 oxidation state occur in complexes.

All three metals of the Iron Family have a strong magnetic property. This property is commonly known as *ferromagnetism* because of the unusual extent to which it is possessed by iron. Cobalt is strongly magnetic, while nickel is the least magnetic of the three. The ferromagnetic nature of these metals is thought to be related to the similar spin orientations of their unpaired 3*d*

Table 26-7
ELECTRON CONFIGURATIONS OF THE IRON FAMILY

Sublevel	1*s*	2*s*	2*p*	3*s*	3*p*	3*d*	4*s*
Maximum number of electrons	2	2	6	2	6	10	2
iron	2	2	6	2	6	:....	:
cobalt	2	2	6	2	6	::...	:
nickel	2	2	6	2	6	:::..	:

electrons. Table 26-7 shows that atoms of iron have 4 unpaired 3*d* electrons, cobalt 3, and nickel 2.

Each spinning electron acts like a tiny magnet. Electron pairs are formed by two electrons spinning in opposite directions. The electron magnetisms of such a pair of electrons neutralize each other. In Iron Family metals, groups of atoms may be aligned and form small magnetized regions called *domains.* Ordinarily, magnetic domains within the metallic crystals point in random directions. In this way, they cancel one another so that the net magnetism is zero. A piece of iron becomes magnetized when an outside force aligns the domains in the same direction. Figure 26-6 illustrates the mechanism of electron spin for paired and unpaired electrons.

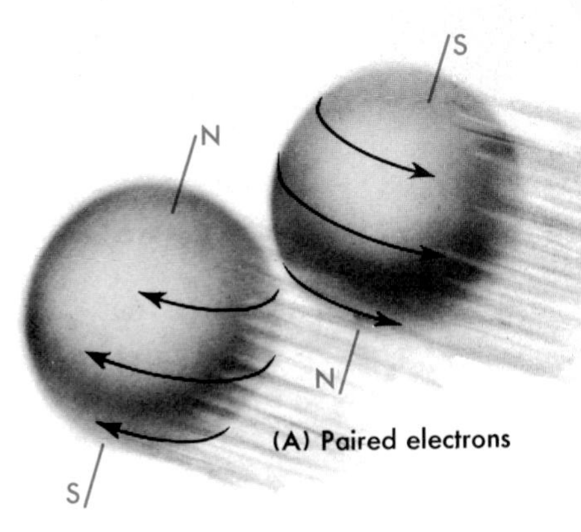

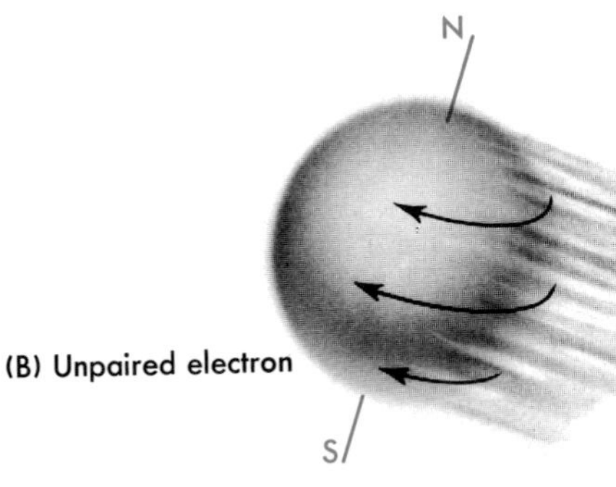

Figure 26-6. Magnetism in matter stems basically from the spin of electrons.

26.9 Occurrence of iron Iron is the fourth element in abundance by mass in the earth's crust. Nearly 5% of the earth's crust is iron. It is the second most abundant metal, following only aluminum. Meteors are known to contain iron. This fact and the known magnetic nature of the earth itself suggest that the earth's core may consist mainly of iron.

Unfortunately, much of the iron in the crust of the earth cannot be removed profitably. Therefore, only minerals from which iron can be profitably recovered by practical methods are considered iron ores.

Great deposits of *hematite,* Fe_2O_3, once existed in the Lake Superior region of the United States. They were the major sources of iron in this country for many years. These rich ore deposits have been depleted, and the iron industry now depends on medium- and low-grade ores, the most abundant of which is *taconite.*

Taconite has an iron content of roughly 25% to 50% in a chemically complex mixture. The main iron minerals present are hematite, Fe_2O_3, and magnetite, Fe_3O_4. The rest of the ore is rock. Modern blast furnaces for reducing iron ore to iron require ores containing well above 60% iron. Thus the raw ores must be concentrated by removing much of the waste materials. This is done at the mines before the ores are transported to the blast furnaces or smelters.

Taconite ores are first crushed and pulverized. They are then concentrated by a variety of complex methods. The concentrated ores are hardened into pellets for shipment to the smelters. Ore concentrates containing 90% iron are produced by the chemical removal of oxygen during the pelletizing treatment as shown in Figure 26-7.

26.10 The blast furnace Iron ore is converted to iron in a giant structure called a blast furnace. The charge placed in the blast furnace consists of iron ore concentrate, coke, and a flux, which causes mineral impurities in the ore to melt more readily. The

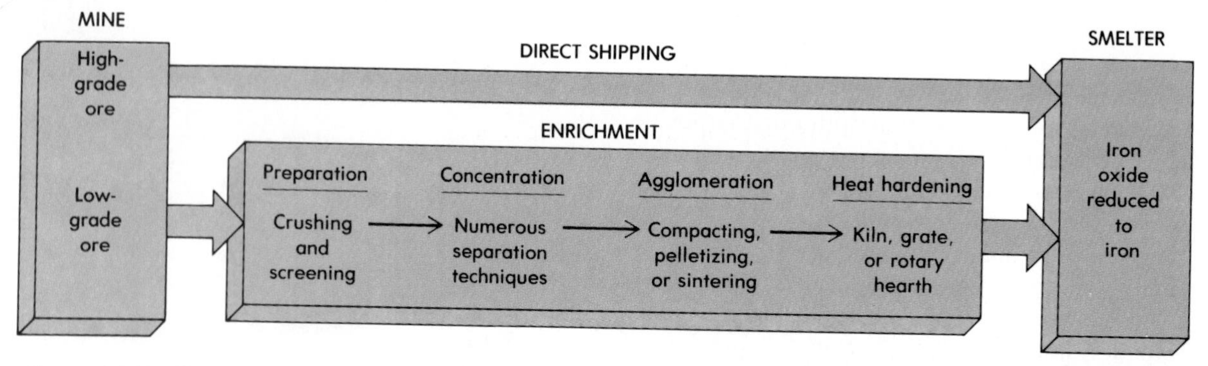

Figure 26-7. Enrichment of low-grade iron ores. Today nearly all iron ore mined in the United States is concentrated before shipment to a blast furnace.

proper proportions for the charge are calculated by analyzing the raw materials. Usually the flux is limestone, because silica or sand is the most common impurity in the iron ore. Some iron ores contain limestone as an impurity; in such cases, the flux added is sand. See Figure 26-8. A blast of hot air, sometimes enriched with oxygen, is forced into the base of the furnace.

Iron oxide is reduced in the blast furnace to iron. The earthy impurities are removed as slag. Coke is required for the first function and limestone for the second. The products of the blast furnace are *pig iron, slag,* and *flue gas.*

The actual chemical changes that occur are complex and somewhat unclear. The coke is ignited by the blast of hot air, and some of it burns to form carbon dioxide.

$$C(s) + O_2(g) \rightarrow CO_2(g)$$

When the oxygen is blown in, the carbon dioxide comes in contact with other pieces of hot coke and is reduced to carbon monoxide gas.

$$CO_2(g) + C(s) \rightarrow 2CO(g)$$

The carbon monoxide thus formed is actually the reducing agent that reduces the iron oxide to metallic iron.

$$Fe_2O_3 + 3CO \rightarrow 2Fe(l) + 3CO_2(g)$$

This reduction probably occurs in steps as the temperature increases toward the bottom of the furnace. Possible steps are

$$Fe_2O_3 \rightarrow Fe_3O_4 \rightarrow FeO \rightarrow Fe$$

The white-hot liquid iron collects in the bottom of the furnace. Every four or five hours it is tapped off. It may be cast in molds as *pig iron* or converted directly to steel.

In the middle region of the furnace, the limestone decomposes to calcium oxide and carbon dioxide.

$$CaCO_3(s) \rightarrow CaO(s) + CO_2(g)$$

The calcium oxide combines with silica to form a calcium silicate slag. This slag melts more readily than silica.

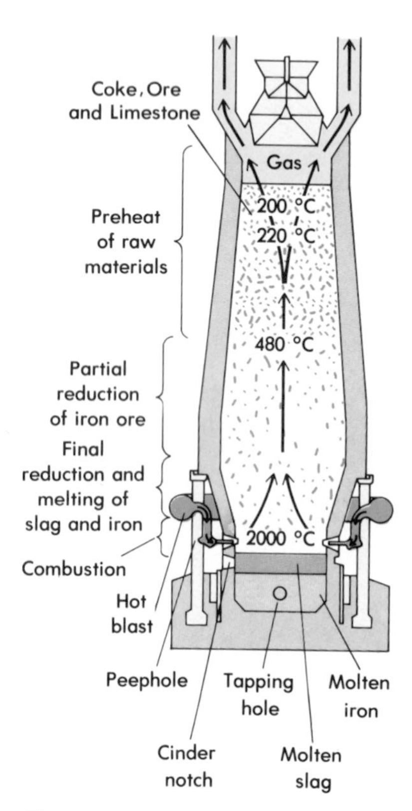

Figure 26-8. A sectional view of a blast furnace.

$$CaO + SiO_2 \rightarrow CaSiO_3(l)$$

This glassy slag also collects at the bottom of the furnace. Since it has a much lower density than liquid iron, it floats on top of the iron. It prevents the reoxidation of the iron. The melted slag is tapped off every few hours. Usually it is thrown away.

26.11 Steel production The relatively high carbon content of iron recovered from the blast furnace makes it very hard and brittle. Two other impurities are phosphorus and sulfur. The phosphorus makes pig iron brittle at low temperatures. The sulfur makes it brittle at high temperatures.

The conversion of iron to steel is *essentially a purification process in which impurities are removed by oxidation.* This purification process is carried out in a furnace at a high temperature. Scrap steel that is being recycled is added to the charge of iron. Iron ore may be added to supply oxygen for oxidizing the impurities. Oxygen gas is blown in for very rapid oxidation. Limestone or lime is included in the charge for forming a slag with nongaseous oxides. Near the end of the process, selected alloying substances are added. By using different kinds and quantities of these alloys, the steel is given different desired properties.

The impurities in the pig iron are oxidized in the following way:

$$3C + Fe_2O_3 \rightarrow 3CO(g) + 2Fe$$

$$3Mn + Fe_2O_3 \rightarrow 3MnO + 2Fe$$

$$12P + 10Fe_2O_3 \rightarrow 3P_4O_{10} + 20Fe$$

$$3Si + 2Fe_2O_3 \rightarrow 3SiO_2 + 4Fe$$

$$3S + 2Fe_2O_3 \rightarrow 3SO_2(g) + 4Fe$$

Carbon and sulfur escape as gases. The limestone flux decomposes as in the blast furnace.

$$CaCO_3 \rightarrow CaO + CO_2(g)$$

Calcium oxide and the oxides of the other impurities react to form slag.

$$P_4O_{10} + 6CaO \rightarrow 2Ca_3(PO_4)_2$$

$$SiO_2 + CaO \rightarrow CaSiO_3$$

$$MnO + SiO_2 \rightarrow MnSiO_3$$

These three reactions are between acidic nonmetallic oxides and basic metallic oxides.

26.12 Pure iron Pure iron is a metal that is seldom seen. It is silver-white, soft, ductile, tough, and does not tarnish readily. It melts at 1535 °C. Commercial iron contains carbon and other impurities that alter its properties. Cast iron melts at approximately 1150 °C. All forms of iron corrode, or rust, in moist air, so it is not a self-protective metal. The rust that forms is brittle and

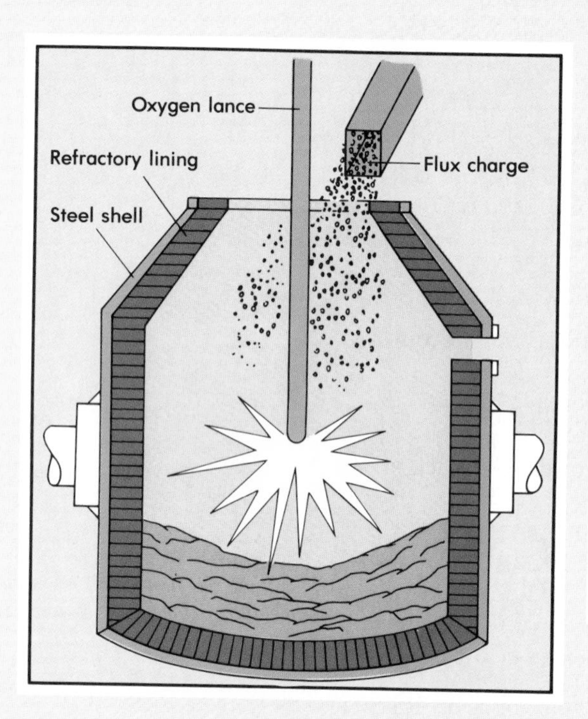

Steelmaking by the basic oxygen process. The furnace receives a charge of scrap steel (upper left) and a charge of molten iron from a blast furnace (upper right). Oxygen is blown into the furnace, and flux is added (lower left). Molten steel is released from a tap hole (lower right) and transferred to a ladle. Alloying substances will then be added according to precise specifications.

scales off, exposing the metal underneath to corrosion. Iron does not rust in dry air or in water that is free of dissolved oxygen.

Iron is only moderately active chemically. Even so, it corrodes more extensively than more active metals such as zinc and aluminum. These metals form oxide coatings that adhere to the metal surface and protect it from further corrosion. The rusting of iron is a complicated (and not fully understood) electrochemical process that involves water, air, and carbon dioxide.

Water containing CO_2 in solution is acidic. It reacts with iron to form Fe^{++} ions. The iron(II) ions are oxidized to the iron(III) state as hydrated Fe_2O_3 (rust) by oxygen. Iron is made rust resistant in several ways: (1) by alloying it with chromium, copper, or nickel; (2) by treating it to form a coating of Fe_3O_4, which adheres to and protects the surface; (3) by painting it, which protects the surface as long as the paint adheres and is not chipped or scratched; (4) by dip-coating it with a self-protective metal (galvanizing with zinc); (5) by plating it with nickel or chromium; and (6) by coating it with a material like porcelain.

Dilute acids generally react readily with iron. Strong hydroxides do not react with it. Concentrated nitric acid does not react with iron. In fact, dipping iron into concentrated nitric acid makes the iron *passive,* or *inactive,* with other chemicals. Concentrated sulfuric acid has little effect on iron.

Figure 26-9. Workman inspects pipelines at iron oxide plant in West Virginia.

26.13 Three oxides of iron Of the three oxides of iron, *iron(II) oxide,* FeO, is of little importance. It oxidizes rapidly when exposed to air to form *iron(III) oxide,* Fe_2O_3. This oxide is the important ore of iron. It is used as a cheap red paint pigment. It is also used for grinding and polishing glass lenses and mirrors. *Magnetic iron oxide,* Fe_3O_4, is an important ore. It is composed of Fe_2O_3 and FeO. Thus it may be considered to be iron(II, III) oxide, and its formula can be written $FeO \cdot Fe_2O_3$.

26.14 Reactions of the Fe^{++} ion Hydrated iron(II) sulfate, $FeSO_4 \cdot 7H_2O$, is the most useful compound of iron in the +2 oxidation state. It is used as a reducing agent and in medicine for iron tonics. Iron(II) sulfate can be prepared by the action of dilute sulfuric acid on iron. The crystalline hydrate loses water of hydration when exposed to air and turns brown because of oxidation. Iron(II) sulfate in solution is gradually oxidized to the iron(III) state by dissolved oxygen. The brown precipitate of basic iron(III) sulfate that forms is evidence of this change:

$$4FeSO_4 + O_2 + 2H_2O \rightarrow 4Fe(OH)SO_4(s)$$

The Fe^{++} ions can be kept in the reduced state by making the solution acidic with sulfuric acid and adding pieces of iron. Hydrated iron(II) ammonium sulfate, $Fe(NH_4)_2(SO_4)_2 \cdot 6H_2O$, is a better source of Fe^{++} ions in the laboratory because it is stable in contact with air.

Iron(II) salts are readily oxidized to iron(III) salts by the corresponding acid and an oxidizing agent. In the case of the nitrate, nitric acid meets both requirements.

$$3Fe(NO_3)_2 + 4HNO_3 \rightarrow 3Fe(NO_3)_3 + NO(g) + 2H_2O$$

26.15 Reactions of the Fe^{+++} ion Hydrated iron(III) chloride, $FeCl_3 \cdot 6H_2O$, is the most useful compound of iron in the +3 oxidation state. It is used as a mordant for dyeing cloth and as an oxidizing agent. The yellow crystalline hydrate is deliquescent. The hydrated Fe^{+++} ion, $Fe(H_2O)_6^{+++}$, is pale violet in color. This color usually is not seen, however, because of hydrolysis, which yields hydroxide complexes that are yellow-brown in color. The hydrolysis of Fe^{+++} ions in water solutions of its salts gives solutions that are acidic.

A mordant is used to "fix" or lock a dye color into a fabric.

***Deliquescence:** the property of certain substances to take up water from the air. See Section 12.19.*

$$Fe^{+++} + 2H_2O \rightleftarrows FeOH^{++} + H_3O^+$$

$$FeOH^{++} + 2H_2O \rightleftarrows Fe(OH)_2^+ + H_3O^+$$

$$Fe(OH)_2^+ + 2H_2O \rightleftarrows Fe(OH)_3 + H_3O^+$$

The hydrolysis is extensive when it occurs in boiling water. A blood-red colloidal suspension of iron(III) hydroxide is formed. This colloidal suspension can be produced by adding a few drops of $FeCl_3$ solution to a flask of boiling water.

Iron(III) ions are removed from solution by adding a solution containing hydroxide ions. A red-brown jelly-like precipitate of iron(III) hydroxide is formed.

$$Fe^{+++} + 3OH^- \rightarrow Fe(OH)_3(s)$$

By evaporating the water, red Fe_2O_3 remains.

26.16 Tests for iron ions Potassium hexacyanoferrate(II), $K_4Fe(CN)_6$ (also called potassium ferrocyanide), is a light yellow crystalline salt. It contains the complex hexacyanoferrate(II) ion (ferrocyanide ion), $Fe(CN)_6^{----}$. The iron is in the +2 oxidation state. The hexacyanoferrate(II) ion forms when an excess of cyanide ions is added to a solution of an iron(II) salt.

***Caution:** Solutions containing cyanide ions are deadly poisons. Safety precautions should be followed when handling these solutions.*

$$6CN^- + Fe^{++} \rightarrow Fe(CN)_6^{----}$$

Suppose KCN is used as the source of CN^- ions and $FeCl_2$ as the source of Fe^{++} ions. The empirical equation is

$$6KCN + FeCl_2 \rightarrow K_4Fe(CN)_6 + 2KCl$$

The iron of the $Fe(CN)_6^{----}$ ion can be oxidized by chlorine to the +3 state. This reaction forms the hexacyanoferrate(III) ion (ferricyanide ion) $Fe(CN)_6^{---}$.

*(**II**) following hexacyanoferrate tells you it is the Fe^{++} ion in the complex. (**III**) tells you it is the Fe^{+++} ion.*

$$2Fe(CN)_6^{----} + Cl_2 \rightarrow 2Fe(CN)_6^{---} + 2Cl^-$$

As the iron is oxidized from the +2 oxidation state to the +3 state, the chlorine is reduced from the 0 state to the −1 state.

Using the potassium salt as the source of the $Fe(CN)_6^{----}$ ions, the empirical equation is

$$2K_4Fe(CN)_6 + Cl_2 \rightarrow 2K_3Fe(CN)_6 + 2KCl$$

$K_3Fe(CN)_6$, potassium hexacyanoferrate(III) (known also as potassium ferricyanide), is a dark red crystalline salt.

Intense colors are observed in most compounds that have an element present in *two different oxidation states.* When iron(II) ions and hexacyanoferrate(III) ions are mixed, and when iron(III) ions and hexacyanoferrate(II) ions are mixed, *the same intense blue substance is formed.*

Before chemists learned that both blue hexacyanoferrate precipitates were the same substance, the one formed with Fe^{++} ions was called Turnbull's blue, and the one formed with Fe^{+++} ions was called Prussian blue.

$$Fe^{++} + K^+ + Fe(CN)_6^{---} + H_2O \rightarrow K\overset{+2}{Fe}\overset{+3}{Fe}(CN)_6 \cdot H_2O(s)$$

and

$$Fe^{+++} + K^+ + Fe(CN)_6^{----} + H_2O \rightarrow K\overset{+3}{Fe}\overset{+2}{Fe}(CN)_6 \cdot H_2O(s)$$

In both reactions, the precipitate contains iron in the +2 and +3 oxidation states, and the same intense blue color is seen.

Iron(II) ions, Fe^{++}, and hexacyanoferrate(II) ions, $Fe(CN)_6^{----}$, form a white precipitate, $K_2FeFe(CN)_6$. Note that both Fe ions are in the same +2 oxidation state. This precipitate remains white if the oxidation of Fe^{++} ions is prevented. Of course, on exposure to air, it begins to turn blue. Iron(III) ions, Fe^{+++}, and hexacyanoferrate(III) ions, $Fe(CN)_6^{---}$, give a brown solution. Here both Fe ions are in the same +3 oxidation state. The reactions given in the two preceding equations provide ways to detect the presence of iron in each of its two oxidation states.

1. *Test for the Fe^{++} ion.* Suppose a few drops of $K_3Fe(CN)_6$ solution is added to a solution of iron(II) sulfate. The characteristic intense blue precipitate $KFeFe(CN)_6 \cdot H_2O$ forms. Two-thirds of the potassium ions and the sulfate ions are merely spectator ions. The net ionic reaction is

$$Fe^{++} + K^+ + Fe(CN)_6^{---} + H_2O \rightarrow KFeFe(CN)_6 \cdot H_2O(s)$$

The formation of a blue precipitate when $K_3Fe(CN)_6$ is added to an unknown solution serves to identify the Fe^{++} ion.

2. *Test for the Fe^{+++} ion.* Suppose a few drops of $K_4Fe(CN)_6$ solution are added to a solution of iron(III) chloride. The characteristic blue precipitate $KFeFe(CN)_6 \cdot H_2O$, is formed. The net reaction is

$$Fe^{+++} + K^+ + Fe(CN)_6^{----} + H_2O \rightarrow KFeFe(CN)_6 \cdot H_2O(s)$$

The formation of a blue precipitate when $K_4Fe(CN)_6$ is added to an unknown solution serves to identify the Fe^{+++} ion.

Blueprints: *The basis for the blueprinting process is a photochemical reaction that reduces Fe^{+++} ions to Fe^{++} ions, which react with hexacyanoferrate(III) ions to form the blue color in the paper. Blueprint paper is first treated with a solution of iron(III) ammonium citrate and potassium hexacyanoferrate(III) and allowed to dry in the dark. A drawing on tracing paper is laid over the blueprint paper and exposed to light. Where light strikes the paper, Fe^{+++} ions are reduced to Fe^{++} ions. The paper is dipped into water and the Fe^{++} ions react with the hexacyanoferrate(III) ions to form the blue color in the paper. Where drawing lines covered the paper, no reduction of Fe^{+++} ions occurs and the paper rinses out white. Thus a blueprint has white lines on blue paper.*

Potassium thiocyanate, KSCN, provides another good test for the Fe^{+++} ion. It is often used to confirm the $K_4Fe(CN)_6$ test. A blood-red solution results from the formation of the complex $FeSCN^{++}$ ion.

THE COPPER FAMILY

26.17 Members of the Copper Family The Copper Family subgroup of transition metals consists of *copper, silver,* and *gold.* All three metals appear below hydrogen in the electrochemical series. They are not easily oxidized and often occur in nature in the free, or *native,* state. Because of their pleasing appearance, durability, and scarcity, these metals have been highly valued since the time of their discovery. All have been used in ornamental objects and coins throughout history.

All soluble compounds of copper, silver, and gold are very toxic.

The atoms of copper, silver, and gold have a single electron in their highest energy levels. Thus they often form compounds in which they exhibit the +1 oxidation state, and thereby resemble the Group I metals of the Sodium Family.

Each metal of the Copper Family has 18 electrons in the next-to-highest energy level. The *d* electrons in this next-to-highest energy level have energies that differ only slightly from the energy of the outer *s* electron. Thus one or two of these *d* electrons can be removed with relative ease. For this reason, copper and gold often form compounds in which they exhibit the +2 and +3 oxidation states respectively. In silver, the +2 oxidation state is reached only under extreme oxidizing conditions.

Copper, silver, and gold are very dense, ductile, and malleable. They are classed as heavy metals along with other transition metals in the central region of the periodic table. Some important properties of each metal are shown in Table 26-8.

26.18 Copper and its recovery Copper, alloyed with tin in the form of bronze, has been in use for over 5000 years. Native copper deposits lie deep underground and are difficult to mine.

Sulfide ores of copper yield most of the supply of this metal. *Chalcocite,* Cu_2S, *chalcopyrite,* $CuFeS_2$, and *bornite,* Cu_3FeS_3, are

Table 26-8
THE COPPER FAMILY

Element	Atomic number	Atomic weight	Electron configuration	Oxidation numbers	Melting point (°C)	Boiling point (°C)	Density (g/cm^3)
copper	29	63.546	2,8,18,1	+1, +2	1083.4	2567	8.96
silver	47	107.8682	2,8,18,18,1	+1	961.9	2212	10.5
gold	79	196.9665	2,8,18,32,18,1	+1, +3	1064.4	2807	19.3

the major sulfide ores. *Malachite*, $Cu_2(OH)_2CO_3$, and *azurite*, $Cu_3(OH)_2(CO_3)_2$, are basic carbonates of copper. Malachite is a rich green, and azurite is a deep blue. Besides serving as ores of copper, fine specimens of these minerals are sometimes polished for use as ornaments or in making jewelry.

The carbonate ores of copper are washed with dilute sulfuric acid, forming a solution of copper(II) sulfate. The copper is then recovered by electrolysis. High-grade carbonate ores are heated in air to convert them to copper(II) oxide. The oxide is then reduced with coke, yielding metallic copper.

The sulfide ores are usually low-grade and require concentrating before they can be refined profitably. The concentration is accomplished by *oil-flotation*. See Figure 26-10. Earthy impurities are wetted by water, and the ore is wetted by oil. Air is blown into the mixture to form a froth. The oil-wetted ore floats to the surface in the froth. This treatment changes the concentration of the ore from about 2% copper to as high as 30% copper. Metallic copper is recovered from the concentrated ore by a process shown diagrammatically in Figure 26-11.

The concentrated ore is partially roasted to form a mixture of Cu_2S, FeS, FeO, and SiO_2. This mixture is known as *calcine*. The roasting process, using oxygen-enriched air, yields high quality sulfur dioxide. This gas is converted to sulfuric acid, as described in Section 29.13. Calcine is fused with limestone in a furnace. Part of the iron is removed as a silicate slag. The rest of the iron, together with the copper, forms a mixture of sulfides known as *matte*. Copper matte is processed in a reverberatory furnace, and the end product contains about 40% copper.

The melted matte is further refined in a converter supplied with oxygen-enriched air as the oxidizing agent. Sulfur from the sulfides, as well as impurities of arsenic and antimony, are

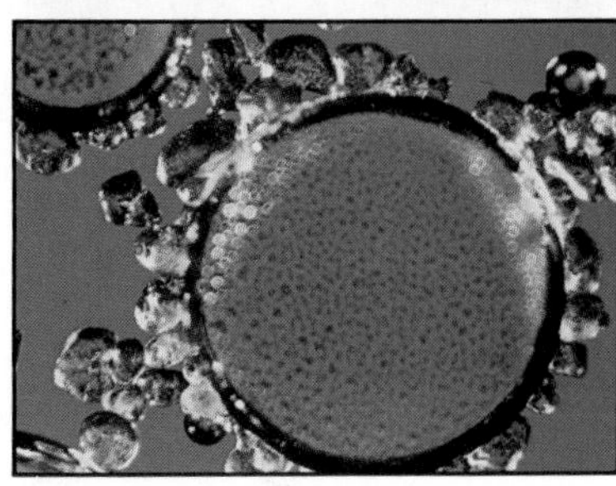

Figure 26-10. (Top photo) The oil-flotation (froth flotation) process for concentrating copper ore. The particles of ore are carried to the surface by air bubbles in the froth. (Bottom photo) Photomicrograph of a flotation bubble from the copper (froth flotation) process.

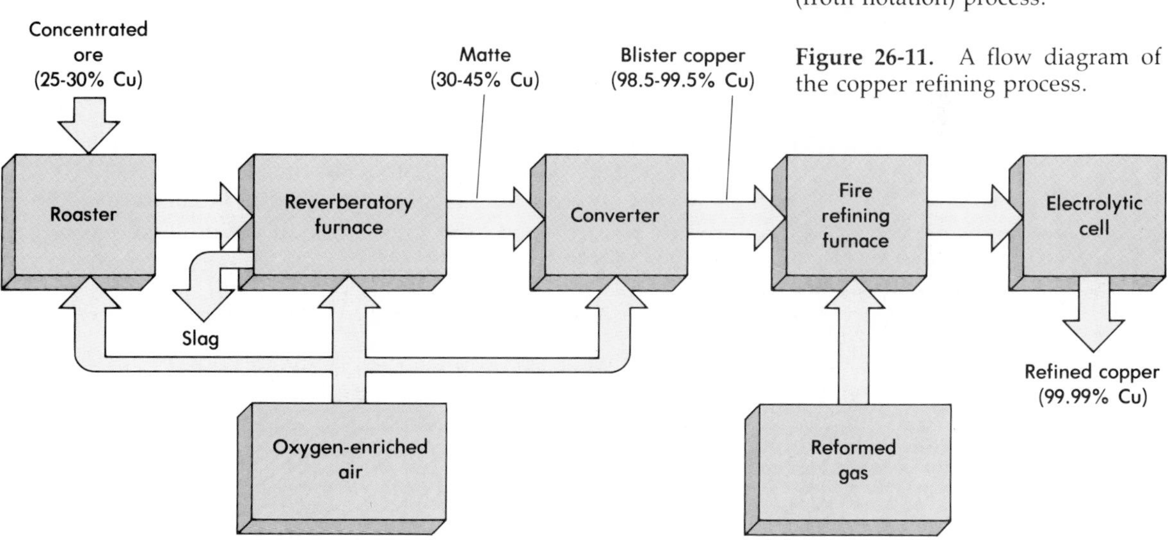

Figure 26-11. A flow diagram of the copper refining process.

removed as volatile oxides. Most of the iron is removed as slag. Some of the copper(I) sulfide is converted into copper(I) oxide. The copper(I) oxide then reacts with more copper(I) sulfide, forming metallic copper and sulfur dioxide. The following equations show the chemical reactions involved in this process.

$$2Cu_2S + 3O_2 \rightarrow 2Cu_2O + 2SO_2(g)$$
$$2Cu_2O + Cu_2S \rightarrow 6Cu + SO_2(g)$$

The molten copper is cast as *blister copper* of 98.5 to 99.5% purity. As the copper cools, dissolved gases escape and form blisters, hence the name. Impurities remaining are iron, silver, gold, and sometimes zinc. In the fourth step of the copper-refining process, blister copper is further purified in a fire-refining furnace.

26.19 Electrolytic refining of copper Unrefined copper contains fairly large amounts of silver and gold. Thus the cost of its refining is offset by the recovery of these precious metals. Copper is largely used for making electric conductors, and very small amounts of impurities greatly increase the electric resistance.

In electrolytic refining, shown in Figure 26-12, sheets of pure copper are used as the negative electrodes in electrolytic cells. Large plates of *impure* copper are used as the positive electrodes. The electrolyte is a solution of copper(II) sulfate in sulfuric acid. A direct current at low voltage is used to operate the cell. During the electrolysis, copper and the other metals in the positive electrode that are above copper in the electrochemical series are oxidized. They enter the solution as ions.

$$Cu \rightarrow Cu^{++} + 2e^-$$
$$Fe \rightarrow Fe^{++} + 2e^-$$
$$Zn \rightarrow Zn^{++} + 2e^-$$

At the low voltages used, the less active silver and gold are not oxidized and so do not go into solution. As the positive electrode is used up, they fall to the bottom of the cell as a sludge and are recovered easily.

You might expect the various positive ions of the electrolyte to be reduced at the negative electrode. However, H_3O^+ ions, Fe^{++} ions, and Zn^{++} ions all require higher voltages than Cu^{++} ions for reduction. At the low potential maintained across the cell, only Cu^{++} ions are reduced at the negative electrode.

$$Cu^{++} + 2e^- \rightarrow Cu$$

Of all the metals present, only copper plates out on the negative electrode. Electrolytic copper is 99.99% pure.

Figure 26-12. Electrolytic refining of copper. In these large electrolytic cells, positive electrodes of impure copper yield Cu^{++} ions in solution. These Cu^{++} ions are plated out on pure copper negative electrodes. Electrolytically refined copper is 99.99% pure.

26.20 Properties of copper Copper is a soft, ductile, malleable, red metal with a density of 8.96 g/cm^3. It is second to silver in electric conductivity.

Heated in air, copper forms a black coating of copper(II) oxide, CuO. Metallic copper and most copper compounds color a Bunsen flame green.

Copper forms copper(II) salts that dissociate in water to give blue solutions. The color is characteristic of the hydrated copper(II) ion, $Cu(H_2O)_4^{++}$. Adding an excess of ammonia to solutions containing this ion produces the deeper blue complex ion, $Cu(NH_3)_4^{++}$. Aqueous solutions of copper(II) salts are weakly acidic because of the mild hydrolysis of the Cu^{++} ion.

$$Cu(H_2O)_4^{++} + H_2O \rightarrow Cu(H_2O)_3OH^+ + H_3O^+$$

An excess of sulfur vapor forms a blue-black coating of copper(I) sulfide on hot copper. In moist air, copper tarnishes and forms a protective coating. This coating is a green basic carbonate, $Cu_2(OH)_2CO_3$. Sulfur dioxide in the air may also combine with copper. If so, a green basic sulfate, $Cu_4(OH)_6SO_4$, is produced. The green color seen on copper roofs is caused by the formation of these compounds.

Figure 26-13. Copper mining pit in Bisbee, Arizona.

Copper(II) compounds are much more common than copper(I) compounds. Copper(II) oxide is used to change alternating current to direct current. It is also used as an oxidizing agent in chemical laboratories.

Hydrated copper(II) sulfate, $CuSO_4 \cdot 5H_2O$, called *blue vitriol,* is an important copper compound. It is used to kill algae in reservoirs and to make certain pesticides. It is also used in electroplating and in preparing other copper compounds.

Because copper stands below hydrogen in the activity series, it does not replace hydrogen from acids. Thus it is not acted on by nonoxidizing acids such as hydrochloric and dilute sulfuric except very slowly when oxygen is present. The oxidizing acids, such as nitric and hot concentrated sulfuric, react vigorously with copper. Such reactions produce the corresponding copper(II) salts.

26.21 Tests for the Cu^{++} ion A dilute solution of a copper(II) salt changes to a very deep blue color when an excess of ammonia is added. This color change is caused by the formation of complex $Cu(NH_3)_4^{++}$ ions.

$$Cu^{++} + 4NH_3 \rightarrow Cu(NH_3)_4^{++}$$

The addition of $K_4Fe(CN)_6$ to a solution containing Cu^{++} ions produces a red precipitate of $Cu_2Fe(CN)_6$. This precipitate is copper(II) hexacyanoferrate(II), also known as copper(II) ferrocyanide. If copper is present in a borax bead formed in an *oxidizing flame,* a clear blue color appears on cooling. The hot bead is green. A bead formed in a *reducing flame* is colorless when hot and an opaque red when cool. The borax-bead test is described in Section 27.11. The test bead for copper is shown in Figure 27-11.

Figure 26-14. Copper penny blanks ready to be coined at the mint.

SUMMARY

The transition metals consist of ten subgroups interposed between Group II and Group III in the periodic table. They are in this position because they represent an interruption of the regular increase in the number of highest energy-level electrons from Group I to Group VIII in the long periods.

Chemical properties of transition metals are varied. Electrons of the next-to-highest *d* sublevel and the highest energy level may become involved in the formation of compounds. Most of the transition metals show different oxidation states in their reactions. They have strong tendencies to form complex ions. Many of their compounds are paramagnetic and are colored.

With minor apparent irregularities, the transition metals have two highest energy-level electrons. In the first row of transition elements, chromium and copper have one 4*s* electron. The availability of 3*d* electrons tends to increase the number of oxidation states up to manganese in the middle of the row, after which the number of common oxidation states for each element decreases.

The color of the compounds of transition elements depends to some extent on the ions or polyatomic groups associated with the element. The colors result from the ease with which *d* electrons can be excited by absorption of discrete quantities of light energy.

Transition metals form many complex ions. A transition element cation is the central ion in the complex. Certain ionic or molecular species coordinate with the central ion to form the complex ion. The coordinating atom of the species must have unshared electron pairs that the central ion can share. Formation of complex ions in a solution will increase the solubility of the metallic ions that form the complex.

The Iron Family consists of iron, cobalt, and nickel. These metals are ferromagnetic. All are present in nature in combined form. Purified iron is used as a construction metal in the form of steel. By alloying other metals with steel, a variety of structural characteristics can be obtained. The metals of the Iron Family are recovered from their ores by reduction methods.

Iron forms compounds in two oxidation states, +2 and +3. Iron(II) ions are easily oxidized to the iron(III) state. Iron(II) compounds are used as reducing agents. [Iron(III) compounds are used as oxidizing agents.] The iron(III) ion hydrolyzes in water solutions of its salts. Its hydroxide is very sparingly soluble and is precipitated by the addition of aqueous ammonia to solutions of iron(III) salts. Both the Fe^{++} ion and the Fe^{+++} ion can be identified by precipitation reactions with potassium hexacyanoferrate(III) and potassium hexacyanoferrate(II), respectively. The color of the intense blue precipitate results from iron's being present in both the +2 and +3 oxidation states.

The Copper Family consists of copper, silver, and gold. All are found as native metals in nature. Most copper used in commerce and industry is recovered from its sulfide ores by reduction and electrolytic refining. All soluble compounds of copper, silver, and gold are very toxic. The copper(II) ions can be identified by the formation of colored complex ions.

VOCABULARY

actinide series
blast furnace
complex ion
coordination number
copper matte
domains
ferromagnetism
hematite
instability constant
lanthanide series
mordant
noble metals
paramagnetism
pig iron
polyatomic ion
slag
taconite
transition elements

QUESTIONS

Group A

1. What structural similarity determines the metallic character of transition elements? *A2b*
2. Copper(II) sulfate crystals are blue. To what color does the light energy absorbed by the crystals correspond? *A1, A2i*
3. Chemically, what happens to iron *(a)* in a blast furnace; *(b)* during steel making? *A3i, A3j*
4. List five points of similarity between copper, silver, and gold. *A2m, A3a, A3r*
5. Why does the copper trim on roofs frequently acquire a green surface? *A3t*
6. What is the oxidation number of manganese in *(a)* Mn; *(b)* MnO_2; *(c)* $MnSO_4 \cdot 4H_2O$; *(d)* $KMnO_4$; *(e)* $BaMnO_4$? *A2e, A2f*

Group B

7. Why were some metals used in very early times, while many of the other metals were obtained and used only within the last century? *A1*
8. Distinguish between a polyatomic ion and a complex ion. *A1*
9. When potassium hexacyanoferrate(II) is added to a solution of iron(II) sulfate, a white precipitate forms that gradually turns blue. Explain the color change. *A3q, A3u*
10. How can you detect an iron(II) and an iron(III) compound if both are present in the same solution? *A3q, A4g*
11. Why does it usually pay to refine copper by electrolysis? *A2u, A4h*

PROBLEMS

Group A

1. How much iron(II) chloride can be produced by adding 165 g of iron to an excess of hydrochloric acid?
2. How much iron(III) chloride can be prepared from the iron(II) chloride in Problem 1 if more hydrochloric acid is added and air is blown through the solution?
3. What is the percentage of iron in a sample of limonite, $2Fe_2O_3 \cdot 3H_2O$?
4. How many grams of silver nitrate can be obtained by adding $10\bar{0}$ g of pure silver to an excess of nitric acid?
5. A sample of hematite ore contains Fe_2O_3 87.0%, silica 8.0%, moisture 4.0%, other impurities 1.0%. What is the percentage of iron in the ore?
6. What will be the loss in mass when 1.0×10^6 metric tons of the ore in Problem 5 are heated to $20\bar{0}$ °C?

Group B

7. How much limestone will be needed to combine with the silica in 1.0×10^6 metric tons of the ore described in Problem 5?
8. *(a)* How much carbon monoxide is required to reduce 1.0×10^6 metric tons of the ore in Problem 5? *(b)* How much coke must be supplied to meet this requirement? (Assume the coke to be $10\bar{0}$% carbon.)
9. Iron(II) sulfate is oxidized to iron(III) sulfate in the presence of sulfuric acid using nitric acid as the oxidizing agent. Nitrogen monoxide and water are also formed. Write the balanced formula equation.
10. Silver reacts with dilute nitric acid to form silver nitrate, water, and nitrogen monoxide. Write the balanced formula equation.
11. Copper reacts with hot concentrated sulfuric acid to form copper(II) sulfate, sulfur dioxide, and water. Write the equation.

Irène Joliot-Curie (French), daughter of Marie Curie, and her husband Frédéric Joliot received the 1935 Nobel Prize in chemistry for their synthesis of new radioactive elements. They produced the first artificial radioactivity when they bombarded aluminum with alpha particles from polonium. The product was the then unknown radioisotope phosphorus-30, which emitted positrons and decayed rapidly to stable silicon. They later produced other artificial radioelements, many of which are now important as medical tracers or used in radiotherapy, such as cobalt-60.

Aluminum and the Metalloids

27

GOALS

In this chapter you will gain an understanding of: • the nature of the seven elements known as metalloids • the important properties, uses, and compounds of aluminum • the natural occurrence and commercial preparation of aluminum • the major properties, uses, and compounds of boron, silicon, arsenic, and antimony

27.1 The nature of metalloids Some elements are neither distinctly metallic nor distinctly nonmetallic. Their properties are intermediate between those of metals and nonmetals. As a group, these elements are called *metalloids* or *semimetals.* In the periodic table, they occupy a diagonal region from the upper center toward the lower right. See Figure 27-1.

The metalloids are the elements *boron, silicon, germanium, arsenic, antimony, tellurium,* and *polonium.* Although aluminum is not included in the metalloids, it is included in this chapter because of its unique position in the periodic table relative to the metalloids.

Elemental aluminum is distinctly metallic and is recognized by the familiar properties of metals. But aluminum forms the negative aluminate ion, $Al(OH)_4^-$. Its hydroxide, $Al(OH)_3$, is amphoteric. The oxide of aluminum is ionic, yet the hydride is a covalent network compound. As a metal, aluminum is so resistant to oxidation that it can be used as cooking ware on hot stove burners. Yet it is easily oxidized in sodium hydroxide solution to yield hydrogen and sodium aluminate. These characteristics tend to place aluminum among the metalloids.

The amphoterism of aluminum hydroxide is discussed in detail in Section 15.9.

Boron, silicon, arsenic, and antimony are the typical metalloids. Table 27-1 lists some of their important properties, together with those of aluminum and the other metalloids.

Germanium is a moderately rare element. With the development of the transistor, germanium became important as a semiconductor material. It is chemically similar to silicon, the element above it in Group IV. Germanium is more metallic than arsenic just to its right in Period Four. The major oxidation state of germanium is +4.

Semiconductor: *A substance with an electric conductivity between that of a metal and an insulator.*

Tellurium is a member of Group VI in the periodic table. As expected, the metal-like characteristics of Group VI elements

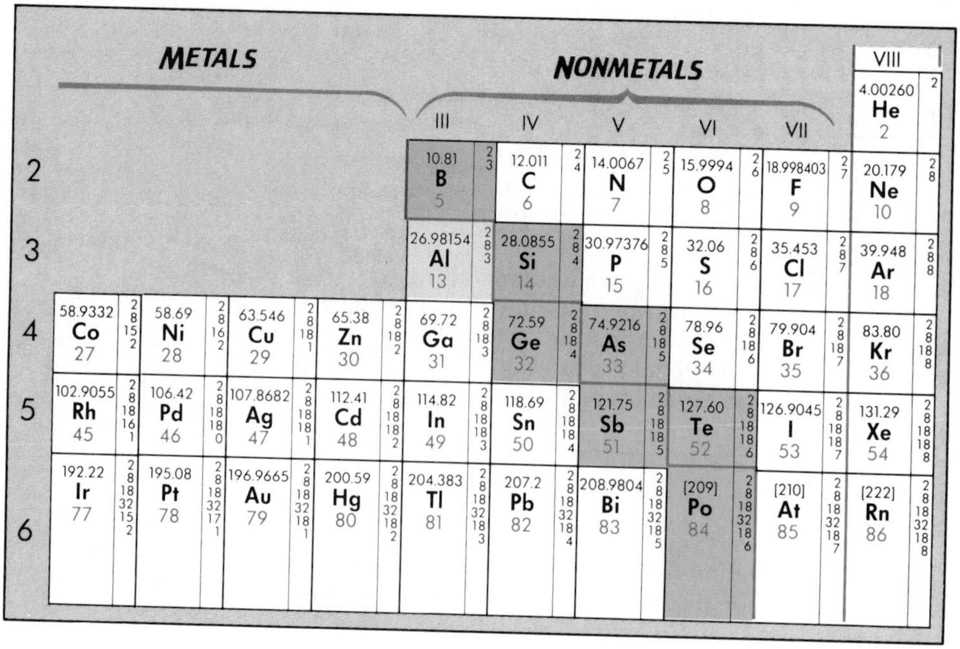

METALS NONMETALS

					III	IV	V	VI	VII	VIII
										4.00260 He 2 (2)
2					10.81 B 5 (2,3)	12.011 C 6 (2,4)	14.0067 N 7 (2,5)	15.9994 O 8 (2,6)	18.998403 F 9 (2,7)	20.179 Ne 10 (2,8)
3					26.98154 Al 13 (2,8,3)	28.0855 Si 14 (2,8,4)	30.97376 P 15 (2,8,5)	32.06 S 16 (2,8,6)	35.453 Cl 17 (2,8,7)	39.948 Ar 18 (2,8,8)
4	58.9332 Co 27 (2,8,15,2)	58.69 Ni 28 (2,8,16,2)	63.546 Cu 29 (2,8,18,1)	65.38 Zn 30 (2,8,18,2)	69.72 Ga 31 (2,8,18,3)	72.59 Ge 32 (2,8,18,4)	74.9216 As 33 (2,8,18,5)	78.96 Se 34 (2,8,18,6)	79.904 Br 35 (2,8,18,7)	83.80 Kr 36 (2,8,18,8)
5	102.9055 Rh 45 (2,8,18,16,1)	106.42 Pd 46 (2,8,18,18,0)	107.8682 Ag 47 (2,8,18,18,1)	112.41 Cd 48 (2,8,18,18,2)	114.82 In 49 (2,8,18,18,3)	118.69 Sn 50 (2,8,18,18,4)	121.75 Sb 51 (2,8,18,18,5)	127.60 Te 52 (2,8,18,18,6)	126.9045 I 53 (2,8,18,18,7)	131.29 Xe 54 (2,8,18,18,8)
6	192.22 Ir 77 (2,8,18,32,15,2)	195.08 Pt 78 (2,8,18,32,17,1)	196.9665 Au 79 (2,8,18,32,18,1)	200.59 Hg 80 (2,8,18,32,18,2)	204.383 Tl 81 (2,8,18,32,18,3)	207.2 Pb 82 (2,8,18,32,18,4)	208.9804 Bi 83 (2,8,18,32,18,5)	[209] Po 84 (2,8,18,32,18,6)	[210] At 85 (2,8,18,32,18,7)	[222] Rn 86 (2,8,18,32,18,8)

Figure 27-1. The metalloids are the elements *boron, silicon, germanium, arsenic, antimony, tellurium,* and *polonium.* Although *aluminum* is generally considered a metal, it has some properties that relate it to the metalloids.

Amorphous: *Without any form, and noncrystalline.*

increase down the group. Tellurium is more metallic (or less nonmetallic) than selenium above it. It is less metallic (or more nonmetallic) than polonium below it. This gradation in properties down the group is illustrated by the change in odor of their hydrogen compounds. Hydrogen oxide (water) is odorless. Hydrogen sulfide has the offensive odor of rotten eggs. The odor of hydrogen selenide is even more offensive, and that of hydrogen telluride is the foulest of them all.

Although it can be amorphous, tellurium is more commonly a brittle, silvery, metal-like crystalline substance. It is classed as a semiconductor, and its chemistry is typically metalloidal. It

Table 27-1
PROPERTIES OF ALUMINUM AND METALLOIDS

Element	*Atomic number*	*Electron configuration*	*Oxidation states*	*Melting point* (°C)	*Boiling point* (°C)	*Density* (g/cm³)	*Atomic radius* (Å)	*First ionization energy* (kcal/mole)
boron	5	2,3	+3	2079	2550	2.34	0.82	191
aluminum	13	2,8,3	+3	660	2467	2.70	1.18	138
silicon	14	2,8,4	+2, +4, −4	1410	2355	2.33	1.11	188
germanium	32	2,8,18,4	+2, +4, −4	937	2830	5.32	1.22	182
arsenic	33	2,8,18,5	+3, +5, −3	sublimes		5.73	1.20	226
antimony	51	2,8,18,18,5	+3, +5, −3	631	1750	6.69	1.40	199
tellurium	52	2,8,18,18,6	+2, +4, +6, −2	450	990	6.24	1.36	208
polonium	84	2,8,18,32,18,6	+4, +6	254	962	9.32	1.46	196

appears with oxygen in both tellurite (TeO_3^{--}) and tellurate (TeO_4^{--}) ions. In these ions, tellurium shows the +4 and +6 oxidation states, respectively. Tellurium combines covalently with oxygen and the halogens, in which the +2, +4, and +6 oxidation states are observed. It forms tellurides (−2 oxidation state) with such elements as gold, hydrogen, and lead. In fact, tellurium is the only element combined with gold in nature.

Polonium is a radioactive element. It was discovered by Pierre and Marie Curie in 1898, just prior to their discovery of radium. Polonium is so rare in nature that little is known of its chemistry. It appears to be more metallic than tellurium.

ALUMINUM

27.2 Aluminum as a light metal Aluminum, atomic number 13, is the second member of Group III. This group is headed by boron and includes gallium, indium, and thallium. All are typically metallic except boron, which is classed as a metalloid. The chemistry of boron differs from that of aluminum and the other Group III elements mainly because of the small size of the boron atom. Its chemistry resembles that of silicon and germanium more than it does that of aluminum. Much of the chemistry of aluminum is similar to that of its corresponding Group II metal, magnesium.

Except in the United States, aluminum is spelled ***aluminium*** *and pronounced* ***al-yuh-min-ee-um.***

Aluminum is a low-density metal; it is about one-third as dense as steel. The pure metal is used in chemical processes, in

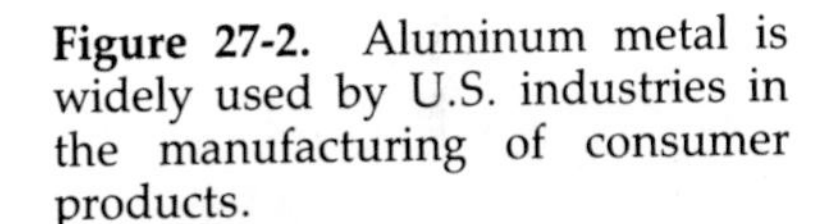

Figure 27-2. Aluminum metal is widely used by U.S. industries in the manufacturing of consumer products.

Figure 27-3. Bauxite being mined at the world's largest bauxite mine located in Weipa, North Queensland, Australia.

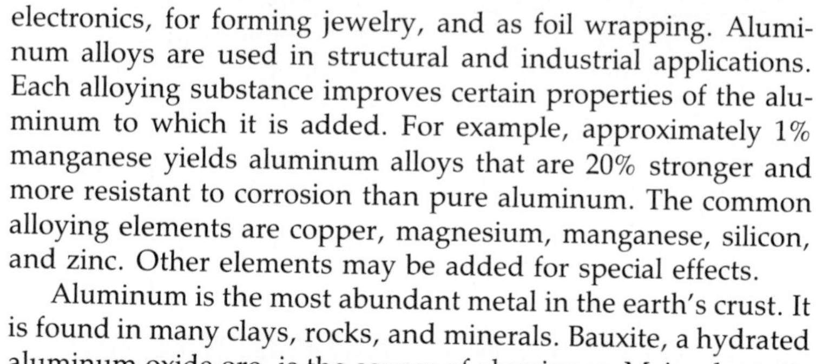
electronics, for forming jewelry, and as foil wrapping. Aluminum alloys are used in structural and industrial applications. Each alloying substance improves certain properties of the aluminum to which it is added. For example, approximately 1% manganese yields aluminum alloys that are 20% stronger and more resistant to corrosion than pure aluminum. The common alloying elements are copper, magnesium, manganese, silicon, and zinc. Other elements may be added for special effects.

Aluminum is the most abundant metal in the earth's crust. It is found in many clays, rocks, and minerals. Bauxite, a hydrated aluminum oxide ore, is the source of aluminum. Major deposits are located throughout the tropical and semitropical regions of the earth. Bauxite containing 40–60% aluminum oxide is required for present aluminum-recovery technology. Major foreign sources are Australia, Jamaica, Surinam, and Guyana. Bauxite is also mined in Georgia, Alabama, Tennessee, and Arkansas. Anhydrous aluminum oxide is called alumina.

Figure 27-4. Aluminum recovery. Electrolytic reduction cells convert refined white bauxite powder (aluminum oxide) into molten aluminum metal.

27.3 Recovery of aluminum Aluminum is extracted by electrolyzing anhydrous aluminum oxide (refined bauxite) dissolved in molten cryolite, Na_3AlF_6. The process requires a temperature slightly below 1000 °C.

In the electrolytic cell, an iron box lined with graphite serves as the negative electrode, graphite rods serve as the positive electrode, and molten cryolite-aluminum oxide is the electrolyte. The cryolite is melted in the cell, and the aluminum oxide dissolves as it is added to the molten cryolite. The operating temperature of the cell is above the melting point of aluminum (660 °C). Thus the aluminum metal at the bottom of the cell is in the liquid phase and is easily drawn off. See Figure 27-4.

The electrode reactions are complex and are not understood completely. Aluminum is reduced at the negative electrode. The positive electrode is gradually converted to carbon dioxide. This fact suggests that oxygen is formed at the positive electrode by oxidation of the O^{--} ion. The following equations for the reaction mechanism may be overly simple. But, they serve to summarize the oxidation-reduction processes in the cell.

negative electrode: $4Al^{+++} + 12e^- \rightarrow 4Al$

positive electrode: $6O^{--} \rightarrow 3O_2 + 12e^-$

$$3C + 3O_2 \rightarrow 3CO_2(g)$$

Figure 27-5. Hot aluminum being carefully poured into molds.

27.4 Properties of aluminum Aluminum has a density of 2.70 g/cm^3 and a melting point of 660 °C. It is ductile and malleable but is not as tenacious as brass, copper, or steel. Only silver, copper, and gold are better conductors of electricity. Aluminum can be welded, cast, or spun and can be soldered only by using a special solder.

Aluminum is a very active metal. The surface is always covered with a thin layer of aluminum oxide. This oxide layer is not affected by air or moisture; aluminum is a self-protective metal. At high temperatures, the metal combines vigorously with oxygen, releasing intense light and heat. Photoflash lamps contain aluminum foil or fine wire in an atmosphere of oxygen.

Ductile: *Capable of being drawn into wire.*

Malleable: *Capable of being rolled into sheets.*

Tenacious: *Resists being pulled apart.*

Aluminum is a very good reducing agent but is not as active as the Group I and Group II metals.

$$Al \rightarrow Al^{+++} + 3e^-$$

The Al^{+++} ion is quite small and carries a large positive charge. The ion hydrates vigorously in water solution and is usually written as the hydrated ion, $Al(H_2O)_6^{+++}$. Water solutions of aluminum salts are generally acidic because of the hydrolysis of $Al(H_2O)_6^{+++}$ ions.

$$Al(H_2O)_6^{+++} + H_2O \rightarrow Al(H_2O)_5OH^{++} + H_3O^+$$

Water molecules are amphoteric but are very weak proton donors or acceptors. Water molecules that hydrate the Al^{+++} ion give up protons more readily, however. This increased activity results from the repulsion effect of the highly positive Al^{+++} ion. In the hydrolysis shown above, a water molecule succeeds in removing one proton from the $Al(H_2O)_6^{+++}$ ion.

Hydrochloric acid reacts with aluminum to form aluminum chloride and release hydrogen. The net ionic equation is

$$2Al(s) + 6H_3O^+ + 6H_2O \rightarrow 2Al(H_2O)_6^{+++} + 3H_2(g)$$

Nitric acid does not react readily with aluminum because of aluminum's protective oxide layer.

In basic solutions, aluminum forms aluminate ions, $Al(OH)_4^-$, and releases hydrogen.

$$2Al(s) + 2OH^- + 6H_2O \rightarrow 2Al(OH)_4^- + 3H_2(g)$$

If the base is sodium hydroxide, sodium aluminate, $NaAl(OH)_4$, is the soluble product and hydrogen is given up as a gas.

$$2Al(s) + 2NaOH + 6H_2O \rightarrow 2NaAl(OH)_4 + 3H_2(g)$$

The oxide surface coating protects aluminum under normal atmospheric conditions. It must be removed before the metal underneath can react with hydronium ions or with hydroxide ions. Hydroxide ions dissolve this oxide layer more readily than do hydronium ions. This explains why aluminum reacts more readily with hydroxides than with acids.

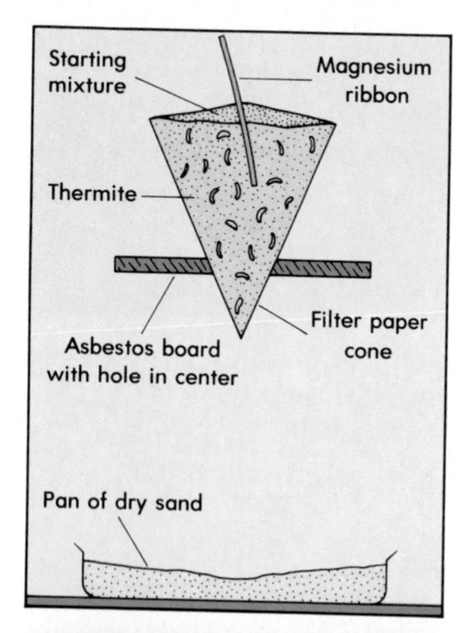

27.5 The thermite reaction When a mixture of powdered aluminum and an oxidizing agent such as iron(III) oxide is ignited, the aluminum reduces the oxide to the free metal. This reduction is rapid and violent and yields a tremendous amount of heat. The sudden release of this heat energy produces temperatures from 3000 to 3500 °C, enough to melt the iron. Such a reaction between aluminum and the oxide of a less active metal is called the *thermite reaction*. The mixture of powdered aluminum and iron(III) oxide is called *thermite*.

The formation of aluminum oxide is strongly exothermic. The reaction releases 399 kilocalories of heat per mole of aluminum oxide formed. The heat of formation of iron(III) oxide is 196 kilocalories per mole. In the thermite reaction, the amount of heat released per mole of aluminum oxide formed equals the difference between these values. The net thermite reaction is considered to be the sum of these two separate reactions.

$$2Al + \tfrac{3}{2}O_2 \rightarrow Al_2O_3 \qquad \Delta H = -399 \text{ kcal}$$

$$Fe_2O_3 \rightarrow 2Fe + \tfrac{3}{2}O_2 \qquad \Delta H = +196 \text{ kcal}$$

$$2Al + Fe_2O_3 \rightarrow 2Fe + Al_2O_3 \qquad \Delta H = -203 \text{ kcal}$$

It is not practical to use aluminum to reduce cheaper metals. However, the thermite reaction is often used to produce small quantities of carbon-free metal. A more important use of this reaction is to reduce metallic oxides that are not readily reduced with carbon. Chromium, manganese, titanium, tungsten, and molybdenum can be recovered from their oxides by the thermite reaction. All these metals are used in making alloy steels.

The very high temperature produced by the thermite reaction makes it useful in welding. Large steel parts, such as propeller shafts and rudder posts on ships or the crankshafts of heavy machinery, are repaired by thermite welding. See Figure 27-6.

Figure 27-6. The very high temperature produced by the thermite reaction makes it useful in welding large steel parts, as in shipbuilding.

27.6 Uses of aluminum oxide *Bauxite*, the chief ore of aluminum is the oxide Al_2O_3. *Corundum* and *emery* are also natural oxides of this metal. Both are useful as abrasives. Emery is the abrasive of emery boards, emery paper, emery cloth, and emery grinding wheels.

Anhydrous aluminum oxide, usually a white powder called alumina, is a very inert substance having an extremely high melting point of 2072 °C. It is used as crucibles for high-temperature reactions.

Pure crystalline alumina, Al_2O_3, is clear and colorless. The *ruby* and *sapphire* gemstones are composed of aluminum oxide contaminated with traces of certain impurities. The presence of a trace quantity of chromium imparts the brilliant ruby-red color to a gem-quality crystal of Al_2O_3, yielding one of the most valuable gemstones found in nature whereas traces of titanium yield the blue sapphire.

Synthetic rubies and sapphires are produced by fusing aluminum oxide with the required trace contaminants in the flame of an oxyhydrogen blowtorch. As synthetic gemstones, they are used in jewelry. See Figure 27-7. Because of the hardness of aluminum oxide, these synthetic crystals are also used as bearings (jewels) in watches and other precision instruments and as dies for drawing wires.

Figure 27-7. Both synthetic and natural sapphires (top) and rubies (bottom) are composed of aluminum oxide, Al_2O_3. The blue color of a sapphire is due to a trace of titanium. The red color of a ruby is due to a trace of chromium.

Scientists are finding new ways to strengthen and stiffen structural materials. One important development involves monocrystalline strands or fibers of one substance embedded in and held in place by some other material. An aluminum rod containing very hard, strong sapphire "whiskers" is quite unlike ordinary soft, ductile aluminum. Such a structure is six times stronger than aluminum and twice as stiff. These *fiber composites* enable engineers to greatly improve the strength-to-weight ratio of structural materials. See Figure 27-8.

Alundum is an oxide of aluminum made by melting bauxite. It is used for making grinding wheels and other abrasives. It is also found in crucibles, funnels, tubing, and other pieces of laboratory equipment.

Figure 27-8. Cross section of a fiber composite magnified 245 times. The experimental composite shown here consists of sapphire whiskers in a silver matrix.

27.7 The alums Alums are common compounds that usually contain aluminum, and they have the general type formula $M^+M^{+++}(SO_4)_2 \cdot 12H_2O$. The M^+ can be any one of several monopositive ions but is usually Na^+, K^+, or NH_4^+. The M^{+++} can be any one of a number of tripositive ions such as Al^{+++}, Cr^{+++}, Co^{+++}, Fe^{+++}, or Mn^{+++}. Potassium aluminum sulfate, $KAl(SO_4)_2 \cdot 12H_2O$, is the most common alum. It is easy to see from this formula why alums are sometimes called double sulfates.

Figure 27-9. Boron, a metalloid, is best known as a constituent of borax which is being mined in the above photo.

All alums have the same crystal structure. This crystal-structure requirement may explain why some monopositive and tripositive ions do not form alums. The alum crystals are hydrated structures in which six water molecules are coordinated to each metallic ion. This coordination number of six for both the monopositive and tripositive ions may restrict the kinds of metallic ions that can form alums.

Alums are used for water purification, as mordants, and in paper manufacture. $NaAl(SO_4)_2 \cdot 12H_2O$ is used as the acid compound in one type of baking powder.

METALLOIDS

27.8 Boron as a metalloid The first member of a periodic group often has properties different from those of the rest of the group. This is true because the outer electrons of its atoms are shielded from the nucleus only by the two electrons of the first energy level. The first member of Group III, boron, is a metalloid; all other Group III elements are metals. Boron also has the highest electronegativity of any element in Group III. Boron atoms are small, with an atomic radius of only 0.82 Å. Their valence electrons are tightly bound, giving boron a high ionization energy for a Group III element. The properties of boron indicate that it forms only covalent bonds with other atoms. At low temperatures boron is a poor conductor of electricity. As the temperature is raised, its electrons have more kinetic energy, and its conductivity increases. This is typical of a *semiconductor.*

Figure 27-10. An alcohol flame has green edges when boric acid is present. This serves as a qualitative test for boric acid.

27.9 Occurrence of boron Boron is not found as the free element. It can be isolated in fairly pure form by reducing boron trichloride with hydrogen at a high temperature. Elemental boron is important in monocrystalline fiber research. Boron filaments, embedded in and held in place by an epoxy plastic, form a very strong and stiff structural material.

Colemanite is a hydrated borate of calcium with the formula $Ca_2B_6O_{11} \cdot 5H_2O$. It is found in the desert regions of California and Nevada. Sodium tetraborate, $Na_2B_4O_7 \cdot 4H_2O$, is found as the mineral *kernite,* also in California. The salt brines of Searles Lake, California, yield most of the commercial supply of boron compounds.

27.10 Useful compounds of boron Boron carbide, B_4C, known as *Norbide,* is an extremely hard abrasive. It is made by combining boron with carbon in an electric furnace. Boron nitride, BN, has a soft, slippery structure similar to graphite.

Under very high pressure it acquires a tetrahedral structure similar to diamond. This form of boron nitride, called *Borazon,* has a hardness second only to diamond. It is used industrially for cutting, grinding, and polishing. Alloys of boron with iron or manganese are used to increase the hardness of steel.

Boric acid, H_3BO_3, can be prepared by adding sulfuric acid to a concentrated solution of sodium tetraborate in water.

$$B_4O_7^{--} + 2H_3O^+ \rightarrow H_2B_4O_7 + 2H_2O$$
$$H_2B_4O_7 + 5H_2O \rightarrow 4H_3BO_3$$
$$\overline{B_4O_7^{--} + 2H_3O^+ + 3H_2O \rightarrow 4H_3BO_3}$$

The acid is moderately soluble and separates as colorless, lustrous scales. It is a weak acid and is used as a mild antiseptic.

Recall that an acid and an alcohol react to form an ester. Sulfuric acid catalyzes the reaction. Suppose a small quantity of methanol and a few drops of dilute sulfuric acid are added to a sample of boric acid. The solution is warmed, and the methanol vapors ignited. The flame has green edges, as shown in Figure 27-10. The alcohol vapor contains the volatile ester $(CH_3)_3BO_3$, which is responsible for the green flame color.

$$3CH_3OH + H_3BO_3 \xrightarrow{H_2SO_4} (CH_3)_3BO_3(g) + 3H_2O$$

This green-flame test indicates the presence of boric acid, or more specifically, the borate ion.

Borax is sodium tetraborate, $Na_2B_4O_7 \cdot 10H_2O$. It is used alone and in washing powders as a water softener. Because it dissolves metallic oxides and leaves a clean metallic surface, it is also used in welding metals.

$$B_4O_7^{--} + 5O^{--} \rightarrow 4BO_3^{---}$$

The borates of certain metals are used in making glazes and enamels. Large amounts of boron compounds are used to make borosilicate glass.

27.11 Borax bead tests Powdered borax held in a burner flame on a platinum wire loop swells and then fuses into a clear, glasslike bead. This bead can be used to identify certain metals. The bead is contaminated with a tiny speck of metal or metallic compound and heated again in the oxidizing flame. The metallic oxide formed fuses with the bead. Certain metals give characteristic transparent colors to the borax bead. Such colors serve to identify the metal involved. For example, cobalt colors the bead *blue;* chromium produces a *green* bead; and nickel yields a *brown* bead. Other metals that can be identified by means of the borax bead test are manganese, copper, and iron. The borax bead colors are shown in Figure 27-11.

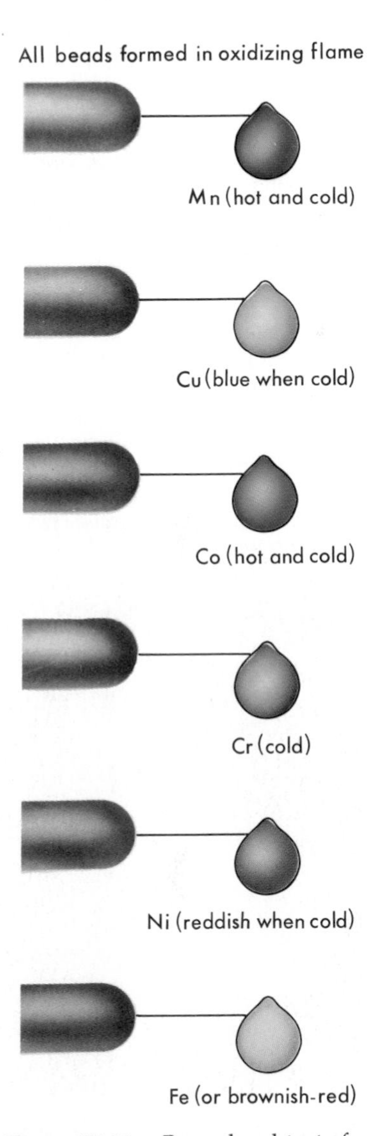

Figure 27-11. Borax bead tests for certain metals. Color is due to a minute trace of a certain metal.

27.12 Silicon as a metalloid Silicon atoms have four valence electrons. Silicon crystallizes with a tetrahedral bond arrangement similar to that of carbon atoms in diamond. Atoms of silicon also have small atomic radii and tightly held electrons. Thus their ionization energy and electronegativity are fairly high. Silicon is a metalloid. It forms bonds with other elements that are essentially covalent. The electric conductivity of silicon is similar to that of boron; it, too, is a semiconductor. Unlike carbon, silicon forms only single bonds. It forms silicon-oxygen bonds more readily than silicon-silicon or silicon-hydrogen bonds. However, much of its chemistry is similar to that of carbon. Silicon has much the same role in mineral chemistry as carbon has in organic chemistry.

Boron and silicon atoms have roughly similar small radii. This similarity allows them to be substituted for one another in glass, even though boron is a member of Group III and silicon of Group IV.

Because the relative size of atoms frequently defines properties, elements in diagonal positions in the periodic table (such as boron-silicon-arsenic) often have similar properties. In general, atom size decreases to the right from Groups I to VIII. It increases down each group. This balancing effect results in atoms of similar size along diagonals in the periodic table.

Figure 27-12. (Left photo) Silicon, the second most abundant element. Its abundance is due to the wide distribution in nature of silicon dioxide (quartz rocks) and many different silicates. It is often used as a semiconductor in integrated circuits (right photo).

27.13 Silicones Silicon resembles carbon in the ability of its atoms to form chains. Unlike carbon chains, which are formed

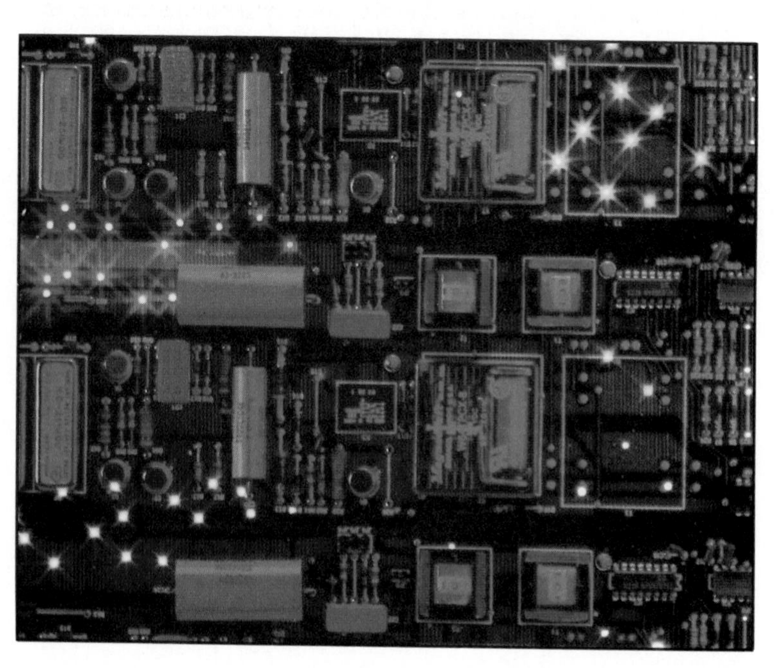

by carbon-carbon bonds, silicon chains are usually formed by silicon-oxygen bonds. That is, silicon atoms are bound into chains by other atoms such as oxygen. A group of synthetic compounds called *silicones* has the silicon chain bound together with oxygen atoms. Hydrocarbon groups are attached to the silicon atoms. Thus the silicones are part organic and part inorganic. By using different hydrocarbon groups, a variety of silicones can be produced. One silicone chain has the structure

$$\left[\begin{array}{ccccc} & H & & H & \\ & | & & | & \\ H- & C & -H \quad H- & C & -H \\ & | & & | & \\ - & Si & -O- & Si & -O- \\ & | & & | & \\ H- & C & -H \quad H- & C & -H \\ & | & & | & \\ & H & & H & \end{array}\right]_x$$

The silicones are not greatly affected by heat. They have very good electric insulating properties and are water repellent. Water does not penetrate cloth treated with a silicone. Some silicones have the character of oils or greases and can be used as lubricants. Silicone varnishes are used to coat wires for the windings of electric motors. This insulation permits the electric motor to operate at high temperatures without developing short circuits. Silicones are used in automobile and furniture polishes.

27.14 Properties and uses of arsenic Metallic arsenic is a brittle, gray solid. Chemically, arsenic may act as a metal and form oxides and chlorides. It may also act as a nonmetal and form acids. When heated in moist air, arsenic sublimes and forms As_2O_3, which has the odor of garlic. Metallic arsenic has few commercial uses except as a hardening agent in certain alloys. A trace of arsenic is used to improve the hardness and sphericity of lead shots. Arsenic and its compounds are poisonous.

CAUTION: Although elemental arsenic is not as toxic as commonly assumed, its dusts and vapors and most of its compounds are very poisonous.

27.15 Properties and uses of antimony Antimony is a dense, brittle, silver-white metalloid with a bright metallic luster. It is less active than arsenic. When strongly heated in air, antimony forms a white oxide Sb_4O_6.

An alloy of lead and antimony is used to reduce friction between the surfaces of moving parts in machinery. Another alloy of lead and antimony is used for the plates in storage batteries. This alloy is stronger and more resistant to acids than lead alone. Tartar emetic, or potassium antimonyl tartrate, $KSbOC_4H_4O_6$, is used as a mordant in the dyeing of cotton goods.

CAUTION: The soluble compounds of antimony are almost as toxic as those of arsenic.

SUMMARY

Metalloids have properties intermediate between those of metals and nonmetals. They occupy a diagonal region of the periodic table from upper center toward lower right. Metalloids include boron, silicon, germanium, arsenic, antimony, tellurium, and polonium.

Aluminum is distinctly metallic in many of its properties. It is always covered with an oxide surface coating that protects it from corrosion by air and moisture. Aluminum is a self-protective metal. Certain properties tend to place aluminum among the metalloids. It forms negative aluminate ions and some covalent halides, and its hydroxide is amphoteric. Its oxide is ionic, and its hydride is a covalent network compound.

Aluminum is recovered by electrolysis of the oxide dissolved in molten cryolite. It is a good reducing agent. Aluminum is a low-density structural metal when alloyed with selected elements. Powdered aluminum mixed with iron(III) oxide and ignited reduces the oxide to the free metal in a rapid, violent combustion reaction. This is the thermite reaction. The aluminum-oxygen bond is a very strong bond. Some natural or synthetic oxides of aluminum are very hard and are used as abrasives. Alums are double sulfates with the general formula $M^{+}M^{+++}(SO_4)_2 \cdot 12H_2O$. The most common alum is $KAl(SO_4)_2 \cdot 12H_2O$.

Boron has typical metalloidal properties. It is a semiconductor. The element can be recovered from its trichloride by reduction with hydrogen. Boron carbide and boron nitride are extremely hard abrasives. Borax, sodium tetraborate, is the most common boron compound. It is used as a cleaning agent and a welding flux. Boric acid is prepared by reacting borax with aqueous sulfuric acid. When present in an alcohol, boric acid imparts a green color to the alcohol flame. Borax bead tests are used to identify cobalt, chromium, nickel, manganese, copper, and iron.

Silicon forms strong bonds with oxygen. Silicon-oxygen lattice chains are similar to carbon-carbon chains. Much of the chemistry of the element silicon is similar to that of carbon. Silicon chains bound together with oxygen atoms and having hydrocarbon groups attached to the silicon atoms are called silicones. Some silicones have properties similar to organic oils and greases.

Arsenic and antimony are used in alloys. Arsenic is a hardening agent for lead. Antimony and lead form "antifriction" alloys and battery-plate alloys.

VOCABULARY

alumina
bauxite
borax
borax bead test
corundum
cryolite
emery
fiber composite
kernite
metalloid
semiconductor
thermite reaction

QUESTIONS

GROUP A

1. Why is aluminum, a familiar structural metal, included in the study of metalloids? *A2b*
2. In what materials does aluminum occur in nature? *A3b*
3. What are the important physical properties of aluminum? *A2a*
4. *(a)* What is the chemical composition of corundum and emery? *(b)* Write the formulas for these substances. *(c)* What is their important use? *A1a, A3b*
5. Why must aluminum oxide be dissolved in molten cryolite before it can be decomposed by electricity? *A2a*
6. Why are certain metallic oxides reduced with aluminum rather than with carbon? *A2g*
7. Write the chemical formulas for four different alums. *A6a*

8. What is the main source of boron compounds in the United States? *A3e*
9. Why can borax be used to prepare metals for welding? *A4f*
10. How does silicon rank in abundance among the elements?
11. Why must extreme care be used in handling arsenic and its compounds? *A2a*

GROUP B

12. What reaction occurs when aluminum is placed in: *(a)* hydrochloric acid solution; *(b)* sodium hydroxide solution? *(c)* Why is there no reaction in neutral water? *A4c*
13. Write equations to show the net positive and negative electrode reactions during the electrolysis of aluminum oxide. *A4a*
14. Why is bauxite imported from the West Indies and South America when almost any clay bank in the United States contains aluminum? *A3e*
15. What geographic conditions affect the location of plants for the production of aluminum from purified bauxite? *A2e*
16. Describe the properties characteristic of metalloids such as boron and silicon in terms of *(a)* atomic radius; *(b)* ionization energy; *(c)* electronegativity; *(d)* types of bonds formed; *(e)* electric conductivity. *A2a*
17. Explain the test for boric acid. *A3f*
18. When a red crystalline compound was tested by means of a borax bead, the bead turned blue. What metal was probably present? *A7*
19. Borazon, a form of boron nitride, and diamond have similar crystal structures. Is there any relationship between this fact and the similarity of their hardness? Explain. *A3b*
20. *(a)* What is the principal form of boric acid in water solution: molecular or ionic? *(b)* What evidence supports your answer? *A3f*
21. *(a)* What is a silicone? *(b)* What are some of the important uses for silicones? *A2j*
22. Why would you expect silicones to be water repellent? *A3b*

PROBLEMS

GROUP A

1. How much aluminum and how much iron(III) oxide must be used in a thermite mixture to produce 10.0 kg of iron for a welding job?
2. What is the percentage of aluminum in sodium alum that crystallizes with 12 molecules of water of hydration?
3. How many liters of hydrogen can be prepared by the reaction of 50.0 g of aluminum metal and $10\bar{0}$ g of sodium hydroxide in solution?
4. Calculate the percentage of boron in colemanite, $Ca_2B_6O_{11} \cdot 5H_2O$.
5. How many grams of $SbCl_3$ can be prepared by the reaction of 10.0 g of antimony with chlorine?

GROUP B

6. A compound contains 96.2% arsenic and 3.85% hydrogen. Its vapor is found to have a density of 3.48 g/L. What is the molecular formula of the compound?
7. Boric acid, H_3BO_3, is produced when sulfuric acid is added to a water solution of borax, $Na_2B_4O_7$. How much boric acid can be prepared from 5.00 g of borax?
8. Tellurium (at. no. 52) appears before iodine (at. no. 53) in the periodic table, yet the atomic weight of tellurium is higher than that of iodine. Using the following data, show why this is so.

Isotope	*Atomic mass*	*Distribution*
Te-120	119.904 *u*	0.09%
Te-122	121.903 *u*	2.46%
Te-123	122.904 *u*	0.87%
Te-124	123.903 *u*	4.61%
Te-125	124.904 *u*	6.99%
Te-126	125.903 *u*	18.71%
Te-128	127.905 *u*	31.79%
Te-130	129.906 *u*	34.48%
I-127	126.904 *u*	100.00%

Fritz Haber (German) received the Nobel Prize in chemistry in 1918 for synthesizing ammonia from its elements, nitrogen and hydrogen, a reaction that is normally reversible. Despite expert opinions to the contrary, Haber planned to carry out the reaction under high pressure and high temperature in order to shift the equilibrium to favor ammonia formation. When he used uranium or osmium as a catalyst, he was able to synthesize ammonia at 500–600 °C and 200 atm. The success of Haber's process led to the development of other high-pressure reactions.

Nitrogen and Its Compounds

28

GOALS

In this chapter you will gain an understanding of: • the occurrence and preparation of nitrogen • the properties and uses of nitrogen • natural and artificial methods of nitrogen fixation • the preparation, properties, and uses of ammonia and nitric acid

NITROGEN

28.1 Occurrence of nitrogen About four-fifths of the volume of the earth's atmosphere is elemental nitrogen. Its presence on the earth is directly related to the action of bacteria during the nitrogen cycle. Although Earth is the only planet of the solar system where nitrogen is a significant component of the atmosphere, Saturn's moon Titan is believed to have a dense atmosphere containing 82–94% nitrogen.

Combined nitrogen is widely distributed on the earth. It is found in the proteins of both plants and animals. Natural deposits of potassium nitrate and sodium nitrate are used as raw materials for producing other nitrogen compounds.

28.2 Preparation of nitrogen The commercial method for preparing nitrogen is fractional distillation of liquid air. Small amounts of liquid air were first produced in France in 1877. Today it is made in large amounts as a first step in separating the various gases of the atmosphere. The physical principles involved in liquefying gases were discussed in Section 12.7.

Liquefying a gas involves compressing the gas and removing the heat of compression. Then the cool, compressed gas is allowed to expand without absorbing external energy.

Liquid air resembles water in appearance. Under ordinary atmospheric pressure, liquid air boils at a temperature of about −190 °C. Because liquid air is mainly a mixture of liquid nitrogen and liquid oxygen, its boiling temperature is not constant. Liquid nitrogen boils at −195.8 °C, while liquid oxygen boils at −183.0 °C. The lower boiling point of the nitrogen causes it to separate in the first portions that evaporate. Fractional distillation of liquid air is also used for producing oxygen and the noble gases (except helium).

Nitrogen ranks second behind sulfuric acid among chemicals in the quantity produced in the United States. Production in 1983 was well over 19 million metric tons.

28.3 Physical properties of nitrogen Nitrogen is a colorless, odorless, and tasteless gas. It is slightly less dense than air and only slightly soluble in water. Its density indicates that its molecules are diatomic, N_2. Nitrogen condenses to a colorless liquid at −195.8 °C and freezes to a white solid at −209.9 °C.

28.4 Chemical properties of nitrogen Nitrogen atoms have the electron configuration $1s^22s^22p^3$. Nitrogen atoms share $2p$ electrons to form nonpolar diatomic molecules having a triple covalent bond.

$$:N:::N:$$

This triple covalent bond is very strong. Even at 3000 °C, nitrogen molecules do not decompose measurably. The nitrogen molecule bond energy is very large, 225 kcal/mole. This explains why elemental nitrogen is rather inactive. It combines with other elements only with difficulty. Many nitrogen compounds have positive heats of formation.

The electron-dot symbol for a nitride ion is $:\ddot{\underset{..}{N}}:^{---}$.

At a high temperature, nitrogen combines directly with such metals as magnesium, titanium, and aluminum. Nitrides are formed in these reactions.

Nitrogen does not burn in oxygen. However, when nitrogen and oxygen are passed through an electric arc, nitrogen monoxide, NO, is formed. Similarly, NO is formed when lightning passes through the air.

Oxides of nitrogen are also formed during the high-temperature and high-pressure combustion in automobile engines. When catalyzed by the sun's ultraviolet radiation, these oxides of nitrogen can react with unburned hydrocarbons from automobiles. In urban areas with poor natural air circulation, the result is a buildup of an irritating cloud of photochemical smog. Ozone and a variety of organic compounds are produced in such a smog.

By the use of a catalyst, nitrogen can be made to combine with hydrogen at a practical rate. Ammonia, NH_3, is formed in this reaction.

28.5 Uses of elemental nitrogen An important use of pure nitrogen is in the commercial preparation of ammonia. Other uses of pure nitrogen are based on its inactivity.

Substances burn rapidly in pure oxygen, but nitrogen does not support combustion. In the air, therefore, nitrogen serves as a diluting agent and lowers the rate of combustion. Its inactivity makes pure nitrogen useful in food processing, metallurgical operations, chemical production, petroleum recovery, and the manufacture of electronic devices. As a "blanket" atmosphere, it prevents unwanted oxidation in these processes. It is similarly used in the chemical, petroleum, and paint industries to prevent

The use of nitrogen in petroleum recovery is similar to the use of carbon dioxide. See Section 17.19(3).

fires or explosions. Electric lamps are filled with a mixture of nitrogen and argon.

Liquid nitrogen is used for freezing and cooling foods to about −100 °C for preservation during storage and transportation. It is also used for the freezing and storage of whole blood for transfusions and in the preservation of bone marrow cells and other body parts. Liquid nitrogen is used in cryosurgery, the removal of cell tissue by freezing.

Figure 28-1. The nodules on these soybean roots are the result of infection by nitrogen-fixing bacteria.

28.6 Nitrogen fixation All living things contain nitrogen compounds. The nitrogen in these compounds is called *combined* or *fixed* nitrogen. *Any process that converts elemental nitrogen to nitrogen compounds is called* ***nitrogen fixation.*** Such processes are important because nitrogen compounds in the soil are necessary for plant growth. There are both *natural* and *artificial* methods of nitrogen fixation.

1. *One natural method* is to grow crops that restore nitrogen compounds to the soil. Most crops rapidly remove such compounds from the soil. On the other hand, certain plants called *legumes* actually restore large amounts of nitrogen compounds. These plants include beans and peas. They have small swellings or *nodules* on their roots. Organisms known as ***nitrogen-fixing bacteria*** grow in these nodules. See Figure 28-1.

In alkaline soil, the nitrogen-fixing bacteria take free nitrogen from the air and convert it to nitrogen compounds. If these plants are plowed under or if their roots are left to decay, the soil is enriched by nitrogen compounds.

Certain blue-green algae also have nitrogen-fixing ability.

2. Another *natural method* of nitrogen fixation occurs during electric storms. Lightning supplies energy which enables some of the nitrogen and oxygen of the air to combine. An oxide of nitrogen is formed. After a series of changes, nitrogen compounds are washed down into the soil in rain.

3. The most important *artificial method* of nitrogen fixation involves making ammonia from a mixture of nitrogen and hydrogen. This ammonia is then oxidized to nitric acid. The nitric acid, in turn, is converted to nitrates suitable for fertilizer.

AMMONIA AND AMMONIUM COMPOUNDS

28.7 Preparation of ammonia 1. *By decomposing ammonium compounds.* In the laboratory, ammonia is prepared by heating a mixture of a moderately strong hydroxide and an ammonium compound. Usually, calcium hydroxide and ammonium chloride are used.

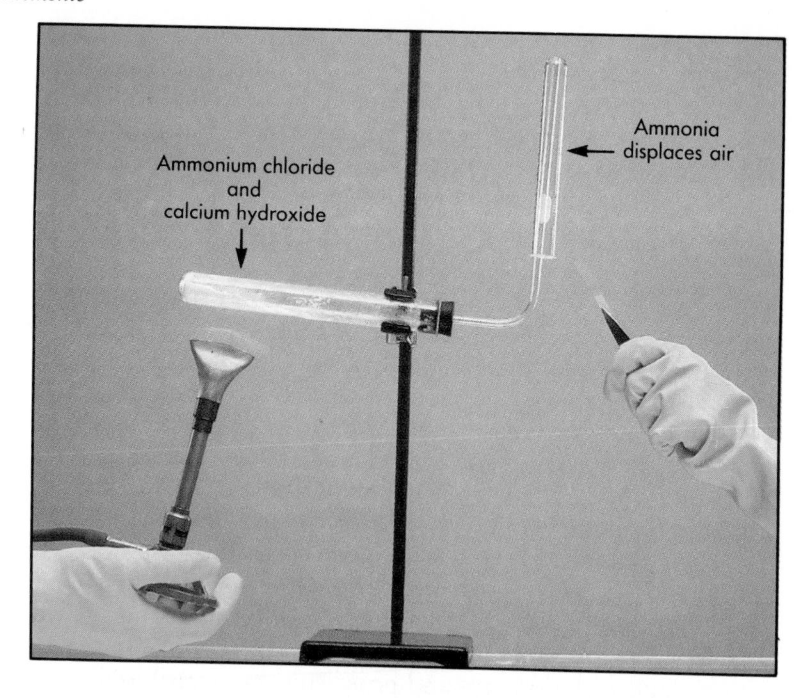

Figure 28-2. When the mixture of ammonium chloride and calcium hydroxide in the test tube is heated, ammonia is given off. Ammonia is collected by downward displacement of air because it is less dense than air and very soluble in water.

The density of ammonia is 0.77 g/L at STP; the density of air is 1.29 g/L at STP. Gases that are soluble in water and less dense than air can be collected by downward displacement of air.

$$Ca(OH)_2(s) + 2NH_4Cl(s) \rightarrow CaCl_2(s) + 2NH_3(g) + 2H_2O(g)$$

The mixture is heated in a test tube fitted with an L-shaped delivery tube, as shown in Figure 28-2. Ammonia is so soluble in water that it cannot be collected by water displacement. Instead, it is collected by downward displacement of air in an inverted container. In what way would moist red litmus paper show when the test tube is full?

2. *By destructive distillation of bituminous coal.* When bituminous coal is heated in a closed container without air, ammonia is one of the gaseous products.

The Haber process is described in Section 21.4.

The nitrogen used in the Haber process comes from the air; the hydrogen comes from the methane in natural gas. See Section 9.17(4).

3. *By the Haber process.* Chemists long ago discovered that ammonia is formed when an electric spark passes through a mixture of nitrogen and hydrogen. But, the reaction is reversible:

$$N_2(g) + 3H_2(g) \rightleftarrows 2NH_3(g) + 22 \text{ kcal}$$

Only a very small percentage of ammonia is produced at equilibrium. The problem of increasing that percentage was solved in 1913 by Fritz Haber (1868–1934), a German chemist.

The reaction between nitrogen and hydrogen is exothermic. Higher temperatures increase the rate at which the molecules of nitrogen and hydrogen react. But, *higher temperatures shift the equilibrium toward the left.* On the other hand, four volumes of reactants produce only two volumes of products. Thus, *increased pressure shifts the equilibrium toward the right.* Haber used a catalyst to increase the speed of reaction. He used a temperature of about 600 °C and a pressure of about 200 atmospheres. These conditions resulted in a yield of about 8% ammonia.

Today, the yield from the Haber process has been raised to about 40% to 60%. This was done by using pressures as high as 1000 atmospheres and an improved catalyst. This catalyst is a mixture of porous iron and the oxides of potassium and aluminum. Its use enables the reaction to proceed at a satisfactory rate at the lower temperature of 400 °–550 °C. The pressures and temperatures used in different Haber-process plants vary widely.

The ammonia is produced in special chrome-vanadium steel converters designed to withstand the tremendous pressure. It is separated from the unreacted nitrogen and hydrogen by being dissolved in water or by being cooled until it liquefies. The uncombined gases are returned to the converters and exposed again to the action of the catalyst.

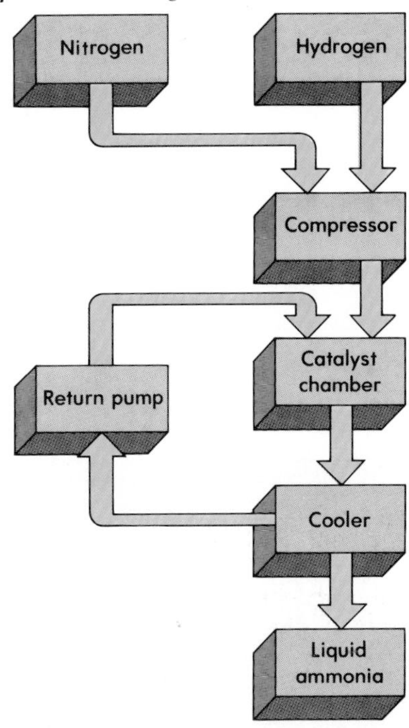

Figure 28-3. A flow diagram of the Haber process. Ammonia gas produced in the catalyst chamber is condensed into a liquid in the cooler. The uncombined nitrogen and hydrogen are recirculated through the catalyst chamber.

28.8 Physical properties of ammonia Ammonia is a colorless gas with a characteristic, strong odor. It is less dense than air and is easily liquefied when cooled to the proper temperature. Liquid ammonia, which boils at −33 °C at atmospheric pressure, is sold in steel cylinders. The melting point of ammonia is −78 °C.

An important property of ammonia is its great solubility in water. One liter of water at 20 °C dissolves about 700 liters of ammonia. At 0 °C about 1100 volumes of ammonia can be dissolved in one volume of water.

The electron-dot formula for ammonia is

$$\begin{array}{c} \mathrm{H} : \overset{\circ\circ}{\mathrm{N}} : \mathrm{H} \\ \mathrm{H} \end{array}$$

The hydrogen atoms bond covalently to the nitrogen atom. In this bonding, the 1*s* hydrogen electrons occupy the half-filled 2*p* nitrogen orbitals. The hydrogen nuclei in ammonia molecules form an equilateral triangle 1.6 Å on a side. The nitrogen nucleus is 0.38 Å vertically above the midpoint of this triangle. This pyramid structure is evidence of sp^3 hybridization in the molecule. See Figure 6-12 on page 134.

For a compound with such a simple molecular structure and low molecular weight, ammonia has a high boiling point and a very high melting point. In Chapter 12, you learned that water shows these properties to an even greater degree. The high melting and boiling points of both ammonia and water are evidence of hydrogen bonding between their molecules in the solid and liquid phases. Ammonia molecules are polar because the three hydrogen atoms are not symmetrically bonded to the nitrogen atom. Hydrogen atoms from one ammonia molecule form hydrogen bonds with the nitrogen atom in adjacent ammonia molecules. The high solubility of ammonia in water is a further indication of the polar nature of water and ammonia molecules.

Ammonia's high heat of vaporization is also evidence of its hydrogen bonding. Much energy is required to break hydrogen

Figure 28-4. A plant in which ammonia is made from atmospheric nitrogen and natural gas.

bonds between molecules and to separate the molecules when ammonia boils.

Liquid ammonia shows many of the solvent properties of water. Many salts dissolve and dissociate in liquid ammonia.

The reaction of ammonia and water is described in Chapter 15. The structure of the ammonium ion is given in text Section 6.23. Formation of complex ions containing ammonia is explained in text Section 26.5.

28.9 Chemical properties of ammonia Gaseous ammonia does not support combustion. It does burn in air or oxygen.

$$4NH_3 + 3O_2 \rightarrow 2N_2 + 6H_2O$$

At ordinary temperatures ammonia is a stable compound. However, it decomposes into nitrogen and hydrogen at high temperatures. When ammonia is dissolved in water, most of the ammonia forms a simple solution. A very small part of the ammonia, however, reacts with water and ionizes.

Figure 28-5. When used as a fertilizer, ammonia may be applied directly to the soil.

28.10 Uses of ammonia and ammonium compounds 1. *As fertilizers.* Ammonia can be applied directly as a fertilizer. It is also used to make large quantities of urea, $(NH_2)_2CO$, and diammonium hydrogen phosphate, $(NH_4)_2HPO_4$, which are widely used fertilizers.

2. *As a cleaning agent.* Ammonia-water solution makes a good cleaning agent. It is weakly basic, emulsifies grease, and leaves no residue to be wiped up.

3. *As a refrigerant.* Ammonia is used as a refrigerant in frozen food production and storage plants.

4. *For making other compounds.* Great quantities of ammonia are oxidized to make nitric acid, as explained in Section 28.11. Ammonia is also used in producing nylon and one type of rayon, and as a catalyst in making several types of plastic. Ammonia is used to make sulfa drugs, vitamins, and drugs for treating the tropical disease malaria. It is also used as a neutralizing agent in the petroleum industry. In the rubber industry, it prevents the rubber in latex from separating out during shipment.

Nitric Acid

28.11 Preparation of nitric acid Two methods are commonly used to prepare this important acid.

1. *From nitrates.* Small amounts of nitric acid are prepared in the laboratory by heating a nitrate with concentrated sulfuric acid. The reaction is carried out in a glass-stoppered retort because nitric acid oxidizes rubber stoppers and also rubber connectors.

$$NaNO_3(s) + H_2SO_4(aq) \rightarrow NaHSO_4(aq) + HNO_3(g)$$

The nitric acid vapor is condensed in the side arm of the retort

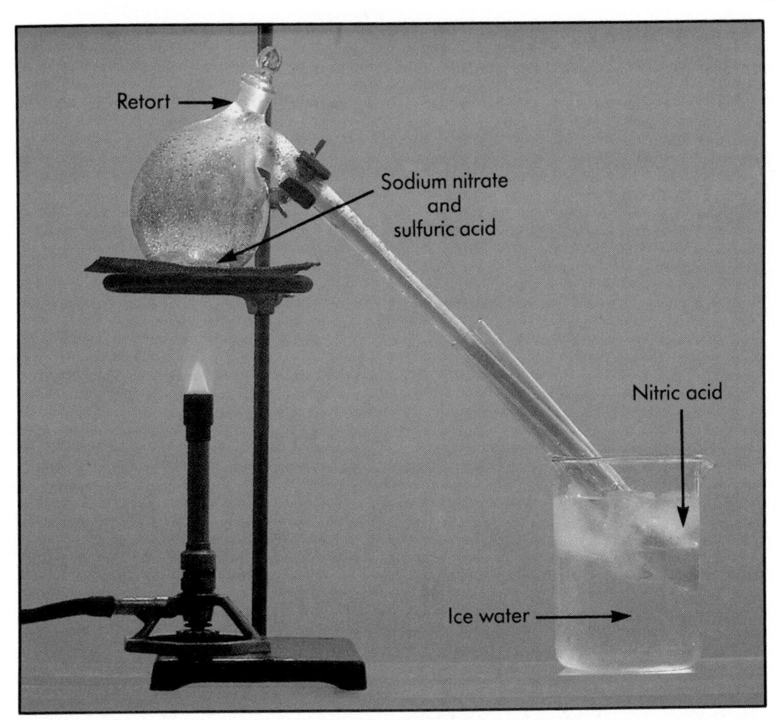

Figure 28-6. Nitric acid may be prepared in the laboratory by the action of sulfuric acid on sodium nitrate. Some reddish-brown nitrogen dioxide is also formed because of the decomposition of some nitric acid. It fills the retort and colors the nitric acid product. Pure nitric acid is colorless.

and collected in the receiver. The reddish-brown gas nitrogen dioxide, NO_2, is produced because some of the HNO_3 formed does decompose. Nitrogen dioxide appears in the retort and contaminates the nitric acid.

2. *From ammonia.* Wilhelm Ostwald (1853–1932), a German chemist, discovered how to oxidize ammonia to nitric acid with a catalyst. In the Ostwald process, a mixture of ammonia and air is heated to a temperature of 600 °C. It is then passed through a tube containing platinum gauze, which serves as the catalyst. On the surface of the platinum, the ammonia is oxidized to colorless nitrogen monoxide, NO.

$$4NH_3 + 5O_2 \rightarrow 4NO + 6H_2O$$

This reaction is exothermic and raises the temperature of the mixture of gases to about 900 °C. Then more air is mixed with colorless nitrogen monoxide to oxidize it to reddish-brown nitrogen dioxide, NO_2.

$$2NO + O_2 \rightarrow 2NO_2$$

The nitrogen dioxide is cooled and absorbed in water, forming nitric acid.

$$3NO_2(g) + H_2O(l) \rightarrow 2HNO_3(aq) + NO(g)$$

The nitrogen monoxide produced is recycled to oxidize it to nitrogen dioxide and absorb it in water.

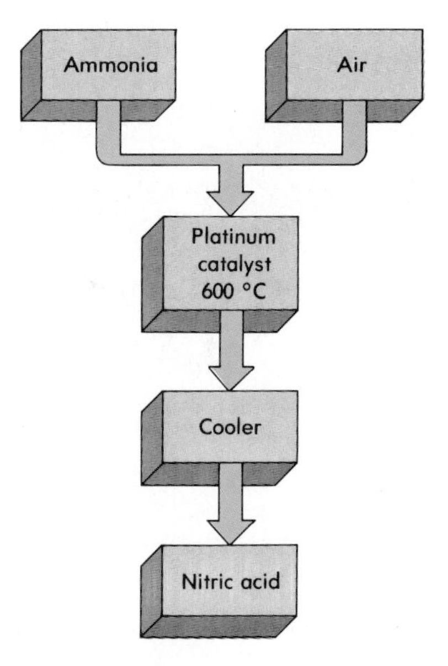

Figure 28-7. Flow diagram of the Ostwald process for the oxidation of ammonia into nitric acid.

Figure 28-8. The reactions of concentrated nitric acid with copper and zinc. What is the reddish-brown gas? What salt is in solution in each beaker? The gaseous products are being drawn off by an exhaust system not shown in the photograph. These reactions should be carried out under a hood or where there is similar provision for removing the product gases because they are poisonous.

Figure 28-9. Nitric acid is used in the manufacturing processes of dyes.

28.12 Physical properties of nitric acid Pure HNO_3 is a colorless liquid, about 1.5 times as dense as water. It fumes in moist air and boils at 83 °C. Pure HNO_3 is unstable. For this reason, commercial, concentrated nitric acid is a 70% solution of HNO_3 in water that has a density of 1.42 g/L. The solution boils at approximately 120 °C.

28.13 Chemical properties of nitric acid 1. *Stability.* Concentrated nitric acid is not very stable. When boiled, or even when exposed to sunlight, it decomposes to some extent. Nitrogen dioxide, water, and oxygen are the decomposition products.

$$4HNO_3 \rightarrow 4NO_2 + 2H_2O + O_2$$

2. *Acid properties. Dilute* nitric acid has the usual properties of acids. It reacts with metals, metallic oxides, and metallic hydroxides, forming salts known as *nitrates.*

3. *As an oxidizing agent.* Nitric acid is a powerful oxidizing agent. It can react as an oxidizing agent in a variety of ways. The concentration of the acid, the activity of the reducing agent mixed with it, and the reaction temperature determine what products are formed. Under ordinary conditions, moderately dilute nitric acid is reduced to nitrogen monoxide. If concentrated nitric acid is reduced, nitrogen dioxide is the product.

4. *Action with metals.* Nitric acid is such a vigorous oxidizing agent that hydrogen gas is *not* produced in significant amounts when the acid is added to common metals. *Very dilute nitric acid,* however, does react with the active metal magnesium to produce hydrogen and a nitrate.

In reactions with less active metals such as zinc and copper, the hydrogen appears in the water product. In these reactions, the nitrogen of the nitric acid is reduced. Copper reacts with cold, dilute nitric acid as shown by this equation:

$$3Cu(s) + 8HNO_3(aq) \rightarrow 3Cu(NO_3)_2(aq) + 2NO(g) + 4H_2O(l)$$

With concentrated nitric acid, copper reacts as follows:

$$Cu(s) + 4HNO_3(aq) \rightarrow Cu(NO_3)_2(aq) + 2NO_2(g) + 2H_2O(l)$$

Nitric acid does not react with gold or platinum because of the stability of these metals. Very concentrated nitric acid reacts with aluminum and iron extremely slowly. The reactions are probably slowed by the formation of semiprotective surface coatings on the metal.

28.14 Uses of nitric acid 1. *For making fertilizers.* About 65% of the nitric acid produced in the United States is used in the manufacture of fertilizers. Ammonium nitrate is the most important nitrate so used.

2. *For making explosives.* Many explosives are made directly or

indirectly from nitric acid. The acid itself is not an explosive. However, some of its compounds form the most violent explosives known. Among these are nitroglycerin, smokeless powders, and TNT (trinitrotoluene).

3. *For making dyes.* Nitric acid reacts with several products obtained from coal tar to form *nitro compounds.* One of these coal tar products is benzene. Benzene reacts with nitric acid to form nitrobenzene, $C_6H_5NO_2$. (See Section 18.21.) Aniline, $C_6H_5NH_2$, is used in making many different dyes. It is made by reducing nitrobenzene with hydrogen.

4. *For making plastics.* Cotton consists mainly of cellulose, $(C_6H_{10}O_5)_n$. When treated with a mixture of nitric acid and sulfuric acid, cellulose forms nitrocellulose plastics. Celluloid, pyroxylins, and many other products are made from nitrocellulose plastics.

SUMMARY

About four-fifths by volume of the earth's atmosphere is free nitrogen. Nitrogen is obtained commercially from the air by fractional distillation. Nitrogen is colorless, odorless, tasteless, slightly less dense than air, and only slightly soluble in water. Elemental nitrogen is rather inactive. Because of this inactivity, nitrogen is used as a blanketing atmosphere to prevent unwanted oxidation. Nitrogen does not readily combine with other elements. The compounds it does form are usually unstable. Nitrogen does not burn, but nitrogen and oxygen do combine to form several oxides of nitrogen. Nitrogen and hydrogen combine to form ammonia under proper conditions. The conversion of free nitrogen to nitrogen compounds is called nitrogen fixation.

In the laboratory, ammonia is prepared by heating calcium hydroxide and an ammonium compound. Using the Haber process, ammonia is made industrially from nitrogen of the air, and hydrogen prepared from natural gas. Ammonia is a colorless gas with a characteristic, strong odor. It is less dense than air, easily liquefied, and very soluble in water. Ammonium ions act like metallic ions; they combine with negative ions to form salts. Ammonia and ammonium compounds are used as fertilizers. Ammonia-water solution is a good cleaning agent. Ammonia is also used as a refrigerant and for making medicinal drugs such as sulfa drugs and certain vitamins.

Nitric acid is prepared in the laboratory by heating a mixture of sodium nitrate and concentrated sulfuric acid. Commercially, nitric acid is prepared by the catalytic oxidation of ammonia, using the Ostwald process. Nitric acid is a dense, colorless liquid that fumes in moist air. Pure nitric acid is not very stable. Concentrated nitric acid is a powerful oxidizing agent. Nitric acid reacts with metals, metallic oxides, and metallic hydroxides. Its salts are called nitrates. Nitric acid is used for making fertilizers, explosives, dyes, and plastics.

VOCABULARY

ammonia
blanket atmosphere
fixed nitrogen
Haber process
legume
nitrate
nitride
nitrogen fixation
nitrogen-fixing bacteria
nodule
Ostwald process
photochemical smog

QUESTIONS

Group A

1. *(a)* Where on Earth do large quantities of elemental nitrogen occur? *(b)* Where else in the solar system are large quantities of elemental nitrogen found? *A3a*
2. In what kinds of compounds does combined nitrogen occur naturally on Earth? *A3a*
3. What is the boiling point *(a)* of liquid nitrogen; *(b)* of liquid oxygen? *(c)* Which boiling point is lower? *(d)* Of what practical advantage is this difference in boiling points in the fractional distillation of liquid air? *A2a*
4. List five important physical properties of nitrogen. *A2b*
5. Several oxides of nitrogen are found in automobile engine exhaust. What is the effect of these oxides on the atmosphere? *A2d*
6. Define *nitrogen fixation.* *A1*
7. There are various methods of nitrogen fixation. Briefly describe *(a)* two natural methods; *(b)* one artificial method. *A2e*
8. In the Haber process *(a)* why is high pressure used; *(b)* why is a catalyst used? *A2f*
9. *(a)* What is the shape of an ammonia molecule? *(b)* What type of bonding holds the atoms together in an ammonia molecule? *A2c*
10. Why are ammonia and water two of the ingredients in solutions used as glass cleaners? *A2b, A3c*
11. Ammonia is used in the manufacture of substances other than nitric acid and fertilizer. List five of these substances. *A3c*
12. Why is a glass-stoppered retort used for the laboratory preparation of nitric acid rather than a rubber-stoppered flask and rubber delivery tube? *A2a, A2b*
13. What properties of the substances involved enable the reaction between sodium nitrate and sulfuric acid to run to completion? *A2a, A2b*
14. There are three chemical reactions that occur in sequence when nitric acid is produced from ammonia. Write balanced formula equations for these reactions. *A4d*
15. What is the impurity frequently found in concentrated nitric acid that makes it appear yellow? *A2b, A3e*
16. In addition to the production of fertilizers, name three broad areas of chemical manufacturing in which nitric acid is used. *A3c*

Group B

17. *(a)* What effect does compressing a gas have on its temperature? *(b)* Why? *A2a*
18. *(a)* What effect does expansion have on the temperature of a gas? *(b)* Why? *A2a*
19. What structural feature of its molecules accounts for nitrogen's inactivity? *A2b, A2c*
20. Give three significantly different uses for nitrogen. For each use describe the physical or chemical property of nitrogen that makes each possible. *A2b, A2c*
21. Under what soil condition are nitrogen-fixing bacteria most effective? *A2e*
22. Why do midwestern growers alternate crops of corn and soybeans on their fields in successive years? *A2e*
23. *(a)* Write the balanced equation for the reaction between calcium hydroxide and ammonium chloride that produces ammonia. *(b)* Write the net ionic equation for this reaction. *(c)* How does the net ionic equation for this reaction compare with the net ionic equation for the reaction between potassium hydroxide and ammonium phosphate? Between sodium hydroxide and ammonium sulfate? *A2a, A4a*
24. *(a)* What is the boiling point of ammonia? *(b)* From Table 18-2, what is the boiling point of methane? *(c)* Explain the difference in these boiling points even though ammonia molecules and methane molecules have nearly the same molecular weight. *A2b*
25. Account for the difference in solubility in water of ammonia and methane. Consult Appendix Table 11 for the necessary data. *A2b*
26. What solvent and complexing properties does ammonia have? *A2b*

27. Zinc can be used with dilute hydrochloric acid or dilute sulfuric acid in the laboratory preparation of hydrogen, Section 9.17. Why can zinc not be used with dilute nitric acid? *A2b*
28. The equation for the reaction of copper and dilute nitric acid indicates that colorless nitrogen monoxide gas is one of the products. Yet when this reaction is carried out in an evaporating dish, dense reddish-brown nitrogen dioxide gas flows over the rim of the dish. Explain. *A3e*
29. What is the oxidation number of nitrogen in *(a)* NH_3, *(b)* N_2H_4, *(c)* N_2, *(d)* NO, *(e)* HNO_2, *(f)* NO_2, *(g)* HNO_3?
30. Balance the following oxidation-reduction equations:
    (a) $Cu(s) + HNO_3(aq) \rightarrow Cu(NO_3)_2(aq) + NO(g) + H_2O(l)$
    (b) $Mg(s) + HNO_3(aq) \rightarrow Mg(NO_3)_2(aq) + N_2(g) + H_2O(l)$
    (c) $Zn(s) + HNO_3(aq) \rightarrow Zn(NO_3)_2(aq) + NH_4NO_3(aq) + H_2O(l)$
    (d) $CuS(s) + HNO_3(aq) \rightarrow Cu(NO_3)_2(aq) + NO_2(g) + S(s) + H_2O(l)$
    (e) $HCl(aq) + HNO_3(aq) \rightarrow NOCl(g) + Cl_2(g) + H_2O(l)$

PROBLEMS

Group A

1. A quantity, $15\bar{0}$ L, of ammonia is to be prepared. How many liters of nitrogen and hydrogen are required? *A4b*
2. From the density of nitrogen (Appendix Table 9) and its atomic weight (Appendix Table 5), show that nitrogen molecules are diatomic.
3. How many grams of ammonia can be produced by the reaction of 25.0 g of ammonium chloride and excess calcium hydroxide? *A4a*
4. Calculate the percentage of nitrogen in *(a)* ammonia, NH_3; *(b)* urea, $(NH_2)_2CO$; and *(c)* diammonium hydrogen phosphate, $(NH_4)_2HPO_4$.
5. What mass, in grams, of sodium nitrate is required for the preparation of 12 g of pure HNO_3 that is needed for a laboratory experiment? *A4a*

Group B

6. What volume of nitrogen, in liters at STP, can be prepared from a mixture of $1\bar{0}$ g of NH_4Cl and $1\bar{0}$ g of $NaNO_2$ in solution? The equation for the reaction is: $NH_4Cl(aq) + NaNO_2(aq) \rightarrow N_2(g) + NaCl(aq) + 2H_2O(l)$.
7. How many grams of HNO_3 can be prepared from 50.0 g of KNO_3 that is 95.0% pure? *A4a*
8. *(a)* What mass of copper(II) nitrate can be prepared from 135 g of copper by reaction with nitric acid? *(b)* How many liters of nitrogen monoxide at STP are also produced? *A4f*
9. Ammonia burns in oxygen, forming nitrogen and water. The equation for the combustion of one mole of ammonia is: $NH_3(g) + \frac{3}{4}O_2(g) \rightarrow \frac{1}{2}N_2(g) + \frac{3}{2}H_2O(l)$. *(a)* Calculate the heat of reaction. *(b)* Is the reaction endothermic or exothermic? Obtain data from Appendix Table 14.
10. In the Ostwald process, ammonia mixed with air is oxidized to nitrogen monoxide and water vapor. *(a)* Write the balanced equation for this oxidation of one mole of ammonia. *(b)* Calculate the heat of reaction. Obtain data from Appendix Table 14. *A4d*

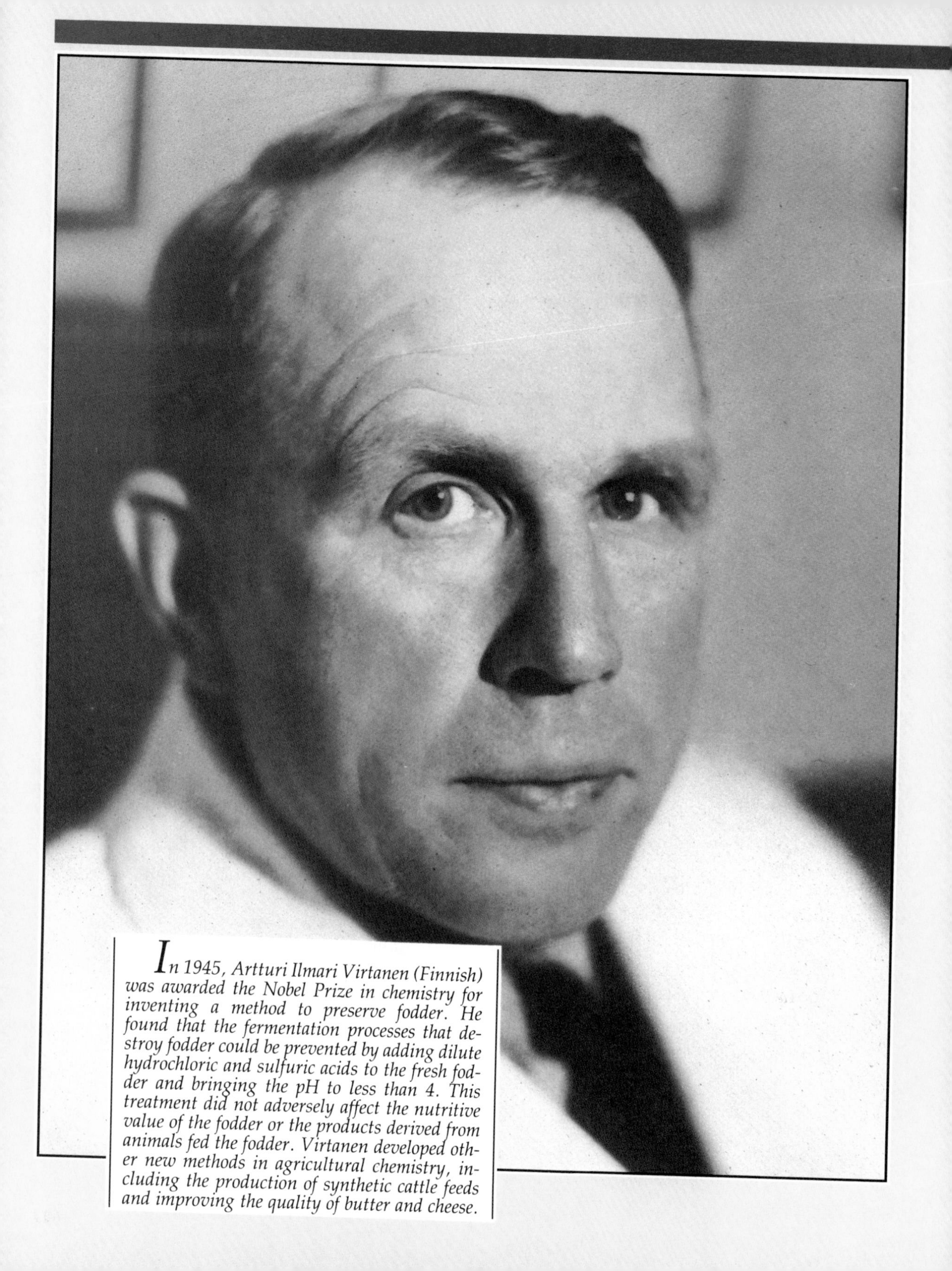

In 1945, Artturi Ilmari Virtanen (Finnish) was awarded the Nobel Prize in chemistry for inventing a method to preserve fodder. He found that the fermentation processes that destroy fodder could be prevented by adding dilute hydrochloric and sulfuric acids to the fresh fodder and bringing the pH to less than 4. This treatment did not adversely affect the nutritive value of the fodder or the products derived from animals fed the fodder. Virtanen developed other new methods in agricultural chemistry, including the production of synthetic cattle feeds and improving the quality of butter and cheese.

Sulfur and Its Compounds

GOALS

In this chapter you will gain an understanding of: • the occurrence and production of sulfur • the allotropic forms of sulfur • the physical and chemical properties of sulfur • the uses of sulfur • the important compounds of sulfur: sulfur dioxide and sulfuric acid • acid rain

SULFUR

29.1 Occurrence of sulfur Sulfur is one of the elements known since ancient times. It occurs in nature as the free element or combined with other elements in sulfides and sulfates. In the United States, huge deposits of nearly pure sulfur occur between 150 and 600 meters underground in Texas and Louisiana, near the Gulf of Mexico.

The Voyager mission has discovered at least nine active volcanoes on Jupiter's moon Io. The gas vented from these volcanoes is believed to contain sulfur. The colors of Io's surface can be explained by the colors of various forms of sulfur.

29.2 The production of sulfur The sulfur beds in Texas and Louisiana are as much as 60 meters thick. Between the surface of the ground and the sulfur, there is usually a layer of quicksand. This makes it difficult to sink a shaft and mine the sulfur by ordinary methods.

The American chemist Herman Frasch (1851–1914) developed a method of obtaining the sulfur without sinking a shaft. This method uses a complex system of pipes. Superheated water (170 °C, under pressure) is pumped into the sulfur deposit. The hot water melts the sulfur, which is then forced to the surface by compressed air. See Figure 29-1.

About half of the elemental sulfur produced in the United States is now obtained from pyrite (FeS_2) and from the purification of coke oven gas, smelter gases, petroleum, or natural gas as a result of air pollution reduction.

In 1983, the total United States production of elemental sulfur was over 9 million metric tons.

29.3 Physical properties of sulfur Common sulfur is a yellow, *odorless* solid. It is practically insoluble in water and is twice

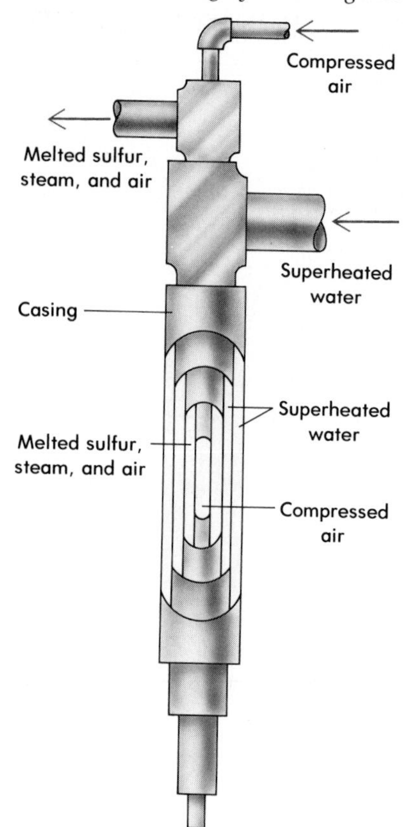

Figure 29-1. The system of concentric pipes used in the Frasch process of mining sulfur.

as dense as water. It dissolves readily in carbon disulfide and less readily in carbon tetrachloride.

Sulfur melts at a temperature of 112.8 °C, forming a pale yellow liquid that flows easily. When heated to a higher temperature, it becomes more *viscous* (thicker), instead of flowing more easily as liquids usually do. At a temperature above about 160 °C, melted sulfur becomes so thick that it hardly flows at all. As the temperature rises, the color changes from a light yellow to a reddish-brown, and then almost to black. Near its boiling point the liquid again flows freely. Sulfur boils at 445 °C. This unusual behavior is caused by the properties of different allotropes of liquid sulfur, allotropes formed at different temperatures.

29.4 Allotropes of sulfur Sulfur exists in several different solid and liquid allotropic forms. These forms are produced by different arrangements of groups of sulfur atoms.

Allotropy is explained in Sections 9.11 and 17.3.

1. *Rhombic sulfur.* This form of solid sulfur is stable at ordinary temperatures. It consists of eight-membered, puckered rings of sulfur atoms. The sulfur atoms are connected in these rings by single covalent bonds. See Figure 29-2B. Crystals of rhombic sulfur are prepared by dissolving sulfur in carbon disulfide and then allowing the solvent to evaporate slowly. The density of rhombic sulfur is 2.07 g/cm^3.

2. *Monoclinic sulfur.* Sulfur can also be crystallized in the form of long needlelike monoclinic crystals. This allotropic form is prepared by first melting some sulfur in a crucible at as low a temperature as possible. Next the sulfur is allowed to cool slowly until a crust begins to form. The crust is broken, and the remaining liquid sulfur is poured off. A mass of monoclinic crystals is then found lining the walls of the crucible. Heat energy must be added to produce this type of sulfur. When such crystals cool below 95 °C, they gradually change back into the rhombic form. In monoclinic sulfur, eight-membered rings of sulfur atoms are arranged in a monoclinic crystal pattern. The density of monoclinic sulfur is 1.96 g/cm^3.

3. *λ-sulfur. (Lambda-sulfur.)* This is the liquid allotropic form of sulfur produced at temperatures just above the melting point. It flows easily and has a straw-yellow color. It is believed to consist of eight-membered rings of sulfur atoms. The rounded shape of these S_8 molecules enables them to roll over one another easily, giving this form of sulfur its fluidity.

4. *μ-sulfur. (Mu-sulfur).* If λ-sulfur is heated to about 160 °C, it darkens to a reddish and then almost black liquid. The melted sulfur becomes so viscous that it does not flow. It is thought that the heating gives enough energy to the sulfur atoms to break some of the eight-membered rings. When a ring of sulfur atoms breaks open, the sulfur atoms on either side of the break are each

Figure 29-2. (A) Part of a chain of sulfur atoms as found in μ-sulfur (mu-sulfur). (B) The structure of S_8 molecules of sulfur.

left with an unshared electron. These sulfur atoms form bonds with similar sulfur atoms from other open rings. In this way, long chains of sulfur atoms are formed. These chains are another allotropic form of sulfur, μ-sulfur. The dark color of μ-sulfur arises from the greater absorption of light by electrons from the broken ring structure. In μ-sulfur these electrons are free to migrate along the chain structure.

The high viscosity of μ-sulfur is explained by the tangling of the sulfur-atom chains. As the temperature is raised, however, these chains break up into smaller groups of atoms. The mass then flows more easily. The color becomes still darker because more free electrons, which absorb more light, are produced by the breaking of chains.

Sulfur vapor, produced when sulfur boils at 445 °C, also consists of S_8 molecules. If sulfur vapor is heated to a higher temperature, these molecules gradually dissociate into S_6 and then S_2 molecules. Monatomic molecules of sulfur are produced at very high temperatures.

5. *Amorphous sulfur*. Amorphous sulfur is a rubbery, plastic mass made by pouring boiling sulfur into cold water. It is dark brown or even black in color and is elastic like rubber. At the boiling point of sulfur, the long tangled chains of μ-sulfur have largely broken down, and the sulfur is fluid again. Eight-membered rings of sulfur atoms and chains are in equilibrium. The S_8 rings are evaporating. When this boiling mixture is suddenly cooled, the chains of μ-sulfur have no time to re-form into rings. Instead, amorphous sulfur is produced. A mass of amorphous sulfur soon loses its elasticity and becomes hard and brittle. For amorphous sulfur, at room temperature, the successive changes into allotropic forms occur in reverse order; eventually the amorphous sulfur again becomes the S_8 arrangement of stable rhombic sulfur. Amorphous sulfur is insoluble in carbon disulfide.

Figure 29-3. The sudden cooling of μ-sulfur produces the allotropic amorphous sulfur.

29.5 Chemical properties of sulfur At room temperature, sulfur is not very active chemically. When heated, sulfur combines with oxygen to produce sulfur dioxide.

$$S(s) + O_2(g) \rightarrow SO_2(g)$$

Traces of sulfur trioxide, SO_3, also form when sulfur burns in the air. Sulfur can be made to combine with nonmetals such as hydrogen, carbon, and chlorine. However, such compounds are formed with difficulty and are not very stable.

The formulas SO_3, SO_2, and H_2S indicate that sulfur can have oxidation numbers of +6 or +4 when combined with oxygen and −2 when combined with hydrogen. Electron-dot formulas for these compounds are shown below.

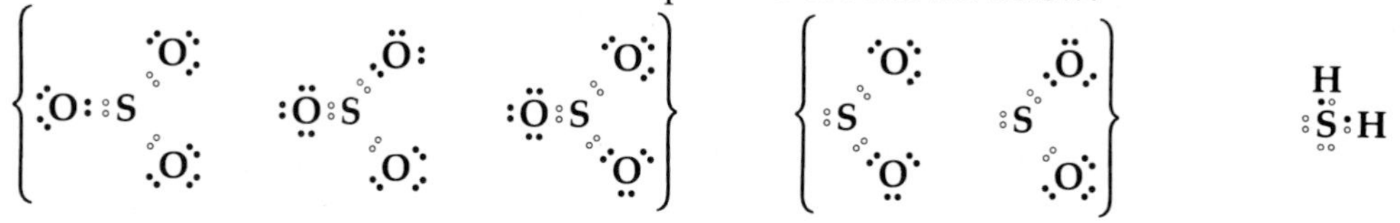

sulfur trioxide **sulfur dioxide** **hydrogen sulfide**

The actual molecules of sulfur trioxide and sulfur dioxide are resonance hybrids of the possible structures given.

Sulfur combines directly with all metals except gold and platinum. Powdered zinc and sulfur combine vigorously. The heat produced when iron filings and sulfur unite causes the whole mass to glow red hot. Copper unites with the vapor of boiling sulfur to form copper(I) sulfide.

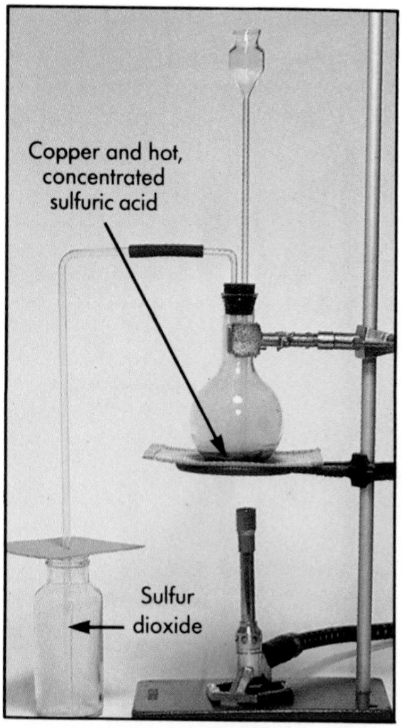

29.6 Uses of sulfur About 90% of the sulfur produced is used to make sulfuric acid. Sulfur is also used in making sulfur dioxide, carbon disulfide, and other sulfur compounds. Matches, fireworks, and black gunpowder all contain either sulfur or sulfur compounds. Sulfur is used in the preparation of certain dyes, medicines, and fungicides. It is also used in the vulcanization of rubber. (See Section 18.26.) New uses for sulfur are in a sulfur concrete and a sulfur asphalt for road building.

Oxides of Sulfur

29.7 Occurrence of sulfur dioxide Traces of sulfur dioxide get into the air from several sources. Sulfur dioxide occurs in some volcanic gases and in some mineral waters. Coal and fuel oil may contain sulfur as an impurity. As these fuels are burned, the sulfur burns to sulfur dioxide. Coal and fuel oil of low sulfur content produce less sulfur dioxide air pollution than similar fuels of high sulfur content.

The roasting of sulfide ores converts the sulfur in the ore to sulfur dioxide. In modern smelting plants, this sulfur dioxide is converted to sulfuric acid.

29.8 Preparation of sulfur dioxide *1. By burning sulfur.* The simplest way to prepare sulfur dioxide is to burn sulfur in air or in pure oxygen.

$$S(s) + O_2(g) \rightarrow SO_2(g)$$

Figure 29-4. Sulfur dioxide can be prepared in the laboratory by reducing hot, concentrated sulfuric acid with copper. The gas is collected by upward displacement of air. This preparation should be performed under a hood to prevent the escape of sulfur dioxide with its suffocating, choking odor.

2. *By roasting sulfides.* Huge quantities of sulfur dioxide are produced by roasting sulfide ores. The roasting of zinc sulfide ore is typical.

$$2ZnS(s) + 3O_2(g) \rightarrow 2ZnO(s) + 2SO_2(g)$$

Sulfur dioxide is a by-product in this operation.

3. *By the reduction of sulfuric acid.* In one laboratory method of preparing this gas, copper is heated with concentrated sulfuric acid. See Figure 29-4.

$$Cu(s) + 2H_2SO_4(aq) \rightarrow CuSO_4(aq) + 2H_2O(l) + SO_2(g)$$

4. *By the decomposition of sulfites.* In this second laboratory method, sulfur dioxide is formed by the action of a strong acid on a sulfite. See Figure 29-5.

$$Na_2SO_3(aq) + H_2SO_4(aq) \rightarrow Na_2SO_4(aq) + H_2O(l) + SO_2(g)$$

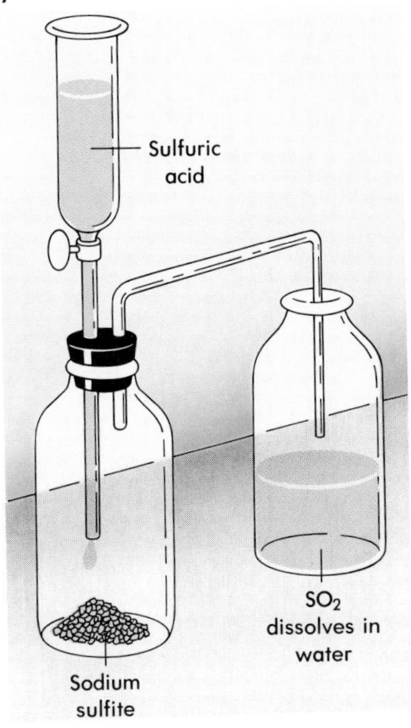

Figure 29-5. An acid added to a sulfite releases sulfur dioxide. Here the sulfur dioxide is being passed into a wide-mouth bottle partly filled with water. The sulfur dioxide will dissolve in the water. This preparation should be performed under a hood.

29.9 Physical properties of sulfur dioxide Pure sulfur dioxide is a colorless gas with a suffocating, choking odor. It is more than twice as dense as air and is very soluble in water. It is one of the easiest gases to liquefy, becoming liquid at room temperature under a pressure of about three atmospheres.

29.10 Chemical properties of sulfur dioxide 1. *It is an acid anhydride.* Sulfur dioxide dissolves in water and also reacts with the water.

$$SO_2(g) + 2H_2O(l) \rightleftharpoons H_3O^+(aq) + HSO_3^-(aq)$$

$$HSO_3^-(aq) + H_2O(l) \rightleftharpoons H_3O^+(aq) + SO_3^{--}(aq)$$

These reactions partly account for the high solubility of sulfur dioxide in water. The solution is commonly called "sulfurous acid" although no significant amount of H_2SO_3 exists in the solution. The water solution of sulfur dioxide is weakly acid. It turns litmus paper red, neutralizes hydroxides, and forms hydrogen sulfites and sulfites. If exposed to the air, the water solution of sulfur dioxide reacts slowly with oxygen to form sulfuric acid.

2. *It is a stable gas.* Sulfur dioxide does not burn. With a suitable catalyst and at a high temperature, it can be oxidized to sulfur trioxide.

$$2SO_2(g) + O_2(g) \rightleftharpoons 2SO_3(g)$$

29.11 Uses for sulfur dioxide 1. *For making sulfuric acid.* In the chemical industry, great quantities of sulfur dioxide are oxidized to sulfur trioxide. The sulfur trioxide is then combined with water, forming sulfuric acid. (See Section 29.13.)

2. *As a preservative.* Dried fruits such as apricots and prunes are treated with sulfur dioxide, which acts as a preservative.

Figure 29-6. Oxides of sulfur are used in the manufacturing processes of paper pulp.

3. *For bleaching.* The weakly acid water solution of sulfur dioxide does not harm the fibers of wool, silk, straw, or paper. Thus, it can be used to bleach these materials. It is believed that the colored compounds in these materials are converted to colorless sulfites.

4. *In preparing paper pulp.* Sulfur dioxide in water reacts with limestone to form calcium hydrogen sulfite, $Ca(HSO_3)_2$. Wood chips are heated in calcium hydrogen sulfite solution as a first step in making paper. The hot solution dissolves the lignin, which binds the cellulose fibers of wood together. The cellulose fibers are left unchanged and are processed to form paper.

29.12 Acid rain Sulfur dioxide is produced when sulfur-containing coal is burned or when sulfide ores are smelted. The waste gases of newer industrial plants are treated to remove sulfur dioxide. But millions of metric tons still enter the atmosphere annually. Sulfur dioxide reacts with oxygen and water vapor in the air to form sulfuric acid.

Oxides of nitrogen, principally nitrogen monoxide, NO, and nitrogen dioxide, NO_2, get into the air as combustion products of coal and automobile engines (as described in Section 28.4).

Figure 29-7. This map shows the variation in pH of the rainfall over North America. The most acid rainfall occurs in the northeast.

Nitrogen monoxide reacts with oxygen in the air to form nitrogen dioxide. Nitrogen dioxide reacts with water vapor in the air to form nitric acid.

Oxide and acid pollutants of both sulfur and nitrogen can be carried many miles through the atmosphere by the prevailing winds.

The pH of normal rainwater is about 5.0. This slight acidity is caused by dissolved carbon dioxide and organic acids of natural origin. Carbon dioxide reacts with water to form hydronium ions and hydrogen carbonate ions, as explained in Section 17.18. Sulfuric acid and nitric acid are both very soluble in water and readily dissolve in falling raindrops. The pH of such rainwater ranges from 4.5 to 4.0, but in extreme cases can be as low as 1.4. Rainwater having a pH below about 5.0 is called *acid rain.* Acid rain may fall many miles away from the sources of sulfur dioxide and oxides of nitrogen. Where it falls, acid rain may be harmful to plant and animal life. More needs to be learned about the effects of acid rain and about methods to minimize the release of sulfur dioxide and oxides of nitrogen into the atmosphere.

Sulfuric Acid

29.13 Preparation of sulfuric acid Sulfuric acid is produced in the United States today by the contact process. In this process, sulfur dioxide is prepared by burning sulfur or by roasting iron pyrite, FeS_2. Impurities that might combine with and ruin the catalyst used in the process are removed from the sulfur dioxide gas. The purified sulfur dioxide is mixed with air and passed through heated iron pipes. These pipes contain the catalyst, usually divanadium pentoxide, V_2O_5. This close "contact" of the sulfur dioxide and the catalyst gives the *contact process* its name. Sulfur dioxide and oxygen of the air are both adsorbed on the surface of the catalyst. There they react to form sulfur trioxide. See Figure 29-8.

The sulfur trioxide is then dissolved in approximately 98% sulfuric acid. Sulfur trioxide combines readily and smoothly with sulfuric acid, forming pyrosulfuric acid, $H_2S_2O_7$.

$$SO_3(g) + H_2SO_4(l) \rightarrow H_2S_2O_7(l)$$

When diluted with water, pyrosulfuric acid yields sulfuric acid.

$$H_2S_2O_7(l) + H_2O(l) \rightarrow 2H_2SO_4(l)$$

Very pure, highly concentrated sulfuric acid is produced by the contact process.

In terms of tonnage, sulfuric acid ranks first among the substances produced by the United States chemical industry. In 1983, over 31 million metric tons was prepared.

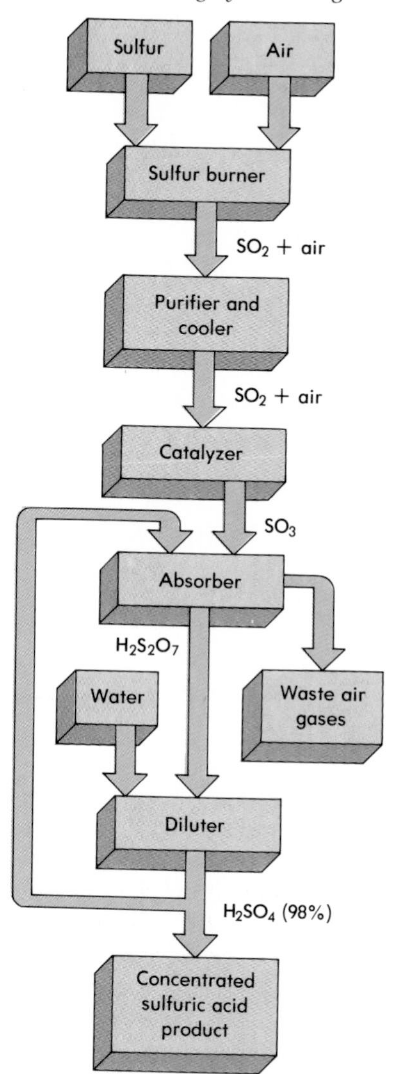

Figure 29-8. A flow diagram of the contact process for manufacturing sulfuric acid.

29.14 Physical properties of sulfuric acid Concentrated sulfuric acid is a dense, oily, colorless liquid. It contains about 2% water, has a density of about 1.84 g/mL, and a boiling point of 338 °C.

When sulfuric acid is added to water, a large amount of heat is released as the hydrates $H_2SO_4 \cdot H_2O$ and $H_2SO_4 \cdot 2H_2O$ are formed.

CAUTION: *Never add water to sulfuric acid.*

29.15 Chemical properties of sulfuric acid 1. *Its acid properties.* Sulfuric acid is a diprotic acid. It ionizes in dilute water solution in two stages:

$$H_2SO_4 + H_2O \rightleftarrows H_3O^+ + HSO_4^-$$

$$HSO_4^- + H_2O \rightleftarrows H_3O^+ + SO_4^{--} \qquad K_i = 1.26 \times 10^{-2}$$

At 25 °C, 0.1-*M* H_2SO_4 is completely ionized in the first stage and about 10% ionized in the second stage. Sulfuric acid reacts with hydroxides to form hydrogen sulfates and sulfates. It reacts with metals and with the oxides of metals. Dilute sulfuric acid is more highly ionized than cold, concentrated sulfuric acid. Thus the dilute acid reacts more vigorously than the cold, concentrated acid with metals above hydrogen in the oxidizing and reducing agents series. See Table 22-3 for this series.

2. *Its oxidizing properties.* Hot, concentrated sulfuric acid is a vigorous oxidizing agent. The sulfur is reduced from the +6 oxidation state to the +4 or −2 oxidation state. The extent of the reduction depends on the strength of the acid and on the reducing agent used.

3. *Its dehydrating properties.* The strong attraction of sulfuric acid for water makes it a very good *dehydrating* agent. Gases that do not react with sulfuric acid can be dried by bubbling them through the concentrated acid. Sulfuric acid is such an active dehydrating agent that it takes hydrogen and oxygen *directly and in the proportion found in water* from certain substances. For example, it does so with cellulose, $(C_6H_{10}O_5)_n$, and sucrose, $C_{12}H_{22}O_{11}$, leaving the carbon uncombined. The equation for this process with sucrose is

$$C_{12}H_{22}O_{11} + 11H_2SO_4 \rightarrow 12C + 11H_2SO_4 \cdot H_2O$$

In some commercial chemical processes, water is formed as a by-product. The production of nitroglycerin, $C_3H_5(NO_3)_3$, is such a process.

$$C_3H_5(OH)_3 + 3HNO_3 \rightarrow C_3H_5(NO_3)_3 + 3H_2O$$

The sulfuric acid acts as a dehydrating agent. It absorbs the water as fast as it is formed and maintains the reaction rate.

CAUTION: *Sulfuric acid burns the flesh severely.* The burns are the

The dehydration of formic acid by sulfuric acid to prepare carbon monoxide is described in Section 17.21.

result of its dehydrating action on the skin. Handling sulfuric acid requires constant care to prevent its contact with the skin.

29.16 Uses of sulfuric acid Calcium phosphate, $Ca_3(PO_4)_2$, is found in great quantities in Florida and Tennessee. It is treated with sulfuric acid to make it more soluble. The soluble form is used as *superphosphate* fertilizer. About 70% of the sulfuric acid produced in the United States is used to make fertilizers.

Sulfuric acid is also used in making phosphoric and other acids, various sulfates, and many chemicals. The iron and steel industries use sulfuric acid to remove oxides from the surface of iron or steel before the metal is plated or coated with an enamel. In petroleum refining, sulfuric acid is used to remove certain organic impurities.

Sulfuric acid serves as a dehydrating agent in the production of smokeless powder and nitroglycerin. It is used in making photographic film, nitrocellulose plastics, and rayon. It is useful in producing paints and pigments, cellophane, and thousands of other commercial articles. The electrolyte in lead storage batteries is dilute sulfuric acid.

SUMMARY

Sulfur occurs in nature as the free element or combined with other elements in sulfides and sulfates. The Frasch process is used to mine sulfur in Texas and Louisiana. Elemental sulfur is also obtained from pyrite and from the purification of coke oven gas, smelter gases, petroleum, and natural gas.

Ordinary sulfur is a yellow, odorless solid that is practically insoluble in water. It is soluble in carbon disulfide and in carbon tetrachloride. Sulfur exists in several allotropic forms. The solid allotropes are rhombic, monoclinic, and amorphous sulfur. The liquid allotropes are lambda- and mu-sulfur. Sulfur is not very active chemically at room temperature. When heated, sulfur combines with oxygen to form sulfur dioxide. Zinc, iron, and copper unite with sulfur at elevated temperatures to form sulfides. Sulfur is used for making sulfur dioxide, carbon disulfide, sulfuric acid, and other sulfur compounds.

Sulfur dioxide is produced by (1) burning sulfur, (2) roasting sulfides, (3) the reduction of sulfuric acid, and (4) the decomposition of sulfites. It is a dense, suffocating gas that is easily liquefied and extremely soluble in water. Water solutions of sulfur dioxide are weakly acidic. Sulfur dioxide does not burn and is a fairly stable compound. With a suitable catalyst, sulfur dioxide can be oxidized to sulfur trioxide.

Sulfur dioxide is used for making sulfuric acid and sulfites. It is also used for bleaching, as a preservative, and in preparing paper pulp.

Acid rain is rain with a pH below about 5.0. Oxides of sulfur and nitrogen react with oxygen and water vapor in the air to form sulfuric acid and nitric acid, the chief pollutants in acid rain.

Sulfuric acid is made from sulfur dioxide, oxygen, and water by the contact process. It is a dense, oily, colorless liquid that is soluble in water in all proportions. In dilute form, it acts as an acid. When hot and concentrated, it is a vigorous oxidizing agent. It is also a good dehydrating agent. Sulfuric acid is one of the most important chemicals used in industry. Some of its uses are (1) in making fertilizer, (2) in making phosphoric acid, (3) in cleaning metals, (4) in petroleum refining, (5) in producing explosives, and (6) as the electrolyte in lead storage batteries.

VOCABULARY

acid rain
amorphous sulfur
contact process
dehydrating agent
fluidity
Frasch process
fungicide
lambda-sulfur
monoclinic sulfur
mu-sulfur
pyrite
rhombic sulfur
roasting
superheated water
superphosphate
viscous

QUESTIONS

Group A

1. In what forms does sulfur occur in nature? *A2a*
2. What is the location of the deposits of elemental sulfur in the United States? *A2a*
3. In the Frasch process, what is the purpose of *(a)* the superheated water? *(b)* the compressed air? *A2b*
4. In addition to Frasch process sulfur, what other sources of sulfur are important in the United States? *A2b*
5. Describe the odor of *(a)* sulfur; *(b)* sulfur dioxide. *A2c*
6. Define *(a) pyrite; (b) viscous; (c) allotrope; (d) fluidity; (e) amorphous.* *A1*
7. A pupil prepared some nearly black amorphous sulfur in the laboratory. When it was examined the following week, it had become brittle and much lighter in color. Explain. *A2d*
8. List five uses for elemental sulfur. *A2g*
9. Sulfur dioxide is an important air pollutant. Where does it come from? *A2a*
10. Write balanced formula equations for two commercial methods of preparing sulfur dioxide. *A3a*
11. *(a)* In the laboratory preparation of sulfur dioxide, what method of collecting the gas is used? *(b)* What properties of sulfur dioxide are considered in making this decision? *A2e*
12. What is the most important use of sulfur dioxide? *A2g*
13. Write balanced formula equations to show the formation from a water solution of sulfur dioxide and sodium hydroxide of *(a)* sodium hydrogen sulfite; *(b)* sodium sulfite. *A5b*
14. *(a)* What is the pH of normal rainwater? *(b)* Why is the pH not 7.0? *(c)* What is *acid rain? (d)* What are the principal causes of acid rain? *(e)* Why is it believed to be undesirable? *A1, A2h*
15. *(a)* What process is used for the manufacture of sulfuric acid in the United States? *(b)* How did the process get its name? *A2e*
16. How is sulfuric acid diluted safely? *A2c*
17. Give three general chemical properties of sulfuric acid. *A2f*
18. What is the most important use for sulfuric acid in the United States? *A2g*
19. For what purpose is sulfuric acid used in the iron and steel industries? *A2g*
20. Why are both nitric acid and sulfuric acid used in making nitroglycerin? *A2f*

Group B

21. Where in the solar system, besides Earth, is there extensive volcanic activity involving sulfur and sulfur compounds? *A2a*
22. Why was the Frasch process for obtaining sulfur developed? *A2b, A4*
23. *(a)* What is *superheated water*? *(b)* Why must the water used in the Frasch process be superheated? *A1, A2b*
24. *(a)* Give the molecular formula for rhombic sulfur. *(b)* Why do you suppose this molecular formula is not usually used in formula equations? *A2d*
25. As sulfur is heated from the melting point to the boiling point, it undergoes changes in color and fluidity. Describe these changes and explain why they occur. *A2d*
26. Is the change from rhombic sulfur to monoclinic sulfur exothermic or endothermic? Explain. *A2d*

27. Write balanced net ionic equations for two laboratory preparations of sulfur dioxide. *A3a*

28. *(a)* Define *resonance.* *(b)* Draw electron-dot formulas to demonstrate the resonance structure of sulfur dioxide. *A1, A5a*

29. Is sulfur dioxide easy or difficult to liquefy? Explain. *A2c*

30. *(a)* Using data from Appendix Table 11, compare the solubility in water of CO_2, NO_2, and SO_2. *(b)* What do these solubilities indicate about the extent to which these oxides react with water? *A2f*

31. What uses are there for sulfur dioxide other than the manufacture of sulfuric acid? *A2g*

32. *(a)* How is the nitric acid in acid rain formed? *(b)* How is the sulfuric acid formed? *A2h*

33. What is being done now about the environmental problem of acid rain? *A2h*

34. Give two reasons why boiling concentrated sulfuric acid burns the flesh. *A2c, A2f*

35. Balance the following oxidation-reduction equations:

(a) $H_2S(g) + SO_2(g) \rightarrow S(s) + H_2O(l)$

(b) $I_2(aq) + Na_2SO_3(aq) + H_2O(l) \rightarrow NaI(aq) + H_2SO_4(aq)$

(c) $Hg(l) + H_2SO_4(aq) \rightarrow HgSO_4(aq) + SO_2(g) + H_2O(l)$

(d) $Cu_2S(s) + O_2(g) \rightarrow Cu_2O(s) + SO_2(g)$

(e) $SO_2(g) + H_2O(l) + KMnO_4(aq) \rightarrow MnSO_4(aq) + K_2SO_4(aq) + H_2SO_4(aq)$

PROBLEMS

Group A

1. A railroad gondola car has a capacity of 49.4 m^3. How many metric tons of sulfur does it contain when loaded to capacity? Obtain density data from Appendix Table 16. *A2c*

2. Common sulfides of Period Three elements are Na_2S, MgS, Al_2S_3, SiS_2, P_4S_{10}, and SCl_2. What is the percentage of ionic character of the bonds in each of these compounds?

3. Calculate the percentage composition of H_2SO_4.

4. A compound of lead and sulfur contains 86.6% lead. Determine the empirical formula of the compound.

5. How many kilograms of sulfur dioxide can be produced by burning 2.5 kg of pure sulfur? *A3a*

6. A lead smelter processes $45\bar{0}$ metric tons of zinc sulfide, ZnS, each day. If no sulfur dioxide is lost, how many kilograms of sulfuric acid could be made in the plant daily? *A3a*

7. How many grams of sodium sulfite are needed to produce 5.00 L of sulfur dioxide at STP by reaction with sulfuric acid? *A3a*

Group B

8. Natural sulfur consists of 95.018% S-32, 31.97207 *u*; 0.750% S-33, 32.97146 *u*; 4.215% S-34, 33.96786 *u*; 0.017% S-36, 35.96709 *u*. Calculate the atomic weight of sulfur.

9. How many kilograms of sulfuric acid can be prepared from 50.0 kg of sulfur that is 99.5% pure? *A3a, A3b*

10. *(a)* If 70.0 kg of scrap iron is added to a large vat of dilute sulfuric acid, how many kilograms of iron(II) sulfate can be produced? *(b)* How many kilograms of 95% sulfuric acid are required?

11. A laboratory technician has on hand $10\bar{0}$ g of copper but only $10\bar{0}$ g of H_2SO_4. How many liters of sulfur dioxide at STP can the technician prepare? *A3a*

12. How many liters of sulfur dioxide at 25 °C and $74\bar{0}$ mm pressure can be produced by roasting $150\bar{0}$ kg of iron pyrite, FeS_2? *A3a*

13. A 0.01-*N* solution of sulfuric acid has a pH of 2.1. Calculate $[H_3O^+]$ and $[OH^-]$.

14. For barium sulfate, $K_{sp} = 1.1 \times 10^{-10}$. If a barium chloride solution is 0.01 *M*, what is the smallest sulfate ion concentration that can be detected by precipitation?

The 1906 Nobel Prize in chemistry was given to Henri Moissan (French) for isolating pure fluorine and developing the electric furnace. He was able to isolate fluorine, a very reactive gas, by electrolyzing a solution of potassium fluoride and hydrofluoric acid. He then investigated fluorine's properties and its reactions with other elements. The Moissan furnace, developed in 1892, utilized an electric arc to obtain temperatures over 3500 °C. It was used to prepare many new compounds and vaporize others. Moissan also claimed to have synthesized diamonds.

The Halogen Family

30

GOALS

In this chapter you will gain an understanding of: • the electron configurations of the halogens: their similarity and differences • the variation in properties of the halogens and its relationship to their atomic structure • the occurrence, preparation, properties, and uses of fluorine, chlorine, bromine, and iodine • the preparation, properties, and uses of hydrogen chloride • the properties and uses of halogen compounds

30.1 The Halogen Family: Group VII of the Periodic Table

The members of the Halogen Family are the colorful, active, nonmetallic elements fluorine, chlorine, bromine, iodine, and astatine. Figure 30-1 shows samples of the first four of these elements. Table 30-1 gives some data about them.

The atoms of each halogen have seven electrons in the highest energy level. The addition of one electron to a *halogen* atom converts it to a *halide* ion having an outer octet of electrons. Using a chlorine atom as an example,

$$:\ddot{\underset{\cdot}{Cl}}: + e^- \rightarrow :\ddot{\underset{\cdot\circ}{Cl}}:^-$$

The halogens are all active elements that have high electronegativities. They are very rarely found free in nature. In the elemental state they exist as covalent, diatomic molecules.

Fluorine has the smallest atoms in the Halogen Family. It has a strong attraction for electrons. Fluorine is the most highly electronegative element and the most active nonmetal. Because of these properties, fluorine cannot be prepared from its compounds by chemical reduction. It must be prepared by electrolysis.

The other halogens, with increasingly larger atoms, are less electronegative than fluorine. As a result, the smaller, lighter halogens can replace and oxidize the larger, heavier halogens from compounds. Astatine is a synthetic radioactive halogen.

Reactions involving the replacement of halogens are described in text Section 8.6(4).

Table 30-1 clearly shows the regular change in properties that occurs in this family. This change in properties proceeds from the smallest and lightest atom to the largest and heaviest. Refer to Figures 5-7, 5-10, and 6-14 while studying Table 30-1. These figures provide ionization energy, electron affinity, and electronegativity data for members of the Halogen Family.

Each of the halogens combines with hydrogen. The great

electronegativity difference between hydrogen and fluorine explains why hydrogen fluoride molecules are so polar that they associate by hydrogen bonding. The remaining hydrogen halides have smaller electronegativity differences and do not show this property. All hydrogen halides are colorless gases that are ionized in water solution. Hydrofluoric acid is a weak acid because hydronium ions are strongly hydrogen bonded to fluoride ions. The other binary halogen acids are highly dissociated, strong acids.

Each of the halogens forms ionic salts with metals. Hence the name *halogen,* which means "salt producer."

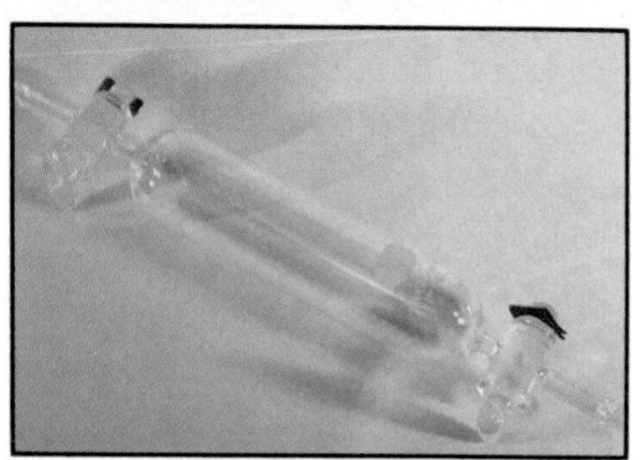

Figure 30-1. The halogens are colorful elements. Seen above, fluorine is a pale yellow gas; chlorine, a greenish-yellow gas; bromine, a reddish-brown liquid; iodine, a bluish-black crystalline solid.

FLUORINE

30.2 Preparation and properties of fluorine Fluorine is prepared by electrolyzing a mixture of potassium fluoride and hydrogen fluoride. A steel and Monel metal electrolytic cell with a carbon positive electrode is used. The fluoride coating that forms on these metals protects them from further reaction.

Fluorine is the most active nonmetallic element. It unites with hydrogen explosively, even in the dark. It forms compounds with all elements except helium, neon, and argon. There are no known positive oxidation states of fluorine. It forms salts known as *fluorides.* Fluorine reacts with gold and platinum slowly. Special carbon steel containers are used to transport fluorine. These containers become coated with iron fluoride, which resists further action.

30.3 Usefulness of fluorine compounds The mineral fluorspar, CaF_2, is often used in preparing most fluorine compounds. Sodium fluoride is used as a poison for destroying roaches, rats, and other pests. A trace of sodium fluoride, or the less expensive sodium hexafluorosilicate, Na_2SiF_6, is added to drinking water in many areas to help prevent tooth decay. The fluoride ion concentration is carefully regulated to be 1 ppm. Fluorides have also been added to some toothpastes.

One of the Freons, dichlorodifluoromethane, CCl_2F_2, is used as a refrigerant. It is odorless, nonflammable, and nontoxic. Its critical temperature is 111.7 °C and critical pressure is 39.4 atm. In producing aluminum, melted cryolite, $AlF_3 \cdot 3NaF$, is used as a solvent for aluminum oxide. Uranium is changed to uranium hexafluoride gas, UF_6, in the process for separating the uranium isotopes.

Fluorine combines with the noble gases krypton, xenon, and radon. Krypton difluoride, KrF_2, is prepared by passing electricity through a krypton-fluorine mixture. This process is carried out at the temperature of liquid nitrogen. KrF_2 is a volatile white solid that decomposes slowly at room temperature.

Table 30-1
THE HALOGEN FAMILY

Element	Atomic number	Atomic weight	Electron configuration	Principal oxidation number	Melting point (°C)	Boiling point (°C)	Color	Density, 0 °C	Atomic radius (Å)	Ionic radius (Å)
fluorine	9	18.998403	2,7	−1	−219.6	−188.1	pale yellow gas	1.696 g/L	0.68	1.33
chlorine	17	35.453	2,8,7	−1	−101.0	−34.6	greenish-yellow gas	3.214 g/L	0.99	1.81
bromine	35	79.904	2,8,18,7	−1	−7.2	58.78	reddish-brown liquid	3.12 g/mL	1.14	1.96
iodine	53	126.9045	2,8,18,18,7	−1	113.5	184.35	bluish-black crystals	4.93 g/mL	1.33	2.20
astatine	85	21$\bar{0}$	2,8,18,32,18,7	−1	302	337			1.45	

Three fluorides of xenon, XeF_2, XeF_4, and XeF_6, have been made. All are colorless crystalline solids at ordinary temperatures. XeF_6 is the most highly reactive. Each of the three compounds reacts with hydrogen to produce elemental xenon and hydrogen fluoride. With water, XeF_2 produces xenon, oxygen, and hydrogen fluoride. The other two fluorides react with water to yield xenon trioxide, XeO_3, a white, highly explosive solid.

Hydrofluoric acid, HF, is used as a catalyst in producing high-octane gasoline. It is also used in making synthetic cryolite for aluminum production. For many years hydrofluoric acid has been used for etching glass. Glassware is given a frosty appearance by exposing it to hydrogen fluoride fumes.

Figure 30-2. Crystals of xenon tetrafluoride.

CHLORINE

30.4 Wide occurrence of compounds Chlorine is rarely found uncombined in nature. Elemental chlorine is found in small amounts in some volcanic gases. Chlorides of sodium, potassium, and magnesium are fairly abundant. Common table salt, sodium chloride, is a widely distributed compound. It is found in sea water, in salt brines underground, and in rock salt deposits. Sodium chloride is the commercial source of chlorine.

30.5 Preparation of chlorine The element chlorine was first isolated in 1774 by Carl Wilhelm Scheele, the codiscoverer of oxygen. The preparation of elemental chlorine involves the oxidation of chloride ions. Strong oxidizing agents are required.

1. *By the electrolysis of sodium chloride.* Chlorine is most often prepared by the electrolysis of a saturated water solution of sodium chloride. At this concentration, hydrogen from the

water is released at the negative electrode, and chlorine is set free at the positive electrode. The hydrogen and chlorine gases are kept separate from each other and from the solution by asbestos partitions. The sodium and hydroxide ions remaining in the solution are recovered as sodium hydroxide.

$$2NaCl(aq) + 2H_2O(l) \xrightarrow{(electricity)} 2NaOH(aq) + H_2(g) + Cl_2(g)$$

2. *By the oxidation of hydrogen chloride.* This method involves heating a mixture of manganese dioxide and concentrated hydrochloric acid. The manganese oxidizes half of the chloride ions in the reacting HCl to chlorine atoms. Manganese is reduced during the reaction from the +4 oxidation state to the +2 state.

$$MnO_2(s) + 4HCl(aq) \rightarrow MnCl_2(aq) + 2H_2O(l) + Cl_2(g)$$

This was the method used by Scheele in first preparing chlorine. It is a useful laboratory preparation.

An alternate process involves heating a mixture of manganese dioxide, sodium chloride, and sulfuric acid. Refer to Figure 30-3.

$$2NaCl(s) + 2H_2SO_4(aq) + MnO_2(s) \rightarrow Na_2SO_4(aq) + MnSO_4(aq) + 2H_2O(l) + Cl_2(g)$$

3. *By the action of hydrochloric acid on calcium hypochlorite.* Hydrochloric acid added drop by drop to calcium hypochlorite powder releases chlorine and forms calcium chloride and water.

$$4HCl(aq) + Ca(ClO)_2(s) \rightarrow CaCl_2(aq) + 2Cl_2(g) + 2H_2O(l)$$

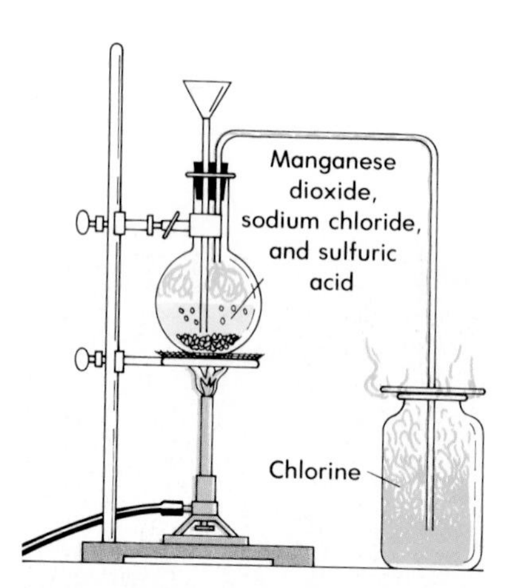

Figure 30-3. One method of preparing chlorine in the laboratory is by heating a mixture of manganese dioxide, sodium chloride, and sulfuric acid. Chlorine is collected by upward displacement of air. The preparation should be performed under a hood.

30.6 Physical properties of chlorine At room temperature, chlorine is a greenish-yellow gas with a disagreeable, suffocating odor. It is about 2.5 times as dense as air. It is moderately soluble in water, forming a pale greenish-yellow solution. Chlorine is easily liquefied and is usually marketed in steel cylinders.

When inhaled in small quantities, chlorine affects the mucous membranes of the nose and throat. If inhaled in larger quantities, chlorine is so poisonous that it may cause death.

30.7 Chemical properties of chlorine The third energy level of a chlorine atom contains seven electrons. There are many reactions by which chlorine atoms can acquire an additional electron and complete the third-energy-level octet.

1. *Action with metals.* All metals react with chlorine directly. For example, when powdered antimony is sprinkled into a jar of moist chlorine, the two elements combine spontaneously, emitting a shower of sparks. Antimony trichloride is formed. See Figure 22-1.

$$2Sb(s) + 3Cl_2(g) \rightarrow 2SbCl_3(s)$$

In a similar manner, hot metallic sodium burns in chlorine to form sodium chloride. See Figure 6-2. Chlorine combines directly with such metals as copper, iron, zinc, and arsenic, if they are heated slightly.

2. *Action with hydrogen.* If hydrogen and chlorine are mixed in the dark, no reaction occurs. But such a mixture explodes violently if it is heated or exposed to sunlight. The heat or sunlight provides the activation energy. A jet of hydrogen that is burning in air will continue to burn if it is inserted into a bottle of chlorine.

$$H_2(g) + Cl_2(g) \rightarrow 2HCl(g)$$

Chlorine does not support the combustion of wood or paper. A paraffin candle, however, continues to burn in chlorine with a smoky flame, as shown in Figure 30-4. In this reaction the hydrogen of the paraffin combines with the chlorine, forming hydrogen chloride. The carbon is left uncombined.

Figure 30-4. Chlorine has such a great attraction for hydrogen that it can take hydrogen from some of its compounds. A paraffin candle will burn in chlorine. The hydrogen of the paraffin combines with the chlorine to form hydrogen chloride. The carbon is left uncombined as the smoke.

3. *Action with water.* A freshly prepared solution of chlorine in water has a greenish-yellow color. If it stands in sunlight for a few days, both the color and the strong chlorine odor will disappear. The chlorine combines with the water to form hypochlorous acid and hydrochloric acid. Hypochlorous acid is unstable and decomposes into hydrochloric acid and oxygen.

$$2H_2O(l) + 2Cl_2(g) \rightarrow 2HClO(aq) + 2HCl(aq)$$
$$\searrow 2HCl(aq) + O_2(g)$$

The instability of hypochlorous acid makes chlorine water a good oxidizing agent.

30.8 Uses of chlorine 1. *For bleaching.* Bleaching solution is usually a solution of sodium hypochlorite. It is made by reacting liquid chlorine with sodium hydroxide solution. **CAUTION:** Chlorine destroys silk or wool fibers. *Never use commercial bleaches containing hypochlorites on silk or wool.* Chlorine is used as a bleaching agent by the pulp and paper industry.

2. *As a disinfectant.* Since moist chlorine is a good oxidizing agent, it destroys bacteria. Large quantities of chlorinated lime, $Ca(ClO)Cl$, are used as a disinfectant.

In city water systems, billions of gallons of water are treated with chlorine to kill disease-producing bacteria. The water in swimming pools is usually treated with chlorine. As part of the sewage purification process, chlorine is sometimes used to kill bacteria in sewage.

Figure 30-5. Chlorine is added to the water in swimming pools to maintain the bacterial content at a safe level.

3. *For making compounds.* Because chlorine combines directly with many substances, it is used to produce a variety of compounds. Among these are aluminum chloride, Al_2Cl_6, used as a catalyst; carbon tetrachloride, CCl_4, a solvent; 1,1,2-trichloroethene, $CHCl{=}CCl_2$, a cleaning agent; and polyvinyl chloride, the polymer of vinyl chloride, $CH_2{=}CHCl$, a plastic.

30.9 Preparation of hydrogen chloride In the laboratory, hydrogen chloride can be prepared by treating sodium chloride with concentrated sulfuric acid.

$$NaCl(s) + H_2SO_4(l) \rightarrow NaHSO_4(s) + HCl(g)$$

This same reaction, carried out at a higher temperature, is used commercially. If more NaCl is used at this higher temperature, a second molecule of HCl can be produced per molecule of H_2SO_4.

$$2NaCl(s) + H_2SO_4(l) \rightarrow Na_2SO_4(s) + 2HCl(g)$$

Another commercial preparation involves the direct union of hydrogen and chlorine. Both elements are obtained by the electrolysis of concentrated sodium chloride solution. Refer to Section 30.5.

By far the most important commercial source of hydrogen chloride results from the formation of HCl, as a by-product of chlorination processes. Some of these processes are the chlorination of hydrocarbons described in Sections 18.12 and 18.21.

Hydrogen chloride dissolved in pure water is sold under the name of hydrochloric acid.

30.10 Physical properties of hydrogen chloride Hydrogen chloride is a colorless gas with a sharp, penetrating odor. It is denser than air and extremely soluble in water. One volume of water at 0 °C dissolves more than 500 volumes of hydrogen chloride at standard pressure. Hydrogen chloride makes a fog with moist air. It is so soluble that it condenses water vapor from the air into tiny drops of hydrochloric acid.

Figure 30-6. Large quantities of hydrochloric acid are used for cleaning metals.

30.11 Chemical properties of hydrogen chloride Hydrogen chloride is a stable compound that does not burn. Some vigorous oxidizing agents react with it to form water and chlorine. The water solution of hydrogen chloride is a strong acid known as *hydrochloric acid*. Concentrated hydrochloric acid contains about 38% hydrogen chloride by weight and is about 1.2 times as dense as water. Hydrochloric acid reacts with metals above hydrogen in the activity series and with the oxides of metals. It neutralizes hydroxides, forming salts and water.

30.12 Uses of hydrochloric acid Hydrochloric acid is used in preparing certain chlorides and in cleaning metals. This cleaning involves removing oxides and other forms of tarnish.

Some hydrochloric acid is essential in the process of digestion. The concentration of hydrochloric acid in gastric juice is 0.16 *M*.

Bromine

30.13 Bromine from bromides In the laboratory, bromine can be prepared by using manganese dioxide, sulfuric acid, and sodium bromide.

$$2NaBr(s) + MnO_2(s) + 2H_2SO_4(aq) \rightarrow Na_2SO_4(aq) + MnSO_4(aq) + 2H_2O(l) + Br_2(g)$$

This is similar to the preparation of chlorine in Figure 30-7.

The commercial production of bromine from salt-well brines depends on the ability of chlorine to displace bromide ions from solution. Chlorine displaces bromide ions because it is more highly electronegative than bromine.

$$2Br^-(aq) + Cl_2(g) \rightarrow 2Cl^-(aq) + Br_2(l)$$

30.14 Physical properties of bromine Bromine is a dark, reddish-brown liquid that is about three times as dense as water. It evaporates readily, forming a vapor that burns the eyes and throat and has a very disagreeable odor. Bromine is moderately soluble in water. The reddish-brown water solution used in the laboratory is known as bromine water. Bromine dissolves readily in carbon tetrachloride and carbon disulfide and in water solutions of bromides.

CAUTION: *Use great care in handling bromine.* It burns the flesh and forms wounds that heal slowly.

30.15 Chemical properties of bromine Bromine unites directly with hydrogen to form hydrogen bromide, as shown in Figure 30-8. It combines with most metals to form bromides. When it is moist, bromine is a good bleaching agent. Its water solution is a strong oxidizing agent that forms hydrobromic acid and oxygen in sunlight.

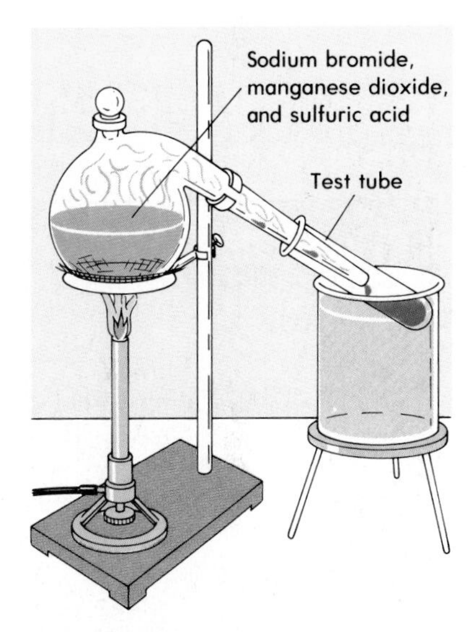

Figure 30-7. Bromine can be prepared in the laboratory by heating a mixture of sodium bromide, manganese dioxide, and sulfuric acid in a glass-stoppered retort. This preparation should be performed under a hood.

30.16 Uses of bromine compounds Complex bromine-containing organic compounds are added to polyurethane and polystyrene plastics to make them flame retardant. Methyl bromide, CH_3Br, is a useful soil fumigant. Improved crop yields result from such treatment. Ethylene bromide, $C_2H_4Br_2$, increases the efficiency of lead tetraethyl in antiknock gasoline. Ethylene bromide was formerly used to control insects in grain. This use has been stopped because ethylene bromide is believed to cause cancer.

Silver bromide, AgBr, is a yellowish solid. It is highly photosensitive (sensitive to light rays) and is widely used in making photographic film. A solution of calcium bromide, $CaBr_2$, is used as a high-density fluid in oil-well drilling. The bromides of sodium and potassium are used in medicine as sedatives. Such medicines should not be used unless prescribed by a physician. Certain bromine compounds may be used as disinfectants.

Figure 30-8. Hydrogen burns in an atmosphere of bromine vapor. Hydrogen bromide is formed.

IODINE

30.17 Preparation of iodine The laboratory preparation of iodine is similar to that of chlorine and bromine. An iodide is heated with manganese dioxide and sulfuric acid.

$$2NaI(s) + MnO_2(s) + 2H_2SO_4(aq) \rightarrow Na_2SO_4(aq) + MnSO_4(aq) + 2H_2O(l) + I_2(g)$$

The iodine is driven off as a vapor. It can be condensed as a solid on the walls of a cold dish or beaker, as shown in Figure 30-9.

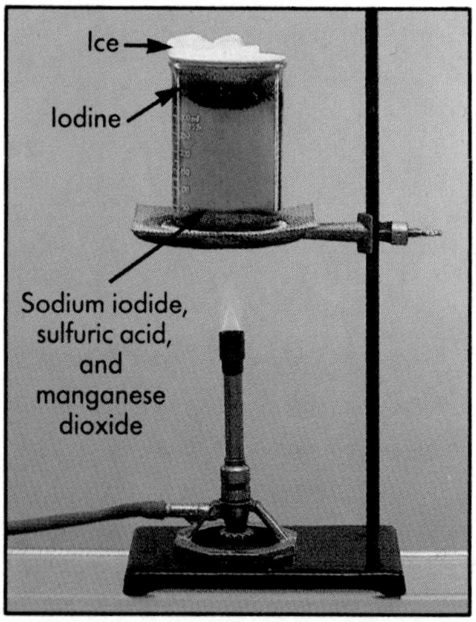

Figure 30-9. Iodine is prepared in the laboratory by heating sodium iodide and manganese dioxide with sulfuric acid.

30.18 Physical properties of iodine Iodine is a bluish-black crystalline solid. When heated, it sublimes (vaporizes without melting) and produces a violet-colored vapor. The odor of this vapor is irritating, resembling that of chlorine.

Iodine is very slightly soluble in water. It is much more soluble in water solutions of sodium or potassium iodide. With these solutions it forms complex I_3^- ions. It dissolves readily in alcohol, forming a dark-brown solution. It is very soluble in carbon disulfide and carbon tetrachloride, giving a rich violet color.

30.19 Chemical properties of iodine Iodine is active chemically, though less so than either bromine or chlorine. It combines with metals to form iodides. It can also form compounds with all nonmetals except sulfur, selenium, and the noble gases. Iodine reacts with organic compounds just as chlorine and bromine do. But since the carbon-iodine bond is not strong, many of the compounds formed decompose readily.

30.20 Uses of iodine Iodine is most useful as a disinfectant. Iodine complexed (loosely bonded) with certain organic compounds is used as an antiseptic on the skin. Detergents with which iodine is complexed make good combination cleaning-and-sanitizing agents. Iodine and some iodine compounds are used as catalysts in certain organic reactions.

30.21 Uses of iodides Silver iodide is a light-sensitive compound used in photographic film. Potassium iodide is added to table salt to provide the iodine necessary for proper nutrition.

SUMMARY

The Halogen Family consists of the highly electronegative elements fluorine, chlorine, bromine, iodine, and astatine. The atoms of each halogen have seven electrons in the highest energy level. Very rarely found free in nature, the elements exist as covalent, diatomic molecules.

Fluorine is prepared by the electrolysis of a mixture of potassium fluoride and hydrogen fluoride. The other three common halogens are prepared in the laboratory by oxidizing their binary acids with manganese dioxide. Commercially, chlorine is made by the electrolysis of sodium chloride solution.

Hydrogen chloride is prepared by treating a salt of the acid with sulfuric acid. Each of the hydrogen halides is a colorless gas that is ionized in water solution.

Fluorine compounds are used (1) as refrigerants, (2) to help prevent tooth decay, (3) in aluminum production, (4) for separating uranium isotopes, (5) in gasoline production, and (6) for etching glass. Chlorine is used for bleaching, disinfecting, and making chlorine-containing compounds. Bromine compounds are used (1) to make plastics flame retardant, (2) in agriculture, (3) in antiknock gasoline, (4) in oil-well drilling, (5) as disinfectants, (6) in medicine, and (7) in photography. Iodine is used as a disinfectant and in organic synthesis. Iodides are used in photography and as nutrients.

VOCABULARY

antiseptic
brine
chlorinated lime
detergent
disinfectant
halide
halogen
hydrobromic acid
hydrochloric acid
hydrofluoric acid
Monel metal

QUESTIONS

GROUP A

1. Define: *(a) halogen; (b) halogen atom; (c) halogen molecule; (d) halide ion.* *A1*
2. *(a)* In what form are the halogens found in nature, as atoms, molecules, or ions? *(b)* Why? *A2a, A3c*
3. As one goes down the Halogen Family on the periodic table, how do the members vary in *(a)* atomic radius; *(b)* ionic radius; *(c)* first ionization energy; *(d)* electron affinity; *(e)* electronegativity; *(f)* melting point; *(g)* boiling point; *(h)* density; *(i)* principal oxidation number? *A2a*
4. *(a)* What kind of container is used for fluorine? *(b)* What chemical property makes this use possible? *A2d*

5. What is the beneficial effect of drinking water having a $[F^-]$ of 1 ppm? *A3a*
6. What properties make dichlorodifluoromethane useful as a refrigerant? *A3a*
7. List three uses of hydrofluoric acid. *A3a*
8. *(a)* What compound is the commercial source of chlorine? *(b)* For what other element is this compound the commercial source? *A2e, A3c*
9. What effects does chlorine have on the body? *A2d*
10. *(a)* For which does chlorine have greater attraction, carbon or hydrogen? *(b)* What experimental evidence can you give to support your answer? *A2d*
11. What are the three general commercial uses of chlorine? *A3a*
12. *(a)* List the physical and chemical properties of hydrogen chloride that must be considered in choosing a method of collecting this gas in the laboratory. *(b)* Which method of collection would you choose for a gas with this combination of properties? *A2e*
13. *(a)* Describe an industrial use for hydrochloric acid. *(b)* What is the function of hydrochloric acid in the human body? *A2b*
14. *(a)* Write the ionic equation for the reaction involved in extracting bromine from salt-well brines. *(b)* What type of reaction is this? *(c)* Why does this reaction occur? *A4d*
15. Describe the important physical properties of bromine. *A2c*
16. Write formulas for two organic and two inorganic bromine compounds and give a use for each. *A3a*
17. What phase change does iodine undergo that the other halogens do not? *A1, A2c*
18. Write an equation using electron-dot symbols for the formation of an iodide ion from an iodine atom. *A5*
19. Define the following terms: *(a) Monel metal; (b) bleaching; (c) chlorinated lime; (d) disinfectant; (e) antiseptic; (f) detergent.* *A1*
20. Write balanced formula equations for the following: *A2d, A4b*
    (a) strontium + fluorine →
    (b) potassium + chlorine →
    (c) silver + bromine →
    (d) nickel + iodine →

Group B

21. If a reaction occurs, complete and balance the following ionic equations. *A2d*
    (a) $Cl^- + F_2 \rightarrow$
    (b) $F^- + Br_2 \rightarrow$
    (c) $I^- + F_2 \rightarrow$
    (d) $Br^- + Cl_2 \rightarrow$
    (e) $Cl^- + I_2 \rightarrow$
    (f) $Br^- + I_2 \rightarrow$
22. All the halogens except fluorine exhibit positive oxidation states. Why is fluorine different? *A2a, A2d*
23. On the basis of atomic or molecular structure, explain the variations with increasing atomic number, if any, the halogens show in *(a)* ionic radius; *(b)* first ionization energy; *(c)* melting point; *(d)* electronegativity; *(e)* principal oxidation number. *A2a*
24. Water reacts with xenon difluoride and yields xenon, oxygen, and hydrogen fluoride. *(a)* Write an equation for this reaction. *(b)* Assign oxidation numbers to each element and balance the equation by a method used to balance oxidation-reduction equations. *A3b*
25. In the commercial electrolysis of sodium chloride solution, the products chlorine, hydrogen, and sodium hydroxide are kept separated from each other. Why? *A2e*
26. Write the balanced formula equation for the lab preparation of chlorine by the action of hydrochloric acid on calcium hypochlorite. Assign oxidation numbers. Tell what is oxidized and what is reduced. *A4a*
27. *(a)* What is the color of freshly prepared chlorine water? *(b)* Why? *(c)* Does the color change as chlorine water is exposed to sunlight? *(d)* Explain. *A2d*
28. Write the equation for the laboratory preparation of bromine from *(a)* manganese dioxide and hydrobromic acid; *(b)* manganese dioxide, sodium bromide, and sulfuric acid. *(c)* Assign oxidation numbers in both equations and tell which element is oxidized and which element is reduced. *(d)* How do these reactions compare? *A4d*
29. Compare the colors of *(a)* solid iodine; *(b)* iodine in alcohol; *(c)* iodine in carbon tetrachloride; *(d)* iodine vapor. *A2c, A2d*

30. Hydrogen forms binary compounds with each of the four common halogens. *(a)* Write the formulas for these compounds. *(b)* How do the hydrogen-halogen bonds in these compounds vary in polarity? (Use electronegativity data to support your answer.) *(c)* Each hydrogen halide reacts with water. Write equations for the reactions. *A2b*

31. The chemical reactions between water molecules and molecules of the hydrogen halides are reversible. *(a)* Qualitatively, at equilibrium, what are the relative concentrations of the particles involved? *(b)* What does this indicate about the relative stability of the hydrogen halide molecules compared with the stability of the ions that can be formed from them? *A2b*

32. Sodium chloride and magnesium chloride are ionic compounds, while aluminum chloride is molecular. Why? *A2d*

33. In 1983, unusually high levels of organic bromide gases were reported in samples of Arctic air. At that time the sources were not known, but industrial pollution and marine microorganisms were suspected. There was also speculation on the relative effects of bromine and chlorine compounds on stratospheric ozone. What progress is now being made on this environmental problem?

34. The improved health that results from using chlorine to kill disease-producing bacteria in municipal water supply systems has been recognized for many years. But chlorine can also react with natural substances in water to produce chloroform, $CHCl_3$, which causes cancer in laboratory animals. In what ways can scientists deal constructively with this dilemma?

PROBLEMS

Group A

1. *(a)* Using density data (Appendix Table 16), how many moles of Cl_2 are there in 5.00 L of the gas at STP? *(b)* How many Cl_2 molecules is this?

2. What is the percentage of bromine in ethylene bromide, $C_2H_4Br_2$?

3. Chlorine reacts with calcium hydroxide to produce bleaching powder, $Ca(ClO)Cl$, and water. *(a)* What mass of calcium hydroxide is required for making $20\bar{0}$ g of bleaching powder? *(b)* What mass of chlorine is also required?

4. How many liters of chlorine at STP can be obtained from 375 g of sodium chloride by electrolysis? *A4a*

5. How many grams of zinc chloride can be produced from 0.500 L of chlorine at STP? *A4b*

Group B

6. Natural chlorine consists of only the two isotopes Cl-35, 34.96885 *u*, and Cl-37, 36.96590 *u*. If the atomic weight is 35.453, what are the percentages of these two isotopes in naturally occurring chlorine?

7. Calculate the heat of reaction for $\frac{1}{2}Cl_2 + HBr \rightarrow HCl + \frac{1}{2}Br_2$ from *(a)* bond energy data (Table 6-4); *(b)* heat of formation data (Appendix Table 14).

8. A laboratory experiment requires five $25\bar{0}$-mL bottles of chlorine, measured at 22 °C and $73\bar{0}$ mm pressure. What volume of 38% hydrochloric acid (density 1.20 g/mL) and what mass of manganese dioxide will be required? *A4a*

9. Using the ideal gas equation (Section 11.12), calculate the number of moles of Cl_2 in 5.00 L of the gas at STP. Why does your answer differ from that obtained in Problem 1?

10. A solution contains 62.5 g of $CaCl_2$ in $50\bar{0}$ g of water. What is the theoretical freezing point of the water of the solution?

11. How many grams of hydrogen chloride can be obtained when $50\bar{0}$ g of 95% sulfuric acid reacts with $50\bar{0}$ g of sodium chloride? *A4c*

Nikolai N. Semenov (Russian), along with Sir Cyril Hinshelwood (British), was awarded the Nobel Prize in chemistry in 1956 for work on the kinetics of chemical reactions. Starting with studies on phosphorus oxidation, he determined that many reactions proceed by a chain reaction mechanism, initiated by the formation of unstable and very reactive free radicals. Semenov used radioactive isotopes as tracers while studying the oxidations of cyclohexane and n-decane, and this helped him to determine the stable intermediate products of the reactions and their mechanisms.

Radioactivity

31

GOALS

In this chapter you will gain an understanding of: • the discovery of radioactivity • the properties of radioactive nuclides: half-life • the three kinds of particles and rays given off by radioactive elements • transmutations and writing nuclear equations • radioactive dating • nuclear stability: nuclear mass defect and nuclear binding energy • the types of nuclear reactions: radioactive decay, nuclear disintegration, fission, fusion • particle accelerators • the production of transuranium elements • the uses of artificial radioactive nuclides • the action in a nuclear reactor • natural and controlled fusion reactions

Natural Radioactivity

31.1 Discovery of radioactivity In 1896 Henri Becquerel (bek-*rel*) (1852–1908) was studying the properties of uranium compounds. He was particularly interested in their ability to *fluoresce* (give off visible light after being exposed to sunlight). By accident Becquerel found that, whether they fluoresce or not, all uranium compounds give off invisible rays. He discovered that these rays penetrate the lightproof covering of a photographic plate and affect the film as if it had been exposed to light rays directly. Substances that give off such invisible rays are *radioactive*, and the property is called *radioactivity*. ***Radioactivity*** *is the spontaneous breakdown of an unstable nucleus with the release of particles and rays.*

31.2 Discovery of polonium and radium Becquerel was very interested in the source of radioactivity. At his suggestion, Pierre (1859–1906) and Marie (1867–1934) Curie began to investigate the properties of uranium and its ores. They soon learned that uranium and uranium compounds are only mildly radioactive. They also discovered that one uranium ore, pitchblende, has four times the amount of radioactivity expected on the basis of its uranium content.

In 1898 the Curies discovered two new radioactive metallic elements in pitchblende. These elements, *polonium* and *radium*, accounted for the high radioactivity of pitchblende. Radium is more than 1,000,000 times as radioactive as the same mass of uranium. Polonium is 5000 times as radioactive as the same mass of radium.

Table 31-1
PROPERTIES OF SOME RADIOACTIVE NUCLIDES

Nuclide	Electron configuration	Principal oxidation numbers	Melting point (°C)	Boiling point (°C)	Color	Density	Atomic radius (Å)	Ionic radius (Å)	Decay type	Half-life
U-238	2,8,18,32,21,9,2	+4, +6	1132	3818	silv metal	19.0 g/mL	1.42	0.93(+4)	α	4.468×10^9 y
Th-232	2,8,18,32,18,10,2	+4	1750	4790	silv metal	11.7 g/mL	1.65	0.99(+4)	α	1.41×10^{10} y
Ra-226	2,8,18,32,18,8,2	+2	700	1140	silv metal	5(?) g/mL	2.20	1.40(+2)	α	1.60×10^3 y
Rn-222	2,8,18,32,18,8	0	−71	−61.8	colorless gas	9.73 g/L	1.43(?)		α	3.8235 d
Po-210	2,8,18,32,18,6	−2, +2, +4, +6	254	962	silv metal	9.32 g/mL	1.46	2.30(−2)	α	138.38 d

Polonium is the heaviest member of Group VI, the oxygen family. It is a silvery metal, confirming the trend from nonmetallic to metallic properties with increasing atomic weight shown by families of active elements at the right of the periodic table. Radium, also a silvery metal, is the heaviest member of Group II, the calcium family. Its chemical properties are similar to those of barium. Some properties of the most abundant nuclides of these radioactive elements are given in Table 31-1.

Radium is always found in uranium ores. However, it never occurs in such ores in a greater proportion than 1 part of radium to 3×10^6 parts of uranium. The reason for this proportion will be explained in Section 31.5. The production of radium bromide (its usual commercial form) from uranium ore is a long, difficult, and costly procedure. Radium was produced commercially in Europe, North America, and Australia from 1904 until 1960. Today, there still remains an ample supply for its limited medicinal and industrial uses.

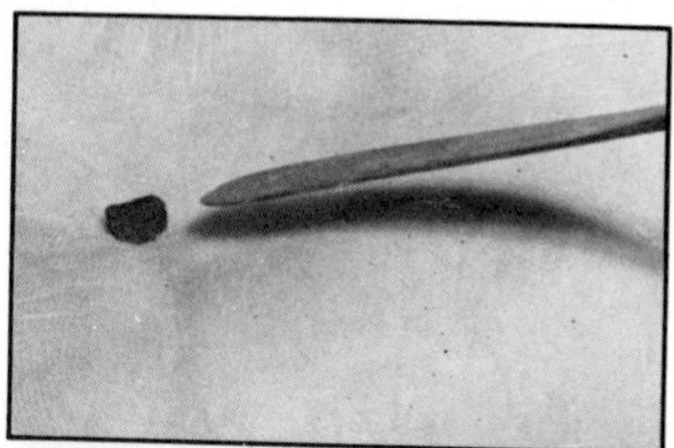

Figure 31-1. (Top) A fragment of metallic uranium, one of the radioactive elements. (Bottom) A photograph produced when radiation from the same fragment of uranium penetrated the light-tight wrappings of a photographic plate.

31.3 Properties of radioactive nuclides Of the elements known to Becquerel in 1896, testing by the Curies showed that only uranium and thorium were radioactive. The Curies discovered two more radioactive elements, polonium and radium. Since that time, many other natural radioactive nuclides have been identified. Some of these are listed in Table 31-2. All nuclides of the elements beyond bismuth (at. no. 83) in the periodic table are radioactive. However, only polonium (at. no. 84), radon (at. no. 86), radium (at. no. 88), actinium (at. no. 89), thorium (at. no. 90), protactinium (at. no. 91), and uranium (at. no. 92) have any natural radioactive nuclides. The remainder of the elements beyond bismuth have only radioactive nuclides that have been artificially produced.

Because of their radioactivity, these nuclides have several unusual properties. Some of these properties are:

1. *They affect the light-sensitive emulsion on a photographic film.* Photographic film can be wrapped in heavy black paper and stored in the dark. This will protect it from ordinary light. How-

Table 31-2
REPRESENTATIVE NATURAL RADIOACTIVE NUCLIDES WITH ATOMIC NUMBERS UP TO 83

Nuclide	Abundance in natural element (%)	Half-life (years)	Nuclide	Abundance in natural element (%)	Half-life (years)
$^{40}_{19}K$	0.0118	1.3×10^{9}	$^{130}_{52}Te$	34.49	8×10^{20}
$^{48}_{20}Ca$	0.185	$> 10^{18}$	$^{180}_{74}W$	0.135	$> 1.1 \times 10^{15}$
$^{64}_{30}Zn$	48.89	$> 8 \times 10^{15}$	$^{182}_{74}W$	26.4	$> 2 \times 10^{17}$
$^{70}_{30}Zn$	0.62	$> 10^{15}$	$^{183}_{74}W$	14.4	$> 1.1 \times 10^{17}$
$^{87}_{37}Rb$	27.85	4.8×10^{10}	$^{186}_{74}W$	28.4	$> 6 \times 10^{15}$
$^{113}_{48}Cd$	12.26	$> 1.3 \times 10^{15}$	$^{190}_{78}Pt$	0.0127	6.9×10^{11}
$^{116}_{48}Cd$	7.58	$> 10^{17}$	$^{192}_{78}Pt$	0.78	10^{15}
$^{124}_{50}Sn$	5.98	$> 2 \times 10^{17}$	$^{198}_{78}Pt$	7.19	$> 10^{15}$
$^{123}_{51}Sb$	42.75	$> 1.3 \times 10^{16}$	$^{196}_{80}Hg$	0.146	$> 1 \times 10^{14}$
$^{123}_{52}Te$	0.87	1.2×10^{13}	$^{209}_{83}Bi$	100	$> 2 \times 10^{18}$

ever, radiations from radioactive nuclides penetrate the wrapping. They affect the film in the same way that light does when the film is exposed. When the film is developed, a black spot shows up on the negative where the invisible radiation struck it. The rays from radioactive nuclides penetrate paper, wood, flesh, and *thin* sheets of metal.

2. *They produce an electric charge in the surrounding air.* The radiation from radioactive nuclides ionizes the molecules of the gases in the air surrounding it. These ionized molecules conduct electric charges away from the knob of a charged electroscope, thus discharging it. In early studies of radioactivity, the activity of a radium compound was measured by the rate at which it discharged an electroscope. Today, this method is of historic interest only. Now, the radiation given off by radioactive nuclides is used to ionize the low pressure gas in the tube of a Geiger counter. Electricity thus passes through the tube for an instant. The passage of electricity is registered as a "click" in a set of earphones.

3. *They produce fluorescence with certain other compounds.* A small quantity of radium bromide added to zinc sulfide causes the sulfide to glow. Since the glow is visible in the dark, the mixture has been used in making luminous paint. A radioactive promethium compound is now used instead of radium bromide.

4. *Their radiations have special physiological effects.* The radiation from radium can destroy the germinating power of seed. It can kill bacteria or animals. People who work with radium may be severely burned by the rays it emits. Such burns heal slowly and can be fatal. However, controlled radiations from a variety of radioactive nuclides are used in the treatment of cancer and certain skin diseases.

Figure 31-2. In early studies of radioactivity, the activity of a radioactive material was measured by the rate at which it discharged a charged goldleaf electroscope, as shown in this modern version. The material to be tested is placed in the hinged drawer at the bottom of the scope and the measurement is then read.

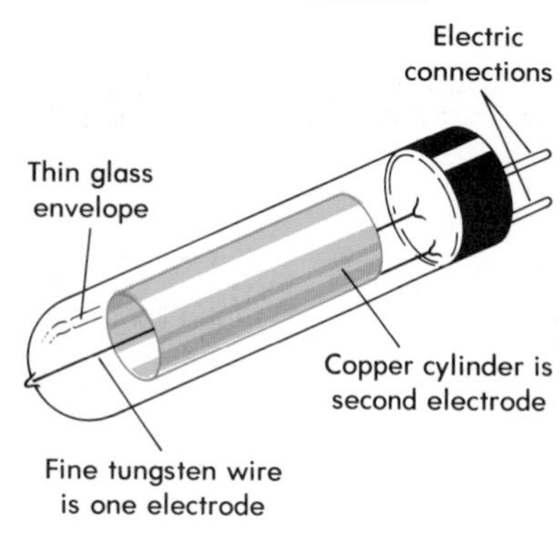

Figure 31-3. A diagram showing the construction of a Geiger-Müller counter tube. Radiation passing through the tube ionizes the gas it contains and enables current to flow.

5. *They undergo radioactive decay.* The atoms of all radioactive nuclides steadily decay into simpler atoms as they release radiation. For example, one-half of any number of radium-226 atoms decays into simpler atoms in 1602 years. One-half of what remains, or one-fourth of the original atoms, decays in the next 1602 years. One-half of what is left, or one-eighth of the original atoms, decays in the next 1602 years, and so on. This period of 1602 years is called the *half-life of radium-226*. ***Half-life** is the length of time during which half of a given number of atoms of a radioactive nuclide decays.* Each radioactive nuclide has its own half-life.

The half-life of a nuclide varies slightly depending on its chemical form. For example, beryllium-7 decays at very slightly different rates when it is elemental, in BeO, in HCl(aq) solution as Be^{++}, or in solid $Be_2P_2O_7$. The decay rate is also affected by large changes in the environment. Some of these changes are very low temperature, high pressure, or large variations in a surrounding electric field. Somewhat larger alteration of the half-life of certain nuclides has been found when the decaying nucleus and its valence electrons interact.

Practice Problems

1. How many years will be needed for the decay of 15/16 of a given amount of radium-226? *ans.* 6408 years
2. The half-life of radon-222 is 3.823 days. After what time will only one-fourth of a given amount of radon remain? *ans.* 7.646 days
3. The half-life of polonium-210 is 138.4 days. What fraction remains after 415.2 days? *ans.* 1/8

31.4 Nature of the radiation The radiation given off by radioactive nuclides can be separated into three different kinds of particles and rays.

1. *The α (alpha) particles are helium nuclei.* Their mass is nearly four times that of a protium atom. They have a +2 charge and move at speeds that are approximately one-tenth the speed of light. Because of their relatively low speed, they have low penetrating ability. A thin sheet of aluminum foil or a sheet of paper stops them. However, they burn flesh and ionize air easily.

Electrons are described in Section 3.4.

2. *The β (beta) particles are electrons.* They travel at speeds close to the speed of light, with penetrating ability about 100 times greater than that of alpha particles.

Electromagnetic radiation is explained in Section 4.2.

3. *The γ (gamma) rays are high-energy electromagnetic waves.* They are the same kind of radiation as visible light, but are of much shorter wavelength and higher frequency. Gamma rays are produced when nuclear particles undergo transitions in nuclear energy levels. They are the most penetrating of the radiations given off by radioactive nuclides. Alpha and beta particles

are seldom, if ever, given off simultaneously from the same nucleus. Gamma rays, however, are often produced along with either alpha or beta particles.

Figure 31-4 shows the effect of a powerful magnetic field on the complex radiation given off by a small particle of radioactive material. The field is perpendicular to the plane of the paper. The heavy alpha particles are slightly deflected in one direction. The lighter beta particles are deflected more sharply in the opposite direction. The gamma rays, being uncharged, are not affected by the magnet.

Radioactive nuclides such as uranium-238, radium-226, and polonium-210 decay spontaneously, yielding energy. A long series of experiments has shown that this energy results from the decay of their nuclei. Alpha and beta particles are the products of such nuclear decay. Certain heavy nuclei break down spontaneously into simpler and lighter nuclei, releasing enormous quantities of energy.

At first it was believed that radioactive nuclides did not lose mass and would give off energy forever. However, more careful investigation proved that radioactive materials do lose mass slowly. The presence of electrons and helium nuclei among the radiations is evidence for the loss of mass.

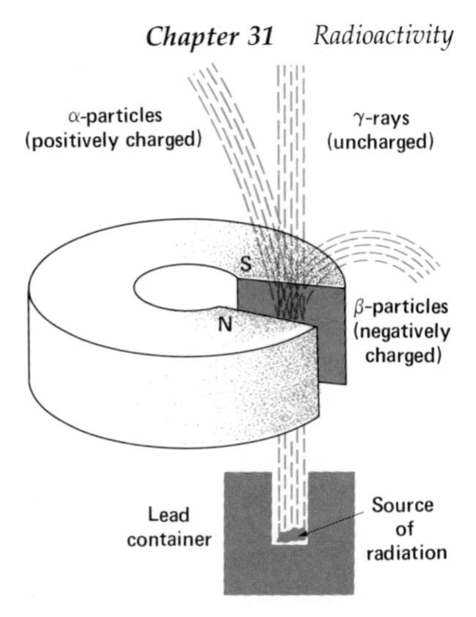

Figure 31-4. The effect of a magnet on the different types of radiations. The north pole of the magnet is toward the reader, and the south pole is away from the reader.

31.5 A series of related radioactive nuclides All naturally occurring radioactive nuclides with atomic numbers greater than 83 belong to one of three series of related nuclides. The heaviest nuclide of each series is called the *parent* nuclide. The parent nuclides are uranium-238, uranium-235, and thorium-232. The decay series of uranium-238 contains radium-226 and is traced in the following example. The various nuclear changes are charted in Figure 31-5.

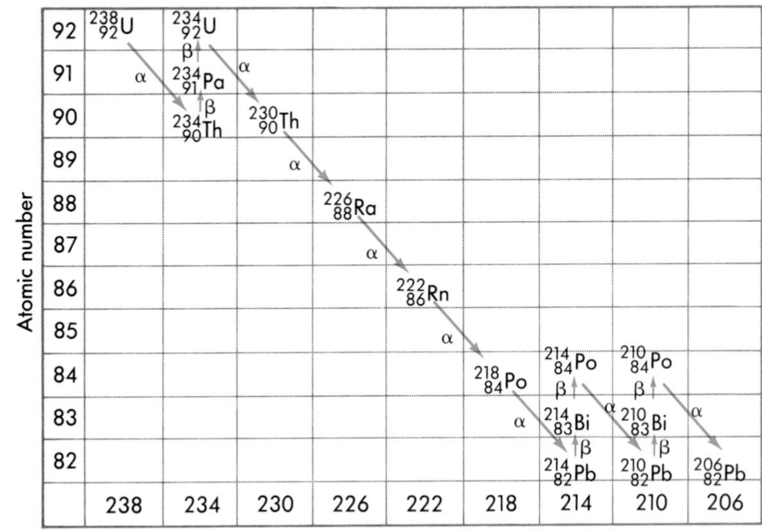

Figure 31-5. The parent nuclide of the uranium decay series is $^{238}_{92}$U. The final nuclide of the series is $^{206}_{82}$Pb.

The nucleus of a uranium-238 atom contains 92 protons (the atomic number of uranium is 92). It has a mass number (number of protons + number of neutrons) of 238. As this nucleus decays, it emits an alpha particle, which becomes an atom of helium when its positive charge is neutralized. An alpha particle has a mass number of 4. Since it contains two protons, it has an atomic number of 2. The remainder of the uranium nucleus thus has an atomic number of 90 and a mass number of 234. This *daughter* nuclide is an isotope of thorium. See the left portion of Figure 31-6. A *transmutation reaction* has taken place. *A* ***transmutation*** *is a change in the identity of a nucleus because of a change in the number of its protons.* The *nuclear equation* for this *transmutation reaction* can be written as

The various nuclides produced by the sequential decay of a parent *nuclide are called* daughter *nuclides.*

$$^{238}_{92}\mathbf{U} \rightarrow {}^{234}_{90}\mathbf{Th} + {}^{4}_{2}\mathbf{He}$$

Since the above equation is a nuclear equation, only nuclei are represented. The superscript is the mass number. The subscript is the atomic number. Alpha particles are represented as helium nuclei, $^{4}_{2}$He. The total of the mass numbers on the left side of the equation must equal the total of the mass numbers on the right side of the equation. The total of the atomic numbers on the left side must equal the total of the atomic numbers on the right.

A physical change is one in which certain physical properties of a substance change, yet its identifying properties remain unchanged.
A chemical change is one in which different substances with new properties are formed by a rearrangement of atoms.
A nuclear change is one in which new substances are formed by changes in the identity of atoms.
See Sections 2.11, 2.12, and 2.15.

The half-life of $^{234}_{90}$Th is about 24 days. It decays by giving off beta particles. The loss of a beta particle from a nucleus increases the number of positive charges in the nucleus (the atomic number) by one. The beta particle is believed to be formed by the change of a neutron into a proton and beta particle (electron). Since the mass of the lost beta particle is so small that it may be neglected, the mass number of the resulting nuclide stays the same. See the right portion of Figure 31-6.

$$^{234}_{90}\mathbf{Th} \rightarrow {}^{234}_{91}\mathbf{Pa} + {}^{0}_{-1}\mathbf{e}$$

The symbol $^{0}_{-1}$e represents an electron with an atomic number of −1 and a mass number of 0. $^{234}_{91}$Pa is an isotope of protac-

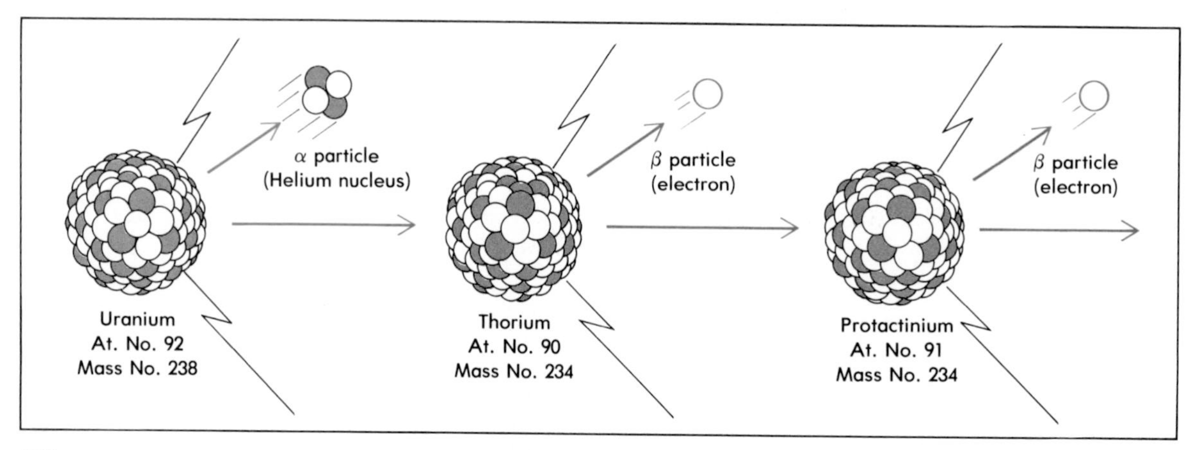

Figure 31-6. This diagram shows successive alpha and beta particle emissions in the decay of $^{238}_{92}$U.

tinium. This nuclide decays by releasing beta particles to produce $^{234}_{92}U$.

$$^{234}_{91}\mathrm{Pa} \rightarrow {}^{234}_{92}\mathrm{U} + {}^{0}_{-1}\mathrm{e}$$

The $^{234}_{92}U$ nuclide decays by giving off alpha particles.

$$^{234}_{92}\mathrm{U} \rightarrow {}^{230}_{90}\mathrm{Th} + {}^{4}_{2}\mathrm{He}$$

The resulting isotope of thorium also emits alpha particles, forming radium-226.

$$^{230}_{90}\mathrm{Th} \rightarrow {}^{226}_{88}\mathrm{Ra} + {}^{4}_{2}\mathrm{He}$$

Now you can see why ores of uranium contain radium. Radium is one of the products of the decay of uranium atoms. The half-lives of $^{238}_{92}U$ and $^{226}_{88}Ra$ determine the proportion of uranium atoms to radium atoms in uranium ores.

The decay of $^{226}_{88}Ra$ proceeds according to the chart shown in Figure 31-5. The $^{226}_{88}Ra$ nuclide decays by giving off alpha particles, forming radon-222. The nuclear equation is

$$^{226}_{88}\mathrm{Ra} \rightarrow {}^{222}_{86}\mathrm{Rn} + {}^{4}_{2}\mathrm{He}$$

Radon-222 is a radioactive noble gas. It is collected in tubes and used for the treatment of disease. Variations in the amount of radon in soil gases near geologic faults may help in predicting earthquakes. The accumulation of radon in airtight energy-efficient homes may be a health hazard. The source of the radon is the decay of uranium, which is widely distributed in the earth's crust.

The $^{222}_{86}Rn$ nuclei are unstable and have a half-life of about four days. They decay by giving off alpha particles.

$$^{222}_{86}\mathrm{Rn} \rightarrow {}^{218}_{84}\mathrm{Po} + {}^{4}_{2}\mathrm{He}$$

The remaining atomic number and mass number changes shown on the decay chart are also explained in terms of the particles given off. When it loses alpha particles, $^{210}_{84}Po$ forms $^{206}_{82}Pb$. This is a stable, nonradioactive isotope of lead. Thus a series of spontaneous transmutations begins with $^{238}_{92}U$, passes through $^{226}_{88}Ra$, and ends with the formation of stable $^{206}_{82}Pb$.

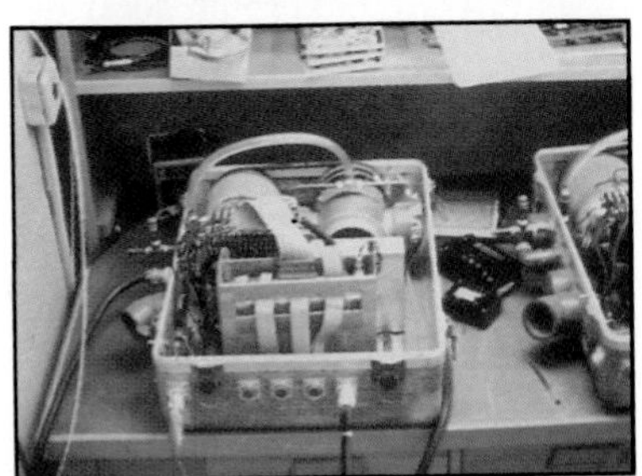

Figure 31-7. The variation in the amount of radon in soil gases near geologic faults helps in predicting earthquakes. The monitoring station shown in the photos measures and records the amount of radon and on signal transmits these data to a central laboratory where they are analyzed.

Practice Problems

Complete the following nuclear equations:

1. $^{218}_{84}\mathrm{Po} \rightarrow \ldots\ldots + {}^{4}_{2}\mathrm{He}$
2. $^{214}_{82}\mathrm{Pb} \rightarrow {}^{214}_{83}\mathrm{Bi} + \ldots\ldots$
3. $^{214}_{83}\mathrm{Bi} \rightarrow \ldots\ldots + \ldots\ldots$
4. $^{214}_{84}\mathrm{Po} \rightarrow \ldots\ldots + \ldots\ldots$

31.6 Applications of natural radioactivity The age of any mineral containing radioactive substances can be estimated with a fair degree of accuracy. Such an estimate is based on the fact

Figure 31-8. Computer display of high-energy cosmic ray event in the IBM underground proton decay detector, Ohio.

that radioactive substances decay at known rates. It is also assumed that these rates have not changed during the existence of the mineral and that there has been no gain or loss of parent or daughter nuclides except by radioactive decay. The mineral is analyzed to determine the amount of long-lived parent nuclide and the amounts of shorter-lived daughter nuclides in the sample. Then, by calculation, scientists can determine how long it must have taken for these amounts of daughter nuclides to have been produced. This time is assumed to be the age of the mineral. By this method, the oldest known minerals on earth, found in western Australia in 1983, have been estimated to be between 4.1 and 4.2 billion years old. Dust from sites of moon landings has been found to be about 4.6 billion years old. The ages of moon rocks range from 3.2 to 4.6 billion years.

The age of more recent potassium-containing minerals, 50 thousand to 50 million years old, is determined quite accurately by the proportion of potassium to argon they contain. Some nuclei of $^{40}_{19}K$ decay by capturing an orbital electron to form $^{40}_{18}Ar$. So, over time, the proportion of argon to potassium in the mineral increases and is used to establish the mineral's age.

Some carbon atoms involved in the oxygen–carbon dioxide cycle of living plants and animals are radioactive. Radioactive $^{14}_{6}C$ is continuously being produced from $^{14}_{7}N$ atoms in the atmosphere. This change is brought about by the action of *cosmic rays*. (Cosmic rays are protons and other nuclei of very high energy. These particles come to the earth from outer space.) When living things die, the oxygen–carbon dioxide cycle ceases to continue in them. They no longer replace carbon atoms in their cells with other carbon atoms. Thus, the level of radioactivity produced by the radioactive carbon in a given amount of nonliving material slowly diminishes.

Carbon from a wooden beam taken from the tomb of an Egyptian pharaoh yields about half the radiation of carbon in living trees. The half-life of a $^{14}_{6}C$ atom is about 5730 years. Thus the age of dead wood with half the radioactivity of living wood is about 5730 years. The use of $^{14}_{6}C$ dating has also been applied to the study of bone proteins. Dating by this method can presently be used with objects up to about 50,000 years old. However, more sensitive measuring techniques are under development.

Artificial Radioactivity

31.7 Stability of a nucleus On the atomic mass scale the isotope of carbon with six protons and six neutrons in its nucleus is defined as having an *atomic mass* of exactly 12 u. On this scale, a $^{4}_{2}He$ nucleus has a mass of 4.0015 u. The mass of a proton is 1.0073 u, and the mass of a neutron is 1.0087 u. A $^{4}_{2}He$ nucleus contains two protons and two neutrons. Thus you might

The atomic mass scale based on carbon-12 is described in Section 3.13.

expect its mass to be the combined mass of these four particles, 4.0320 u. [2(1.0073 u) + 2(1.0087 u) = 4.0320 u.] Note, however, that there is a *difference* of 0.0305 u between the measured mass, 4.0015 u, and the calculated mass, 4.0320 u, of a ^4_2He nucleus. This difference in mass is called the *nuclear mass defect. The **nuclear mass defect** is the difference between the mass of a nucleus and the sum of the masses of its constituent particles.* The mass defect, converted into energy units by using Einstein's equation, $E = mc^2$, is the energy released when a nucleus is formed from the particles that compose it. This energy is generally referred to as the *binding energy. The **nuclear binding energy** is the energy released when a nucleus is formed from its constituent particles.* This energy must be supplied to a nucleus to separate it into its constituent particles.

Calculations of binding energies of the atoms of the elements show that the lightest and the heaviest elements have the smallest binding energies per nuclear particle. Elements having intermediate atomic weights have the greatest binding energies per nuclear particle. See Figure 31-10. The elements with the greatest binding energies per nuclear particle are those with the most stable nuclei. Therefore, the nuclei of the lightest and heaviest elements are less stable than the nuclei of elements having intermediate atomic weights.

The stability of atomic nuclei is affected by the ratio of the neutrons to protons that compose them. *Among atoms having low atomic numbers,* the most stable nuclei are those whose neutron-to-proton ratio is 1:1. Nuclei with a greater number of neutrons than protons have lower binding energies and are less stable. Many properties of nuclear particles indicate that energy levels

Figure 31-9. Based on radiocarbon-determined growth rates, creosote clones are certainly among the oldest *living things* known and may have persisted continuously since the first seedlings established in the Mojave Desert at the close of the Wisconsin glaciation. King Clone, pictured above, is an elliptical creosote clone about 23 meters long and estimated to be 11,700 years of age. It is located in Johnson Valley, California.

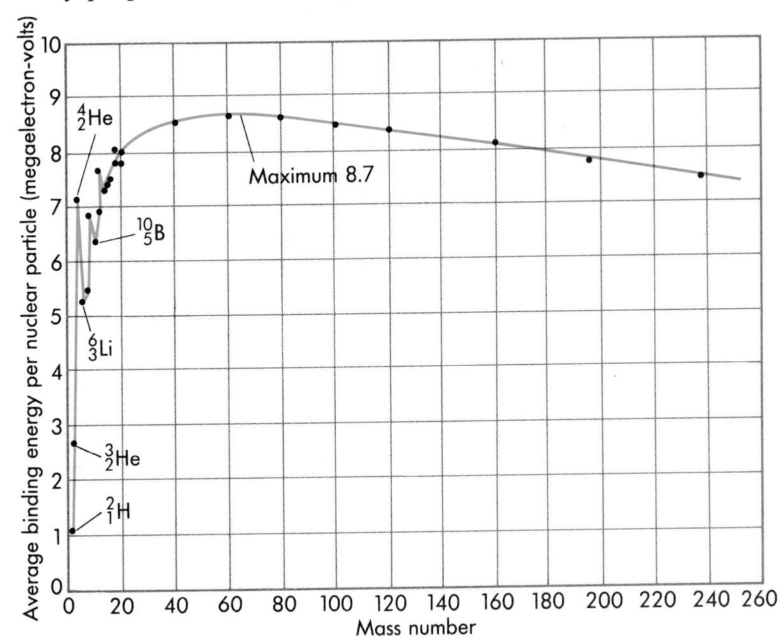

Figure 31-10. This graph shows the relationship between binding energy per nuclear particle and mass number.

exist *within* the atomic nucleus. In nuclei with an equal number of neutrons and protons, the particles apparently occupy the lowest energy levels in the nucleus. In this way, they give it stability. However, in low-atomic-number nuclei that contain an excess of neutrons over protons, some of the neutrons seem to occupy higher energy levels. This reduces the binding energy and consequently lowers the stability of the nucleus.

As the atomic number increases, stable nuclei have a neutron-to-proton ratio increasingly greater than 1:1. For example, $^{127}_{53}I$, the stable isotope of iodine, has a $\frac{\text{neutron}}{\text{proton}}$ ratio of $\frac{74}{53}$ (about 1.40:1). The stable end product of the uranium-238 decay series is $^{206}_{82}Pb$. It has a $\frac{\text{neutron}}{\text{proton}}$ ratio of $\frac{124}{82}$ (about 1.51:1).

The stability of a nucleus *also* depends on the even-odd relationship of the number of protons and neutrons. Most stable nuclei, 157 of them, have even numbers of both protons and neutrons. There are 55 stable nuclei having an even number of protons and an odd number of neutrons; 50 stable nuclei have an odd number of protons and an even number of neutrons. Only four stable nuclei having odd numbers of both protons and neutrons are known.

Because of the difference in stability of different nuclei, there are four types of nuclear reactions. In each type a small amount of the mass of the reactants is converted into energy, forming products of greater stability.

1. A nucleus undergoes ***radioactive decay.*** The nucleus releases an alpha or beta particle and gamma rays, forming a slightly lighter, more stable nucleus. The process of radioactive decay was explained in Sections 31.4 and 31.5.
2. A nucleus is bombarded with alpha particles, protons, deuterons (deuterium nuclei, $^{2}_{1}H$), neutrons, or other particles. The now unstable nucleus emits a proton or a neutron and becomes more stable. This process is ***nuclear disintegration.***
3. A very heavy nucleus splits to form medium-weight nuclei. This process is known as ***fission.***
4. Light-weight nuclei combine to form heavier, more stable nuclei. This process is known as ***fusion.***

Figure 31-11. "M-2 Telemanipulator" which is being developed by Oak Ridge National Laboratory for use in remote maintenance of future nuclear fuel reprocessing plants. This robotics technology is also applicable to deep-sea and space exploration, bomb disposal, and handling toxic materials.

31.8 Stable nuclei from radioactive decay The release of an alpha particle from a radioactive nucleus such as $^{238}_{92}U$ decreases the mass of the nucleus. The resulting lighter nucleus has higher binding energy per nuclear particle. The release of an alpha particle decreases the number of protons and neutrons in a nucleus *equally and also by an even number.*

Beta particles are released when neutrons change into protons, as during the decay of $^{234}_{90}Th$. This change lowers the neutron-to-proton ratio toward the value found for stable nuclei of the same mass number. Both alpha-particle release and beta-particle release yield a product nucleus that is more stable than the original nucleus. Refer again to Section 31.5 and Figure 31-5.

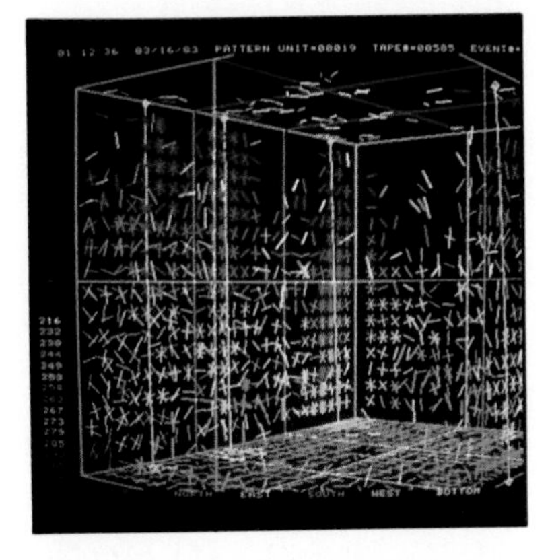

Figure 31-12. Computer display of a four muon event in an underground proton decay detector. Muons are subatomic particles that exist in positive and negative forms. They have a mass about 207 times that of an electron. They decay into electrons and neutrinos.

31.9 The first artificial nuclear disintegration After scientists discovered how uranium and radium undergo natural decay and transmutation, they worked to produce artificial transmutations. They had to find a way to add protons to a nucleus of an atom of an element, converting it to the nucleus of an atom of a different element. In 1919, Rutherford produced the first artificial nuclear disintegration. His method involved bombarding nitrogen with alpha particles from radium. He obtained protons (hydrogen nuclei) and a stable isotope of oxygen. See Figure 31-13. This nuclear disintegration is represented by the following equation:

$$^{14}_{7}\mathrm{N} + ^{4}_{2}\mathrm{He} \rightarrow ^{17}_{8}\mathrm{O} + ^{1}_{1}\mathrm{H}$$

31.10 Proofs of Einstein's equation In 1932, the two English scientists J. D. Cockcroft (1897–1967) and E. T. S. Walton (b. 1903) experimentally proved Einstein's equation, $E = mc^2$. They bombarded lithium with high-speed protons. Alpha particles and an enormous amount of energy were produced.

$$^{7}_{3}\mathrm{Li} + ^{1}_{1}\mathrm{H} \rightarrow ^{4}_{2}\mathrm{He} + ^{4}_{2}\mathrm{He} + \text{energy}$$

There is a loss of matter in this reaction. One lithium nucleus (mass 7.0144 u) was hit by a proton (mass 1.0073 u). These particles formed two alpha particles (helium nuclei), each having a mass of 4.0015 u. Calculation shows that there is a loss of 0.0187 u. $[(7.0144\ u + 1.0073\ u) - 2(4.0015\ u)]$. Cockcroft and Walton found that the energy released very nearly equaled that predicted by Einstein for such a loss in mass. Additional experiments have further supported Einstein's equation.

31.11 Neutron emission in some nuclear disintegrations You have already learned that neutrons were discovered by Chadwick in 1932. He first detected them in an experiment that involved bombarding beryllium with alpha particles.

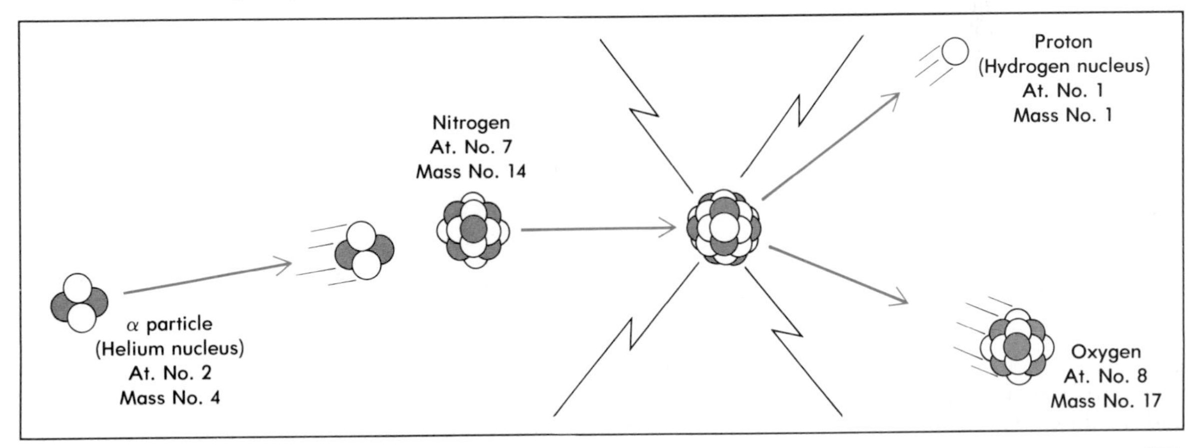

Figure 31-13. This diagram shows the historic nuclear disintegration performed by Rutherford.

$$^{9}_{4}Be + ^{4}_{2}He \rightarrow ^{12}_{6}C + ^{1}_{0}n$$

The symbol for a neutron is $^{1}_{0}n$. This symbol indicates a particle with zero atomic number (no protons) and a mass number of 1. The reaction described above proved that neutrons were a second type of particle in the nuclei of atoms.

Neutrons are described in Section 3.5.

31.12 The cyclotron and other particle accelerators Radium was used as a natural source of alpha particles in many early experiments. However, radium is not very efficient in producing nuclear changes. As a result, scientists sought more effective ways of producing high-energy particles for bombarding nuclei. This search resulted in the development of many large electric devices for accelerating charged particles.

The *cyclotron* was invented by E. O. Lawrence (1901–1958) of the University of California. It consists of a cylindrical box placed between the poles of a huge electromagnet. Air is pumped out of the box until a high vacuum is produced. The "bullets" used to bombard nuclei are usually protons (protium nuclei) or deuterons (deuterium nuclei). They enter the cylindrical box through its center.

Inside the box are two hollow, D-shaped electrodes called *dees*. These dees are connected through an oscillator to a source of very high voltage. When the cyclotron is in operation, the oscillator reverses the electric charge on the dees very rapidly. The combined effects of the high-voltage alternating potential and the electromagnetic field cause the protons or deuterons inside to move in a spiral course. They move faster and faster as they near the outside of the box, gaining more and more energy. When they reach the outer rim of the box, they are deflected toward the target. The energy of particles accelerated in a simple cyclotron can reach 15,000,000 electron-volts. This is the energy an electron would have if it were accelerated across a potential difference of 1.5×10^{7} volts. By studying the fragments of atoms formed by bombardment, scientists have learned a great deal about atomic structure. They have also discovered much about the products formed when atoms disintegrate.

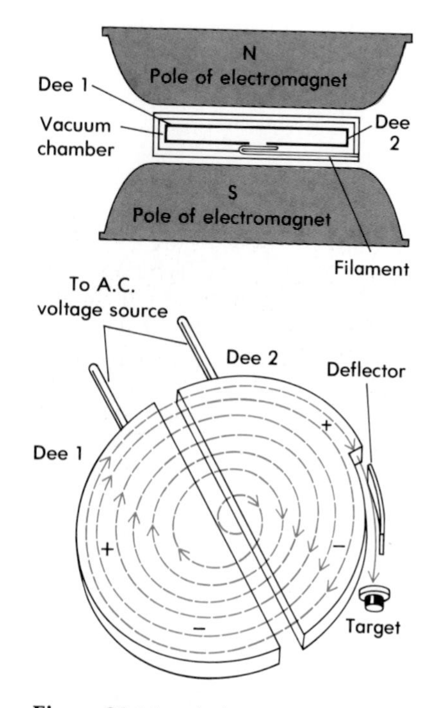

Figure 31-14. A diagram of the cyclotron used to produce "atomic bullets" of very high energy.

Other machines for bombarding atomic nuclei are the *synchrotron* and the *linear accelerator*. The synchrotron works much like the cyclotron. By varying both the oscillating voltage and the magnetic field, particles are accelerated in a narrow circular path rather than in a spiral. A synchrotron can give an energy of 800 billion electron-volts to the protons it accelerates. In the linear accelerator, the particles travel in a straight line. They are accelerated by passage through many stages of potential difference. The acceleration of electrons for bombardment or for production of high-energy X rays was formerly carried out in a circular accelerator called a *betatron*. Today synchrotrons and linear accelerators are better devices for producing electron acceleration.

31.13 Neutrons as "bullets" Before the discovery of neutrons in 1932, alpha particles and protons were used to study atomic nuclei. But alpha particles and protons are charged particles. It requires great quantities of energy to "fire" these charged "bullets" into a nucleus. Their positive charge causes them to be repelled by the positive nuclear charge. The various kinds of particle accelerators were developed to give charged "bullets" enough energy to overcome this repelling force.

When accelerated positive particles strike a target material, usually lithium or beryllium, neutrons are produced. Neutrons have no charge. Thus there is no repelling force, and they can easily penetrate the nucleus of an atom. Some fast neutrons may go through an atom without causing any change in it. Other fast neutrons may cause nuclear disintegration. Slow neutrons, on the other hand, are sometimes trapped by a nucleus. This nucleus then becomes unstable and may break apart. Fast neutrons are slowed down by passage through materials composed of elements of low atomic weight. Examples are deuterium oxide or graphite.

Figure 31-15. Giant magnet for acceleration experiment at Fermi National Accelerator Laboratory.

31.14 Artificial elements from neutron bombardment The $^{238}_{92}U$ nuclide is the most plentiful isotope of uranium. When hit by slow neutrons, a $^{238}_{92}U$ nucleus may capture a neutron. This capture produces the nucleus of an atom of an unstable isotope of uranium, $^{239}_{92}U$. This nucleus emits a beta particle and, in so doing, becomes the nucleus of an atom of an artificial radioactive element, neptunium. Neptunium has atomic number 93. The nuclide formed has the symbol $^{239}_{93}Np$.

$$^{238}_{92}U + ^{1}_{0}n \rightarrow ^{239}_{92}U$$

$$^{239}_{92}U \rightarrow ^{239}_{93}Np + ^{0}_{-1}e$$

Neptunium is itself an unstable element. The nucleus of a neptunium atom gives off a beta particle. This change produces the nucleus of an atom of still another artificial element, plutonium, atomic number 94. This nuclide has the symbol $^{239}_{94}Pu$.

$$^{239}_{93}Np \rightarrow ^{239}_{94}Pu + ^{0}_{-1}e$$

Neptunium and plutonium were the first artificial *transuranium* elements. ***Transuranium elements*** are those with more than 92 protons in their nuclei. As this is written, 18 artificially prepared transuranium elements have been reported. In addition to neptunium and plutonium, there are americium, curium, berkelium, californium, einsteinium, fermium, mendelevium, nobelium, and lawrencium. Elements 104 through 109 have only systematic names. All these elements are prepared by bombarding the nuclei of uranium or other complex atoms with neutrons, alpha particles, or other "nuclear bullets." See Table 31-3.

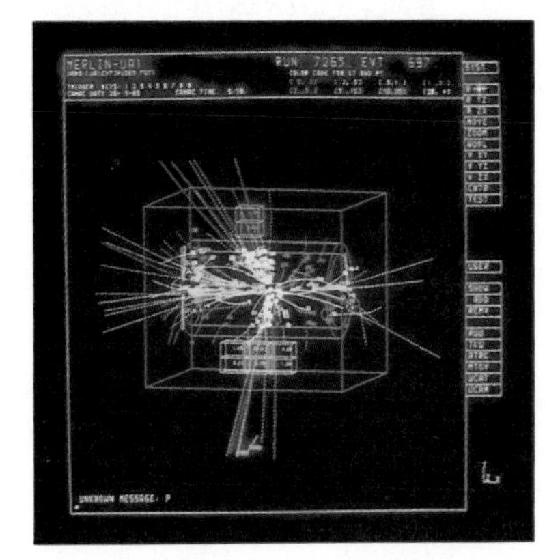

Figure 31-16. Two-jet event in UA1 detector, CERN. Two jets of blue, purple, and green tracks produced by quarks. What's a quark? Quarks are even smaller particles that make up subatomic particles.

Table 31-3
REACTIONS FOR THE FIRST PREPARATION OF TRANSURANIUM ELEMENTS

Atomic number	*Name*	*Symbol*	*Nuclear reaction*
93	neptunium	Np	$^{238}_{92}U + ^{1}_{0}n \rightarrow ^{239}_{93}Np + ^{0}_{-1}e$
94	plutonium	Pu	$^{238}_{92}U + ^{2}_{1}H \rightarrow ^{238}_{93}Np + 2^{1}_{0}n$
			$^{238}_{93}Np \rightarrow ^{238}_{94}Pu + ^{0}_{-1}e$
95	americium	Am	$^{239}_{94}Pu + 2^{1}_{0}n \rightarrow ^{241}_{95}Am + ^{0}_{-1}e$
96	curium	Cm	$^{239}_{94}Pu + ^{4}_{2}He \rightarrow ^{242}_{96}Cm + ^{1}_{0}n$
97	berkelium	Bk	$^{241}_{95}Am + ^{4}_{2}He \rightarrow ^{243}_{97}Bk + 2^{1}_{0}n$
98	californium	Cf	$^{242}_{96}Cm + ^{4}_{2}He \rightarrow ^{245}_{98}Cf + ^{1}_{0}n$
99	einsteinium	Es	$^{238}_{92}U + 15^{1}_{0}n \rightarrow ^{253}_{99}Es + 7^{0}_{-1}e$
100	fermium	Fm	$^{238}_{92}U + 17^{1}_{0}n \rightarrow ^{255}_{100}Fm + 8^{0}_{-1}e$
101	mendelevium	Md	$^{253}_{99}Es + ^{4}_{2}He \rightarrow ^{256}_{101}Md + ^{1}_{0}n$
102	nobelium	No	$^{246}_{96}Cm + ^{12}_{6}C \rightarrow ^{254}_{102}No + 4^{1}_{0}n$
103	lawrencium	Lr	$^{252}_{98}Cf + ^{10}_{5}B \rightarrow ^{258}_{103}Lr + 4^{1}_{0}n$
104	unnilquadium (USSR)	Unq	$^{242}_{94}Pu + ^{22}_{10}Ne \rightarrow ^{260}_{104}Unq + 4^{1}_{0}n$
104	unnilquadium (US)	Unq	$^{249}_{98}Cf + ^{12}_{6}C \rightarrow ^{257}_{104}Unq + 4^{1}_{0}n$
105	unnilpentium (US)	Unp	$^{249}_{98}Cf + ^{15}_{7}N \rightarrow ^{260}_{105}Unp + 4^{1}_{0}n$
106	unnilhexium (US)	Unh	$^{249}_{98}Cf + ^{18}_{8}O \rightarrow ^{263}_{106}Unh + 4^{1}_{0}n$
107	unnilseptium (USSR)	Uns	$^{209}_{83}Bi + ^{54}_{24}Cr \rightarrow ^{261}_{107}Uns + 2^{1}_{0}n$
108	unniloctium (W. Germany)	Uno	$^{208}_{82}Pb + ^{58}_{26}Fe \rightarrow ^{265}_{108}Uno + ^{1}_{0}n$
109	unnilennium (W. Germany)	Une	$^{209}_{83}Bi + ^{58}_{26}Fe \rightarrow ^{266}_{109}Une + ^{1}_{0}n$

31.15 Artificial radioactive atoms In 1934, Madame Curie's daughter Irène (1897–1956) and her husband Frédéric Joliot (1900–1958) discovered that stable atoms can be made radioactive by artificial means. This occurs when they are bombarded with deuterons or neutrons. Radioactive isotopes of all the elements have been prepared. For example, radioactive $^{60}_{27}Co$ can be produced from natural nonradioactive $^{59}_{27}Co$ by slow-neutron bombardment. The nuclear equation is

$$^{59}_{27}Co + ^{1}_{0}n \rightarrow ^{60}_{27}Co$$

Radiation from $^{60}_{27}Co$ consists of beta particles and gamma rays.

Radioactive $^{32}_{15}P$ is prepared by bombardment of $^{32}_{16}S$ with slow neutrons.

$$^{32}_{16}S + ^{1}_{0}n \rightarrow ^{32}_{15}P + ^{1}_{1}H$$

The radiation from $^{32}_{15}P$ consists of only beta particles.

Radioactive phosphorus, radioactive cobalt, and some other radioactive nuclides are used to treat certain forms of cancer. Radioactive drugs are used for diagnostic purposes and for test purposes on blood and tissue samples. The gamma radiation from $^{60}_{27}Co$ can be used to preserve food. It also kills bacteria that spoil food and insects that infest food. Many radioactive nuclides are used as *tracers.* By using them, scientists can determine the course of chemical reactions, the cleaning ability of deter-

Figure 31-17. The experimental area of the UNILAC (Universal Linear Accelerator) near Darmstadt, Germany, where elements 108 and 109 were produced.

gents, the wearing ability of various products, the efficiency of fertilizers, the flow of fluids through pipelines, and the movement of sand along sea coasts. Many new radioactive nuclides are made by slow-neutron bombardment in the nuclear reactor at Oak Ridge, Tennessee.

31.16 Fission of uranium The element uranium exists as three naturally occurring isotopes: $^{238}_{92}U$, $^{235}_{92}U$, and $^{234}_{92}U$. Most uranium is the nuclide $^{238}_{92}U$. Only 0.7% of natural uranium is $^{235}_{92}U$. The nuclide $^{234}_{92}U$ occurs in only the slightest traces. You have learned that transuranium elements can be produced when $^{238}_{92}U$ is bombarded with slow neutrons. However, when $^{235}_{92}U$ is bombarded with slow neutrons, each atom may capture one of the neutrons. This extra neutron in the nucleus makes it very unstable. Instead of giving off an alpha or beta particle, as in other radioactive changes, the nucleus splits into medium-weight parts. Neutrons are usually produced during this *fission*. There is a small loss of mass, which appears as a great amount of energy. One equation for the fission of $^{235}_{92}U$ is

$$^{235}_{92}\mathrm{U} + {}^{1}_{0}\mathrm{n} \rightarrow {}^{138}_{56}\mathrm{Ba} + {}^{95}_{36}\mathrm{Kr} + 3{}^{1}_{0}\mathrm{n} + \text{energy}$$

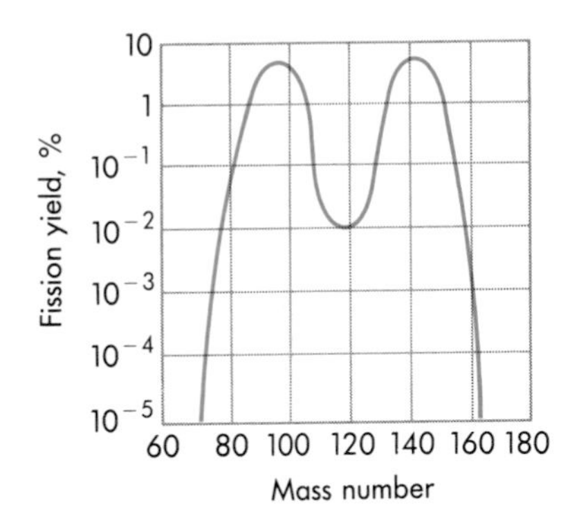

Figure 31-18. This graph shows the yield of various nuclides that are produced by the slow-neutron fission of uranium-235. The products vary in mass number from 72 to 161. The most probable products have mass numbers of 95 and 138. Observe the low probability of fission products of nearly equal mass numbers.

The atomic mass of $^{235}_{92}U$ is slightly greater than 235 *u*. The atomic masses of the unstable barium and krypton isotopes are slightly less than 138 *u* and 95 *u*, respectively. Thus the masses of the reactants and the masses of the products are not equal. Instead, about 0.2 *u* of mass is converted to energy for each uranium atom undergoing fission. Plutonium, made from $^{238}_{92}U$, also undergoes

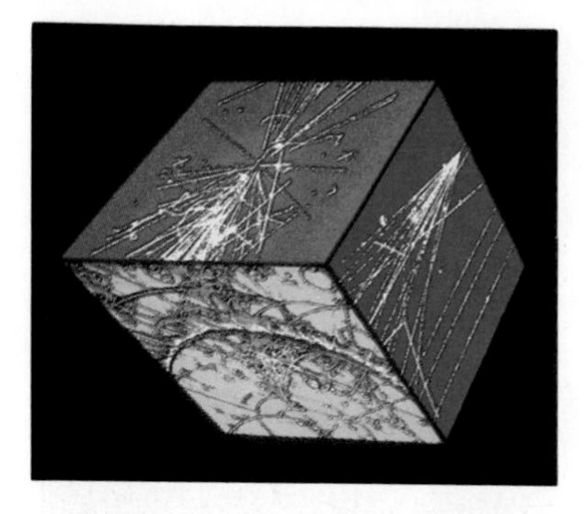

Figure 31-19. Cube of artificially colored bubble chamber photos, showing the tracks of subatomic particles.

fission to produce more neutrons when bombarded with slow neutrons.

Uranium atoms exist in very small amounts in many minerals. In the time since their formation, some $^{238}_{92}U$ atoms have undergone spontaneous fission. These fissions have left tracks in certain mineral crystals and in glassy materials. These tracks can be made visible under a microscope by etching. The number of tracks in a given area and the amount of $^{238}_{92}U$ in the specimen are used to determine the age of the crystal or the time since a glassy material was last heated to a high temperature. This method, called fission-track dating, can establish the age of materials from a few decades old to as old as the solar system.

31.17 Nuclear chain reaction *A* ***chain reaction*** *is one in which the material or energy that starts the reaction is also one of the products.* The fissions of $^{235}_{92}U$ and $^{239}_{94}Pu$ can produce chain reactions. One neutron causes the fission of one $^{235}_{92}U$ nucleus. Two or three neutrons given off when this fission occurs can cause the fission of other $^{235}_{92}U$ nuclei. Again neutrons are emitted. These can cause the fission of still other $^{235}_{92}U$ nuclei. This is a chain reaction. It continues until all the $^{235}_{92}U$ atoms have split or until the neutrons fail to strike $^{235}_{92}U$ nuclei.

31.18 Action in a nuclear reactor A ***nuclear reactor*** *is a device in which the controlled fission of radioactive material produces new radioactive substances and energy.* One of the earliest nuclear reactors was built at Oak Ridge, Tennessee, in 1943. This reactor uses natural uranium. It has a lattice-type structure with blocks of graphite forming the framework. Spaced between the blocks of graphite are rods of uranium, encased in aluminum cans for pro-

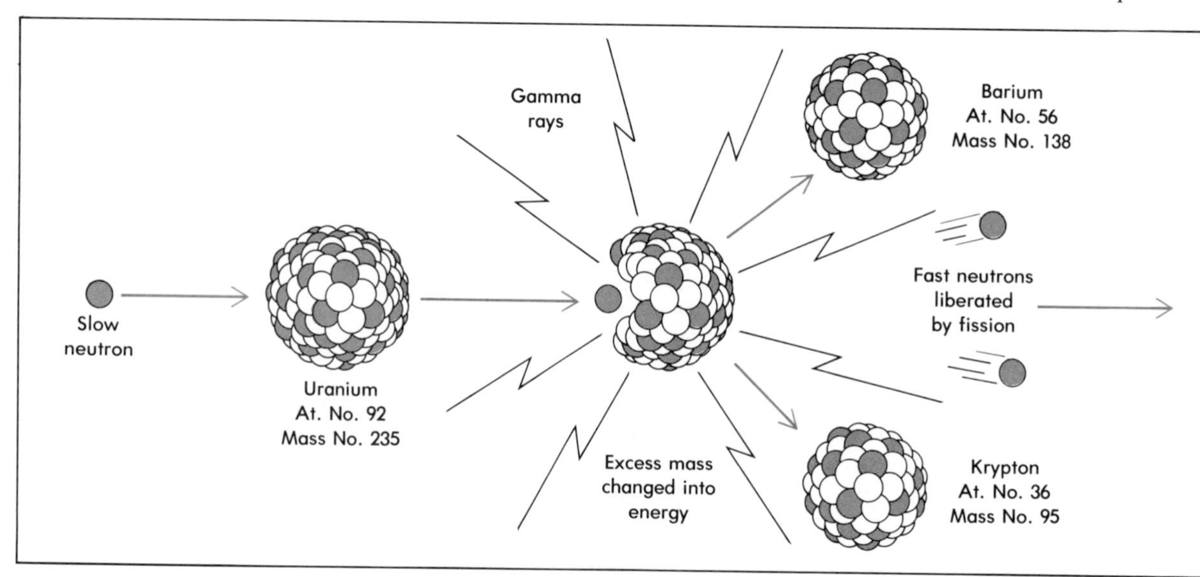

tection. *Control rods* of neutron-absorbing boron steel are inserted into the lattice to limit the number of free neutrons. The reactor is air-cooled.

The rods of uranium or uranium oxide are the *nuclear fuel* for the reactor. The energy released in the reactor comes from changes in the uranium nuclei. Graphite is said to be the *moderator* because it slows down the fast neutrons produced by fission. By doing so, it makes them more readily captured by a nucleus and thus more effective for producing additional nuclear changes. The mass of uranium in such a reactor is important. Enough uranium must be present to provide the number of neutrons needed to sustain a chain reaction. This quantity of uranium is called the ***critical mass.***

Two types of reactions occur in the fuel in such a reactor. Neutrons cause $^{235}_{92}U$ nuclei to undergo fission. The fast neutrons from this fission are slowed down as they pass through the graphite. Some strike other $^{235}_{92}U$ nuclei and continue the chain reaction. Other neutrons strike $^{238}_{92}U$ nuclei, starting the changes that finally produce plutonium. Great quantities of heat energy are released. For this reason the reactor must be cooled continuously by blowing air through tubes in the lattice. The rate of the reaction is controlled by the insertion or removal of the neutron-absorbing control rods. This type of reactor is now used to produce radioactive isotopes.

In nuclear power plants, the reactor is the source of heat energy. Pressurized water is both the moderator and the coolant. The heat from the reactor absorbed by the pressurized water is used to produce steam. This steam turns the turbines, which drive the electric generators. Present problems with nuclear power plant development include questions about location, and

Figure 31-20. Nuclear plant near San Clemente, California.

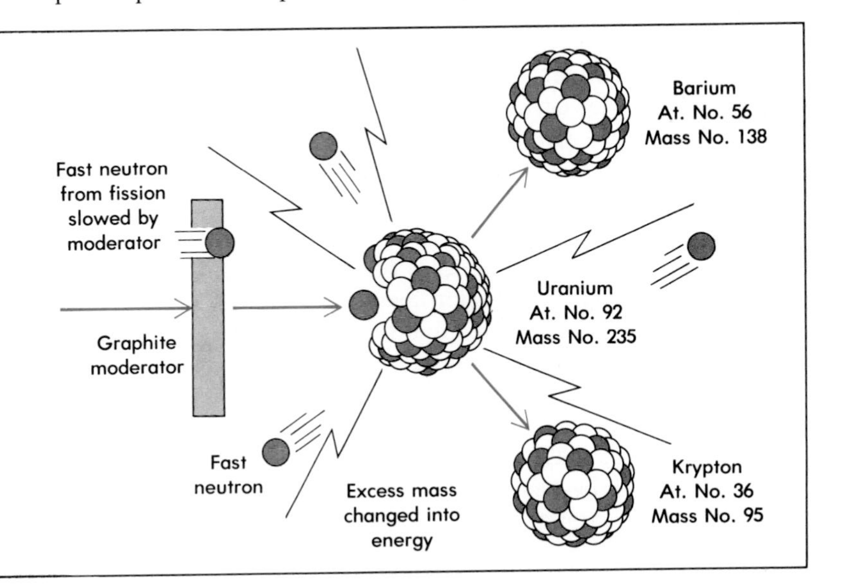

Figure 31-21. Neutrons from fission of a $^{235}_{92}U$ nucleus, when slowed down by a carbon moderator, can cause fission in a second $^{235}_{92}U$ nucleus. This process makes a chain reaction possible.

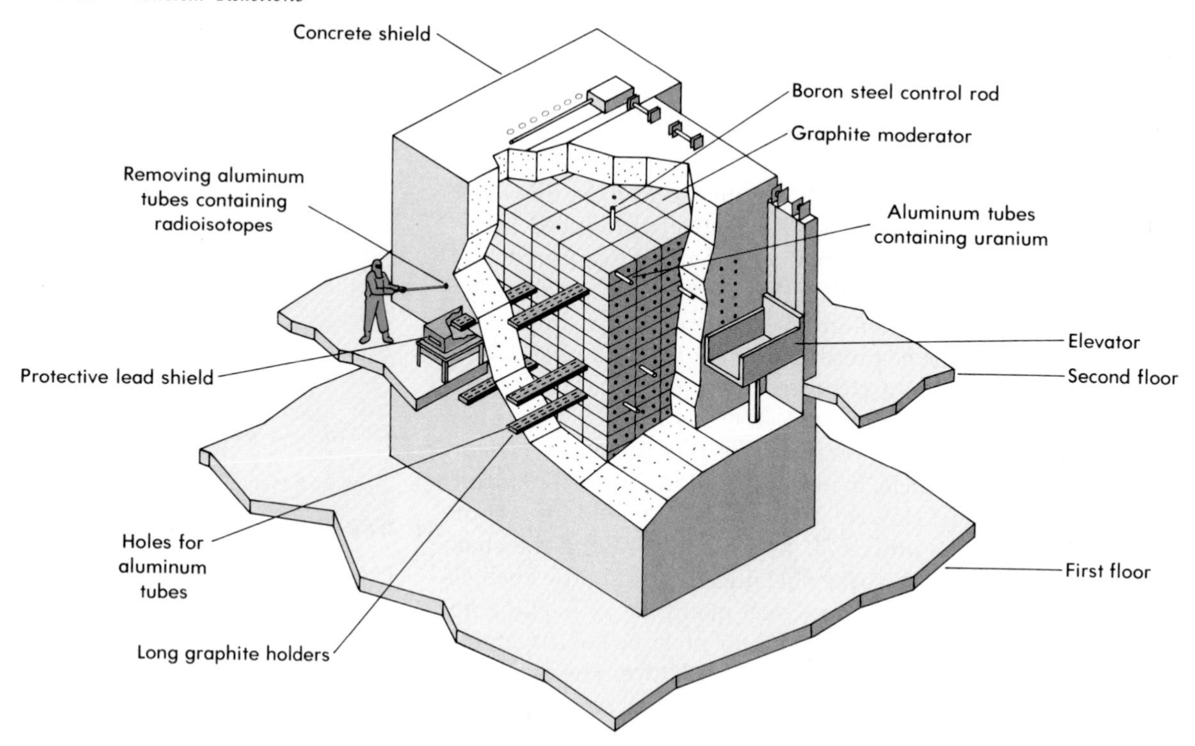

Figure 31-22. A cutaway view of the Oak Ridge reactor.

Figure 31-23. Glass pellet containing deuterium and tritium: The target contains over 100 atmospheres of deuterium-tritium fuel ready to be fused by the heat from a powerful laser.

environmental requirements, safety of operation, and plant construction costs. The procurement and enrichment of uranium, as well as methods for storing or reprocessing used nuclear reactor fuel, are also being studied.

31.19 Fusion reactions Nuclear stability can be increased by combining light-weight nuclei into heavier nuclei. This process is defined as ***fusion.***

Fusion reactions are the source of the sun's energy. It is believed that there are two series of such reactions going on in the sun. One series occurs at the very hot center of the sun, and the other takes place in the cooler outer portion of the sun. These two reactions proceed by different pathways. However, their net effect is the combination of four hydrogen nuclei into a helium nucleus. A loss of mass occurs, and a tremendous amount of energy is released. More energy is released per gram of fuel in a fusion reaction than in a fission reaction.

Current research indicates that fusion reactions may be controlled. Scientists are attempting to find ways to confine the nuclear fuel, usually ionized deuterium and tritium, and at the same time make it hot enough and dense enough long enough for fusion to occur. Magnetic fields are usually used to confine the ionized fuel because, at the initial temperature required, about 10^8 °K, no known material container will do. The "particle density × confinement time" product is about $\frac{10^{14}\ \text{particles}}{\text{cm}^3} \times \text{s}$.

This can be interpreted as requiring a density of $\frac{10^{14}\ \text{particles}}{\text{cm}^3}$ for 1 s or some equivalent, such as a density of $\frac{10^{18}\ \text{particles}}{\text{cm}^3}$ for 10^{-4} s. Experiments using high-powered laser light or beams of particles, such as protons or lithium ions, as ways to start the fusion reaction are being conducted. If fusion reactions are to be a practical source of energy, the energy given off must be greater than the energy required to heat and compress the ionized fuel. Such reactions may someday produce useful energy that can be converted to electricity.

SUMMARY

Radioactivity is the spontaneous breakdown of an unstable atomic nucleus with the release of particles and rays.

Radium was discovered by Pierre and Marie Curie in 1898. It is a very radioactive element always found in uranium ores. Radium resembles barium in its chemical properties. Radioactive nuclides and their compounds have several unusual properties: (1) they affect the light-sensitive emulsion on a photographic film, (2) they produce an electric charge in the surrounding air, (3) they produce fluorescence with certain other compounds, (4) their radiations have special physiological effects, (5) they undergo radioactive decay. The half-life of a radioactive nuclide is the length of time that it takes for half of a given number of atoms of the nuclide to decay.

The radiation given off by radioactive nuclides consists of three different kinds of particles and rays: (1) alpha particles, which are helium nuclei; (2) beta particles, which are electrons; and (3) gamma rays, which are high-energy X rays. The emission of these particles from the nuclei of radioactive nuclides causes the nuclides to decay into simpler ones.

The age of certain minerals and of carbon-containing materials can be estimated by the amounts of radioactive nuclides they contain.

The difference between the sum of the masses of the separate particles making up a nucleus and the actual mass of a nucleus is the nuclear mass defect.

There are four types of reactions that nuclei undergo to become more stable: (1) radioactive decay, (2) nuclear disintegration, (3) fission, and (4) fusion.

A nuclear reactor is a device in which the controlled fission of radioactive material produces new radioactive substances and heat energy, which may be used to generate electricity.

A fusion reaction is one in which light-weight nuclei are combined into heavier nuclei. Fusion reactions produce the sun's heat and light.

VOCABULARY

alpha particle
beta particle
chain reaction
control rod
cosmic ray
critical mass
cyclotron
daughter nuclide
deuteron
fission
fluoresce
fusion
gamma ray
half-life
linear accelerator
moderator
nuclear binding energy
nuclear disintegration
nuclear equation
nuclear mass defect
nuclear reactor
parent nuclide
radioactive decay
radioactive tracer
radioactivity
synchrotron
transmutation
transuranium element

QUESTIONS

Group A

1. *(a)* What is radioactivity? *(b)* When and by whom was radioactivity discovered? *(c)* How was the discovery made? *A1, A5*
2. *(a)* Why did Marie and Pierre Curie suspect that there was a radioactive element other than uranium in pitchblende? *(b)* What did their subsequent discoveries show? *A3a, A5*
3. Compare the radioactivity of uranium, radium, and polonium. *A3a*
4. Define *half-life*. *A1*
5. What is the source of the alpha or beta particles emitted by a radioactive nuclide? *A2c*
6. *(a)* What practical use was formerly made of the fluorescence produced in zinc sulfide by a radium compound? *(b)* Why is a promethium-147 compound now used instead of a radium compound? Pm-147 is a beta-particle emitter with a half-life of 2.62 years. *A2a*
7. What change in atomic number, identity, and mass number occurs when a radioactive nuclide gives off an alpha particle? *A2c, A2d*
8. What change in atomic number, identity, and mass number occurs when a radioactive nuclide gives off a beta particle? *A2c, A2d*
9. *(a)* Radon-222 has a half-life of less than four days, yet it is continually found in very low concentrations in the earth's crust, surface waters, and atmosphere. Why? *(b)* In what ways is radon useful? *(c)* In what way may radon be harmful? *A2d, A3a*
10. *(a)* Describe how the age of a radioactive mineral is estimated. *(b)* What assumptions are necessary? *A2e*
11. Name and briefly describe the four types of nuclear reactions that produce more stable nuclei. *A2h*
12. Write the nuclear equation for the reaction by which Chadwick discovered neutrons. *A4b*
13. Neutrons are more effective than protons or alpha particles for bombarding atomic nuclei. Why? *A2j*
14. What may happen to a neutron that is fired at the nucleus of an atom? *A2j*
15. List five different uses of artificially radioactive nuclides. *A3c*
16. Name the naturally occurring isotopes of uranium and indicate their relative abundance. *A3a*
17. How is the fission of a ${}^{235}_{92}U$ nucleus produced? *A2h*
18. Define: *(a) chain reaction; (b) nuclear reactor; (c) critical mass.* *A1*
19. What factors are currently affecting the construction of nuclear power plants?
20. Describe the reaction that produces the sun's energy. *A2l*

Group B

21. Where on the periodic table are most of the radioactive nuclides located? *A3d*
22. Why can the radiation from a radioactive material affect photographic film, even though the film is well wrapped in black paper? *A2a*
23. Explain how a radioactive material can change the discharge rate of a charged electroscope. *A2b*
24. What instrument for studying radioactivity has replaced the electroscope? *A2b*
25. *(a)* What conditions cause variations in the decay rate of a radioactive material? *(b)* Are these variations slight, moderate, or great? *A3b*
26. Make a chart that compares the following properties of alpha particles and beta particles: identity, mass number, charge, speed, and penetrating ability. *A2c*
27. *(a)* What are gamma rays? *(b)* How do scientists believe they are produced? *(c)* How does their penetrating ability compare with that of alpha particles and beta particles? *A2c*
28. Write the nuclear equation for the release of an alpha particle by ${}^{210}_{84}Po$. *A4a, A4b*
29. Write the nuclear equation for the release of a beta particle by ${}^{210}_{82}Pb$. *A4a, A4b*

30. The parent nuclide of the thorium decay series is $^{232}_{90}Th$. The first four decays are: alpha particle emission, beta particle emission, beta particle emission, alpha particle emission. Write the nuclear equations for this series of emissions. *A4b*
31. *(a)* How does binding energy per nuclear particle vary with mass number? *(b)* How does binding energy per nuclear particle affect the stability of a nucleus? *A2g*
32. Describe two ways in which the number of protons and the number of neutrons in a nucleus affect its stability. *A2g*
33. Explain how each type of nuclear reaction produces more stable nuclei. *A2h*
34. *(a)* Who produced the first artificial nuclear disintegration? *(b)* How many years ago? *(c)* Write the equation for this reaction. *A2h, A5*
35. *(a)* How was Einstein's equation for the relationship between matter and energy, $E = mc^2$, proved to be correct? *(b)* Give the names and nationality of the scientists responsible for the proof. *A2h, A5*
36. *(a)* What is the nature of the path of the accelerated particles in a cyclotron? *(b)* What causes them to take this path? *A2i*
37. *(a)* Describe the path of the accelerated particles in a synchrotron; *(b)* in a linear accelerator. *A1*
38. Explain the nuclear changes that occur when $^{239}_{94}Pu$ is produced from $^{238}_{92}U$. *A2h*
39. Assume that element 110, ununnilium, can be prepared by a reaction similar to those used in preparing unniloctium and unnilennium. *(a)* If $^{209}_{83}Bi$ is used as a target nucleus, what projectile nucleus might be used? *(b)* If $^{58}_{26}Fe$ is the projectile nucleus, what target nucleus might be used? *(c)* Has the preparation of element 110 been announced? *(d)* If so, what reaction was actually used? (Appendix Table 2 may be helpful.) *A2h*
40. *(a)* How are artificially radioactive nuclides prepared? *(b)* Write a nuclear equation to show the preparation of such a nuclide. *(c)* What use is made of the above nuclide? *A3c, A4b*
41. How does the fission of $^{235}_{92}U$ produce a chain reaction? *A2h*
42. How is a uranium-graphite reactor constructed? *A2k*
43. What problems must be overcome before energy-producing controlled fusion reactions are a reality? *A2l*
44. To what extent has the use of radiation replaced the use of chemical fumigants and insecticides in food preservation?
45. What progress is being made in devising environmentally safe methods for storing radioactive wastes?

PROBLEMS

Group A

1. The half-life of thorium-227 is 18.72 days. How many days are required for three-fourths of a given amount to decay? *B1*
2. The half-life of $^{234}_{91}Pa$ is 6.75 hours. How much of a given amount remains after 27.0 hours? *B1*
3. After 4806 years, how much of an original 0.250 g of $^{226}_{88}Ra$ remains? *B1*
4. The mass of a $^{7}_{3}Li$ nucleus is 7.01436 *u*. Calculate the nuclear mass defect. The atomic masses of nuclear particles are given in Section 3.13. *B2*
5. Calculate the nuclear mass defect of $^{20}_{10}Ne$ if its nucleus has a mass of 19.98695 *u*. *B2*

Group B

6. The half-life of $^{222}_{86}Rn$ is 3.823 days. What was the original mass of $^{222}_{86}Rn$ if 0.0500 g remains after 7.646 days? *B1*
7. The half-life of $^{239}_{94}Pu$ is 24,100 years. Of an original mass of $10\bar{0}$ g, approximately how much remains after 10^5 years? *B1*
8. Calculate the nuclear mass defect per nuclear particle for $^{7}_{3}Li$ using the answer to Problem 4. Convert the mass in *u* to binding energy in megaelectron-volts by using the relationship 1 *u* = 931 Mev. *A2f, B2*
9. Calculate the binding energy per nuclear particle of $^{238}_{92}U$ in Mev if the atomic mass of its nucleus is 238.0003 *u*. *A2f, B2*

APPENDIX

Table 1 METRIC SYSTEM PREFIXES					
Factor	*Prefix*	*Symbol*	*Factor*	*Prefix*	*Symbol*
10^{12}	tera	T	10^{-1}	deci	d
10^{9}	giga	G	10^{-2}	centi	c
10^{6}	mega	M	10^{-3}	milli	m
10^{3}	kilo	k	10^{-6}	micro	μ (mu)
10^{2}	hecto	h	10^{-9}	nano	n
10^{1}	deka	da	10^{-12}	pico	p

Table 2
ISOTOPES OF SOME ELEMENTS

(Naturally occurring nonradioactive isotopes are given in bold type. Naturally occurring radioactive isotopes are boldface italics in color. All other radioactive isotopes are in italics. Naturally occurring isotopes are listed in order of their abundance. All other isotopes are listed in order of length of half-life.)

Element	*Mass numbers of isotopes*
H	**1, 2,** *3*
He	**4, 3,** *6, 8*
Li	**7, 6,** *8, 9*
Be	**9,** *10, 7, 11, 6*
B	**11, 10,** *8*
C	**12, 13,** *14, 11, 10, 15, 16, 9*
N	**14, 15,** *13, 16, 17, 18*
O	**16, 18, 17,** *15, 14, 19, 20*
F	**19,** *18, 17, 20, 21, 22*
Ne	**20, 22, 21,** *24, 23, 19, 18, 17*
Na	**23,** *22, 24, 25, 21, 26, 20*
Mg	**24, 26, 25,** *28, 27, 23, 20, 21*
Al	**27,** *26, 29, 28, 25, 30, 24*
Si	**28, 29, 30,** *32, 31, 27, 26, 25*
P	**31,** *33, 32, 30, 34, 29, 28*
S	**32, 34, 33, 36,** *35, 38, 37, 31, 30, 29*
Cl	**35, 37,** *36, 39, 38, 40, 33, 34, 32*
Ar	**40, 36, 38,** *39, 42, 37, 41, 35, 33*
K	**39, 41,** ***40,*** *43, 42, 44, 45, 38, 47, 37*
Ca	**40, 44, 42,** ***48,*** **43, 46,** *41, 45, 47, 49, 50, 39, 38, 37*
Cr	**52, 53, 50, 54,** *51, 48, 49, 56, 55, 46*
Fe	**56, 54, 57, 58,** *60, 55, 59, 52, 53, 61*
Co	**59,** *60, 57, 56, 58, 55, 61, 62, 63, 54*
Ni	**58, 60, 62, 61, 64,** *59, 63, 56, 66, 57, 65, 67*
Cu	**63, 65,** *67, 64, 61, 60, 62, 58, 66, 59, 68, 57*
Zn	***64,*** **66, 68, 67,** ***70,*** *65, 72, 62, 69, 63, 71, 60, 61*
Br	**79, 81,** *77, 82, 76, 83, 75, 74, 84, 80, 78, 85, 87, 86, 88, 89, 90*
Sr	**88, 86, 87, 84,** *90, 85, 89, 82, 91, 83, 92, 80, 81, 93, 94, 95*
Ag	**107, 109,** *105, 111, 113, 112, 104, 103, 106, 115, 102, 116, 108, 117, 110, 114*
Sn	**120, 118, 116, 119, 117,** ***124,*** **122, 112, 114, 115,** *126, 123, 113, 125, 121, 110, 127, 128, 111, 109, 108, 129, 131, 130, 132*
I	**127,** *129, 125, 126, 131, 124, 133, 123, 130, 135, 132, 121, 120, 134, 128, 119, 118, 117, 122, 136, 137, 138, 139*

Table 2
ISOTOPES OF SOME ELEMENTS (cont'd)

Element	Mass numbers of isotopes
Ba	**138, 137, 136, 135, 134, 130, 132,** 133, 140, 131, 128, 129, 126, 139, 141, 142, 127, 125, 123, 143
W	**184,** 186, 182, 183, 180, 181, 185, 188, 178, 187, 176, 177, 179, 175, 174, 173, 189
Pt	**195, 194, 196,** 198, 192, 190, 193, 188, 191, 197, 200, 189, 186, 187, 185, 199, 184, 183, 182, 201, 181, 180, 179, 178, 177, 176, 175, 174
Pb	**208, 206, 207, 204,** 205, 202, 210, 203, 200, 212, 201, 209, 198, 199, 197, 196, 211, 214, 195, 194, 213
Bi	209, 208, 207, 205, 206, 210, 203, 204, 201, 202, 212, 213, 200, 199, 214, 197, 215, 211, 198
Po	209, 208, 210, 206, 207, 204, 205, 202, 203, 201, 200, 199, 218, 198, 197, 217, 196, 195, 211, 194, 216
Rn	222, 211, 210, 224, 223, 209, 221, 212, 208, 207, 206, 205, 204, 220, 203, 202, 219
Ra	226, 228, 225, 223, 224, 230, 227, 213, 222, 221
U	238, 235, 236, 234, 233, 232, 230, 237, 231, 240, 229, 239, 228, 227
Np	237, 236, 235, 234, 239, 238, 236, 240, 231, 233, 241, 232, 230, 229
Pu	244, 242, 239, 240, 238, 241, 236, 237, 246, 245, 234, 243, 232, 235, 233
Am	243, 241, 240, 242, 239, 244, 245, 238, 237, 246, 247
Cm	247, 248, 250, 245, 246, 243, 244, 242, 241, 240, 239, 238, 249
Bk	247, 249, 245, 246, 248, 243, 244, 250, 251
Cf	251, 249, 250, 252, 248, 254, 253, 246, 247, 245, 244, 243, 242
Es	252, 254, 255, 253, 251, 250, 249, 248, 246, 247, 245
Fm	257, 253, 252, 255, 251, 254, 256, 250, 249, 248, 247, 245, 246, 258
Md	258, 257, 256, 255, 254, 252, 251, 250, 249, 248
No	259, 255, 253, 254, 257, 256, 252, 251, 258
Lr	260, 256, 255, 254, 259, 258, 257, 253
Unq	261, 257, 259, 260, 258
Unp	262, 263, 261, 260
Unh	263
Uns	261?
Uno	265
Une	266?

Table 3
PHYSICAL CONSTANTS

Quantity	Symbol	Value
atomic mass unit	u	$1.6605655 \times 10^{-24}$ g
Avogadro number	N_A	6.022045×10^{23}/mole
electron rest mass	m_e	9.109534×10^{-28} g
mechanical equivalent of heat	J	4.1868 J/cal
molar gas constant	R	8.20568×10^{-2} L atm/mole K
molar volume of ideal gas at STP	V_m	22.4136 L/mole
neutron rest mass	m_n	$1.6749543 \times 10^{-24}$ g
normal boiling point of water	T_b	373.15 K = 100.0 °C
normal freezing point of water	T_f	273.15 K = 0.00 °C
Planck's constant	h	6.626176×10^{-34} J s
proton rest mass	m_p	$1.6726485 \times 10^{-24}$ g
speed of light in vacuum	c	2.99792458×10^{8} m/s
temperature of triple point of water		273.16 K = 0.01 °C

Table 4
THE ELEMENTS—THEIR SYMBOLS, ATOMIC NUMBERS, AND ATOMIC WEIGHTS

The more common elements are printed in color.

Name of element	Symbol	Atomic number	Atomic weight
actinium	Ac	89	227.0278
aluminum	Al	13	26.98154
americium	Am	95	[243]
antimony	Sb	51	121.75
argon	Ar	18	39.948
arsenic	As	33	74.9216
astatine	At	85	[210]
barium	Ba	56	137.33
berkelium	Bk	97	[247]
beryllium	Be	4	9.01218
bismuth	Bi	83	208.9804
boron	B	5	10.81
bromine	Br	35	79.904
cadmium	Cd	48	112.41
calcium	Ca	20	40.08
californium	Cf	98	[251]
carbon	C	6	12.011
cerium	Ce	58	140.12
cesium	Cs	55	132.9054
chlorine	Cl	17	35.453
chromium	Cr	24	51.996
cobalt	Co	27	58.9332
copper	Cu	29	63.546
curium	Cm	96	[247]
dysprosium	Dy	66	162.50
einsteinium	Es	99	[252]
erbium	Er	68	167.26
europium	Eu	63	151.96
fermium	Fm	100	[257]
fluorine	F	9	18.998403
francium	Fr	87	[223]
gadolinium	Gd	64	157.25
gallium	Ga	31	69.72
germanium	Ge	32	72.59
gold	Au	79	196.9665
hafnium	Hf	72	178.49
helium	He	2	4.00260
holmium	Ho	67	164.9304
hydrogen	H	1	1.00794
indium	In	49	114.82
iodine	I	53	126.9045
iridium	Ir	77	192.22
iron	Fe	26	55.847
krypton	Kr	36	83.80
lanthanum	La	57	138.9055
lawrencium	Lr	103	[260]
lead	Pb	82	207.2
lithium	Li	3	6.941
lutetium	Lu	71	174.967
magnesium	Mg	12	24.305
manganese	Mn	25	54.9380
mendelevium	Md	101	[258]
mercury	Hg	80	200.59
molybdenum	Mo	42	95.94
neodymium	Nd	60	144.24
neon	Ne	10	20.179
neptunium	Np	93	237.0482
nickel	Ni	28	58.69
niobium	Nb	41	92.9064
nitrogen	N	7	14.0067
nobelium	No	102	[259]
osmium	Os	76	190.2
oxygen	O	8	15.9994
palladium	Pd	46	106.42
phosphorus	P	15	30.97376
platinum	Pt	78	195.08
plutonium	Pu	94	[244]
polonium	Po	84	[209]
potassium	K	19	39.0983
praseodymium	Pr	59	140.9077
promethium	Pm	61	[145]
protactinium	Pa	91	231.0359
radium	Ra	88	226.0254
radon	Rn	86	[222]
rhenium	Re	75	186.207
rhodium	Rh	45	102.9055
rubidium	Rb	37	85.4678
ruthenium	Ru	44	101.07
samarium	Sm	62	150.36
scandium	Sc	21	44.9559
selenium	Se	34	78.96
silicon	Si	14	28.0855
silver	Ag	47	107.8682
sodium	Na	11	22.98977
strontium	Sr	38	87.62
sulfur	S	16	32.06
tantalum	Ta	73	180.9479
technetium	Tc	43	[98]
tellurium	Te	52	127.60
terbium	Tb	65	158.9254
thallium	Tl	81	204.383
thorium	Th	90	232.0381
thulium	Tm	69	168.9342
tin	Sn	50	118.69
titanium	Ti	22	47.88
tungsten	W	74	183.85
unnilennium	Une	109	[266?]
unnilhexium	Unh	106	[263]
unniloctium	Uno	108	[265]
unnilpentium	Unp	105	[262]
unnilquadium	Unq	104	[261]
unnilseptium	Uns	107	[261?]
uranium	U	92	238.0289
vanadium	V	23	50.9415
xenon	Xe	54	131.29
ytterbium	Yb	70	173.04
yttrium	Y	39	88.9059
zinc	Zn	30	65.38
zirconium	Zr	40	91.22

A value given in brackets denotes the mass number of the isotope of longest known half-life. The atomic weights of most of these elements are believed to have an error no greater than ±1 in the last digit given.

Table 5
COMMON ELEMENTS

Name	Symbol	Approx. at. wt.	Common ox. nos.	Name	Symbol	Approx. at. wt.	Common ox. nos.
aluminum	Al	27.0	+3	magnesium	Mg	24.3	+2
antimony	Sb	121.8	+3, +5	manganese	Mn	54.9	+2, +4, +7
arsenic	As	74.9	+3, +5	mercury	Hg	200.6	+1, +2
barium	Ba	137.3	+2	nickel	Ni	58.7	+2
bismuth	Bi	209.0	+3	nitrogen	N	14.0	−3, +3, +5
bromine	Br	79.9	−1, +5	oxygen	O	16.0	−2
calcium	Ca	40.1	+2	phosphorus	P	31.0	+3, +5
carbon	C	12.0	+2, +4	platinum	Pt	195.1	+2, +4
chlorine	Cl	35.5	−1, +5, +7	potassium	K	39.1	+1
chromium	Cr	52.0	+2, +3, +6	silicon	Si	28.1	+4
cobalt	Co	58.9	+2, +3	silver	Ag	107.9	+1
copper	Cu	63.5	+1, +2	sodium	Na	23.0	+1
fluorine	F	19.0	−1	strontium	Sr	87.6	+2
gold	Au	197.0	+1, +3	sulfur	S	32.1	−2, +4, +6
hydrogen	H	1.0	−1, +1	tin	Sn	118.7	+2, +4
iodine	I	126.9	−1, +5	titanium	Ti	47.9	+3, +4
iron	Fe	55.8	+2, +3	tungsten	W	183.8	+6
lead	Pb	207.2	+2, +4	zinc	Zn	65.4	+2

Table 6
COMMON IONS AND THEIR CHARGES

Name	Symbol	Charge	Name	Symbol	Charge
aluminum	Al^{+++}	+3	lead(II)	Pb^{++}	+2
ammonium	NH_4^+	+1	magnesium	Mg^{++}	+2
barium	Ba^{++}	+2	mercury(I)	Hg_2^{++}	+2
calcium	Ca^{++}	+2	mercury(II)	Hg^{++}	+2
chromium(III)	Cr^{+++}	+3	nickel(II)	Ni^{++}	+2
cobalt(II)	Co^{++}	+2	potassium	K^+	+1
copper(I)	Cu^+	+1	silver	Ag^+	+1
copper(II)	Cu^{++}	+2	sodium	Na^+	+1
hydronium	H_3O^+	+1	tin(II)	Sn^{++}	+2
iron(II)	Fe^{++}	+2	tin(IV)	Sn^{++++}	+4
iron(III)	Fe^{+++}	+3	zinc	Zn^{++}	+2
acetate	$C_2H_3O_2^-$	−1	hydrogen sulfate	HSO_4^-	−1
bromide	Br^-	−1	hydroxide	OH^-	−1
carbonate	CO_3^{--}	−2	hypochlorite	ClO^-	−1
chlorate	ClO_3^-	−1	iodide	I^-	−1
chloride	Cl^-	−1	nitrate	NO_3^-	−1
chlorite	ClO_2^-	−1	nitrite	NO_2^-	−1
chromate	CrO_4^{--}	−2	oxide	O^{--}	−2
cyanide	CN^-	−1	perchlorate	ClO_4^-	−1
dichromate	$Cr_2O_7^{--}$	−2	permanganate	MnO_4^-	−1
fluoride	F^-	−1	peroxide	O_2^{--}	−2
hexacyanoferrate(II)	$Fe(CN)_6^{----}$	−4	phosphate	PO_4^{---}	−3
hexacyanoferrate(III)	$Fe(CN)_6^{---}$	−3	sulfate	SO_4^{--}	−2
hydride	H^-	−1	sulfide	S^{--}	−2
hydrogen carbonate	HCO_3^-	−1	sulfite	SO_3^{--}	−2

Table 7
ELECTRON ARRANGEMENT OF THE ELEMENTS

	Sublevels	1*s*	2*s*	2*p*	3*s*	3*p*	3*d*	4*s*	4*p*	4*d*	4*f*	5*s*	5*p*	5*d*	5*f*	6*s*	6*p*	6*d*	6*f*	7*s*
1	hydrogen	1																		
2	helium	2																		
3	lithium	2	1																	
4	beryllium	2	2																	
5	boron	2	2	1																
6	carbon	2	2	2																
7	nitrogen	2	2	3																
8	oxygen	2	2	4																
9	fluorine	2	2	5																
10	neon	2	2	6																
11	sodium	2	2	6	1															
12	magnesium	2	2	6	2															
13	aluminum	2	2	6	2	1														
14	silicon	2	2	6	2	2														
15	phosphorus	2	2	6	2	3														
16	sulfur	2	2	6	2	4														
17	chlorine	2	2	6	2	5														
18	argon	2	2	6	2	6														
19	potassium	2	2	6	2	6		1												
20	calcium	2	2	6	2	6		2												
21	scandium	2	2	6	2	6	1	2												
22	titanium	2	2	6	2	6	2	2												
23	vanadium	2	2	6	2	6	3	2												
24	chromium	2	2	6	2	6	5	1												
25	manganese	2	2	6	2	6	5	2												
26	iron	2	2	6	2	6	6	2												
27	cobalt	2	2	6	2	6	7	2												
28	nickel	2	2	6	2	6	8	2												
29	copper	2	2	6	2	6	10	1												
30	zinc	2	2	6	2	6	10	2												
31	gallium	2	2	6	2	6	10	2	1											
32	germanium	2	2	6	2	6	10	2	2											
33	arsenic	2	2	6	2	6	10	2	3											
34	selenium	2	2	6	2	6	10	2	4											
35	bromine	2	2	6	2	6	10	2	5											
36	krypton	2	2	6	2	6	10	2	6											
37	rubidium	2	2	6	2	6	10	2	6			1								
38	strontium	2	2	6	2	6	10	2	6			2								
39	yttrium	2	2	6	2	6	10	2	6	1		2								
40	zirconium	2	2	6	2	6	10	2	6	2		2								
41	niobium	2	2	6	2	6	10	2	6	4		1								
42	molybdenum	2	2	6	2	6	10	2	6	5		1								
43	technetium	2	2	6	2	6	10	2	6	5		2								
44	ruthenium	2	2	6	2	6	10	2	6	7		1								
45	rhodium	2	2	6	2	6	10	2	6	8		1								
46	palladium	2	2	6	2	6	10	2	6	10										
47	silver	2	2	6	2	6	10	2	6	10		1								
48	cadmium	2	2	6	2	6	10	2	6	10		2								
49	indium	2	2	6	2	6	10	2	6	10		2	1							
50	tin	2	2	6	2	6	10	2	6	10		2	2							
51	antimony	2	2	6	2	6	10	2	6	10		2	3							
52	tellurium	2	2	6	2	6	10	2	6	10		2	4							
53	iodine	2	2	6	2	6	10	2	6	10		2	5							
54	xenon	2	2	6	2	6	10	2	6	10		2	6							

Table 7
ELECTRON ARRANGEMENT OF THE ELEMENTS (cont'd)

	Sublevels	1*s*	2*s*	2*p*	3*s*	3*p*	3*d*	4*s*	4*p*	4*d*	4*f*	5*s*	5*p*	5*d*	5*f*	6*s*	6*p*	6*d*	6*f*	7*s*
55	cesium	2	2	6	2	6	10	2	6	10		2	6			1				
56	barium	2	2	6	2	6	10	2	6	10		2	6			2				
57	lanthanum	2	2	6	2	6	10	2	6	10		2	6	1		2				
58	cerium	2	2	6	2	6	10	2	6	10	2	2	6			2				
59	praseodymium	2	2	6	2	6	10	2	6	10	3	2	6			2				
60	neodymium	2	2	6	2	6	10	2	6	10	4	2	6			2				
61	promethium	2	2	6	2	6	10	2	6	10	5	2	6			2				
62	samarium	2	2	6	2	6	10	2	6	10	6	2	6			2				
63	europium	2	2	6	2	6	10	2	6	10	7	2	6			2				
64	gadolinium	2	2	6	2	6	10	2	6	10	7	2	6	1		2				
65	terbium	2	2	6	2	6	10	2	6	10	9	2	6			2				
66	dysprosium	2	2	6	2	6	10	2	6	10	10	2	6			2				
67	holmium	2	2	6	2	6	10	2	6	10	11	2	6			2				
68	erbium	2	2	6	2	6	10	2	6	10	12	2	6			2				
69	thulium	2	2	6	2	6	10	2	6	10	13	2	6			2				
70	ytterbium	2	2	6	2	6	10	2	6	10	14	2	6			2				
71	lutetium	2	2	6	2	6	10	2	6	10	14	2	6	1		2				
72	hafnium	2	2	6	2	6	10	2	6	10	14	2	6	2		2				
73	tantalum	2	2	6	2	6	10	2	6	10	14	2	6	3		2				
74	tungsten	2	2	6	2	6	10	2	6	10	14	2	6	4		2				
75	rhenium	2	2	6	2	6	10	2	6	10	14	2	6	5		2				
76	osmium	2	2	6	2	6	10	2	6	10	14	2	6	6		2				
77	iridium	2	2	6	2	6	10	2	6	10	14	2	6	7		2				
78	platinum	2	2	6	2	6	10	2	6	10	14	2	6	9		1				
79	gold	2	2	6	2	6	10	2	6	10	14	2	6	10		1				
80	mercury	2	2	6	2	6	10	2	6	10	14	2	6	10		2				
81	thallium	2	2	6	2	6	10	2	6	10	14	2	6	10		2	1			
82	lead	2	2	6	2	6	10	2	6	10	14	2	6	10		2	2			
83	bismuth	2	2	6	2	6	10	2	6	10	14	2	6	10		2	3			
84	polonium	2	2	6	2	6	10	2	6	10	14	2	6	10		2	4			
85	astatine	2	2	6	2	6	10	2	6	10	14	2	6	10		2	5			
86	radon	2	2	6	2	6	10	2	6	10	14	2	6	10		2	6			
87	francium	2	2	6	2	6	10	2	6	10	14	2	6	10		2	6			1
88	radium	2	2	6	2	6	10	2	6	10	14	2	6	10		2	6			2
89	actinium	2	2	6	2	6	10	2	6	10	14	2	6	10		2	6	1		2
90	thorium	2	2	6	2	6	10	2	6	10	14	2	6	10		2	6	2		2
91	protactinium	2	2	6	2	6	10	2	6	10	14	2	6	10	2	2	6	1		2
92	uranium	2	2	6	2	6	10	2	6	10	14	2	6	10	3	2	6	1		2
93	neptunium	2	2	6	2	6	10	2	6	10	14	2	6	10	4	2	6	1		2
94	plutonium	2	2	6	2	6	10	2	6	10	14	2	6	10	6	2	6			2
95	americium	2	2	6	2	6	10	2	6	10	14	2	6	10	7	2	6			2
96	curium	2	2	6	2	6	10	2	6	10	14	2	6	10	7	2	6	1		2
97	berkelium	2	2	6	2	6	10	2	6	10	14	2	6	10	9	2	6			2
98	californium	2	2	6	2	6	10	2	6	10	14	2	6	10	10	2	6			2
99	einsteinium	2	2	6	2	6	10	2	6	10	14	2	6	10	11	2	6			2
100	fermium	2	2	6	2	6	10	2	6	10	14	2	6	10	12	2	6			2
101	mendelevium	2	2	6	2	6	10	2	6	10	14	2	6	10	13	2	6			2
102	nobelium	2	2	6	2	6	10	2	6	10	14	2	6	10	14	2	6			2
103	lawrencium	2	2	6	2	6	10	2	6	10	14	2	6	10	14	2	6	1		2
104	unnilquadium	2	2	6	2	6	10	2	6	10	14	2	6	10	14	2	6	2		2?
105	unnilpentium	2	2	6	2	6	10	2	6	10	14	2	6	10	14	2	6	3		2?
106	unnilhexium	2	2	6	2	6	10	2	6	10	14	2	6	10	14	2	6	4		2?
107	unnilseptium	2	2	6	2	6	10	2	6	10	14	2	6	10	14	2	6	5		2?
108	unniloctium	2	2	6	2	6	10	2	6	10	14	2	6	10	14	2	6	6		2?
109	unnilennium	2	2	6	2	6	10	2	6	10	14	2	6	10	14	2	6	7		2?

Table 8
WATER-VAPOR PRESSURE

Temperature (°C)	Pressure (mm Hg)	Temperature (°C)	Pressure (mm Hg)	Temperature (°C)	Pressure (mm Hg)
0.0	4.6	19.5	17.0	27.0	26.7
5.0	6.5	20.0	17.5	28.0	28.3
10.0	9.2	20.5	18.1	29.0	30.0
12.5	10.9	21.0	18.6	30.0	31.8
15.0	12.8	21.5	19.2	35.0	42.2
15.5	13.2	22.0	19.8	40.0	55.3
16.0	13.6	22.5	20.4	50.0	92.5
16.5	14.1	23.0	21.1	60.0	149.4
17.0	14.5	23.5	21.7	70.0	233.7
17.5	15.0	24.0	22.4	80.0	355.1
18.0	15.5	24.5	23.1	90.0	525.8
18.5	16.0	25.0	23.8	95.0	633.9
19.0	16.5	26.0	25.2	100.0	760.0

Table 9
DENSITY OF GASES AT STP

Gas	Density (g/L)	Gas	Density (g/L)
air, dry	1.2929	hydrogen	0.0899
ammonia	0.771	hydrogen chloride	1.639
carbon dioxide	1.977	hydrogen sulfide	1.539
carbon monoxide	1.250	methane	0.716
chlorine	3.214	nitrogen	1.251
dinitrogen monoxide	1.977	nitrogen monoxide	1.340
ethyne (acetylene)	1.171	oxygen	1.429
helium	0.1785	sulfur dioxide	2.927

Table 10
DENSITY OF WATER

Temperature (°C)	Density (g/mL)	Temperature (°C)	Density (g/mL)
0	0.99987	15	0.99913
1	0.99993	20	0.99823
2	0.99997	25	0.99707
3	0.99999	30	0.99567
4	1.00000	40	0.99224
5	0.99999	50	0.98807
6	0.99997	60	0.98324
7	0.99993	70	0.97781
8	0.99988	80	0.97183
9	0.99981	90	0.96534
10	0.99973	100	0.95838

Table 11
SOLUBILITY OF GASES IN WATER

Volume of gas (reduced to STP) that can be dissolved in 1 volume of water at the temperature (°C) indicated.

Gas	0°	10°	20°	60°
air	0.02918	0.02284	0.01868	0.01216
ammonia	1130	870	680	200
carbon dioxide	1.713	1.194	0.878	0.359
carbon monoxide	0.03537	0.02816	0.02319	0.01488
chlorine	4.54	3.148	2.299	1.023
hydrogen	0.02148	0.01955	0.01819	0.01600
hydrogen chloride	512	475	442	339
hydrogen sulfide	4.670	3.399	2.582	1.190
methane	0.05563	0.04177	0.03308	0.01954
nitrogen	0.02354	0.01861	0.01545	0.01023
nitrogen dioxide	0.07381	0.05709	0.04706	0.02954
oxygen	0.04889	0.03802	0.03102	0.01946
sulfur dioxide	79.789	56.647	39.374	—

Table 12
SOLUBILITY CHART

S = soluble in water. A = soluble in acids, insoluble in water. P = partially soluble in water, soluble in dilute acids. I = insoluble in dilute acids and in water. a = slightly soluble in acids, insoluble in water. d = decomposes in water.

	acetate	bromide	carbonate	chlorate	chloride	chromate	hydroxide	iodide	nitrate	oxide	phosphate	silicate	sulfate	sulfide
aluminum	S	S	—	S	S	—	A	S	S	a	A	I	S	d
ammonium	S	S	S	S	S	S	—	S	S	—	S	—	S	S
barium	S	S	P	S	S	A	S	S	S	S	A	S	a	d
calcium	S	S	P	S	S	S	P	S	S	P	P	P	P	P
copper(II)	S	S	—	S	S	—	A	—	S	A	A	A	S	A
hydrogen	S	S	—	S	S	—	—	S	S	—	S	I	S	S
iron(II)	S	S	P	S	S	—	A	S	S	A	A	—	S	A
iron(III)	S	S	—	S	S	A	A	S	S	A	P	—	P	d
lead(II)	S	S	A	S	S	A	P	P	S	P	A	A	P	A
magnesium	S	S	P	S	S	S	A	S	S	A	P	A	S	d
manganese(II)	S	S	P	S	S	—	A	S	S	A	P	I	S	A
mercury(I)	P	A	A	S	a	P	—	A	S	A	A	—	P	I
mercury(II)	S	S	—	S	S	P	A	P	S	P	A	—	d	I
potassium	S	S	S	S	S	S	S	S	S	S	S	S	S	S
silver	P	a	A	S	a	P	—	I	S	P	A	—	P	A
sodium	S	S	S	S	S	S	S	S	S	S	S	S	S	S
strontium	S	S	P	S	S	P	S	S	S	S	A	A	P	S
tin(II)	d	S	—	S	S	A	A	S	d	A	A	—	S	A
tin(IV)	S	S	—	—	S	S	P	d	—	A	—	—	S	A
zinc	S	S	P	S	S	P	A	S	S	P	A	A	S	A

Table 15
HEAT OF COMBUSTION

ΔH_c = heat of combustion of the given substance. All values of ΔH_c are expressed as kcal/mole of substance oxidized to $H_2O(l)$ and/or $CO_2(g)$ at constant pressure and 25 °C. s = solid, l = liquid, g = gas.

Substance	*Formula*	*Phase*	ΔH_c
hydrogen	H_2	g	−68.32
graphite	C	s	−94.05
carbon monoxide	CO	g	−67.64
methane	CH_4	g	−212.79
ethane	C_2H_6	g	−372.81
propane	C_3H_8	g	−530.57
butane	C_4H_{10}	g	−687.98
pentane	C_5H_{12}	g	−845.16
hexane	C_6H_{14}	l	−995.01
heptane	C_7H_{16}	l	−1149.9
octane	C_8H_{18}	l	−1302.7
ethene (ethylene)	C_2H_4	g	−337.23
propene (propylene)	C_3H_6	g	−490.2
ethyne (acetylene)	C_2H_2	g	−310.61
benzene	C_6H_6	l	−780.96
toluene	C_7H_8	l	−934.2
naphthalene	$C_{10}H_8$	s	−1231.8
anthracene	$C_{14}H_{10}$	s	−1712.0
methanol	CH_3OH	l	−173.64
ethanol	C_2H_5OH	l	−326.68
ether	$(C_2H_5)_2O$	l	−657.52
formaldehyde	CH_2O	g	−136.42
glucose	$C_6H_{12}O_6$	s	−669.94
sucrose	$C_{12}H_{22}O_{11}$	s	−1348.2

Table 16
PROPERTIES OF COMMON ELEMENTS

Name	Form/color at room temperature	Density (g/cm³)	Melting point (°C)	Boiling point (°C)	Common oxidation numbers
aluminum	silv metal	2.70	660.4	2467	+3
antimony	silv metal	6.69	630.7	1750	+3, +5
argon	colorless gas	1.784*	−189.2	−185.7	0
arsenic	gray metal	5.73	817 (28 atm)	613 (sublimes)	+3, +5
barium	silv metal	3.5	725	1640	+2
beryllium	gray metal	1.85	1280	2970	+2
bismuth	silv metal	9.75	271.3	1560	+3
boron	blk solid	2.34	2079	2550 (sublimes)	+3
bromine	red-br liquid	3.12	−7.2	58.8	−1, +5
calcium	silv metal	1.55	839	1484	+2
carbon	diamond	3.51	3700	4200	+2, +4
	graphite	2.26	3620 (sublimes)	4200	
chlorine	grn-yel gas	3.214*	−101.0	−34.6	−1, +5, +7
chromium	silv metal	7.19	1860	2672	+2, +3, +6
cobalt	silv metal	8.9	1495	2870	+2, +3
copper	red metal	8.96	1083.4	2567	+1, +2
fluorine	yel gas	1.696*	−219.6	−188.1	−1
germanium	gray metalloid	5.32	937.4	2830	+4
gold	yel metal	19.3	1064.4	2807	+1, +3
helium	colorless gas	0.1785*	−272.2 (26 atm)	−268.9	0
hydrogen	colorless gas	0.08988*	−259.1	−252.9	−1, +1
iodine	bl-blk solid	4.93	113.5	184.4	−1, +5
iron	silv metal	7.87	1535	2750	+2, +3
lead	silv metal	11.4	327.5	1740	+2, +4
lithium	silv metal	0.534	180.5	1347	+1
magnesium	silv metal	1.74	649	1090	+2
manganese	silv metal	7.3	1244	1962	+2, +4, +7
mercury	silv liquid	13.5	−38.8	356.6	+1, +2
neon	colorless gas	0.8999*	−248.7	−246.0	0
nickel	silv metal	8.90	1453	2732	+2
nitrogen	colorless gas	1.251*	−209.9	−195.8	−3, +3, +5
oxygen	colorless gas	1.429*	−218.4	−183.0	−2
phosphorus	yel solid	1.82	44.1	280	+3, +5
platinum	silv metal	21.4	1772	3800	+2, +4
plutonium	silv metal	19.8	641	3232	+3, +4, +5, +6
potassium	silv metal	0.862	63.6	774	+1
radium	silv metal	5(?)	700	1140	+2
radon	colorless gas	9.73*	−71	−61.8	0
silicon	gray solid	2.33	1410	2355	+4
silver	silv metal	10.5	961.9	2212	+1
sodium	silv metal	0.971	97.8	882.9	+1
strontium	silv metal	2.54	769	1384	+2
sulfur	yel solid	2.07	112.8	444.7	−2, +4, +6
tin	silv metal	7.31	232.0	2270	+2, +4
titanium	silv metal	4.54	1660	3287	+3, +4
tungsten	gray metal	19.3	3410	5660	+6
uranium	silv metal	19.0	1132	3818	+4, +6
xenon	colorless gas	5.89*	−111.9	−107	0
zinc	silv metal	7.13	419.6	907	+2

*Densities of gases are given in grams/liter at STP.

Table 17
FOUR-PLACE LOGARITHMS OF NUMBERS

n	0	1	2	3	4	5	6	7	8	9
10	0000	0043	0086	0128	0170	0212	0253	0294	0334	0374
11	0414	0453	0492	0531	0569	0607	0645	0682	0719	0755
12	0792	0828	0864	0899	0934	0969	1004	1038	1072	1106
13	1139	1173	1206	1239	1271	1303	1335	1367	1399	1430
14	1461	1492	1523	1553	1584	1614	1644	1673	1703	1732
15	1761	1790	1818	1847	1875	1903	1931	1959	1987	2014
16	2041	2068	2095	2122	2148	2175	2201	2227	2253	2279
17	2304	2330	2355	2380	2405	2430	2455	2480	2504	2529
18	2553	2577	2601	2625	2648	2672	2695	2718	2742	2765
19	2788	2810	2833	2856	2878	2900	2923	2945	2967	2989
20	3010	3032	3054	3075	3096	3118	3139	3160	3181	3201
21	3222	3243	3263	3284	3304	3324	3345	3365	3385	3404
22	3424	3444	3464	3483	3502	3522	3541	3560	3579	3598
23	3617	3636	3655	3674	3692	3711	3729	3747	3766	3784
24	3802	3820	3838	3856	3874	3892	3909	3927	3945	3962
25	3979	3997	4014	4031	4048	4065	4082	4099	4116	4133
26	4150	4166	4183	4200	4216	4232	4249	4265	4281	4298
27	4314	4330	4346	4362	4378	4393	4409	4425	4440	4456
28	4472	4487	4502	4518	4533	4548	4564	4579	4594	4609
29	4624	4639	4654	4669	4683	4698	4713	4728	4742	4757
30	4771	4786	4800	4814	4829	4843	4857	4871	4886	4900
31	4914	4928	4942	4955	4969	4983	4997	5011	5024	5038
32	5051	5065	5079	5092	5105	5119	5132	5145	5159	5172
33	5185	5198	5211	5224	5237	5250	5263	5276	5289	5302
34	5315	5328	5340	5353	5366	5378	5391	5403	5416	5428
35	5441	5453	5465	5478	5490	5502	5514	5527	5539	5551
36	5563	5575	5587	5599	5611	5623	5635	5647	5658	5670
37	5682	5694	5705	5717	5729	5740	5752	5763	5775	5786
38	5798	5809	5821	5832	5843	5855	5866	5877	5888	5899
39	5911	5922	5933	5944	5955	5966	5977	5988	5999	6010
40	6021	6031	6042	6053	6064	6075	6085	6096	6107	6117
41	6128	6138	6149	6160	6170	6180	6191	6201	6212	6222
42	6232	6243	6253	6263	6274	6284	6294	6304	6314	6325
43	6335	6345	6355	6365	6375	6385	6395	6405	6415	6425
44	6435	6444	6454	6464	6474	6484	6493	6503	6513	6522
45	6532	6542	6551	6561	6571	6580	6590	6599	6609	6618
46	6628	6637	6646	6656	6665	6675	6684	6693	6702	6712
47	6721	6730	6739	6749	6758	6767	6776	6785	6794	6803
48	6812	6821	6830	6839	6848	6857	6866	6875	6884	6893
49	6902	6911	6920	6928	6937	6946	6955	6964	6972	6981
50	6990	6998	7007	7016	7024	7033	7042	7050	7059	7067
51	7076	7084	7093	7101	7110	7118	7126	7135	7143	7152
52	7160	7168	7177	7185	7193	7202	7210	7218	7226	7235
53	7243	7251	7259	7267	7275	7284	7292	7300	7308	7316
54	7324	7332	7340	7348	7356	7364	7372	7380	7388	7396

Table 17
FOUR-PLACE LOGARITHMS OF NUMBERS (cont'd)

n	0	1	2	3	4	5	6	7	8	9
55	7404	7412	7419	7427	7435	7443	7451	7459	7466	7474
56	7482	7490	7497	7505	7513	7520	7528	7536	7543	7551
57	7559	7566	7574	7582	7589	7597	7604	7612	7619	7627
58	7634	7642	7649	7657	7664	7672	7679	7686	7694	7701
59	7709	7716	7723	7731	7738	7745	7752	7760	7767	7774
60	7782	7789	7796	7803	7810	7818	7825	7832	7839	7846
61	7853	7860	7868	7875	7882	7889	7896	7903	7910	7917
62	7924	7931	7938	7945	7952	7959	7966	7973	7980	7987
63	7993	8000	8007	8014	8021	8028	8035	8041	8048	8055
64	8062	8069	8075	8082	8089	8096	8102	8109	8116	8122
65	8129	8136	8142	8149	8156	8162	8169	8176	8182	8189
66	8195	8202	8209	8215	8222	8228	8235	8241	8248	8254
67	8261	8267	8274	8280	8287	8293	8299	8306	8312	8319
68	8325	8331	8338	8344	8351	8357	8363	8370	8376	8382
69	8388	8395	8401	8407	8414	8420	8426	8432	8439	8445
70	8451	8457	8463	8470	8476	8482	8488	8494	8500	8506
71	8513	8519	8525	8531	8537	8543	8549	8555	8561	8567
72	8573	8579	8585	8591	8597	8603	8609	8615	8621	8627
73	8633	8639	8645	8651	8657	8663	8669	8675	8681	8686
74	8692	8698	8704	8710	8716	8722	8727	8733	8739	8745
75	8751	8756	8762	8768	8774	8779	8785	8791	8797	8802
76	8808	8814	8820	8825	8831	8837	8842	8848	8854	8859
77	8865	8871	8876	8882	8887	8893	8899	8904	8910	8915
78	8921	8927	8932	8938	8943	8949	8954	8960	8965	8971
79	8976	8982	8987	8993	8998	9004	9009	9015	9020	9025
80	9031	9036	9042	9047	9053	9058	9063	9069	9074	9079
81	9085	9090	9096	9101	9106	9112	9117	9122	9128	9133
82	9138	9143	9149	9154	9159	9165	9170	9175	9180	9186
83	9191	9196	9201	9206	9212	9217	9222	9227	9232	9238
84	9243	9248	9253	9258	9263	9269	9274	9279	9284	9289
85	9294	9299	9304	9309	9315	9320	9325	9330	9335	9340
86	9345	9350	9355	9360	9365	9370	9375	9380	9385	9390
87	9395	9400	9405	9410	9415	9420	9425	9430	9435	9440
88	9445	9450	9455	9460	9465	9469	9474	9479	9484	9489
89	9494	9499	9504	9509	9513	9518	9523	9528	9533	9538
90	9542	9547	9552	9557	9562	9566	9571	9576	9581	9586
91	9590	9595	9600	9605	9609	9614	9619	9624	9628	9633
92	9638	9643	9647	9652	9657	9661	9666	9671	9675	9680
93	9685	9689	9694	9699	9703	9708	9713	9717	9722	9727
94	9731	9736	9741	9745	9750	9754	9759	9763	9768	9773
95	9777	9782	9786	9791	9795	9800	9805	9809	9814	9818
96	9823	9827	9832	9836	9841	9845	9850	9854	9859	9863
97	9868	9872	9877	9881	9886	9890	9894	9899	9903	9908
98	9912	9917	9921	9926	9930	9934	9939	9943	9948	9952
99	9956	9961	9965	9969	9974	9978	9983	9987	9991	9996

GLOSSARY

absolute zero The lowest possible temperature, 0 K or −273.15 °C.
absorption The taking up of one material through the entire mass of another.
accuracy The nearness of a measurement to its accepted value.
acid (1) A substance that increases the hydronium-ion concentration of its aqueous solution. (2) A proton donor. (3) An electron-pair acceptor.
acid anhydride An oxide that reacts with water to form an acid, or that is formed by the removal of water from an acid.
acid rain Rainwater having a pH below about 5.0.
actinide series Rare-earth elements of the seventh period following radium, in which the transitional inner building of the 6*d* sublevel is interrupted by the inner building of the 5*f* sublevel.
activated complex The transitional structure resulting from an effective collision of reactant particles.
activation energy Energy required to transform reactants into an activated complex.
activity series A table of metals or nonmetals arranged in order of descending activities.
adsorption The concentration of a gas, liquid, or solid on the surface of a liquid or solid with which it is in contact.
alcohol A compound containing a hydrocarbon group and one or more —OH (hydroxyl) groups.
aldehyde A compound that has a hydrocarbon group and one or more $-\mathrm{C}(=\mathrm{O})\mathrm{H}$ (formyl) groups.
alkadiene A straight- or branched-chain hydrocarbon with two double covalent bonds between carbon atoms in each molecule.
alkali metal An element of Group I of the periodic table.
alkaline-earth metal An element of Group II of the periodic table.
alkane A straight- or branched-chain hydrocarbon in which the carbon atoms are connected by only single covalent bonds; a member of the paraffin series.
alkene A straight- or branched-chain hydrocarbon in which two carbon atoms in each molecule are connected by a double covalent bond; a member of the olefin series.
alkylation The combining of simple hydrocarbons with unsaturated hydrocarbons by heat in the presence of a catalyst.
alkyl group A group derived from an alkane by the loss of a hydrogen atom; frequently symbolized by R—.
alkyl halide An alkane in which a halogen atom is substituted for a hydrogen atom.
alkyne A straight- or branched-chain hydrocarbon in which two carbon atoms in each molecule are connected by a triple covalent bond; a member of the acetylene series.
allotrope One of the two or more different forms of an element in the same physical phase.
allotropy The existence of an element in two or more forms in the same physical phase.
alloy A material composed of two or more metals.
alpha particle A helium nucleus emitted from the nucleus of a radioactive nuclide.
alum A double salt of the type $M^{+}M^{+++}(SO_4)_2 \cdot 12H_2O$, $KAl(SO_4)_2 \cdot 12H_2O$ being the most common.
alumina Anhydrous aluminum oxide, Al_2O_3.
amorphous Having neither definite form nor structure.
amphoteric Capable of acting as either an acid or a base.
amyl group The C_5H_{11}—group.
angstrom A unit of linear measure; 1×10^{-8} cm.
anhydrous Without water of crystallization.
anion An ion attracted to the anode of an electrolytic cell; a negative ion.
anion hydrolysis Hydrolysis reaction in which an anion base accepts a proton from a water molecule, increasing the OH^- ion concentration of the solution.
anode In electrochemistry, the electrode at which oxidation occurs.
antifriction alloy An alloy that reduces friction.
antiseptic A substance that checks the growth or action of microorganisms on or in the body.
aqueous In water; watery.
aqueous acid A water solution having acid properties due to the nature of the solute present.
aqueous base A water solution having basic properties due to the nature of the solute present.
aqueous solution A solution in which water is the solvent.
aromatic hydrocarbon A hydrocarbon having a resonance structure sometimes represented by alternating single and double covalent bonds in six-membered carbon rings.
atom The smallest unit of an element that can exist either alone or in combination with atoms of the same or different elements.
atomic mass The mass of a nuclide expressed in atomic mass units.
atomic mass unit A unit of mass that is exactly 1/12 the mass of a carbon-12 atom; $1.6605655 \times 10^{-24}$g.
atomic number The number of protons in the nucleus of an atom.
atomic theory A theory that includes information about the structure and properties of atoms, the kinds of compounds they form, and the properties of these compounds. It also includes information about the mass, volume, and energy relationships in reactions between atoms.
atomic weight The ratio of the average atomic mass of an element to 1/12 of the mass of an atom of carbon-12.
autooxidation Self-oxidizing and reducing. Redox process in which the substance acts both as the oxidizing agent and reducing agent.

Avogadro number The number of carbon-12 atoms in exactly 12 grams of this nuclide; 6.022045×10^{23}.

baking soda Sodium hydrogen carbonate, $NaHCO_3$.
base (1) A substance that increases the hydroxide-ion concentration of its aqueous solution. (2) A proton acceptor. (3) An electron-pair donor.
basic anhydride An oxide that reacts with water to form a solution containing OH^- ions.
bauxite A hydrated aluminum oxide ore.
beta particle An electron emitted from the nucleus of a radioactive nuclide.
binary compound A compound consisting of only two elements.
binding energy See *nuclear binding energy*.
biomass The collective term for living materials (or matter derived from living things) that could be used to provide energy.
blast furnace A tall, cylindrical chamber in which iron oxide is reduced using coke, limestone, and a blast of hot air.
blister copper Crude copper as refined in a converter.
boiling point The temperature at which the equilibrium vapor pressure of a liquid is equal to the prevailing atmospheric pressure.
bond energy The energy required to break chemical bonds and form neutral atoms.
borax Sodium tetraborate, $Na_2B_4O_7 \cdot 10H_2O$.
borax-bead test An identification test for certain metals whose oxides impart characteristic colors to borax beads when fused with them.
borazon A crystalline form of boron nitride, BN, having about the same hardness as diamond.
bright-line spectrum A spectrum consisting of a series of bright lines, which have frequencies characteristic of the atoms present.
buffer A substance that, when added to a solution, causes a resistance to any change in pH.
buffered solution A solution containing a relatively high concentration of a buffer salt, which tends to maintain a constant pH.

calcine A partially roasted copper ore.
calorie A unit of heat; the heat required to raise the temperature of one gram of water through one Celsius degree.
calorimeter An apparatus for measuring heats of reactions.
carbonyl group The $\backslash\!\!/$C=O group.
carboxyl group The —C(=O)—O—H group.
catalyst A substance or combination of substances that accelerates a chemical reaction without itself being used up.
catalytic cracking The breaking up of large molecules into smaller ones by using a catalyst at high temperature.
cathode In electrochemistry, the electrode at which reduction occurs.
cation An ion attracted to the cathode of an electrolytic cell; a positive ion.
cation hydrolysis Hydrolysis reaction in which a cation acid donates a proton to a water molecule, increasing the H_3O^+ ion concentration of the solution.
caustic (1) Capable of converting some types of animal and vegetable matter into soluble materials by chemical action. (2) A substance with such properties.
Celsius temperature Temperature on the Celsius scale, which has two fixed points, the freezing point and the boiling point of water, 0° and 100° respectively.
centi- Metric prefix meaning 0.01.
centigrade scale The Celsius temperature scale.
chain reaction A reaction in which the material or energy that starts the reaction is also one of the products.
chemical bond The linkage between atoms produced by the transfer or sharing of electrons.
chemical change A change in which new substances with new properties are formed.
chemical equilibrium The state of balance attained in a reversible chemical reaction in which the rates of the opposing reactions are equal.
chemical formula A shorthand method of representing the composition of a substance by using chemical symbols and numerical subscripts.
chemical kinetics The branch of chemistry that deals with reaction rates and reaction mechanisms.
chemical properties Those properties that pertain to the behavior of a material in changes in which its identity is altered.
chemical symbol Either a single capital letter or a capital letter and a small letter used together that serves as an abbreviation for (1) an element; (2) an atom of an element; and (3) a mole of atoms of an element.
chemistry The science dealing with the structure and composition of substances, the changes in composition, and the mechanisms by which these changes occur.
coalescence The act of coalescing, or joining together.
colligative property A property of a system that is determined by the number of particles present in the system, but is independent of the nature of the particles themselves.
colloidal state A state of subdivision of matter ranging between the dimensions of ordinary molecules and microscopic particles.
colloidal suspension A two-phase system having dispersed particles suspended in a dispersing medium.
combining weight The weight of any element that combines with a fixed weight of a particular element—oxygen, for example.
combustion Any chemical action that occurs so rapidly that both noticeable heat and light are produced.
common-ion effect The decrease in ionization of a weak electrolyte by the addition of a salt having an ion common to the solution of the electrolyte.

complementary colors Two colors that, when combined, yield white light.
complex ion An ionic species composed of a central metal ion combined with a specific number of polar molecules or ions.
composition reaction A chemical reaction in which two or more substances combine to form a more complex substance.
compound A substance that can be decomposed into two or more simpler substances by ordinary chemical means.
concentrated Containing a relatively large amount of solute.
condensation The process of converting a gas into a liquid or a solid.
condensation temperature The lowest temperature at which a substance can exist as a gas at atmospheric pressure.
conjugate acid The species formed when a base acquires a proton.
conjugate base The species that remains after an acid has donated a proton.
control rod A rod of neutron-absorbing material used to regulate the reaction in a nuclear reactor.
coordination number The number of molecules or ions covalently bonded to (coordinated with) a central ion.
corrosive (1) Capable of irritating, altering, or gradually destroying by surface action. (2) A substance with such properties.
cosmic ray A proton or other nucleus of very high energy coming to the earth from outer space.
covalent bonding Bonding in which atoms share electrons.
covalent molecular crystal A crystal consisting of molecules arranged in a systematic order.
covalent network crystal A crystal consisting of an array of atoms that share electrons with their neighboring atoms to form a giant, compact, interlocking structure.
cracking A process of breaking up complex organic molecules by the action of heat and usually a catalyst.
critical mass The amount of radioactive material required to sustain a chain reaction.
critical pressure The pressure required to liquefy a gas at its critical temperature.
critical temperature The highest temperature at which it is possible to liquefy a gas with any amount of pressure.
critical volume The volume occupied by one mole of a gas at its critical temperature and critical pressure.
crystal A homogeneous portion of a substance bounded by plane surfaces making definite angles with each other, giving a regular geometric form.
crystal lattice The pattern of points that describes the arrangement of particles in a crystal structure.
cyclotron An electromagnetic device for accelerating protons or deuterons in a spiral path.

daughter nuclide A nuclide that is the product of the sequential radioactive decay of a parent nuclide.
decomposition reaction A chemical reaction in which one substance breaks down to form two or more simpler substances.
dehydrating agent A substance that removes water from a material.
dehydration The removal of oxygen and hydrogen atoms in the form of water from a substance.
deliquescence The property of certain substances to take up water from the air to form a solution.
denatured alcohol Ethanol to which poisonous materials have been added so it is unfit for drinking.
density The mass per unit volume of a material.
destructive distillation The process of decomposing materials by heating them in a closed container without access to air or oxygen.
deuterium The isotope of hydrogen having one proton and one neutron in the nucleus; hydrogen-2.
deuteron The nucleus of deuterium, consisting of one proton and one neutron.
diatomic Consisting of two atoms.
diffusion The process of spreading out spontaneously to occupy a space uniformly; the intermingling of the particles of substances.
dilute Containing a relatively small amount of solute.
dimer A compound formed by two simpler molecules or radicals.
dimeric Capable of forming twofold polymers.
dimerization A chemical reaction in which two simple molecules or radicals combine to form a more complex molecule.
dipole A polar molecule, one region of which is positive and the other region negative.
dipole-dipole attraction A type of van der Waals force that is the attraction between the oppositely charged portions of neighboring polar molecules.
diprotic Pertaining to an acid capable of donating two protons per molecule.
dispersion interaction A type of van der Waals force dependent on the number of electrons in the interacting molecules and the tightness with which they are held.
dissociation The separation of the ions from the crystals of an ionic compound during the solution process.
distillation The process of evaporation followed by condensation of the vapors in a separate vessel.
domain Small magnetized regions formed by groups of properly aligned atoms of ferromagnetic substances.
ductile Capable of being drawn into a wire.

effervescence The rapid evolution of a gas from a liquid in which it is dissolved.
efflorescence The property of hydrated crystals to lose water of crystallization when exposed to the air.
elastic collision A collision in which there is no net loss of energy.
electrochemical Pertaining to spontaneous oxidation-

reduction reactions used as a source of electric energy.

electrochemical cell A system of electrodes and electrolyte by which a spontaneous oxidation-reduction reaction can be used as a source of electric current.

electrochemical reaction A spontaneous oxidation-reduction reaction in which chemical energy is transformed into electric energy.

electrode A conductor used to establish electric contact with a nonmetallic part of a circuit.

electrode potential The potential difference between an electrode and its solution in a half-reaction.

electrolysis (1) The separation of a compound into simpler substances by an electric current. (2) The process by which an electric current is used to drive an oxidation-reduction reaction.

electrolyte A substance whose aqueous solution conducts an electric current.

electrolytic Pertaining to driven oxidation-reduction reactions that utilize electric energy from an external source.

electrolytic cell A system of electrodes and electrolyte by which an electric current is used to drive an oxidation-reduction reaction.

electrolytic reaction A driven oxidation-reduction reaction in which electric energy is converted to chemical energy.

electromagnetic radiation A form of energy, such as light, X rays, or radio waves, which travels through space as waves at the rate of 3.00×10^8 m/s.

electron A negatively charged pointlike particle found in an atom. It has $1/1837$ of the mass of the simplest type of hydrogen atom.

electron affinity The energy change that occurs when an electron is acquired by a neutral atom.

electron cloud The part of an atom outside the nucleus in which the electrons may most probably be found.

electron configuration The arrangement of electrons in an atom that is in its ground state.

electronegativity The property of an atom of attracting the shared electrons that form a bond between it and another atom.

electron pair Two electrons of opposite spin in the same orbital.

electron-volt The energy required to move an electron across a potential difference of one volt.

electroplating An electrolytic process by which a metal is deposited on a surface.

electroscope A device for determining the presence of electric charge.

element A substance that cannot be further decomposed by ordinary chemical means; a substance in which all the atoms have the same number of protons.

empirical formula A chemical formula that denotes the constituent elements of a substance and the simplest whole-number ratio of atoms of each.

endothermic Pertaining to a process that occurs with the absorption of energy.

end point The point of completion in a titration process.

energy The capacity for doing work.

energy level A region about the nucleus of an atom in which electrons move.

enthalpy The heat content of a system at constant pressure.

enthalpy change A measure of the quantity of heat exchanged by a system and its surroundings (at constant pressure).

entropy That property which describes the disorder of a system.

enzyme A catalyst produced by living cells.

equilibrium A dynamic state in which two opposing processes take place at the same time and at the same rate.

equilibrium constant The ratio of the product of the concentrations of the substances produced at equilibrium to the product of the concentrations of reactants, each concentration raised to that power which is the coefficient of the substance in the chemical equation.

equilibrium vapor pressure The pressure exerted by a vapor in equilibrium with its liquid.

equivalence point The theoretical end point in a titration process. That point in a titration process at which equivalent quantities of reactants are present.

equivalent (1) The mass in grams of a reactant that contains, replaces, or reacts with (directly or indirectly) the Avogadro number of hydrogen atoms. (2) The mass in grams of a reactant that acquires or supplies the Avogadro number of electrons.

ester A compound formed by the reaction between an acid and an alcohol.

esterification The process of producing an ester by the reaction of an acid with an alcohol.

ether An organic oxide.

eudiometer A gas-measuring tube.

evaporation The escape of molecules from the surface of liquids and solids.

excited atom An atom that has absorbed a photon.

exothermic Pertaining to a process that occurs with the liberation of energy.

external phase The dispersing medium of a colloidal suspension.

fat An ester of glycerol and long-carbon-chain acids having a plant or animal origin.

fermentation A chemical change produced by the action of an enzyme.

ferromagnetism The property of certain metals whereby they are strongly attracted by a magnet.

fission The breakup of a very heavy nucleus into medium-weight nuclei.

fluoresce To give off visible light after exposure to sunlight.

flux A material used to promote the melting of minerals.

formula A shorthand method of representing the composition of substances by using chemical symbols and numerical subscripts.

formula equation A concise, symbolized statement of a chemical change.

formula weight The sum of the atomic weights of all the atoms represented in the chemical formula.

formyl group The 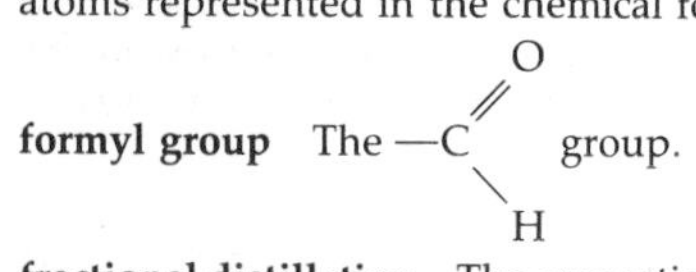 group.

fractional distillation The separation of the components of a mixture that have different boiling points by carefully controlled vaporization.

free energy The function of the state of a reaction system that assesses the tendencies toward lowest energy and highest entropy at a given temperature.

free-energy change The net driving force of a reaction system.

freezing The process of converting a liquid into a solid.

Friedel-Crafts reaction A substitution reaction, catalyzed by aluminum chloride, in which an alkyl group replaces a hydrogen atom of a benzene ring.

fuel A material that is burned to provide heat.

fuse To melt; to change to the liquid phase by heating.

fusion (1) The act or process of liquefying by heat; melting. (2) The combination of light-weight nuclei to form heavier, more stable nuclei.

galvanize To coat iron or steel with zinc.

gamma ray A high-energy electromagnetic wave emitted from the nucleus of a radioactive nuclide.

gas The phase of matter characterized by neither a definite volume nor a definite shape.

gas constant The value of the quotient pV/nT; 0.082057 L atm/mole K.

generator In chemistry, the vessel in which a reaction occurs to produce a desired gaseous product.

gram A metric unit of mass equal to one-thousandth of the standard kilogram.

gram-atomic weight The atomic weight of an element expressed in grams; contains one mole of atoms of the element.

gram-formula weight (1) The mass of a substance in grams equal to its formula weight. (2) The mass of one mole of the substance.

gram-molecular weight (1) The mass of a molecular substance in grams equal to its molecular weight. (2) The mass of one mole of molecules of the substance.

ground state The most stable state of an atom.

group A vertical column of elements in the periodic table.

half-cell The portion of a voltaic cell consisting of an electrode immersed in a solution of its ions.

half-life The length of time during which half of a given number of atoms of a radioactive nuclide decays.

half-reaction The reaction at an electrode in a half-cell of a voltaic cell.

halide (1) A binary compound of a halogen with a less electronegative element or group of elements. (2) Fluoride, chloride, bromide, iodide, or astatide.

halogen The name given to the family of elements having seven valence electrons.

halogenation A chemical reaction by which a halogen is introduced into a compound by addition or substitution.

hard water Water containing ions such as calcium or magnesium that form precipitates with soap.

heat energy The energy transferred between two systems that is associated exclusively with the difference in temperature between the two systems.

heat of combustion The heat of reaction released by the complete combustion of one mole of a substance.

heat of formation The heat released or absorbed in a composition reaction.

heat of reaction The quantity of heat evolved or absorbed during a chemical reaction.

heat of solution The difference between the heat content of a solution and the heat contents of its components.

heterogeneous Having parts with different properties.

heterogeneous catalyst A catalyst introduced into a reaction system in a different phase from that of the reactants.

heterogeneous reaction A reaction system in which reactants and products are present in different phases.

hexagonal A crystalline system in which three equilateral axes intersect at angles of 60° and with a vertical axis of variable length at right angles to the equilateral axes.

homogeneous Having similar properties throughout.

homogeneous catalyst A catalyst introduced into a reaction system in the same phase as all reactants and products.

homogeneous reaction A reaction system in which all reactants and products are in the same phase.

homologous series A series of similar compounds in which adjacent members differ by a constant unit.

hybridization The combining of two or more orbitals of nearly the same energy into new orbitals of equal energy.

hydrate A crystallized substance that contains water of crystallization.

hydrated ion An ion of a solute to which molecules of water are attached.

hydration (1) The attachment of water molecules to particles of a solute. (2) The solvation process in which water is the solvent. (3) The addition of hydrogen and oxygen atoms to a substance in the proportion in which they occur in water.

hydride A compound consisting of hydrogen and one other less electronegative element.

hydrocarbon A compound containing hydrogen and carbon.

hydrogenation The chemical addition of hydrogen to a material.

hydrogen bond A weak chemical bond between a hydrogen atom in one polar molecule and a very electronegative atom in a second polar molecule.

hydrolysis An acid-base reaction between water and ions of a dissolved salt.

hydrolysis constant The equilibrium constant of a reversible reaction between an ion of a dissolved salt and water.

hydronium ion A hydrated proton; the H_3O^+ ion.

hygroscopic Absorbing and retaining moisture from the atmosphere.
hypothesis A possible or tentative explanation.

ideal gas An imaginary gas whose behavior is described by the gas laws.
immiscible Not capable of being mixed.
indicator A substance that changes in color on the passage from acidity to alkalinity, or the reverse.
inertia Resistance of matter to change in position or motion.
inhibitor A substance that hinders catalytic action.
inorganic Pertaining to materials that are not hydrocarbons or their derivatives.
insoluble (1) Not soluble. (2) So sparingly soluble as to be considered not soluble in the usual sense.
instability constant The equilibrium constant for complex-ion equilibria.
internal phase The dispersed particles of a colloidal suspension.
ion An atom or group of atoms that has either a net positive or negative charge resulting from unequal numbers of positively charged protons and negatively charged electrons.
ion-exchange resin A resin that can exchange hydronium ions for positive ions, or one that can exchange hydroxide ions for negative ions.
ionic bonding Bonding in which one or more electrons are transferred from one atom to another.
ionic crystal A crystal consisting of ions arranged in a regular pattern.
ionic equilibrium The state of balance attained in a reversible ionization reaction between un-ionized molecules in solution and their hydrated ions.
ionic reaction A chemical reaction in which ions in solution combine to form a product that leaves the reaction environment.
ionization The formation of ions from polar solute molecules by the action of the solvent.
ionization constant The equilibrium constant of a reversible reaction by which ions are produced from molecules.
ionization energy The energy required to remove an electron from an atom.
isomer One of two or more compounds having the same molecular formula but different structures.
isotope One of two or more forms of atoms with the same atomic number but different atomic masses.

Kelvin temperature Temperature on the Kelvin scale, which is numerically 273 higher than that on the Celsius scale.
kernel The portion of an atom excluding the valence electrons.
ketone An organic compound that contains the $\rangle C{=}O$ (carbonyl) group.
kilo- Metric prefix meaning 1000.
kilocalorie The quantity of heat required to raise the temperature of one kilogram of water through one Celsius degree.
kinetic energy Energy of motion.
kinetic theory A theory that explains the properties of gases, liquids, and solids in terms of the forces between the particles of matter and the energy these particles possess.

lanthanide series Rare-earth elements of the sixth period following barium, in which the transitional inner building of the 5*d* sublevel is interrupted by the inner building of the 4*f* sublevel.
law A generalization that describes behavior in nature.
lichen (1) Two plants, an alga and a fungus, living symbiotically. (2) The source of litmus, an acid-base indicator.
limewater A water solution of calcium hydroxide.
linear accelerator A particle accelerator in which the particles travel in a straight line through many stages of potential difference.
liquid The phase of matter characterized by a definite volume but an indefinite shape.
liter One cubic decimeter.
litmus A dye extracted from lichens used as an acid-base indicator.
lye A commercial grade of either sodium hydroxide or potassium hydroxide.

magnetic quantum number The quantum number that indicates the position of the orbital about the three axes in space.
malleable Capable of being shaped by hammering or rolling.
mass The quantity of matter that a body possesses; a measure of the inertia of a body.
mass action equation See *equilibrium constant.*
mass defect See *nuclear mass defect.*
mass number (1) The whole number closest to the atomic mass of an atom. (2) The sum of the number of protons and neutrons in the nucleus of an atom.
matte A partially refined copper ore consisting of a mixture of the sulfides of iron and copper.
matter Anything that occupies space and has mass.
melting The process of converting a solid into a liquid.
melting point The temperature at which a solid changes into a liquid.
metal One of a class of elements that shows a luster, is a good conductor of heat and electricity, and is electropositive.
metallic crystal A crystal lattice consisting of positive ions surrounded by a cloud of valence electrons.
metalloid An element having certain properties characteristic of a metal, but which is generally classed as a nonmetal.
metamorphic Pertains to rocks that have undergone a change in form due to heat or pressure.
meter The metric unit of length.
metric system A decimal system of measurement.
milli- Metric prefix meaning 0.001.
miscible Capable of being mixed.

mixture A material composed of two or more substances, each of which retains its own characteristic properties.
moderator A material that slows down neutrons.
molal boiling-point constant The boiling-point elevation of a solvent in a 1-molal solution of a nonvolatile, molecular solute in the solvent.
molal freezing-point constant The freezing-point depression of a solvent in a 1-molal solution of a molecular solute in the solvent.
molality The concentration of a solution expressed in moles of solute per 1000 grams of solvent.
molal solution A solution containing one mole of solute per 1000 grams of solvent.
molar heat of formation The heat of reaction released or absorbed when one mole of a compound is formed from its elements.
molar heat of fusion The heat energy required to melt one mole of solid at its melting point.
molarity The concentration of a solution expressed in moles of solute per liter of solution.
molar solution A solution containing one mole of solute per liter of solution.
molar volume The volume in liters of one mole of a gas at STP, taken as 22.4 liters for ordinary gases, 22.414 liters for the ideal gas.
mole The amount of substance containing the Avogadro number of any kind of chemical unit. In practice, the gram-atomic weight of an element represented as monatomic; the gram-molecular weight of a molecular substance; the gram-formula weight of a nonmolecular substance; and the gram-ionic weight of an ion.
molecular formula A chemical formula that denotes the constituent elements of a molecular substance and the number of atoms of each element composing one molecule.
molecular weight The formula weight of a molecular substance.
molecule The smallest chemical unit of a substance that is capable of stable independent existence.
monatomic Consisting of one atom.
Monel metal An alloy of nickel and copper that is highly resistant to corrosion.
monomer A simple molecule, or single unit, of a polymer.
monoprotic Pertaining to an acid capable of donating one proton per molecule.
mordant A substance that, by combining with a dye, produces a fast color in a textile fiber.

natural gas A mixture of hydrocarbon gases and vapors found in porous formations in the earth's crust.
neutralization The reaction between hydronium ions and hydroxide ions to form water.
neutron A neutral particle found in the nucleus of an atom and that has about the same mass as a proton.
nitration A chemical reaction by which an $—NO_2$ group is introduced into an organic compound.
nitride A compound of nitrogen and a less electronegative element.
nitrogen fixation The process of converting elemental nitrogen into nitrogen compounds.
noble metal A metal that shows little chemical activity, especially toward oxygen.
nonelectrolyte A substance whose water solution does not conduct an electric current appreciably.
nonmetal One of a class of elements that is usually a poor conductor of heat and electricity and is electronegative.
nonpolar covalent bond A covalent bond in which there is an equal attraction for the shared electrons and a resulting balanced distribution of charge.
nonpolar molecule A molecule with all nonpolar bonds or with uniformly spaced, like polar bonds that has a uniform exterior electron distribution.
normality The concentration of a solution expressed in equivalents of solute per liter of solution.
normal solution A solution containing one equivalent of solute per liter of solution.
nuclear binding energy The energy released when a nucleus is formed from its constituent particles.
nuclear change Formation of a new substance through changes in the identity of the atoms involved.
nuclear disintegration The emission of a proton or neutron from a nucleus as a result of bombarding the nucleus with alpha particles, protons, deuterons, neutrons, etc.
nuclear equation An equation representing changes in the nuclei of atoms.
nuclear mass defect The difference between the mass of a nucleus and the sum of the masses of its constituent particles.
nuclear reactor A device in which the controlled fission of radioactive material produces new radioactive substances and energy.
nucleus The positively charged, dense central part of an atom.
nuclide A variety of atom as determined by the number of protons and number of neutrons in its nucleus.

octet The group of eight electrons filling the *s* and *p* orbitals of the highest-numbered energy level of an atom.
oil A liquid ester of glycerol and long-carbon-chain acids having a plant or animal origin.
orbital A highly probable location about a nucleus where an electron may be found.
orbital quantum number The quantum number that indicates the shape of an orbital.
ore A mineral containing an element that can be extracted profitably.
organic Pertaining to carbon compounds, particularly hydrocarbons and their derivatives.
orthorhombic A crystalline system in which there are three unequal axes at right angles.
oxidation A chemical reaction in which an element attains a more positive oxidation state.
oxidation number A signed number, assigned to an element according to a set of rules, which designates its oxidation state.

oxidation-reduction reaction Any chemical process in which there is a simultaneous attainment of a more positive oxidation state by one element and a more negative oxidation state by an associated element.
oxidation state See *oxidation number.*
oxide A compound consisting of oxygen and usually one other element in which oxygen has an oxidation number of −2.
oxidizing agent The substance that is reduced in an oxidation-reduction reaction.
oxyacid An acid containing hydrogen, oxygen, and a third element.
oxygen-carbon dioxide cycle The combination of photosynthesis and the various natural and artificial methods of producing atmospheric carbon dioxide.
ozone An allotropic form of oxygen containing three atoms per molecule.

paramagnetism The property of a substance whereby it is weakly attracted into a magnetic field.
parent nuclide The heaviest, most complex, naturally occurring nuclide in a decay series of radioactive nuclides.
partial pressure The pressure each gas of a gaseous mixture would exert if it alone were present.
period A horizontal row of elements in the periodic table.
periodic table A tabular arrangement of the chemical elements based on their atomic structure.
permanent hardness Hardness in water caused by the sulfates of calcium or magnesium, which can be removed by precipitation or ion-exchange methods.
permutit A synthetic zeolite, used in softening water.
petroleum A liquid mixture of hydrocarbons obtained from beneath the surface of the ground.
pH Hydronium ion index; the common logarithm of the reciprocal of the hydronium-ion concentration.
phenyl group The C_6H_5— group.
photon A quantum (unit) of electromagnetic radiation energy.
photosynthesis The process by which plants produce carbohydrates and oxygen with the aid of sunlight, using carbon dioxide and water as the raw materials and chlorophyll as the catalyst.
physical change A change in which the identifying properties of a substance remain unchanged.
physical equilibrium A dynamic state in which two opposing physical changes occur at equal rates in the same system.
physical properties Those properties that can be determined without causing a change in the identity of a material.
pig iron Iron recovered from a blast furnace.
polar covalent bond A covalent bond in which there is an unequal attraction for the shared electrons and a resulting unbalanced distribution of charge.
polar molecule A molecule containing one or more nonuniformly arranged polar covalent bonds and having a nonuniform exterior electron distribution.
polyatomic ion A charged group of covalently bonded atoms.
polymer A compound formed by two or more simpler molecules or radicals with repeating structural units.
polymeric Capable of forming a polymer.
polymerization A chemical reaction in which two or more simple molecules combine to form more complex molecules that contain repeating structural units of the original molecule.
potential energy Energy of position.
precipitate (1) A substance, usually a solid, that separates from a solution as a result of some physical or chemical change. (2) To produce such a substance.
precipitation The separation of a solid from a solution.
precision The agreement between the numerical values of two or more measurements made in the same way; the reproducibility of measured data.
pressure Force per unit area.
principal quantum number The quantum number that indicates the most probable distance of an orbital from the nucleus of an atom.
product An element or compound resulting from a chemical reaction.
promoter A substance that increases the activity of a catalyst when introduced in trace quantities.
protium The isotope of hydrogen having one proton and no neutrons in the nucleus; hydrogen-1.
protolysis Proton-transfer reactions.
proton A positively charged particle found in the nucleus of an atom. It has $1836/1837$ of the mass of the simplest type of hydrogen atom.
proton acceptor A base according to the Brønsted system.
proton donor An acid according to the Brønsted system.

quantum numbers The numbers that describe the distance from the nucleus, the shape, and the position with respect to the three axes in space of an orbital, as well as the direction of spin of the electron(s) in each orbital.
quicklime Calcium oxide, CaO; also called lime.

radioactive Having the property of radioactivity.
radioactive decay A radioactive change in which a nucleus emits a particle and rays, forming a slightly lighter, more stable nucleus.
radioactive tracer A radioactive nuclide introduced in small quantities to determine the behavior of chemically similar nonradioactive nuclides in various physical or chemical changes.
radioactivity The spontaneous breakdown of an unstable atomic nucleus with the release of particles and rays.
rare-earth element An element that usually differs in electronic configuration from that of next lower or higher atomic number only in the number of *f* electrons in the second-from-highest energy level.
rate-determining step The slowest of a sequence of steps along a reaction pathway.
rate law An equation that relates the reaction rate and concentrations of reactants.

reactant An element or compound entering into a chemical reaction.
reaction mechanism The pathway of a chemical reaction; the sequence of steps by which a reaction occurs.
reaction rate A measure of the amount of reactants converted to products per unit of time.
reagent A substance, usually in solution, used to react with another substance (or mixture of substances) to dissolve, precipitate, oxidize, or reduce it (or them) for analysis or use.
redox Pertaining to oxidation-reduction reactions.
reducing agent The substance that is oxidized in an oxidation-reduction reaction.
reducing atmosphere An atmosphere containing hydrogen or other reducing gases.
reduction A chemical reaction in which an element attains a more negative oxidation state.
refractory Any material that is difficult or slow to melt or corrode, and resists the action of heat.
replacement reaction A chemical reaction in which one substance is displaced from its compound by another substance.
resonance The bonding situation in substances whose bond properties cannot be satisfactorily represented by any single formula that uses the electron-dot notation system and keeps the octet rule.
resonance hybrid A substance whose properties show that its structure is intermediate between several electron-dot structures.
respiration The process by which a plant or animal absorbs oxygen and gives off products of oxidation in the tissues, especially carbon dioxide.
reversible reaction A chemical reaction in which the products re-form the original reactants under suitable conditions.
roasting Heating in the presence of air.

salt A compound composed of the positive ions of an aqueous base and the negative ions of an aqueous acid.
saponification The process of making a soap by the hydrolysis of a fat with a strong hydroxide.
saturated (1) Pertaining to a solution in which the concentration of solute is the maximum possible under existing conditions. (2) Pertaining to an organic compound that has only single covalent bonds between carbon atoms.
saturated solution A solution in which the dissolved and undissolved solutes are in equilibrium.
sedimentary Pertains to rocks formed from sediment that has been deposited in layers.
self-protective metal A metal that forms a nonporous, nonscaling coat of tarnish.
semiconductor A substance with an electric conductivity between that of a metal and an insulator.
significant figures The digits in a measurement that represent the number of units counted with reasonable assurance.
silicone One of a group of compounds containing a chain of alternate silicon and oxygen atoms, with hydrocarbon groups attached to the silicon atoms.
simplest formula See *empirical formula.*
slag An easily melted product of the reaction between the flux and the impurities of an ore.
slaked lime Calcium hydroxide, $Ca(OH)_2$; also called hydrated lime.
slaking The addition of water to lime, CaO, to produce hydrated (slaked) lime, $Ca(OH)_2$.
soda ash Anhydrous sodium carbonate, Na_2CO_3.
soft water (1) Water that lathers readily with soap. (2) Water that is free of hardening agents or water from which these agents have been removed.
solid The phase of matter characterized by a definite shape.
solubility The amount of a solute dissolved in a given amount of solvent at equilibrium, under specified conditions.
solubility-product constant The product of the molar concentrations of the ions of a sparingly soluble substance in a saturated solution, each concentration raised to the appropriate power.
soluble Capable of being dissolved.
solute The dissolved substance in a solution.
solution A homogeneous mixture of two or more substances, the composition of which may be varied within definite limits.
solution equilibrium The physical state attained in which the opposing processes of dissolving and crystallizing of a solute occur at equal rates.
solvation The clustering of solvent particles about the particles of solute.
solvent The dissolving medium in a solution.
spectator ion An ion in a reaction system that takes no part in the chemical action.
spectroscope An optical instrument consisting of a collimator tube, a glass prism, and a telescope, used for producing and viewing spectra.
spectrum The pattern of colors formed by passing light through a prism.
spin quantum number The quantum number that indicates the direction of spin of an electron.
stable compound A compound that is not decomposed easily.
standard boiling point The temperature at which the equilibrium vapor pressure of a liquid is equal to the standard atmospheric pressure, 760 mm of mercury.
standard molar heat of vaporization The heat energy required to vaporize one mole of liquid at its standard boiling point.
standard pressure The pressure exerted by a column of mercury exactly 760 mm high at 0 °C.
standard solution A solution that contains a definite concentration of solute, which is known precisely.
standard temperature 0 °Celsius.
stoichiometry Pertaining to the numerical relationships of elements and compounds and the mathematical proportions of reactants and products in chemical reactions.
STP The abbreviation for "standard temperature and pressure."

structural formula A formula that indicates the kind, number, arrangement, and valence bonds of the atoms in a molecule.
sublimation The change of phase from a solid to a vapor.
sublime To pass from the solid to the gaseous phase without liquefying.
subscript A number written below and to the side of a symbol. If at the left, it represents the atomic number; if at the right, it represents the number of atoms of the element.
substance A homogeneous material consisting of one particular kind of matter.
substitution A reaction in which one or more atoms are substituted for hydrogen atoms in a hydrocarbon.
substitution product A compound in which various atoms or groups have been substituted for one or more atoms.
sulfation A chemical reaction in which a compound is treated with sulfuric acid, replacing a hydrogen atom bonded to an oxygen atom with an $—SO_2OH$ group.
sulfonation A chemical reaction in which an organic compound is treated with sulfuric acid, replacing a hydrogen atom bonded to a carbon atom with an $—SO_2OH$ group.
superheated water Water heated under pressure to a temperature above its normal boiling point.
supersaturated Pertaining to a solution that contains an amount of solute in excess of that normally possible under existing conditions.
synchrotron A particle accelerator in which particles move in a circular path due to the varying of the oscillating voltage and the magnetic field.
synthetic Artificial.

temperature A measure of the ability of a system to transfer heat to, or acquire heat from, other systems.
temporary hardness Hardness in water caused by the presence of hydrogen carbonates of calcium or magnesium, which can be removed by boiling.
tensile strength The resistance of a material to being pulled apart.
theory A plausible explanation of an observed phenomenon that is confirmed by experiments designed to test predictions based upon the explanation.
thermal cracking The breaking up of large molecules into smaller ones by the use of high temperature.
thermal equilibrium The condition in which all objects in an isolated system are at the same temperature.
thermite reaction The reaction by which a metal is prepared from its oxide by reduction with aluminum.
titration The process by which the capacity of a solution of unknown concentration to combine with one of known concentration is measured.
transition Pertaining to subgroups of elements characterized by the belated filling of the next-to-highest energy level of the atoms.
transition element An element that usually differs in electronic configuration from that of the next lower or higher atomic number only in the number of *d* electrons in the next-to-highest energy level.
transition interval The pH range over which the color change of an indicator occurs.
transmutation reaction A reaction in which the nucleus of an atom undergoes a change in the number of its protons and, consequently, in its identity.
transuranium element An element with an atomic number higher than uranium, which has atomic number 92.
triclinic A crystalline system in which there are three unequal axes and three unequal oblique intersections.
trigonal See *rhombohedral*.
triple point of water The single temperature and pressure condition at which water exists in all three phases at equilibrium.
triprotic Pertaining to an acid capable of donating three protons per molecule.
tritium The isotope of hydrogen having one proton and two neutrons in the nucleus; hydrogen-3.

unit cell The smallest portion of the crystal lattice that exhibits the pattern of the lattice structure.
unsaturated Pertaining to an organic compound with one or more double or triple covalent bonds between carbon atoms in each molecule.
unstable compound A compound that is decomposed easily.

valence electron One of the electrons in an incomplete highest energy level of an atom.
van der Waals forces Forces of attraction between molecules.
vapor A gas at a temperature that is below its critical temperature.
vapor pressure Pressure due to the vapor of confined liquids and solids.
vinyl group The $CH_2{=}CH—$ group.
viscous Slow-flowing.
volatile Easily vaporized.
voltaic cell (1) An electrochemical cell arranged to deliver an electric current to an external circuit. (2) An electric cell that converts chemical energy to electric energy.
vulcanization The heating of rubber with other materials to improve its properties.

waft (1) To cause a slight puff of a gas (or odor) to move lightly by the nostrils. (2) The act of wafting an odor.
water gas A fuel gas containing mainly CO and H_2 made by blowing a blast of steam through a bed of red-hot coke.
water of crystallization Water that has united with some compounds as they crystallize from solution.
weight The measure of the earth's gravitational attraction for a body.

X rays Electromagnetic radiations of high frequency and short wavelength.

zeolite A natural mineral, sodium silico-aluminate, used to soften water.

INDEX

A

B

C

D

E

F

G

H

I

J

K

L

M

N

O

P

Q

R

S

T

U